Dr. M. Gunasekaran
Department of Biology
Fisk University
Nashville, TN 37203, U.S.A.

Microbiology: Essentials and Applications

LARRY McKANE
California State Polytechnic University, Pomona

JUDY KANDEL
California State University, Fullerton

MICRO-
BIOLOGY

Essentials
and
Applications

McGraw-Hill Book Company

New York | St. Louis | San Francisco | Auckland
Bogotá | Hamburg | Johannesburg | London | Madrid
Mexico | Montreal | New Delhi | Panama | Paris
São Paulo | Singapore | Sydney | Tokyo | Toronto

MICROBIOLOGY:
Essentials
and Applications

1234567890VNHVNH898765

ISBN 0-07-045125-7

This book was set in Palatino by Black Dot, Inc. (ECU).
The editors were Mary Jane Martin and Jo Satloff;
the designer was Nicholas Krenitsky;
the production supervisor was Joe Campanella.
The drawings were done by Fine Line Illustrations, Inc.
Von Hoffmann Press, Inc., was printer and binder.

Cover photographs courtesy of N. Furjanick, Nikon Corporation,
R. Williams, J. Higbee, and R. Marchino.

See Illustration Credits on pages 739–743. Copyrights included on this page by reference.

Library of Congress Cataloging in Publication Data

McKane, Larry.
    Microbiology: essentials and applications.

    Includes index.
    1. Microbiology. I. Kandel, Judy. II. Title.
[DNLM: 1. Microbiology. QW 4 M478m]
QR41.2.M38 1985   576   84-12538
ISBN 0-07-045125-7

Contents

x

CONTENTS

Preface

This book is intended for use by all undergraduate students, including those who plan to pursue careers in fields other than microbiology. We have therefore tried to write a book that is stimulating and enjoyable for students majoring in any aspect of science and the health-related fields. We feel this is best accomplished by following guidelines of clarity and brevity. Writing a short book is in many ways more difficult than writing a long one. The decisions to omit some material traditionally found in microbiology texts were often agonizing (it hurts to throw out one's microbiological pets). These decisions were carefully considered to avoid sacrificing the academic concepts necessary for thorough understanding and for practical implementation of microbiological principles. We tried to find a balance that will challenge the student without intimidating, thereby reducing anxiety and increasing learning efficiency. We hope this book generates sufficient interest that students will continue to informally study microorganisms long after they finish this class.

Our goals are to not only convey an important body of knowledge, but to share with the reader our fascination and excitement about microorganisms and the science that studies them. In this way we hope we can help students develop respect for microorganisms, for their vast number of beneficial activities as well as for their enormous potential to cause human suffering. This book should help students realize that, regardless of their career plans, microorganisms will continue to affect their lives in many notable ways. We have tried to create a book that helps students better recognize the problems created by the presence of microbes (as well as the solutions to these problems), and to personally modify their behavior as the result of intelligent consideration. Knowing the microbiological alternatives helps people make good decisions.

Such decision making is especially critical for students of health care. We hope to make these students realize that microbiology must be considered more than coursework. Effective medical professionals let their knowledge of microorganisms determine their behavior in clinical settings. The student (and teacher) should be aware of the potential role of hospital personnel in the transmission of infectious diseases in the hospital. Lives depend on the knowledge and attitudes these students are acquiring right now in their introductory microbiology classes. Inadequate quality control in medical microbiology curricula often leads to grave consequences. The overriding goal of this text is to help medical professionals avoid becoming unwitting participants in the spread of disease.

Our presentation strategy is to initially guide students through the fundamentals of microbiology. In doing so, we decided not to defer the presentation of practical applications to the latter chapters, but rather discuss them in the same sections employed for presenting the conceptual information. This makes it easier for sudents to connect theory and practice. In the early chapters the reader receives an overview of the microbial world and an introduction to microscopy, laboratory cultivation, and other techniques for studying the microbes. The specifics of bacterial anatomy, physiology, reproduction, and growth are followed by chapters on the fungi, algae, protozoa, and viruses. The chemistry of life is presented in two chapters located late in the book's first half (Chapters 10 and 11). This way, students with no chemistry background won't have to surmount this hurdle until they have enough microbiological information to relate molecular concepts to principles they already understand. The same is true of Chapter 12 on microbial genetics, which cannot be fully appreciated before learning the biological importance of enzymes and metabolic reactions.

At this stage in the book, the student has acquired sufficient information about microbial fundamentals to understand fully how these processes may be intentionally impaired by physical and chemical agents for controlling microorganisms. A separate chapter is devoted to antibiotics and chemotherapeutic agents.

Except for the final chapter on microbial biotechnology, the remainder of the book is devoted to discussing the microbes as agents of infectious disease and the body's response to these intruders. We have attempted to paint an integrated picture of the infectious disease process, emphasizing the host-parasite relationship and the factors that distinguish colonization from infection, and infection from disease. Students become acquainted with the dynamic equilibrium between humans and potentially pathogenic agents and with the properties of host and parasite that are important in the development and clinical manifestations of infectious disease.

Separate chapters are devoted to concepts of disease pathogenesis, epidemiology, and clinical specimen collection. This unique triad is strategically located just before the chapters discussing the specific infectious diseases. These diseases are discussed according to the pathogen's portals of entry into the body. In this way the reader can better associate each disease with the most effective approaches for preventing the spread of infection to uninfected persons.

The book's final chapter deals with the modern "industrial revolution" created by our increasing ability to harness the activities of microorganisms. Microbial biotechnology is providing new sources of important compounds, of food that may help relieve the world's hunger problem, and of modern weapons for fighting disease and human suffering.

Students will never fully appreciate the importance of microbiology, however, unless they are motivated to read about the microbes. For many, the subject itself will be of sufficient interest to keep them reading. We have incorporated several other features to capture the interest of students who may be either bored or intimidated by the sciences. Our diagrams are designed to suggest motion and draw attention to the important elements in the proper viewing sequence. We feel the use of our bold arrows helps change abstract concepts into easily comprehended physical images. To provide additional motivation, each chapter is supplemented with relevant asides enclosed in color boxes. These boxes enliven the text and better enable students to understand and appreciate the more human aspects of microbiology. Many of these boxed asides are provocative, others are historical, and still others help readers realize the personal intimacy of their own relationships with microorganisms.

Each chapter is preceded by an outline of the chapter contents and is followed by an overview, a list of key terms, and several questions that direct attention to the most important information.

To help reduce the continually growing incidence of nosocomial (hospital-acquired) infections, we have devoted a full chapter to this complex problem. Frequent references to these dangers are made during the discussions of concepts underlying growth, transmission, pathogenesis, epidemiology, and control of microorganisms and infectious disease. Although the importance of medical asepsis is stressed, the nosocomial infection problem cannot be solved by technique alone. This can be achieved only by an awareness of microorganisms and their influence on health and disease.

We would like to express our thanks for the many useful comments and suggestions provided by colleagues who reviewed this text during the course of its development, especially to Lucia Anderson, Queensborough Community College; David Campbell, St. Louis Community College, Meramec; Albert Canaris, University of Texas, El Paso; Garry J. Ciskowski, Indiana University of Pennsylvania; Elaine B. McClanahan, Crafton Hills College; Carolyn P. Eau Claire, Hackensack Medical Center; Cindy Erwin, City College of San Francisco; Richard Gross, Motlow State Community College; Joan Handley, the University of Kansas; John Hendry; Dean Hoganson, Drake University; Robert Janssen, The University of Arizona; Virginia C. Kelley, Auburn University; Leah Koditschek, Montclair State College; John Lancaster, the University of Oklahoma at Norman; William Lester, Humboldt State University; Lynne McFarland, University of Washington; William O'Dell, the University of Nebraska at Omaha; Rosemarie Palmer, Pennsylvania State University; Loy Pike, Indiana University at South Bend; Robert Quackenbush, the University of South Dakota; Mary Lee Richeson, Indiana University–Purdue University at Fort Wayne; Warren S. Silver, University of South Florida; William Tidwell, San Jose State University; Susan Turner, University of Portland; Curtis Williams, State University of New York at Purchase; Fred Williams, Iowa State University; Marshal A. Yokell, Middlesex Community College; and Betty Yost, Orange Coast College.

Larry McKane
Judy Kandel

Microbiology: Essentials and Applications

The World of Microbiology

1

Our world is populated by invisible creatures too small to be seen with the unaided eye. These life forms, the **microorganisms**, or **microbes**, may be seen only by magnifying their image with a microscope. Despite their small size, the effects of microbes on humans, and on the world in general, are critical for maintaining life on earth. It is therefore impossible, indeed undesirable, for people to avoid microorganisms or their influences. We should, however, understand the activities of microbes as well as their potential for enhancing or diminishing the quality of our lives. Such knowledge is instrumental for controlling microorganisms, minimizing their harmful effects, and maximizing their beneficial activities.

A few microorganisms are capable of growing in or on the human body. Some of these microbes are responsible for infectious disease, one of the great burdens people throughout history have had to endure. Until relatively recently, people were helpless when such diseases as smallpox, diphtheria, and typhoid swept through their communities. Controlling infectious diseases has saved millions of human lives. The cornerstone of this remarkable accomplishment is **microbiology**, the study of microorganisms.

BEGINNINGS OF MICROBIOLOGY

In 1674 Antoni van Leeuwenhoek, an amateur lens grinder of extraordinary skill and patience, looked through his simple microscope and discovered a new world (Fig. 1-1). The existence of this universe of microscopic inhabitants had been suspected by only a few insightful scientists. The "animalcules," as Leeuwenhoek called them in a series of letters to the Royal Society of London, were found in his mouth, stagnant water, and foods. Leeuwenhoek observed microbes in samples from nearly every environment he investigated. The animalcules seemed to be everywhere.

2

FIGURE 1-1

(*a*) Antoni van Leeuwenhoek with his simple microscope in hand. (*b*) Drawings from Leeuwenhoek's notebook of some of the microbes he observed.

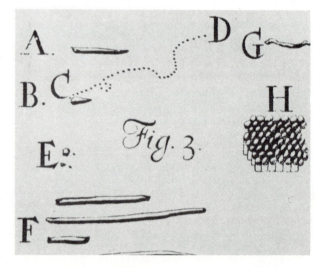

The Spontaneous Generation Controversy

Leeuwenhoek's discovery of microorganisms explains why many foods and drinks spoil. After sufficient time they simply become overgrown by microbes. Microbial proliferation is often evidenced by the cloudiness of a once-clear liquid. The origin of the microorganisms, however, was the subject of a scientific controversy that was finally resolved 200 years after Leeuwenhoek's findings. Some people suggested that nonliving substances were converted into living organisms; in other words, they believed in the **spontaneous generation** of life from nonliving materials. Opponents of this explanation supported **biogenesis**, a theory that all organisms arise only from other living organisms, and insisted that overgrown food has been "seeded" with at least one **viable** (living) parent microbe.

Spontaneous generation appeared to be disproved when it was demonstrated that beef broth that had been boiled to kill all the microbes in it remained **sterile** (free from all living organisms) as long as the container was plugged with a solid stopper. In response to criticism that air was necessary for spontaneous generation to occur, cotton plugs were substituted for solid stoppers. (Cotton filters remove suspended particles from air—microbes are among the suspended particles.) Broth that was protected from suspended particles remained sterile, demonstrating that the broth itself could not give rise to living organisms. Removing the cotton plug allowed microorganisms to enter and subsequently overgrow the liquid within 18 hours. Thus, boiling did not destroy the ability of the broth to support microbial growth. Biogenesis seemed to be the only logical explanation for such a phenomenon.

Many proponents of spontaneous generation, however, remained unconvinced. The controversy was finally resolved in 1861 by the powerful logic of Louis Pasteur. He designed swan-necked flasks that allowed the introduction of fresh, unaltered air but trapped dust particles and microorganisms in the curved neck, thereby eliminating the need for the cotton plug (Fig. 1-2). Broth that had been boiled in these flasks remained sterile. Thus, Pasteur proved that neither broth nor air could spontaneously produce microorganisms, since sterility was preserved in the presence of both. Furthermore, when the flask was tipped, allowing the liquid to flow into the neck, visible microbial growth developed within hours. He thus demonstrated that the agents responsible for spoiling the broth were the microbes trapped in the flask's neck.

In spite of his convincing experiments, a few scientists continued to reject biogenesis, and for good reason—their own experiments still failed to confirm Pasteur's claim. These investigators boiled and sealed liquid infusions of hay rather than beef broth. The hay infusions often became overgrown with bacteria, even when protected from all external sources of microbes. The conflicting results were due to the types of bacteria found in hay but rarely in beef. These bacteria form **endospores**, protective structures that are among the most heat-resistant forms of life on earth. Many endospores can survive several hours of boiling and then, at lower, more hospitable temperatures, germinate to an actively growing stage. Thus, the

FIGURE 1-2

Swan-necked Pasteur flask.
Dust settles in the neck of
the flask, trapping particles
and microorganisms while
allowing air to reach the
nutrient medium. The
broth remains clear and
sterile. If the flask is tilted
so the broth flows into the
neck and reaches the
trapped particles, the liquid
becomes cloudy within
hours, indicative of
microbial proliferation.

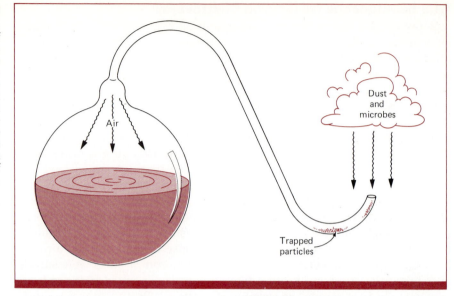

4

THE WORLD OF
MICROBIOLOGY

boiled hay infusions were actually inoculated from within by the surviving endospores, resulting in overgrowth with no apparent source of microbial parents. Spontaneous generation was finally disproved with the discovery of the endospore by John Tyndall.

MICROBES

Microbes are grouped into six major categories: (1) protozoa, (2) microscopic algae, (3) fungi (yeasts and molds), (4) bacteria, (5) cyanobacteria, and (6) viruses. Although most are unicellular (single-celled) organisms, some, like the molds and many algae, are multicellular. **Bacteria** and **cyanobacteria** possess the least complex cellular structure. The cell structure of **protozoa** is similar to that of animals. The cells of **algae** are similar to those of plants, whereas **fungi** resemble plant cells that lack chlorophyll and thus have lost that ability to photosynthesize. In spite of these resemblances, however, microorganisms are considered neither plant nor animal. **Viruses** lack cellular structure common to other microbes and are referred to as noncellular particles. They are classically considered to be the simplest of all microbes.

Some typical members of each group of microorganisms are pictured in Figure 1-3.

MICROBES IN THE ENVIRONMENT

Microorganisms exist virtually everywhere. They are in our food, in the water we use for drinking and bathing, on our utensils, and on our bedsheets. The air we breathe carries a wide variety of microbes. Their varied nature allows some types to survive in many unlikely environments. Microbes have been found in the air miles

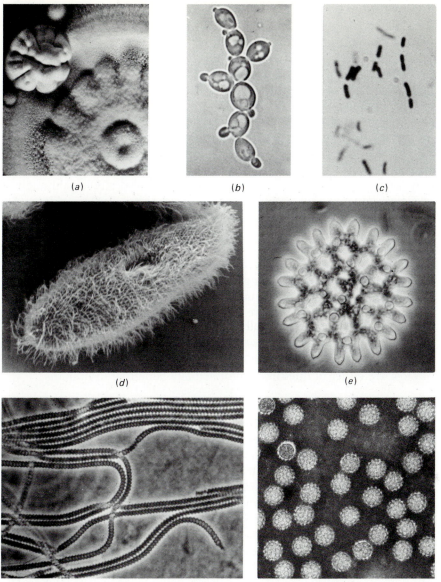

FIGURE 1-3

Some examples of common microorganisms: (a) mold; (b) baker's yeast; (c) bacteria; (d) protozoan; (e) algae; (f) cyanobacteria; (g) viruses.

(a)   (b)   (c)

(d)   (e)

(f)   (g)

above the earth. Other microbes thrive in natural hot springs at a temperature of 90°C, and at least one type grows well at 250°C. Some bacteria can grow in sulfuric acid, others in distilled or deionized water, using minute amounts of nutrients dissolved from air.

Microorganisms inhabit the surfaces of living human and animal bodies and grow abundantly in the mouth and the intestinal tract. One-third the dry weight of human feces is bacteria. Yet most persons are unaware of the presence of billions of microorganisms on their bodies because the microbes ordinarily stimulate no physiological response and cause no disease. Such harmless microorganisms comprise the **normal flora**, those microorganisms which live on the human body in a normally harmonious relationship with their host.

It is much easier to list the types of environments devoid of microorganisms because only a few such places exist on earth. The interior of the healthy human body usually contains no microbes. ("Interior" does not include the alimentary tract or upper respiratory tract, which are extensions of the external environment.) The cerebrospinal fluid must be free of microorganisms for a person to remain healthy. Similarly, the blood and tissue fluids of a healthy person are microbe-free, except for transient microorganisms introduced by trauma such as cutting the skin or accidentally biting the tongue. Urine in the bladder is also sterile, although it becomes contaminated with normal flora organisms during voiding. All the internal organs of a healthy person are free of microbial growth.

MICROBIAL ACTIVITIES

Life requires biochemical activity. All living organisms must be able to change the chemicals in their environment into forms that can be incorporated into cellular material. The chemical changes performed by an organism are collectively known as **metabolism**. The multitude of influences that microbes exert on their environment are direct reflections of their diverse metabolic activities. Considering the widespread distribution of microbes, it is fortunate that the metabolic processes of most are either nondetrimental or actually beneficial to humans.

Beneficial Activities

Without microorganisms all life on earth would perish. Microorganisms decompose dead organic matter into simple nutrients that can be used by plants and other photosynthetic organisms. As decomposers, microbes are critical to the **biogeochemical cycle**, the flow of nutrient compounds through the world's food network (Fig. 1-4). Simple inorganic nutrients are assimilated into complex organic compounds by photosynthetic organisms, which are the ultimate source of food for all the larger animal consumers. Ultimately these nutrients are trapped in the bodies of dead animals and plants. Microorganisms recycle these nutrients by decomposing the complex constituents of dead organisms back into simple chemical components, making them once again available to photosynthetic organisms. This recycling process allows the earth, with its limited supply of nutrients, to sustain a continuum of life.

Some bacteria and cyanobacteria contribute to soil fertility by changing atmospheric nitrogen into a biologically useful form by a process called **nitrogen fixation**. Although 80 percent of the earth's atmosphere is molecular nitrogen (N_2), the overwhelming majority of organisms cannot use nitrogen in this form. Nitrogen-fixing microorganisms convert molecular nitrogen to ammonia, which can then be converted to the organic form (proteins and nucleic acids) by plants. Certain plants called *legumes* (clover, alfalfa, and soybeans, for example) establish a mutually beneficial relation-

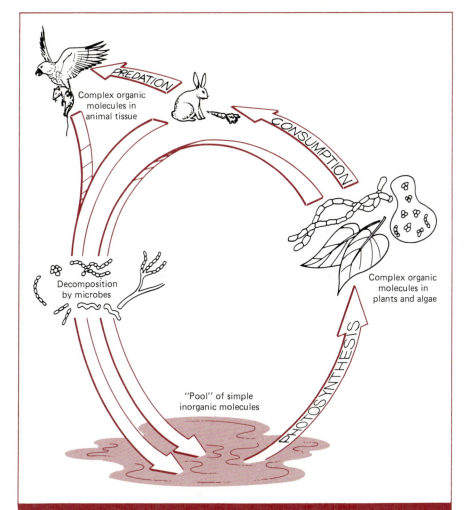

Complex organic molecules in animal tissue

PREDATION

CONSUMPTION

Decomposition by microbes

Complex organic molecules in plants and algae

"Pool" of simple inorganic molecules

PHOTOSYNTHESIS

FIGURE 1·4

Biogeochemical cycle: Elements used as nutrients (carbon, nitrogen, etc.) become incorporated into complex organic molecules of the tissues in plants and subsequent consumers. Microbial decomposers recycle the nutrients by converting organic wastes and dead organisms back to simple inorganic compounds, making them once more available to photosynthetic organisms.

ship with nitrogen-fixing bacteria housed in nodules on their roots. Because more nitrogen is fixed than the host plants can use, legumes restore soil fertility to land that has been exhausted of its usable nitrogen.

Our meat supplies are also dependent on microorganisms. Cattle and sheep, for example, are incapable of digesting cellulose, the bulk of their grass diet. Cellulose-digesting microorganisms, however, live in a special stomach chamber called a *rumen*. These microbes help the cattle to acquire nutrients from foods that the animal would otherwise be unable to digest.

Some of our foods are actually by-products of microbial growth. Cheese, for example, is produced by the growth of microorganisms in milk. The type and flavor of the cheese is largely determined by the microorganism used for processing. The blue mold *Penicillium roqueforti*, for example, provides both the flavor and color of Roquefort cheese. Yogurt results from the growth of bacteria in milk. The leavening of bread is accomplished by *Saccharomyces cerevisiae* (baker's yeast), which produces carbon dioxide as a waste product. This gas is trapped as tiny bubbles that cause the dough to rise. This yeast is also responsible for the production of alcoholic beverages. Other microbes ferment cabbage to sauerkraut. Still others change soybean extract to soy sauce. Even chocolate is the result of the action of yeasts on the cocoa bean prior to roasting.

Our energy needs may also be partially supplied by microbial activity. Microbial methane generators are already in limited use to convert manure to combustible fuel for powering vehicles and heaters. By converting our wastes into usable energy, biological fuel generators may someday be used to supply power to residential communities and industry.

Microorganisms can be tiny biological factories that produce commercially useful products such as dietary supplements, food stabilizers, solvents, flavorings, meat tenderizers, and products for treating disease. The field of industrial microbiology is growing rapidly. Perhaps the most promising of these endeavors is *genetic engineering* of microorganisms. Microbiologists have developed the methodology for isolating **genes**, segments of chromosomes that contain information directing a cell to synthesize certain chemical compounds. Human genes may be introduced into bacteria or yeast cells, which may then synthesize the compound specified by the newly acquired genetic information. In this fashion, products such as insulin and other human hormones that were traditionally difficult or impossible to obtain can be inexpensively produced by genetically altered microbes growing in large quantities. These tiny microbial factories will ultimately provide an inexpensive source of many biological compounds.

The door to commercial genetic engineering was opened in June 1980 when the United States Supreme Court made it legal to patent life forms. Entrepreneurs may now own the organisms they create by genetic manipulation. The commercial production of extraordinary new products will likely be encouraged by this legal precedent.

One of the frequently overlooked medical benefits provided by microorganisms is their role as normal flora on the human body. Microbial competition between our established flora and **pathogens** (microbes that cause disease) usually prevents infection. Many pathogens lose the competition for the limited space and nutrients, or are inhibited by antimicrobial chemicals produced by members of the normal flora. Such microbial antagonism helps prevent upper respiratory tract infections, pneumonia, gastrointestinal invasion, vaginal infections, and even gonorrhea in women.

Detrimental Activities

Unfortunately, the consequences of microbial proliferation are not always beneficial. The detrimental activities of microbes include decomposition of useful materials, spoilage of food or water, and diseases of plants and animals. There is scarcely a material that is not subject to microbial deterioration. Paint, rubber, insulation on electrical wires, textiles, and metals can be destroyed by the effect of microbial metabolism. Microbes are even responsible for decomposition of wood by termites. This insect can utilize wood only because cellulose-digesting protozoa inhabit its gastrointestinal tract.

Foods are especially susceptible to microbial decomposition because they are rich in nutrients that support the growth of microbes. Contaminated foods are often inedible and may also be the source of human disease if the contaminating organisms are pathogens or toxin producers. Water can be rendered undrinkable or dangerous if it contains pathogens or microbes that produce toxins or substances with unpleasant odors or tastes. Control measures such as **pasteurization** (see color box) can reduce the incidence of both spoilage and disease.

In addition to spoilage, our food supplies are also reduced by infectious diseases of living plants and animals. For example, the fungus that caused the potato blight of the mid-1800s destroyed the major food crop of Ireland and was responsible for the starvation deaths of over a million people. Similarly, animal diseases caused by microbes, such as anthrax in cattle, have resulted in major losses of food.

MICROBES AND HUMAN DISEASE

Even before 1876, when microbes were first proved to be the cause of infectious human diseases, they were suspected of being responsible for many illnesses. This seemed most apparent in hospitals, where diseases were often spread from one patient to another. This pattern of disease transmission was noticed by Ignaz Semmelweiss, a Hungarian physician who, in 1847, was alarmed by the

BATTLING THE SPOILERS

■ Louis Pasteur's renown as a scientist was partially established by his role in disproving spontaneous generation. He became a national hero, however, when he saved the economy of France by developing a technique for controlling the growth of undesirable microorganisms in wine. In the mid-1800s, the French wine industry began to produce diseased wines of undrinkable quality. Theodor Schwann had already proved that yeast cells were responsible for alcohol fermentation in the production of beer, wine, and spirits from fruit and grain juices. Pasteur demonstrated that the quality of wine depended on the type of microorganism in the grape mash. To produce wine of predictably high quality, undesirable microbes that produce poor or undrinkable wine had to be eliminated and the grape mash then inoculated with the desired yeasts. Although boiling destroys the contaminated microbes it also destroys the quality of the grape mash. Pasteur found that gentle heating at 60°C for 30 minutes kills the great majority of microorganisms with no adverse influence on the mash. The decontaminated product is then inoculated with a superior strain of yeast. This controlled heating process, now called **pasteurization** in honor of its inventor, is still used by the food and dairy industry to kill pathogens and retard spoilage without damaging the product.

FIGURE 1-5
Joseph Lister supervising
the use of phenol to
prevent wound infection in
the hospital. The
instrument operated by the
person to the right sprays a
fine mist of carbolic acid
into the air, producing an
"antiseptic" environment.

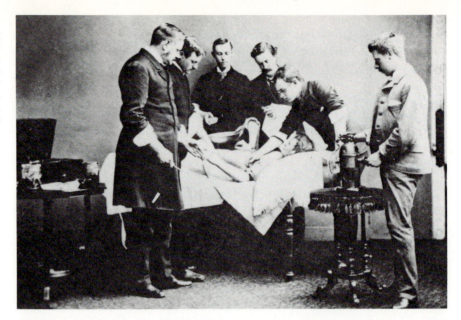

high incidence of childbed fever, fatal infections among women who had just had babies. He observed that deliveries were often performed by doctors who moved from one mother to the next with no intervening sanitary precautions such as hand washing. In fact, physicians often went directly from postmortem dissections to the maternity ward. Semmelweiss suspected that the agent causing childbed fever was transmitted on the hands of these doctors. He therefore instituted mandatory hand washing in a chlorine solution for all physicians prior to performing each delivery. The incidence of childbed fever dropped from 50 to 1 percent. Similar successes were achieved by Joseph Lister in 1865 against wound and postsurgical infections. He recommended that all instruments and dressings, as well as the wounds themselves, be treated with carbolic acid (phenol), an antimicrobial agent (Fig. 1-5). These triumphs not only improved health conditions within hospitals but also provided support for the **germ theory of infectious disease**, which attributes many illnesses to the growth of microorganisms in or on the body.

The Germ Theory of Disease

No microbe was unequivocally proved to be the cause of human disease until 1876 when Robert Koch showed that *Bacillus anthracis* causes anthrax, a potentially fatal disease acquired from livestock. Koch applied a set of criteria for conclusively demonstrating the **etiology** (specific cause) of an infectious disease. Collectively known as **Koch's postulates**, the four criteria are:

1. The suspected microorganism must always be found in diseased but never in healthy individuals.

2. The microorganism must be cultivated in a pure culture (one free of all other types of microbes) on an artificial laboratory **medium** (a nutrient material that supports cell growth).

3. Pure cultures of the microorganism must cause the same disease when inoculated into susceptible experimental animals.

4. The same organisms must be reisolated from experimentally infected animals.

Although viruses and a few other microbes cannot be cultured in artificial media, Koch's postulates have been valuable in determining the cause of most infectious diseases.

Fulfillment of Koch's postulates requires pure cultures of microorganisms. Most natural environments, however, are populated by many types of microbes. Koch devised an easy method for obtaining pure cultures (Fig. 1-6*a*). He spread material containing microbes over the surface of a *solid medium,* using a technique that isolates individual cells from each other. Each isolated cell proliferates and gives rise to a **colony**, a visible aggregate of billions of cells, all descended from the same original parent organism (Fig. 1-6*b*). All the cells in an isolated colony are therefore virtually identical. Thus, the colony represents a pure culture.

Immunization

The best approach for combating infectious disease is *prophylaxis,* that is, disease prevention. Among the most powerful prophylactic weapons against pathogens are those that enhance people's resistance to disease. During the course of some infectious diseases, the body develops

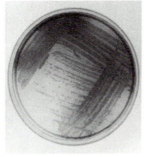

(a) (b)

FIGURE 1-6

(*a*) Robert Koch with the basic tools of the microbiologist. The petri dishes contain a solid medium and the flask contains broth, a liquid medium. (*b*) Surface of solid medium (nutrient agar) inoculated with bacteria and incubated for 24 hours. An isolated colony may be selected for obtaining pure cultures.

FIGURE 1-7

Edward Jenner administers the first vaccine against infectious disease. His procedure will protect this child from smallpox.

immunity, resistance to further attacks by the same pathogen. Thus a person suffers from some illnesses such as measles only once. Immunity can be clinically induced by introducing harmless variants of pathogens into the body, which consequently develops resistance without acquiring the disease. This procedure, known as **vaccination** (*vacca*, cow), was first used successfully in 1798 by Edward Jenner, who utilized cowpox virus to protect against smallpox. Jenner observed that persons who contracted cowpox (a mild disease characterized by sores on the hands) never developed smallpox. He proved that inoculating people with pus from cowpox lesions safely protected them from smallpox (Fig. 1-7).

In spite of his success, Jenner never understood the mechanism by which his vaccine provided protection. We now know that the success of the vaccine was due to similarities in the physical and chemical structure between smallpox and cowpox viruses. Immunity develops because the body produces specific blood substances, called **antibodies**, in response to the presence of foreign agents such as viruses or bacteria. Antibodies react with the specific agent that stimulated their production and often prevent it from damaging the host (Fig. 1-8).

Vaccination became a useful tool for preventing other diseases when Louis Pasteur (who had his fingers in many scientific pies) developed methods for converting pathogens into harmless variants that could induce immunity. Using this method he developed successful vaccines against anthrax and against rabies. Subsequent vaccines have dramatically reduced the incidence of diphtheria, polio, whooping cough, tetanus, mumps, and measles.

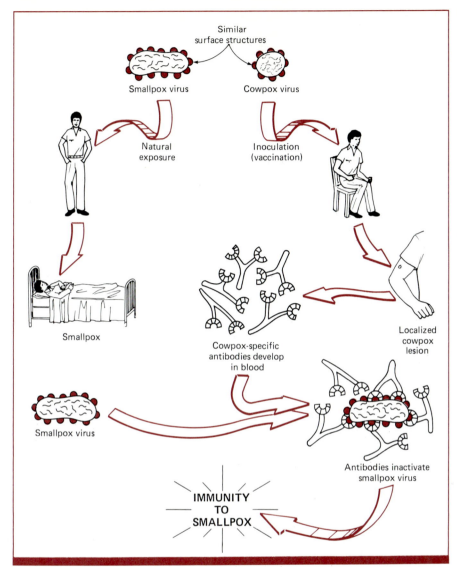

FIGURE 1-8

Using cowpox virus to vaccinate against smallpox. Nonimmune persons acquire smallpox by exposure to the smallpox virus shed from persons with the disease. Previous exposure to cowpox virus stimulates the person to produce cowpox-specific antibodies. Because of coincidental similarities between surface structures of cowpox and smallpox viruses, the vaccinated person becomes immune to smallpox as well as cowpox. (See Chapter 16 for structure of antibodies.)

Similar surface structures

Smallpox virus

Cowpox virus

Natural exposure

Inoculation (vaccination)

Smallpox

Cowpox-specific antibodies develop in blood

Localized cowpox lesion

Smallpox virus

Antibodies inactivate smallpox virus

IMMUNITY TO SMALLPOX

13

MICROBES AND HUMAN DISEASE

Chemotherapy

Some diseases can strike the same person many times because infection fails to induce immunity. Malaria and some sexually transmitted diseases are examples of such recurrent illnesses. Fortunately many diseases can be cured by **chemotherapy**, the use of chemicals that selectively inhibit or kill pathogens without killing the patient. These chemicals can therefore be introduced into infected persons to treat their diseases.

Many chemotherapeutic agents are **antibiotics**, chemicals produced by one type of microorganism, that interfere with specific biological processes of other types of microbes. Alexander Fleming's accidental discovery of penicillin in 1928 opened the door to the antibiotic era of medicine. Fleming observed that a contaminating mold growing on a solid medium somehow inhibited the growth of the bacteria that he had inoculated on the medium. Years later other investigators purified this inhibitory chemical from mold cultures. Penicillin became the first of many effective antibiotics for treating infectious disease. Millions of human lives have been saved by antibiotic therapy, and the search continues for new and more effective antimicrobial agents.

Epidemiology

An important field within microbiology is **epidemiology**, an investigative science that provides information about the factors and conditions that contribute to the occurrence of diseases. Epidemiologists monitor the occurrence of each illness in a population and the number of persons dying from each disease. This information helps authorities identify potential health hazards in the community so that appropriate precautions may be instituted. For example, an increase in the incidence of measles suggests the need for a more extensive immunization (vaccination) program. Epidemiologists also identify sources of disease outbreaks. Once identified, measures may be instituted to control the source of infection, thereby preventing the spread of disease.

The development of microbiology has produced dramatic progress in the war against human diseases (Fig. 1-9). Citizens of developed countries no longer need fear the onslaught of crippling or killer epidemics such as polio, diphtheria, whooping cough, typhoid fever, plague, yellow fever, or cholera. Yet outbreaks of these diseases may still occur if control measures become inadequate; thus these diseases are not completely defeated. The first total microbiological victory was achieved in 1980 when smallpox, once a disease that scarred or killed one-quarter of the world's population, became the first disease declared by the World Health Organization to be successfully eradicated from the earth.

Despite these advances, infectious diseases such as tuberculosis, hepatitis, and pneumonia continue to cause human suffering and death. In addition, diseases are still acquired in hospitals at an alarming rate. Their frequency of occurrence is a startling 5 to 10 percent, or as many as one in every ten hospitalized patients. Such hospital-acquired diseases are called

FIGURE 1-9

A few of the scientific accomplishments that fostered the development of microbiology. The foundations were laid by scientists interested in finding answers to four basic questions: (1) Does life spontaneously arise from inanimate substances? (2) How can fermentation be controlled to enhance the quality of wine and beer? (3) What are the molecular mechanisms of inheritance? and (4) What is the nature of infectious disease? The accomplishments chronicled here are only the tip of the microbiological iceberg. Modern developments are discussed throughout the book.

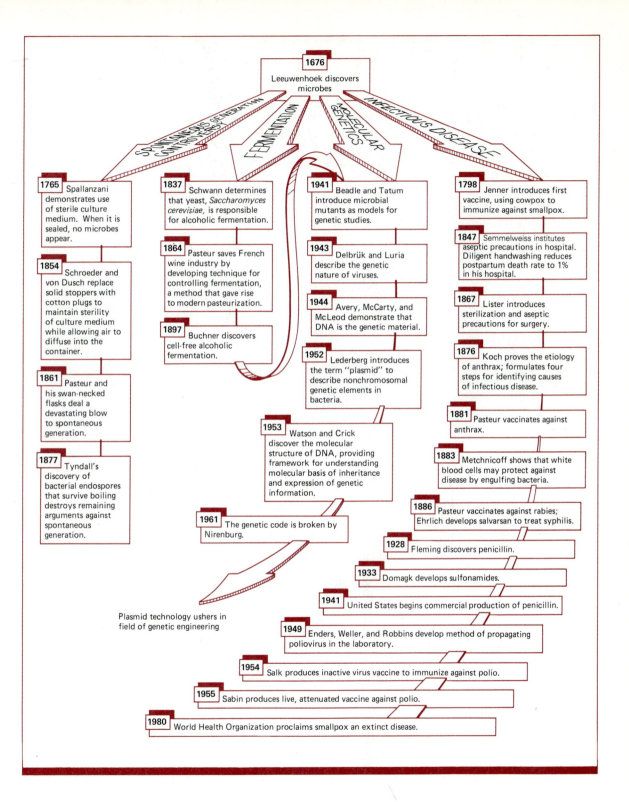

1676 Leeuwenhoek discovers microbes

SPONTANEOUS GENERATION CONTROVERSY FERMENTATION MOLECULAR GENETICS INFECTIOUS DISEASE

1765 Spallanzani demonstrates use of sterile culture medium. When it is sealed, no microbes appear.

1854 Schroeder and von Dusch replace solid stoppers with cotton plugs to maintain sterility of culture medium while allowing air to diffuse into the container.

1861 Pasteur and his swan-necked flasks deal a devastating blow to spontaneous generation.

1877 Tyndall's discovery of bacterial endospores that survive boiling destroys remaining arguments against spontaneous generation.

1837 Schwann determines that yeast, *Saccharomyces cerevisiae,* is responsible for alcoholic fermentation.

1864 Pasteur saves French wine industry by developing technique for controlling fermentation, a method that gave rise to modern pasteurization.

1897 Buchner discovers cell-free alcoholic fermentation.

1941 Beadle and Tatum introduce microbial mutants as models for genetic studies.

1943 Delbrük and Luria describe the genetic nature of viruses.

1944 Avery, McCarty, and McLeod demonstrate that DNA is the genetic material.

1952 Lederberg introduces the term "plasmid" to describe nonchromosomal genetic elements in bacteria.

1953 Watson and Crick discover the molecular structure of DNA, providing framework for understanding molecular basis of inheritance and expression of genetic information.

1961 The genetic code is broken by Nirenburg.

Plasmid technology ushers in field of genetic engineering

1798 Jenner introduces first vaccine, using cowpox to immunize against smallpox.

1847 Semmelweiss institutes aseptic precautions in hospital. Diligent handwashing reduces postpartum death rate to 1% in his hospital.

1867 Lister introduces sterilization and aseptic precautions for surgery.

1876 Koch proves the etiology of anthrax; formulates four steps for identifying causes of infectious disease.

1881 Pasteur vaccinates against anthrax.

1883 Metchnicoff shows that white blood cells may protect against disease by engulfing bacteria.

1886 Pasteur vaccinates against rabies; Ehrlich develops salvarsan to treat syphilis.

1928 Fleming discovers penicillin.

1933 Domagk develops sulfonamides.

1941 United States begins commercial production of penicillin.

1949 Enders, Weller, and Robbins develop method of propagating poliovirus in the laboratory.

1954 Salk produces inactive virus vaccine to immunize against polio.

1955 Sabin produces live, attenuated vaccine against polio.

1980 World Health Organization proclaims smallpox an extinct disease.

DEDICATED MICROBIOLOGISTS

■ Many scientists who contributed to microbiology's development were motivated by a curiosity that compelled them to investigate and understand as much as possible. Their efforts sometimes appear extreme. For example, Spallanzani, the man who introduced the use of sterile culture media, was also interested in the functions of the human body. To document the digestive processes, he filled a hollow wooden cage with various foods. He then repeatedly swallowed the cage, which was attached to a line, and removed it periodically to observe the changes that occurred in the stomach.

For these early microbiologists, such total involvement in their work occasionally seemed to transcend their instincts for self-preservation. Louis Pasteur routinely flamed his hands to avoid contaminating his cultures (he claimed it could be done safely). Dr. Jesse Lazear intentionally infected himself with yellow fever during Walter Reed's successful effort to determine how this disease was transmitted. He died before realizing the success of the project. Similarly, another curious investigator inoculated himself with contagious pus from a person with gonorrhea to better understand and combat this disease. Unfortunately he selected a patient who was coinfected with undiagnosed syphilis. The unfortunate scientist accidentally infected himself with a disease that eventually claimed his life.

Several microbiologists have made such ultimate sacrifices in their quests to understand the nature of infectious diseases. The microorganism that causes epidemic typhus killed both of the men who discovered and characterized it. Rocky Mountain spotted fever fought savagely against its investigators, killing five of them before being understood and controlled. Such laboratory-related fatalities illustrate the dedication of early microbiologists. Today equally dedicated researchers enjoy many safeguards because of the work and sacrifices of the pioneer microbiologists.

---

nosocomial infections. Control of these diseases depends on an expanded knowledge of microorganisms and their activities, as well as on the human body's response to them.

OVERVIEW

The development of microbiology over the last 100 years has enabled us to control many formerly dreaded epidemic diseases. The pioneers of microbiology provided the tools that are still the cornerstones of a science that continues to revolutionize the world.

Microbiology has practical and theoretical consequences beyond its immediate application to medicine. Microorganisms are essential to the recycling of nutrients and are fundamental elements in the food chain. They can live in association with animals and plants and are often vital to the survival of their hosts. Microbes can be controlled to produce important foods, commercially valuable chemicals, and substances that can combat disease. Thus, while they are sometimes the agents of human misery, they are also sustainers of life.

Among the many professional fields touched by microbiology, medicine is perhaps the most directly affected. Health care professionals must always consider the effects of microorganisms that inhabit the hospital, clinic, community, and bodies of both the patient and practitioner. The hands of the unaware, unprepared, or careless medical professional can disseminate potential pathogens that may claim a patient's life.

The majority of persons seeking the aid of a physician do so because of infectious disease. Although most cases are mild and self-limiting, others are life-threatening. Prevention and treatment of infectious disease depend on the collection and application of microbiological information. A thorough understanding of microorganisms and of their activities and effects on the human host is essential to quality patient care.

KEY WORDS

microorganism

microbe

microbiology

spontaneous generation

biogenesis

viable cell

sterile culture

endospore

bacterium

cyanobacterium

protozoan

alga

fungus

virus

normal flora

metabolism

biogeochemical cycle

nitrogen fixation

gene

pathogen

pasteurization

germ theory of disease

etiology

Koch's postulates

medium

colony

immunity

vaccination

antibody

chemotherapy

antibiotic

epidemiology

nosocomial infection

REVIEW QUESTIONS

1. Define (a) normal flora; (b) pathogen; (c) nosocomial infection.

2. What is the importance of microorganisms in (a) biogeochemical cycles? (b) the pharmaceutical industry? (c) the preparation of vaccines?

3. List four ways each in which microbial activities are (a) beneficial and (b) detrimental to humans.

4. Describe a prophylactic method employed to decrease the occurrence of infectious disease.

5. Identify the major contribution of each of the following men:

A. van Leeuwenhoek
L. Pasteur
R. Koch
J. Lister
E. Jenner
A. Fleming

Introduction to the Microbes

2

The fundamental unit of life for all organisms (except viruses) is the cell. Even such a highly evolved, complex organism as a human being is actually an intricately coordinated aggregate of single cells. Microorganisms are generally unicellular, each microbial cell performing all the functions required to maintain itself and propagate. These biological functions are performed by mechanisms that are strikingly similar to those of individual cells that compose the bodies of plants and animals. All living cells, in fact, share a biochemical likeness. They use many of the same chemical constituents to build their structural components, the same genetic code to transmit hereditary information from parent to offspring, and similar, often identical, metabolic processes. Like all cells, microorganisms reflect this unity of life at the cellular and subcellular levels. On the other hand, there are many distinct properties among microbes that provide morphological, physiological, and behavioral diversity.

EUCARYOTES AND PROCARYOTES

At one time, it was commonly assumed that the structural arrangements within bacteria were the same as those in the larger cells of plants and animals. In the 1950s, however, the electron microscope revealed striking differences between bacteria and the cells of other organisms. These physical differences formed the basis for classifying cells into two categories, eucaryotic cells and procaryotic cells. The most outstanding feature observed in the **eucaryotic cell** (*eu*, true; *caryote*, nucleus) is the presence of a true nucleus, surrounded by a membrane which separates it from the cytoplasmic contents of the cell (Fig. 2-1). The eucaryotic group includes the cells of plants, animals, protozoa, algae, and fungi. **Procaryotes** (*pro*, before), on the other hand, possess no

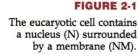
FIGURE 2-1

The eucaryotic cell contains a nucleus (N) surrounded by a membrane (NM).

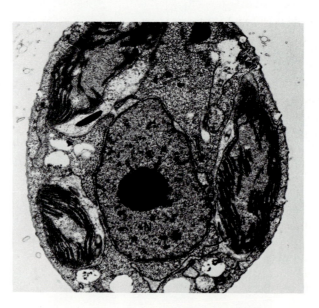

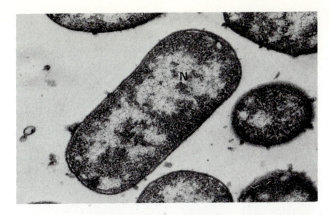

FIGURE 2·2

The procaryotic cell contains a nucleoid (N). No membrane separates this region from the rest of the cell.

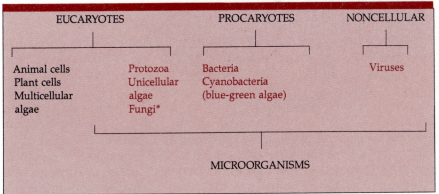

EUCARYOTES		PROCARYOTES	NONCELLULAR
Animal cells	Protozoa	Bacteria	Viruses
Plant cells	Unicellular	Cyanobacteria	
Multicellular	algae	(blue-green algae)	
algae	Fungi*		

MICROORGANISMS

*Although some fungi are large enough to view with the naked eye, most are microscopic and are considered microorganisms.

TABLE 2·1

A Spectrum of the World's Organisms. The microorganisms are printed in color. The eucaryotes include macroorganisms, such as plants and animals, and some of the microorganisms (protozoa, algae, fungi). The procaryotes, on the other hand, are all microbes. Because of their noncellular structure, viruses are considered neither eucaryotic nor procaryotic and are placed in a separate category.

membrane-bound nucleus. Instead, their hereditary information is suspended in a portion of the cytoplasm called the nucleoid of nuclear region (Fig. 2-2). The procaryotic group consists of bacteria and cyanobacteria. Eucaryotes, procaryotes, and viruses are shown in Table 2-1.

Beyond this nuclear distinction, eucaryotic and procaryotic cells exhibit other important structural and chemical differences. Many of the morphological and chemical components unique to procaryotes contribute to the ability of certain microorganisms to cause disease. Differences between eucaryotic and procaryotic cells also provide the basis for treating diseases caused by bacteria. Such practical considerations will be evident in the following discussion on cell structure and function.

THE EUCARYOTIC CELL

Eucaryotic cell organization is common to the fungi, algae, and protozoa, as well as to all higher multicellular organisms, including humans. The cell is filled with a jellylike fluid called the **cytoplasm**. The cytoplasm of eucaryotic cells is in constant motion, a phenomenon termed *cytoplasmic streaming*. Proteins and dissolved nutri-

FIGURE 2-3

The eucaryotic cell: Thin section of the fungus *Cryptococcus neoformans.* Note the nuclear membrane (NM), cell wall (CW), cell membrane (CM), mitochondrion (M), ribosomes (R), vacuole (V), and endoplasmic reticulum (ER).

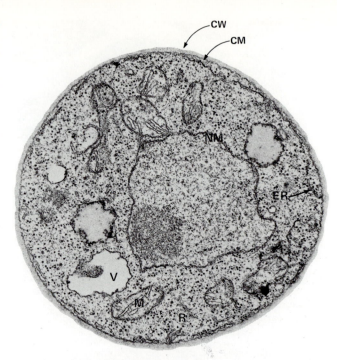

ents are transported through the cytoplasm as a result of this movement. The cytoplasm of all eucaryotic cells contains **organelles**, intracellular structures bounded by membranes that segregate their contents from the cytoplasm. Membrane-bound organelles are chemically and physically adapted to accomplish specialized tasks, such as generating usable energy. These organelles include the nucleus, mitochondria, vacuoles, and, in photosynthetic cells, the chloroplasts. In addition, the cytoplasm of eucaryotic cells is laced with an elaborate membrane network called the endoplasmic reticulum. These structures are all visible in the fungal cell shown in Figure 2-3.

The Nucleus

The **nucleus** is separated from the rest of the cell by the **nuclear membrane** and contains **chromosomes**, structures responsible for determining a cell's characteristics and for transmitting those properties to offspring. It is generally the largest organelle in the eucaryotic cell. All eucaryotic cells contain more than one chromosome (in contrast to procaryotes, which have only one chromosome). The genetic material of chromosomes is **deoxyribonucleic acid (DNA)**. In addition to chromosomes, the nucleus contains a smaller body called the *nucleolus,* which manufactures structural components used in protein synthesis. Differences in the appearance of the nucleus, especially in the chromosomes and nucleolus, help to identify certain pathogenic protozoa and are instrumental in the diagnosis of disease. For example, the protozoan that causes dysentery can be distinguished from harmless intestinal protozoa by microscopic examination.

The Endoplasmic Reticulum

The **endoplasmic reticulum** is an internal membrane network that extends throughout the cytoplasm from the outer cell membrane to the nucleus, where it is continuous with the nuclear membrane. It serves to increase the membrane surface area of the cell, enabling materials to reach all parts of the large eucaryotic cell. The endoplasmic reticulum also contains enzymes that perform essential reactions. Without cytoplasmic streaming and the endoplasmic reticulum, the eucaryotic cell would have to be much smaller (Fig. 2-4).

Ribosomes

The membranes of the endoplasmic reticulum may be covered with **ribosomes**, the sites where proteins are synthesized. Ribosomes are also found in the cytoplasm, not associated with any membrane structure. These free ribosomes produce most of the protein used within the cell, whereas the ribosomes associated with the endoplasmic reticulum synthesize proteins that are released from the cell. Both types of ribosomes are

FIGURE 2-4

The maximum size of a cell is limited by its surface-to-volume ratio. As cells get larger, they have less surface area per volume of cytoplasm. The 10 smaller cells on the right have the same combined volume as does the larger cell on the left, but with about twice the total surface area. The inner recesses of the larger cell are too far from the surface for nutrients to reach. The smaller cells don't have this problem.

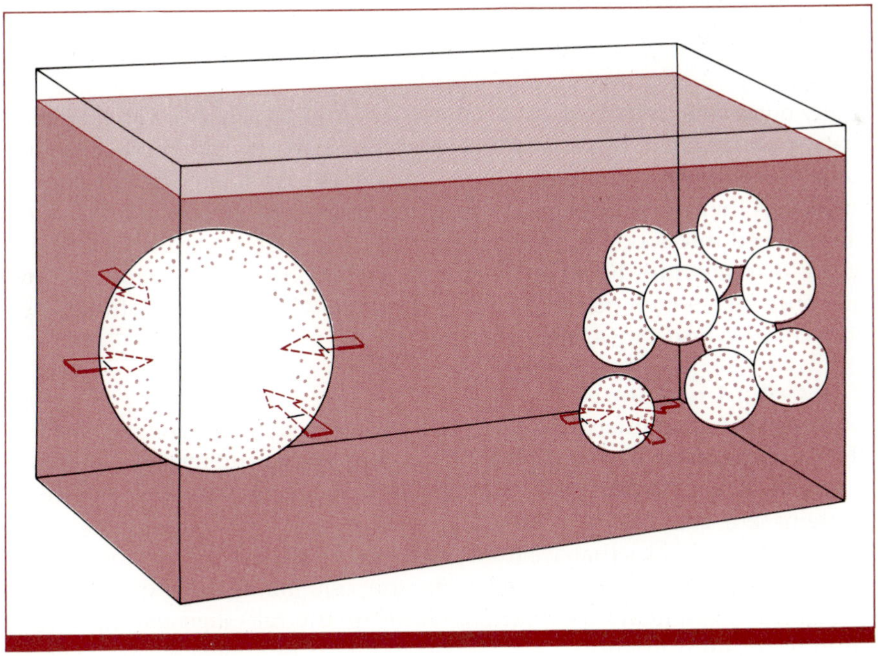

referred to as 80 S particles. This is simply an indication of their size as determined by their rate of sedimentation when centrifuged at extremely high speeds. Heavier particles sediment faster and therefore are assigned a higher S value (called the *sedimentation coefficient*). For example, 80 S particles are heavier than 70 S particles.

Mitochondria

Most of the energy for cell functions is produced in organelles called **mitochondria** (singular, mitochondrion). Energy is generated in mitochondria by **respiration**, a process that releases energy from food molecules, usually using molecular oxygen and special enzymes bound to the mitochondrion's membranes or dissolved in its fluid matrix. (The details of respiration are presented in Chapter 11.) Mitochondria contain their own DNA and ribosomes. Unlike that found in the cell's nucleus, the DNA in mitochondria is a single, small, circular piece. The mitochondrial ribosomes are somewhat smaller than the cytoplasmic ribosomes. Mitochondria use their DNA and ribosomes to manufacture many of their own protein constituents independently of the cell nucleus. Consequently they are capable of replicating within the cell and may increase in number in response to an increase in the cell's demand for energy.

Chloroplasts

Photosynthetic eucaryotic cells contain membrane-bound organelles called **chloroplasts**. Chloroplast membranes contain chlorophyll, a green pigment used for **photosynthesis**, the process by which cells convert energy of sunlight into chemical energy needed for biological functions. Like mitochondria, chloroplasts have their own DNA and ribosomes and are also self-replicating within cytoplasm.

Vacuoles

Many eucaryotic cells form intracellular sacs called **vacuoles**. For example, many protozoa feed by *phagocytosis*, engulfing particles into food vacuoles. Other sacs in the cytoplasm, called **lysosomes**, are filled with **enzymes**, molecules that promote chemical reactions in the cell. Lysosomal enzymes are released into food vacuoles where they help digest engulfed particles. Many of our white blood cells protect us by phagocytic engulfment of microbes that enter our body. The engulfed prey is enzymatically destroyed after lysosomes fuse with the phagocytic vacuole.

The Golgi Complex

Lysosomes are manufactured by a stack of flattened membrane disks called the **Golgi complex**. This cytoplasmic structure stores and modifies proteins it receives from the ribosome-coated endoplasmic reticulum. Some of these proteins are enzymes that the Golgi encases in lysosomes. Others are packaged in vacuoles that are transported to the surface membrane, where they are eliminated or secreted from the cell. Many of these secreted proteins are essential for *extracellular* (outside the cell) digestive processes used to obtain nutrients.

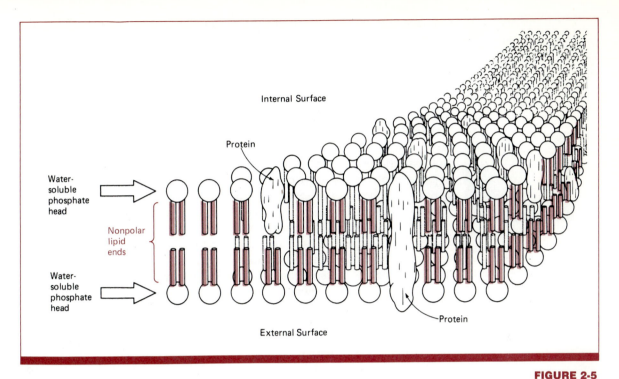

Internal Surface

Protein

Water-
soluble
phosphate
head

Nonpolar
lipid
ends

Water-
soluble
phosphate
head

External Surface

Protein

FIGURE 2-5
The cell membrane is composed of a bilayer of phospholipid molecules each having a hydrophilic (water-seeking) and hydrophobic (water-avoiding) end. The hydrophobic ends are always in the interior where they are protected from water. The proteins dispersed in the membrane have multiple functions, including transport of molecules across this barrier.

Surface Layers

Directly surrounding the cytoplasmic contents in all cells is the **cell membrane** (also called the **plasma membrane**). In addition to physically separating the cell from its environment, the cell membrane determines which molecules may travel between the external medium and the interior of the cell, a process known as **selective transport**. Because of this selectivity, cells may accumulate nutrients from the environment and dispose of toxic waste products.

Membranes of both eucaryotes and procaryotes have the same basic structure (Fig. 2-5). They are composed of a double layer of *phospholipids* (a combination of lipid and phosphate), with proteins dispersed throughout the structure. The proteins are generally responsible for mediating special membrane functions, especially transporting substances into and out of the cell.

Eucaryotic membranes typically contain compounds called *sterols.* Cholesterol, for example, is commonly found in human and animal cell membranes, whereas ergosterol is predominant in the membranes of fungi. Therefore, drugs that can selectively damage ergosterol-containing membranes without harming cholesterol-containing membranes are useful in treating human diseases caused by fungi.

The cell membrane is the outermost layer in most protozoa. In fungi and most algae, however, a **cell wall** always surrounds the membrane. The wall confers rigidity and shape to the cell. In addition, the cell wall protects

25

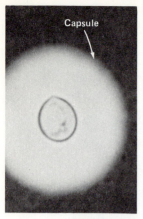

FIGURE 2-6

Cells of the yeast *Cryptococcus neoformans* have been treated with india ink in order to show the surrounding capsule.

the cell by limiting the expansion of the elastic cell membrane. Consequently, under conditions where large volumes of water tend to enter the cell, **lysis** (bursting) of the cell is less likely.

Most eucaryotic cell walls are composed of some type of *polysaccharide*, a large molecule composed of repeating sugar subunits. The polysaccharide in plants and most algae is cellulose. Most fungi, on the other hand, have cell walls made of polysaccharides other than cellulose.

In a few species, the cell wall may be surrounded by an additional layer, the **capsule**. Under most environmental conditions, its presence is not essential for the cell's survival, but under some circumstances the capsule protects the microorganism from destruction. For example, the fungus *Cryptococcus neoformans* uses its capsule to escape the defenses of an infected person. The capsule prevents access to the fungal cell surface by the protective white blood cells that normally defend the body against infection. The encapsulated fungus may then become established in the lung and invade the central nervous system, causing a fatal brain infection. The large encapsulated cells can be seen by microscopically examining stained clinical specimens (Fig. 2-6). Nonencapsulated *Cryptococcus neoformans* cells are readily engulfed by phagocytes and are incapable of causing human disease.

Structures for Motility

Eucaryotic cells move by a number of different mechanisms. Some microbes, such as an amoeba (a protozoan), move by sending out extensions of their surface, called **pseudopods**, (*pseudo*, false; *pod*, foot). The rest of the cell then flows toward the tip of the extended pseudopod. (Human white blood cells also move in this manner.) Other eucaryotic microbes depend on flexible appendages, flagella or cilia, for motility. **Flagella** (singular, flagellum) are long filaments that whip back and forth, propelling the cell

FIGURE 2-7

(*a*) The characteristic whipping motion of cilia moves the protozoan. (*b*) The structure of cilia and eucaryotic flagella is characterized by 10 pairs of tubes that traverse the length of the appendage.

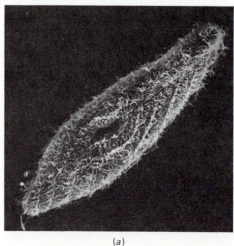

(a)

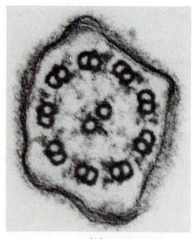

(b)

forward. **Cilia** (singular, cilium) are shorter than flagella and generally more numerous, but are otherwise identical in morphology and chemical composition. They function as tiny oars (Fig. 2-7*a*). Eucaryotic flagella and cilia contain 10 pairs of microtubules that extend down the length of the appendage (Fig. 2-7*b*). Each pair of microtubules slides back and forth next to the neighboring pair. This sliding motion bends the flagellum or cilium, producing the propulsive motion.

EUCARYOTIC MICROBES

Distinctive features separate eucaryotic microorganisms into three categories—fungi, protozoa, and algae.

Fungi

The outstanding features that characterize most **fungi** (singular, fungus) are the presence of cell walls, the lack of motility, and the absence of photosynthesis. Because fungi cannot derive energy from sunlight, they must depend on an external source of organic compounds to provide the energy necessary for survival. Although these nutrients are usually obtained from dead organisms, some fungi can utilize living tissue as a food source, often causing disease in the process.

Some fungi develop as single cells, called **yeasts** (Fig. 2-8*a*), whereas others produce networks of filaments characteristic of **molds** (Fig. 2-8*b*). A few fungi are *dimorphic* (*di*, two; *morph*, form), existing as either yeasts or molds depending on environmental conditions. For example, a pathogenic fungus may grow as a mold in the soil and as yeast cells in the infected human body. Fungi cause some of the most persistent and disfiguring diseases still prevalent throughout the world (including the United States). Common diseases caused by fungi include ringworm infections, yeast infections of the mouth and vagina, and several serious systemic diseases.

Among the most familiar fungi are the mushrooms, filamentous organisms that produce large reproductive structures. These and other fungi are discussed in more detail in Chapter 7.

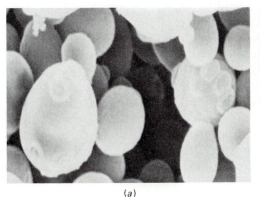 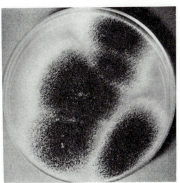

(*a*) (*b*)

FIGURE 2-8
(*a*) *Saccharomyces*, the yeast used in bread and beer production. (*b*) *Aspergillus*, a mold commonly found in air.

FIGURE 2-9
Stentor, a ciliated
protozoan.

Protozoa

These unicellular eucaryotic organisms are nonphotosynthetic and lack a cell wall. Most **protozoa** (singular, protozoan) are extremely motile, moving about by the action of pseudopods, flagella, or cilia (Fig. 2-9). Protozoa are responsible for many diseases of widespread occurrence. For example, malaria is a protozoan disease that affects more people than any other infectious disease. The agents of trypanosomiasis ("sleeping sickness") and amebic dysentery are also protozoa. This ubiquitous group of organisms will be discussed in Chapter 8.

Algae

Algae (singular, alga) are a diverse group of organisms ranging from microscopic single cells to large multicellular seaweeds (Fig. 2-10). They may be either motile or nonmotile and most possess a cell wall. With one exception, algae are photosynthetic and are easily recognized by the presence of chloroplasts within the cytoplasm. The chloroplasts enable the algae to synthesize their own organic material from carbon dioxide and water. Molecular oxygen is a by-product of photosynthesis. Through photosynthesis algae generate almost half of the earth's atmospheric oxygen.

Because they require sunlight for photosynthesis, most algae cannot infect people. The one exception is *Prototheca,* an alga that has apparently lost its photosynthetic ability and is responsible for the rare disease prototheosis. Other algae may release toxic products and contaminate the waters in which they grow. These toxins can contribute to diseases of fish, animals, and people. Algae are discussed more fully in Chapter 8.

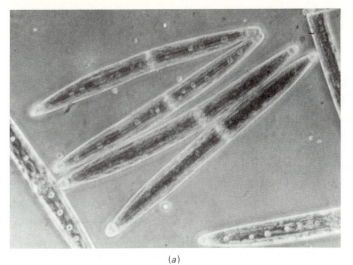

(a)

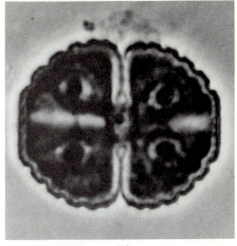

(b)

FIGURE 2-10
(a) *Cosmarium* and
(b) *Closterium* illustrate the
diversity among the algae.

29

THE
PROCARYOTIC CELL

THE PROCARYOTIC CELL

Typical procaryotic cells are approximately one–twenty-fifth the volume of eucaryotic cells. The primary feature that distinguishes procaryotes from eucaryotes, however, is the absence of a nucleus, mitochondria, and other membrane-bound organelles. Although structurally dissimilar to the eucaryotic cells, procaryotes still perform most of the same functions. They must generate usable energy, eliminate wastes, and reproduce. The manner in which some procaryotic cells achieve these ends, however, is unique to this category of microorganisms. For example, some procaryotes can use chemical energy extracted from inorganic nitrogen or hydrogen sulfide, a process no eucaryote can duplicate.

The procaryotic cell is structurally less complex than the eucaryotic cell. All procaryotes possess cytoplasm, a nucleoid, a cell membrane, and ribosomes (Fig. 2-11). A number of additional characteristics are shared by the majority of procaryotic cells. Although briefly described here, the details of these procaryotic structures will be discussed in Chapter 4.

Nucleoid The nuclear region of procaryotes is not separated from the

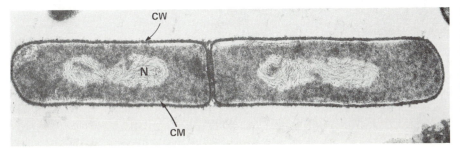

FIGURE 2-11
The procaryotic cell: Thin section of the bacterium *Bacillus subtilis*. Note cell wall (CW), cytoplasmic membrane (CM), and nucleoid (N). The cytoplasm is filled with ribosomes, which cause it to stain dark.

cytoplasm by a nuclear membrane. It is seen in the electron microscope as an area lighter than the cytoplasmic contents. The **nucleoid** contains a single circular chromosome of DNA.

Cell Membrane   The procaryote's cell membrane must assume many of the biological functions performed by membrane-bound organelles in eucaryotic cells. For example, enzymes for generating energy by respiration are integrated into the cell membrane since mitochondria are absent. Some procaryotes possess **mesosomes**, invaginations (infoldings) of the cell membrane into the cytoplasm. These increase the membrane surface area without a corresponding increase in cell size, thereby allowing the cell to absorb more nutrients and to better eliminate wastes. Mesosomes may also function in photosynthesis, respiration, and cell division.

Ribosomes   As with eucaryotes, all protein synthesis in procaryotes takes place at the ribosomes. Ribosomes are distributed throughout the cytoplasm; they are also attached to membrane surfaces. Procaryotic ribosomes differ from their eucaryotic counterparts in at least two ways: procaryotic ribosomes are smaller, having a sedimentation coefficient of 70 S and their function can be inhibited by chemicals that do not affect the 80 S ribosomes of eucaryotes.

Cell Wall   With few exceptions, the procaryotic cell is surrounded by a cell wall. Although its function is similar to that of the eucaryotic cell wall, its chemical composition is very different and is one of the distinguishing features between procaryotes and eucaryotes. Rather than the polysaccharides of eucaryotic cell walls, the procaryotic cell wall contains **peptidoglycan**, a substance found nowhere else in nature.

TABLE 2-2

Comparison of Eucaryotes and Procaryotes

	Eucaryotes (True nucleus)	Procaryotes (Nucleoid)
Nuclear Region		
Membrane-bound	Yes	No
Nucleolus	Yes	No
Number of chromosomes	>1	1
Cytoplasm		
Cytoplasmic streaming	Yes	No
Membrane-bound organelles (mitochondria, chloroplasts, vacuoles)		
Golgi complex	Yes	No
Endoplasmic reticulum	Yes	No
Mesosomes	No	Yes
Ribosomes	80 S	70 S
Surface Layers		
Cytoplasmic membrane	Yes	Yes
Sterols in membrane	Yes	Rare or absent
Peptidoglycan in cell wall	No	Yes
Flagella, if present	Contain microtubules	Composed of flagellin
Size of Typical Cell, diameter	2–25 μm	0.3–2 μm

Flagella Many procaryotes possess flagella which are structurally and chemically different from those of eucaryotes. They are much thinner and are composed of a *flagellin*, a noncontractile protein. Procaryotic flagella propel the cell by rotating rather than whipping.

The features that distinguish procaryotic and eucaryotic cells are summarized in Table 2-2.

PROCARYOTIC MICROBES

Procaryotic cells are divided into two major groups, cyanobacteria and bacteria. **Cyanobacteria** (formerly called blue-green algae) are procaryotes that perform oxygen-evolving photosynthesis in a manner similar to eucaryotic algae and plants. Some cyanobacteria transform molecular nitrogen (N_2) into a form usable by plants, thereby introducing this essential nutrient into the food chain. Many of these microbes grow as filaments of cells, as illustrated by *Anabaena* in Figure 2-12. Although they are not pathogenic, some cyanobacteria grow in close association with a living host (see color box, p. 32).

All procaryotes that are not cyanobacteria are classified as **bacteria**. They are among the earth's most abundant and diverse organisms. Because of the metabolic diversity among bacteria, these organisms are found in virtually every habitat, even some formerly believed to be incapable of supporting life. Bacteria grow in such extreme environments as thermal vents at 250°C in the ocean floor, acid runoff from mines, and many regions devoid of oxygen. A few bacteria are photosynthetic, but

FIGURE 2-12

The cyanobacterium *Anabaena* is typically arranged in chains of cells. Nitrogen fixation occurs in specialized structures called heterocysts (arrow).

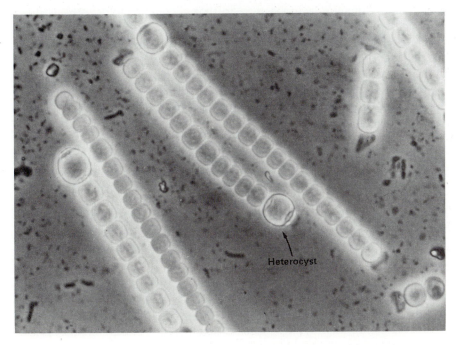

Heterocyst

THE GREEN POLAR BEARS

■ For several years visitors to a few zoos in the United States and Europe have observed an unusual group of animals—green polar bears. Although the bears were white in their arctic habitat, in captivity they acquired a greenish cast to their coats (see color plate). The pigment is more noticeable in the warmer months and was at first believed to be algae growing on the hair surfaces. Microscopic inspection of the bears' hairs, however, revealed an unexpected finding.

The green color is indeed due to photosynthetic microbes, but inside the hair, not on its surface. The bears' outer "guard" hairs are hollow, and it is in this hollow shaft that the organisms have taken up residence. Microscopic examination and analysis of the procaryote's pigments revealed the photosynthetic freeloader to belong to the Chroococcales, a family of cyanobacteria. This "blue-green alga" apparently gains access to the hollow space when the end of the hair breaks off, exposing the inner chamber.

The hairs of polar bears are especially susceptible to such colonization because they lack pigment, permitting the penetration of light into the hollow compartment, which also provides protection for the microbe. Such a relationship has been reported in no other animal and only in captive specimens of the polar bear. In the wild the temperatures are too cold to encourage growth of the cyanobacteria.

Aside from cosmetic considerations, the microbes have no detrimental effects on the colonized bears. Zoo officials are nonetheless eager to display white polar bears and are experimenting with different environmental factors to discourage the microbe's growth. San Diego Zoo scientists have successfully controlled the problem by increasing water salinity.

---

these differ from the cyanobacteria in that bacterial photosynthesis does not release oxygen. Most bacteria, however, are nonphotosynthetic. Some of these require only simple inorganic chemicals for growth. Most derive nutrients by decomposing organic materials. Some bacteria are so efficient at this that the low concentrations of nutrients in distilled water can support their growth. A representative bacterium is shown in Figure 2-13.

Bacteria are currently divided into major groups based on morphological and physiological properties. Many of these properties determine the ability of an organism to cause disease. Bacteria have caused devastating epidemics of human suffering such as the plague (black death) that killed one-quarter of the population of Europe in the fourteenth century. Although bacteria continue to be a major cause of disease, their many other biological activities should not be overlooked. Because of the importance of bacteria, much of this book has been devoted to discussing their biology, their impact on people, and methods for controlling their activities.

ARCHAEBACTERIA

Scientists are discovering a growing number of cells possessing unique characteristics that upset the traditional eucaryote-procaryote dichotomy. Although classified as procaryotes, these cells, called archaebacteria, appear to be fundamentally different from typical bacteria or cyanobacteria. In fact, they represent a cell type that seems to be

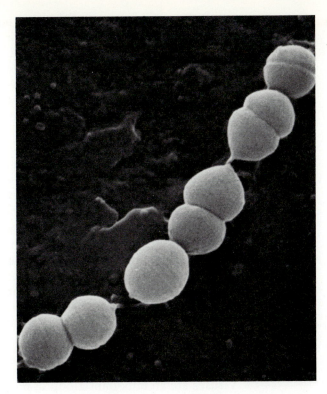

neither eucaryotic nor eubacterial (true bacteria). The **archaebacteria** have the following unique combination of traits:

Procaryotic Traits

- They are about 1 micrometer (μm) in diameter, the size of typical procaryotes.

- They lack membrane-bound organelles.

- They have nuclear bodies (nucleoids) rather than true, membrane-bound nuclei.

- Their ribosomes are 70 S, the size of those found in typical procaryotes.

Eucaryotic Traits

- Their cell walls completely lack peptidoglycan.

- Their protein synthesis machinery is sensitive to inhibitors that typically affect only eucaryotes and is resistant to many inhibitors that affect procaryotes.

- Some of their proteins, pigments, and biochemical processes closely resemble those found in eucaryotic cells.

Archaebacteria also possess unique traits found in neither eucaryotes nor procaryotes. For example, their membranes contain lipids with ether, a chemical uncharacteristic of either of the traditional cell types.

By 1983 archaebacteria had been found among four groups of bacteria, all of which thrive in environments that would be extremely hostile to other forms of life. These four groups are the extreme halophiles (salt lovers), two genera of thermoacidophiles (heat and acid lovers), and methane-producing bacteria.

This atypical group of microbes reveals new information about the early history of life. Conventional evolutionary theory proposes that the ancestor of all cells was an ancient procaryote that generated two lines of descent, the procaryotes and eucaryotes. The archaebacteria may be a living example of the "missing link" in the development of eucaryotes from procaryotes. Most investigators, however, believe that the universal ancestor was a *progenote*, a cell much simpler than the simplest of present-day procaryotes. From this progenote three separate lines of cellular evolution developed: archaebacteria, procaryotes, and eucaryotes. Each is believed to be the product of its own branch of the evolutionary tree. If this is true, these unique cells contain information that will provide a new understanding of how life originated and evolved on this planet.

VIRUSES

Viruses are an important group of microbes that are neither procaryotic nor eucaryotic because they have no cell structure. Some are composed solely of nucleic acid surrounded by a protective protein shell. They have no cytoplasm, no internal organelles, and no capacity to synthesize their own protein or to produce energy. Viruses are **obligate intracellular parasites**, organisms that require the biological machinery of a host cell for virtually every function needed for their reproduction and survival.

Viruses cause infectious diseases of humans, plants, and animals. Even bacteria are susceptible to virus infections. Viruses are so small they can only be observed with an electron microscope (Fig. 2-14). They are the subject of Chapter 9.

MICROBIAL DIVERSITY AND ANTIMICROBIAL THERAPY

The biological differences between procaryotes and eucaryotes provide a basis for treating diseases caused by bacteria. An essential property of chemical agents used for antimicrobial therapy inside the human body is **selective toxicity**, the ability to destroy invading procaryotic cells while leaving the eucaryotic host cells unharmed. Drugs that selectively interfere with processes that occur solely in procaryotes may be able to subdue a bacterial pathogen while having no effect on the eucaryotic cells of the human body. Such antimicrobial drugs are powerful weapons against infectious disease.

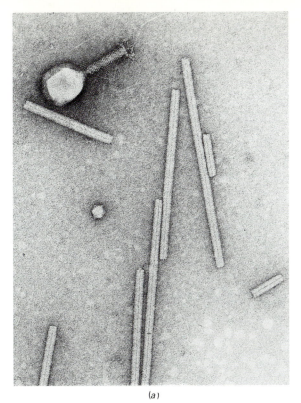

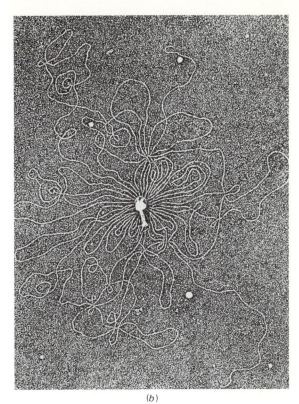

(a)

(b)

Selective toxicity is perhaps best represented by penicillins and cepha-losporins. These antibiotics prevent the synthesis of the peptidoglycan layer of the bacterial cell wall, a structure with no chemical equivalent in the eucaryotic human cell. Without their cell walls, most bacteria are so fragile they have little chance of survival. Other therapeutically effective antibacterial agents selectively inhibit procaryotic ribosomes (see color box).

Most antibacterial agents have no effect on eucaryotic pathogens. In addition, since eucaryotic pathogens, such as fungi and protozoa, have the same basic cell structure as human cells, few drugs can kill or inhibit them without having adverse effects on human cells. The few drugs that are used to treat diseases caused by fungi and protozoa possess some toxicity to the human host as well. Some have very serious side effects and are employed only in life-threatening situations. Diseases caused by viruses are even more difficult to treat. Once these obligate intracellular parasites become an integral part of the infected cell, it is nearly impossible to therapeutically destroy them without also killing the host cells.

Regardless of the microbe, however, many properties distinguish it from human cells. A more thorough knowledge of the structure and function of each organism increases the likelihood of developing safer, more effective drugs against a wider range of pathogens. Herein lies one of the great promises in the continuing battle against infectious disease.

FIGURE 2-14

(a) The three different virus types in this electron micrograph vary in shape but are remarkably similar in their chemical composition. Each of these viruses is composed of a protein coat surrounding a piece of genetic material. (b) A single disrupted virus that has released its nucleic acid from its protein coat.

MITOCHONDRIA: ANCIENT BACTERIA?

■ A number of remarkable similarities exist between procaryotic cells and the mitochondria of eucaryotes. Mitochondria, for example, are approximately the same size as bacteria and contain a circular piece of DNA similar to the bacterial chromosome. Mitochondria possess their own protein synthesis machinery, including small ribosomes that appear more procaryotic than eucaryotic. Mitochondria reproduce themselves by a process resembling bacterial cell division rather than being produced by the eucaryotic cell itself. These similarities suggest that the mitochondria may have developed from an ancient procaryotic ancestor which entered and lived within the cytoplasm of a larger cell, establishing an *endosymbiotic* relationship. The original procaryote was probably very efficient at generating usable energy by respiration. After millions of years of such coexistence, the procaryote lost the capacity for an independent existence and developed into the organelle responsible for supplying the cell with energy.

The similarity between the procaryotes and mitochondria has far more importance than as simply an interesting theory of evolution. It may explain why human cells are damaged by some antibacterial drugs. These adverse side effects may be due to inhibition of mitochondrial processes that are similar to their procaryotic counterparts. For example, drugs such as tetracycline and erythromycin, which interfere with bacterial protein synthesis, also inhibit mitochondrial ribosomes, although to a lesser degree. Similarities between mitochondria and bacteria impose safety limitations that must be evaluated when choosing the most effective drug for treatment.

OVERVIEW

All cells perform similar processes in order to survive and reproduce. The specialized cellular structures that perform many of these processes, however, provide the fundamental differences between eucaryotic and procaryotic cells. These two basic cell types are distinguished by their microscopic anatomy. The eucaryotic cell possesses a cell membrane and several membrane-bound organelles, including the nucleus, mitochondria, vacuoles, and, in photosynthetic organisms, chloroplasts. Eucaryotes also contain internal membrane systems such as the endoplasmic reticulum. In addition, eucaryotes possess 80 S ribosomes, where protein synthesis occurs. In some but not all eucaryotes, the cell membrane is surrounded by a cell wall or capsule. Some are capable of moving about by forming pseudopods or by the action of flagella or cilia. Algae, fungi, and protozoa are eucaryotic microorganisms. The cells of plants and animals are also eucaryotic.

Bacteria and cyanobacteria, the procaryotes, also possess cell membranes, but lack membrane-bound organelles. They generally possess a cell wall, containing peptidoglycan, exterior to the cell membrane. Procaryotic ribosomes are 70 S, smaller than those of eucaryotes. Rather than a nucleus, the procaryotic nuclear region is the nucleoid, a single circular chromosome suspended in the cytoplasm without being surrounded by a nuclear membrane.

Viruses are noncellular microbes so small that they are observable only with an electron microscope. They are obligate intracellular parasites composed of nucleic acid surrounded by a protective protein coat.

Structural and biochemical differences among eucaryotes, procaryotes, and viruses are important aids for identifying these microbes in the laboratory. These unique features are also the targets in the search for effective antimicrobial drugs that possess high selective toxicity, the ability to interfere with the growth of pathogens while causing few adverse side effects for the host.

KEY WORDS

eucaryote

procaryote

cytoplasm

organelle

nucleus

nuclear membrane

chromosome

DNA (deoxyribonucleic acid)

endoplasmic reticulum

ribosome

mitochondrion

respiration

chloroplast

photosynthesis

vacuole

lysosome

enzyme

Golgi complex

cell membrane

selective transport

cell wall

lysis

capsule

pseudopod

flagellum

cilium

fungus

yeast

mold

protozoan

alga

nucleoid

mesosome

peptidoglycan

cyanobacterium

bacterium

archaebacterium

virus

obligate intracellular parasite

selective toxicity

REVIEW QUESTIONS

1. List and describe the structure and functions of the following parts of a eucaryotic cell:
   (a) Nucleus
   (b) Mitochondria
   (c) Endoplasmic reticulum
   (d) Chloroplasts

2. Compare the functions of the cell membranes of eucaryotic and procaryotic cells.

3. List the identifying characteristics of the fungi, algae, and protozoa.

4. Describe three structural and three chemical differences between eucaryotes and procaryotes.

5. What characteristics of viruses make them obligate intracellular parasites?

6. Discuss one mechanism of selective toxicity. Why is this an important consideration of chemotherapy?

INTRODUCTION
TO THE MICROBES

Investigating the Microbial World

3

Because microorganisms are too small to be seen with the naked eye, the microscope has become the tool most often associated with microbiology. Microscopic observation, however, is only one method of detecting these organisms. Formation of **colonies** (visible aggregates of microorganisms) on laboratory culture media, for example, allows observation and characterization of microbes without the aid of any instrument. Other laboratory techniques enable us to indirectly detect and characterize microorganisms by demonstrating evidence of microbial biochemical activity or a specific immunological reaction. Many laboratory procedures are needed to identify species of yeast or bacteria because their microscopic anatomy alone provides very little detail. Nonetheless, the microscope is usually the first tool the microbiologist uses in the systematic characterization of a microbe. Without this instrument a microbe's anatomy, indeed the microbes themselves, would remain hidden.

MICROSCOPIC OBSERVATIONS

The earliest microscopes possessed a single lens, usually in a hand-held instrument. These simple microscopes magnified objects 200 to 300 times, thereby providing the opportunity to view objects with diameters below the lower limits of human vision (see color box). With these instruments, however, only the gross form of cells could be discerned. Modern instruments are **compound microscopes**, that is, they contain at least two lenses. The second lens magnifies the image from the first lens, and consequently the size of the observed image depends on the magnification of both lenses. For example, combining a 10× lens with a 45× lens produces an image 450 times larger

THE FATHER OF MICROSCOPY

■ Although Antoni van Leeuwenhoek is remembered as the first microbiologist, he was by trade a cloth merchant. More important, he was a man devoted to his hobby—his passion—of grinding fine lenses. He did not invent his microscopes with the intention of discovering the microbial world, but rather to solve such mysteries as "the cause of the pungency of pepper on the tongue." His mastery at grinding lenses and of microscopic technique allowed him to describe bacteria, protozoa, and other microorganisms in such detail that we can identify many of his microbes from his recorded information.

Focusing the instrument was the most difficult aspect of using his microscope, and Leeuwenhoek found it easier to build a new microscope than to change specimens. Consequently, he produced hundreds of scopes, many of which still survive in museums, some containing the original specimen placed there by Leeuwenhoek.

Leeuwenhoek developed a very effective technique for using his instruments. Modern microbiologists are still not sure how he could have seen what he accurately described at the magnification attainable with his microscopes. It was proposed in 1976, 300 years after his initial observations, that Leeuwenhoek may have used an indirect source of illumination, creating an effect similar to that of the darkfield microscope.

than the object being examined. Objects too small to be seen with a single lens are often observable through the compound microscope.

Theoretically, the size of an image can be increased by additional lenses. **Magnification**, the process of enlarging an object's image, is only useful, however, if the details of the enlarged image are clearly visible. *Effective magnification* depends on the microscope's **resolving power**, the ability to distinguish two adjacent objects as separate and distinct images rather than as a single blurred image. The lens system of the human eye, for example, has a resolving power of approximately 0.2 mm. This means that the eye cannot see anything that is smaller than 0.2 mm (the approximate width of a point of a needle). Two objects, each of which is large enough to be visible, will appear to be touching each other if the space separating them is less than 0.2 mm. Since the goal of microscopy is to increase the degree of detail that can be observed, the microscope must maximize both magnification and resolving power.

The greatest *useful magnification* of a microscope is that which makes the smallest visible particles clearly resolvable. The maximum useful magnification is limited by the resolving power of the microscope, since increases in magnification without corresponding increases in resolving power will only produce a larger blurred image.

The resolving power of a microscope is largely dependent on the wavelength of the beam used for illumination. Shorter wavelengths give better resolution. Ultraviolet light or high-voltage electron beams are two sources of illumination with wavelengths shorter than those of visible light. Both are used in the construction of high-resolution instruments (Fig. 3-1). In addition to wavelength, the resolving power of the light microscope is largely dependent on the optical quality of the lenses themselves.

FIGURE 3-1

Comparison of the images obtained from the (a) light and (b) electron microscopes. Both pictures are at the same magnification (1000×). This magnification is the upper limit for the light microscope but is very low magnification for the electron microscope because of its greater resolving power.

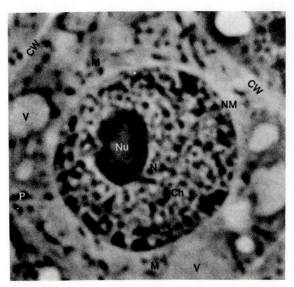

(a)

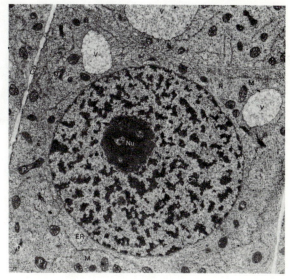

(b)

The Brightfield Microscope

The most commonly used microscope for general laboratory observations is the standard **brightfield microscope** (Fig. 3-2a). This instrument uses a direct light source (light shines directly on the specimen), and the entire field of view is illuminated. The final magnified image is produced by a

FIGURE 3-2

(a) The brightfield microscope: (1) The lamp is the source of illumination; (2) the iris diaphragm controls the amount of light entering the objective lens; (3) the condenser focuses the available light on the specimen; (4) the objective lenses magnify the specimen; (5) the ocular lens further magnifies the image produced by the objective lens, usually about 10×; (6) the focusing knobs provide coarse and fine adjustment.
(b) Spherical bacteria as seen with a brightfield microscope. Only details of shape and arrangement are observable.

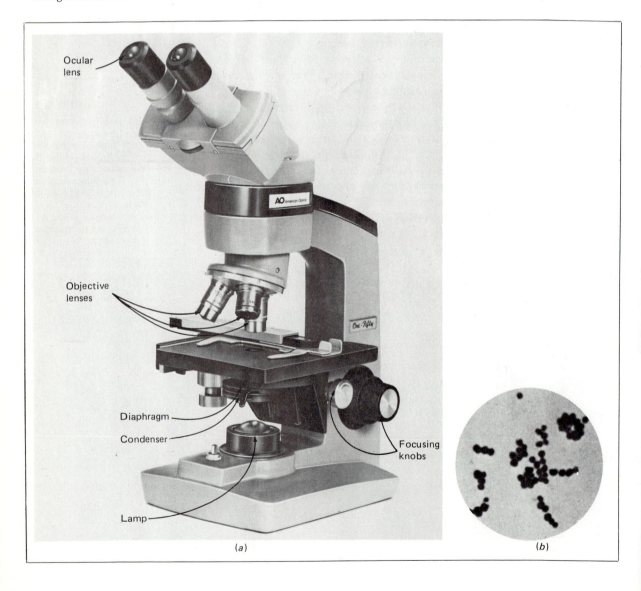

Ocular lens

Objective lenses

Diaphragm

Condenser

Focusing knobs

Lamp

(a)

(b)

series of two lenses, the **objective lens** located close to the object being examined (the specimen) and the **ocular lens** (eyepiece) through which the image is viewed. Most brightfield microscopes are equipped with at least three different objective lenses on a rotating nose piece. Typically these provide magnification powers of 10, 45, and 100X. The enlarged image produced by the objective lens is further magnified by the ocular lens, usually an additional tenfold. Thus the greatest magnification achievable by most standard brightfield microscopes is 1000X (10 x 100). The resolution of such a scope is 0.00027 mm, about one-thousandth the thickness of a human hair. Most microorganisms, except viruses and some very thin bacteria, can be clearly resolved by the brightfield microscope (Fig. 3-2*b*).

The brightfield microscope is most often used as an aid in counting cells and in the laboratory identification of microorganisms. Microscopic observations of cell morphology and reactions of cellular constituents to various stains are discussed later in this chapter.

Some unstained living cells large enough to be seen in the microscope remain invisible if there is not enough contrast between the specimen and its background. Therefore, contrast is as critical a factor as resolution when making microscopic observations. Although contrast can be artificially increased by staining the specimen, these procedures usually kill the specimen, making it impossible to view the activities of living cells. To view living unstained cells, special optical devices that increase specimen contrast may be attached to standard brightfield microscopes. Two such commonly used instruments are the darkfield microscope and the phase-contrast microscope.

The Darkfield Microscope

The **darkfield microscope** is equipped with a condenser that blocks light from entering the objective lens directly from the light source. The result is a completely dark field of view. Light that passes through the specimen is scattered, and some of it is deflected into the objective lens (Fig. 3-3*a*). The specimen thus appears brightly illuminated on a dark background. A bright object in the dark is much easier to see than a dark object in bright light. The darkfield microscope can therefore detect organisms that are too thin to be seen in the brightfield microscope, for example, the organism that causes syphilis (Fig. 3-3*b*). Because it cannot detect color, the darkfield microscope is not used for examining stained specimens. It is excellent, however, for viewing living unstained cells.

The Phase-Contrast Microscope

Living cells in liquid are best viewed with a **phase-contrast microscope**. This instrument highlights details between structures with dissimilar densities, such as a cell and its surrounding medium. The optics of the phase-contrast microscope detect these density differences and translate them into patterns of shadows and light (Fig. 3-4). Different organelles within a cell also have different densities. The phase-contrast microscope therefore has the advantage of revealing subcellular anatomy without

FIGURE 3-3

Darkfield microscopy:
(a) Because of the darkfield
ring in the condenser, the
only light that reaches the
observer's eye is that
scattered by the specimen.
(b) *Treponema pallidum*, too
slender to observe with the
brightfield microscope, is
easily seen with the
darkfield microscope.

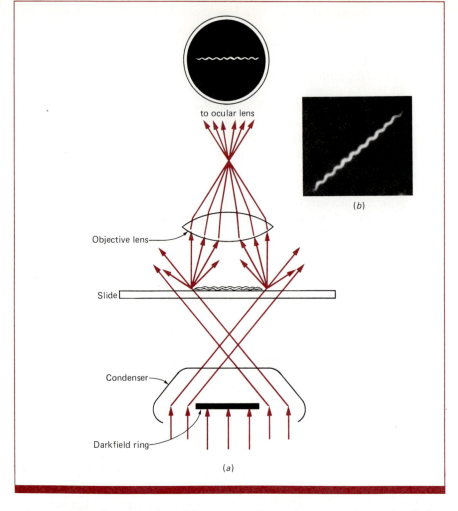

INVESTIGATING THE
MICROBIAL WORLD

using structural stains. In addition, motility, phagocytosis, and cellular activities may be more clearly observed than with the standard brightfield microscope.

Both the darkfield and phase-contrast microscopes have limitations. Neither microscope provides greater resolution or magnification than the standard brightfield instrument. More important, however, neither instrument can be used for determining the color of stained specimens. Because important diagnostic information is obtained from ascertaining the way a microorganism reacts to certain stains, neither darkfield nor phase-contrast can replace the standard brightfield microscope in the clinical or research laboratory.

The Fluorescence Microscope

This instrument uses ultraviolet (uv) light as the source of illumination. Although invisible to the human eye, ultraviolet light causes some microbes to fluoresce colorfully against a dark background. The most

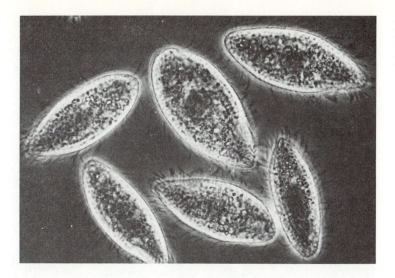

FIGURE 3-4

Phase-contrast microscopy: Observation of these living protozoa by phase-contrast microscopy reveals organelles as well as movement by the beating of cilia.

widespread uses of fluorescence microscopy, however, employ dyes that glow when exposed to ultraviolet light. For example, *Mycobacterium tuberculosis* in sputum can be identified by its characteristic yellow fluorescence after staining with the fluorescent dye auramine. Since few other bacteria contain the waxy surface layer that combines with this dye, this information is useful in the diagnosis of tuberculosis.

Fluorescence microscopy also aids in the rapid identification of other microorganisms. This is accomplished by the **fluorescent antibody technique**. A fluorescent dye is chemically attached to an antibody molecule which can recognize and combine with a specific infectious agent. For example, fluorescent-labeled antibodies can be mixed with a sample of tissue taken from a patient suspected of being infected with this microbe. If the microbe is present in the patient's tissue, the fluorescent antibodies will attach to it. This tissue will thus appear to glow when exposed to ultraviolet light (Fig. 3-5). If the microorganism is absent, the antibodies fail to coat the tissue and the specimen appears dark. The fluorescent antibody technique is one of the most useful techniques available for the rapid diagnosis of Legionnaires' disease, syphilis, rabies, and herpesvirus infections.

The Electron Microscope

The electron microscope uses a beam of electrons instead of light. Because a beam of electrons has a much shorter wavelength than light, electron microscopes dramatically increase resolving power and therefore the upper limit of effective magnification. Magnetic lenses focus the electron beam much the same way that glass lenses focus light. The image formed by the electrons cannot be viewed directly and is projected onto a viewing screen (similar to a television screen) or onto photographic film to be permanently recorded as a photograph.

There are two types of electron microscopes: (1) the **transmission electron microscope** and (2) the **scanning electron microscope**.

FIGURE 3-5

Fluorescence microscopy: The organism *Legionella pneumophila* combines with a specific antibody that has been labeled with a fluorescent stain. The organism becomes coated with the labeled antibody and fluoresces when exposed to ultraviolet light. Microbes other than *L. pneumophila* fail to react with the specific antibody, remain uncoated, and do not fluoresce.

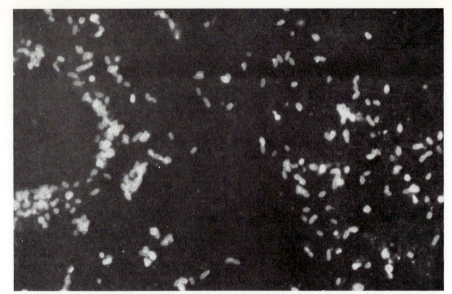

46

INVESTIGATING THE
MICROBIAL WORLD

The Transmission Electron Microscope (TEM)   The useful magnification of the transmission electron microscope is 1000 times greater than that possible with conventional brightfield microscopes. The TEM can produce an image 1 million times the actual size. The introduction of the TEM allowed the first observations of viruses and other particles too small to be detected by light microscopes. Similarly, the fine details of intracellular structures are revealed, providing new understanding of the fundamental life processes of all cells.

FIGURE 3-6

Transmission electron microscopy: The bacterium has been stained to increase contrast between specimen and background.

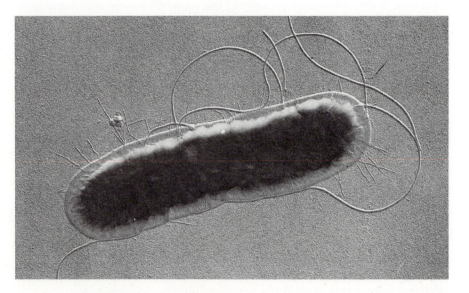

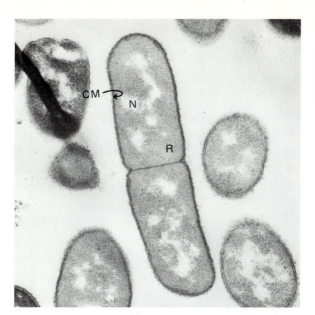

FIGURE 3-7

Transmission electron microscopy: Thin section of bacterium reveals details of cell ultrastructure such as ribosomes (R), nucleoid (N), and cell membrane (CM).

Most specimens appear transparent in the TEM and are virtually invisible unless stained to increase contrast between the specimen and the supporting background (Fig. 3-6). Heavy metals that reflect electrons, most commonly tungsten, uranium, or lead, are used for this purpose. In addition, because electrons have poor penetrating power, internal cellular structures can best be viewed in a thin slice cut from the cell. Using a special cutting instrument called an *ultramicrotome,* a single bacterial cell can be sliced into several hundred sections (Fig. 3-7). Although some recently developed techniques reduce the need for sectioning, the ultramicrotome is still standard equipment in TEM facilities.

The Scanning Electron Microscope (SEM) The more recently developed scanning electron microscope reveals surface details of whole specimens in striking three-dimensional relief (Fig. 3-8). This instrument uses a moving electron beam to scan the specimen's surface, generating pictures with depth. Although only the surface can be observed, scanning electron microscopy has revealed unique information about the form and function of many microbes.

Neither SEM nor TEM allows the observation of living specimens. Furthermore, the harsh treatment of specimens in preparing them for viewing often introduces **artifacts** (artificial features not possessed by the living organism) such as shrinkage and distortion of the specimen. Interpretation of electron micrographs, therefore, requires skilled individuals capable of distinguishing artifacts from natural cellular structures.

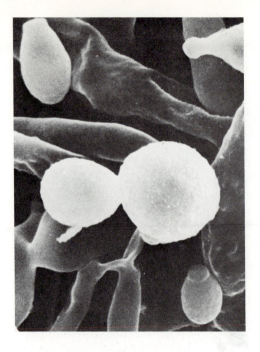

A comparison of the properties of each of the five commonly used microscopes is provided in Table 3-1.

Microscopic Dimensions

Since the average bacterial cell is approximately 0.001 mm in diameter, the dimensions of microorganisms and of their component parts are more conveniently expressed in units smaller than millimeters. The **micrometer** (μm) is one-millionth (10^{-6}) of a meter (one-thousandth of a millimeter). This allows us to express the diameter of the average bacterial cell as 1 μm. For even smaller structures, such as viruses, the nanometer and angstrom are often employed. The **nanometer** (nm) is one-billionth (10^{-9}) of a meter (one-millionth of a millimeter). The **angstrom** is one-tenth of a nanometer, or one-ten-billionth (10^{-10}) of a meter. For ease of reference, these units are summarized in Table 3-2.

Tremendous size diversity exists among the microorganisms (Fig. 3-9). For example, the 27-nm poliovirus, one of the smallest viruses, is about one-thousandth the diameter of the average eucaryotic cell. This size difference is similar to the height difference between a housefly and an elephant. Size is one of the properties of a microorganism useful for determining its identity in the laboratory.

MICROBIOLOGIC STAINS

Viewed in the microscope, most bacteria are transparent and difficult to see. Specimens are therefore routinely stained to increase visibility and to reveal additional information to help

TABLE 3-1
Comparison of the Five Types of Microscopes

Microscope	Maximum Magnification	Common Uses	Advantages	Disadvantages
Standard brightfield	1000×	Observing stained specimens Counting microbes	Easy to use Relatively inexpensive Allows staining reactions to be interpreted Readily available	Lack of contrast Inability to resolve viruses and some very thin bacteria Most intracellular structures must be stained to be observable Introduction of artifacts during staining procedures
Darkfield	1000×	Detecting unstained microbes not easily observable with brightfield microscopy	Allows living specimens to be viewed Reduces artifacts	Inability to evaluate staining reactions Subcellular detail not readily observable
Phase-contrast	1000×	Observing living unstained cells Observing intracellular structures	Allows observation of intracellular biological activity Enhances subcellular details	Inability to evaluate staining reactions
Fluorescence	1000×	Identifying microorganisms Detecting specific infectious agents in tissue Detecting immunological reactions	Allows very rapid identification of infectious agents	Limited to viewing specimens that naturally fluoresce or that are stained with a fluorescent dye
Electron	1,000,000×	Observing fine structure of cells and viruses Diagnosing certain viral diseases and cancers Detecting certain giant molecules	Permits viewing of objects too small to be detected by light microscopy	Specimen must be killed Harsh preparation procedures increase the probability of artifacts in the specimen Very expensive

TABLE 3-2
Units of Measurement

Unit	Fraction of a Meter	Examples of Objects Measured with These Units
Centimeter (cm)	10^{-2}	Whole chicken egg (~7 cm long)
Millimeter (mm)	10^{-3}	Flea (~1 mm long)
Micrometer (μm)	10^{-6}	Typical bacterium (1 μm diameter)
Nanometer (nm)	10^{-9}	Poliovirus (27 nm)
Angstrom (Å)	10^{-10}	Cell membrane thickness (10 Å)

identify microbes. There are two categories of staining procedures, simple and differential.

Simple stains employ a single dye, most commonly methylene blue, crystal violet, or fuchsin. Most cells and most structures within each cell

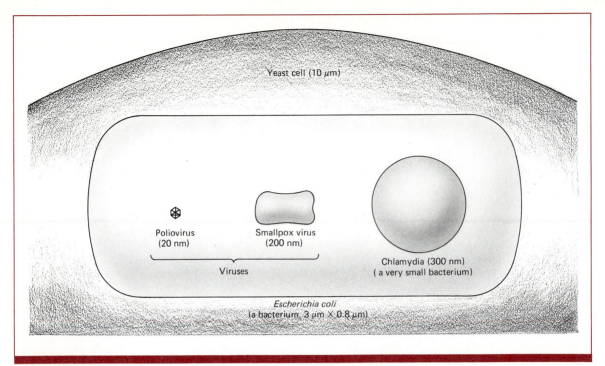

Yeast cell (10 μm)

Poliovirus
(20 nm)

Smallpox virus
(200 nm)

Viruses

Chlamydia (300 nm)
(a very small bacterium)

Escherichia coli
(a bacterium, 3 μm X 0.8 μm)

FIGURE 3-9

Size relationships among
some single-celled
microorganisms. (The
shapes are stylized for
diagrammatic purposes.)

will stain the same color. Simple stains, therefore, do little more than
reveal characteristics of size, shape, and cell arrangement.

Differential stains, which require more than one dye, distinguish
between structures within a cell or types of cells by staining them different
colors. Some differential stains color specific microbial structures, such as
flagella, capsules, and spores. The structure stained depends on the type
of dye and the procedure employed. Another differential stain, the **Gram
stain**, divides bacteria into two categories, the purple gram-positive cells
and the red gram-negative cells. The Gram stain is the single most
important stain in the clinical microbiology laboratory. The differential
acid-fast stain is also routinely performed whenever tuberculosis or leprosy
is suspected.

The Gram Stain

The Gram stain was developed by Dr. Hans Christian Gram in 1884. The
procedure requires four steps:

1. *Crystal violet* is added as a primary stain. It colors the cytoplasm of all
cells purple.

2. *Iodine* is then used as a *mordant,* an agent that binds the dye to the cell
and helps resist decolorization. It combines with crystal violet to form an
insoluble complex inside the cell.

3. A *decolorizing agent* (alcohol or a mixture of acetone and alcohol) is then added. The purple dye complex is retained by gram-positive organisms but is readily removed from gram-negative cells. Gram-negative cells will therefore be colorless at this stage.

4. The red dye *safranin* is applied as a counterstain. This stains gram-negative bacteria red while gram-positive cells remain purple.

The Gram stain of clinical material taken directly from an infected patient can rapidly provide valuable diagnostic information about the microorganism(s) that is causing the disease. This information is a guide to determining what subsequent tests are needed to identify the bacteria. For example, identifying gram-positive cocci (spherical cells) or gram-positive bacilli (rod-shaped cells) requires a different battery of tests than are used for identifying the gram-negative rods.

Gram stains of clinical specimens are especially important in determining the most effective antibiotic for critically ill patients who require immediate therapy. Penicillin, for example, is more effective against most gram-positive bacteria than against most gram-negative bacteria. Penicillin would be of little benefit to the patient with an infection caused by most gram-negative organisms. Gram staining clinical specimens, however, provides only a preliminary indication of the identity of the **etiological agent** (the organism causing the disease). Although it is an important first step a Gram stain does not serve as a substitute for isolating the suspected microorganism and determining its biochemical characteristics and its antibiotic susceptibilities.

The Acid-Fast Stain

The **acid-fast stain** is used to identify members of the genus *Mycobacterium*. The etiological agents of tuberculosis and leprosy are both mycobacteria. Because mycobacteria contain high levels of lipid material, they are difficult to stain by standard procedures. Hot carbol fuchsin, however, will penetrate the bacteria and stain them red. Unlike most other bacteria, the mycobacteria are difficult to decolorize once they are stained. Their resistance to decolorization by an acid-alcohol mixture explains the term "acid-fast." Following decolorization, methylene blue is added as a counterstain to make the unstained non-acid-fast organisms visible. Mycobacteria remain red, whereas non-acid-fast bacteria stain blue. The acid-fast stain is a rapid method for indicating the presence of mycobacteria in specimens such as sputum (material expectorated from a person's lower air passages) that may contain a large number of non-acid-fast normal flora.

Table 3-3 summarizes some common bacteriological staining procedures.

CULTURE TECHNIQUES

The morphological simplicity of most microbes prevents their identification by microscopic examination alone. Usually they must be isolated and grown in a **pure culture**, one that

TABLE 3-3

Common Biologic Stains
Used in Bacteriology

Reagents	Appearance of Stained Cells	Purpose
Simple stain Methylene blue or Crystal violet or Carbol fuchsin	Blue Purple Red	A simple stain used for a variety of microorganisms. Most commonly used to determine the presence of microorganisms, for example, in urine or tissue specimens from patient; the Gram stain, however, is better for this purpose.
Gram stain 1. Crystal violet 2. Gram's iodine 3. Decolorizer 4. Counterstain	Gram-positive cells, purple; gram-negative cells, red (Occasionally in older cultures gram-positive cells will stain as gram-negative. Older gram-negative cells, however, do not show a tendency to stain as gram-positive.)	A differential stain used as the first step in determining characteristic properties of any bacteria. Gram-positive bacteria retain the crystal-violet dye after decolorization and appear deep blue (or purple). Gram-negative bacteria fail to retain the dye and are counterstained red by safranin.
Acid-fast stain 1. Hot carbol fuchsin 2. Acid-alcohol 3. Methylene blue	Acid-fast cells, red; non-acid-fast cells, blue	Acid-fast bacilli, such as the agents of tuberculosis and leprosy, resist staining unless heat is applied to allow the stain to penetrate. Once stained, acid-fast bacteria resist decolorization and remain red. Other bacteria are destained with the acid-alcohol and counterstained with methylene blue.
Endospore stain 1. Hot malachite green 2. Water 3. Safranin	Endospores stain green; rest of the cell stains pink	A differential stain used to detect and characterize bacterial endospores, structures resistant to staining by traditional techniques

contains a single species. Pure cultures can then be used to determine properties such as biochemical and immunological characteristics. Pure cultures are also needed for other microbiological applications, such as harvesting useful compounds produced by microorganisms. Isolation of microbes in pure culture requires aseptic technique and special methods of inoculation.

Aseptic Techniques

Aseptic techniques are procedures that allow us to handle materials without introducing microbial contaminants from air, water, hands, or other nonsterile sources. It is usually impossible to determine which microbes are contaminants and which are from the specimen. In addition, the contaminants may multiply faster than the microbe of interest and are capable of overgrowing a sample. In improperly handled clinical specimens contaminants may obscure the diagnosis and result in incorrect treatment.

To prevent microbial contamination, one must (1) minimize the exposure of the sample to nonsterile environments, such as normal air, and (2) avoid contact between the sample and all nonsterile objects and surfaces. For example, tubes and plates should be kept closed and should be opened only when necessary. In addition, equipment for transfer (such as wire loops or forceps) should be sterilized by flaming. Strict adherence to aseptic technique will prevent contamination and reduce the time required to achieve valid results.

Inoculation Techniques

Microbes may be isolated in pure cultures by inoculating solid media in a manner that results in the development of single colonies. Since all the cells in a single isolated colony are descendants of one organism, they are identical. Obtaining isolated colonies requires the use of a solid medium, since cells in a liquid cannot be kept separate. Solid media are easily produced by adding the solidifying agent, agar, to a nutrient solution. **Agar** is an extract of seaweed that melts at 100°C and then solidifies the liquid in which it is dissolved when the temperature drops below 45°C. The melted medium is usually poured into *petri dishes*, where it is allowed to solidify. The surface of agar media is ideal for the growth of isolated colonies.

There are so many microbes in most specimens, however, that direct inoculation of a solid medium would result in overgrowth of the entire surface rather than isolated colonies. The sample must therefore be processed in such a way that cells are deposited on the medium far enough from each other that well-isolated colonies will develop. This may be accomplished by preparing streak plates, pour plates, or spread plates.

Streak Plates Organisms may be inoculated onto the surface of a plate with a wire loop or swab. A series of streaks spreads the inoculum over a large surface area until very few cells are left on the loop. Several patterns for streaking have proved effective for obtaining isolated colonies, but the most commonly employed technique is the *quadrant streak method* illustrated in Figure 3-10. These plates have four distinct areas of inoculation. The first quadrant may be inoculated with a loop or a swab. (Swabs are usually used to collect clinical specimens from the throat, eye, vaginal tract, and exudates from wounds or lesions.) The remaining three quadrants must each be streaked with a sterile instrument such as a flamed loop or fresh swab.

All streaking techniques depend on spreading the inoculum over a

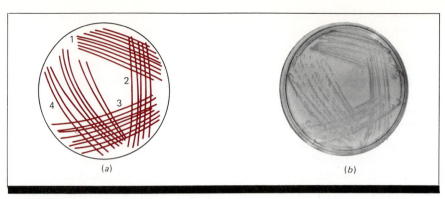

FIGURE 3-10

The quadrant streak method is the most commonly employed technique for obtaining isolated colonies.
(*a*) Quadrants are streaked successively through a portion of the previously streaked quadrant following sterilization of the loop. The first quadrant may be inoculated with a swab, after which a loop is used. (*b*) A properly prepared streak plate reveals isolated colonies after 24-hour incubation.

large surface area so that the organisms fall off the loop a single cell at a time. It is often difficult to isolate an organism in a single attempt, especially if there are many competing contaminants in the specimen. A fresh plate may be streaked using material from the area of the streak plate that appears to be richest in the desired organism.

Pour Plates and Spread Plates   Another way to obtain isolated colonies is to inoculate agar media with a sample that has been diluted in sterile liquid. In the pour-plate technique a sample of the liquid, usually 1 ml, is added to melted agar at 45°C. The pour-plate technique generates many colonies that are embedded in agar. Spread plates are prepared by spreading a small volume of the diluted sample across the surface of a solid medium by using a bent glass rod. In this case, all colonies grow on the agar's surface. Pour plates and spread plates are generally employed for counting microbes. Because of its convenience, the streak technique is usually employed for isolating bacteria from clinical specimens.

IDENTIFICATION TECHNIQUES

Culture Characteristics

When growing on suitable solid media, most microbes form visible aggregates called *colonies.* The appearance of these colonies provides additional clues as to the identity of microorganisms. Cultural characteristics that contribute to identification are size, shape, pigmentation, and texture of the colonies. (See color plates.)

Biochemical Tests

A wide range of microbial activities can be determined by biochemical tests. These characteristics can be used to differentiate between microbes that appear morphologically identical. For example, different types of microbes can usually be distinguished by the characteristic pattern of nutrients each can utilize. In addition, microbes produce biochemicals such as acids, alcohols, gases, or specific enzymes, the detection of which may aid in identifying the organism. A series of biochemical tests can provide a microbial "fingerprint."

Biochemical activities of microbes are discussed in more detail in Chapters 10 and 11.

Immunological Tests

Immunological reactions can be used to identify microorganisms by determining whether the microbe can combine with specific antibodies. Antibodies can be obtained from the blood of animals that have been vaccinated (immunized) by injecting them with microbes or specific microbial components. These antibodies usually react only with the same

FIGURE 3-11

Example of an immunological test for identifying unknown microbes. The suspected organism is mixed with antibodies prepared against specific microbes. Since antibodies usually react only with the corresponding microbes, a positive immunological reaction can identify the organism. In this test, fluorescence identifies the organism as *Treponema pallidum*, the bacterium that causes syphilis. (In actual testing only the antibody against *T. pallidum* would be used.)

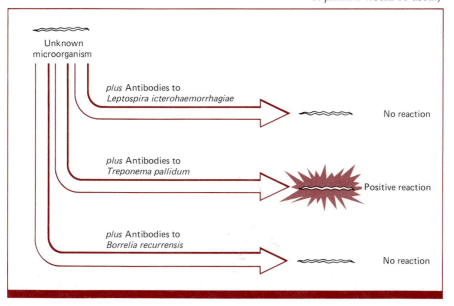

type of organism that was used to stimulate their production; thus, reaction with a specific antibody is evidence of a microbe's identity. Immunological tests are particularly useful because they are rapid and can identify pathogens that cannot be grown on laboratory media. For example, *Treponema pallidum*, the bacterium that causes syphilis, is most often diagnosed by a fluorescent antibody test that uses fluorescence as the indicator of reaction (Fig. 3-11). Immunity and its applications for identifying microbes are discussed in Chapter 16.

OVERVIEW

A fundamental tool for studying microorganisms is the microscope. Magnification, however, is only useful if the detail of the enlarged image is clearly discernible. The useful magnifying capacity of a microscope is limited by its resolving power, its ability to create a clear image. Substituting an electron beam for visible light boosts resolution and magnification to 1000 times that achievable by light microscopes. The introduction of the electron microscope provided the first look at previously unseen members of the microbial world, such as viruses.

Some variations on the conventional light microscope are designed to enhance contrast rather than to increase useful magnification. Darkfield and phase-contrast microscopes reduce the need for staining the cells, thereby allowing detailed observation of living microorganisms. The fluorescence microscope provides an especially rapid means for identifying microorganisms by detecting natural fluorescence in certain microbes or by allowing the observation of specific immunological reactions between a microbe and a fluorescent-labeled antibody. The common drawback to all such variations on the standard brightfield microscope is the inability to distinguish the colors of differentially stained cells. The gram-stain reaction, for example, can be interpreted only with a standard brightfield microscope. Since such information is usually necessary for identification, the basic brightfield microscope has a permanent place in most microbiology laboratories.

Laboratory methods for detecting and characterizing microorganisms usually begin with microscopic examination. This generally means determination of gram-stain reaction and morphological features. Whereas many eucaryotic microbes may be microscopically identified, procaryotes lack the morphological complexity needed for microscopic determination of identity. Further characterization usually requires isolating the microbe in pure culture. This may be accomplished by the streak-plate, pour-plate, or spread-plate techniques. Additional microbial properties can be determined by biochemical and immunological tests.

colony	micrometer (μm)
compound microscope	nanometer (nm)
magnification	angstrom (Å)
resolving power	simple stain
brightfield microscope	differential stain
objective lens	Gram stain
ocular lens	etiological agent
darkfield microscope	acid-fast stain
phase-contrast microscope	pure culture
fluorescence microscope	aseptic technique
fluorescent antibody technique	agar
transmission electron microscope	streak plate
scanning electron microscope	pour plate
artifact	spread plate

REVIEW QUESTIONS

1. Why is it impractical to increase magnification beyond the resolving power of a microscope?

2. What is the advantage of the compound microscope over the simple microscope? Why are light microscopes limited to two lens systems?

3. What are the specific advantages and disadvantages of each of the following types of microscopes?
    (a) Brightfield
    (b) Phase-contrast
    (c) Fluorescence
    (d) Scanning electron
    (e) Transmission electron

4. Differentiate between simple stains and two categories of differential stains. Give an example of each.

5. Express the following measurements in millimeters.
    (a) 200 μm
    (b) 17 m
    (c) 75 nm
    (d) 2 cm
    (e) 155 Å

6. Describe four laboratory procedures for determining the identification of a bacterium. What is the importance of identifying bacteria?

7. Why is the acid-fast stain used for diagnosing tuberculosis and leprosy?

8. Describe three techniques used for isolating microbes in a pure culture.

9. What is meant by aseptic technique?

Structure and Function of Bacteria

4

Many of our most dreaded diseases are caused by bacteria, the most common group of procaryotes. The structures and biological activities of bacteria enable them to infect and damage the body. Yet structures and functions unique to bacteria may be sensitive to antibiotics that inhibit or destroy essential properties of procaryotic cells that are not shared by eucaryotes. As a result, many diseases that killed our ancestors are easily cured today.

Unlike the larger features of eucaryotic cells, bacterial structures are difficult or impossible to distinguish using light microscopy. The finer details of subcellular structure are revealed best by high-resolution electron microscopy.

MORPHOLOGY

The general shape of individual bacterial cells is usually discernible with the brightfield microscope. Bacteria are differentiated into major categories based on such microscopic observations. These

FIGURE 4-1

Characteristic morphology of (a) cocci, (b) bacilli, and (c) spirals.

(a)

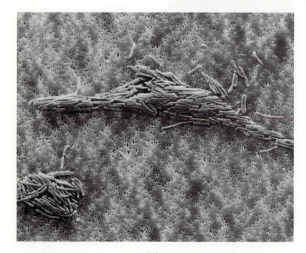

(b)

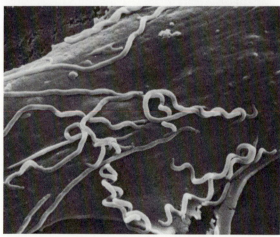

(c)

groupings reflect morphological properties such as (1) cell shape, (2) cell size, (3) staining characteristics, and (4) the manner in which similar cells are arranged.

Cell Shape and Arrangement

Most bacteria have a defined shape that falls into one of three morphological categories (Fig. 4-1): (1) spherical **cocci** (singular, coccus) (2) rod-shaped **bacilli** (singular, bacillus), and (3) **spiral** organisms. In addition, some bacteria are **filamentous**; they tend to form long strands composed of many cells (Fig. 4-2a). In these cases, an occasional cell may be seen after it breaks away from a longer filament. A few bacteria change their shapes and are called **pleomorphic** (*pleo,* more; *morph,* form) (Fig. 4-2b).

Coccus   A coccus is a spherical organism normally ranging between 0.4 and 2 μm in diameter. Cocci generally appear in groups formed by the incomplete separation of cells during the reproductive process. The arrangement of the cells in the group is often indicative of an organism's identity (Fig. 4-3). For example, *Neisseria gonorrhoeae,* the causative agent of gonorrhea occurs as pairs of cells called *diplococci.* Bacteria that characteristically grow in long chains are called *streptococci* (*strep,* thread). Still other cocci form packets of four cells, called *tetrads,* or of eight cells in a cubical arrangement. Spherical bacteria in grapelike clusters often belong to the genus *Staphylococcus.* (*Staphyle* is the Greek word meaning a bunch of grapes.) Members of this genus may be part of the normal skin flora as well as important agents of human disease such as abscesses and toxic shock syndrome.

Although cell arrangement can give some clue as to the identity of an organism, under certain conditions bacteria fail to grow in their characteristic patterns. For example, variations in culture media, growth temperature, and specimen preparation often influence the arrangement observed.

FIGURE 4-2

(a) Early growth of a filamentous bacterium, *Actinomyces naeslundii.* (b) *Mycoplasma* bacteria do not maintain a defined shape.

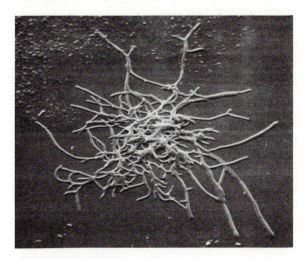

(a)

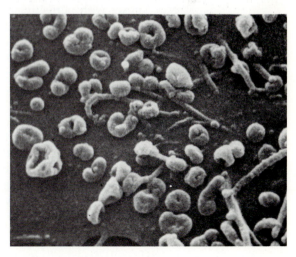

(b)

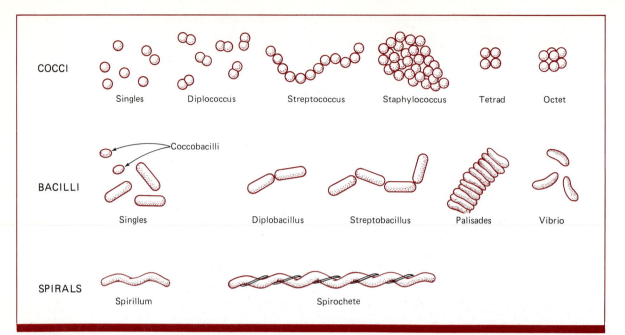

COCCI

Singles Diplococcus Streptococcus Staphylococcus Tetrad Octet

BACILLI

Coccobacilli

Singles Diplobacillus Streptobacillus Palisades Vibrio

SPIRALS

Spirillum Spirochete

FIGURE 4-3
Cell shapes and
arrangements.

Thus, microscopic observations of cellular arrangement may be of limited diagnostic value.

Bacillus Bacilli are rod-shaped organisms usually ranging between 1 and 10 μm in length. Some bacilli are characteristically long and slender. Others are so short and stumpy they appear ovoid and are referred to as **coccobacilli**. Some bacilli are curved into a form resembling a comma. These cells are called **vibrios**, as in *Vibrio cholera*, a pathogen that causes potentially fatal gastrointestinal disease. Some bacteria are flat rectangular boxes with perfectly straight edges (see color box).

Bacilli usually grow as single cells, although pairs *(diplobacilli)* or chains *(streptobacilli)* are occasionally observed. Some organisms, for example, *Corynebacterium diphtheriae*, bend at the point of division following reproduction, resulting in a palisade arrangement resembling a picket fence and angular patterns that look like Chinese letters.

Spiral Spiral organisms are less common than cocci or bacilli and include the agent of syphilis. They are divided into two groups, **spirilla** (singular, spirillum) and **spirochetes**. These two groups of microbes are very similar in shape, but spirochetes are flexible whereas spirilla are rigid. While they can be clearly recognized by their shape, some of these organisms are far too thin to be seen with the standard brightfield microscope and are therefore usually observed by darkfield microscopy. Spiral organisms usually occur singly.

LIVING BOXES

■ The shape of a newly discovered bacterium has captured the interest of microbiologists. The unique microbe was first described in 1980 after its discovery in a natural salt pond along the shore of the Red Sea. Its discovery introduces a new morphological category to the traditional coccus-bacillus-spiral triad of bacterial shapes. The new microbes are called "square bacteria."

The square bacteria are really flat rectangular boxes with perfectly straight edges and sharp 90° angles at each corner. Although they are several micrometers in length and width, their thickness is a uniform 0.25 μm, just at the limits of resolution for the brightfield microscope. Each bacterium is a thin flexible sheet that may be bent and twisted by external forces, much like a thin sheet of flexible plastic. Their surfaces are unusually smooth. Smaller cells are usually perfectly square with dimensions of 2 × 2 μm. Larger cells are rectangles about twice as long as they are wide (4 × 2 μm). After division, these cells may apparently remain

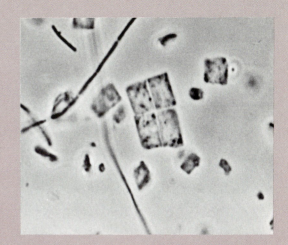

attached to each other, producing the larger sheets of squares shown in the figure.

Square bacteria originally remained undescribed because, although detected by microscopy, they were not recognized as cells because of their unusual shapes. This shape provides the basis for naming these microbes *Arcula*, from the Latin word for small box.

CELL STRUCTURE

In procaryotes a single structure may accomplish functions performed by several different membrane-bound organelles in eucaryotes. This reduces the structural complexity of procaryotes. All bacteria possess a nucleoid, ribosomes, and a cell membrane. Most bacteria also have a cell wall. Additional structures include a capsule, cytoplasmic inclusions, and surface appendages (Fig. 4-4).

Surface Layers

Although more than 1500 species of bacteria have been described, only about 100 are primarily human pathogens. The factors contributing to many biological activities, including their ability to initiate or promote disease, are often directly related to the exposed surface structures and appendages of the bacterial cell. The eucaryotic host often defends against microbial invasion by neutralizing these surface bacterial components.

In all cells cytoplasmic contents are separated from the environment by the cell membrane. In addition, almost all procaryotic cells are sur-

FIGURE 4-4

(*a*) Diagrammatic representation of the major structural components of bacteria. The structures in color are not found in all cells and are usually not essential for growth. (*b*) A cross section of a typical rod-shaped bacterium.

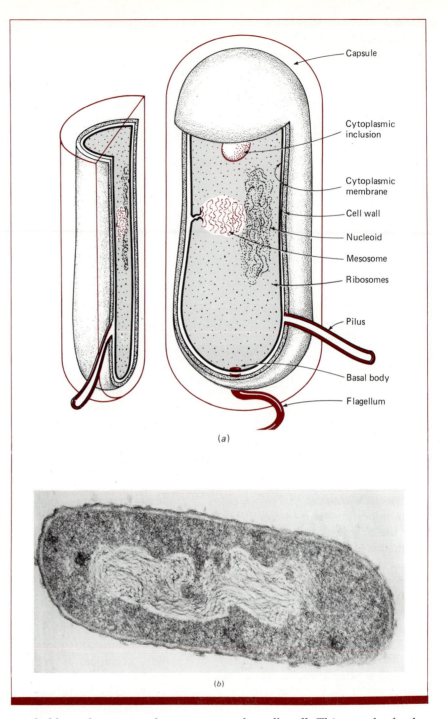

rounded by at least one other structure, the cell wall. This may be further enveloped by a capsule or slime layer.

The Cell Membrane   In order to survive, the cell must remain separate from its environment. This separation is accomplished by the **cell membrane**. Major disruptions in this barrier result in the spilling of the cytoplasm from the cell and the death of the organism.

The cell membrane in bacteria is a phospholipid-protein bilayer similar to that present in eucaryotic cells (see Fig. 2-3). The major difference is that there are no sterols in the cell membranes of most procaryotes.

The cell membrane is the site of many functions that are accomplished by specialized internal organelles in eucaryotes. These include:

1. Transport of molecules in and out of the cell

2. Secretion of extracellular enzymes

3. Respiration and photosynthesis

4. Regulation of reproduction

5. Cell wall synthesis

MEMBRANE TRANSPORT Organisms must selectively regulate the transport of nutrients into and waste products out of the cell. Usually each type of molecule requires specific receptors that can recognize it and move it across the membrane. This is the function of many of the membrane proteins. The presence or absence of the corresponding receptor proteins in the cell membrane determines which molecules can be transported. When this process requires energy, it is known as **active transport**.

Some small molecules move across the membrane by a process of simple **diffusion**, the tendency of molecules to move from areas where they are in higher concentration to areas of lower concentration. Diffusion requires no energy expenditure by the cell. When the diffusing molecule is water, the process is termed **osmosis**. Since water freely passes through the membrane, its uptake or loss depends on its concentration in the environment relative to that in the cytoplasm and on the available space inside the cell. A bacterium can accumulate water only until its internal volume increases to the limitation imposed by the nonexpandable cell wall.

SECRETION OF EXTRACELLULAR ENZYMES Before they can be transported across the cell membrane, larger food molecules must be broken down outside the cell into smaller subunits. Such extracellular digestion is mediated by enzymes released from the bacteria into their fluid environment. Secretion of extracellular enzymes is another function of the cell membrane.

RESPIRATION AND PHOTOSYNTHESIS The functions accomplished by eucaryotic mitochondria (energy generation) and chloroplasts (photosynthesis) are performed by the cell membrane in procaryotes. For example, those proteins located on the internal membranes of the eucaryotic organelles are embedded in the cell membrane of the procaryote. The greater the membrane area in the cell, the more respiration and/or photosynthesis can occur.

In some bacteria the membrane invaginates into the cytoplasm, forming a **mesosome**. Mesosomes significantly increase membrane surface area, allowing the cell greater activity in respiration and active transport. Mesosomes also play a role in cell reproduction. In photosynthetic bacteria another invaginated membrane structure, the **chromatophore**, contains enzymes and pigments that perform functions similar to those of the eucaryotic chloroplast.

REPRODUCTION Production of bacterial progeny is partially regulated by the cell membrane. Specific proteins in the membrane attach to the replicating DNA and separate the duplicated chromosomes from each other. In addition, the cytoplasm of the two daughter cells may be physically separated from each other by formation of a septum (cross wall). In some bacteria, chromosome separation and cross-wall formation are performed by mesosomes (Fig. 4-5). In others, these tasks are accomplished by the noninvaginated cell membrane.

CELL WALL SYNTHESIS Molecules needed for constructing and repairing the cell wall are synthesized in the cytoplasm and transported outside the cell. Specialized membrane proteins transport these subunits across the cell membrane and assemble them into the growing cell wall.

The Cell Wall The bacterial **cell wall** is the structure that immediately surrounds the cell membrane. The shape of the cell is imposed by the

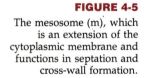

66

FIGURE 4-5

The mesosome (m), which is an extension of the cytoplasmic membrane and functions in septation and cross-wall formation.

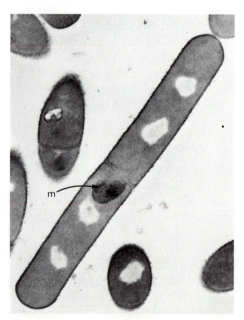

shape of the cell wall. (Without their cell walls, all bacteria assume a spherical shape.) Thus the cell wall allows the existence of larger bacteria by forming them into cylindrical shapes that reduce the distance between the center of the cell and the external environment. Since procaryotes lack cytoplasmic streaming, large spherical bacteria cannot exist.

The most important function of the cell wall is to physically protect the cell. This protection is necessary because of the susceptibility of the cell membrane to physical or osmotic lysis (disruption). In many aqueous environments bacteria will accumulate water due to osmosis. Water influx increases cell size and would burst the cell (as an overinflated balloon) if not for the physical restriction of this expansion imposed by the cell wall. The cell wall in some bacteria is strong enough to withstand 25 atm (atmospheres) of internal osmotic pressure.

A bacterium from which the wall has been completely removed, usually by enzymatic digestion, is called a **protoplast**. These cells are

FIGURE 4-6

Water tends to rush into cells when the medium is more dilute than the cell's cytoplasm. (a) The cell wall prevents cell swelling due to such internal water gain. Since the wall does not expand, eventually the internal pressure becomes great enough to oppose further accumulation of water and the cell survives. (b) Protoplasts of the same cell, however, have no cell wall to protect them. These cells expand and eventually lyse because of osmotic influx of water.

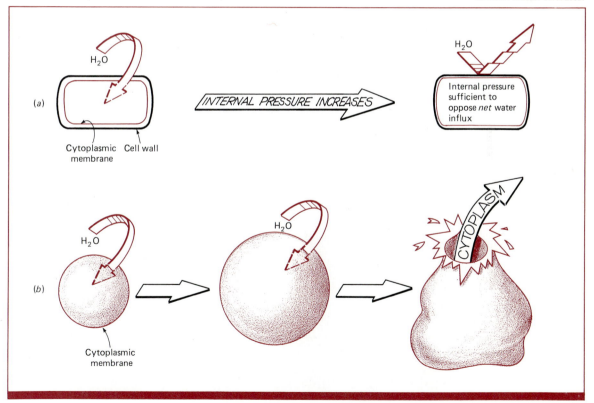

poorly protected and will lyse in environments that encourage uncontrolled influx of water (Fig. 4-6). We can prevent protoplast lysis by balancing the exchange of water across the cell membrane. Such a balance is accomplished by suspending protoplasts in solutions containing high concentrations of salts or sugars. Protoplasts are also susceptible to lysis by physical trauma. Even shaking a broth culture of protoplasts is enough to destroy many cells.

All protoplasts assume a more or less spherical shape regardless of the shape of the bacteria from which they were derived (Fig. 4-7). The empty cell wall fragment from which the protoplast escaped still retains the shape characteristic of the original cell.

The cell wall is a complex structure composed of several substances. Its amazing strength is due primarily to **peptidoglycan** (also known as murein and mucopeptide), a substance found only in procaryotes. Peptidoglycan is an enormous molecule composed of amino acids and sugars (*peptide*, amino acid; *glycan*, sugar). The sugars *N-acetylglucosamine (NAG)* and *N-acetylmuramic acid (NAM)* alternate to form long parallel chains (Fig. 4-8a). Attached to the NAM molecules are short side chains of four or five amino acids. A single layer of peptidoglycan is a network of adjacent sugar chains bound together through the amino acid side chains, thus making a cross-linked structure that covers the entire cell. The cross-linking may be a direct connection of two side chains (Fig. 4-8b) or may require an additional chain of five amino acids (Fig. 4-8c).

Some bacteria have several layers of peptidoglycan. Each layer increases the strength of the cell wall and is cross-linked to the adjacent layers, forming a gigantic molecular network. The strength of the individual layers is largely determined by the extent of cross-linking between the backbones. (See the discussion of gram-positive and gram-negative cell walls below.)

Interference with a bacterium's ability to produce peptidoglycan results in a defective cell wall, and the bacterium usually will not survive. The penicillin and cephalosporin antibiotics specifically prevent pepti-

FIGURE 4-7
This spherical protoplast was derived from a bacillus. Note that the flagella remain attached to the cell, although they are no longer capable of motility.

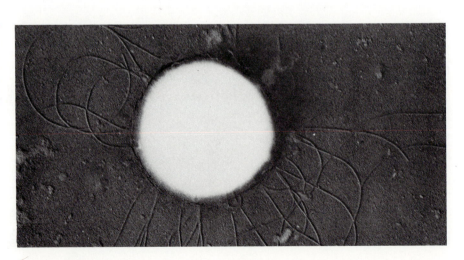

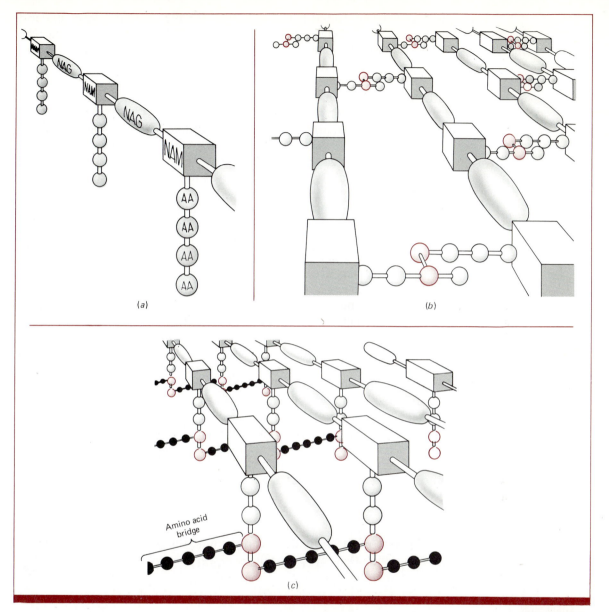

FIGURE 4-8

The bacterial cell wall: (*a*) Structure of the peptidoglycan backbone. NAM = *N*-acetylmuramic acid; NAG = *N*-acetylglucosamine; AA = amino acid. (*b*) Representation of the peptidoglycan cross-linked structure in gram-negative bacteria. The peptidoglycan layer is usually one or two layers thick. The amino acid chains are either free or linked directly to one another. This cross bridging is much less extensive than in gram-positive cell walls. (*c*) Representation of the peptidoglycan cross-linked structure in gram-positive bacteria. Multiple layers of peptidoglycan form a very thick covering around the cell. Within each layer, backbones are extensively cross-linked by amino acid bridges which connect side chain amino acids.

doglycan synthesis and induce cell lysis. Since no peptidoglycan is found in human cells (which have no cell walls), these antibiotics can be introduced into infected persons to selectively destroy bacteria without risking a similar effect on the human host.

GRAM-POSITIVE BACTERIA   The gram-positive cell wall is composed primarily of several layers of peptidoglycan (Fig. 4-9). The thickness of their peptidoglycan armor makes gram-positive bacteria especially resistant to osmotic lysis. Extensive cross-linking within each layer also contributes to cell-wall strength.

In addition to peptidoglycan, most gram-positive walls contain **teichoic acids**, large molecules composed of repeating units of sugars and phosphates. They are found only in the cell walls and cell membranes of gram-positive bacteria. Teichoic acids give the cell surface a negative charge, which may be important in determining the types of substances attracted to and ultimately transported into the cell.

Other auxiliary molecules associated with the cell wall impart properties to the outer surface of gram-positive cells that influence their ability to cause human disease. For example, the cell-wall-associated M protein of *Streptococcus pyogenes* effectively prevents its engulfment by protective white blood cells, thereby aiding the ability of this bacterium to cause disease. (*Streptococcus pyogenes* causes more types of disease than any other etiological agent known. These diseases include strep throat, rheumatic fever, scarlet fever, impetigo, and serious kidney disorders.)

Teichoic acids and cell-wall-associated proteins are the major surface antigens of the gram-positive cell wall. **Antigens** are molecules that stimulate the host immune system to make *antibodies*. These antibodies specifically react with antigens on the bacterial surface, often assisting the host in eliminating the invading microbe. The antibodies recognize and react only with the type of antigen that stimulated their production. Because different types of cells have distinct surface antigens, antibodies against one type of bacterium will not likely protect against invasion by another species.

In spite of these auxiliary compounds, the gram-positive cell wall is readily penetrated by penicillin and cephalosporin antibiotics, which

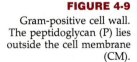

FIGURE 4-9

Gram-positive cell wall. The peptidoglycan (P) lies outside the cell membrane (CM).

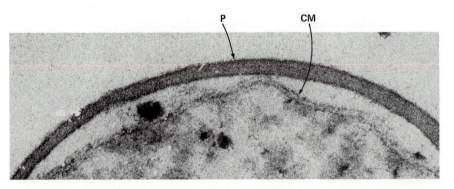

(a)

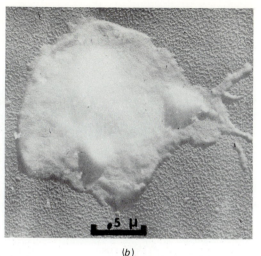

(b)

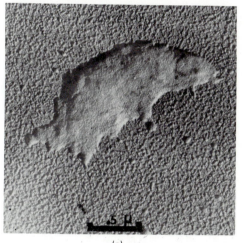

(c)

(d)

FIGURE 4-10
Lysozyme destroys the peptidoglycan layer, which results in the gradual digestion of the gram-positive cell wall.

prevent peptidoglycan synthesis, and *lysozyme,* an enzyme that digests the chemical bond between NAG and NAM. Lysozyme destroys the peptidoglycan, leaving cells that have no protective cell walls (Fig. 4-10). Lysozyme is found in tears, saliva, and other secretions as a natural defense against bacterial diseases.

GRAM-NEGATIVE BACTERIA The gram-negative cell wall is structurally more fragile than its gram-positive counterpart. Its peptidoglycan is much thinner, accounting for only 1 to 2 percent of the dry weight of the cell (compared to 20 percent in gram-positive bacteria). The outer surface of the peptidoglycan is covered by a series of layers (Fig. 4-11). Together these layers provide a protective coating around the cell that resists penetration by some potentially toxic chemicals.

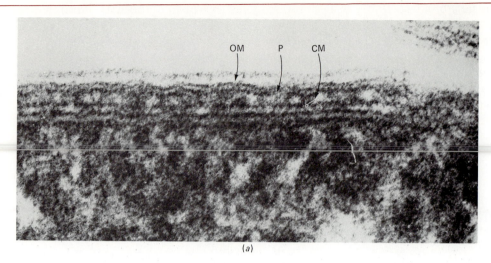

OM   P   CM

(a)

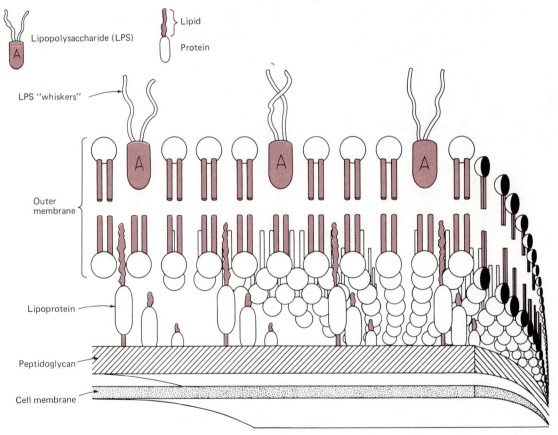

Lipopolysaccharide (LPS)

Lipid

Protein

LPS "whiskers"

Outer membrane

Lipoprotein

Peptidoglycan

Cell membrane

(b)

The first of these layers is **lipoprotein**. One end of lipoprotein is directly attached to peptidoglycan. The other end extends into the **outer membrane**, a lipid-protein bilayer similar to the cell membrane. This outer membrane protects gram-negative bacteria from the effects of anti-peptidoglycan chemicals such as penicillin and lysozyme. These antibacterial agents fail to penetrate this membrane and consequently are unable to either attack the peptidoglycan layer (lysozyme) or interfere with its synthesis (penicillin). Determining the Gram reaction of a pathogen is therefore important for selecting the best antibiotic to treat serious infectious diseases when there is no time to wait for laboratory cultures and identification. The use of penicillin against gram-negative organisms will usually be of limited or no value. (Important exceptions are *Neisseria gonorrhoeae* and *Neisseria meningitidis,* the gram-negative cocci that cause gonorrhea and meningitis. Unique protein components in the outer membrane of *Neisseria* allow penicillin to penetrate to its site of action and inhibit the synthesis of peptidoglycan.)

The surface of the outer membrane contains molecules of **lipopolysaccharide (LPS)**. LPS consists of a molecule called *lipid A* covalently linked to a polysaccharide (a large molecule composed of repeating sugar subunits). LPS has important medical consequences. When a gram-negative bacterium is destroyed within the human body, the lipid A portion of LPS is released from the disrupted cell wall and elicits toxic reactions in the host. These reactions include fever, diarrhea, and potentially fatal shock. Because of its ability to injure a host and because it is an integral part of the bacterial cell wall, lipid A is commonly referred to as **endotoxin** (see color box). The presence of endotoxin enables gram-negative pathogens to produce symptoms of disease that are rarely provoked by infection with gram-positive bacteria.

The polysaccharides that extend outward from the lipid A are the outermost molecules of the cell wall and are major surface antigens of the gram-negative bacterial cell. Antibodies directed against one gram-negative bacterium usually do not protect against another gram-negative species.

CELL WALLS AND GRAM STAINING   The difference in staining behavior between gram-positive and gram-negative cells is due to differences in the physical structure of their cell walls. Although the exact mechanism of differentiation is still unclear, it is believed to be due to dehydration of the thick peptidoglycan layer of gram-positive cells by alcohol, closing the

FIGURE 4-11

Gram-negative cell wall: (*a*) A series of layers that are visible here are the cytoplasmic membrane (CM), peptidoglycan (P), and outer membrane (OM). (*b*) Diagrammatic representation of the gram-negative cell surface. (See text for discussion.) Drawing is not to scale; the outer membranes have been enlarged to show structural detail.

DANGEROUS GHOSTS OF DEAD BACTERIA

■ Dead gram-negative bacteria retain their ability to injure or kill people. This dangerous "ghost effect" is due to residual endotoxin, which retains its biological activity when exposed to many conditions employed to kill bacteria. This alarming fact is often evident when fever develops shortly after sterile solutions or objects such as needle catheters or prosthetic devices are introduced into the body. Even though such solutions and objects are sterilized to kill all microbial contaminants, some of these sterilization procedures fail to remove residual cell-wall debris. For example, the activity of endotoxin survives steam heating at 120°C for 1 hour. This condition far exceeds that required to kill any type of bacteria.

Even small amounts of residual endotoxin can elicit symptoms. Typical adults will develop fever within 3 hours after receiving as few as 1 million killed bacteria. This is about 300 bacteria per milliliter of blood, a tiny number when considering that bacteria may typically achieve population sizes of 1 billion cells per milliliter.

To prevent such occurrences, all commercially prepared objects and substances to be introduced into the body are examined by the manufacturer for evidence of endotoxin contamination. Two techniques are commonly employed for gathering this evidence. One assay procedure detects the pyrogenic (fever-inducing) activity of endotoxin. Laboratory rabbits are fitted with rectal thermometers and injected with the test solution. (Solid objects must be rinsed in a pyrogen-free sterile solution, which is then tested for eluted endotoxin.) Three hours later the rabbits' thermometers are examined for evidence of fever. Although the rabbit-assay test (also called the *pyrogen test*) can detect endotoxin concentrations as small as 0.1 ng/ml, it requires a large facility for housing the many rabbits needed.

A more convenient and sensitive assay procedure was developed after the 1968 discovery that amoebocyte blood cells from horseshoe crabs (*Limulus polyphemus*) contained a substance that visibly gelled within 1 hour when mixed with very small amounts of endotoxin. This test, called the *limulus amoebocyte lysate assay (LAL)*, has replaced the pyrogen assay procedure in many labs (an advance hailed by rabbits). In addition to convenience, it is faster and sensitive enough to detect the endotoxin in as few as 300 bacteria. Since the amoebocyte lysate is now commercially available, collecting and bleeding your own crabs is no longer necessary. Some chemicals such as albumen (a protein found in serum), however, interfere with the limulus assay. Samples containing these substances must be assayed by another method, such as the pyrogen test.

pores in the wall and trapping the crystal-violet–iodine complex inside the cell. Thus, gram-positive cells resist decolorization by alcohol. The thinner, more porous peptidoglycan layer in gram-negative cell walls fails to retain the dye complex. The solvent dissolves the outer layer, penetrates the cell, and washes away the crystal violet.

Physical disruption or damage to the gram-positive wall decreases its ability to retain the crystal violet, making a gram-positive cell appear to be gram-negative. For example, removal of the cell wall of a gram-positive bacterium produces a gram-negative protoplast. Because of cell wall deterioration, "old" cultures of gram-positive bacteria will stain as gram-negative bacteria. Gram-negative cells, however, always remain gram-negative, regardless of age.

WALL-DEFICIENT VARIANTS Only one group of bacteria, the **mycoplasmas**, naturally exist without cell walls. The mycoplasmas are protected from osmotic lysis by the presence of sterols in the cell membrane and by adopting a parasitic existence in the osmotically favorable environment of a eucaryotic host. A second group of bacteria, found in marine environments, have unusual cell walls devoid of peptidoglycan. These bacteria can exist without the rigidity provided by a peptidoglycan-rich wall because the salt concentration in seawater prevents influx of water into the cell, thereby eliminating the danger of osmotic lysis.

Wall-defective or wall-deficient forms are known to develop among other procaryotes, for example, in the presence of penicillin or lysozyme. If the bacterium is situated in an osmotically favorable environment, the resulting protoplast may survive. Such an environment may be found within the human body, particularly in such areas as pus-filled wounds. These wall-less forms are called **L forms** ("L" for the Lister Institute in London where they were first discovered). Most L forms will resynthesize their walls once the antibiotic is removed. Some, however, are believed to permanently lose the capacity to produce cell walls. Because of their osmotic fragility, they are difficult to isolate or identify in most laboratories. While their role in disease is still unclear, it is possible that L forms are responsible for some infectious diseases for which no organisms can be cultured. L forms may survive antibiotic therapy because of their lack of susceptibility to anti-peptidoglycan agents. For example, relapses of infection after completion of penicillin therapy may result from the resumed growth of these surviving L forms.

The major properties of the bacterial cell wall can be summarized as follows:

1. The cell wall provides protection from osmotic lysis.

2. It gives rigidity and shape to the cell.

3. It provides protection from some antibiotics and destructive chemicals (gram-negative).

4. It determines differences in Gram-stain reaction.

5. It possesses surface antigens.

6. It may elicit certain toxic symptoms of diseases caused by gram-negative bacteria.

7. Its synthesis can be inhibited by some antibiotics.

8. It provides the support necessary for propulsion by flagella. (See Flagella later in this chapter.)

TABLE 4-1

Comparison of Gram-
Positive and Gram-
Negative Cell Walls

	Gram-Positive	Gram-Negative
Peptidoglycan	Very thick layer, extensively cross-linked	Thin layer with few cross-links
Auxiliary compounds	Teichoic acids; proteins	Lipoprotein; outer membrane; lipopolysaccharide (endotoxin)
Resistance to osmotic pressure	More resistant (25 atm)	Less resistant (5 atm)
Penicillin sensitivity	Usually sensitive	Usually insensitive
Response to lysozyme	Digested cell wall	Resistant cell wall

The features that distinguish the cell walls of gram-positive and gram-negative bacteria are presented in Table 4-1.

The Capsule, the Slime Layer, and the Glycocalyx Most bacteria extrude some material that collects outside the cell wall to form an additional surface layer. When this layer adheres to the surface of the cell and is clearly differentiated from the environment, it is called a **capsule**. It can be microscopically detected after differential staining using a procedure that stains the cell and the background darker than the capsule itself (Fig. 4-12). Surface layers that are loosely distributed around the cell and diffuse into the medium are referred to as **slime layers**. A different structure, the **glycocalyx**, is a tangled mass of thin polysaccharide fibers that extends from the bacterial surface (Fig. 4-13). The capsule stain does not reveal the presence of the glycocalyx, which can only be seen by electron microscopy.

In some bacteria, these layers help initiate infectious diseases ranging

FIGURE 4-12

A thick polysaccharide
capsule surrounds *Klebsiella
pneumoniae.*

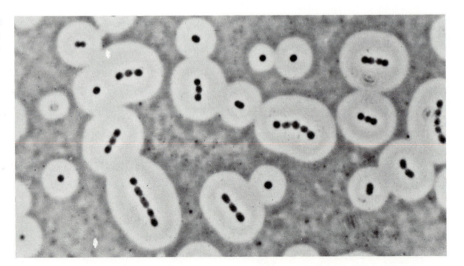

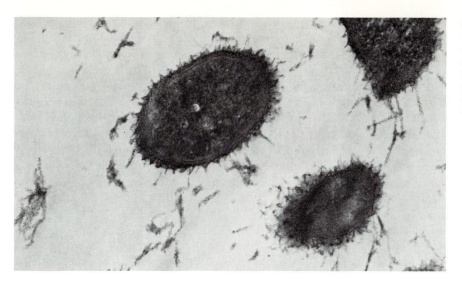

in severity from localized dental caries to life-threatening pneumonias. Host antibodies that bind to the surface layer help eliminate these bacteria from an infected person.

When present, the capsule or glycocalyx is the outermost surface of the cell. As such they may play several major roles:

■ A thick capsule may protect the cell from dehydration. It does not, however, prevent the passage of water-soluble nutrients and waste products.

■ The capsule protects some pathogens from engulfment and destruction by the body's white blood cells, thereby contributing to an organism's ability to cause disease. This is the major factor in the ability of *Streptococcus pneumoniae* to survive the pulmonary defenses and establish respiratory disease. Pathogenic and harmless strains of *Streptococcus pneumoniae* differ solely by the presence or absence of the capsule. Encapsulated strains can escape host defenses and cause fatal pulmonary infection. Nonencapsulated strains are eliminated by phagocytosis. Other bacteria which may possess antiphagocytic capsules include *Clostridium perfringens* (which causes gas gangrene), *Bacillus anthracis* (anthrax), *Klebsiella pneumoniae* (pneumonia), and *Haemophilus influenzae* (meningitis).

Unlike the capsule, the glycolcalyx is unable to protect an individual cell from destruction. Some bacteria, however, adhere to each other in groups or microcolonies "cemented" together by the glycocalyx. An attacking white blood cell may be confronted with a bacterial aggregate that is too large to engulf.

■ The polysaccharide of the glyocalyx is responsible for allowing some bacteria to attach to surfaces. The ability to adhere is often crucial to the development of some diseases. Many pathogens in the urinary or alimentary tracts must attach to tissue surfaces in order to establish infection. Without some protection against the flushing action of urine or the motility of the fluids in the intestines, these organisms would be constantly washed away.

The development of dental caries depends on attachment of the bacterium *Streptococcus mutans* to the tooth. This bacterium, a normal resident in the mouth, forms an elaborate glycocalyx when supplied with the sugar, sucrose. As the polysaccharide layer thickens, more bacteria become entrapped and anchored against the tooth surface where their growth and metabolism ultimately decay the tooth. Eliminating or reducing the sucrose in the diet reduces glycocalyx formation, thereby opposing bacterial adherence to teeth and retarding the development of dental caries.

The chemical composition of capsules is genetically determined for each species that produces them. Most capsules are polysaccharides or polysaccharide-protein complexes. *Bacillus anthracis* secretes a capsule consisting solely of polypeptide (short fragments of protein). The specific nature of the capsule can be used to help identify bacteria. For example, when antibodies react with the particular capsule material that stimulated their formation, they give the capsule a swollen appearance. This response, called the *quellung reaction,* can be observed microscopically and used to identify specific bacteria that have reacted with the antibody.

Even though a bacterium is capable of producing a capsule or

FIGURE 4-14

Influence of sucrose on glycocalyx production by *Streptococcus mutans*. The organism was cultured in two flasks containing media that differed only in the sugar supplement. In the presence of sucrose, glycocalyx cements the bacteria to the glass (inverted flask on the right). Without sucrose, the bacterium produces no glycocalyx and fails to attach to the surface of the inverted flask on the left.

glycocalyx, the organism may remain naked. Environmental and nutritional factors dictate whether these surface layers will be produced. For example, sucrose is important in promoting glycocalyx formation by cariogenic bacteria (those that promote tooth decay). Although *Streptococcus mutans* grows perfectly well in the absence of sucrose, it will not form the glycocalyx without it (Fig. 4-14). Environmental influences are further illustrated by the observation that pathogens freshly isolated from infected humans are more likely to be encapsulated than the same organisms after cultivation on laboratory media. The presence of the capsule often gives colonies of fresh isolates an especially moist and slimy appearance.

The Cytoplasm

Few morphologically distinct components can be found within the cytoplasm of most procaryotic cells. All bacteria possess a relatively transparent region of chromosomal material called the nucleoid. The remainder of the cytoplasm contains large numbers of ribosomes which give the cytoplasm a dark appearance when viewed by transmission electron microscopy. In some microbes, cytoplasmic inclusions are microscopically evident. These structures store excess nutrients. Their presence is dependent on environmental conditions. The remainder of the cytoplasm is composed of water, enzymes, and small molecules.

The Nucleoid   The information for bacterial functions is housed on a single circular molecule of DNA. Most cells have only one copy of the chromosome. Before cell division, however, the chromosome will duplicate. The copies are then distributed to the progeny cells and separated by the formation of the septum (cross wall). In some cells, two distinct nuclear regions may be seen because septum formation occurs later than chromosome reproduction.

The size of the chromosome varies according to species. The smallest bacteria, the mycoplasmas, contain the smallest DNAs, directing the synthesis of less than 1000 cellular products. The larger procaryotic organisms, such as cyanobacteria, make over 10 times as many products. *Escherichia coli*, a microbe commonly isolated from the gastrointestinal tract and probably the most studied bacterium, contains a chromosome that houses information for the production of approximately 3000 products.

Additional genetic information may be found on **plasmids** in many bacteria. These are small, circular pieces of DNA that can replicate independently of the chromosome. They are usually less than one-hundredth the size of the chromosome. Because of their small size these components cannot be seen within the cell's cytoplasm when viewed by electron microscopy.

Plasmid DNA may give a bacterium the capacity to synthesize new products. Although plasmids are not usually essential for bacterial growth, they may carry information providing the cell with resistance to antibiotics, with the ability to produce toxins, or with the ability to produce surface appendages essential for attachment and establishment of infection.

Ribosomes Proteins are constructed at the ribosomes. These structures, which occupy much of the cytoplasmic volume, are composed of protein and nucleic acid. Bacterial ribosomes are smaller than eucaryote ribosomes. In addition to size, chemical differences exist between procaryotic and eucaryotic ribosomes, making them, like the cell wall, selective targets for antibiotic action. Streptomycin, for example, binds to bacterial ribosomes and alters their ability to function correctly in protein synthesis. Eucaryotic ribosomes of an infected person's cells continue their activity unaffected. (The toxic side effects of streptomycin are due to another property and are discussed in Chapter 14.)

Inclusions A number of bacteria characteristically store deposits of nutrient materials, usually phosphate, sulfur, carbohydrate, or fat, in structures called **cytoplasmic inclusions** or **granules**. Most cells have the capacity to store only one kind of material, producing a characteristic inclusion which can be stained and microscopically observed. These stained inclusions may help identify the organism. *Corynebacterium diphtheriae*, for example, stores its reserves of inorganic phosphate in granules known as *metachromatic* or *volutin granules* (when stained with methylene blue, the granules turn red). *Yersinia pestis*, the bacillus that causes plague, also stores phosphate, an observation useful in diagnosis.

Appendages

Several structures project through the cell wall to form surface appendages. The most commonly observed bacterial appendages are flagella, axial filaments, and pili.

Flagella Many genera of bacteria move by means of flagella. The location and number of flagella on a cell, as well as the number of waves of an individual flagellum, vary according to bacterial species. These three factors, however, are constant for any given species.

Some organisms have only a single flagellum, an arrangement called *monotrichous* flagellation (*mono*, one; *trichous*, hair). Some bacteria are *lophotrichous*, possessing many flagella arranged in tufts or clusters at one end, or *amphitrichous*, having flagella at both ends of the cell, either singly or in tufts. Since the flagella are on the ends of the cell in these three arrangements, they are referred to as **polar flagella**. If flagella are distributed around the entire cell surface, the cell is said to be **peritrichous**. Some of these arrangements are depicted in Figure 4-15.

Flagella are only about 20 nm in diameter, too thin to be resolved by light microscopes unless special staining procedures are used. The flagella stain employs a chemical that precipitates onto the appendage and increases its apparent diameter to resolvable dimensions.

Bacterial flagella, unlike their flexible eucaryotic counterparts, are somewhat rigid and do not whip back and forth. Instead they spin much

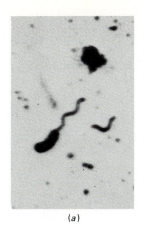
(a)

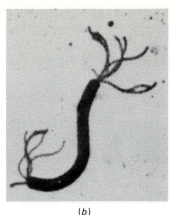
(b)

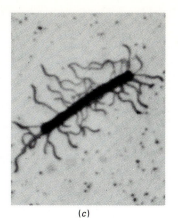
(c)

FIGURE 4-15

The arrangements of flagella on bacterial cells: (a) monotrichous, polar; (b) amphitrichous, polar; and (c) peritrichous.

like the propeller of a boat, using the method described in Figure 4-16. Flagellated bacteria are capable of very rapid movement. *Pseudomonas aeruginosa*, for example, can travel 37 times its cell length in a second. This is equivalent to a 6-foot person swimming at approximately 150 miles per hour. The direction of movement is often determined by **chemotaxis**, the ability to move toward chemical attractants, such as nutrients, and away from repellants such as poisons. Flagella are rarely observed among the cocci.

FIGURE 4-16

Semirigid bacterial flagellum propels the cell by spinning like a propeller. The flagellum shaft rotates in a series of rings anchored to the cell wall. The terminal ring, shown in color, is attached to the flagellum shaft and may be part of the "motor" that drives the flagellum. Rotation of the lower ring spins the flagellum.

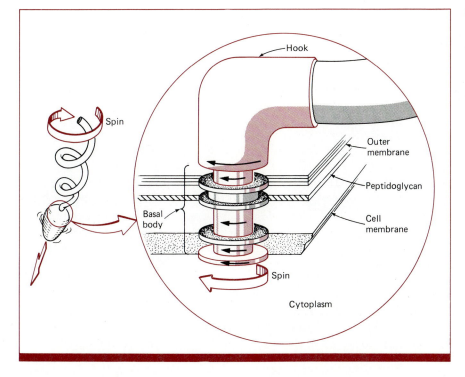

Hook

Spin

Outer membrane

Peptidoglycan

Basal body

Cell membrane

Spin

Cytoplasm

FIGURE 4-17

The axial filament of *Treponema pallidum* is composed of flagella fibrils that arise at each end of the cell and extend halfway down the organism, winding around its surface. The flagella are covered by the cell wall but can be seen if the wall is removed, as in this preparation.

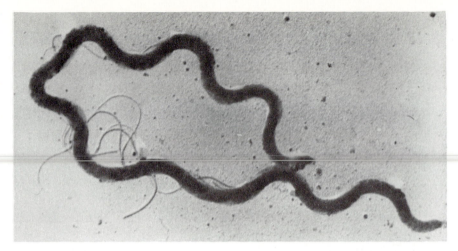

Axial Filaments Spirochetes move by means of specialized structures called **axial filaments** (Fig. 4-17). This structure is composed of two groups of fibers which originate at opposite ends of the cell and overlap in the middle. Structurally and chemically the fibers of the axial filament are similar to flagella and are sometimes called endoflagella (*endo*, within). Unlike flagella, however, they are covered by the spirochete's cell wall. The cell moves by rotating around its longitudinal axis and by flexing and bending along its length.

Pili Pili (singular, pilus) are hollow tubes that extend from the cell. Unlike flagella, pili play no role in motility. They are straighter, shorter, and thinner than flagella and can be observed only by electron microscopy (Fig. 4-18). They are found only on certain species of gram-negative

82

STRUCTURE AND
FUNCTION OF BACTERIA

FIGURE 4-18

Electron micrograph of *Klebsiella pneumoniae* showing numerous pili distributed over the cell surface.

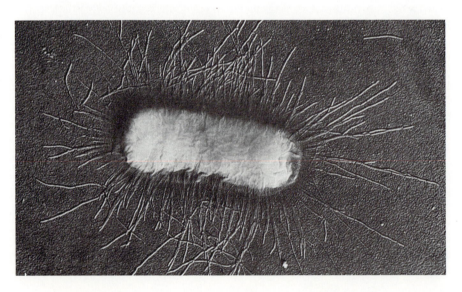

bacteria, usually in large numbers. Pili may differ in appearance (generally in length) or function. Their known functions include:

1. Conjugation between bacteria and subsequent transfer of genetic information (copies of chromosomes or plasmids). Conjugation requires a special pilus called the *F pilus* (F, fertility). After attachment of the F pilus DNA is transferred. Genetic transfer by conjugation is discussed in detail in Chapter 12.

2. Attachment to surfaces such as the tissues of an infected person. Thus pili, like the capsule and glycocalyx, contribute to the establishment of infection by binding bacteria to cell surfaces in areas where they would otherwise be eliminated by movement of body fluids. For example, certain pathogenic variants of *Escherichia coli* which produce a severe diarrhea possess pili (unlike their nonpathogenic counterparts). They remain attached to the intestinal wall rather than being removed by the movement of the gastrointestinal contents. *Neisseria gonorrhoeae* withstand the flushing action of urine by adhering to the urethral canal in a similar fashion; thus pili are critical in the development of gonorrhea.

Endospores

A number of bacteria can change into dormant structures that exhibit no detectable metabolism and do not grow or reproduce. These structures, called **endospores**, are remarkably resistant to heat, radiation, chemicals, and other agents that are typically lethal to the organism. Thus, endospores may protect the cell from many adverse conditions.

Endospore formation is not a means of reproduction since usually there is no increase in cell numbers during the spore cycle (Fig. 4-19). A single bacterium forms a single spore by a process called **sporulation**. The actively growing vegetative state is restored by **germination**. A bacterium may exist for indefinite periods as either a vegetative cell or an endospore.

Sporulation appears to be a response initiated by depletion of an essential nutrient in the bacterial environment. This is a signal to the growing cells that conditions are becoming unfavorable for growth and survival. The spore requires several hours to be produced, and must be completely formed before it encounters such adverse physical conditions as heat, cold, or radiation. Otherwise it will not survive.

The spore develops within the vegetative cell, thus the name endospore. During sporulation, the cell membrane encloses a section of the cytoplasm containing the entire bacterial chromosome, some ribosomes, and other soluble cytoplasmic materials. These components are needed to produce a new vegetative cell when the spore later germinates. A series of relatively impermeable layers develops around and protects this core from damage by toxic chemicals. During sporulation the cell loses most of its water. The dehydrated state of the spore may account for its resistance to heat, since proteins are less susceptible to thermal inactivation in the absence of water.

FIGURE 4-19

Stages in the spore cycle.

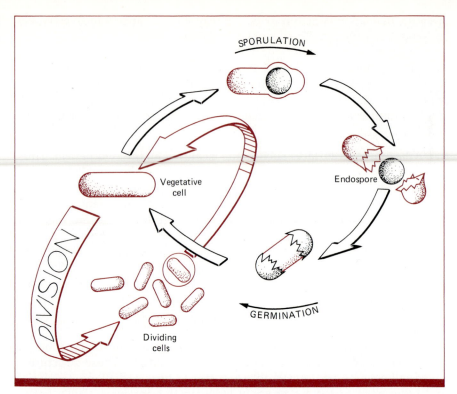

STRUCTURE AND
FUNCTION OF BACTERIA

Despite the fact that mature spores are metabolically inert, they are able to respond quickly to changes in their environment, returning within minutes to the vegetative state. During germination the cell takes up water and enlarges. At the same time the spore coats disintegrate and the vegetative cell emerges (Fig. 4-20). These changes are accompanied by the loss of the spore's resistant properties.

Endospores remain uncolored following Gram staining and may be seen as clear areas within stained cells. The spores can be differentially stained by using special procedures that help dyes penetrate the spore wall. The spore may then be observed as a colored structure when viewed with a standard light microscope. The position of a spore within its parent cell is characteristic of the species and may aid in the microscopic identification of the bacterium. The position may be central, subterminal, or terminal. Additional diagnostic evidence can be obtained by observing the shape of the spore and whether it causes the parent cell to swell (Fig. 4-21).

Endospores are only formed by members of a few genera of bacteria. The majority of spore formers are rod-shaped organisms, members of the genera *Bacillus* and *Clostridium*, most of which are nonpathogenic inhabitants of the soil. The pathogenic spore formers include the agents of tetanus (*Clostridium tetani*), gas gangrene (*Clostridium perfringens*), botulism (*Clostridium botulinum*), and anthrax (*Bacillus anthracis*) (see color box).

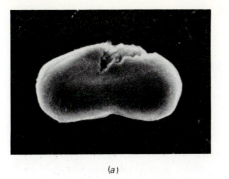

(a)

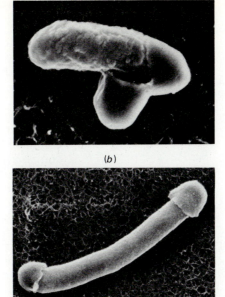

(b)

(c)

85

CELL STRUCTURE

Endospores are probably the most resistant form of life. For example, many can survive in boiling water at 100°C for several hours. Temperatures above 120°C for 10 to 15 minutes are needed to kill the more heat-resistant types. Spores of the bacterium that causes botulism, a fatal food poisoning, can survive in foods that have been subjected to insufficient heating, such as boiling. They may later germinate into vegetative cells that contaminate the food with lethal botulinum toxin. Another potentially fatal disease,

FIGURE 4-21
The appearance of bacterial endospores within their parent cell.

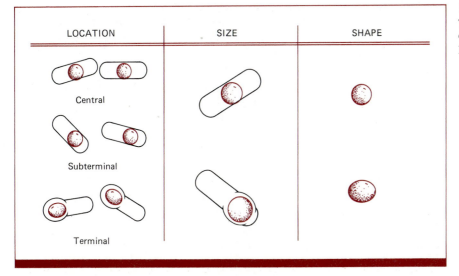

IMMORTAL SEEDS OF DEATH

◼ In addition to being one of earth's most heat-resistant forms of life, bacterial endospores may also be the oldest living microbes on this planet. Most studies on endospore longevity have found that their viability is retained for at least 100 years. Many investigators of longevity chose to study endospores of *Bacillus anthracis*, the bacterium that causes anthrax in people and animals. In 1982, however, their research findings were paled by discovery of an unrecorded disaster in the Soviet Union.

Evidence of the event was unearthed during an archeological excavation at the site of a seventh-century settlement. The village was abandoned and its homes burned 13 centuries ago when it was apparently struck by an epidemic of anthrax. Microbiologists examining the site discovered viable anthrax spores. These spores had lain dormant underground from A.D. 700 until they were eventually disturbed by modern-day archeologists. Viable endospores were even found in closed pots that were buried with the village. These pots contained living organisms 1300 years old.

This finding is probably disturbing to residents in England who have found themselves haunted by an action that occurred during World War II. Fearing the Germans may be developing deadly bacteria into weapons of war, British authorities conducted a series of tests to determine if anthrax spores would survive when dropped and dispersed by bombs. The tests were conducted on Gruinard Island, an uninhabited strip of land 600 yards off the coast of Scotland.

Unfortunately, the experiments were a complete success. Today the island is still contaminated with deadly anthrax endospores, although they lie 6 feet below the ground where, presumably, they are harmless if undisturbed. In October 1981, however, British officials received shocking news that the spores were being disturbed. They found a package containing 10 pounds of dirt removed from Gruinard Island that had been placed near the British laboratories that developed these biological weapons. The dirt, which contained viable anthrax spores, had been collected by members of Operation Dark Harvest, a group conducting a campaign of environmental terrorism in retaliation against those who seeded the island. They claim to have removed 300 pounds of contaminated soil from the island and threaten to distribute these "seeds of death" throughout Britain until the government decontaminates the island. Authorities believe that 280 pounds of contaminated earth are still hidden somewhere in England—280 pounds of earth containing these seemingly immortal seeds of death.

tetanus, may occur when endospores are introduced into human tissue. *Clostridium tetani* spores, which remain dormant in soil, can germinate in the dead tissues of a deep wound where the vegetative cells grow and secrete a toxin capable of causing paralysis and death.

OVERVIEW

Morphological properties—cell size, cell shape, and cell arrangement—are used to differentiate bacteria into major categories. Most bacteria have one of three characteristic cell shapes: coccus, bacillus, or spiral. A few bacteria are either filamentous or pleomorphic.

The components essential to all bacterial cells are the nucleoid, ribosomes, and cell membrane. In bacteria, the cell membrane performs many of the functions accomplished by the specialized organelles of the eucaryote. Internal extensions of the cell membrane, mesosomes and chromatophores, are present in some bacteria. These extensions increase the surface area of the membrane, thereby enhancing its ability to function in respiration, reproduction, and photosynthesis. With few exceptions, a rigid cell wall composed of peptidoglycan surrounds and protects the cell membrane in bacteria. The cell wall also determines the cell shape and the Gram stain reaction. In gram-positive bacteria, the cell wall contains few chemical constituents other than peptidoglycan. The cell wall of gram-negative bacteria contains an outer membrane, which influences the permeability of the cell surface, lipopolysaccharides with endotoxin activity, lipoproteins, and peptidoglycan.

Other structures found in many bacteria are not essential for growth. These are flagella and axial filaments (the most common organelles of motility), pili (mediators of conjugation and attachment to surfaces), protective capsules, and the adhesive glycocalyx. In addition, some bacteria may contain cytoplasmic inclusions, where deposits of excess nutrients are stored. The vegetative cells of several bacteria are able to develop into dormant, resistant structures called endospores. Vegetative growth resumes when the spores germinate in a favorable environment.

KEY WORDS

coccus	protoplast
bacillus	peptidoglycan
spiral	teichoic acid
filamentous bacterium	antigen
pleomorphic bacterium	lipoprotein
coccobacillus	outer membrane
vibrio	lipopolysaccharide (LPS)
spirillum	endotoxin
spirochete	mycoplasma
cell membrane	L form
active transport	capsule
diffusion	slime layer
osmosis	glycocalyx
mesosome	plasmid
chromatophore	cytoplasmic inclusion
cell wall	granule

polar flagellum   pilus

peritrichous flagella   endospore

chemotaxis   sporulation

axial filament   germination

REVIEW QUESTIONS

1. Bacteria characteristically appear in specific shapes and arrangements. Draw and label a simple diagram of common bacterial shapes and arrangements.

2. Describe the structure and function of the following parts of a typical procaryote cell:
   - (a) Cell membrane
   - (b) Nucleoid
   - (c) Inclusions
   - (d) Ribosomes
   - (e) Flagella
   - (f) Pili

3. Compare and contrast the cell wall structure of gram-positive and gram-negative cells.

4. What is the medical significance of
   - (a) Organisms that form endospores?
   - (b) Plasmids?
   - (c) The lipopolysaccharide layer of the gram-negative cell wall?

5. What role does the glycocalyx, capsule, or slime layer play in certain disease processes?

6. What role do pili play in human disease?

7. How do spores differ from vegetative cells?

8. Membranous structures play a vital role in eucaryotes. How do bacteria compensate for the lack of internal membranous organelles?

88

STRUCTURE AND
FUNCTION OF BACTERIA

Bacteria: Systematics

CLASSIFICATION OF PROCARYOTES

BERGEY'S MANUAL
Cyanobacteria
Bacteria
Rickettsias / Chlamydias / Mycoplasmas

OVERVIEW

5

The systematic classification and naming of organisms provides biologists a common basis for communication. The name *Staphylococcus aureus* or *Escherichia coli*, for example, brings a distinct set of characteristics to a microbiologist's mind, as does the name Harry Truman or Winston Churchill to the mind of a historian. In addition, an orderly arrangement of classification establishes rules that can be used to differentiate among microbes. The practical value of classifying microorganisms, therefore, is to provide a method for determining the identity of unknown microbes, such as suspected pathogens isolated from infected patients.

Like all organisms, procaryotes are named according to a system of **binomial nomenclature**; that is, each organism is assigned a name composed of two words, one for its genus and another for the species. Organisms that comprise a species are presumed to be most genetically related. Groups of species that are similar are assigned to a single genus. The genus name appears first and the first letter is always capitalized; the species name, however, is never capitalized. Both names are always underlined (or italicized). Ideally, an organism's name conveys some information about the organism's properties (see color box). For example, *Staphylococcus aureus* was named for its shape (*coccus*, seed) and arrangement (*staphylo*, grapelike clusters) and for the gold color of its colony (*aureus*, gold). Another member of the genus, *Staphylococcus epidermidis* is named for its natural habitat, the skin.

CLASSIFICATION OF PROCARYOTES

Classification of eucaryotes into species is generally based on the ability of two organisms to sexually produce fertile offspring with each other and not with members of other species. This criterion is not applicable to procaryotes, which do not sexually reproduce (bacterial conjugation is not a reproductive process). Classification of procaryotes is generally based on a number of arbitrary criteria which are believed to indicate whether organisms are genetically related to one another. Such criteria include Gram stain reaction, morphology, the ability to produce endospores, and mechanisms of metabolism, motility, and reproduction. These criteria are not necessarily an accurate reflection of genetic relatedness. However, because most of these morphological and biochemical properties are relatively easy to determine in the laboratory, classification based on these criteria has proven to be practical for identifying bacteria isolated from clinical and environmental sources.

The most important criterion currently influencing **taxonomy**, the science of classification, is similarity in chromosomal DNA. DNA similarity is the ultimate indicator of genetic relatedness. Procaryotic cells are easiest to analyze for DNA similarities because they have only one chromosome. Even with procaryotes, however, DNA relatedness is not as easy to measure as most morphological and biochemical properties, and analysis is therefore usually performed only in research laboratories.

Three criteria for genetic relatedness are measured in the research

■ Most bacteria were discovered and named before strict rules for bacterial nomenclature were developed. Many names honor the microbiologist who first described the organism. For example, the genera *Escherichia*, *Salmonella*, *Shigella*, and *Neisseria* derive their names from the microbiologists Escherich, Salmon, Shiga, and Neisser, respectively. Similarly *Rickettsia prowazekii*, the cause of epidemic typhus, is named for two microbiologists, Ricketts and Prowazek, both of whom became infected with this pathogen and died while studying it. Other names describe properties characteristic of the bacterium. *Bacillus* (rod), *Treponema* (spiral), *Clostridium* (spindle-shaped), and *Mycobacterium* (funguslike bacterium), for example, refer to the morphology of the organism. *Proteus*, a motile bacterium that spreads over the surface of solid media and gives rise to a variety of colonial forms, is named after a Greek god who could change his shape. *Clostridium botulinum*, the etiologic agent of botulism, derives its species name from the Latin word *botulus*, meaning sausage. Early outbreaks of this food-borne illness in Europe were traced to the ingestion of homemade sausages contaminated with the toxin-producing bacteria.

In 1976, a new bacterium was isolated as the agent of a sometimes fatal pulmonary infection. The first recognized victims were people attending a convention of the American Legion in Philadelphia. The organism was named *Legionella pneumophila* (*pneumo*, lung; *phile*, lover), commemorating the first recognized victims and their disease.

laboratory. The first and easiest approach is to compare the *size* of the entire chromosome. Clearly, cells whose DNAs differ greatly in size are not related. Similarity in chromosome size, however, does not prove that the organisms are related, since the size similarities may be coincidental. DNAs of similar size must be further compared by a second and stricter criterion, chemical composition of the DNA. In such studies, the amounts of two DNA components, guanine (G) and cytosine (C), are measured, yielding a value called the $G + C$ *content*. Organisms that are related have similar or identical $G + C$ contents. Not all organisms with similar $G + C$ content, however, are related. To rule out coincidence these DNAs must be subjected to the third and strictest test for genetic relatedness, **DNA homology**. This test measures similarities in the sequential order in which the molecular building blocks of DNA are arranged. The degree of DNA homology is the best indicator of genetic relatedness.

New information on the details of microbial relatedness at the molecular level is evaluated by an international judicial commission. One of the tasks of the commission is to define how genetically similar two organisms must be for assignment to the same genus and species. Organisms that are about 40 to 60 percent related according to DNA homology are considered members of a single genus. Those organisms that are more than 70 percent related are assigned to a single species. The natural variability within a species (organisms may be 30 percent unrelated) may be reflected by immunological and biochemical differences. In some species, these differences are used to further subdivide the members of the species into subgroups called **types** or **strains**. If an organism is too different to be assigned to any existing genus or species, it is declared a new organism and given its own name.

DNA homology has shown that some organisms formerly believed to be similar enough to belong to the same genus are actually genetically dissimilar. Conversely, some organisms assigned to different genera have been shown to belong to a single genus. For example, genetic data indicate that members of the genus *Shigella* (the agents of dysentery) are closely related to *Escherichia coli* and should be reclassified as different types of the same species. This change in nomenclature has not yet been made, however, in order to avoid confusion within the medical community. Measurements of DNA were also used to show that the agent of Legionnaires' disease was a newly discovered bacterium, genetically unrelated to any previously known.

Decisions of the judicial commission are published in the *International Journal of Systematic Bacteriology*, which appears four times a year. Their decisions on classification are also incorporated into *Bergey's Manual of Determinative Bacteriology*.

BERGEY'S MANUAL

This manual has been a practical guide for the identification of bacteria since it was first published in 1923. Its classification scheme is based on available genetic data, and it relates the genetic data to those morphological and biochemical tests that are simple to perform in the laboratory. The 1974 (eighth) edition of *Bergey's Manual of Determinative Bacteriology* separates the procaryotes into two divisions, the bacteria and the cyanobacteria. **Cyanobacteria** are photosynthetic procaryotes that produce oxygen as an end product of photosynthesis. All other procaryotes are classified as **bacteria**.

The first of four volumes of the ninth edition of *Bergey's Manual* was released in 1983. This edition separates procaryotes into four divisions, based on the nature of the cell wall. These divisions are Graculicutes (*graculo*, thin; *cutes*, skin), gram-negative bacteria; Firmicutes (*firmi*, strong), gram-positive bacteria; Tenericutes (*tener*, soft), wall-less bacteria; and Mendosicutes (*mendos*, faulty), bacteria possessing cell walls that lack peptidoglycan. For now, the eighth edition continues to be the major taxonomic instrument used by bacteriologists.

Cyanobacteria

Cyanobacteria (also referred to as blue-green algae or blue-green bacteria) contain chlorophylls and accessory pigments similar to those in algae or green plants (the photosynthetic eucaryotes). In addition, the photosynthetic processes of cyanobacteria and photosynthetic eucaryotes are biochemically alike. These two factors differentiate cyanobacteria from the second class of photosynthetic procaryotes, the photosynthetic bacteria.

Cyanobacteria may exist as single cells or as colonies or chains of cells (Fig. 5-1). The internal cell structure of cyanobacteria is more complex than that of bacteria because the photosynthetic apparatus of the former consists of many membrane layers that are connected to the cell mem-

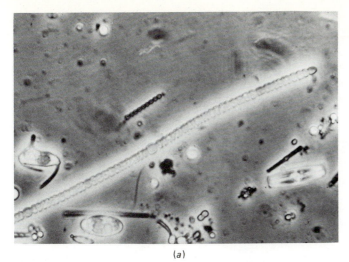

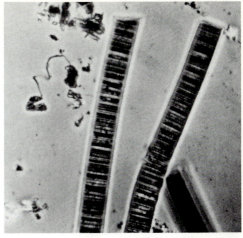

(a)

(b)

FIGURE 5-1
Filamentous cyanobacteria:
(a) *Anabaena* and (b)
Oscillatoria.

brane. Some cyanobacteria produce specialized structures called *heterocysts* that enable the cell to fix atmospheric nitrogen into a form usable by plants and animals. Such activities introduce nitrogen into the food chain.

Cyanobacteria are abundant in aquatic and soil habitats, including such extreme environments as hot springs and saline waters. Although they may produce toxic and odorous compounds, none of the cyanobacteria are known to cause human infectious diseases.

Bacteria

The eighth edition of *Bergey's Manual* divides the bacteria into 19 groups (called parts) distinguished by such properties as Gram stain reaction, morphology, the ability to produce endospores, and mechanisms of metabolism, motility, and reproduction (Table 5-1). The vast majority of medically important bacteria are found in 13 of the parts. Identification of a pathogen from an infected patient usually begins by determining its Gram stain reaction and shape. This information limits the possible choices for the identity of a pathogen to members of only one or two of the 19 parts. Within each part, genera and species are differentiated by more detailed morphological and biochemical tests (Table 5-2). For example, spirochetes (part 5) may differ in size, number of turns in the spiral, and ability to grow in the absence of oxygen. These criteria differentiate among pathogens belonging to the genera *Treponema* (agents of syphilis and yaws), *Borrelia* (agent of relapsing fever), and *Leptospira* (agent of leptospirosis).

Clinical laboratories spend much time identifying bacteria described in part 8. Many of these gram-negative bacilli are important pathogens; others are members of the normal human flora or environmental contaminants. They are all *facultative anaerobes*; that is, they can grow in the presence or absence of oxygen. Bacteria in this part include species of *Salmonella*, *Shigella*, *Escherichia*, *Proteus*, *Yersinia*, *Vibrio*, and *Haemophilus*. Within this part, bacteria that possess peritrichous flagella have been

TABLE 5-1

Outline of Bacteria Classification According to *Bergey's Manual**

Part	Distinguishing Features	Some Medically Important Genera	Representative Diseases
1. Phototrophic bacteria	Photosynthetic without the production of oxygen		
2. Gliding bacteria	Capable of slow, gliding movement but lacking locomotor organelles		
3. Sheathed bacteria	Covered with a sheath		
4. Budding or appendaged bacteria	Reproduce by budding and/or produce appendages (other than flagella or pili)		
5. Spirochetes	Flexible, coiled morphology; motility by axial filament	*Treponema* *Borrelia* *Leptospira*	Syphilis Relapsing fever Leptospirosis
6. Spiral and curved bacteria	Rigid, coiled morphology	*Spirillum* *Campylobacter*	Ratbite fever Diarrhea
7. Gram-negative aerobic rods and cocci	Capable of growth only in presence of oxygen *a.* Rods motile by polar flagella *b.* Coccobacilli of uncertain affiliation	*a. Pseudomonas* *b. Brucella* *Bordetella* *Francisella*	Wound infections Undulant fever Whooping cough Tularemia
8. Gram-negative facultatively anaerobic rods	Capable of growth in presence or absence of oxygen *a.* Motile by peritrichous flagella	*a. Escherichia* *Shigella* *Salmonella* *Klebsiella* *Yersinia*	Intestinal and urinary tract infections Dysentery Food poisoning, typhoid fever Pneumonia Plague
	b. Motile by polar flagella *c.* Uncertain affiliation	*b. Vibrio* *c. Haemophilus*	Cholera Meningitis
9. Gram-negative anaerobic bacteria	Capable of growth only in absence of oxygen; nonspore formers; pleomorphic	*Bacteroides*	Abscesses
10. Gram-negative cocci and coccobacilli	Generally occurring as diplococci; nonflagellated	*Neisseria* *Acinetobacter*	Gonorrhea, meningitis Meningitis
11. Gram-negative anaerobic cocci	Capable of growth only in absence of oxygen	*Veillonella*	Endocarditis
12. Gram-negative chemolithotrophic bacteria	Derive energy from inorganic compounds		
13. Methane-producing bacteria	Obtain energy for growth via formation of methane; anaerobic		
14. Gram-positive cocci	Nonspore formers	*Staphylococcus* *Streptococcus*	Boils, food poisoning, pneumonia Scarlet fever, impetigo, pneumonia
15. Endospore-forming rods and cocci	*a.* Obligate aerobes *b.* Obligate anaerobes	*a. Bacillus* *b. Clostridium*	Anthrax Tetanus, botulism, gas gangrene

TABLE 5·1 (*continued*)
Outline of Bacteria Classification According to *Bergey's Manual**

Part	Distinguishing Features	Some Medically Important Genera	Representative Diseases
16. Gram-positive asporogenous rods	Nonspore formers	*Listeria*	Meningitis
17. Actinomycetes and related organisms	Rods, pleomorphic rods, or filaments; some acid-fast	*Actinomyces Corynebacterium Mycobacterium Streptomyces†*	Lumpy jaw Diphtheria Tuberculosis, leprosy
18. Rickettsias	Obligate intracellular parasites of eucaryotic cells	*Rickettsia Chlamydia*	Typhus, spotted fever Trachoma, urethritis
19. Mycoplasmas	Lack cell wall; highly pleomorphic	*Mycoplasma Ureaplasma*	Pneumonia Urethritis

*J. G. Holt (ed.), *The Shorter Bergey's Manual of Determinative Bacteriology*, 8th ed., Williams & Wilkens Co., Baltimore, 1974.
†Important antibiotic producers.

assigned to the family Enterobacteriaceae (Fig. 5-2). Because all members appear to be virtually identical, practical identification of Enterobacteriaceae is based primarily on biochemical characteristics that are easily determined in the laboratory.

Two parts consist of atypical bacteria. These are the obligate intracellular parasites (rickettsias and chlamydias, part 18) and the mycoplasmas, bacteria that lack cell walls (part 19).

TABLE 5·2
An Example of a Key to Distinguish among Genera

> **Key Characteristics of the Genera of Spirochetes**
> 1. Size, shape of end, and degree of coiling of cells
> 2. Number of turns
> 3. Oxygen requirement
> 4. Habitat
> 5. Presence of inclusions and bundles of axial filaments as seen with phase-contrast microscopy
> 6. Catalase and oxidase biochemical reactions
> 7. Pathogenicity

Source: J. G. Holt (ed.) *The Shorter Bergey's Manual of Determinative Bacteriology*, 8th ed., Williams & Wilkens Co., Baltimore, 1974.

Rickettsias   Rickettsias are small gram-negative pleomorphic rods that possess all the structural and chemical features of typical bacteria. Unlike most bacteria, however, they cannot be cultivated on inanimate (cell-free) media. (The one exception is *Rochalimaea quintana,* the agent of trench fever.) They are obligate intracellular parasites that require living host cells for propagation. Within the eucaryotic cell, the nutritionally rich cytoplasm provides the microbe with material needed for growth (Fig. 5-3). Rickettsias are grown in the laboratory in embryonated chicken eggs or in animal cells cultured in test tubes, bottles, or petri dishes. The rickettsias are naturally found in the host cells of a wide variety of arthropods and

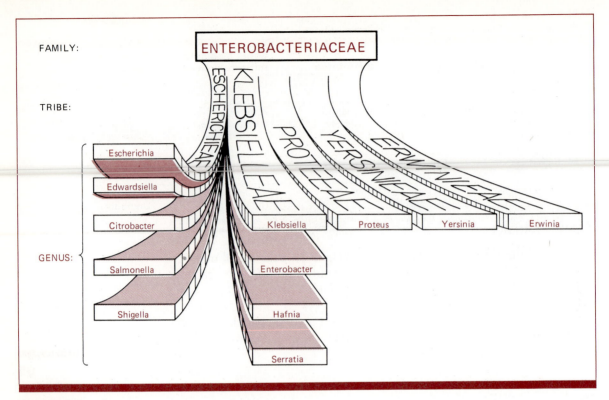

FIGURE 5-2
The family *Enterobacteriaceae* is divided into five tribes based primarily upon genetic data (G + C content) and major biochemical distinctions. Within each tribe assignment to genus and species is based on DNA homology if the information is available. Laboratory identification depends upon biochemical tests that reflect only about 10 percent of the information in the bacterium's DNA.

animals. The arthropod may be the primary host or a vector transmitting the rickettsias between animal hosts. In humans, rickettsias may cause typhus, Rocky Mountain spotted fever, or other severe diseases.

Chlamydias   Chlamydias are another group of obligate intracellular parasites that are classified in part 18 with the rickettsias. These procaryotes cannot grow outside a host cell because they lack critical enzymes for generating compounds needed for energy transfers. Thus chlamydias, like rickettsias, utilize the products of the eucaryotic host cell system. Although classified with rickettsias in *Bergey's Manual*, chlamydias differ from rickettsias in their mode of reproduction. Rickettsias, as most other bacteria, reproduce by binary fission, a process of simple cell division by which one parent cell produces two daughters of equal size. Chlamydias,

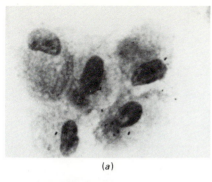

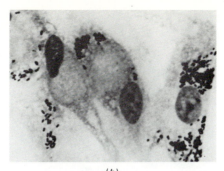

(a)   (b)

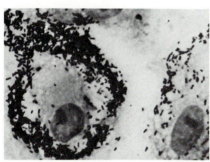

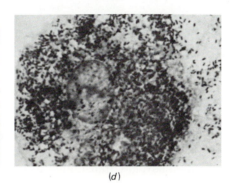

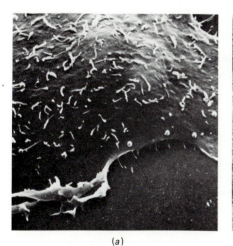

(c)   (d)

FIGURE 5-3

Rickettsia rickettsii proliferating within a host cell. The microbe is dividing by binary fission. After 2 hours (*a*) only a few bacteria are present. After 6 days (*d*) the engorged cell releases the infectious progeny.

on the other hand, undergo a unique cycle of developmental changes. Outside a host cell the organisms exist as dense sporelike cells, the *elementary bodies* (Fig. 5-4*a*). These are engulfed through phagocytosis by a eucaryotic cell. Intracellularly, the elementary bodies develop into larger *reticulate bodies*, which multiply by binary fission (Fig. 5-4*b*). The reticulate bodies are the only metabolically active form of chlamydias. (They cannot, however, make their own energy transfer compounds.) Progeny reticulate

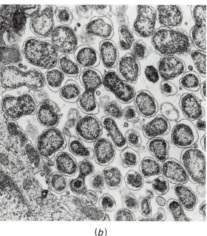

FIGURE 5-4

(*a*) Elementary bodies of *Chlamydia psittaci* attached to the surface of a host cell. (*b*) *Chlamydia psittaci* reproducing within the host cell. These reticulate bodies are dividing by binary fission.

(a)   (b)

FIGURE 5-5

Mycoplasmas are characterized by the absence of a defined shape. Note the tapered ends of many of the cells.

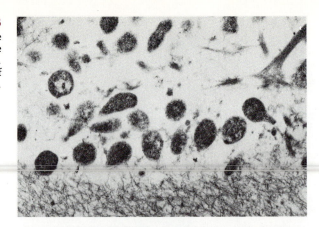

bodies ultimately reorganize into new elementary bodies, which are released from the host cell and are able to initiate another infection cycle.

There are only two species of chlamydias. *Chlamydiae psittaci* causes psittacosis, a form of human pneumonia. *Chlamydiae trachomatis* causes trachoma (currently the world's leading cause of blindness), as well as several genital and respiratory infections. *C. trachomatis* is also one of the causes of nongonococcal urethritis (NGU), considered to be the most common sexually transmitted disease in the western world.

Mycoplasmas Mycoplasmas are distinguished from other bacteria by the absence of a cell wall. The lack of a rigid cell wall has four major consequences.

1. Mycoplasmas do not assume a defined and clearly recognizable shape. Instead they are highly pleomorphic organisms showing variations in size and shape (Fig. 5-5).

2. They are gram-negative organisms.

3. They are resistant to penicillin and cephalosporins, antibiotics that function by preventing cell wall formation.

4. Mycoplasmas must be maintained on special media that protect them from osmotic lysis. On agar they produce characteristic colonies with a "fried egg" appearance (Fig. 5-6).

FIGURE 5-6

The "fried egg" appearance is a characteristic of colonies of mycoplasmas.

Many members of the genus *Mycoplasma* are unique among the procaryotes because they have sterols in their plasma membrane. Sterols are thought to confer some rigidity and protection to the membranes of these wall-less forms.

Mycoplasmas are best known as pathogens of "nonbacterial" pneumonia and genitourinary tract infections such as nongonococcal urethritis.

OVERVIEW

Systematic classification of procaryotes establishes a practical method to identify microbes and to assign a meaningful name to each organism. Whenever possible, classification is based on DNA studies. Routine bacterial identification, however, is usually based on morphological and biochemical tests that reflect the genetic composition of the organism. Such information is compiled in a practical guide called *Bergey's Manual*.

The procaryotes are divided into the cyanobacteria and bacteria. Bacteria are assigned to one of 19 parts based on morphological and biochemical properties. Further division of each part into genera and species is based on more elaborate tests. The vast majority of the bacteria are free-living organisms with typical procaryote features. Two groups are atypical: mycoplasmas lack a cell wall and rickettsias and chlamydias are obligate intracellular parasites.

KEY WORDS

binomial nomenclature	cyanobacterium
taxonomy	bacterium
DNA homology	rickettsia
type	chlamydia
strain	mycoplasma
Bergey's Manual	

REVIEW QUESTIONS

1. List five major characteristics currently used in the classification of bacteria.

2. Describe the three criteria used to measure genetic (DNA) relatedness.

3. What is *Bergey's Manual of Determinative Bacteriology?*

4. What differentiates the following from typical procaryotes?
   (a) Mycoplasmas
   (b) Rickettsias
   (c) Chlamydias

5. Why is the Gram stain an important initial step in the diagnosis of infectious disease?

BACTERIA: SYSTEMATICS
AND NOMENCLATURE

Bacteria: Growth and Laboratory Cultivation

Microorganisms are active creatures. In an accommodating environment they are constantly metabolizing, growing in both size and numbers. Simply stated, microbes are metabolic machines with an explosive potential for multiplication. Sometimes we can use the growth of microbes to our advantage by utilizing their metabolic activities or harvesting their many valuable by-products. At other times, however, we are victims of their harmful activities. For example, a single *Streptococcus* accidentally inoculated through a cut in the human skin can quickly increase in numbers within the body to levels sufficient to cause severe fever or even death. Fortunately other bacteria produce antibiotics which can be harvested and used to fight infectious disease. One major role of microbiology is to study and manipulate microbial growth so we may better take advantage of microbes' beneficial properties and minimize their harmful effects.

The production of valuable chemotherapeutic agents is but one of the beneficial consequences of microbial growth. Other useful materials manufactured by microbes include foods, beverages, alcohol, acetone, and acetic acid. In addition, the ability to grow microorganisms in the laboratory is essential for studying the basic characteristics of microbes, as well as for establishing an accurate diagnosis of most infectious diseases and for selecting the most effective antimicrobial therapy. Some pathogens grow faster under laboratory conditions than in the body and can be rapidly identified. This gives the clinician the added time advantage that is needed to halt the disease process before irreversible injury is done. However, if conditions for isolating and culturing a pathogen are undefined or unsuccessful, diagnosis may be complicated or impossible. For example, in 1976 when the sudden emergence of Legionnaires' disease killed 29 members of the American Legion attending a convention in Philadelphia, attempts to determine the cause of this illness were frustrated by the failure to isolate a pathogen from the victims. Meanwhile antibiotic therapy was a "hit-or-miss" affair, since accurate choices must be based on tests that utilize the isolated pathogen. Only after 5 months had passed and several hundred cases had been reported were the attempts at isolation successful. Identification of the growth requirements of the organism finally facilitated the laboratory diagnosis of the disease and dramatically reduced the death rate. Our ability to control many such infectious diseases depends on our understanding of microbial reproduction and growth (see color box).

REPRODUCTION AND EXPONENTIAL GROWTH

Bacterial growth usually refers to reproduction—increases in population size—rather than to enlargement of individual cells. Bacteria can reproduce by a variety of mechanisms. The actinomycetes form long filaments which split into living fragments by the process of *filamentation* and breakage. Each fragment is an individual cell capable of giving rise to a new filament. The streptomycetes develop *spores* at their tips. Once released, these spores can give rise to a new colony of

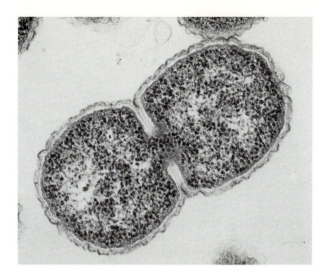

FIGURE 6-1
Neisseria gonorrhoeae
multiplying by binary
fission.

microorganisms. A few bacteria are capable of *budding*, producing new, smaller cells from the surface of a single organism. The majority of bacterial cells, however, reproduce by **binary fission**, a process whereby one cell divides into two daughter cells of equal size (Fig. 6-1).

The Potential of Binary Fission

Each time a cell divides by binary fission it forms a new **generation** of cells. Each cell of the new generation is capable of further division by fission. A

MICROBIAL GROWTH AND PREVENTION OF DISEASE

■ Although microorganisms can be agents of annoying and sometimes fatal disease, they also contribute to the arsenal against disease. In fact, some pathogens can even be used against themselves. For example, the essential component of a vaccine is a microbe (or one of its products) that has been altered so it can no longer cause disease. When introduced into the body the modified agent stimulates immunity against that disease. In this way whooping cough, a dangerous disease in children, has been successfully combated by vaccinating people with killed preparations of the organism. The vaccine against pneumococcal pneumonia is an extract of the microbe's capsule. Vaccines developed against diphtheria and tetanus are inactivated derivatives of the disease-producing toxins. These toxins are elaborated during laboratory cultivation of the pathogenic agents.

Microbes with potential to serve as major food sources may someday also contribute to disease prevention. Since malnourishment reduces resistance to disease, new and economical alternatives for nutrition, typified by *single-cell protein*, can help prevent infectious diseases. This food would be produced by the industrial cultivation of large quantities of nutritionally rich bacteria, fungi, or algae which can be economically grown on waste materials. The distribution of single-cell protein to developing countries can help reduce tragic outbreaks of cholera, an infectious disease that rarely kills well-nourished persons.

FIGURE 6-2

Exponential growth in a
bacterial population with a
doubling time of 30
minutes. Population
growth may be expressed
as number of bacteria or as
logarithm of the number of
bacteria.

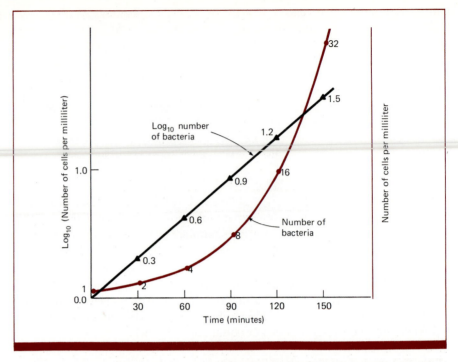

BACTERIA: GROWTH AND
LABORATORY CULTIVATION

single cell, after one division, gives rise to two cells. The next division finds these two cells becoming four (2×2, or 2^2), then four cells become eight (4×2, or 2^3), and so on. Each generation has twice the number of cells as the preceding generation. Therefore, during active growth, the population size is increasing at an exponential rate (Fig. 6-2). The number of bacteria that are present at any time depends on the size of the initial population and the number of generations that have been produced. Mathematically, this is expressed by the formula

$$B_f = B_i \times 2^n$$

where B_f is the final number of bacteria, B_i is the initial population size, and n is the number of generations. For example, if we start with a single cell, the first five doublings (generations) will yield 2^5, or 32, progeny. Because of the exponential nature of microbial growth, these numbers increase rapidly with each additional generation. Thus after another five doublings there are 2^{10}, or over 1000 cells and after 20 generations, over 1 million bacteria have been produced from a single cell.

The amount of time it takes for a population to reproduce is called the **doubling time**, or **generation time**. During this time one cell may be dividing into two, or 1000 cells into 2000. The doubling time is usually constant for each organism as long as physical and chemical conditions do not change. For example, *Escherichia coli* has a generation time of less than 30 minutes under ideal conditions in the laboratory. On the other hand, in the gut, where nutritional conditions are not optimal or where competing microorganisms exist, it may require as much as 12 hours to produce a single generation. Other microorganisms multiply even more slowly. *Mycobacterium tuberculosis* has a generation time of 12 hours on laboratory media, and requires up to 6 weeks before producing a visible colony. *Treponema pallidum* has a generation time of 30 hours in rabbit testes; *Mycobacterium leprae* reproduces once every 13 days in the armadillo. Neither of the last two microbes can be cultured on laboratory media; they are propagated only in living tissue.

The initial concentration of microbes introduced into a person's body influences the number of microbes present after a period of time. To illustrate, suppose that 1000 of the potentially pathogenic organisms released in an unrestrained sneeze are transmitted into the respiratory tract of a susceptible host. This may be well above the **infectious dose**, the number of organisms needed to initiate infection in the host. Covering the mouth during sneezing may limit this number to 100, a difference of only 900 organisms for the host defenses to eliminate. It takes time, however, to mobilize host defenses, time during which the invading pathogens may be actively reproducing. After one generation, sometimes only a half hour, the microbes from the unrestrained sneeze will have produced 2000 progeny, compared to 200 generated following the controlled sneeze. With more exponential growth, the difference between the numbers of organisms that the host defenses must eliminate becomes huge. Such growth dynamics also emphasize the importance of rapid diagnosis and treatment of infectious diseases, since delays may allow pathogens to proliferate to uncontrollable concentrations.

The short generation times of most of the commonly encountered microorganisms facilitate the rapid production of colonies composed of billions of cells in less than a day after inoculation on a solid growth

FIGURE 6-3
(*a*) A developing colony of *Staphylococcus aureus* as seen by scanning electron microscopy. (*b*) Macroscopic appearance of *Staphylococcus aureus* colonies.

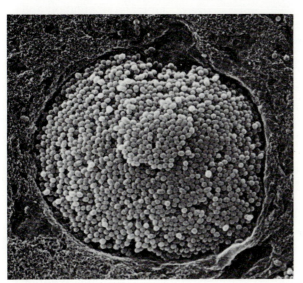

(a)

(b)

medium (Fig. 6-3). This burst of growth, however, is far short of the growth potential of the organisms. A microbe with a doubling time of 30 minutes, two generations per hour, has the theoretical capacity to produce 2^{48} — or 281,474,976,710,656 — progeny in 24 hours. The volume of this number of bacteria alone would be approximately one liter. Continued exponential growth for one or two more days would yield an astounding microbial population having a mass greater than that of the earth. Obviously, microbial growth on agar or in liquid does not continue unchecked. Microbes do not come pouring out over the sides of the petri dish nor do they crowd together in a flask, creating a solid mass of cells. Despite their capacity to reproduce, there are some natural limitations on microbial growth. These limitations are illustrated by the kinetics of bacterial growth in a **batch culture**, a closed vessel or system containing a nutrient medium to which nothing is added or removed during the growth of the culture.

Bacterial Growth in Batch Culture

The dynamics of bacterial growth can be observed by inoculating bacteria into a liquid culture medium and measuring population sizes at regular time intervals. When these measurements are plotted on a graph, the resulting **growth curve** shows four distinct phases of growth: (1) lag phase, (2) logarithmic phase, (3) stationary phase, and (4) death phase. A typical growth curve is shown in Figure 6-4.

The Lag Phase When microbes encounter a new environment, they do not begin to multiply immediately but require a period of adjustment. During this **lag phase** there is no detectable increase in cell number. Despite the fact that no reproduction is occurring, the microbes are

106

BACTERIA: GROWTH AND
LABORATORY CULTIVATION

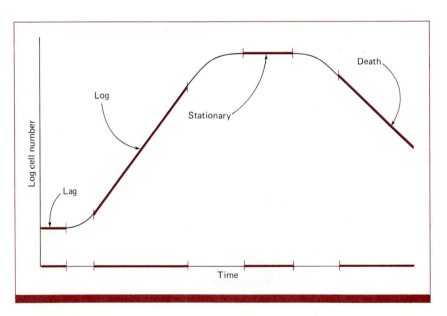

FIGURE 6-4

The bacterial growth curve. The actual length of each phase varies with the organism and with environmental conditions.

metabolically active, adapting to their new environment before they begin to increase their numbers by cell division. If the cells are old or have been dormant, essential constituents that have become depleted or damaged are replaced during the lag phase. Cells that have been transferred from a nutritionally rich environment to one that is nutritionally poor need time to synthesize enzymes for the manufacture of materials that were previously supplied in the growth medium. Cells unable to adapt to the requirements of a new medium will not grow. On the other hand, the lag phase may be short or absent altogether if healthy, actively multiplying microbes are inoculated into the same type of medium as that in which they were already growing.

The Logarithmic Phase   The lag phase ends when cells have adapted to their new surroundings and begin to divide. When all the cells in the culture are actively dividing, the culture is in the **exponential**, or **logarithmic (log)**, **phase** of growth. (The number of bacterial cells during exponential growth is recorded as a logarithm to represent exponential changes.) The **growth rate**, the number of doublings per hour, is influenced by genetic factors that determine the speed at which materials from the culture medium can be absorbed and assimilated by each type of organism. In addition, growth rate is strongly influenced by the physical environment and the composition of the culture medium. During logarithmic growth, most members of the population are at their metabolic peak and are therefore most sensitive to inhibition by agents that disrupt metabolic functions.

Microbes continue to multiply at a logarithmic rate until either of two factors terminates the log phase: (1) the supply of an essential nutrient is exhausted or (2) toxic products from microbial metabolism accumulate to inhibitory levels. Fewer and fewer cells divide and the culture begins the transition into the stationary phase.

The Stationary Phase   During the **stationary phase**, there is no net increase in population size. Constant cell numbers are maintained because the number of cells being produced is equivalent to the numbers that are dying. More important, the microbes gradually stop multiplying and switch to the lowest levels of metabolism and energy expenditure that will allow them to remain viable. At this stage a large part of the population is in a "suspended" state—no longer dividing but still able to return to active growth if conditions should once again become favorable. Cells in the stationary phase are usually less susceptible to antibiotics or other agents that inhibit metabolic processes because these processes are temporarily inactive. As the stationary phase continues, cells begin to age as more of the cellular reserve materials are utilized for growth and maintenance. Some microbes undergo protective changes, such as formation of endospores, that make them even more resistant to damage by adverse physical and chemical agents.

The Death Phase Eventually, those microbes that cannot maintain the dormant state will die. For bacteria, death is defined as the inability to reproduce when transferred to a fresh medium in ideal conditions. Some organisms contain enzymes called *autolysins* which lyse the cells as they age. Death, like growth, occurs logarithmically.

Continuous Cultures

Batch cultures eventually die because of nutritional depletion or accumulation of toxic by-products. Research and industry, however, often require bacterial cultures that can be maintained indefinitely in the logarithmic phase of growth. This can be accomplished by the use of **continuous cultures**. These are open systems in which microbial populations are maintained in exponential growth by continuing to supply fresh nutrients to the incubation vessel while simultaneously removing toxic wastes and excess microorganisms. A continuous culture can be maintained in the laboratory by using a *chemostat* (Fig. 6-5a). The continuous flow from the growth vessel produces a steady, readily available supply of actively multiplying organisms and of any metabolic by-products produced during exponential growth.

Continuous culture environments are common in nature. For example, the human gastrointestinal tract supports a population of well-fed normal flora microbes. A person's diet serves as a constantly renewed source of nutrition to these resident bacteria, which can multiply to

FIGURE 6-5

(*a*) The chemostat, a continuous culture apparatus, maintains logarithmic growth by the constant replacement of used media with fresh media. (*b*) Like the chemostat, the gastrointestinal tract maintains organisms in the logarithmic phase of growth.

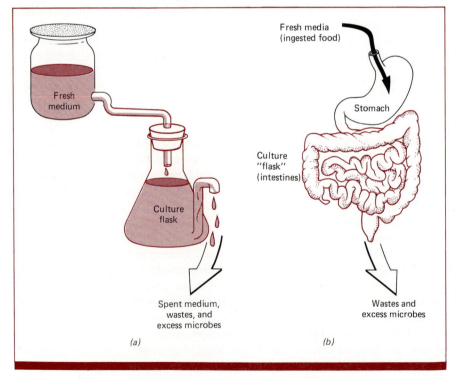

FIGURE 6-5
(*a*) The chemostat, a continuous culture apparatus, maintains logarithmic growth by the constant replacement of used media with fresh media. (*b*) Like the chemostat, the gastrointestinal tract maintains organisms in the logarithmic phase of growth.

concentrations greater than 10^{10} organisms per gram of intestinal contents. These organisms are in a constant state of reproduction, since fresh nutrients are supplied and gastrointestinal contents are flushed out of the body with the feces (Fig. 6-5b).

REQUIREMENTS FOR MICROBIAL GROWTH

To grow, all microbes must have suitable physical and chemical conditions, nutrients, and freedom from interference by antagonistic competitors. In natural habitats, conditions are rarely conducive to *optimal microbial growth,* multiplication at the maximum growth rate (the shortest possible generation time). In the laboratory, however, reproductive rates can be increased or decreased by manipulating nutrients and physical conditions.

Nutritional Requirements

To survive, every cell on earth needs (1) moisture, (2) a source of energy to fuel the demands of cell work, and (3) chemical compounds to supply the building blocks from which the cell is composed. If an organism is to be successfully cultivated in the laboratory, these needs must be satisfied by the culture medium.

The majority of microorganisms have simple nutritional needs; they can be cultivated with little difficulty on culture media containing sugars, water, and salts to provide essential elements. Some microbes, on the other hand, are **fastidious**, requiring unusual nutrients. These needs make it more difficult and sometimes impossible to grow them in the laboratory. The isolation and identification of fastidious pathogens may, therefore, present a problem to the clinical microbiologist.

Moisture Most organisms have difficulty growing or even surviving in environments where water is scarce. Outside the cell, water carries dissolved particles that are to be transported into the cell. Water also serves as the solvent in which the cell's internal biochemical reactions occur and as the medium for elimination of soluble waste materials. Pathogenic organisms are generally bathed in an abundance of water within the tissues of their living host, and their survival outside the host often depends on their ability to resist drying. *Treponema pallidum* and *Neisseria gonorrhoeae,* for example, are so sensitive that they die within 20 seconds after drying. Direct contact is therefore necessary for their transmission from one host to another. In clinical specimens microorganisms must be protected from drying by placement in special liquid media for transport to the laboratory. Alternatively, the sample may be inoculated directly onto solid media at the time of collection. (The addition of agar to a liquid medium does not reduce its water content. It merely solidifies it.)

Many organisms, on the other hand, become metabolically inactive when desiccated but resume growth when moisture is restored. They may survive for long periods in the dry environments of soil or dust or on inanimate objects. Dormant microbes remain viable in dehydrated foods and can resume growth when the product is rehydrated. In fact, desiccation is one of the most common methods of preserving bacterial cultures in the lab. The viability of dried cultures can be maintained for years.

Energy and Carbon Sources Growth, movement, metabolism, protein synthesis, and many other essential cell processes demand energy. Furthermore, most of these processes require a source of carbon, one of the major building blocks for constructing cell material. All organisms can be usefully divided into one of two categories according to their energy source. The **phototrophs** derive energy from sunlight through photosynthesis. The **chemotrophs**, on the other hand, depend on energy they harvest by breaking molecular bonds. Chemical energy sources are usually organic compounds (molecules containing carbon and hydrogen) such as sugars and amino acids. Some bacteria, however, derive energy from inorganic compounds, especially those containing nitrogen, iron, or sulfur.

Organisms are further classified according to the nature of the carbon source. The **autotrophs** use inorganic carbon, in the form of carbon dioxide (CO_2), as their sole source of carbon, assimilating it into complex organic compounds of which the cell is made. **Heterotrophs**, on the other hand, require a supply of carbon in the form of organic molecules.

Bacteriologists usually combine these terms to characterize an organism's nutritional needs according to both its carbon and its energy sources. This scheme produces four categories: *photoautotrophs, photoheterotrophs, chemoautotrophs,* and *chemoheterotrophs* (Table 6-1). The majority of bacteria, including all pathogens, are chemoheterotrophs, obtaining both their energy and carbon from organic compounds.

Essential Elements In addition to carbon, all cells need supplies of hydrogen, oxygen, nitrogen, phosphorus, and sulfur. Hydrogen and oxygen, along with carbon, are essential for the synthesis of most organic compounds. Phosphorus is needed for nucleic acids, sulfur for proteins, and nitrogen for nucleic acids and proteins. Some chemotrophs use the same compounds as sources of both energy and essential elements.

TABLE 6-1
Nutritional Categories among Microorganisms (Based on Carbon and Energy Sources)

Category	Energy Source	Carbon Source	Representative Microbe
Photoautotroph	Light	CO_2	Algae; photosynthetic bacteria
Photoheterotroph	Light	Organic compounds	Photosynthetic bacteria
Chemoautotroph	Inorganic compounds	CO_2	Sulfur, iron, and ammonia-oxidizing bacteria; several types of methane-producing bacteria
Chemoheterotroph*	Organic compounds	Organic compounds	Protozoa; fungi; most bacteria

*Chemoheterotrophs usually use the same compounds for both carbon and energy.

Metals are also essential to the cell but they are required in relatively small quantities. Potassium, magnesium, calcium, and iron are used as components in some critical cell structures or to assist in certain enzyme reactions. Several other metals, called *trace elements,* are required in such minute concentrations that the levels normally found in the water used to prepare media in the laboratory are usually sufficient to meet the microbes' demands. Molybdenum, copper, cobalt, and zinc are examples of trace elements.

Organic Growth Factors All cells also require *amino acids* for manufacturing proteins, *purines* and *pyrimidines* for making nucleic acids, and *vitamins,* which assist in many enzyme-mediated reactions. Organisms that cannot synthesize these organic growth requirements must be supplied with them in the culture medium. Some bacteria manufacture all these components within the cell and do not need an external source. Such organisms have simple nutritional requirements and can grow on a medium that contains only a source of carbon, nitrogen, energy, water, and other essential elements. This is called a *minimal medium* because it supports only the growth of microbes having minimal needs. Fastidious microbes fail to grow on minimal media.

Physical Requirements

Microorganisms need more than nutrients to proliferate. Environmental conditions must be within a suitable range. Temperature, pH, oxygen, and osmotic pressure all influence survival and growth. Light is an additional factor essential for phototrophs. The range of tolerance for these conditions largely determines where an organism will be found in nature. The human body itself can be considered a series of microenvironments, each of which has the potential to harbor a limited spectrum of microorganisms. The physical and nutritional requirements of the more fastidious organisms, especially obligate parasites, are usually as complex as those of their hosts. These organisms are often difficult to culture in artificial environments.

Temperature All organisms can be characterized by the range of temperatures within which they grow. Within this range is an optimum temperature at which the growth rate is maximal (Fig. 6-6). Bacteria may be placed into one of three categories according to the temperature range in which they grow best:

1. **Thermophiles** are organisms whose optimum temperatures are above 40°C. Habitats from which thermophiles may be isolated include hot springs, tropical soils, compost piles, hot water heaters, hot tubs, and thermal vents in the ocean floor (see color box on page 113).

2. **Mesophiles** prefer temperatures between 20 and 40°C. Human pathogens are mesophiles, adapted to our 37°C body temperature, which, even with fever, rarely exceeds 40°C.

FIGURE 6-6

Categorization of bacteria
according to range of
temperatures that allow
growth.

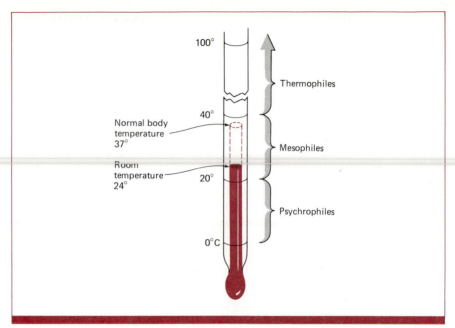

3. **Psychrophiles** (also called cryophiles) thrive at temperatures below 20°C, some down to −7°C. They are all capable of growing at 0°C. These organisms proliferate in deep ocean waters, arctic and antarctic habitats, and in refrigerated or frozen foods.

Some organisms are capable of multiplying throughout a wider range of temperatures than indicated by these categories. A few mesophiles can grow at the elevated temperatures found in foods kept warm on heating tables and in electric slow cookers. This phenomenon is responsible for many cases of food-borne illness associated with catered meals and cafeteria-style restaurants.

Low temperatures generally retard cell growth without killing microbes. Even freezing fails to kill most microorganisms in food or medical supplies. Once permissive temperatures are restored, growth resumes. In fact, rapid freezing such as in liquid nitrogen at temperatures of −196°C is a common mechanism for preservation of viable cultures. Slow freezing, on the other hand, especially repeated freeze-thawing, generates large ice crystals that can cut the cell membrane, causing cell death in even the heartiest bacteria. Some fragile microbes, however, such as *Neisseria* species are cold-sensitive and die when merely refrigerated. Refrigerating specimens for diagnosing gonorrhea or meningitis may therefore result in misdiagnosis.

As temperatures increase, enzyme-mediated reactions occur faster. Bacteria consequently grow more rapidly in warmer environments. Above a critical temperature, however, heat denatures protein, irreversibly dam-

NATURE'S LAWBREAKERS

■ The laws of nature place unavoidable restrictions on the activities of the earth's organisms. One of the most fundamental of these laws concerns heat: "Thou shall not grow in temperatures above 100°C." The boiling point of water has long been believed to be the upper thermal limit for growth. Penalties for breaking the law are severe — cytoplasm boils, proteins denature, DNA melts into separate strands, and lipids run into a nonfunctional mess. In short, violators suffer the ultimate punishment. This law was slightly bent in 1982 with the discovery of thermophiles growing at 105°C in the hot waters streaming from thermal vents in the ocean floor. Less than a year later, the law was shattered by a discovery that staggers scientific imagination. Another bacterium was brought up from the hot submarine floor, this one recklessly proliferating at 250°C (482°F) — well above the temperature required to ignite this page and burn your book.

How can this biological outlaw escape the consequences of its disobedience? Part of the answer is found in the depth at which the microbe lives. The enormous pressure exerted by 2500 m of seawater protects the cytoplasm from boiling. At this pressure (265 atm) water does not boil until it reaches 460°C (860°F). Proteins denature when heat energy exceeds the intrinsic forces that maintain the bonds determining the molecules' shapes. A few extra chemical bonds could theoretically stabilize protein to withstand terrific thermal influences. In addition, the proteins of extreme thermophiles possess at least five amino acids of unknown identity, although it is not apparent whether these increase thermostability. DNA could theoretically be stabilized by special proteins that maintain its configuration. The molecular structure of the membrane of extreme thermophiles is substantially different from more cosmopolitan bacteria and eucaryotes. These microbes are probably archaebacteria (see p. 32) whose membranes are composed of long-chained ether lipids that span from one surface to the other, forming thermostable structures.

It is clear that some of nature's "rules" are not really laws, but merely the product of conventional biological wisdom. It is also clear that the upper thermal limit of life has yet to be defined. The recent discovery of these submarine bacteria has led to speculation that life may exist within areas traditionally believed to be sterile because of their elevated temperatures. The earth's crust, for example, and even other planets may be populated by extreme thermophiles. These microbes may also change our views of the origin of life on earth. The evolution of living forms from chemical aggregates could have proceeded much more quickly in the superheated milieu of the primitive planet. Thus, the first organisms may have appeared much earlier than traditionally speculated, even before the earth's crust cooled.

aging molecules essential to the cell's survival. Thus, unlike cold, which induces protective dormancy, heat actually kills cells. The difference of even a few degrees may be crucial. High fevers, only 3 or 4°C above normal body temperatures, are sometimes sufficient to inactivate an invading pathogen. Medieval remedies based on this observation often included artificial inducement of fever and were occasionally successful if the pathogens were exceptionally heat-sensitive. Unfortunately most pathogens are not killed by the heat of fever unless the temperature is high enough to cause similar heat-mediated damage to the infected person. Fevers above 41°C (105°F) can cause permanent brain damage and death.

pH The measure of the relative acidity or alkalinity of a solution is called pH. pH is expressed as a number from 0 to 14. Pure water is neither acidic nor basic and has a pH of 7 (neutral). Acidic substances have pH values less than 7, whereas numbers above 7 indicate increasing alkalinity. The number is an expression of the concentration of hydrogen ions (H^+) in the solution—the higher the concentration, the more acidic the solution. The value is the negative exponent of the H^+ concentration. For example, when $[H^+] = 0.00001$ (10^{-5}) M the pH is 5. A more acidic solution at 0.01 (10^{-2}) M has a lower pH value (pH = 2).

A microbe's responses to acidity and alkalinity are similar to those observed with heat—each organism has a minimum, optimum, and maximum pH for growth. Most bacteria grow best at a neutral pH of 7, with a normal range between 6 and 8, and cannot survive at pH values below 4. The waste products of normal microbial metabolism are often acids or bases which can rapidly lower or elevate pH to intolerable levels. Therefore, laboratory culture media often contain **buffers**, chemical substances that tend to resist changes in the pH even when acid or alkali is added.

A variety of environments with broadly different pH levels exist within the human body. Most microbes cannot survive the extremely low pH in the human stomach. Only pathogens that can tolerate prolonged exposures to strong acid can successfully establish infection after entering the body through the gastrointestinal tract. The acidic pH of the adult's vagina and skin also helps protect against infection.

Molecular Oxygen Microbes also differ in their response to molecular oxygen. Some organisms, called **aerobes**, require the presence of molecular oxygen for metabolism. Other organisms function in either the presence or absence of molecular oxygen. These **facultative anaerobes** use a less efficient alternative mechanism for energy production in *anaerobic* (oxygen-free) environments.

A third group, the **microaerophiles**, appears to require some oxygen, but too much or too little is detrimental to them. Many also require an elevated concentration of CO_2. Microaerophilic conditions are common in many tissues of the human body, or they may be established and maintained by the metabolism of neighboring, oxygen-utilizing organisms. In the laboratory, microaerophiles are easily cultured in a *candle jar*, a container into which a lit candle is introduced before sealing the airtight lid. The flame burns until extinguished by oxygen deprivation, creating a CO_2-rich, oxygen-poor atmosphere. Special media or incubators may also be used to culture microaerophiles.

A fourth group of bacteria utilize no molecular oxygen, and some may even be killed by its presence. These **obligate anaerobes** reproduce in oxygen-free environments—for example, the dead tissue in deep wounds. Although some anaerobes are *aerotolerant* (they can survive but do not grow in the presence of oxygen), many others (the *strict anaerobes*) are killed by even the briefest exposure to oxygen. In all cells the presence of

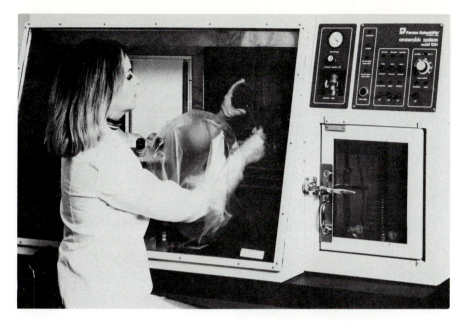

molecular oxygen results in the production of the lethal by-products hydrogen peroxide and superoxide (an unstable charged atom of oxygen), which quickly oxidize and destroy cytoplasm. Aerobic and facultative organisms are equipped with enzymes for immediately eliminating these lethal compounds before cytoplasmic damage can occur. Strict anaerobes, however, possess no enzymes to protect against oxygen's by-products and are incapable of surviving exposure to even trace amounts of oxygen.

Reducing agents are chemicals that react with and inactivate molecular oxygen before it can do cell damage. Inclusion of the reducing agents cysteine, thioglycollate, or hemin in culture media is one way to grow obligate anaerobes. Special equipment is also helpful to assure that these oxygen-sensitive microorganisms are protected during incubation. The most reliable, but unfortunately the most expensive, systems are *anaerobic glove boxes*. In these completely enclosed units (Fig. 6-7) oxygen is replaced by nitrogen or another relatively inert gas. This arrangement allows manipulations and incubations to be performed in an oxygen-free atmosphere. Fortunately, many clinically important anaerobes appear to be able to tolerate oxygen for the amount of time required for laboratory manipulations. They are then incubated in anaerobic jars (Fig. 6-8). These are airtight containers from which oxygen has been removed, either by pumping out the air or by enclosing chemicals that generate hydrogen and combine it with free oxygen to form water.

Anaerobes and microaerophiles are distributed throughout the environment in soil, sediments, sewage, and food. They also comprise part of a person's normal flora, inhabiting the vagina, urethra, mouth, intestinal tract, and even the conjunctiva of the eye. In the colon they may

FIGURE 6-8

The GasPak anaerobic jar contains a package of chemicals that generates hydrogen gas when water is added. A chemical catalyst removes the oxygen by complexing it with the excess hydrogen gas, forming water. An indicator strip should be included to verify the existence of anaerobic conditions.

116

BACTERIA: GROWTH AND
LABORATORY CULTIVATION

outnumber the aerobes 1000 to 1. Some are serious pathogens found in blood, abscesses, intraabdominal infections, and wounds. The pathogens that cause tetanus and gas gangrene grow in oxygen-poor areas of the body. Botulism is caused by ingesting the toxin of the obligate anaerobe *Clostridium botulinum*, usually produced during growth in the oxygen-free environment of canned food. Laboratory cultivation for diagnoses of these diseases depends on culturing the pathogens in anaerobic conditions.

When the pathogen's oxygen requirements are unknown, incubating multiple specimens in aerobic, microaerophilic, and anaerobic conditions helps assure their isolation (Fig. 6-9). Facultative anaerobes will be able to grow in all oxygen concentrations.

Carbon Dioxide   Some important pathogens grow best in the elevated CO_2 environments of candle jars used for generating microaerophilic conditions. Alternatively, a special incubator may be employed that maintains the CO_2 level at 3 to 10 percent. Organisms that require increased levels of CO_2 are called **capneic** organisms. Many human pathogens, notably *Neisseria gonorrhoeae*, grow better when supplied with elevated concentrations of CO_2.

Osmotic Pressure   The internal liquid environment of the bacterial cell is separated from the external environment by the cell membrane and, to a lesser extent, the cell wall. Water moves freely in and out of the cell by

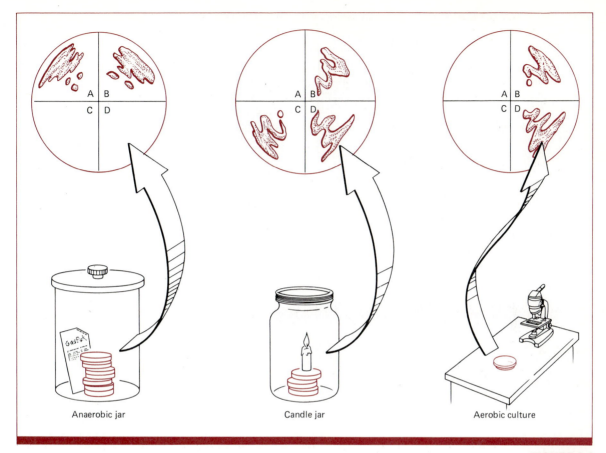

Anaerobic jar   Candle jar   Aerobic culture

FIGURE 6-9

Identification of isolates according to their response to oxygen. Each plate is streaked with the same four organisms (labeled *A*, *B*, *C*, and *D*) and incubated under the conditions indicated.
(*A*) Obligate anaerobe;
(*B*) facultative anaerobe;
(*C*) microaerophile; and
(*D*) obligate aerobe.

osmosis, migrating from dilute areas (pure distilled water, for example) into more concentrated solutions. If the surrounding medium is more dilute than the cytoplasm, water tends to enter the cell, creating pressure inside the cell that forces it to expand. Although such **osmotic pressure** may rupture some cells, bacteria are protected from exploding by the rigid peptidoglycan layer of the cell wall. (See Chap. 4.) Media for growing mycoplasma, protoplasts, or other procaryotes that lack cell walls must contain sugars or salts to prevent osmotic influx of water into the cell. Otherwise, these fragile cells will rapidly lyse.

Most microbes prefer environments within a narrow range of osmotic pressures. Solutions with high concentrations of salt or sugar tend to draw water out of the cell, resulting in dehydration and inhibition of metabolism. Treatment with salts or sugar has been used since ancient times to preserve food. Osmotic dehydration explains the antimicrobial properties of anchovies, beef jerky, preserves, honey, and sweetened condensed milk (see color box on page 118).

Some organisms, however, grow only at high osmotic pressures. For example, marine bacteria grow in moderately high salt concentrations (3.5

ARE YOU WORTH YOUR SALT?

■ This odd expression probably originated during the prerefrigeration days when salt was one of the most valuable of commodities. Its antimicrobial properties helped preserve foods long enough to prevent starvation when sources of fresh nutrition were unavailable. Salt also helped disguise the objectionable taste of rotted food when it was available. During the Roman Empire's domination, campaigning soldiers were paid their wages in salt, considered to be more precious than gold. Today we still earn our wages as salary, a term derived from *sal*, the Latin word for salt.

Many of the foods still popular today were initially developed to preserve quality. The high sugar content of jelly and jam prevents the growth of microbes that cannot tolerate high osmotic environments (hence the name "fruit preserves"). Sweetened condensed milk is also preserved in this manner. Desiccated food spoils more slowly because it lacks the water essential for microbial activity. Beef jerky, dried fruit, powdered milk, freeze-dried food, and Parmesan cheese remain palatable much longer than their hydrated counterparts. Milk can also be preserved longer if the curd is removed from the whey (the liquid portion of curdled milk). The curd is then called cottage cheese. The curd can also be compressed and heated to remove additional moisture and then microbially ripened to produce other types of cheese. The flavor and aroma of cheese are determined by the type of microbe that ferments the curd. Roquefort cheese, for example, is ripened by the blue mold *Penicillium roqueforti*. Cheese is much more resistant to spoilage than the fresh curd. Butter can also be kept much longer than fresh whole milk because of its low water content and its high concentration of antimicrobial fatty acids. Some foods last longer because of their high acid contents. Sauerkraut, pickles, and vinegar are all preserved by their low pH. These techniques have provided many culinary variations while prolonging the edibility of food stuffs. They have certainly proven to be worth their salt.

percent). *Vibrio parahaemolyticus,* a bacterium that can cause food poisoning, is found in coastal waters and may be a natural contaminant of seafood. Organisms that require high concentrations of salt are called **halophiles** (*halo,* salt; *phile,* loving). Other organisms, notably the staphylococci that normally reside on the salty surface of human skin, exhibit *salt tolerance;* they can grow in high salt concentrations, but grow better in normal conditions (0.85% NaCl).

Yeasts and molds are well known for growing on sugar-rich jams and jellies, environments that would dehydrate most bacteria. Because they are already dehydrated, endospores of bacteria such as *Clostridium botulinum* readily survive in the osmotically unfavorable environment of honey. Almost one-half of all cases of infant botulism have been due to feeding babies endospore-contaminated honey.

The high salt content of some laboratory media inhibits the growth of most microbes while allowing a few types to propagate. On media containing up to 10% sodium chloride the salt-tolerant staphylococci proliferate while most other microbes in specimens taken from the body are inhibited. High salt concentrations are also used in media for isolating the halophilic marine vibrios. Media with high sugar content specifically encourage fungal growth while simultaneously inhibiting most bacteria. The most common fungal media contain 2 or 4% sugar compared to the 0.5 to 1% sugar content found in most media used for the cultivation of bacteria.

CULTURE MEDIA

To cultivate organisms in the laboratory, microbiologists must provide the nutritional requirements. Many pathogens are quite fastidious. In the extreme cases of obligate intracellular parasites, viruses, rickettsias, and chlamydias, cultivation on laboratory media is impossible; they can only be grown in living host cells. Most hospital laboratories are not equipped to isolate these pathogens, and specimens are shipped to local or state facilities with the appropriate resources. Because of this, unfortunately, many cases of diseases caused by these microbes go undiagnosed. Other pathogens are not as fussy, but still proliferate only when specific nutritional needs are met. Fortunately, media have been developed to culture most microorganisms. The choice of appropriate media can significantly reduce the time and effort required to characterize microorganisms in the laboratory.

Defined versus Complex Media

When all the specific chemicals in a medium and their concentrations can be identified, it is referred to as a synthetic medium, or a **chemically defined medium**. These media usually contain a source of carbon, energy (often glucose), and salts, plus any needed amino acids and vitamins. Most laboratory media, however, are complex, that is, the exact chemical constituents and their quantities are unknown. A **complex medium** contains a source of energy and carbon, sometimes provided by the sugar glucose, as well as extracts of yeast or beef to supply water-soluble vitamins and salts. Some media contain peptone, digested protein material rich in amino acids that provide nitrogen and carbon. Complex media are easier to prepare than synthetic media and are usually able to support the growth of all but the most fastidious organisms. Table 6-2 compares the ingredients in a synthetic medium and in two complex media commonly employed in the laboratory.

Specialized Media

Some media aid in the transportation and maintenance of organisms; others encourage the isolation of a single group of microorganisms from a specimen containing many microbes. Still others provide information about an organism to help differentiate and identify it. Media may be described according to their function:

1. **Transport media** are used for the temporary storage of specimens while they are being transported to the laboratory for examination. Transport media ideally maintain the viability of all organisms in the specimen without altering their concentrations. Transport media typically

TABLE 6-2

Composition of Several
Commonly Used Media

Medium Components	Principal Purpose
Synthetic Medium	
Minimal agar	*Cultivation of nonfastidious heterotrophs*
Glucose, 2.0 g	Source of energy and carbon
$(NH_4)_2SO_4$, 1.0 g	Source of nitrogen
K_2HPO_4, 7.0 g	Source of phosphorus
$MgSO_4$, 0.5 g	Source of sulfur
Agar, 15 g	Solidifying agent
Water, 1 liter	Source of trace elements
Complex Media	
Nutrient agar	*General cultivation of bacteria*
Beef extract, 3 g	Source of vitamins and salts
Peptone, 5 g	Source of energy, carbon, and nitrogen
Agar, 15 g	Solidifying agent
Water, 1 liter	
Trypticase soy broth	*General cultivation of bacteria*
Trypticase (animal peptone), 17.0 g	Source of nitrogen
Phytone (soy peptone), 3.0 g	Source of nitrogen, carbohydrates, and vitamins
Sodium chloride, 5.0 g	Inorganic salt
K_2HPO_4, 2.5 g	Buffer (maintains pH)
Glucose, 2.5 g	Source of readily utilizable energy and carbon
Water, 1 liter	

contain only buffers and salts. The lack of carbon, nitrogen, and organic growth factors prevents microbial multiplication. Transport media used in the isolation of anaerobes must be free of molecular oxygen.

2. **Enriched media** contain the nutrients required to support the growth of a wide variety of organisms, including some of the more fastidious pathogens. They are commonly used to isolate as many microbes as possible. Blood agar is an enriched medium in which nutritionally rich whole blood supplements the basic nutrients. Chocolate agar is enriched with heat-treated blood, which turns brown and gives the medium the color for which it is named. Serum, yeast extract, and vitamin supplements are also used to enrich media.

3. **Selective media** support the growth of some organisms while specifically inhibiting the growth of others. Selective media expedite rapid isolation and identification of the desired microbes by inhibiting the growth of interfering organisms. For example, when a particular pathogen is suspected, the specimen can be inoculated onto a selective medium that prevents the normal microbial flora contaminants in the material from overgrowing the pathogen. Media that are highly selective permit the use of inocula containing a large number of contaminating organisms. In some selective media, all nutrients have been omitted except those that can be utilized by only a limited group of microbes. Most selective media, however, contain inhibitors that suppress the growth of contaminants.

MacConkey's agar and eosin methylene blue (EMB) agar, for example, contain dyes that inhibit the growth of gram-positive organisms but allow gram-negatives to thrive. Selective media used for the isolation of the salt-tolerant staphylococci contain 7.5% NaCl. Sabouraud's agar may be considered a selective medium for fungi because its pH, usually at 5.6, and its high sugar content favor the growth of molds and yeast and select against bacteria. Other media are made selective by the addition of antibiotics that prevent the growth of sensitive contaminants.

4. **Differential media** contain indicators that distinguish between organisms on the basis of their appearance on the medium. These media allow certain organisms to produce macroscopically distinct colonies (or characteristic zones around colonies) which are helpful in differentiating these organisms from others in the sample. MacConkey's agar, for example, contains one sugar (lactose) and a dye which turns red when the pH drops below 6.8. Any organism that can ferment lactose produces an acid end product that lowers the pH and causes the colony to turn red. Organisms that fail to use lactose produce colorless colonies. This visual characterization helps identify gram-negative gastrointestinal pathogens because they rarely ferment lactose and therefore produce colorless colonies on MacConkey's agar. Most harmless normal flora, on the other hand, readily utilize this sugar and produce red colonies. The dye in MacConkey's agar also inhibits the growth of gram-positive bacteria; this medium is therefore both selective and differential.

Blood agar is perhaps the most commonly used differential medium. Certain organisms have the ability to lyse red blood cells and are thus termed **hemolytic**. If red blood cells are completely lysed, a clear area surrounds the colony. This is called **beta hemolysis**. Partial destruction of red cells yields a green opaque zone that is termed **alpha hemolysis**. Organisms that cannot destroy red blood cells are *gamma hemolytic* (also called nonhemolytic).

5. **Enrichment broths** encourage the growth of a particular organism which would likely be overgrown by competitors in a sample with a complex mixture of microbes. The desired microbe becomes the dominant species because it is better adapted than its competitors to grow in the medium. In other words, after incubation the relative concentration of the desired microorganism has been enriched. This distinguishes enrichment broths from enriched media which contain general nutrient supplements. Enrichment broths are commonly used to process fecal specimens that are suspected of containing *Salmonella* or *Shigella*, two agents of food poisoning. The concentrations of these pathogens in fecal samples are usually low compared to the normal gram-negative intestinal population. The enrichment broths, selenite or gram-negative (GN) broth, hold the normal flora in the lag phase while allowing logarithmic growth of the pathogens. The organisms can then be isolated by streaking on a solid medium to obtain colonies.

The properties of some of the more commonly used media are summarized in Table 6-3.

TABLE 6-3
Characteristics of Some Commonly Employed Media

Medium	Type	Special Ingredients	Characteristic Feature	Useful for Identifying
Blood agar	Enriched and differential	Blood	Grows many fastidious microbes	Streptococci and many others
		Blood	Gives evidence of hemolysis	
MacConkey's agar	Selective and differential	Crystal violet	Inhibits gram-positive bacteria	Enteric gram-negative bacilli
		Bile salts	Inhibits fastidious gram-negative bacteria	
		Neutral red	Lactose fermenters produce red colonies	
Eosin methylene blue (EMB) agar	Selective and differential	Eosin and methylene blue	Inhibits gram-positive and fastidious gram-negative bacteria; lactose fermenters produce red colonies	Enteric gram-negative bacilli
Salmonella-Shigella (SS) agar	Selective	Bile salts Neutral red	Inhibits most other bacteria Lactose fermenters produce red colonies	Salmonella and Shigella
Selenite broth	Enrichment	Sodium selenite	Temporarily inhibits other enteric bacteria	Salmonella
Gram-negative (GN) broth	Enrichment	Sodium deoxycholate	Temporarily inhibits other enteric bacteria	Salmonella and Shigella
Mannitol salt agar	Selective and differential	Sodium chloride	Inhibits gram-negative and most gram-positive bacteria	Staphylococcus
		Phenol red	Mannitol fermenters produce yellow zones around colony	
Chocolate agar	Enriched	Heated blood	Grows many fastidious microbes	*Neisseria*
Thayer-Martin agar	Selective	Antibiotics	Inhibits growth of most other bacteria	*Neisseria*
Chopped meat glucose broth	Selective	Cysteine	Reducing agent removes molecular oxygen, producing anaerobiosis	Anaerobes
Amies medium	Transport	Salts and buffers	Preserves viability without multiplication	

Incubation

No single set of incubation conditions is applicable to all microorganisms. Cultures grow best when incubated in environments similar to the physical conditions of the microbe's natural habitat. Some incubators not only control temperature, but humidity, oxygen, and CO_2 concentrations as well. Cultures should be incubated at their natural temperatures. Isolates of bacteria that normally reside in soil or water are usually incubated at room temperature (25°C), while most clinical isolates are incubated at body temperature (35 to 37°C). Some microbes change their appearance at different temperatures. These cultures should be incubated at temperatures that encourage the development of the desired characteristics.

Serratia marcescens, for example, produces red colonies when grown at 24°C and white colonies at 37°C. Some fungi grow as spore-forming filaments at 24°C and single cells at 37°C. Duplicate cultures must be prepared, one for each temperature, to demonstrate such temperature-sensitive characteristics.

The purity of cultures is usually verified by microscopic observation of morphological properties, and by observing colony characteristics on a streak plate. Contamination is often revealed by the presence of colonies with different characteristics. Variations in shape, size, pigmentation, consistency, elevation, and margin of colonies suggest contamination. Except for infrequent mutations, pure cultures show uniform culture characteristics.

MAINTENANCE AND PRESERVATION TECHNIQUES

Following isolation, the organism is inoculated on plates or agar **slants** (Fig. 6-10) and transferred every few weeks or months to maintain viability. The interval between transfers depends on the storage conditions and on the growth rate of the organisms. Organisms can be transferred to **maintenance media** specifically designed to support low growth rates and extend the culture's life. Refrigeration also slows growth and preserves the culture. It is advisable to maintain two slants, one as the *working culture* to be used as a source of the organism for routine inoculations. The other is saved as the *stock culture* from which new working cultures are prepared. Such a system decreases the probability of contaminating the stock culture. Important organisms may also be preserved by some long-term procedure such as lyophilization (freeze-drying) or freezing in liquid nitrogen.

Lyophilization (freeze-drying) is the rapid dehydration of organisms

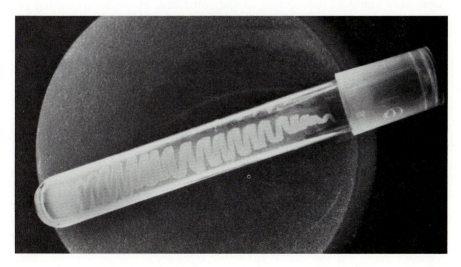

FIGURE 6-10
Bacteria growing on agar slant. Slants of solid media are prepared by keeping tubes tilted as agar solidifies. In a tube a sloped surface provides more area and is easier to streak than a horizontal surface. Slants are preferred to plates for storage of microbes because they require less media and are less readily contaminated.

while they are in a frozen state. This procedure protects most microbes from the damage often associated with water loss. The culture is rapidly frozen at −70°C and then dehydrated by vacuum. While they are still under vacuum, tubes containing the freeze-dried culture are sealed. Because metabolism requires water, the organisms are in a dormant state and can retain viability for years.

Freezing in **liquid nitrogen** at temperatures of −196°C also suspends cell metabolism. Organisms are rapidly frozen in the presence of stabilizing agents, such as glycerol, which prevent the formation of ice crystals that may kill frozen cells. Most organisms survive unchanged for long periods. Because they can resume growth when thawed, frozen or lyophilized cultures pose a biohazard when discarded unless destroyed by incineration, autoclaving, or other sterilizing procedures (see Chap. 13).

MEASURING MICROBIAL CONCENTRATION

In many instances it is important to determine the population density (concentration of microbes) in a sample. For example, the safety of our water and food is ascertained by determining the degree of contamination. The concentration of microorganisms in some clinical specimens often helps determine whether that organism is causing disease or is merely a contaminant from the normal flora.

Microbial concentration can be measured in the laboratory by several methods. The cells can be directly counted, for example under a microscope, or indirectly counted by measuring some property that is proportional to the number of individual organisms present in the sample (Fig. 6-11). Both direct counts and indirect counts require that the organisms be uniformly dispersed throughout the medium. Otherwise the numbers will not accurately represent the true microbial population in the entire sample. For example, microorganisms suspended in a liquid environment will commonly settle to the bottom of the container. Such cultures must be resuspended by swirling the culture prior to removing a small volume for quantification.

Direct Counts

In the direct microscopic count, a known volume of liquid is deposited on a slide and the number of organisms within the volume are counted by examining the slide with the brightfield microscope. *Petroff-Hauser counters* are special slides that are easily and accurately filled to a fixed volume. Since direct counts do not permit differentiation between living and dead cells, they usually yield higher numbers than a count of viable cells. Furthermore, some smaller organisms are difficult to see, particularly in the presence of contaminating debris.

Direct counts may also be accomplished by an electronic particle counter. This instrument automatically counts cells as they are pumped past a beam of light. Another number is registered each time the beam is

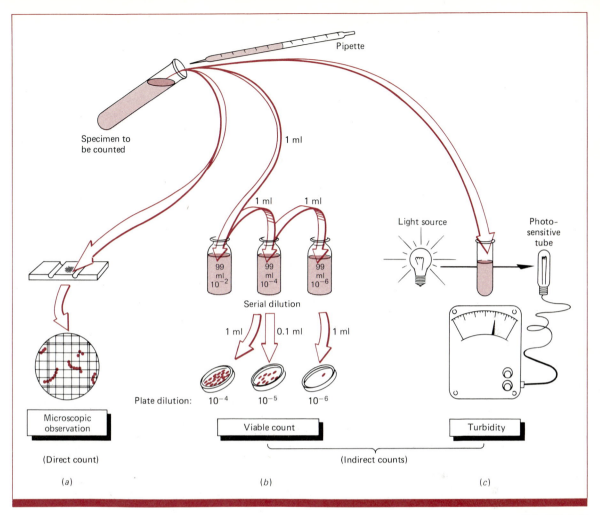

Specimen to be counted

Pipette

1 ml

1 ml   1 ml

99 ml 10^{-2} 99 ml 10^{-4} 99 ml 10^{-6}

Serial dilution

Light source

Photo-sensitive tube

1 ml   0.1 ml   1 ml

Plate dilution: 10^{-4} 10^{-5} 10^{-6}

Microscopic observation

Viable count

Turbidity

(Direct count)

(Indirect counts)

(a)   (b)   (c)

FIGURE 6-11
Methods of counting
microorganisms. (See text
for description.)

interrupted by a cell. Because of the expense of such an instrument, direct counts are usually performed by microscopic examination.

Direct counts are used for immediate estimates of how many microorganisms are present in a solution. This information is valuable for grading the quality of milk, for rapid preliminary diagnosis of urinary tract infections, and for determining the number of particles in a vaccine preparation containing killed cells.

Indirect Counts

Population density may be determined by observing some property that provides indirect evidence of microbial numbers. The two most common indirect methods are the **viable count** (plate count), which detects living organisms by their ability to form colonies on agar surfaces, and **turbidi-**

125

metric measurements, which relate cell number to the turbidity (cloudiness) of a liquid culture.

Viable Counts   The most accurate measurement of the number of *living* microbes is usually the viable count. Theoretically, each viable cell forms a single colony on solid media that can support its growth. Thus, following incubation, the number of colonies on the plate ideally equals the number of cells in the sample inoculated on the agar. In reality, each colony represents a **colony-forming unit** (CFU). The CFU count is identical to the cell number *only* if each colony is descended from a single cell. Some organisms, however, grow in aggregates. For example, staphylococci grow in clusters, streptococci in chains, and actinomycetes and molds in filaments. In these cases, each colony represents an aggregate rather than a single cell, and the number of colony-forming units will be lower than the actual number of microbes in the culture. The viable count is ineffective for counting organisms that form colonies that swarm together, organisms that cannot yet be cultured on artificial media, and organisms that multiply very slowly.

Viable counts depend upon the sample being sufficiently dilute so that well-isolated and readily countable colonies will develop. For example, a milliliter of raw milk applied directly to the plate may contain so many bacteria that colonies overlap and cannot be counted. A series of dilutions, therefore, are usually plated to assure at least one plate with a countable number of distinct colonies (Fig. 6-11b). In addition, all manipulations must be performed aseptically to prevent contamination of the sample, which would indicate a larger number of organisms than were actually in the sample.

On a standard-size petri dish a statistically valid sample contains between 30 and 300 colony-forming units. Fewer than 30 colonies represents an unreliably small sample size. When the colony number exceeds 300, multiple colonies become superimposed on each other and may be mistakenly counted as a single unit.

If the microbial count is too low to generate 30 or more colonies from an undiluted liquid or air sample, the microbes can be concentrated by passing a large volume through a sterile bacteriological membrane filter. Bacteria are too large to pass through the filter's pores and will be retained on the membrane's surface. The microbe-laden filter can be placed on a solid culture medium and incubated until colonies develop (Fig. 6-12).

Turbidimetric Measurements   *Turbidity*, or *optical density*, is the cloudiness of a suspension. The more turbid a suspension, the less light will be transmitted through it. As bacterial cultures grow in broth, the clear liquid medium becomes turbid. Since the turbidity increases as the number of cells increases, this property can be used as an indirect indicator of bacterial concentration. In the laboratory, turbidity is quantified with a spectrophotometer, an instrument that measures the amount of light transmitted directly through a sample. Some inaccuracy is unavoidable in

BACTERIA: GROWTH AND
LABORATORY CULTIVATION

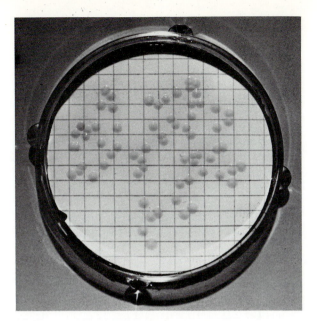

FIGURE 6-12
Bacterial colonies growing on membrane filter. Sample was filtered through the membrane which was subsequently incubated on a solid medium.

turbidimetry since dead cells and contaminating debris also block the direct passage of light.

Turbidity is also useful for standardizing the population densities of bacterial cultures. For example, many clinical laboratory procedures, such as determining antibiotic susceptibilities of an isolated pathogen, require the rapid preparation of suspensions containing a specific concentration of microorganisms. Cultures grown overnight can be adjusted quickly by dilution to the turbidity that represents the desired concentration of cells. Turbidimetric measurements may be converted to values representing cell counts (number of cells per milliliter) by experimentally constructing a **standard curve** (Fig. 6-13). Both turbidity and direct count of samples containing different cell concentrations are determined. The data, when plotted on a graph, usually produce a straight line. This graph is then used to obtain direct counts by simply measuring the turbidity of a culture.

The most commonly employed methods of enumerating microbes are compared in Table 6-4.

Microbiological Assays

The growth of microbes provides one of the most sensitive indicators for determining the presence and concentrations of many chemical compounds. Because these assays use living organisms as indicators, they are known as **bioassays**. Bioassays are the only means of measuring concentrations of the vitamins biotin and B_{12} below 1 $\mu g/ml$. The test bacterium will grow only in media that contain the vitamin. When the vitamin has been consumed, bacterial growth will cease. The amount of microbial growth will therefore be proportional to the concentration of the limiting vitamin

FIGURE 6-13

Construction of standard curve relating turbidity to direct microbial count. Samples A through F are cultures with increasing cellular concentrations. The turbidity and number of colony-forming units of each sample are measured. Each standard curve is applicable only to cultures of one strain grown in the same medium and under the same conditions.

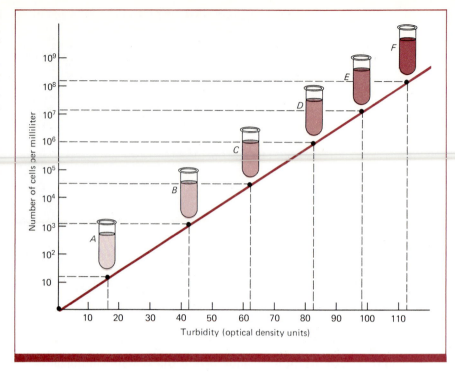

TABLE 6-4 Comparison of Enumeration Methods

| | | Indirect Counts | |
	Direct Counts	Turbidity Measurement	Viable Count
Advantages	Rapid (requires no incubation time)	Rapid	Counts living cells only; dilute solutions can be counted by filtration
Disadvantages	Counts include dead cells; organisms in dilute solution cannot be readily counted by microscopy	Counts all light-scattering material; useless for dilute solution	Requires incubation period; danger of contamination; clumping of cells introduces inaccuracy
Uses	Microbial counts in milk, urine, and vaccines	Microbiological assays; estimation of cell numbers in liquids	Microbial counts in food, soils, liquids, and gases
Instruments	Microscope; particle counter	Spectrophotometer	Incubator; petri dishes; solid media; occasionally filtration apparatus

as long as all other nutrients are supplied in excess. Thus a solution containing 0.2 μg/ml of the vitamin will allow the production of twice as many bacteria as one containing 0.1 μg/ml. A standard curve that relates vitamin concentration to total growth is experimentally constructed. Then by measuring total bacterial growth in a vitamin-free medium supplemented with a standard amount of the substance to be tested, the concentration of vitamin can be determined in any unknown materials, such as vitamin pills, cereals, or other foods. The concentrations of many nutrients listed on food labels are often determined by bioassay.

Microorganisms are also used for evaluating the antimicrobial effectiveness of antibiotics, disinfectants, antiseptics, and sterilizing agents. Antibiotic assays are routinely performed in order to determine the most effective therapy for a patient with infectious disease. Chemical disinfectants and antiseptics are evaluated in order to determine the best concentration for practical use. Whereas nutritional bioassays measure bacterial growth, these tests measure its inhibition. Details of these assays are found in Chapters 13 and 14.

OVERVIEW

A knowledge of bacterial growth contributes to our understanding of many fundamental principles of life, as well as aiding in the prevention, diagnosis, and treatment of infectious diseases. Most bacteria multiply asexually by binary fission and have short generation times. The appropriate nutritional and physical requirements for growth vary with the organism. Optimal culture conditions are rarely supplied in nature but can usually be created in the laboratory by selecting a medium that satisfies the cell's nutritional needs, then incubating at temperatures and atmospheric conditions that best encourage growth. The most effective combination depends on the microbe to be cultured. As long as nutrients are available and toxic waste products do not accumulate, microbes continue to multiply logarithmically, as seen in continuous cultures. In batch cultures, however, nutrient exhaustion or toxic wastes eventually impede growth.

Pure cultures for identification or study may be obtained by using specialized media designed to enhance the growth of specific organisms, inhibit contaminants, or yield reactions characteristic of particular microbes. Cultures may be maintained by a number of methods that preserve microbial viability.

The concentration of microbes in a sample can be ascertained either by direct counts or by indirect methods such as viable counts or turbidity determinations. The choice of method usually depends upon the availability of time and equipment and upon the need for an accurate determination of viable cells.

KEY WORDS

binary fission

generation

doubling time

generation time

infectious dose

batch culture

growth curve

lag phase

logarithmic phase

growth rate

stationary phase

death phase

continuous culture

fastidious microbe

phototroph

chemotroph

autotroph

heterotroph

thermophile

mesophile

psychrophile

buffer

aerobe

facultative anaerobe

microaerophile

obligate anaerobe

capneic organism

osmotic pressure

halophile

synthetic medium

defined medium

complex medium

transport medium

enriched medium

selective medium

differential medium

hemolytic organism

alpha hemolysis

beta hemolysis

enrichment broth

slant

maintenance medium

lyophilization

liquid nitrogen

direct count

indirect count

viable (plate) count

turbidimetric measurement

colony-forming unit

standard curve

bioassay

REVIEW QUESTIONS

1. Draw and label a typical bacterial growth curve.
   (a) What are the major events in each phase of microbial growth?
   (b) During which phase in the growth cycle are organisms most susceptible to destruction by metabolic inhibitors?
   (c) What two factors contribute to the termination of the log phase?

2. Differentiate between:
   (a) Batch cultures and continuous cultures
   (b) Synthetic and complex media
   (c) Obligate aerobes, facultative anaerobes, obligate anaerobes, and microaerophiles
   (d) Enrichment broths and enriched media
   (e) Alpha, beta, and gamma hemolysis

3. Some pathogens are very sensitive to dehydration. How does this affect their mode of transmission?

4. Pathogenic organisms are chemoheterotrophs. Explain.

5. Are human pathogens thermophiles, mesophiles, or psychrophiles? Explain.

6. What is the purpose of each of the following types of media?
   (a) Transport   (b) Enriched
   (c) Selective   (d) Differential

7. What are the advantages of determining microbial numbers by the viable plate-dilution count over the direct cell count? What are the disadvantages?

8. What is a colony-forming unit? Why is this term used instead of cell number?

9. How are bioassays used to determine the concentration of various chemical compounds in products?

10. What types of physical conditions should be taken into consideration when trying to cultivate bacteria?

11. Why are some microbes more difficult to cultivate under laboratory conditions than other microorganisms?

Fungi

7

One of the first microbiological sciences was **mycology**, the study of fungi. The **fungi** (singular, fungus) comprise molds, yeasts, and a group of macroscopic organisms often called the fleshy fungi. They are all eucaryotic nonphotosynthetic organisms enclosed by cell walls, usually made of the polysaccharide chitin. Many fungi are familiar to all of us. Molds that grow on bread, fruit, and cheese; mildew in damp textiles; yeast used in pastry baking and brewing; mushrooms and toadstools are all types of fungi. Some fungi produce antibiotics that we use therapeutically against many bacterial infections. Others, however, affect our lives in adverse ways; for example, they are agents of important infectious diseases in humans. Other fungi may destroy our food resources, as did the agent of the Irish potato blight that caused death by starvation for a million persons. Powerful toxins produced by several fungi can be fatal when ingested.

MORPHOLOGY OF THE FUNGI

Fungi are generally larger than bacteria, their individual cell diameters ranging from 1 to 30 μm. Microscopic fungi exist as either molds or yeasts, or both. The **molds** form large multicellular aggregates of long branching filaments, called **hyphae** (Fig. 7-1a). These tubelike hyphae are responsible for the fluffy appearance of the macroscopic mold colony (Fig. 7-1b). **Yeasts**, on the other hand, are single cells that rarely form filaments (Fig. 7-2a). Yeast colonies are usually characterized by a smooth surface similar to that of many bacteria (Fig. 7-2b).

Fleshy fungi produce large, macroscopically visible reproductive structures, of which mushrooms and toadstools are the best known examples. The fleshy structure, however, represents only part of the organism, most of which grows beneath the soil as microscopic filaments—hyphae. Because the fleshy fungi form hyphae, they will be discussed with the molds.

FIGURE 7-1

(a) Microscopic and (b) macroscopic appearance of typical mold.

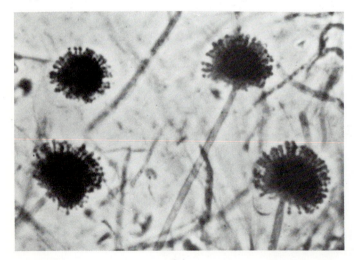

(a)

(b)

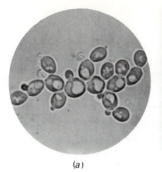 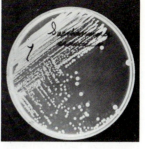

(a)   (b)

FIGURE 7-2
(a) Microscopic and
(b) macroscopic appearance
of typical yeast
(*Saccharomyces cerevisiae*).

Molds

Hyphae   The hyphae of a mold colony grow as an intertwined mass of filaments, collectively called a **mycelium**. Molds are classified and identified partially on the basis of whether the hyphae are septate or nonseptate. **Septate hyphae** are filaments with cross walls, called septa, that partition the hyphae into individual cellular compartments. Most septa have pores that allow migration of cytoplasm and many organelles. In **nonseptate hyphae** there are no physical boundaries to distinguish individual cells in the hyphae (Fig. 7-3).

Most hyphae are **vegetative**, that is, they are actively growing and form the main body of the colony. **Aerial hyphae**, on the other hand, support the reproductive structures and further contribute to the fluffy appearance of the mold colony.

Mold Spores   The reproductive structures elevated at the ends of the aerial hyphae produce specialized cells, called **spores**, each of which is capable of generating a new colony in a favorable environment. Mold spores have the following characteristics:

They are generally produced in large numbers.

They are easily disseminated.

Some are resistant to conditions that would kill the vegetative cell.

Thus, the collective functions of mold spores are threefold: reproduction,

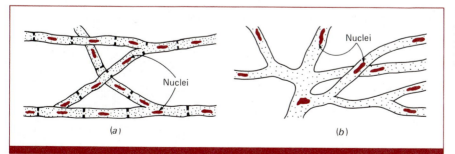

(a)   Nuclei   Nuclei   (b)

FIGURE 7-3
Characteristic (a) septate
hyphae and (b) nonseptate
hyphae (note the absence
of septum between nuclei).

dissemination, and protection of the species against adverse environmental conditions.

Molds may produce spores that are either sexual or asexual. Sexual spores are products of a sexual cycle, an alternation between the **diploid** state (each nucleus contains two sets of chromosomes) and **haploid** state (one set of chromosomes per nucleus). One of the characteristic events in sexual reproduction is *fertilization,* the fusion of two haploid nuclei to form a diploid nucleus. The process by which the number of chromosomes in the diploid nucleus is reduced in half is called *meiosis.* Depending on the organism, either the diploid or haploid cell germinates into a new vegetative fungus. Asexual spores, on the other hand, are produced by the process of simple cell division in the absence of fertilization and meiosis. Many fungi are capable of producing both sexual and asexual spores.

Individual fungal pathogens can be identified by differences in the appearance of spore types and the sporulating structures. Consequently, the presence of characteristic spores in clinical specimens is helpful in diagnosing certain fungal diseases.

FIGURE 7-4
Various sporulating structures and the corresponding asexual spores: (*a*) Clusters of conidia; (*b*) packaged and discharged sporangiospores; (*c*) spherical shlamydospores attached to vegetative cells; and (*d*) arthrospores.

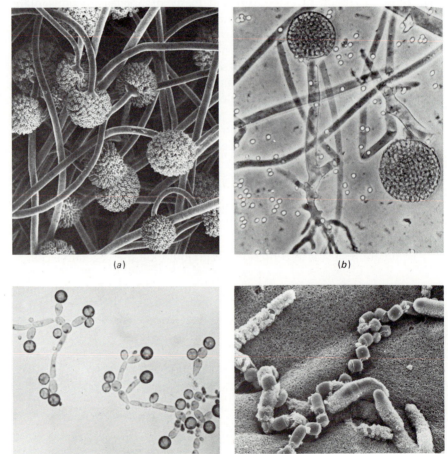

(a)

(b)

(c)

(d)

ASEXUAL SPORES Most fungal diseases are acquired by contact with the asexual spore of the fungus. Many of the fungal spores and structures responsible for the spore production are illustrated in Figure 7-4. Perhaps the most common spore is the **conidium**, (plural, conidia) which forms in clusters at the tips of specialized hyphae (Fig. 7-4*a*). Conidia are not particularly resistant to adverse conditions, but they are lightweight and easily disseminated through the air, enhancing spread of the fungus. The conidia of some fungi can infect people and cause disease.

Sporangiospores are spores contained in saclike structures called **sporangia**, borne at the end of aerial hyphae (Fig. 7-4*b*). The sac erupts and discharges the mature spores. Sporangiospores are normally harmless to people, but can occasionally infect persons with lowered resistance and cause a fatal disease. **Chlamydospores** are formed by the vegetative hypha rather than by special structures (Fig. 7-4*c*). They are resistant to heat, drying, and freezing.

Individual cells within the hyphae occasionally undergo thickening of their cell walls and fragment away from the parent filament, forming **arthrospores** (Fig. 7-4*d*). Like conidia, arthrospores may be dangerous infectious agents for humans. Some fungi produce asexual spores by budding. These buds are referred to as **blastospores**.

SEXUAL SPORES Sexual fungal spores are rarely agents of human disease. Structures that bear sexual spores in fungi are called **fruiting bodies**. The mushroom, for example, is a common fungal fruiting body

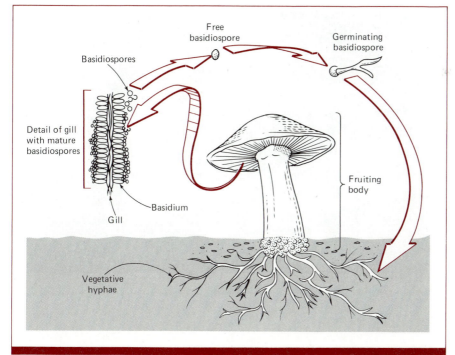

FIGURE 7-6

(a) Diagram of an ascocarp (fruiting body, with ascospores contained in a sac, the ascus). (b) Photo of asci containing ascospores.

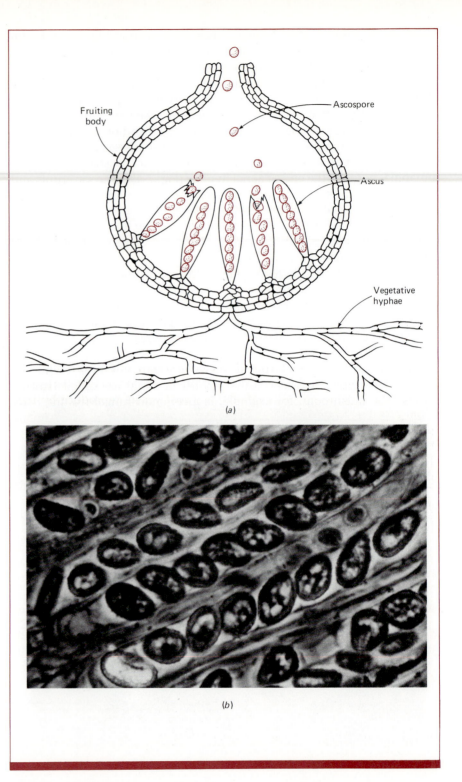

Fruiting body

Ascospore

Ascus

Vegetative hyphae

(a)

(b)

that produces **basidiospores** at the tip of clublike structures, called **basidia** (singular, basidium), which are arranged along the gills of the mushroom (Fig. 7-5). Beneath the mushroom, in the soil, is the mold colony itself, consisting of a mat of intertwined hyphae, sometimes several feet in diameter.

Another sexual spore, the **ascospore,** is contained in a saclike structure called an **ascus** (plural, asci). Several asci are usually contained in a single larger fruiting body (Fig. 7-6). **Zygospores** and **oospores** are two sexual spores that are not associated with well-developed fruiting bodies. Common sexual and asexual spores of fungi are listed in Table 7-1.

Vegetative Reproduction New mold colonies can also be formed from fragments of hyphae. This is called *vegetative reproduction* because it requires no spores. The hyphae fragments simply continue the same growth activities that were going on at the time of fragmentation. The fragmented cells produce a new colony identical to the parent. These fragments do not undergo the morphological differentiation found in the arthrospores with their thickened cell walls.

Yeasts

Yeast cells usually grow as large single cells (5 to 8 μm in diameter, compared to the 1-μm diameter of bacterial cocci), rarely forming filaments. Only a few yeast reproduce by binary fission (as in bacteria), forming two daughter cells of equal size. Most yeasts reproduce by the asexual process of **budding** (Fig. 7-2). The parent cell produces a growing bud which eventually acquires a full complement of cellular components. The bud, also called a blastospore, then breaks away from the larger parent cell and grows to full size. Secondary buds often develop on the daughter bud before it separates from the parent yeast cell. Extreme expressions of the "bud-on-bud" phenomenon are thought to account for the occasional development of chains of elongated yeast cells called *pseudohyphae* (false hyphae), which may be confused with the true hyphae of molds.

Sexual reproduction in yeasts produces four or more thick-walled, dormant haploid spores inside the yeast cell. The cell wall of the parent yeast cell functions as an ascus, a sac containing the spores. The sexual spores of most yeasts are, therefore, ascospores.

Sexual	Asexual
Mold Spores†	
Ascospores	Conidia
Basidiospores	Sporangiospores
Zygospores	Chlamydospores
Oospores	Arthrospores
	Blastospores (buds)
Yeast Spores	
Ascospores	Blastospores (buds)
Blastospores	

*See text for description of each type.
†Includes the fleshy fungi.

TABLE 7-1

Common Asexual and Sexual Spores in Fungi*

Dimorphism

Some fungi grow only in the yeast form, others grow only as molds. Many pathogenic species, however, are **dimorphic**; they can be either molds or yeast, usually depending on the environmental conditions. For example, many fungi grow as molds at room temperatures (24°C), but transform into yeasts at human body temperature (37°C), especially in a moist, blood-enriched medium. Most of the more dangerous human pathogenic fungi are dimorphic, one form (a mold spore) adapted for entry into the human host, the other (a yeast) capable of replication following establishment in the tissues.

For example, in its natural habitat (soil), the dimorphic fungus *Histoplasma capsulatum* is a white mold. The filamentous appearance at this temperature is due to the abundant growth of hyphae that support the formation of conidia (Fig. 7-7a). The mature conidia are readily disseminated by wind, passing animals, and birds, distributing the fungus to new environments where the conidia may germinate and produce new mold colonies. If inhaled by a human, however, the conidia may germinate in the lungs or bronchioles. Here, no hyphae are formed, *Histoplasma capsulatum* growing only as single-celled yeasts incapable of producing conidia at body temperature (Fig. 7-7b). The yeasts spread throughout the lung and may cause tissue damage and clinical lesions resembling those of tuberculosis. In fact, many x-ray findings of "calcified" lesions in human lungs mistakenly identified as tuberculosis are actually histoplasmosis, which is the major cause of human lung calcification in the United States.

Although the yeast stage is responsible for damage to the human body, only the conidia are capable of initially entering the body and surviving the host defenses. The yeast stage cannot cause the initial infection; it is found in the body only because the infectious conidia enter

FIGURE 7-7

Histoplasma capsulatum in (a) filamentous stage (in natural or artificial medium) and (b) yeast stage (in human tissue).

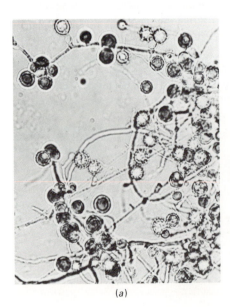
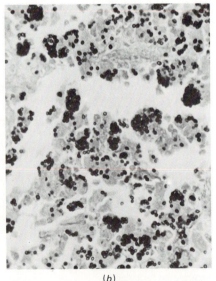

(a)   (b)

the lung and differentiate at body temperature into yeast. A patient with histoplasmosis is incapable of spreading the disease directly to another person because infectious conidia are not produced in the body. Thus, because of dimorphism, histoplasmosis is a noncommunicable disease—it cannot be spread directly from person to person.

CLASSIFICATION OF THE FUNGI

Every medically important fungus belongs to one of four taxonomical groups, distinguished on the basis of spore types, morphology of hyphae, and sexual cycles (see Table 7-2). These groups are the zygomycetes, ascomycetes, basidiomycetes, and deuteromycetes. Except for the deuteromycetes, all fungi are capable of producing sexual spores characteristic of their group. Most fungi, however, usually reproduce by asexual sporulation. Asexual spores are generally the forms responsible for initiating human infections.

Zygomycetes

The **zygomycetes** are characterized by nonseptate hyphae, asexual sporangiospores, and sexual zygospores. The zygomycetes include black bread mold *(Rhizopus nigrificans)* and common soil fungi such as *Mucor*. Ordinarily these fungi are harmless, but they can occasionally cause serious diseases in diabetics and other people with lowered resistance. These diseases are collectively called **mucormycoses** because the causative fungi all belong to the order Mucorales. Mucormycoses are serious, often fatal diseases of the respiratory tract, blood vessels, brain, or other organs.

Ascomycetes

Molds and yeasts that produce ascospores belong to the **ascomycetes**. The filamentous fungi of this class characteristically produce septate hyphae

TABLE 7-2
Distinguishing Characteristics of the Four Medically Important Groups of Fungi

Group	Hyphae	Sexual Spores	Commonly Observed Asexual Spores	Some Medically Important Genera
Zygomycetes	Nonseptate	Zygospores	Sporangiospores	*Mucor* *Rhizopus*
Ascomycetes	Septate	Ascospores	Conidia Arthrospores Blastospores	*Aspergillus* *Histoplasma* *Trichophyton* *Penicillium*
Basidiomycetes	Septate	Basidiospores	Characteristically none	*Cryptococcus* *Amanita* (death angel mushroom)
Deuteromycetes	Septate	None	Conidia Arthrospores Blastospores Chlamydospores	*Candida* *Sporothrix* *Coccidioides*

and several types of asexual spores. They include such interesting and important fungi as *Penicillium notatum*, the mold that produces the antibiotic penicillin. Another mold, *Aspergillus* (one of the most common microbial contaminants), occasionally causes a fatal pulmonary disease (aspergillosis) in people with lowered resistance. Other members of this genus synthesize dangerous poisons called **aflatoxins** when growing on certain foods such as peanuts. In addition to the toxicity, aflatoxins have been found to be carcinogenic (cancer-causing). Many yeasts, notably the economically important members of the genus *Saccharomyces* (brewer's and baker's yeasts) are also members of the ascomycetes.

Basidiomycetes

The **basidiomycetes** characteristically produce sexual basidiospores and septate hyphae. Most members of the basidiomycetes—the mushrooms, toadstools, and puff balls—are considered fleshy fungi; their sexual spores are borne in unusually fleshy fruiting bodies. Asexual spores are uncommon. Although the basidiomycetes cause few human infections, the toxins present in poisonous mushrooms represent a deadly threat to the unwary consumer. Other members of this group, the grain rusts and smuts, are economically important plant pathogens.

Deuteromycetes (Fungi Imperfecti)

Some fungi have no known sexual cycle of reproduction (called the "perfect stage"). These organisms are all grouped with the **deuteromycetes**. Since sex is either nonexistent or undiscovered in these organisms, they are commonly called the "imperfect fungi." The only characteristic shared by all the fungi of this group is the absence of a sexual cycle; their appearance may resemble organisms in any of the other three classes. Most of the deuteromycetes, however, look like ascomycetes, with septate hyphae and similar asexual spores. Members of these imperfect fungi are continually being placed in other groups as their elusive sexual cycles are discovered. For example, the mold that produces the antibiotic penicillin, once classified with the deuteromycetes, is now a member of the ascomycetes. Many important pathogens, however, are still considered fungi imperfecti.

NUTRITION AND CULTIVATION OF THE FUNGI

Most fungi are *saprophytes*, nonparasitic organisms that live on dead organic matter as a source of nutrients. Saprophytic fungi are important decomposers in the biogeochemical cycle. (See Chap. 1.) Unfortunately, they also decompose wood, rubber, and many other useful materials. Each year millions of dollars are spent to retard product deterioration caused by fungi. On the other hand, some metabolic by-products of fungal growth are useful. For example, the yeast *Saccharomyces cerevisiae* produces ethyl (grain) alcohol when glucose is fermented. Soybean protein is converted to soy sauce partially by the action of another fungus, *Aspergillus oryzae*. Fungi help produce bread and cheese. Vitamins, enzymes, organic acids, antiasthmatic drugs, and antibiotics are harvested from fungal cultures. Clearly, the nutrition and cultivation of fungi are of substantial practical importance.

The general nutritional and cultural requirements of fungi differ from those of bacteria. Many fungi, especially molds, grow more slowly than bacteria. Rapidly growing bacterial contaminants are an occasional frustration when culturing fungi. Most fungi, however, grow best at a low pH between 5.0 and 6.0 and can tolerate extremely high sugar concentrations, sometimes greater than 50% sucrose. They can therefore grow on media that would exclude most bacteria. The usual medium employed for isolating fungi is **Sabouraud's agar**, which incorporates all the above cultural and selective conditions. In addition, antibacterial antibiotics are often included in the medium to prevent the growth of the few bacterial contaminants that could otherwise grow on this medium. Definitive identification of most fungi requires additional types of media to stimulate characteristic morphology or reveal biochemical properties. The choice of media depends on the suspected fungus. Fungi should never be incubated anaerobically since all are either aerobes or facultative anaerobes.

Mature mold cultures contain millions of spores that are readily disseminated whenever the culture container is opened. People working with fungi should exercise caution for two reasons: fungal spores are persistent sources of airborne contaminants of other cultures in the laboratory; and serious diseases may result from inhaling fungal spores, especially those of such pathogens as *Histoplasma* and *Coccidioides*. A special laboratory hood should be used for working with these pathogenic fungi.

DISEASES CAUSED BY FUNGI

Of the 150,000 known species of fungi, fewer than 50 have been identified as common human pathogens. Fungal diseases are consequently much less common than bacterial or viral diseases. The disorders they cause are usually slowly progressing and range from very mild painless diseases of the skin to overwhelming invasions of the infected person's entire body.

Diseases caused by fungi are collectively called **mycoses** (singular, mycosis). They are divided into four general categories on the basis of the primary tissue affinity of the pathogen (see Table 7-3):

Superficial mycoses are infections limited to the hair and dead layers of the skin.

Cutaneous mycoses (dermatophytoses or ringworm) affect only the skin, hair, and nails.

Subcutaneous mycoses affect the subcutaneous tissue below the skin and occasionally bone.

Systemic ("deep") **mycoses** infect the internal organs and may spread throughout the host.

Type of Mycosis	Primary Area Affected	Representative Disease
Superficial	Top layers of skin and hair	Tinea versicolor (pityriasis versicolor)
Cutaneous	Skin, hair, and nails	Dermatophytoses (ringworm), including tinea capitis, athlete's foot, and jock itch; cutaneous candidiasis
Subcutaneous	Subcutaneous tissue, bone, and lymphatics	Sporotrichosis; mycetoma
Systemic	Any organ in the body, including the skin	Histoplasmosis; coccidioidomycosis; cryptococcosis; candidiasis; blastomycosis

Since many fungi that cause superficial or cutaneous mycoses are skin parasites by nature, the sources of infection are usually humans or animals. The disease is transmitted directly from an infected person or animal to a susceptible host. The agents of subcutaneous and systemic mycoses, however, are normally saprophytic fungi growing in the soil. These fungi do not need human or animal hosts for survival. Because the primary source of infection is the soil, humans acquire systemic and subcutaneous mycoses only when the soil organisms, usually spores, are either inhaled or introduced into the body through a break in the skin. Direct transmission from person to person is unknown.

Superficial Mycoses

Pathogens in this minor group of diseases cause little damage to their host. They infect only hair and the outermost layers of dead skin and do not invade living tissue. Since they elicit neither a host response nor discomfort, they are primarily cosmetic diseases. *Tinea (pityriasis) versicolor* is the most common superficial mycosis and is characterized by white or tan scaly areas on the trunk of the body (Fig. 7-8). The discoloration is augmented by exposure to sunlight.

Cutaneous Mycoses

Most cutaneous mycoses are caused by a group of fungi called **dermatophytes** (*dermis*, skin; *phytes*, plant). Because of their affinity for tissues rich in the protein keratin, they are parasites of skin, hair, and nails, where they cause diseases collectively known as **dermatophytoses**. These maladies are among the most common and persistent of all human infections. The dermatophytes often grow in a radial pattern on the skin with an elevated margin that gives the appearance of a circular worm beneath the cutaneous layer. This feature is responsible for the common name, "ringworm," although the pathogen is a fungus, not a worm. The specific dermatophytoses are usually named by a double term consisting of the word *tinea* (Latin word meaning "a gnawing worm") followed by a term specifying the area of the body affected. For example, tinea capitis is

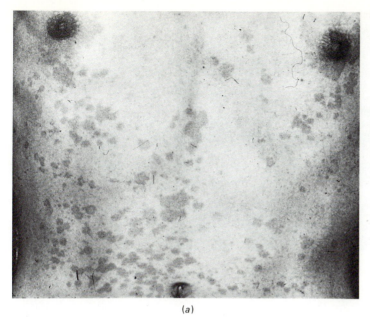

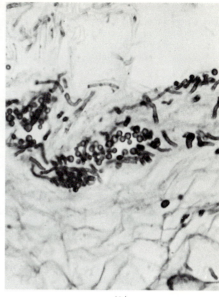

(a)   (b)

ringworm of the scalp and tinea pedis is athlete's foot (Table 7-4). The details of these diseases will be discussed in Chapter 23.

The organisms that cause the dermatophytoses are members of three genera of filamentous fungi, *Epidermophyton, Microsporum,* and *Trichophyton.* Microscopic examination of material from the lesions may reveal hyphae and spores characteristic of the etiological dermatophyte (Fig. 7-9). In addition, the organisms can be cultured on laboratory media for isolation and identification.

Dermatophytoses are rarely life-threatening, but some infections of the feet and toenails, especially by certain species of *Trichophyton,* may be difficult to cure (see color box, p. 146).

Subcutaneous Mycoses

A group of slowly progressing but extremely persistent diseases, the subcutaneous mycoses, are characterized by multiplication of fungi beneath the skin. The organisms are usually introduced by a puncture wound. The most common of these diseases is *sporotrichosis,* caused by a

FIGURE 7-8

Tinea versicolor: (*a*) Clinical appearance; (*b*) direct microscopic examination of stained skin specimen.

Dermatophytosis	Common Name	Target Area
Tinea corporis	Ringworm	Trunk
Tinea capitis	Ringworm	Scalp
Tinea cruris	Jock itch	Groin
Tinea pedis	Athlete's foot	Feet
Tinea barbae	Barber's itch	Bearded facial areas
Tinea unguium	Ringworm	Fingernails and toenails

TABLE 7-4

Common Dermatophytoses and Areas Affected

FIGURE 7-9

Dermatophytoses: (*a*) Tinea capitus and (*b*) microscopic appearance of *Microsporum canis* on hair.

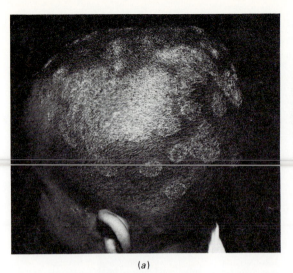

(a)

(b)

RELENTLESS FUNGI OF THE FEET

■ Tinea pedis (athlete's foot) is one of the most common infectious diseases, affecting males and females of every age group and race. The disease is extremely contagious and may be transmitted on contaminated floors or towels. Athlete's foot begins as cracking and scaling between the toes, where conditions are ideal for the growth of the fungus. Once established, this painful disease can become chronic, especially in hot, humid weather, persisting in spite of various antifungal therapies. Athlete's foot can sometimes be eradicated simply by correcting the factors that encourage fungal growth, for example, by preventing the accumulation of moisture between the toes. The use of powders to absorb excess moisture or of shoes that allow evaporation may eliminate tinea pedis in some patients. In some people, however, abrasion from toes rubbing together, coupled with the moisture that continues to accumulate between the toes makes tinea pedis virtually incurable. Many such unfortunate patients suffer from advanced stages of the disease that may prevent walking.

In a radical effort to provide permanent relief from such chronic disease, some physicians have successfully altered the structure of the foot so that fungal growth between the toes is no longer encouraged. For example, surgical removal of the second and fourth toes enlarges the interdigital spaces. This prevents the toes from rubbing together, while discouraging moisture accumulation.

The physician who pioneered the surgery reported that every patient who received the surgery subsequently recovered from persistent tinea pedis. Another surgeon reported that complete elimination of the interdigital spaces was also effective against athlete's foot. The skin on the inner surfaces of each toe was surgically cut and joined to the cut skin of the adjacent toe. Once the toes were sewn together, the fungus never regained its "foothold." Such radical approaches were used as last-resort tactics and will probably never be accepted as a rational option in most cases. The success of the techniques, however, illustrates the important role predisposing factors play in the development of persistent fungal diseases.

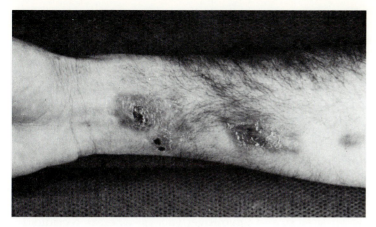

(a)   (b)

dimorphic saprophytic fungus that grows as a mold in soil and as yeast in the human body. When the conidia are inoculated beneath the skin, often by the stab of a thorn, an abscess forms at the site of puncture and spreads along the lymphatic system (Fig. 7-10a). If the organisms reach the central nervous system, meningitis will develop, but this rarely occurs. In the body, the fungus grows as a cigar-shaped yeast which occasionally may be microscopically detected in pus from a lesion. It is far more reliable to culture the fungus using conditions similar to those of its natural environmental habitat. These conditions encourage the growth of the filamentous form of the fungus which can be easily identified for diagnosis.

Systemic Mycoses

The most serious fungal diseases are the systemic mycoses, those that can disseminate to all parts of the body. They are usually difficult to eradicate, and many patients die from deep mycosis even with extensive medical care, especially when the central nervous system is involved. Some information on the four major systemic mycoses in North America is provided in Table 7-5.

Most of the fungi causing deep mycoses are dimorphic. Only the mold form, found in the soil, produces spores capable of infecting humans. The spores of some of these pathogens are so infectious that inhalation of a single spore can initiate pulmonary histoplasmosis or coccidioidomycosis. In most cases these diseases are asymptomatic, but in some individuals the pathogen invades the body. These cases may be fatal.

The fungi that cause histoplasmosis and cryptococcosis are especially prevalent in soil contaminated with fresh feces of bats and birds, usually pigeons or chickens. The disease itself is not found in these animals, but the droppings serve as rich nutrient sources for growth of the fungi in the soil. These enriched environments are often dangerous reservoirs of pathogenic fungi. For example, spelunkers (cave explorers) run a greater risk of acquiring histoplasmosis because of enormous deposits of bat feces in many caves.

FIGURE 7-10

Sporotrichosis: (a) Clinical picture showing progression up the lymphatics; (b) microscopic appearance of *Sporothrix schenkii* in culture.

DISEASES CAUSED BY FUNGI

TABLE 7-5

Major Systemic Mycoses of North America

Disease	Fungus	Appearance In Infected Tissue	Appearance In Culture*	Dimorphic
Coccidioidomycosis	*Coccidioides immitis*	Spherule†	Mold	Yes
Histoplasmosis	*Histoplasma capsulatum*	Yeast	Mold	Yes
Cryptococcosis	*Cryptococcus neoformans*	Yeast	Yeast	Rarely
Blastomycosis	*Blastomyces dermatitidis*	Yeast	Mold	Yes

*Cultures grown at room temperature (25°C).
†A nonfilamentous form of the fungus characterized by a sac filled with spores. When liberated each spore matures into a spherule.

Soil conditions and climate are important factors in determining the geographical distribution of infection. For example, histoplasmosis is most common in the Mississippi and Ohio River valleys, whereas coccidioidomycosis (Fig. 7-11) is endemic in the hot, dry climates of the western and southwestern United States, especially the San Joaquin Valley of California.

Opportunistic Fungi

Some fungi that are normally incapable of producing disease in healthy humans can cause serious and often fatal mycoses in people whose resistance has been lowered. Diseases caused by these **opportunistic fungi** may be either cutaneous or systemic. Many diabetics are susceptible to pulmonary infection by the common saprophytic molds of the genera *Rhizopus* and *Mucor* (both zygomycetes) and *Aspergillus.* Other high-risk groups susceptible to opportunistic fungal infections are alcoholics, leukemia patients, and persons on chemotherapy for treatment of cancer. Perhaps the most common fungal opportunistic mycosis, especially among hospitalized patients, is *candidiasis,* caused by the yeast *Candida albicans.* Unless the predisposing condition can be corrected, many of these opportunistic mycoses may be fatal, even with antibiotic treatment.

One often overlooked factor contributing to the development of opportunistic mycoses is the indiscriminate use of antibiotics. When a person's harmless bacterial flora is disrupted by antibacterial chemotherapy, the area formerly occupied by the displaced bacteria is recolonized by a fungus, which, by its eucaryotic nature, is resistant to antibacterial agents. For example, the yeast *Candida albicans* is usually controlled by the presence of normal flora bacteria, which easily outcompete the fungus for the available space and nutrients. Following destruction of these bacteria, however, several forms of candidiasis may occur, most commonly vaginitis ("yeast infections") and oral thrush.

Mycotoxicoses

Some fungi incapable of causing infectious diseases produce toxic substances that poison a person who ingests them. These poisonous substances are collectively called **mycotoxins** (*myco,* fungus; *toxin,* poison). The most common mycotoxicoses follow ingestion of poisonous mush-

FIGURE 7-11
Coccidioidomycosis:
(*a*) Clinical picture of
cutaneous form;
(*b*) scanning electron
micrograph of the
spherule, the form found
in infected lung tissue;
(*c*) extremely infectious
arthrospores of *Coccidioides
immitis* growing as a mold
in culture.

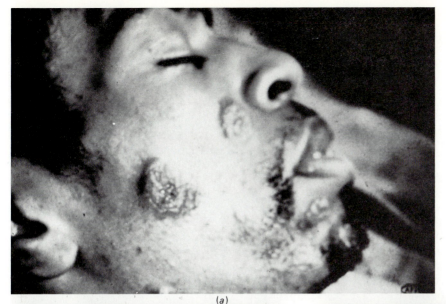

(a)

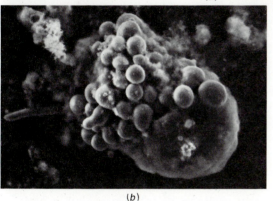

(b)

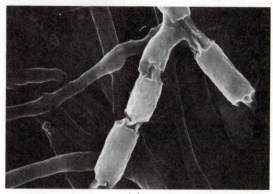

(c)

rooms, such as those in the genus *Amanita* (also called the "death angel"). These mushrooms contain lethal substances that destroy liver cells and excite the nervous system. Other poisonous substances may be produced by fungi growing on grain, nuts, and other agricultural products. For example, *Claviceps purpurea* growing on rye produces a variety of substances, some of which cause ergot poisoning (see color box, p. 150). Aflatoxin poisoning is associated with the growth of some common species of *Aspergillus* and *Penicillium* on nuts and grains. Beer and peanut butter made from contaminated grains and nuts may still contain toxin levels high enough to poison the consumer. (Aflatoxin primarily damages the liver and may also predispose for the development of cancer.) The toxin is not destroyed by cooking. Fortunately, in the United States federal law requires testing of peanuts for evidence of aflatoxin contamination. Other

149

A EUPHORIC DEATH

■ The young physician could scarcely believe what he was seeing. The patient standing before him was holding his detached left arm in his right hand. The doctor's disbelief was compounded by the expression on the patient's face. This man who had less than an hour ago watched his own arm fall from his body was calm and smiling.

The patient would soon be dead of *ergot poisoning*. Earlier he had eaten bread made from grain contaminated with a lethal fungus called the ergot of rye. While growing on grain, the ergot fungus (*Claviceps* sp.), an ascomycete, produces a potent toxin that, when eaten, causes spasms of the smooth muscles that surround the blood vessel. These muscle spasms restrict the blood flow, causing the body's extremities to *necrose*, turn black and die from lack of nourishment. The poison's effects eventually reach more vital organs and kill the victim. One bizarre feature of such cases stems from an additional characteristic of the ergot toxin—it causes powerful hallucinations. Thus, while killing a person, the toxin also distorts the victim's perception, often producing a state of euphoria that contradicts the nightmarish magnitude of the ordeal.

In addition to the toxin, the ergot fungus produces several interesting or medically valuable substances. Lysergic acid is a powerful hallucinogen when chemically converted to lysergic acid diethylamide (LSD). Another group of compounds, the *ergotamines*, induces smooth-muscle contractions, as does the toxin, although much less severe. The ergotamines are useful in the treatment of migraine headaches (which are caused by the expansion of the vessels in the head) or for inducing labor in women with prolonged pregnancy. Because they induce uterine contractions, ergotamines should not be used to treat migraine headaches in pregnant women.

Today the incidence of ergot poisoning in the United States is low because of government-sponsored agricultural inspections and controls of contaminated grain. The use of *fungicides*, chemicals that kill fungi, provides additional protection.

fungal toxins have milder effects. "Drunken breadeaters disease," for example, is a condition in persons who consume bread made from flour contaminated with a fungus that produces an intoxicating substance.

Mycotoxicosis occurs more frequently among domestic animals than humans because animals are more likely to ingest the contaminated food. This often causes substantial losses of livestock. In 1960 100,000 turkeys died after eating ground peanut meal in which *Aspergillus flavus* was growing. It was this outbreak of "turkey X disease" that led to the discovery of aflatoxins.

Most fungal toxins are only produced when the fungus grows in moist environments at relatively high temperatures. Unfortunately, these conditions are often found in silos and grain storage facilities. Most of these substances are toxic in extremely minute quantities, and lethal concentrations are produced long before the grain shows visible evidence of mold growth or spoilage.

Control and Treatment

Fungal diseases are difficult to control by prophylactic measures alone. No effective vaccines against mycoses are available. The transmission of contagious mycoses may be reduced by quickly diagnosing cases and treating them so they are no longer sources of infection. Sanitary conditions and good hygiene can further reduce the incidence, especially of cutaneous mycoses. Many fungal pathogens, however, reside in all types of soil and contact with them is virtually unavoidable. Fortunately, most of these fungi cause overt disease only if encountered in large numbers or if the exposed person is debilitated. For diseases caused by these opportunistic pathogens, identification and correction of the factors that predispose a person to infection are often the most effective therapeutic measures.

Since fungi are eucaryotic, antibacterial agents are useless in eradicating mycoses. The first antibiotics that could inhibit eucaryotic fungal pathogens without having a similar effect on the eucaryotic human host cells became available in 1959. Before that time, disseminated systemic infections by *Coccidioides* and *Cryptococcus* were usually fatal. A group of antibiotics, called *polyenes*, are now used with reasonable success for treating deep mycoses, depending on the specific disease. The most common of these is amphotericin B, which, like the other polyenes, inhibits the function of fungal cell membranes by binding to sterols. Sterols in human cells are different from fungal sterols and have a lower affinity for the polyenes, which are therefore more toxic for fungi. The toxicity, however, is not strictly selective for the fungus. Human red blood cells and kidney cells can also be affected by amphotericin B. This often leads to hemolytic anemia, liver damage, and impaired renal functions. Although previously fatal mycoses may be controlled by amphotericin B, many failures still occur because the dosage must be kept below the level that would be fatally toxic for the human.

Nystatin is another polyene used against less-invasive mycoses, such as candidiasis. This antibiotic acts much like amphotericin B, but is restricted to topical application since it is too toxic for internal use. A third antibiotic, griseofulvin, is effective against the dermatophytes, but it has no therapeutic value in dealing with systemic or subcutaneous mycoses.

A newer group of promising antifungal agents may reduce the mortality of mycoses. These agents, the *imidazoles*, are less toxic than amphotericin B. Some can be used for treating systemic mycoses and others for localized yeast infections. Antifungal therapy is discussed in greater detail in Chapter 14.

OVERVIEW

The fungi are nonphotosynthetic eucaryotic organisms with cell walls. Their members, the molds and yeasts, are generally saprophytes that are essential decomposers in the biogeochemical cycle. A few species are human pathogens capable of causing unusually persistent diseases. These maladies are grouped into four categories according to the anatomical sites where they are most commonly found. An additional group of diseases, the opportunistic mycoses, can infect persons with compromised resistance or persons receiving long-term antibacterial antibiotic therapy. Control of mycoses depends on correcting predisposing factors, reducing exposure to infected persons or to environments that promote infectious spore production, and using effective antifungal antibiotics in patients suffering from fungal disease.

Some fungi are dimorphic, growing as filamentous molds in certain conditions and as unicellular yeasts in others. Molds and yeasts are capable of both asexual and sexual reproduction. The types of sexual spores produced provide the basis for taxonomic assignment of a fungus to one of four groups—the zygomycetes, ascomycetes, basidiomycetes, or deuteromycetes. Of these groups, only the deuteromycetes demonstrate no sexual reproduction.

The fungi are among the most common organisms on earth. Many are important plant pathogens. Other fungi, such as poisonous mushrooms, are highly toxic or fatal when eaten. On the other hand, fungi are our only sources of many valuable products—leavened bread and pastries, some cheeses, soy sauce, wine, beer, and pharmaceuticals such as penicillin.

KEY WORDS

mycology	fruiting body
fungus	basidiospore
mold	basidium
hypha	ascospore
yeast	ascus
mycelium	zygospore
septate hypha	oospore
nonseptate hypha	budding
vegetative hypha	dimorphic fungus
aerial hypha	zygomycete
spore	mucormycosis
diploid state	ascomycete
haploid state	aflatoxin
conidium	basidiomycete
sporangiospore	deuteromycete
sporangium	Sabouraud's agar
chlamydospore	mycosis
arthrospore	superficial mycosis
blastospore	cutaneous mycosis

subcutaneous mycosis   dermatophytosis

systemic mycosis   opportunistic fungus

dermatophyte   mycotoxin

REVIEW QUESTIONS

1. Distinguish between yeasts and molds.

2. What is a fungal spore? How does it differ from a bacterial endo-spore?

3. List and describe five types of asexual fungal spores.

4. Many pathogenic fungi are dimorphic. What does this mean? How does it affect the transmission of the systemic mycoses?

5. List the four medically important groups of fungi and their characteristics.

6. What medium is usually employed for the primary isolation of fungi? What characteristics make it suitable for primary isolation of fungi?

7. Why are dermatophytes restricted to growth on the cutaneous layers of the body?

8. What are some factors that contribute to the development of opportunistic mycosis?

9. Why are mycoses more difficult to treat than bacterial diseases?

10. Name three antibiotics effective against mycotic diseases.

Protozoa and Algae

8

Protozoa were among the active little "animalcules" that fascinated Leeuwenhoek in 1674. These single-cell organisms, along with the **algae**, comprise an important group of eucaryotic microbes. Although relatively few diseases are caused by protozoa in modern western countries, over half the people on earth will suffer from a protozoan infection sometime during their lives, and no single bacterial pathogen causes more human death and suffering than the protozoa that cause malaria. Algae, on the other hand, which rarely cause human disease, benefit all our lives through their photosynthetic activities. The importance of these two groups of eucaryotes cannot be overestimated.

PROTOZOA: NATURAL DISTRIBUTION AND BIOLOGICAL ACTIVITIES

Most of the 40,000 known species of protozoa (singular, protozoan) are found in oceans, fresh water, or damp soil. Free-living (nonparasitic) protozoa are critical links in the food chain, feeding on other microorganisms and subsequently serving as a nutrient source for higher organisms. The majority of these protozoa cannot survive in the human body, where they are either destroyed or expelled. Most of these organisms that enter through the mouth are digested as they pass through the alimentary tract. Some, however, pass through unharmed and may be found alive in the feces.

Other protozoa parasitize the tissues and fluids of animals and humans. Fortunately, most of these interactions are harmless or even beneficial to the host. Cattle, for example, consume many soil protozoa while eating grass as their major food source. These microbes become established in the rumen (the first chamber of the bovine stomach) where they partially digest cellulose, the main constituent of grass. Without these intestinal protozoa, cattle fed on grass would starve. As the protozoa multiply, some are transported through the other stomach chambers and the intestinal tract. Along the route, most of them are digested by the cow and thus serve as a source of protein. The cells that survive are shed in the feces and reseed the soil.

Cellulose-digesting protozoa are also responsible for the termite's ability to utilize wood as food. Soon after they hatch, young termites are infected with protozoa by eating a mixture of cellulose and protozoa regurgitated directly from the alimentary tracts of their parents.

Eight genera of protozoa cause human diseases; these diseases include malaria, genitourinary tract infections, amebic dysentery, and interstitial cell pneumonia of newborns. Protozoan diseases are especially prevalent in tropical and subtropical environments.

Morphology

Protozoa are nonphotosynthetic unicellular organisms with a eucaryotic cell structure. They are further characterized by the absence of cell walls, which distinguishes them from the fungi. They range in length from about

2 μm (one phase of the human *Leishmania* parasites) to 20,000 μm, or just less than an inch. Thus, the smallest protozoa are similar in size to many of the bacteria; the largest are large enough to be seen without the aid of a microscope. Some protozoa, the radiolaria and foraminifera, possess intricate shells that are several centimeters in diameter.

Most protozoa are **polymorphic** (*poly*, many; *morph*, form); they undergo morphological changes during different phases of their life cycles. Some of these protozoa alternate between two forms, the actively feeding trophozoite and the dormant cyst (Fig. 8-1). The **cyst** is a dehydrated, thick-walled, protective form of the organism similar in function to the bacterial endospore. Cysts survive in both dry and aquatic environments, allowing dissemination over large distances. **Trophozoites**, on the other hand, feed and reproduce as long as environmental conditions are favorable. The onset of adverse conditions that would kill the trophozoite may trigger its transformation into a cyst, thereby increasing the chances for species survival. For example, the trophozoite of the soil organism *Naegleria* transforms itself into a cyst when the soil gets too dry. When moisture returns to the soil, the trophozoite emerges from the cyst.

Cyst formation is common among protozoa and is essential to the survival of some pathogenic species. For example, the cyst forms of intestinal parasites survive for long periods in water, soil, or food, as well as in the extreme acid environment of the human stomach. Trophozoite forms of the same organism survive poorly outside the body and, if ingested, are quickly destroyed by stomach acid. Without cysts, the protozoa could not establish infection in the gastrointestinal tract. The cyst is therefore the only infectious stage; trophozoites play no role in transmis-

PROTOZOA: NATURAL
DISTRIBUTION AND
BIOLOGICAL ACTIVITIES

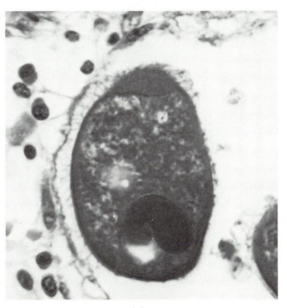
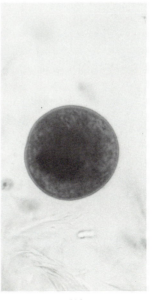

(a)   (b)

FIGURE 8-1
Balantidium coli, a human intestinal pathogen: (*a*) Trophozoite form and (*b*) cyst form.

sion of intestinal illness. Once in the intestine, however, the fragile trophozoite emerges from the cyst and reproduction resumes in the new host (Fig. 8-2).

The cyst state is not formed by all protozoa. Protective cysts are unnecessary if the organisms are directly transmitted between living hosts without exposure to hostile external environments or to the adverse conditions of the stomach. *Trichomonas,* for example, which is transmitted between people by direct sexual contact, forms no cyst. Similarly, no cysts are formed by protozoa that are transmitted by **vectors**, living organisms (usually a biting arthropod) that transfer the protozoa from one vertebrate host to another. Vectors protect the parasites from extreme changes in temperature, moisture, nutrient availability, and other factors in the nonliving environment. Since arthropods are the only vehicles for trans-

FIGURE 8-2

Life cycle of an intestinal amoeba: (1) Cyst in food, water, or on fingers; (2) emergence of trophozoites from cysts in small intestine; (3) colonization of large intestine; trophozoites multiply by binary fission; (4) possible invasion of intestinal wall; (5) encystment; (6, 7) cysts and trophozoites shed into the environment in feces.

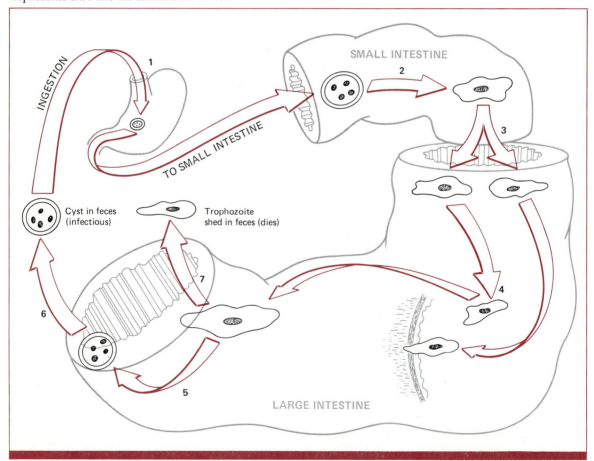

mission of these pathogens, the diseases they cause are geographically restricted to regions that favor the survival and proliferation of their respective vectors. Malaria, for example, is prevalent only in areas populated by the anopheles mosquitos, sleeping sickness only where there are tsetse flies. Control of these diseases depends on eliminating the vector from these regions.

Some protozoa are **pleomorphic**. Their trophozoite form has differing morphologies, often depending on its environment. (The term "pleomorphic" should not be confused with "polymorphic," which denotes the capability of producing both trophozoite and cyst forms.) Organisms with multiple hosts are often pleomorphic, with different trophozoite morphologies in the different host species. The morphology of the trophozoite may also vary in different tissues within the same host. For example, an organism may bear flagella for swimming through body fluids but lose the appendages when invading heart tissue. Identification of characteristic forms within infected tissues is important in the microscopic diagnosis of protozoan illness.

Nutrition and Metabolism

Since protozoa are nonphotosynthetic chemoheterotrophs, they require an external source of food in order to obtain energy and materials for biosynthesis. They obtain their nutrients in one of two ways. Some absorb dissolved nutrients directly through their cell membrane. Parasitic protozoa, for example, absorb their nutrients from the blood and tissues of their host. Others feed by engulfing solid foods, forming intracytoplasmic containers called **food vacuoles**. Amoebas, for example, can take up particulate matter anywhere along their surface by **phagocytosis**. In this process, the cell membrane extends around the extracellular particle until the particle is completely engulfed into a food vacuole (Fig. 8-3).

Ciliated organisms such as *Paramecium* create currents to sweep food particles into a specialized mouthlike structure, the *cytostome*, where they are trapped in food vacuoles. Digestive enzymes stored in **lysosomes** are discharged into the vacuole, where the ingested particle is degraded. The end products of digestion are either metabolized by the cell for energy, used to synthesize structural components, or released as waste products.

Free-living protozoa are usually aerobic. Some parasitic protozoa, however, live in environments where oxygen may be limited; indeed, some intestinal protozoa require low oxygen concentrations. Resident aerobic bacteria help reduce the oxygen level in the intestine and may contribute to host susceptibility to protozoan infection. In rare instances, antibacterial chemotherapeutic agents may cure a person of an intestinal protozoan infection by temporarily suppressing the bacteria on which the protozoa depend.

Reproduction

Most protozoa reproduce by both asexual and sexual processes. Some organisms have no sexual cycle. Others can exhibit sexual cycles but do not

FIGURE 8-3

Phagocytic engulfment of bacteria by an amoeba, with formation of a food vacuole. The bacteria are digested following fusion of the lysosomal membranes with the membrane of the food vacuole.

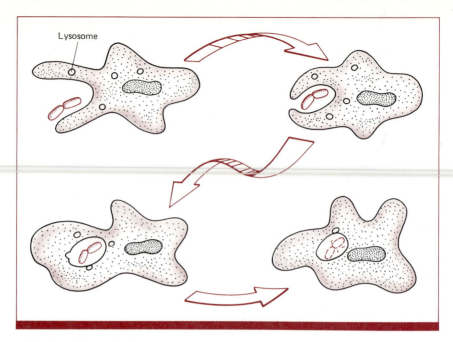

Lysosome

require them for survival. A third group reproduces asexually in one host and sexually in another.

The principal mode of asexual reproduction is **fission**, cell division producing two or more progeny cells. **Binary fission** is the most common. Flagellated protozoa always divide in a longitudinal direction (along their length), whereas protozoa with cilia always divide in a transverse direction. Some protozoa reproduce asexually by the process of **multiple fission**. In this case the nucleus divides many times before partitioning of the cytoplasm. A large number of progeny are thus formed from a single cell. These three variations of fission are depicted in Figure 8-4. A few protozoa multiply by a budding process similar to the budding of yeasts.

Sexual reproduction occurs by fertilization, the fusion of male and female reproductive cells. The fused cell then undergoes division to produce progeny cells. Some protozoa exchange genetic information by **conjugation**. During conjugation, two trophozoites trade copies of their nuclear genetic information while temporarily fused. This creates new combinations of genetic traits which may be more beneficial for survival than those found in the original strains. After separation, the cells reproduce by binary fission.

CLASSIFICATION OF THE PROTOZOA

Protozoa are divided into four major groups according to the type of locomotion used by the trophozoite:

1. **Mastigophora** move by flagella.

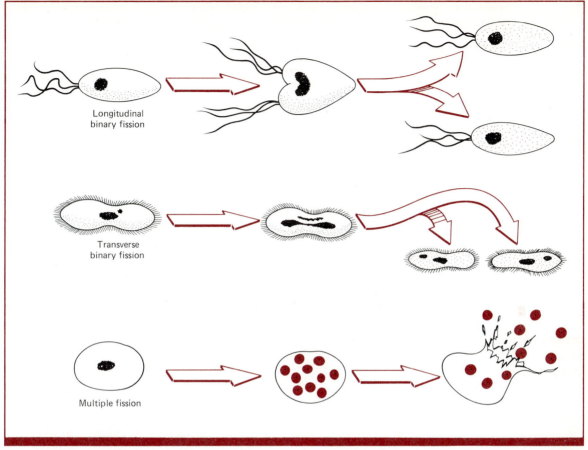

FIGURE 8-4
Reproduction by fission.

2. **Ciliophora** move by cilia.

3. **Sarcodina** move by producing **pseudopods** (*pseudo,* false; *pod,* foot), extensions of cytoplasm from the cell's surface.

4. **Sporozoa** are nonmotile in their mature forms.

Some organisms change mechanisms for motility depending on environmental conditions. For example, *Naegleria* may be amoeboid or flagellated depending on the availability of water. These organisms are assigned to the sarcodina, however, because they live and reproduce primarily as amoebas. Although organisms may display motility during sexual reproduction, they are classified according to the motility of the trophozoite. Further subclassification is based on distinctions in morphology and size.

Human parasites are found among all four classes of protozoa. The major taxonomic groups and their distinguishing characteristics are listed in Table 8-1.

Group	Characteristics	Representative Diseases
Mastigophora	Move by flagella* (or by undulating membrane); reproduce asexually by longitudinal binary fission; sexual reproduction absent	Giardiasis; trichomoniasis; trypanosomiasis; leishmaniasis
Ciliophora	Move and/or feed by cilia*; genetic exchange by conjugation; reproduce asexually by transverse binary fission	Balantidiasis
Sarcodina	Move and feed by pseudopodia*; reproduce asexually by binary fission	Amebic dysentery
Sporozoa	No apparent trophozoite motility;* intracellular parasites; alternate asexual and sexual cycles; many require alternate hosts for completion of life cycle.	Malaria; toxoplasmosis; pneumocystis (interstitial cell pneumonia)

*This property distinguishes these protozoa from the members of the other three subdivisions.

Mastigophora

Mastigophora (*mastigo*, whip; *phora*, bearer) form the largest group of protozoa. All possess flagella in the adult stage of their life cycle. Some flagella propel the cell by pushing or pulling it, other flagella may be used for steering. In some mastigophora, a flagellum is attached to a thin, loose membrane along most of the length of the protozoa, forming a structure called the **undulating membrane** (Fig. 8-5). This structure is characteristic of some dangerous blood and tissue parasites and is believed to facilitate movement of these flagellates through viscous body fluids by amplifying the propulsive forces generated by the flagellum.

Some flagellated organisms produce cysts; others, such as those that

FIGURE 8-5

A typical trypanosome with an undulating membrane.

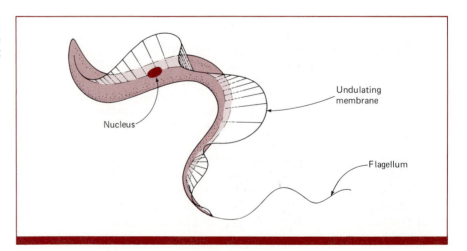

are transmitted by direct contact between hosts or by vectors, have no cyst stage. Flagellates reproduce only by longitudinal fission, an asexual process.

Most species of mastigophora are free-living. Some, however, cause human diseases, the most serious of which are African sleeping sickness (*Trypanosoma gambiense* and *Trypanosoma rhodensiense*) and Chagas's disease (*Trypanosoma cruzi*). Two other parasitic mastigophoras cause less severe but very prevalent diseases. They are *Giardia lamblia,* an intestinal pathogen, and *Trichomonas vaginalis,* an agent of genital infections.

Ciliophora

Ciliophora are characterized by the presence of cilia on the cell surface. Cilia are structurally identical to eucaryotic flagella but are shorter and more numerous. Often, as in paramecia (Fig. 8-6), the entire ciliate body is covered with thousands of these hairlike projections that beat like the oars of a boat in a coordinated fashion to propel the organism. In some organisms cilia are restricted to certain regions or fused together to form tufts called *cirri* (tufts of hair). These structures are able to function as legs that the protozoa use to creep along surfaces. Cilia may also aid in the feeding process. Some ciliophora are attached to solid surfaces such as rocks and are nonmotile. These organisms use their cilia exclusively for obtaining food.

The ciliates are unique among the protozoa in that each cell possesses two types of nuclei. The *macronucleus* is responsible for directing cell growth and asexual reproduction by transverse binary fission. The *micronucleus* contains the genetic information that is exchanged during conjugation.

Balantidium coli, which causes diarrhea, is the only species of ciliophora known to cause human illness.

Sarcodina

Members of sarcodina are commonly known as **amoebas**. They are

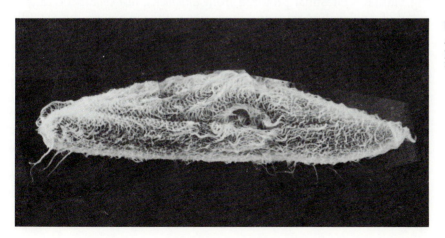

FIGURE 8-6
A typical paramecium. The protozoan is covered with numerous cilia.

FIGURE 8-7

A typical amoeba. This protozoan will change morphology as it moves.

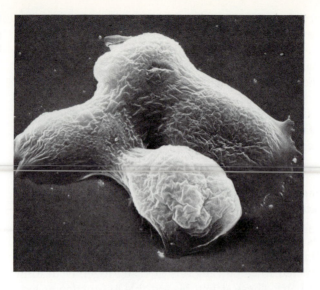

characterized by their ability to change shape by forming pseudopods (Fig. 8-7). The pseudopods are projected in front of the cell and the rest of the cytoplasm flows to the furthest point of extension. The formation of pseudopods allows the amoebas to move and to feed by phagocytosis.

Foraminifera and *radiolaria* are amoebas that have rigid internal or external skeletal structures made of calcium or silica. These organisms move and feed by extending their pseudopods through pores in their shells. The famous White Cliffs of Dover are composed of these protozoan skeletons that, over billions of years, settled to the ocean floor before geological activity thrust the massive deposit above sea level.

Amoeba trophozoites multiply by binary fission. Their cysts often undergo nuclear multiplication without cell division, resulting in several nuclei in a single cyst. The number of nuclei found in a cyst is often characteristic of the species and is diagnostically important for distinguishing harmless intestinal protozoa from *Entamoeba histolytica*, the causative agent of amebic dysentery and the major pathogen among the sarcodina.

Sporozoa

All sporozoa have three things in common: (1) they lack motility in their replicative forms; (2) they are intracellular parasites; and (3) they alternate asexual and sexual reproductive cycles. Some organisms carry out their sexual and asexual cycles in the same host; other sporozoa require multiple hosts. For example, *Toxoplasma gondii* multiplies sexually only in members of the cat family but develops asexually in humans and other mammals, often causing serious disease.

The *Coccidia* are another genus of sporozoa that are animal parasites, particularly of chickens. Entire flocks have been destroyed by these intestinal pathogens. Soil is contaminated by cysts of the organism released in the birds' feces. Young chickens picking food from the ground swallow these cysts and subsequently develop fatal coccidiosis (not to be confused with the human fungal disease, coccidioidomycosis). To prevent large outbreaks, chicken feeds often contain anticoccidial chemicals.

The four species of sporozoa in the genus *Plasmodium*, the etiological agents of malaria, are among the most important of all human pathogens.

DISEASES CAUSED BY PROTOZOA

Pathogenic protozoa generally initiate infection in one of four primary sites: the intestinal tract, the genital tract, the blood and tissues, or the central nervous system. Of these sites, the intestinal tract is the only one in the human body normally populated by harmless protozoa. Isolation of protozoa from any other anatomical location is indicative of disease. Since most pathogenic protozoa are difficult to culture, direct microscopic examination of clinical samples is essential for accurate diagnosis. Pathogenic protozoa are usually detected by examining smears of blood, cerebrospinal fluid, or genital exudates for cysts and trophozoites. Some specimens may have to be concentrated so that there are enough cells to be detected. The specimens are often stained to make distinguishing characteristics microscopically visible.

Intestinal Infections

Three protozoa, the flagellate *Giardia lamblia*, the ciliate *Balantidium coli*, and the amoeba *Entamoeba histolytica*, cause human intestinal infections (Fig. 8-8). The diseases, which are acquired by consuming contaminated food or water, are most prevalent in areas of poor sanitation and inadequate health education. Since 1972 *Giardia lamblia* has caused more outbreaks of water-borne disease in the United States than any other pathogen, affecting over 7000 people. These outbreaks have usually been associated with deficiencies in water treatment systems (see color box, p. 167).

These three organisms have similar life cycles. When ingested, protective cysts survive passage through the stomach. Trophozoites emerge from the cysts in the small intestine, where they produce the alimentary disturbances characteristic of the disease. As the organisms are moved into the more dehydrated regions of the bowel, cyst formation is triggered by the decreasing concentrations of water. Cyst formation is essential to the survival of the organisms, since trophozoites die when passed into the environment in the feces.

Of the three agents, *Entamoeba histolytica* causes the most serious disease, amebic dysentery. The pathogen releases enzymes that break down intestinal tissues (*histolytica*, tissue-lytic). The pathogen can spread and cause fatal abscesses of the liver, lungs, heart, and brain. About 10 percent of the world's population may be carriers of this protozoan, each infected person capable of shedding 300 million cysts per day. In the United States, a million people may be infected with *Entamoeba histolytica*.

E. histolytica is one of the most difficult intestinal pathogens to identify. The distinctive trophozoite and cyst forms can usually be detected

FIGURE 8-8

Agents of intestinal and
genital infections.

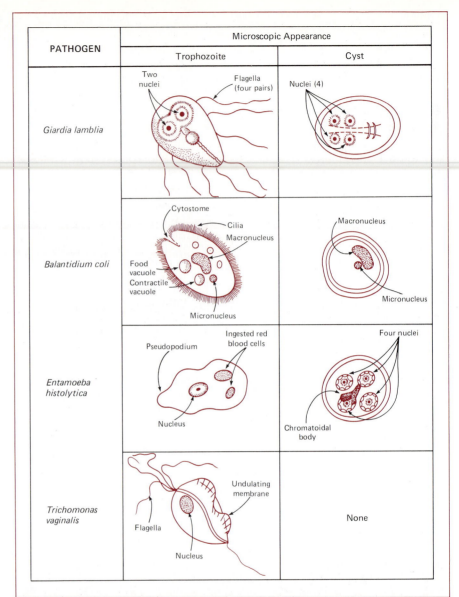

PATHOGEN	Microscopic Appearance	
	Trophozoite	Cyst
Giardia lamblia	Two nuclei — Flagella (four pairs)	Nuclei (4)
Balantidium coli	Cytostome, Cilia, Macronucleus, Food vacuole, Contractile vacuole, Micronucleus	Macronucleus, Micronucleus
Entamoeba histolytica	Pseudopodium, Ingested red blood cells, Nucleus	Four nuclei, Chromatoidal body
Trichomonas vaginalis	Undulating membrane, Flagella, Nucleus	None

by direct microscopic examination of feces. Harmless amoeba found in the
intestinal tract, however, are similar in appearance and must be distinguished from *E. histolytica*. Fortunately, the characteristic number of nuclei
in cysts of *E. histolytica* is four or less, whereas cysts of the commonly
occurring harmless amoeba have eight nuclei. In addition, trophozoites of
the pathogen usually contain red blood cells that have been phagocytosed
following tissue destruction.

"ENJOY AMERICA . . . BUT DON'T DRINK THE WATER"

■ Protozoan infections are often considered to be characteristic of underdeveloped countries where sanitary conditions are poor and insect populations are uncontrolled. Within the United States, however, at least one protozoan disease is alarming public health authorities with its frequency of attack. The pathogen is *Giardia lamblia*, the most common cause of waterborne outbreaks of diarrhea. Victims are usually residents of rural environments or visitors whose vacations are abruptly terminated by the disease.

Giardia cysts are usually removed from municipal water supplies by filtration. Residents are therefore protected from attacks of giardiasis. Outbreaks can usually be traced to contaminated drinking water that has been inadequately treated. Routine chlorination procedures, for example, are totally ineffective against *Giardia* cysts. Defective filtration systems that fail to remove the cysts mechanically are responsible for many incidents of giardiasis.

Giardia cysts may also be found in untreated waters, such as those in streams, rivers, or lakes that are contaminated with animal or human feces. Even those cool, clear waters which seem so pure and refreshing to hikers may be harboring the infectious form of the protozoa and cause "hiker's diarrhea." Vacationers, campers, and hikers who drink from these untreated sources can protect their vacations from ruin only by boiling their drinking water to kill *Giardia* cysts.

Venereal Infections

The flagellated protozoan *Trichomonas vaginalis* is one of the world's most common causes of venereal infections. In some areas of the United States, 50 percent of the women examined show microscopic evidence of *Trichomonas* infection. The disease, *trichomoniasis*, is localized, usually causing vaginitis, a chronic, often irritating inflammation of the vagina. Males are also infected with the organism but are usually asymptomatic. Infection is diagnosed by microscopic detection of the trophozoite in vaginal or urethral discharges. The trophozoite has five flagella, one of which is attached to an undulating membrane (Fig. 8-8).

Trichomonas vaginalis produces no protective cysts and cannot survive outside the host. Its primary mode of transmission is therefore believed to be sexual intercourse. An infected female can transmit the organism to a male, who usually acquires an asymptomatic infection of the urethra. Although the disease can be cured by treatment with the antiprotozoan drug metronidazole (Flagyl), males often receive no treatment because they have no symptoms. These untreated males frequently reinfect women after women have been cured. To prevent continual reoccurrence of trichomoniasis it is important when treating a symptomatic female that her male sex partner(s) be treated concurrently.

Blood Infections

Human blood parasites are nearly always acquired by the bite of an infected insect vector and are generally limited in their geographic distribution to areas where the vectors are prevalent. Occasionally, however, they are transmitted by blood transfusions. The blood is the primary site of infection, but subsequent invasion of the brain, viscera, and other secondary sites may occur. Among the most debilitating protozoan blood infections are trypanosomiasis and malaria.

Trypanosomiasis   African trypanosomiasis is caused by *Trypanosoma gambiense* or *Trypanosoma rhodesiense* (Fig. 8-9). These trypanosomes invade the central nervous system, eventually causing mental deterioration, coma, and death. Both organisms are transmitted by the bite of the tsetse fly, a vector found only in Africa and southern Arabia. A single infected insect may shed the parasite in its saliva for 3 months, spreading the disease to large populations. Each year in Africa there are 250,000 cases of human trypanosomiasis.

Many wild animals in Africa serve as healthy reservoirs of infection for trypanosomes. Although these native animals exhibit few or no symptoms, the acute susceptibility of imported animals, particularly horses and cattle, has prevented their successful introduction into many African regions. Even in many areas where imported livestock have been introduced, periodic outbreaks of trypanosomiasis continue to kill cattle and diminish food resources. The trypanosomes' relentless attacks on human and animal populations impede the development of 4 million square miles of tropical Africa.

American trypanosomiasis, also known as Chagas's disease, is caused by *Trypanosoma cruzi*, commonly found among domestic and wild animals

FIGURE 8-9
Transmission cycle of
African sleeping sickness.

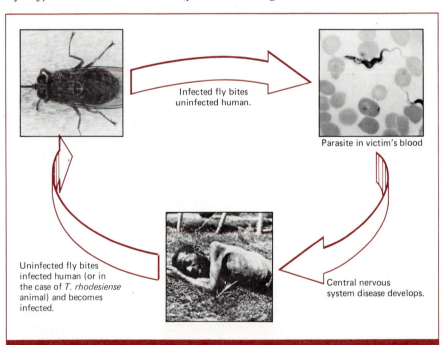

Infected fly bites
uninfected human.

Parasite in victim's blood

Uninfected fly bites
infected human (or in
the case of *T. rhodesiense*
animal) and becomes
infected.

Central nervous
system disease develops.

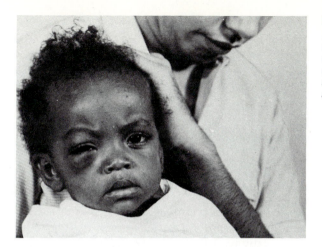

FIGURE 8-10
A doctor examines a young child suffering from an acute form of Chagas's disease. Swollen eyelids are a typical manifestation of acute infection.

in Central and South America. The vectors are reduviid bugs, blood-sucking insects that have an affinity to areas of the body, particularly the head, where blood vessels are close to the body surface. The insects' attraction to the lips accounts for their nickname, the "kissing bug." Another common site of attack is the eyelid (Fig. 8-10). The reduviid bug does not directly inoculate its victims, but deposits feces contaminated with trypanosomes on the skin surface while biting. Scratching the wound facilitates the entry of the protozoa through the bite and into the bloodstream. From the bloodstream, the organisms localize in the heart and central nervous system. The damage they cause is often fatal.

Malaria   It is estimated that there are 100 million cases of malaria annually, making it the most common serious infectious disease of humans. One million of these cases are fatal. The disease occurs only in areas where the female anopheles mosquito exists. This vector has been successfully controlled in most of the United States and other developed countries. However, a large part of the populated world is still at risk of contracting malaria (Fig. 8-11).

The name malaria is derived from two Italian words, *mal* (bad) and *aria* (air). Before the mosquito's role in transmission was known, the disease was blamed on hostile influences in night air. This myth was based on the valid observation that people who stayed indoors after sunset were less likely to contract the disease. (The anopheles mosquito is primarily nocturnal.)

Four members of the genus *Plasmodium* cause malaria in humans: *Plasmodium malariae, P. vivax, P. ovale,* and *P. falciparum.* The disease occurs when any of the four species is present in the saliva of an anopheles mosquito and is injected into the human bloodstream by the insect's bite. The parasites infect the liver and ultimately release progeny back into the bloodstream where red blood cells become infected. The red blood cells lyse, showering the bloodstream with more parasites that infect additional

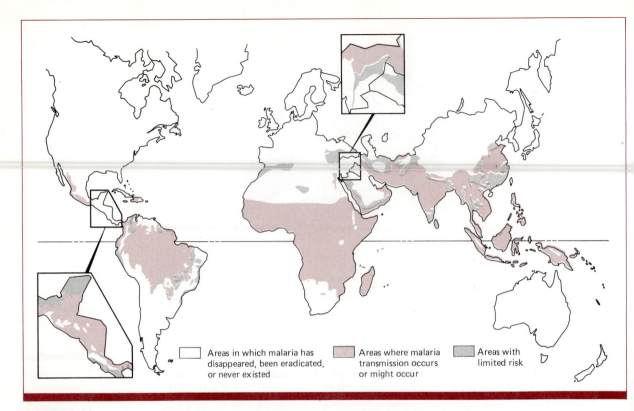

Areas in which malaria has
disappeared, been eradicated,
or never existed

Areas where malaria
transmission occurs
or might occur

Areas with
limited risk

FIGURE 8-11

Areas at risk: Geographic
distribution of malaria,
December 1979.

red blood cells. Rupture of the red blood cells occurs in a synchronized manner, resulting in the periodic episodes of chills and fever that are typical of malaria. (Details of the malaria parasite's life cycle are described in Chap. 23; see Fig. 23-12.)

Infections of the Central Nervous System

Several free-living amoeba are capable of causing meningoencephalitis (an inflammation of the brain and its membrane cover). The two most common causes of amebic meningoencephalitis are *Naegleria fowleri* and several species of *Acanthamoeba*. The disease is contracted while swimming in water containing the parasites, which enter through the nasal cavity and, in a few people, invade the brain. Factors predisposing such unfortunate persons for amebic meningoencephalitis are unknown.

Opportunistic Infections

Some protozoa primarily infect persons with reduced defenses. Two protozoan diseases, toxoplasmosis and interstitial cell pneumonia, are becoming increasingly important opportunistic pathogens, particularly within the hospital environment. Toxoplasmosis, caused by the sporozoan *Toxoplasma gondii*, infects one-third of the world's adults. The organism thrives in several animals and can be transmitted to people in one of two ways. Undercooked or raw meat of animals infected with *Toxoplasma* contains cysts that, after ingestion, release the trophozoites in the intestinal tract. The trophozoites may become widely distributed throughout the body, where they become dormant cysts. The organism may also be acquired when the sexual form of the parasite is ingested. These sexual forms are produced only in cats and are shed in the cats' feces. Human toxoplasmosis is rarely serious unless host resistance is compromised, allowing either an overwhelming primary infection or reactivation of dormant cysts from a previous infection. The disease is also severe when transmitted across an infected mother's placenta, often leading to abortion or serious defects in her fetus.

Another opportunistic pathogen, *Pneumocystis carinii,* causes interstitial cell pneumonia, a frequently fatal pulmonary infection that strikes immunosuppressed or compromised patients and premature infants. This disease is the most frequent cause of death among children with acute lymphoblastic leukemia. *Pneumocystis* infections frequently strike persons with acquired immunodeficiency syndrome (AIDS), an apparently transmissible disorder that impairs host defenses. (AIDS is discussed in Chapter 16.) Very little information is available on the natural source of *Pneumocystis* or its mode of transmission, but it is believed to be acquired by inhalation.

CONTROL AND TREATMENT OF PROTOZOAN DISEASES

The spread and occurrence of most protozoan diseases can be best controlled by drug therapy, vector control, and improvement of sanitary conditions. Despite massive research efforts, by 1984 no effective vaccine against any protozoan has been developed.

Drug Therapy

Because protozoa are eucaryotic parasites, they are difficult to kill selectively without damaging human host cells as well. Many protozoa are intracellular parasites, making them even more difficult to eliminate with drug therapy. Polymorphic protozoa are often hardest to treat because they may go through stages that are resistant to drugs. Nonetheless, some drugs are moderately to very successful against several protozoa. Metronidazole (Flagyl) is used against intestinal amoeba and genital parasites such as trichomoniasis. Chloroquine, primaquine, and pyrimethamine are antimalarial drugs that are used both for treatment and as prophylactic medications for travelers in tropical regions. Unfortunately, because of release of parasites from the liver, where they are protected from chloroquine, relapses may occur after drug therapy has been discontinued. The development of drug resistance in formerly susceptible pathogens further reduces the effectiveness of drug therapy.

Vector Control

Reducing the population of the anopheles mosquito has effectively controlled the incidence of malaria, particularly in temperate environments. Malaria was once common in the southeastern United States, where

600,000 cases were reported in 1914. Since 1950, only 13 cases within the United States have been caused by insect inoculation. (Most cases of malaria that are acquired in the United States are from transfusion of blood harvested from a donor who had the disease.) Vectors can be controlled by using insecticides and by draining swamps to eliminate breeding habitats. The use of mosquito netting and protective clothing limits access of the vector to the body.

Improvement of Sanitary Conditions

The transmission of many protozoan infections can be prevented by proper sanitary conditions. Thorough hand washing, proper cooking of foods, and adequate water treatment facilities are critical in preventing accidental ingestion of pathogenic protozoa.

INTRODUCTION TO THE ALGAE

The **algae** include both unicellular and multicellular eucaryotic organisms capable of photosynthesis. Multicellular algae are distinguished from true vascular plants by their lack of roots, leaves, stems, or vessels for conducting sap and other fluids. The name "algae" is the Latin word for seaweed, referring to the macroscopic members of the group. The study of algae is called *phycology,* derived from the Greek word for seaweed, *phykos.*

Since algae are photosynthetic, they require light and water and are found mainly in damp or aquatic environments. Here they serve as primary producers, the first step in the food chain, harnessing the sun's energy and fixing carbon dioxide into organic materials for metabolism and biosynthesis. Algae produce oxygen as a by-product of photosynthesis and comprise a major source of the earth's free atmospheric oxygen.

Some algae form beneficial associations with other organisms. For example, a **lichen** is a composite organism composed of a fungus and an alga, neither of which could survive alone in the lichen's natural habitat. Such habitats are generally poor in water and organic nutrients. The alga provides organic nutrients by photosynthesis, whereas the fungus is well adapted to gathering and conserving what little water is available. This enables lichens to colonize the surface of granite boulders and other inhospitable habitats. Another alga, *Chorella,* lives within the cytoplasm of a paramecium (a protozoan). The alga releases photosynthesized nutrients to the paramecium while being provided with a relatively safe ride inside the ciliate.

Red alga is the source of agar used for solidifying microbiological culture media (see Chap. 3). Another group of algae, the *diatoms* (Fig. 8-12), are important producers of vitamins A and D. Fish consume the algae and concentrate these vitamins in their livers, from which they can be extracted for human use. Cod-liver oil, for example, is a well-known vitamin supplement. The shells of diatoms are used to prepare abrasive compounds found in such products as toothpaste and silver polish. Large

PROTOZOA AND ALGAE

FIGURE 8-12
Diatoms are characterized by intricate patterns formed by their silica shells.

deposits of diatoms, called diatomaceous earth, are used as water filters in industry as well as in swimming pools and aquariums.

Classification of the Algae

The algae are a large and diverse group of organisms in terms of morphology, size, and chemical composition, properties used to classify the algae into six major subdivisions. These groups and their importance to humans are listed in Table 8-2.

TABLE 8-2
Divisions of the Algae and Their Importance to Humans

Subdivision		Importance
Common Name	**Taxonomic Name**	
Green algae	Chlorophyta	Producers of free oxygen; ancestors of plants; a nonphotosynthetic variant is the agent of protothecosis.
Euglenoids	Euglenophyta	Wall-less algae; possibly evolutionary transition forms between algae and protozoa
Diatoms	Chrysophyta	Producers of free oxygen; involved in vitamin A and D production; diatomaceous earth used in filters and abrasives
Dinoflagellates	Pyrrophyta	Agents of paralytic shellfish poisoning
Red algae	Rhodophyta	Agar production
Brown algae	Phaeophyta	Kelp, a source of food, fertilizer, and alginates

Algae may be unicellular, filamentous, branched, colonial, or leafy, as, for example, in the seaweeds. All photosynthetic algae contain chlorophyll A as the main pigment for trapping the energy of light. Other types of chlorophyll and accessory pigments (carotenoids and phycobilins) may be present, depending on the alga. These assist in photosynthesis and contribute to the characteristic color of the organism. Algae also differ from one another in how the organic products of photosynthesis are stored— either as starches, sugars, fats, oils, or alcohols. Additional criteria for classification include cell-wall composition, chloroplast structure, type of motility, and mode of reproduction.

Algae and Human Disease

Because sunlight is unavailable inside the human body, photosynthetic algae do not cause infectious disease. One alga, *Prototheca*, is unique in that it lacks chlorophyll and therefore the capacity to photosynthesize. It causes the disease *protothecosis*, found principally among people living in the tropics. *Prototheca* is believed to be a variant of the green alga *Chlorella* that has adapted to a heterotrophic existence. Similarities in morphology, mode of reproduction, and several chemical features suggest that *Prototheca* arose from a *Chlorella* that lost its chloroplasts. *Prototheca* can be isolated from water, sewage, and the human gastrointestinal tract. Protothecosis, however, is rare and occurs either in individuals with impaired immune

FIGURE 8-13
A case of protothecosis.

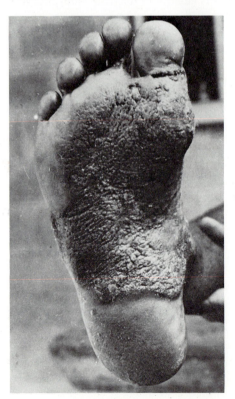

174

systems or in populations with poor nutrition or poor sanitation. The most common manifestation is the formation of skin ulcers (Fig. 8-13).

A more common problem associated with algae is *paralytic shellfish poisoning*, an occasionally fatal condition that follows consumption of shellfish contaminated with toxic algal products. *Gonyaulax* and *Gymnodidium*, both dinoflagellates—algae containing two flagella (Fig. 8-14)—produce a potent nerve toxin that causes paralysis in humans. Shellfish, however, may consume the algae without being affected by their toxin, which becomes harmlessly concentrated in their tissues. Persons who then eat these shellfish may contract paralytic shellfish poisoning.

The danger of paralytic shellfish poisoning increases during seasons when conditions are favorable for algal multiplication. During these times *Gonyaulax* and *Gymnodidium* can increase to concentrations of 50,000 cells per milliliter of water, their pigments imparting a characteristic red color to the waters they inhabit. This phenomenon is known as *red tide* and is indicative of a great abundance of neurotoxin in the shellfish of the region.

Paralytic shellfish poisoning is not always associated with red tide, since algae populations too small to change the water color are still dangerous. Ingestion of 0.5 mg of toxin can be fatal to humans. This amount of toxin is easily acquired in a single meal of shellfish, even after freezing or routine cooking. There is no antidote for the poison; treatment is symptomatic and is usually begun by inducing the patient to vomit in an attempt to reduce to a nonlethal level the amount of toxin-contaminated food remaining in the victim. If treatment is not successful, death usually occurs within 3 to 12 hours.

FIGURE 8-14

Gonyaulax, a toxin-producing dinoflagellate.

It is important in disease prevention that coastal waters and shellfish be strictly monitored to identify the areas where shellfish are most likely to be contaminated. The disease is most commonly associated with shellfish harvested from the coastal waters of the Pacific Northwest and Alaska, as well as of New England. Red tide is also responsible for killing large populations of fish, thereby polluting the water and beaches.

Microscopic algae are not usually visibly detectable in water. Occasionally, however, growth of the organism forms a layer on the water's surface. Such a phenomenon is called an **algal bloom**. Generally such blooms are limited in size by the availability of essential nutrients, specifically nitrates and phosphates, in the water. When the nutrients are exhausted, growth stops. If enough nutrients are present, however, massive algal growths may cover the entire surface of a lake. In many cases, sewage and industrial wastes dumped into waterways provide these nutrients and lead to **eutrophication**, literally, "overfertilization" of the aquatic environment. Algal blooms in these polluted lakes may reach overwhelming dimensions. When these huge mats of algae die, their decomposition by aerobic fungi and bacteria depletes the oxygen supply, suffocating most of the fish and other aquatic organisms. In this way, eutrophication was responsible for the near-death of Lake Erie in the early 1970s. To help reduce eutrophication many states try to prevent disposal of pollutants in their waterways.

OVERVIEW

Protozoa are unicellular, heterotrophic eucaryotes that lack cell walls. Many are critical links in the food chain. The organelles of motility in the mature trophozoite stage provide the basis for classification of protozoa into four groups. The mastigophora use flagella, either free or attached to an undulating membrane, for movement. Ciliophora possess cilia for either movement or feeding (or both). They can also exchange genetic material by conjugation. Sarcodina, the amoebas, move and feed by forming pseudopods. Sporozoa are intracellular parasites with no apparent motility in the trophozoite stage.

Pathogenic protozoa occur in all four groups. Some pathogens exist as trophozoites in the intestinal tract but form cysts before they are released in the feces. The cyst is the infectious stage that survives outside the host. Of the human pathogens, only those acquired by ingestion characteristically form cysts. The trophozoite of venereal pathogens can be transmitted directly from an infected to a susceptible host. No cyst stage is necessary because it is never exposed to environments outside the body. Most blood-borne pathogens also do not form cysts. They alternate between human and arthropod hosts. Protozoan diseases may be controlled by eliminating the arthropod vector or by improving sanitary conditions. Drug therapy is successful against several protozoan diseases.

Algae are photosynthetic eucaryotes which, unlike plants, are not differentiated into roots, leaves, or stems. They may be unicellular organisms or multicellular, such as seaweeds. Algae play important roles as producers of oxygen and as phototrophs, the first step in the food chain. Useful algal products include agar and diatomaceous earth. Other products may be toxic to fish, animals, and humans. Only two human illnesses are caused by algae, paralytic shellfish poisoning and protothecosis.

KEY WORDS

protozoan	multiple fission
alga	conjugation
polymorphic	mastigophora
cyst	ciliophora
trophozoite	sarcodina
vector	pseudopod
pleomorphic	sporozoa
food vacuole	undulating membrane
phagocytosis	amoeba
lysosome	lichen
fission	algal bloom
binary fission	eutrophication

REVIEW QUESTIONS

1. Describe three protozoan structures for motility.

2. Differentiate between the following:
   (a) Trophozoite and cyst
   (b) Pleomorphic and polymorphic
   (c) Binary fission, multiple fission, and budding

3. Describe the life cycle of a pathogenic intestinal protozoa.

4. How does a clinician distinguish between *Entamoeba histolytica* (the cause of amebic dysentery) and the common amoebas normally found in the healthy intestinal tract?

5. Describe two modes of transmission of blood parasites from person to person.

6. Even after effective therapy, recurrences of malaria are common. Compare the reason for these recurrences with that for recurrent trichomoniasis.

7. How are algae distinguished from true vascular plants?

8. Why is red tide a potential danger to people?

9. How are excessive algal blooms promoted by high concentrations of pollutants? What are the ultimate consequences of eutrophication?

PROTOZOA AND ALGAE

Viruses

Compared to cells, viruses are extremely simple in physical and chemical properties. They are noncellular particles that completely lack all structures characteristic of cells and often consist of nothing more than nucleic acid and protein.

The question of whether viruses are living entities has been debated for three-quarters of a century. Despite the heated fervor that this question has generated, finding the answer is not as important as understanding the characteristics of viruses that have created such a debate: in some stages of its life cycle a virus is as inanimate as any complex of crystallized molecules; at other times the same virus behaves as a living entity, demonstrating heredity and reproduction. This apparent paradox is explained by the biological properties of viruses.

PROPERTIES OF VIRUSES

Intracellular Parasitism

Viruses are **obligate intracellular parasites**. They require a host cell to perform nearly every biological function necessary for their survival. They lack the cellular machinery necessary for acquiring nutrients, producing energy, and synthesizing proteins. They are incapable of any independent metabolism, a property that distinguishes them from other obligate intracellular parasites, such as rickettsias and chlamydias. Viruses are the ultimate parasites.

Size

Viruses are among the smallest biological entities, ranging in size from 20 nm to slightly less than 400 nm. Many are only one-hundredth as large as a typical bacterium (Fig. 9-1). Too small to be seen with standard light microscopes, viruses were invisible before the invention of the electron microscope.

FIGURE 9-1
Many small viruses absorbed to the bacterial cell wall.

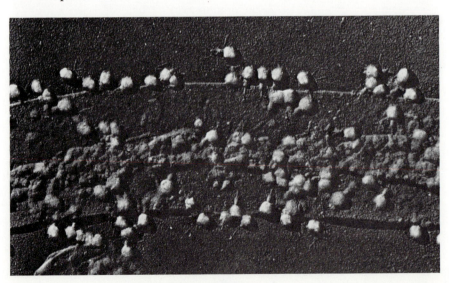

Because of their small size, viruses are *filterable*, that is, they cannot be removed from a liquid by filtration through bacteriological filters. These filters have holes too small to allow the passage of almost all bacteria and are often used for "sterilizing" liquids. Viruses, however, are not retained by such filters because they readily pass through the pores.

Genetic Composition

Viruses are unique in that they possess only a single type of nucleic acid, either deoxyribonucleic acid (DNA) or ribonucleic acid (RNA), but never both. Some viruses use RNA instead of DNA as the genetic material. All microbes other than viruses have DNA for storing genetic information and RNA to translate the stored information into cellular properties (see Chap. 12). In addition, the nucleic acid of some viruses is single-stranded, whereas that of others is double-stranded. Thus, four possible configurations exist: (1) double-stranded DNA, (2) single-stranded DNA, (3) double-stranded RNA, and (4) single-stranded RNA.

Eclipse

The **eclipse phase** is an event unique to the life cycle of all viruses. While reproducing in a host cell, the virus particle disintegrates into its molecular constituents. This is the beginning of viral eclipse. The eclipse phase ends with the reappearance of progeny viruses that are identical to the parent virus. Details of eclipse will be discussed under Virus Replication later in this chapter.

Virus Crystals

Another characteristic unique to viruses demonstrates their inanimate quality. Many can be concentrated into crystals of purified viruses (Fig. 9-2). These crystals resemble those formed by any highly symmetrical molecule, but viruses have the ability to infect cells and propagate.

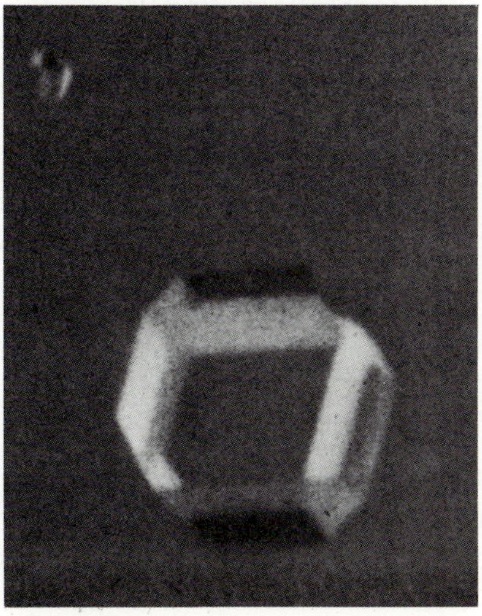

FIGURE 9-2
A purified crystal of poliovirus containing billions of infectious virions. Such crystalline structures are often formed naturally during intracellular development of many viruses. (See Virus Replication.)

TABLE 9-1

Distinguishing
Characteristics of Viruses

1. They are obligate intracellular parasites.
2. They are incapable of independent metabolism.
3. They are filterable through standard bacteriological membrane filters.
4. They possess only a single type of nucleic acid.
5. They have an "eclipse phase" in their life cycle.
6. Some retain infectivity after crystallization.

Those properties that distinguish viruses from all other organisms are listed in Table 9-1.

Host Specificity

Every organism on earth is believed to be susceptible to viral infection. Even bacteria are attacked and killed by viruses. These bacterial viruses are called **bacteriophages** (or simply phages). All viruses are specific for the cells they can infect, and many are restricted to a single host species, or even strains within a species. For example, a bacteriophage that infects one strain of *Staphylococcus aureus* may be incapable of attacking another strain that is identical except for a single molecule on the cell surface. A few viruses infect a broad range of natural hosts. Rabies virus, for example, can be transmitted among different animal hosts as well as to humans. Host specificity determines the types of hosts in which a pathogenic virus can cause infection and disease.

FIGURE 9-3

Diagrammatic representations of viral structure (cross section). (*a*) Nonenveloped virus; (*b*) enveloped virus; (*c*) three-dimensional representation of an enveloped virus showing realistic capsomere arrangement.

Structure and Chemical Composition

The simplest viruses are strands of nucleic acid (DNA or RNA) surrounded by the **capsid**, a protective protein coat. The capsid is made up of repeating

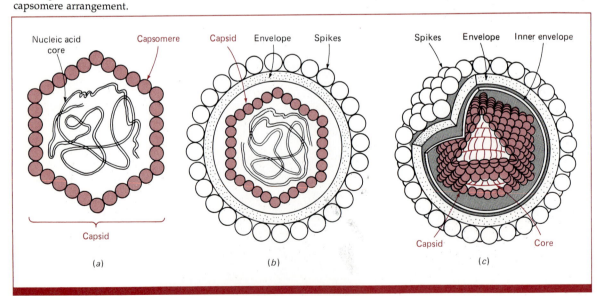

TABLE 9-2

Comparison of Enveloped
and Nonenveloped Viruses

	Enveloped	Nonenveloped
Virion	Nucleic acid, capsid, and envelope	Nucleic acid and capsid
Location of attachment sites	On membrane envelope	On protein capsid
Infectivity of nucleocapsid alone	Noninfectious (requires presence of envelope)	Infectious
Ether-sensitive	Yes	No

units of protein called **capsomeres**. Together, the capsid and the nucleic-acid core form the **nucleocapsid**. Some viruses have an additional outer layer, an **envelope**, that surrounds the nucleocapsid. The envelope is derived from host cell membranes that have been modified by the addition of virus proteins. In some envelopes the proteins appear as projections called *spikes*. The entire infectious virus is called a **virion**. (See Fig. 9-3.)

The surface components of virions allow them to recognize and attach to their susceptible host cells. For nonenveloped viruses host recognition sites are on the nucleocapsid, which is therefore the virion. In enveloped viruses, the recognition sites are found on the envelope; thus their virion consists of the nucleocapsid and the additional membrane envelope. If this envelope is dissolved with a lipid solvent such as ether, the remaining nucleocapsid is incapable of infecting the host cell. Enveloped viruses, therefore, are said to be ether-sensitive. Nonenveloped viruses, on the other hand, are not inactivated by ether since lipid is not part of the virion and the protein capsid is not altered by such solvents. Table 9-2 presents a comparison of the enveloped and nonenveloped viruses. A glossary of terms is provided in Table 9-3.

Morphology

The shape of nonenveloped viruses is determined by the manner in which the capsomeres are arranged around the nucleic acid core. Viruses with *cubic symmetry* are icosahedra, sphere-like structures with 20 triangular sides (Fig. 9-4a and b). Other viruses with *helical symmetry* resemble long rods with the capsomeres arranged around a spiraled coil of nucleic acid (Fig. 9-4c). Envelopes alter the appearance of the virion, often resulting in pleomorphism (having no stable shape) because the outer membrane is

TABLE 9-3

Glossary of Terms Relating
to the Structure of Viruses

Core	The nucleic acid and any associated molecules necessary for its stabilization
Capsid	The protein coat surrounding the nucleic acid core
Capsomere	The repeating protein unit of which the capsid is composed
Nucleocapsid	The nucleic acid core plus the protein coat (capsid)
Envelope	The outer membrane layer which characteristically surrounds certain viruses
Virion	An infectious virus particle

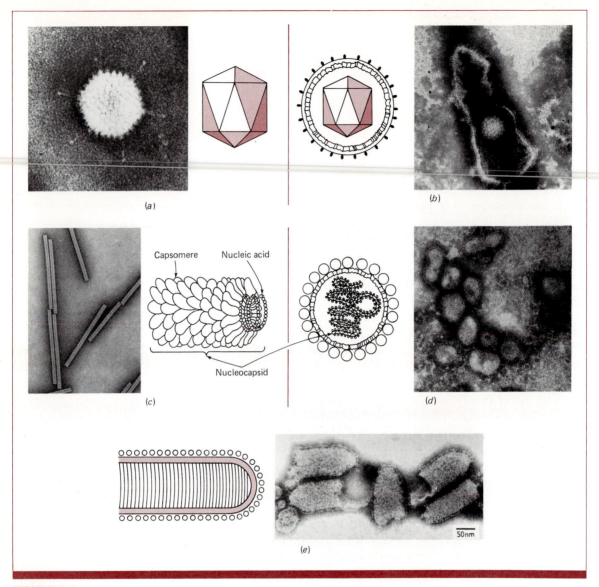

(a)

(b)

Capsomere   Nucleic acid

Nucleocapsid

(c)

(d)

50nm

(e)

FIGURE 9-4

Morphology of viruses
with (a) cubic symmetry
(naked); (b) cubic
symmetry (enveloped);
(c) helical symmetry;
(d) enveloped helix
(pleomorphic);
(e) enveloped helix
(bullet-shaped).

somewhat fluid. The shape of pleomorphic viruses may be determined by the physical forces exerted on the envelope, although the symmetry of the nucleocapsid inside remains unchanged by these influences (Fig. 9-4d). A few enveloped viruses have more stable shapes, such as the peculiar bullet-shaped rabies virus (Fig. 9-4e).

A more complicated viral morphology is demonstrated by some bacteriophages (Fig. 9-5). The nucleic acid is stored in the virus head, analogous to the capsid of viruses with cubic symmetry. The head is attached to a contractile hollow tube called the *tail* that terminates in a *base plate*. Fibers that emanate from the base plate are used by the bacteriophage

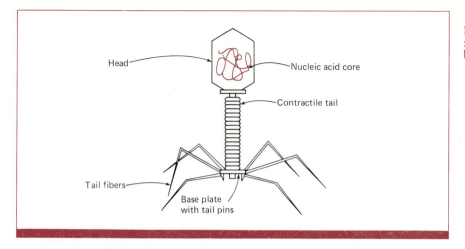

FIGURE 9-5
Diagrammatic
representation of a typical
bacteriophage.

Head

Nucleic acid core

Contractile tail

Tail fibers

Base plate
with tail pins

to recognize and attach to its specific host bacterium. If any of these structures is removed or damaged, the bacteriophage can no longer infect its host bacterium.

Classification of Viruses

Viruses are divided into major categories according to whether they infect animal, plant, or procaryotic cells. Further classification is based primarily on physical and chemical properties and on the range of hosts that can be infected. These properties are usually considered in the following order of importance:

1. Type of nucleic acid (DNA or RNA)

2. Single or double strandedness of the nucleic acid

3. Capsid morphology (size, symmetry, and number of capsomeres)

4. Presence or absence of envelope (thus, sensitivity to inactivation by physical or chemical agents, especially ether)

5. Host range

Although viruses have been assigned names according to the classical binomial system of nomenclature, they are still known by common names. The taxonomy of some viruses that affect animal cells is provided in Table 9-4.

VIRUS REPLICATION

Unlike cellular organisms, viruses cannot reproduce by fission. Because they have no independent metabolism, they must use a host cell to perform all the functions necessary for producing new infectious virions. Viruses efficiently redirect many host-cell metabolic reactions to favor the production of new viral particles rather than new

TABLE 9-4
Classification of Animal Viruses

Type of Nucleic Acid	Strandedness of Nucleic Acid	Morphology of Nucleocapsid	Envelope	Group*	Representative Human Disease
DNA	Single	Icosahedral	No	Parvovirus	
	Double	Icosahedral	No	Papovavirus	Warts
			No	Adenovirus	Respiratory infections
			Yes	Herpesvirus	Chickenpox; cold sores; genital herpes; infectious mononucleosis
		Complex	No	Poxvirus	Smallpox
RNA	Double	Icosahedral	No	Reovirus	Diarrhea
	Single	Icosahedral	No	Picornavirus	Polio; common cold
			Yes	Togavirus	Yellow fever; encephalitis
		Helical	Yes	Orthomyxovirus	Influenza
				Paramyxovirus	Measles; mumps
				Rhabdovirus	Rabies
				Coronavirus	Common cold
				Retrovirus	Possible role in tumor induction
		Unknown	Yes	Arenavirus	Lassa fever

*Additional properties, such as shape and size of the virion, are used to differentiate between groups.

host-cell material, resulting in the eventual release of progeny viruses at the expense of the cell itself.

The reproductive cycles of all viruses consist of five steps: (1) attachment, (2) penetration and uncoating, (3) synthesis of viral components, (4) assembly of viral components, and (5) release of progeny virions. The way these tasks are accomplished depends on the type of virus and on the type of cell infected (animal, plant, or bacterial). The general scheme of viral replication presented in Figure 9-6 is typical of many nonenveloped animal viruses that replicate in the cytoplasm (as opposed to the nucleus). Variations encountered in the replication of other animal viruses, plant viruses, and bacteriophages will be discussed under Viral Adaptations to Different Hosts.

Attachment

The first event in viral infection is attachment of the virus to the surface of a susceptible cell. Attachment follows random collision of the virus and host cell. This step requires the specific interaction of two complementary molecules, the **attachment site** on the surface of the virus and the **receptor site** on the cell's surface (see color box). Since attachment and receptor sites interact in much the same way that two pieces of a puzzle fit together, this step is highly specific. A proper "fit" must exist before attachment can

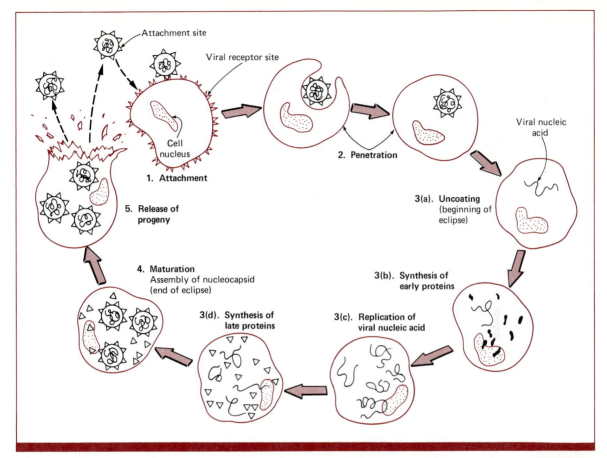

Attachment site

Viral receptor site

Cell nucleus

1. Attachment

2. Penetration

Viral nucleic acid

3(a). Uncoating (beginning of eclipse)

5. Release of progeny

4. Maturation
Assembly of nucleocapsid (end of eclipse)

3(b). Synthesis of early proteins

3(d). Synthesis of late proteins

3(c). Replication of viral nucleic acid

FIGURE 9-6

Replication of a typical nonenveloped virus within the cytoplasm of a susceptible animal cell.

occur. Most viruses, therefore, show an extremely high degree of specificity for their host cells. In many cases, viral infectivity is neutralized by molecules such as antibodies that cover the attachment sites before the virus can adhere to the cell's receptor sites. Formation of such antibodies is one way we develop immunity to viral diseases.

Attachment sites are located in the capsid of nonenveloped viruses and on the viral-specific proteins in the membrane of enveloped viruses.

Penetration and Uncoating

The second step in viral infection is penetration of the host cell. The virions of a few viruses such as poliovirus are altered during attachment so that free nucleic acid is released directly into the cytoplasm. Most enveloped viruses enter the host cell by fusion of the host cell membrane and the viral envelope (Fig. 9-7*a*). This releases the naked nucleocapsid into the cytoplasm. Most nonenveloped and some enveloped viruses are engulfed in an intracellular vacuole by **viropexis**, a process similar to phagocytosis. These nucleocapsids are released into the cytoplasm by one of the mechanisms depicted in Figure 9-7*b* and *c*.

187

THE CELL'S ACHILLES' HEEL

■ It may seem antievolutionary for cells to possess receptor sites that make them sensitive to virus infection. Most of these surface molecules, however, are not suicide sites; that is, their cellular function is not to welcome viruses. These receptor sites normally perform important functions for the cell—for example, the transport of substances across the cell membrane. Viruses, however, have acquired the ability to use these complex surface molecules for attachment, as ports for initiating infection. The cell is faced with a dilemma. Eliminating complex surface molecules would eliminate the ability to perform critical functions and would spell cell death.

The dilemma is well illustrated by a complex component on the surface of *Escherichia coli*, a structure that has been dubbed the "Achilles pore." Less than 20 minutes after a bacteriophage attaches to this site, the bacterium explodes and releases hundreds of viral progeny. But the surface complex also binds iron for transport into the cell. This is the only mechanism for the cell to accumulate intracellular concentrations of this essential element. Like the Achilles pore, virtually any complex surface molecule is a potential site for virus attachment. To be alive is to be vulnerable to viral infection.

Before a virus can reproduce, viral nucleic acid must be released from the capsid. Some viruses are **uncoated** by enzymes in the phagocytic vacuole. These enzymes digest the capsid and free the viral nucleic acid. In most cases, the uncoating process is poorly understood.

Eclipse begins with the uncoating step. The virus is now naked nucleic acid (DNA or RNA) floating freely in the host cell's cytoplasm. No infectious virus can be found in the cell during eclipse. Once free of the protein coat, the viral nucleic acid competes with the cell's chromosomes for control of the biological machinery.

Synthesis and Assembly of Viral Components

One of the first viral-directed acts performed by the cell is the synthesis of **early proteins**. Many of these enzymes interfere with the expression of host-cell genetic information, leaving the viral nucleic acid as the sole

FIGURE 9-7

Entry of viral nucleocapsids into host cells: (*a*) Fusion of viral envelope and cell membrane. Nucleocapsid is released directly into the cytoplasm. (*b*) Viral penetration by endocytosis. The virus envelope absorbs to the cell membrane and the entire virion is engulfed into a vacuole, which then fuses with a lysosome. Lysosomal acids released into the vacuole cause the viral envelope to fuse with the vesicle membrane, releasing the nucleocapsid into the cytoplasm. (*c*) Penetration of nonenveloped virus by envelopment in a temporary vacuole. The vacuole membrane fuses with an internal membrane system (Golgi complex or endoplasmic reticulum), releasing the free nucleocapsid.

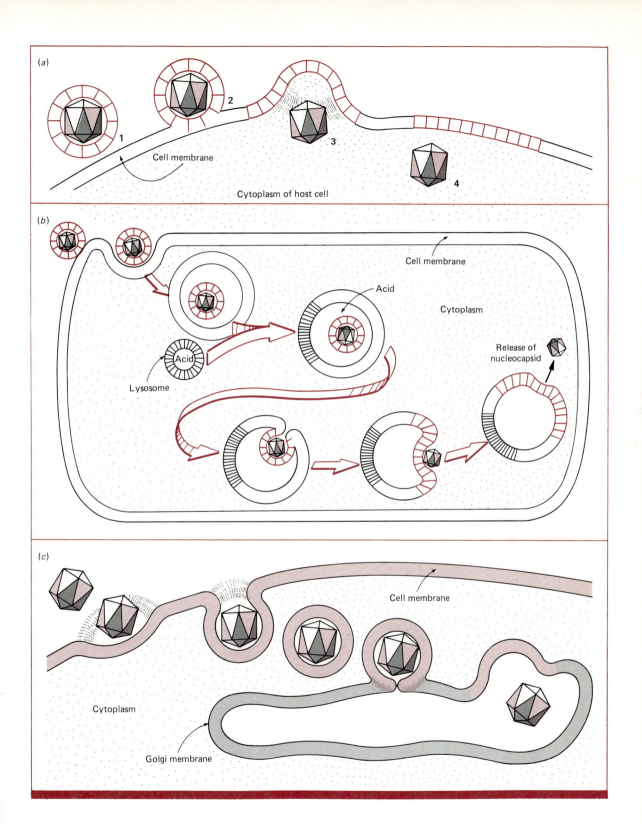

(a)

1   2   3   4

Cell membrane

Cytoplasm of host cell

(b)

Cell membrane

Acid

Cytoplasm

Release of nucleocapsid

Acid

Lysosome

(c)

Cell membrane

Cytoplasm

Golgi membrane

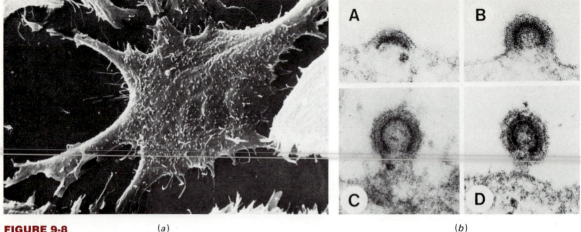

FIGURE 9-8

Enveloped viruses escape the host cell by budding, acquiring their membrane covering in the process. (See text for explanation.) (a) Numerous viruses budding from the surface of an infected cell and (b) sequence showing progeny virus emerging from the surface of an infected cell.

(a)

(b)

director of metabolism; others produce many copies of the viral nucleic acid. After the viral chromosome has been repeatedly duplicated, cellular metabolism is redirected toward the synthesis of **late proteins**. These are the structural components that, along with the viral nucleic acid, assemble into nucleocapsids. For nonenveloped viruses, eclipse ends with the intracellular appearance of these virions.

Release

Some viruses burst their host cells during release. Others leave the host cell intact, escaping by a process that resembles penetration in reverse (Fig. 9-8). Some late viral proteins are incorporated into the host-cell membrane. Each nucleocapsid becomes surrounded by a portion of the modified membrane, forming a "bud" which pinches off as the virion escapes. This event also seals the holes in the host cell's membrane. Because the host cell survives, more viruses are produced. The acquisition of the envelope is the final stage in maturation of these viruses.

Viral Adaptations to Different Hosts

To infect plant cells or bacteria, viruses must get through the protective cell wall of the host. Plant viruses enter through previously damaged cell walls or are introduced by insects and other arthropods that feed on the plant. Progeny virions are then transmitted directly to other cells throughout the plant.

Bacteriophages demonstrate some of the most interesting variations in their replicative cycles, differing primarily in their mode of penetration and release. Figure 9-9 shows the attachment and penetration of one type of bacteriophage. Attachment sites on the tail fibers recognize and fasten to receptor sites on the bacterial cell wall, bringing the base plate on the viral tail into contact and triggering the spring-like contraction of the bacteriophage tail. The nucleic acid is injected directly into the cytoplasm. The empty capsid remains outside the cell so no intracellular uncoating is

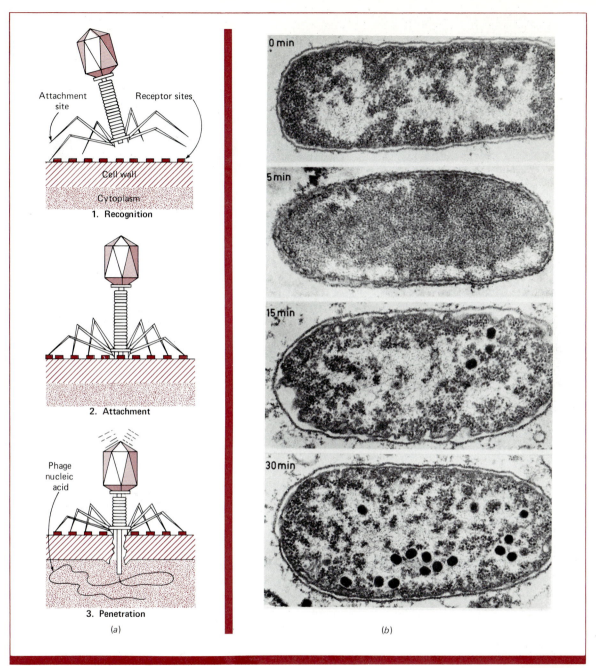

Attachment site
Receptor sites
Cell wall
Cytoplasm
1. Recognition

2. Attachment

Phage nucleic acid
3. Penetration

(a)

0 min

5 min

15 min

30 min

(b)

FIGURE 9-9

T-2 bacteriophage:
(a) Steps in adsorption and penetration
and (b) intracellular development in
host bacterium (*Escherichia coli*).
The phage heads begin to appear
within 15 minutes after infection.

FIGURE 9-10

Lytic release of viruses from host bacterium.

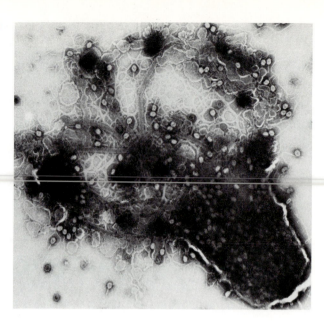

necessary. The free viral nucleic acid inside the cell immediately begins to take control of the cell in much the same way as the typical animal virus previously discussed.

Progeny bacteriophages escape from the host cell after producing an abundant supply of lysozyme late in the replication cycle. This enzyme digests the peptidoglycan layer of the cell wall. The progeny viruses burst from the osmotically fragile cell in an explosive lytic event (Fig. 9-10).

Lysogeny—An Alternative to Lysis

Bacteriophage infection that results in cell lysis and release of progeny virions is called a **lytic cycle**. Certain DNA-containing bacterial viruses, referred to as **temperate bacteriophages**, can infect a cell without producing progeny viruses or damaging the host. These viral-infected cells may reproduce for generations, each cell containing a copy of the bacteriophage chromosome. Because the viral chromosome only replicates when the cell's chromosome replicates, most of these cells never show evidence of infection. These virus-containing bacterial cells are said to be lysogenic.

The establishment of **lysogeny** by temperate bacteriophages occurs by the following mechanism (Fig. 9-11):

1. After penetration, the viral chromosome directs production of repressor proteins that specifically bind to the virus chromosome and turn off replication of viral DNA.

2. The repressed viral DNA then integrates into and becomes a physical part of the host chromosome. The integrated virus DNA is now called a **prophage**.

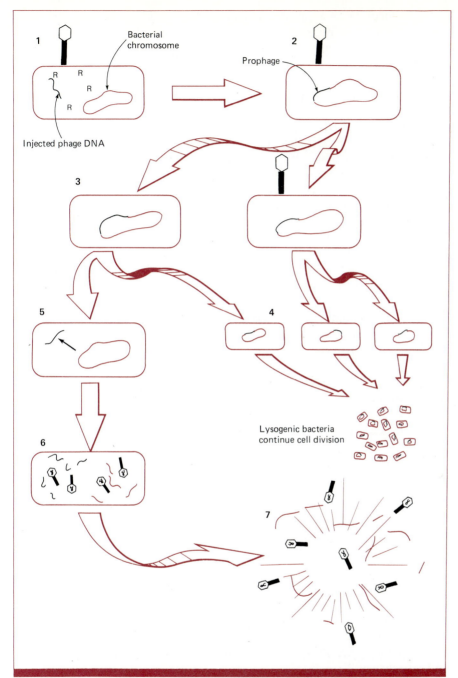

FIGURE 9-11

Temperate bacteriophage infection leading to formation of a lysogenic bacterium. (1) Temperate phage infects the cell and represses its own replication (R, repressor molecules). (2) Phage DNA integrates into host chromosome. (3) Prophage is replicated with host chromosome (all daughter cells possess integrated prophage). (4) Most of the cells continue to divide, showing no evidence of viral infection. (5) Spontaneous induction. In a small percentage of lysogens, prophage DNA excises from the bacterial chromosome. (6) Following excision, phage DNA replicates and progeny temperate phages assemble. (7) The induced cell lyses and infectious temperate phages are released.

3. The virus DNA replicates whenever the bacterial chromosome doubles, so all progeny cells inherit one copy of the prophage in the chromosome and thus carry the potential for producing the temperate bacteriophage.

An occasional lysogenic cell undergoes **induction**, the onset of the productive lytic cycle. Induction occurs whenever the prophage detaches from the bacterial DNA and produces progeny viruses, eventually lysing the host cell. Induction occurs spontaneously, but its frequency is enhanced by irradiation with ultraviolet light or exposure to agents that interfere with DNA replication. Like the parent bacteriophage, viruses produced following induction are also temperate; most will establish lysogeny in the cells they subsequently infect.

The advantage of lysogeny to the virus is reflected by a biological principle: the best-adapted parasite is one that does not harm its host. To do so would destroy the parasite's life support system. The bacterium also benefits from the lysogenic association in that the virus's repressor protein protects the cell from lytic infection by viruses of the same type as the prophage. In addition, bacteriophages may bring in new genes to the cell either by transduction (see Chap. 12) or by lysogenic conversion (below). These new genes may increase the bacterium's ability to survive, for example, by providing it with resistance to antibiotics.

Lysogenic Conversion Some lysogenic cells possess properties not shared by their nonlysogenic counterparts. The acquisition of new properties following lysogeny is called **lysogenic conversion**. These changes are directed by the phage chromosome and are stable as long as the host cell retains the prophage.

The medical significance of lysogenic conversion is illustrated by its role in the pathogenesis of diphtheria. The potentially fatal symptoms of this disease are elicited by a potent exotoxin, a soluble poison produced by the bacteria growing in the victim's throat. Only lysogenic strains of this pathogen can produce the exotoxin and elicit symptoms of diphtheria. Diseases such as scarlet fever and botulism, both caused by bacterial exotoxins, are also believed to be caused exclusively by lysogenic strains of their respective pathogens.

Oncogenesis Lysogeny is a phenomenon found only in bacteria, but an analogous system exists in animal cells infected with viruses that integrate their DNA into the host cell's chromosome. The ability to integrate may be related to **oncogenesis** (the production of cancer) by some of these viruses.

Cancer is a single name applied to a number of diseases characterized by uncontrolled growth of cells in the body. These malignant cells may arise from several types of normal tissue cells. Cancer cells have lost their sensitivity to the signals that inhibit excessive reproduction of normal cells and may metastasize (spread) throughout the body from their original sites of formation.

For some animals, such as chickens and mice, certain cancers have been clearly demonstrated to be caused by *oncogenic viruses*. Viruses have not been proven to cause cancer in humans, however, although their participation has been implicated. The *herpesviruses*, for example, are capable of maintaining their DNA in susceptible host cells. Herpesviruses cause fever blisters in the mouth and on the lips, and they cause the genital lesions of one of the most prevalent venereal diseases in America. Women with genital herpes appear more likely to acquire cancer of the cervix. Some persons infected with another herpesvirus, called *Epstein-Barr (EB) virus* (the agent of infectious mononucleosis), seem more prone to develop certain cancers of the nasopharynx. In addition, Burkitt's lymphoma, a malignant tumor of lymphoid tissue in the facial region, is linked to EB virus infection of black children in Africa.

Another possible candidate for inducing human cancers is a group of RNA viruses frequently isolated from patients with leukemia. These viruses bear striking resemblances to RNA viruses that cause tumors in animals. In their host cells these animal viruses produce DNA copies of their RNA. The DNA integrates into the host cell's chromosomes. Among the newly introduced viral nucleic acid is a genetic segment called an **oncogene**, which is responsible for transforming normal cells to malignant cells. Normal human cells also have a region on their chromosomes comparable to the oncogene. This gene directs the synthesis of a product needed for normal cell growth and development. Abnormal production of the gene's product, however, can lead to transformation to a tumor cell. When viral nucleic acid containing an oncogene is introduced into a cell, the oncogene may be excessively expressed, causing transformation. (Alternatively, some viruses do not bring in oncogenes, but promote the overexpression of the host's "oncogene.") Even in the absence of a virus, however, the cell's oncogene may be triggered to cause transformation when environmental factors, called *carcinogens*, interfere with controls on its expression. Carcinogens, and not viruses, are believed to be the major causes of human cancer, although some oncogenic viruses may work in concert with carcinogens to produce tumors.

DETECTION AND CULTIVATION OF VIRUSES

Although viruses are too small to be seen with the light microscope, they may be detected by electron microscopy. It is often easier or more reliable to employ an indirect method of detecting viruses. Many of these indirect techniques depend on our ability to cultivate viruses in vitro (in laboratory glassware such as petri dishes, test tubes, or flasks). Because they are obligate intracellular parasites, however, viruses cannot be cultivated on inanimate media. They require a living cell system. Bacteriophages, for example, may be propagated on susceptible bacterial cultures. In addition, their effects on host cell cultures are often observable, providing a means of detecting the presence of viruses. Just as a single bacterium generates an isolated colony on a suitable nutrient medium, the progeny of a single virus may create a macroscopically visible effect on a surface of susceptible host cells. For example, when a suspension of virus-sensitive bacteria is spread over the surface of a solid nutrient medium and incubated, a confluent "lawn" of bacterial growth covers the surface of the agar. If bacteriophages are added to the bacteria when they are spread over the surface, each virus that infects and lyses a cell releases progeny viruses that infect and destroy surrounding cells. The process continues until, by the time the bacterial lawn begins to appear, a small zone of clearing will have developed around the original virus-infected cell. Each such zone of clearing, called a **plaque** (Fig. 9-12), represents one bacteriophage applied to the plate.

If too many phages are applied to the seeded medium, all the bacteria will be infected, resulting in confluent lysis and no bacterial growth on the surface. If the virus suspension is diluted, however, the number of isolated plaques that appear can be used for determining the concentration of bacteriophages in the original suspension. This is accomplished, as with the bacterial viable count, by multiplying the dilution factor by the number of plaques on the plate. In this respect, a bacteriophage plaque is analogous to a bacterial colony.

Viruses that infect animal cells may be detected in a similar fashion, using susceptible animal cells as the indicator. **Cell culture**, the growth of animal cells on artificial media, is the most common system for detecting and cultivating animal viruses. Animal cells are grown in a **monolayer**, a uniform layer one cell thick on the inner surface of a bottle or test tube (Fig.

FIGURE 9-12

Bacteriophage plaques in a confluent lawn of susceptible bacteria. The number of viruses in the volume applied to the plate is ascertained by counting the plaques.

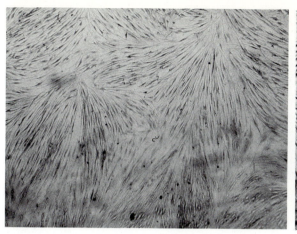
(a)

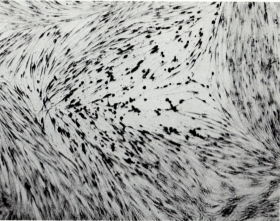
(b)

FIGURE 9-13

Cell culture (100×): (a) The monolayer typical of healthy cells and (b) cytopathic effect of the same cells infected with herpesvirus.

197

DETECTION AND CULTIVATION OF VIRUSES

9-13*a*). A virus which infects a single cell may ultimately form a plaque by a process similar to plaque formation by bacteriophage. Virus infection of cell cultures may induce several other morphological changes—fusion of cells into larger cells (called giant cells), clumping of cells, or development of **inclusion bodies** (intracellular aggregates of developing viruses). These virus-induced changes in cell cultures are referred to as **cytopathic effects (CPE)** and are easily observed with a light microscope.

The appearance of the CPE is often characteristic of the type of virus and is consequently a useful diagnostic aid for identifying viruses isolated from infected patients and grown on cell culture (Fig. 9-13*b*). CPE also develops in the patient's cells during natural infection. Thus many viral diseases are rapidly diagnosed by microscopically observing stained material taken directly from the patient's lesions. For example, intranuclear inclusion bodies (called Cowdry bodies) in epithelial cells are characteristic of herpesvirus infection. In the absence of CPE, some enveloped viruses may be detected by *hemadsorption,* the attachment of red blood cells to the surface of virus-infected cells (Fig. 9-14). This phenomenon is due to the affinity of some viral proteins for red blood cells. These proteins are present in the surface membranes of cells infected with the viruses.

Although cell culture is the preferable method of virus cultivation, some viruses can only be grown in living experimental animals. Other viruses are cultivated on embryonated chicken eggs and are detected when pocks develop on membranes surrounding the embryo. Some viruses, such as the influenza virus, can be detected by their tendency to cause clumping of red blood cells in test tubes, a phenomenon known as *hemagglutination.*

One of the most useful methods of viral detection is the serological reaction between the antigens of the virus and antibody of known viral specificity. For example, if a virus obtained from an infected patient reacts with a herpes-specific antibody, the identity of the virus is confirmed. Most viruses can be detected (and identified) by such immunological tech-

FIGURE 9-14

Detection of measles virus infection by hemadsorption. This phenomenon is observed when red blood cells are added to infected cell cultures.

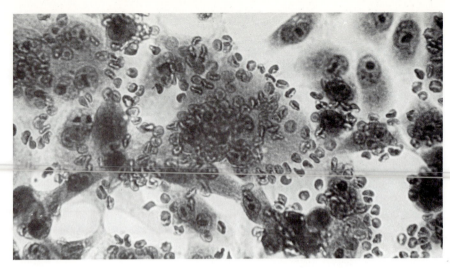

niques. Indicators of positive antibody-virus reaction are discussed later in this chapter (Diagnosis of Viral Diseases).

VIRAL DISEASES

Most pathogenic viruses produce acute or asymptomatic infections that rapidly run their course and stimulate permanent immunity in survivors. The resultant immunity is due to production of immune cells and antibodies that specifically recognize and inhibit subsequent infection by the same types of viruses. This explains why chickenpox, measles, mumps, and several other common childhood diseases occur once in a lifetime. A number of viral diseases, however, pose special medical problems. For example, common colds may be caused by more than 100 distinct strains of *rhinoviruses* (*rhino,* nose). Because immunity is specific, infection with any one strain fails to induce host immunity to the other strains. Another example is the *influenza virus,* which causes recurrent epidemics because of its tendency to periodically alter its surface antigens to a new specificity that is no longer recognized and neutralized by the immune defenses of persons previously exposed to earlier strains of the same virus.

Some viruses escape elimination by the immune response, establishing **latent** (hidden) **infection**. These viruses remain in the host even after disease symptoms disappear and are generally undetectable during these latent periods. The disease may be periodically reactivated by various stimuli. For example, recurrent episodes of Type 1 (oral) herpes and Type 2 (genital) herpes may be triggered by emotional stress, sunburn, menstruation, pregnancy, common colds, or diseases that cause fever (hence the common terms "fever blisters" and "cold sores"). Between episodes the virus silently resides within regional nerve cells.

Latency may also be established in brain tissue following measles infection, leading to the development of a slowly progressive, ultimately fatal neurological disease called subacute sclerosing panencephalitis (*SSPE*). The latent period lasts from 2 to 20 years.

Another group of slowly progressive neurological diseases is caused by a group of agents called **slow viruses**. Kuru and Creutzfeldt-Jakob syndrome, two typical slow-virus diseases of humans, are characterized by incubation periods of months to years (see color box). Once the virus begins to multiply, the disease may progress more rapidly; death is inevitable within 1 year following onset of symptoms. Although their infectious natures and small sizes suggest similarities to viruses, no particles of these slow viruses have been observed, even by electron microscopy. In addition, many of these agents are resistant to physical and chemical treatments that inactivate conventional viruses. Nucleic acid has yet to be detected in preparations of these infectious agents. This has led to the speculation that slow viruses may be *prions*, infectious agents that consist solely of protein (see Chap. 12).

Another group of unusual infectious entities consists solely of a single-stranded circle of RNA unprotected by a protein coat. These are **viroids**. Because they lack capsids, they possess no viral antigens. They were discovered relatively recently (1971) and have all been associated with

KURU: THE LAUGHING DEATH

■ In the mid-1950s an unusual and deadly phenomenon threatened eventual extinction of a small tribe of people in New Guinea. Most of the women and many of the children were being attacked by a fatal neurological disorder that began by causing its victims to giggle uncontrollably. The syndrome baffled American epidemiologist Carleton Gajdusek, who analyzed soil, drinking water, food, even the ashes in the fires in search of the etiological agent and its mode of transmission. After months of inquiry, Gajdusek discovered that the tribe was cannibalistic. As an expression of respect for their dead relatives, the survivors would consume portions of the corpses, including the brain. Years after preparing the brains for cooking, the women and children would begin the fatal giggles.

Kuru, as the disease was named, was subsequently shown by Gajdusek to be caused by a previously undiscovered type of pathogen, called a slow virus because of its 2- to 20-year incubation period. It is transmitted by eating the infected neurological tissue of someone who has died from the laughing death or by cutaneous inoculation of the virus while preparing the brain. The tribe's extinction was avoided when they abandoned their cannibalistic tribute to their dead. Gajdusek's detective work on slow viruses won him the 1976 Nobel Prize in medicine.

Although kuru has been virtually controlled, another slow-virus disease, Creutzfeldt-Jakob disease (CJD) continues to cause invariably fatal neurological infections throughout the world. The transmission of this rare disease is still an enigma, but two cases were caused when contaminated electrodes were inserted into the brain during neurosurgery. Another case followed a corneal transplant from a patient with undiagnosed CJD. CJD is also an occupational hazard among neurosurgeons and neuropathologists. Each year 200 people in the United States die of the disease.

Viral Disease	Major Target Organ(s)	Major Mode of Transmission	Major Portal of Entry
Measles	Respiratory tract and skin	Respiratory droplets	Respiratory tract
Mumps	Respiratory tract, salivary glands, and nervous system	Respiratory droplets and saliva	Respiratory tract
Influenza	Respiratory tract	Respiratory droplets	Respiratory tract
Common cold	Respiratory tract	Respiratory droplets	Respiratory tract
Smallpox	Skin, mucous membrane, liver, spleen, and lungs	Respiratory droplets	Respiratory tract
Chickenpox	Skin and occasionally lungs	Respiratory droplets	Respiratory tract
Herpes simplex Type 1	Buccal mucosa	Contact	Mucous membrane of mouth
Herpex simplex Type 2	Genitals	Sexual contact	Mucous membrane of genitals
Poliomyelitis	Gastrointestinal tract and nervous system	Feces (and fecally contaminated food or water)	Alimentary tract
Rabies	Nervous system	Saliva	Wound, usually an animal bite
Hepatitis A ("infectious hepatitis")	Gastrointestinal tract, liver, and spleen	Feces (and fecally contaminated food, water, or fomites)	Alimentary tract
Hepatitis B ("serum hepatitis")	Liver	Infected blood; occasionally feces, urine, semen, and respiratory secretions	Parenteral, mucous membranes, and alimentary tract

diseases of plants. Many investigators believe that viroids may soon be recovered from animal and human cells also.

Table 9-5 describes some characteristics of many important human diseases caused by viruses.

Tissue Specificity

The disease symptoms that develop following viral infection depend largely on the type of cell infected by the specific virus. Specificity for certain tissue types is undoubtedly related to (1) the presence of sites for viral attachment on the surface of susceptible cells and (2) the virus's ability to replicate in the infected cell. Surface receptors on nerve cells, for example, are complementary to the rabies virus envelope. Rabies virus therefore infects nervous tissue, causing its degeneration and ultimately killing the patient. Influenza ("flu") virus affects only the respiratory tract. ("Intestinal flu" is an inaccurate euphemism for many acute diseases of the gastrointestinal tract. It is not really influenza, but any number of intestinal infections caused by a variety of bacteria or viruses.) Rhinoviruses cause only colds because they can replicate only at the cooler temperatures found in the airways of the upper respiratory tract.

Some pathogenic viruses, on the other hand, have multiple target organs and cause disease symptoms in some or all of the cell types infected. Smallpox virus, for example, affects the skin, mucous membranes, liver, spleen, and lungs.

Diagnosis of Viral Diseases

Since there is no specific chemotherapy for the overwhelming majority of viral diseases, the identification of the pathogen may seem to be relatively unimportant. Rapid diagnosis, however, can spare the patient dangerous and unnecessary antibiotic therapy mistakenly directed against a suspected bacterial infection. It also affords the physician a more accurate prognosis and provides important epidemiological information useful for preventing the further transmission of the disease. For example, isolation and identification of one of the viruses that causes mosquito-borne encephalitis would signal public health officials to initiate measures to eradicate the vector population, perhaps preventing a large-scale epidemic.

Diagnosis of viral diseases depends on a combination of clinical symptomology plus laboratory analysis of infected material from the patient. Viruses in clinical specimens may be detected and identified by the methods described in the previous section on Detection and Cultivation of Viruses. The exact method depends on the virus suspected of being the etiological agent.

Patients suffering from viral infections usually have an elevated concentration of viral-specific antibodies in their blood, especially in the later, convalescent phase of the disease. This rise in antibody level is the result of the body's immune reaction to the presence of the virus. Detection of these circulating antibodies provides an additional diagnostic technique. A patient's serum specifically reacts with a known viral preparation if the individual has been exposed to that virus long enough to produce antibodies against the virus. If a significant increase in the concentration of a patient's antibodies against the virus occurs over a period of 2 weeks or longer, active infection with that virus is indicated. Diagnosis by these serological techniques is discussed in more detail in Chapter 16.

Isolation and cultivation of the virus also contributes to proper diagnosis, although more problems are encountered than with bacteria. Most medical laboratories lack the resources for propagating viruses on living cell cultures, and it is usually necessary to mail clinical specimens to a virology laboratory. In addition, many viruses quickly lose infectivity at room temperature, and clinical specimens must be immediately refrigerated until they can be packed in dry ice and mailed. The whole process requires considerably more time than bacterial cultivation. In addition, very little is known about the viruses that normally inhabit healthy humans, so for some viruses isolation and identification do not automatically establish their role in causing that patient's disease. Their presence, however, provides an important piece of information, which, used with other findings, may help to establish a correct diagnosis.

Resistance and Recovery

Immunity   The immune system can recognize virus-infected cells and destroy them before progeny virions can be formed and released. The skin lesions of chickenpox, for example, are localized areas where viral-infected cells have been lysed by the immune system. In addition, virus infection stimulates the body to produce antibodies that react with viral surface antigens and block their ability to attach to their host cells, thereby neutralizing virus infectivity. Such antibodies are especially effective protection against subsequent attacks by the same type of virus.

Interferon   One of the most important antiviral factors in the human body is a group of related proteins called **interferon**. These proteins are rapidly produced by virus-infected cells and diffuse to other, noninfected cells of the same host. Interferon stimulates these noninfected cells to produce enzymes with antiviral activity that protects the cell from viral multiplication (Fig. 9-15). Interferon usually does not stop the infection in a cell in which viral replication has begun. Consequently, the virus-infected cell is not saved by its own interferon production, but its noninfected neighbor cells do become temporarily resistant to infection by a broad spectrum of viruses in addition to infection by the virus that stimulated interferon production. Interferon also increases the efficiency of the immune system to fight viral infection. It is not clear whether its protective effect is primarily due to enhancement of the immune response or to its direct antiviral activity.

Since its discovery in 1959, interferon has been investigated as a possible chemotherapeutic agent to treat patients with viral diseases. These studies have been hampered by the limited quantities of interferon available. For years human interferon could only be obtained from the white cells of human blood, and only minute quantities were produced and harvested. So much blood was required that, in 1980, the cost of 1 ounce of interferon exceeded 2 billion dollars. Fortunately, new approaches in genetic engineering have provided abundant quantities of pure interferon which are now (1984) being evaluated as possible chemotherapeutic agents. Interferon may become our first broad-spectrum antiviral agent.

Chemotherapy   The most difficult infectious diseases to control are those caused by viruses. The problems stem from the biological nature of the virus, which actually becomes a functioning portion of the infected cell. It is therefore difficult to selectively destroy the virus without killing the host cell as well. In addition, viruses outside the host cell (but still in the human body) are biologically inactive particles and are therefore resistant

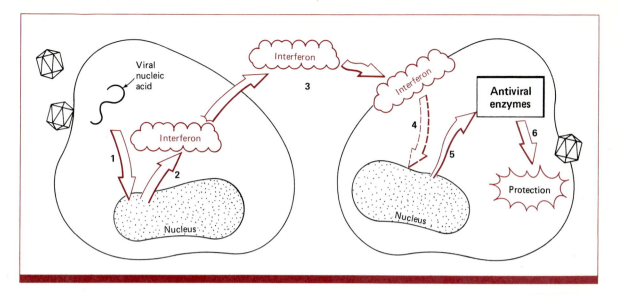

FIGURE 9-15

Antiviral mechanism of interferon: (1) Viral nucleic acid turns on host gene for interferon production. (2) Infected host cell produces interferon. (3) Interferon diffuses out of infected cell and attaches to membrane of uninfected cell. (4) Attachment triggers a signal in the nucleus of the uninfected cell. (5) Signal turns on the production of antiviral enzymes. (6) Antiviral enzymes prevent the synthesis of virus-specific proteins, thereby protecting the uninfected cell from infection.

to metabolic inhibitors. Therefore, *virtually no antibiotics are therapeutically effective against any virus disease.*

Antibiotics administered to patients with influenza, a cold, mumps, fever blisters, or any other self-limiting virus infection do nothing toward curing the disease, but are often given to prevent secondary bacterial infections. This convention is of questionable value, and the practice may actually increase the chance of secondary infection because it disrupts the normal protective bacterial flora which could otherwise compete with pathogenic bacteria (see Microbial Defenses, Chap. 15).

A few chemotherapeutic agents show medical effectiveness, but nearly all of them are limited by their toxic side effects (see Chap. 14).

Control of Virus Diseases

Many virus diseases can be effectively controlled by increasing the population's resistance to infection and preventing exposure to the infectious agent.

People who contract virus diseases and recover are commonly immune to the virus that caused that disease. They are *naturally immunized*. Vaccines have been developed from viruses that are either inactivated ("killed") or attenuated, that is, active but weakened so they are incapable of producing disease. These vaccines, when introduced into a human, expose the person to the virus without the risk of subsequent disease. This exposure stimulates antiviral immunity similar to that acquired by contracting the active disease. These people are *artificially immunized*.

Effective vaccines have been developed for many formerly dreaded virus diseases that are clinically unmanageable once the patient contracts the disease. Vaccines have helped to reduce or eliminate the incidence of smallpox, poliomyelitis, rabies, yellow fever, measles, rubella, and mumps. The effectiveness of such vaccination programs is illustrated by the reduction of congenital rubella (German measles) of the fetus from 20,000 cases reported in 1964 through 1965 to 17 cases reported in 1977.

In addition to vaccination, the prophylactic (preventive) approach to management of virus disease includes a combination of ensuring sanitary conditions, quarantining sick individuals, and controlling animal reservoirs and arthropod vectors. Using these strategies, many formerly dreaded virus diseases are no longer the threats they once were. The most dramatic victory against pathogenic viruses occurred in 1980 when smallpox became the first human disease to be officially eradicated from the earth.

OVERVIEW

Viruses are noncellular biological entities that are too small to be detected by light microscopy. They are obligate intracellular parasites which possess no enzymes for independent metabolism. Viruses contain a single type of nucleic acid and have an eclipse in their replicative cycle; some may be crystallized with no loss of infectivity. Their structure is simple, consisting of nucleic acid, protein, and sometimes a membrane envelope. These components are systematically arranged to form the shape characteristic of the virus. Viral morphology is determined by the symmetry of the nucleocapsid or by the shape of the outer envelope.

The simplicity of viruses is also illustrated by their replication cycles. Following attachment to a host cell and penetration into either the cytoplasm or nucleus, viruses depend entirely on the cell's metabolic machinery for all the events necessary to produce progeny viruses. The host cell is often killed by such infection, but some viruses allow the cell to live and continue to release new virions. The DNA of temperate bacteriophages establishes a stable relationship with the host chromosome, resulting in lysogeny. Some bacterial diseases, such as diphtheria, are caused only by lysogenic bacteria. An analogous mechanism is associated with latent viral diseases and virus-induced cancer in animals (and perhaps humans).

Because of their unique biological properties, pathogenic viruses differ from other infectious entities in the mechanisms by which they produce disease, in the methods by which they are detected and identified, and in their resistance to chemotherapy. No antibiotics are effective in the treatment of viral diseases, and control is largely a matter of prevention.

Many virus diseases have been successfully controlled by preventive measures—vaccination, control of vector populations, isolation of infected individuals, improved sanitary conditions, and control of animal reservoirs of infection.

Natural recovery from viral diseases often results in long-lasting immunity. Recovery is probably due to a combination of neutralizing antibodies, interferon, and protective host cells that can recognize and destroy virus-infected cells.

KEY WORDS

obligate intracellular parasite

eclipse phase

bacteriophage

capsid

capsomere

nucleocapsid

envelope

virion

attachment site

receptor site

viropexis

uncoating

early protein

late protein

lytic cycle

temperate bacteriophage

lysogeny

prophage

induction

lysogenic conversion

oncogenesis

oncogene

plaque

cell culture

monolayer

inclusion body

cytopathic effect (CPE)

latent infection

slow virus

viroid

interferon

REVIEW QUESTIONS

1. Describe the events of infection, replication, and release for
   (a) Nonenveloped animal viruses
   (b) Enveloped viruses
   (c) Bacteriophages

2. Why is it unlikely that a bacteriophage would be enveloped?

3. What is one reason that most viruses are specific for one type of host cell? How does this relate to disease?

4. Some viruses escape the host cell by reversing the penetration process. Explain.

5. Lysogeny is believed to be more common than lytic phage infection. How would a bacteriophage benefit from lysogeny?

6. Why are viruses more difficult to detect than other microbes?

7. Describe how each of the following aids in the detection of viruses:
  (a) Electron microscopy
  (b) Plaque formation
  (c) Cell culture
  (d) Light microscopy
  (e) Serological reactions
  (f) Hemagglutination

The Molecules of Life

10

All organisms, from microbes to humans, are composed entirely of chemicals. The survival of an organism depends on its ability to reorganize the chemicals in its environment into molecules that compose its own cellular materials. The cell-directed chemical reactions that accomplish these changes are called **metabolism**. These metabolic reactions provide energy and raw materials for cell growth and reproduction and create all the characteristic properties of an organism. Metabolism builds the molecules that distinguish one type of organism from all others. Many of these molecules are the most complex substances on earth, and their precise synthesis and utilization require an enormous degree of organization. The study of microbiology is incomplete without an understanding of these biochemicals, the molecules of life.

MACROMOLECULES

The large, highly organized biochemicals required for cell growth and metabolism are referred to as **macromolecules** (*macro*, large). Four major categories of macromolecules are recognized: (1) proteins, (2) polysaccharides, (3) lipids, and (4) nucleic acids. Each type of macromolecule is associated with specific cell functions. Macromolecules are constructed from small molecular subunits, called **monomers**, which are linked together to form large complex molecules. The resulting macromolecule is referred to as a **polymer** (*poly*, many; *meros*, part). Each category of polymer is composed of different types of monomers (Table 10-1). Approximately 40 different types of monomers are known to exist; every organism on earth is constructed from these same 40. How monomers are put together determines the distinct nature of the organism.

The synthesis of a hypothetical polymer is depicted in Figure 10-1. The

208

THE MOLECULES
OF LIFE

TABLE 10–1

Major Macromolecules
Found in Living Systems

Macromolecule	Class of Monomer	Some Major Functions	Examples
Protein	Amino acid	Acts as catalyst for metabolic reactions; provides physical structure	Enzymes Flagellin; pilin
Polysaccharide (a large carbohydrate)	Sugar	Stores energy; provides physical structure	Starch; glycogen; cell walls (cellulose); capsules
Lipid (other than a steroid)	Fatty acid and glycerol	Stores energy; provides membrane structure	Fat; oil; Cell membrane
Nucleic acid	Nucleotides:		
	Deoxyribonucleotide	Carries genetic inheritance	DNA
	Ribonucleotide	Carries expression of genetic information coded in DNA	RNA

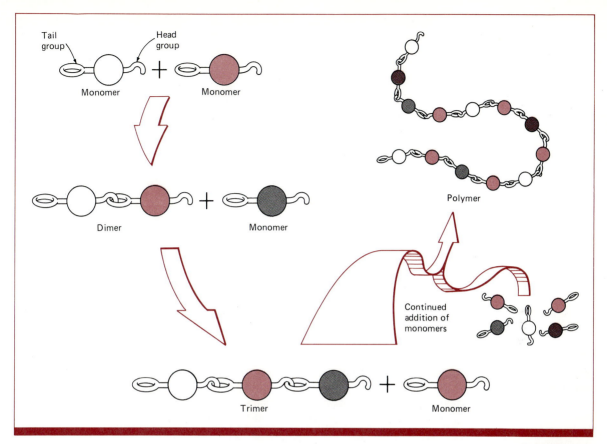

FIGURE 10-1

Synthesis of a hypothetical polymer from monomeric subunits. The "head" group of one monomer reacts with the "tail" group of another, forming a dimer (two subunits). The chain grows with the addition of new monomers. Regardless of chain length, there are free head and tail groups at opposite ends of the chain. The monomers must belong to the same class but need not all be identical.

properties of the resulting macromolecule depend on which monomers are used and their order of arrangement in the chain. Changes in the sequence of monomers may completely alter the properties of the polymer or destroy its physiological function.

Proteins

Proteins are polymers made of *amino acid* monomers. There are approximately 20 naturally occurring amino acids, each characterized by an amino group ($-\overset{\text{H}}{\underset{|}{\text{N}}}-\text{H}$) and a carboxyl group ($-\overset{\text{O}}{\overset{||}{\text{C}}}-\text{OH}$) attached to the same carbon atom. Another chemical group is also attached to this carbon. Its structure varies and distinguishes the 20 amino acids from each other. The amino group of one amino acid may react with the carboxyl group of another, releasing a molecule of water and forming a *peptide bond* that chemically links the two monomers (Fig. 10-2*a*). The bonding process can continue indefinitely since free reactive groups will be available at each end of the chain regardless of the number of amino acids in the sequence.

Proteins perform a myriad of essential functions for a cell. Some comprise the cell's structural matrix, whereas others determine what gets

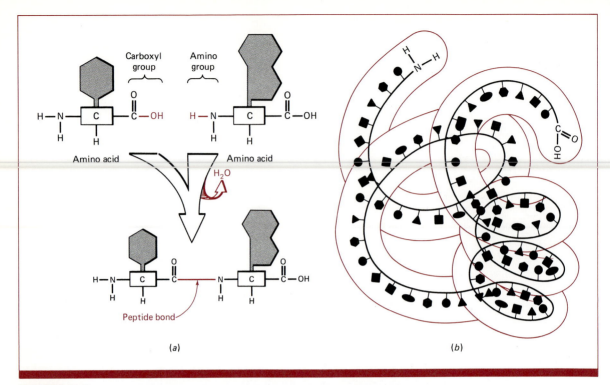

FIGURE 10-2

(a) Formation of a peptide bond between two amino acids. This new linkage replaces the bonds that, in the monomers, held one atom of hydrogen (H) to nitrogen (N) and a hydroxyl group (OH) to the carboxyl end. These molecules react with each other to form water (H₂O). For this reason peptide bond formation is always accompanied by formation of a free water molecule. Notice that a free carboxyl group and an amino group are available at opposite ends, even after coupling. (The energy needed to fuel such chemical bonding is not shown here but is discussed in the next chapter.) (b) The final shape of this linear array of symbols represents how a protein chain may acquire a complex, three-dimensional shape. Each symbol represents an amino acid.

into and out of a cell. Another group of proteins, the *enzymes*, regulate the biochemical activities upon which the cell's existence depends. The great variety among proteins is due to differences in the types of amino acids in the polymer and to the precise sequence in which they are arranged. The amino acid order determines the properties of the protein, primarily by dictating the shape the molecule assumes as the chain folds into its final stable, compacted shape (Fig. 10-2b). In an aqueous environment, for example, the most stable configuration is one that buries **hydrophobic** (water-repelling) amino acids in the center of the protein, placing the **hydrophilic** (water-attracting) amino acids on the polymer's surface in contact with the surrounding water. A number of physical and chemical forces combine to shape each protein into its precise three-dimensional configuration.

The shape of structural proteins is important to their role as building blocks of cell structure. The ability of other proteins, the enzymes, to control metabolic reactions also depends on the shape of the polymer. Thus, the sequence of amino acids in a protein spells out a protein's shape and, therefore, its function.

Polysaccharides

Polysaccharides are polymers composed of simple sugars, such as glucose or fructose. Each sugar subunit is called a *monosaccharide* (*mono*, one; *saccharide*, sugar). Sucrose (table sugar) on the other hand, is a *disaccharide* composed of two simple sugars, glucose and fructose, that are chemically linked together. Sugars, and all compounds constructed from sugar subunits, are called **carbohydrates**.* Thus, polysaccharides are large carbohydrates.

The monomers of polysaccharides are chemically linked to one another by *glycosidic bonds* (Fig. 10-3). Although most polysaccharides are large molecules they usually contain no more than two types of monomers. For example, cellulose, the cell wall material of most higher plants, is com-

---

*Carbohydrates are chemically defined as molecules that contain carbon, hydrogen, and oxygen, with a hydrogen-to-oxygen ratio of 2:1.

FIGURE 10-3

Formation of a glycosidic bond between sugars, resulting in a disaccharide. One molecule of water is removed in the process. As with peptides, each end of the growing carbohydrate chain has a reactive group available for the addition of another monomer, regardless of chain length. Such a polymer could theoretically grow indefinitely. (The energetics of the reaction are not shown in this diagram.)

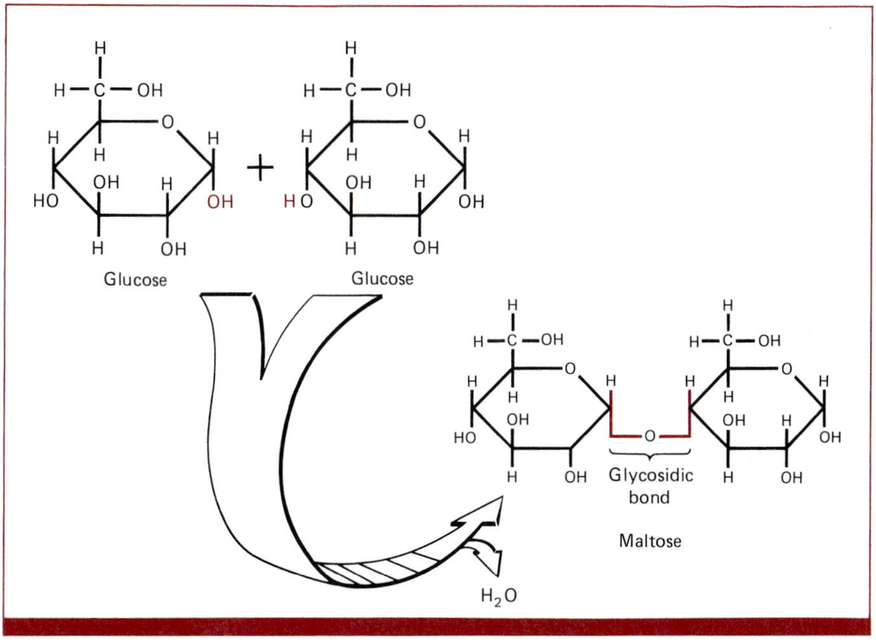

posed entirely of glucose subunits. Cellulose is the primary matrix of wood, cotton, and nearly all plants (see color box).

Many polysaccharides are manufactured by a cell for the sole purpose of storing excess energy. These polymers must be easily digested so the cell can quickly tap their stored energy whenever it is needed. Plants store

CELLULOSE: A VAST UNTAPPED FOOD RESOURCE

■ Cellulose is the most abundant carbon and energy source on earth. Found in trees, grasses, and other plants, cellulose supports the growth of countless chemoheterotrophic organisms, but not humans. People cannot digest cellulose; we have no molecular capacity to release the sugars that comprise the cellulose polymer. Ironically, cellulose is identical in its chemical content to starch, one of our most easily digested polymeric foods. Like cellulose, starch is a polymer composed entirely of glucose molecules. The significant difference is in the way the sugars are fastened to each other.

Starch is an "alpha glucose polymer," which means that one glucose is linked to another by an **alpha glycosidic bond** (see figure). This bond is readily hydrolyzed by amylase, a digestive enzyme found in our saliva and intestinal tracts. Cellulose, however, is a "beta glucose polymer," with glucose molecules joined by **beta glycosidic bonds**. This linkage cannot be broken by animals, including humans, because we lack the necessary digestive proteins. When we eat plants, much of their energy and carbon remain locked in the glucose of undigested cellulose. The polysaccharide passes intact through our digestive tracts, providing needed fiber (bulk) that aids in formation of feces and in defecation but supplies us with nothing nutritional.

We can harvest cellulose's energy and carbon by putting microbes to work. Cellulose-digesting protozoa in the rumen of cattle, for example, allow the animal to survive in grasslands. Other microbes, the decomposers, can recycle the carbon in cellulose of dead plants because they produce the enzyme necessary to do so. Our inability to produce these cellulases, however, has spelled starvation and death for countless millions of people.

A short segment of a starch molecule

A short segment of a cellulose molecule

their energy surpluses in the chemical bonds of *starch*. Humans and other animals, as well as some bacteria, store some of their surplus energy as the polysaccharide *glycogen*. Polysaccharides are also major components of many bacterial structures. Polysaccharides are found in the gram-negative cell wall, bacterial capsules, slime layers, and glycocalyces. Many of these surface polysaccharides are used for bacterial attachment and colonization. For example, *dextran*, a polysaccharide produced by the anaerobic oral streptococci, is a major component of dental plaque, which cements bacteria to tooth surfaces where their acid by-products may deteriorate the enamel and produce dental caries.

Lipids

Fats, oils, waxes, and steroids are all **lipids**, a group of biological compounds whose one common property is their insolubility in water. They are all, however, readily dissolved in hydrophobic organic solvents such as acetone, benzene, ether, and chloroform. This hydrophobic property is essential to the role of some lipids as integral components of membranes and membranous cell structures.

Lipids are also used to store energy. More than twice as much energy can be stored in lipids as in an equivalent amount of proteins or carbohydrates. When energy intake exceeds the needs of the cell, many organisms, from bacteria to humans, will build reserves of fat or oil to store the surplus (Fig. 10-4). Fats and oils consist of three molecules of fatty acid joined to a molecule of glycerol (Fig. 10-5). If the resulting molecule is a liquid at normal room temperature, it is considered an *oil;* if it is a solid, it is called a *fat*.

The lipids in cell membranes are composed of only two fatty acid chains attached to glycerol. The third glycerol carbon is attached to a phosphate group (PO_4^{-3}) that, unlike fatty acids, is soluble in water. These important molecules are called **phospholipids** (Fig. 10-6). Because they have both hydrophobic and hydrophilic ends, phospholipids become spontaneously oriented so the fatty acid "tails" extend away from water. In biological systems, this orientation produces a double layer configuration which allows the hydrophobic tails to be embedded in the interior of the bilayer, protected from exposure to water by the layers of hydrophilic phosphate groups on the exterior. This is the fundamental structure of membranes.

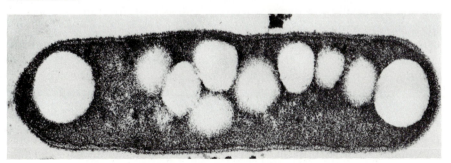

FIGURE 10-4
Granules of poly-beta-hydroxybutyric acid in a bacillus.

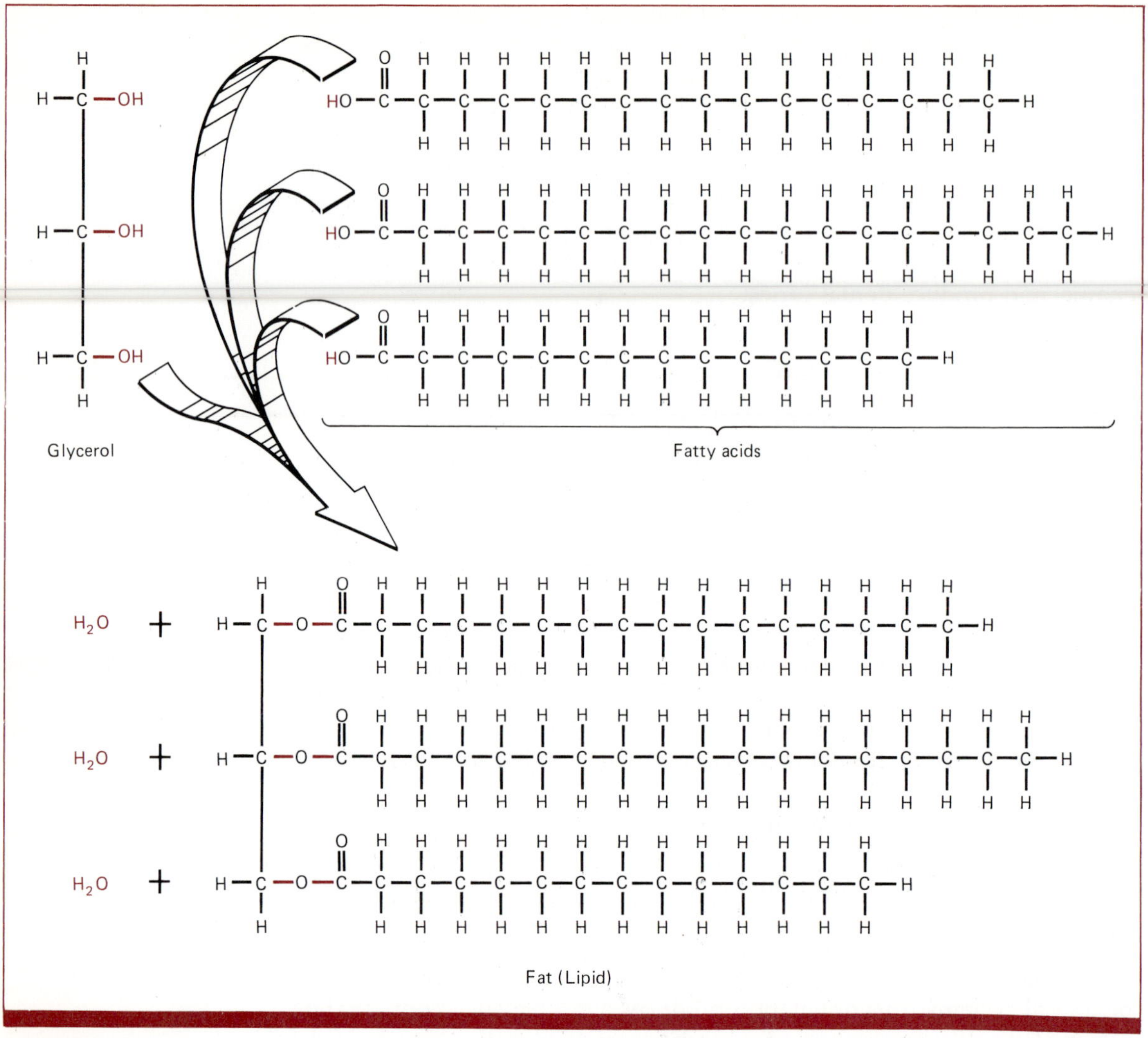

Glycerol

Fatty acids

H_2O +

H_2O +

H_2O +

Fat (Lipid)

FIGURE 10-5

Construction of a typical fat molecule from one glycerol and three long-chain fatty acids. The carboxyl group

$$\overset{O}{\underset{\|}{}}\!-\!C\!-\!OH$$

(—C—OH) of each fatty acid reacts with one of the three hydroxyl groups (—OH) of the glycerol. The reaction is accompanied by the removal of water. (The energetics of the reaction are not shown.)

Another group of important lipids is the *steroids*. These molecules contain no glycerol or fatty acid but rather have a structure that consists of four fused rings (Fig. 10-7). Although steroids are absent from the cell membranes of all procaryotes (except mycoplasmas), they are essential to the structure of eucaryotic membranes, possibly by helping the cell resist osmotic lysis. Cholesterol is a steroid found in the membranes of animal cells. Another steroid, ergosterol, is an important component of the membrane of fungi, but not of human cells, and provides a selective target for some antibiotics used for treating fungal infections (see Chap. 14).

Nucleic Acids (DNA and RNA)

Deoxyribonucleic acid (**DNA**) is a linear polymer that encodes the genetic information that governs the properties and potential activities of a cell.

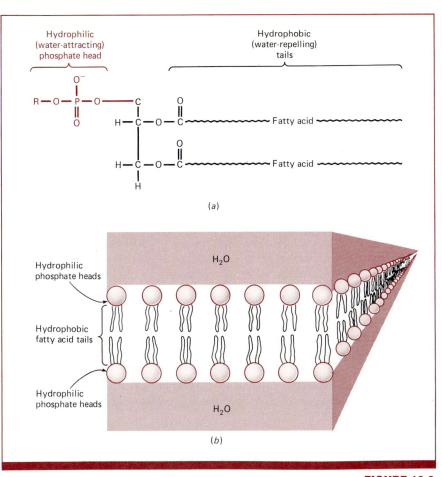

Hydrophilic
(water-attracting)
phosphate head

Hydrophobic
(water-repelling)
tails

$$R-O-\overset{\overset{\displaystyle O^-}{|}}{\underset{\underset{\displaystyle O}{\|}}{P}}-O-C$$

(a)

H₂O

Hydrophilic
phosphate heads

Hydrophobic
fatty acid tails

Hydrophilic
phosphate heads

H₂O

(b)

FIGURE 10-6

Phospholipids are molecules that are both soluble and insoluble in water. (The chains of fatty acids are drawn here as wavy lines in the fat molecule.) (a) Such "molecular schizophrenia" results in a molecule that will automatically orient itself so that its hydrophobic end extends away from water while its hydrophilic end is in contact with water. (b) This orientation accounts for the peculiar sandwich nature of the phospholipid bilayer of cell membranes shown here. (See Chap. 4 for discussion of cell membranes.)

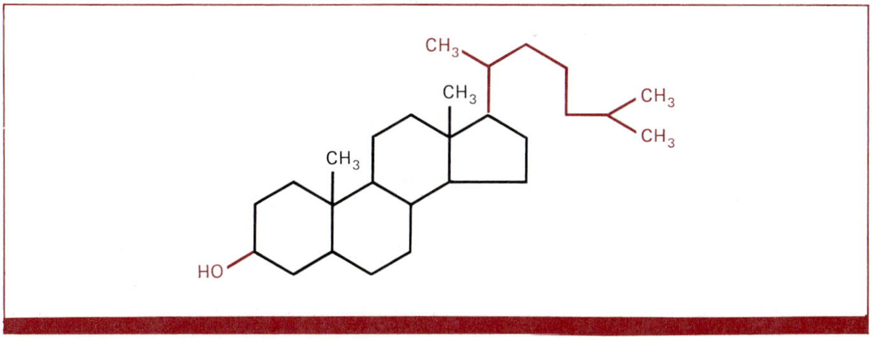

FIGURE 10-7

All steroids share the basic four-ring skeleton shown in black. The colored part of the molecule is unique to this steroid, cholesterol. Although not shown, a carbon atom exists at each angle.

This information is also used to direct the construction of progeny that have biological properties identical to those of the original cell. In other words, DNA is responsible for heredity. In the cell, DNA exists as chromosomes and plasmids. Another class of nucleic acids, **ribonucleic acid (RNA)** "reads" the information encoded in cellular DNA and carries the message to locations where it directs the synthesis of proteins. Such genetically eoded messages are fundamental to all biological systems (Chap. 12). In some viruses, RNA also serves the functions ordinarily performed by DNA.

Nucleic acids are constructed from monomers called **nucleotides**. Each nucleotide contains three major components: a five-carbon sugar, a phosphate group, and a nitrogenous (nitrogen-containing) base (Fig. 10-8). These bases are divided into two categories, single-ringed structures called *pyrimidines* and double-ringed molecules, the *purines*. The monomers of RNA are called ribonucleotides because they always contain the five-carbon sugar ribose. Deoxyribonucleotides (the monomers of DNA) differ from ribonucleotides in that the sugar contains one less oxygen than ribose; hence the five-carbon sugar is called **deoxy**ribose.

Five different nitrogenous bases are found in nucleotides, depending on whether it is a deoxyribonucleotide or a ribonucleotide (Fig. 10-9). The four bases found in DNA are *adenine, thymine, cytosine,* and *guanine.* In RNA the bases are adenine, cytosine, guanine, and *uracil.* Except for the replacement of thymine by uracil, the four bases characteristic of ribonucleotides are the same as those of deoxyribonucleotides. Adenine and guanine are purines, whereas thymine, cytosine, and uracil are pyrimidines.

A single strand of nucleic acid is synthesized by sequentially attaching the sugar group of one nucleotide to the phosphate group of another. Free phosphate groups are available at one end of the chain, with unbonded ribose (or deoxyribose) groups at the other, so the polymer may continue to grow in length. The linear sequence of nucleotides represents the coded information stored in that strand of nucleic acid. Changing the nucleotide order alters the message, just as changing the order of letters in a sentence alters its meaning. With these four "letters" (the four different nucleotides) an infinite number of sequences are possible, creating the potential for an infinite number of unique types of organisms.

In the cell, DNA is usually a double-stranded molecule. The order of

FIGURE 10-8
Skeletal structure of a nucleotide. In RNA nucleotides the sugar is ribose; in DNA it is deoxyribose. Although the nitrogenous base illustrated is a purine, nucleotides may contain a pyrimidine instead (see text for explanation).

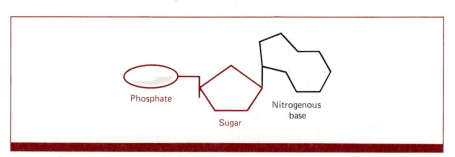

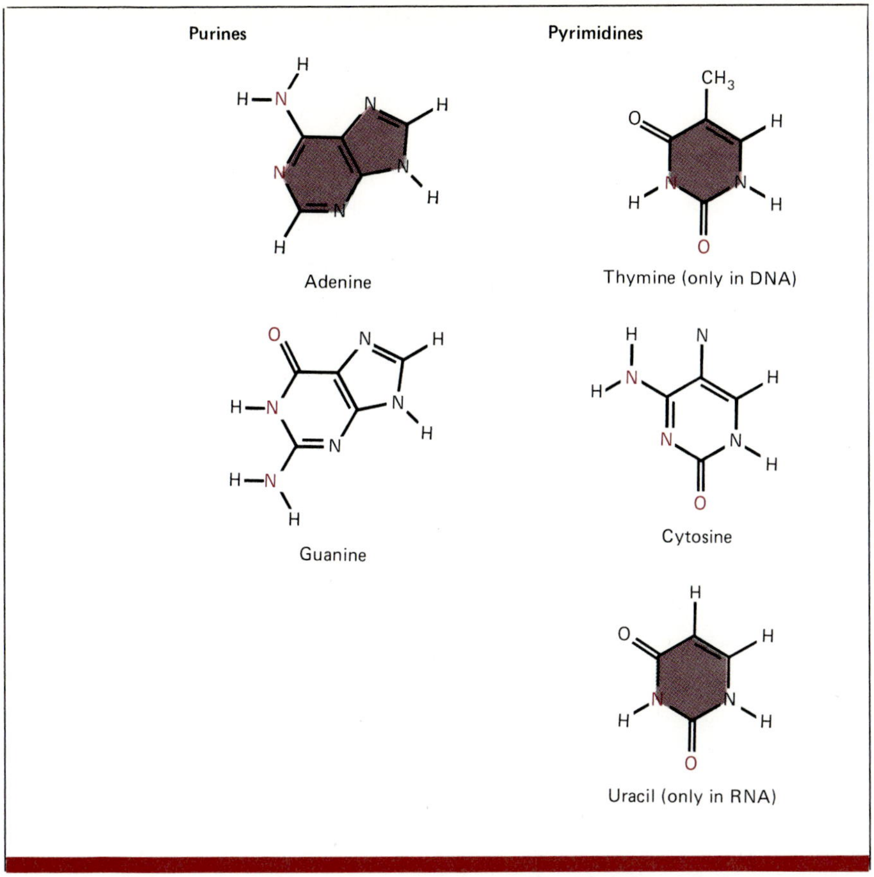

Purines

Adenine

Guanine

Pyrimidines

Thymine (only in DNA)

Cytosine

Uracil (only in RNA)

FIGURE 10-9

The five nitrogenous bases found in nucleotides of DNA and RNA. The nitrogens and oxygens (shown in color) are the points on the nucleotides involved in specific base pairing with another nucleotide (base pairing is explained in the text).

nucleotides in one strand is complementary to the sequence in the other strand, since each guanine (G) pairs only with a cytosine (C) and each adenine (A) only with a thymine (T) (Fig. 10-10). Double-stranded DNA resembles a twisted ladder (Fig. 10-11) in which the "rails" are linear phosphate-sugar linkages and the "steps" are the A-T and G-C base pairs. Internal molecular forces cause the ladder to twist into a helical configuration. For this reason DNA has been dubbed the "double helix." Unlike DNA, RNA molecules are usually single-stranded.

Hydrolysis of Macromolecules

Polymer formation is always accompanied by the liberation of water when bonds are formed between monomers. (See Figs. 10-2, 10-3, and 10-4.) Polymers can be degraded by reversing this biosynthetic process, splitting the bonds between monomers and using water to replace the chemical groups that were lost during bond formation. The degradation of polymers to smaller units therefore consumes water and is called **hydrolysis** (*hydro*, water; *lysis*, split), since water must be used to split the bond. In living systems hydrolysis depends on the activity of specific enzymes. One of the initial processes in the enzymatic digestion of food is the hydrolysis of

FIGURE 10-10

Pairing of the nitrogenous bases in double-stranded DNA. G always pairs with C, and A always pairs with T.

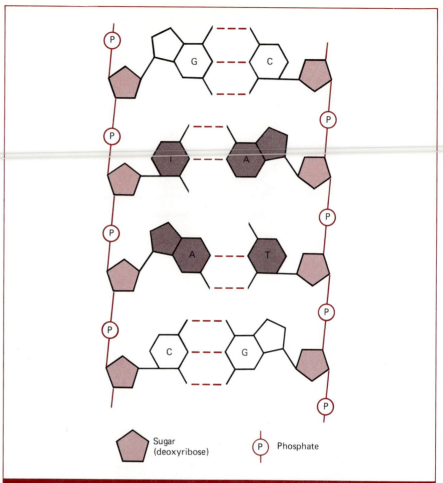

Sugar (deoxyribose)

P Phosphate

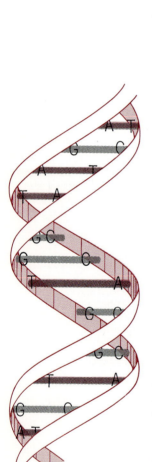

FIGURE 10-11

The DNA double helix. The two strands have complementary sequences.

macromolecules to smaller fragments and monomers. Bacteria hydrolyze macromolecules extracellularly by releasing enzymes into the surrounding medium. When the effects of hydrolytic enzymes can be observed, they may provide laboratory information useful in the identification of microorganisms. For example, a few bacteria can hydrolyze casein, the protein that gives milk its white color. Casein hydrolysis produces a zone of clearing around colonies growing on milk agar (Fig. 10-12).

ENZYMES

Everything an organism is or does is the direct result of metabolic reactions. Metabolism does not occur in a random fashion; rather, each reaction is precisely controlled by some governing factor. This is the role of the enzymes.

An **enzyme** is a biological catalyst, a substance that accelerates the rate of a specific chemical reaction. The presence of an enzyme decreases the

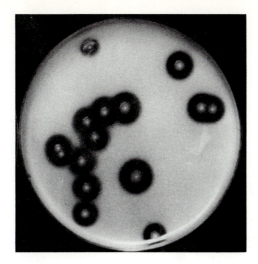

FIGURE 10-12

Pseudomonas growing on milk agar. Hydrolysis of milk protein is evident as a zone of clearing around colonies.

amount of energy needed to initiate a specific chemical reaction. In the cell, certain reactions that would occur spontaneously at very slow rates are speeded up by the action of the appropriate enzymes (Fig. 10-13).

Properties of Enzymes

■ Simple enzymes are composed entirely of protein. Complex enzymes contain additional substances such as sugars or lipids, chemically bonded to the protein. These nonprotein portions are referred to as *prosthetic groups*.

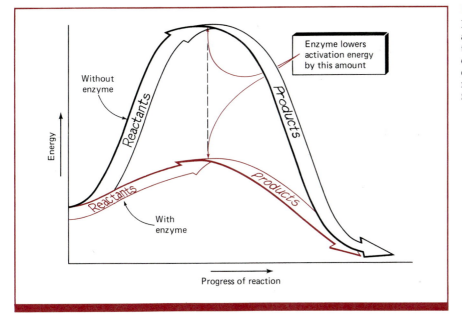

FIGURE 10-13

Enzymes decrease the amount of energy required to initiate a specific chemical reaction. With enzymes, therefore, reactions proceed at much faster rates.

FIGURE 10-14
Enzymatic alteration of
substrate A, forming
products B and C. Notice
that the enzyme is
recycled.

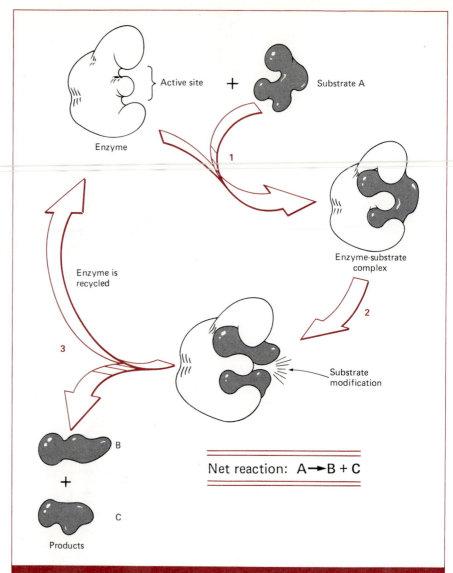

Active site

\+

Substrate A

Enzyme

1

Enzyme-substrate
complex

Enzyme is
recycled

2

3

Substrate
modification

B

\+

C

Products

Net reaction: A → B + C

220

THE MOLECULES
OF LIFE

Enzymes are highly specific in their catalytic action. In general, an enzyme recognizes only one set of *substrates* (the material on which it will react) and converts the substrates to one particular set of *products*. The specificity of an enzyme is due to the complementary three-dimensional shapes of the two molecules. The enzyme contains an area, called the **active site**, that reacts with the substrate(s). The substrate(s) must be able to fit into the active site much the same way a key fits into a lock (Fig. 10-14). If the shapes are not complementary, nothing happens. *This high degree of specificity enables the cell to control most metabolic reactions by regulating the types of enzymes produced.*

TABLE 10–2

Coenzyme activity of
B-complex vitamins

Vitamin	Coenzyme	Types of Reactions Assisted by Coenzyme
Thiamin (B_1)	TTP	Removal of CO_2 from molecules (decarboxylation)
Riboflavin (B_2)	FAD	Hydrogen carrier in energy-generating reactions
Pyridoxine (B_6)	Pyridoxyl phosphate	NH_2 transfer (transamination); decarboxylation of amino acids; removal of H_2O (dehydration) from amino acids
Cobalamin (B_{12})	Cobalamine	Protein and nucleic acid metabolism
Niacin (nicotinic acid)	NAD:NADP	Hydrogen carrier in energy-generating reactions and biosynthesis
Pantothenic acid (B_5)	Coenzyme A	Transfer of small organic molecular fragments in respiration and fatty acid metabolism
Folicin	Folic acid	Single-carbon transfer in nucleic acid and amino acid metabolism
Vitamin H	Biotin	CO_2 transfer

■ Enzymes are remarkably efficient in the reactions they catalyze. Very low concentrations of enzymes cause chemical reactions to proceed at very rapid rates.

■ Enzymes are never consumed in the reactions they catalyze. They always reappear intact after the substrate has been converted to the products and are reused until they eventually deteriorate. This recycling of enzymes accounts for their efficiency.

■ Some enzymes require small molecules to help carry out their catalytic role. Without the molecules such enzymes are incomplete (and inactive). When the helper molecule is a metallic ion, it is called a **cofactor**; if it is an organic molecule, it is referred to as a **coenzyme**. Most microorganisms manufacture their own coenzymes. However, humans, animals, and some fastidious microbes require external sources of many coenzymes. The nutritional role of some vitamins is to provide the precursors for essential coenzymes that an organism cannot synthesize (Table 10-2). Without a source of essential vitamins to provide these coenzymes, an organism suffers malnutrition and dies. Unlike prosthetic groups, cofactors and coenzymes are not permanent parts of the enzyme.

The general properties of enzymes are summarized in Table 10-3.

1. Enzymes are composed of protein.
2. They have a high degree of specificity for substrate.
3. They are unchanged by the reactions they catalyze.
4. They have a high reaction efficiency.
5. They may require coenzymes for activity.
6. They are heat-sensitive.

TABLE 10–3

Summary of Enzyme
Properties

Conditions Affecting Enzyme Activity

Increases in temperature generally boost enzyme activity and accelerate reaction rates. This explains why many organisms grow more rapidly at higher temperatures. Above a critical temperature, however, growth stops. This is the temperature at which enzymes suffer heat damage. They change shape, often lose solubility, and coagulate (in much the same way as does the protein in a cooked egg). This is called protein **denaturation** and results in loss or alteration of enzyme activity. Once an enzyme has been physically damaged, it usually cannot be repaired. Irreversible protein denaturation contributes to the effectiveness of superheated steam as an agent for killing microorganisms. Low temperatures approaching freezing, on the other hand, greatly reduce enzyme activity, but rarely cause physical damage to the enzyme. Microbial growth is therefore arrested by refrigeration, which slows the growth of the microorganism by reducing the rate of enzyme-controlled metabolism. Activity (and growth) is restored once the enzymes return to optimal temperature.

An enzyme can also be denatured by changes in pH. An enzyme that functions optimally at neutral pH (7.0) will usually be inactivated when its environment becomes too alkaline or too acidic. Microorganisms that live in extremely acidic or alkaline environments have developed mechanisms for maintaining a neutral pH in their cytoplasm.

Because the growth of organisms is dependent on the activity of enzymes, any factors that affect enzyme activity will similarly affect growth. Successful cultivation of microorganisms, therefore, requires careful control of temperature and pH to optimize enzyme activity.

Relative concentrations of substrates and products also affect the rate of enzymatically catalyzed reactions. High concentrations of substrate and low levels of product increase the reaction rate, whereas the opposite situation lowers it. Furthermore, the concentration of the enzyme itself plays a role in determining the reaction rate, greater concentrations resulting in increased rates. In fact many metabolic reactions are at least partially regulated by either turning on or shutting off synthesis of the corresponding enzyme. For example, organisms may avoid wasting energy by turning off production of an enzyme when the corresponding substrate is unavailable.

Enzyme Nomenclature

The current international system of naming enzymes calls for adding the suffix "ase" to the end of either the name of the substrate or the name of the reaction catalyzed. An enzyme that accelerates the hydrolysis of protein to amino acids is called a *protease*. Removal of a CO_2 (carboxyl) group, a common event in cellular metabolism, is mediated by a *decarboxylase*. A list of enzymes associated with several types of reactions is provided in Table 10-4.

Enzymologists sometimes employ alternative methods of nomenclature when referring to certain degradative enzymes. Enzymes that break down their substrates are often described by adding the suffix "lytic" to the

TABLE 10–4

Examples of Enzyme
Nomenclature

Name Based on	Enzyme
Substrate	
Lipid	Lipase
Protein	Protease
Nucleic acid	Nuclease
Reaction Catalyzed	
Oxidation	Oxidase
Removal of hydrogen	Dehydrogenase
Transfer of amino group	Transaminase

name of the corresponding substrate. For example, a protease is also called a *proteolytic enzyme*, and a lipase is a *lipolytic enzyme* because it "lyses" lipids. Finally, certain traditional labels for enzymes that were named prior to the evolution of the international nomenclature scheme are still used. Some of the best known are *trypsin* and *pepsin*, both proteolytic enzymes (proteases) found in the mammalian digestive tract, and *lysozyme*, the peptidoglycanase found in body secretions.

Enzymes That Damage the Body

Many pathogenic microorganisms produce enzymes that contribute to their ability to invade the infected body, sometimes with destructive consequences. The potentially fatal symptoms of gas gangrene, for example, are largely the effect of lecithinase, an enzyme produced by the pathogen *Clostridium perfringens*. The substrate for this enzyme is lecithin, a component of human cell membranes. Dissolving cell membranes kills host tissue so it no longer presents an effective solid barrier to microbial invasion of surrounding tissues. The pathogen leaves a trail of enzymatically digested tissue as it invades new body sites. Lecithinase production, which may be detected by growing the pathogen on egg yolk media, has proved to be a valuable laboratory indicator in the diagnosis of gas gangrene (Fig. 10-15).

FIGURE 10-15

(*a*) The effect of lecithinase on egg yolk agar. The halo around the colonies of *Clostridium perfringens* is due to splitting of lecithin in the egg protein. (*b*) The effect of lecithinase in a patient with gas gangrene.

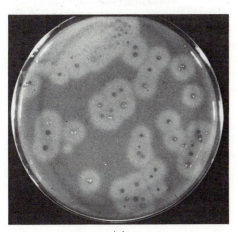

(*a*)

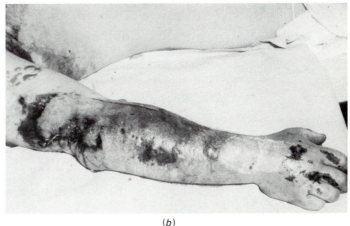

(*b*)

Enzymes that contribute to the ability of a pathogen to injure people do so by one of two mechanisms: (1) direct destruction of host tissue, as seen in gas gangrene, or (2) interruption of essential metabolic processes. The pathogen that causes diphtheria, for example, produces a toxic enzyme that inactivates one of the components needed for protein synthesis by host cells.

In addition to directly injuring the host, microbial enzymes may protect the pathogen by inactivating substances that would otherwise eliminate infection. For example, many species of bacteria produce the enzyme penicillinase, which destroys the antibiotic penicillin. The failure of this antibiotic in treating diseases caused by penicillinase-producing *Staphylococcus aureus* is common.

Chemical Inhibitors of Enzyme Activity

Many reactions essential to microorganisms are not shared by humans, for example, synthesis of bacterial cell walls. An agent that specifically inhibits the enzymes that catalyze these reactions can retard or completely arrest the growth of microorganisms without affecting human metabolism. In other words, these enzymes provide selective targets for antibiotics and chemotherapeutic agents. All effective antibiotics must possess such selective characteristics before they may be introduced safely into a diseased patient.

Many selectively toxic inhibitors are *structural analogs* (similar in shape) of the substrate of the enzyme inhibited. Because of this similarity in shape, the inhibitor actually competes with the substrate for the enzyme's active site (Fig. 10-16). Unlike the substrate however, the inhibitor cannot be converted to products, and, while in the active site, it prevents normal substrate binding. In this tied-up state, the enzyme is inactive. Such **competitive inhibition** of enzymes is characterized by two important properties: (1) It is highly specific for a certain metabolic reaction, that is, only reactions with substrates similar to the inhibitor molecules will be affected; and (2) the effect is reversed once the inhibitor molecule is removed. The inhibitor is constantly dissociating from and rebinding with the enzyme. No structural damage is done to the enzyme. Therefore, inhibition is maintained only as long as the concentration of inhibitor is high enough to successfully compete with the substrate. As inhibitor concentrations decrease, the substrate overcomes the inhibitor, the inhibition is reversed, the enzyme resumes activity, and the organism begins to grow again.

This property of reversibility is an important consideration when using chemotherapeutic agents. Agents that work by competitive inhibition (sulfa drugs, for example) are *microbistatic;* that is, they do not kill microorganisms but merely inhibit cell growth. Most chemotherapeutic agents are used to stop the growth of the pathogenic microorganisms long enough for the body to eradicate the infection using its own defense mechanisms. Thus, if the antibiotic regimen is terminated too soon (such as when the symptoms disappear), the inhibited microbes may resume

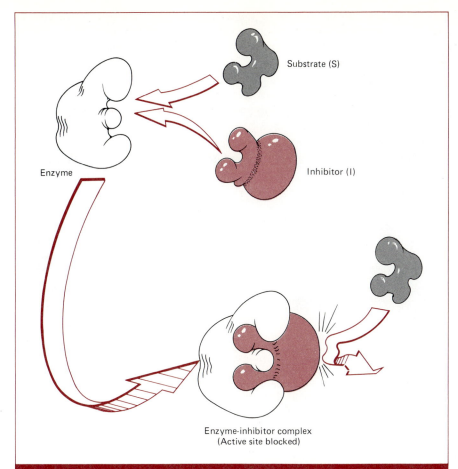

FIGURE 10-16

Competitive (substrate) inhibition of enzyme activity by an inhibitor (I) that is a structural analog of the substrate. The enzyme's active site recognizes both molecules, but the inhibitor cannot be converted to products.

Substrate (S)

Inhibitor (I)

Enzyme

Enzyme-inhibitor complex
(Active site blocked)

growth. The therapeutic role of competitive inhibitors is discussed in more detail in Chapter 14.

Enzymes may be permanently inactivated by the heavy metals lead, silver, mercury, and arsenic. These poisonous substances bear no resemblance to the substrate of the inactivated enzymes, but nonspecifically attach to protein. Some of these inhibitors alter the enzyme's shape so that it is no longer active. Others bind in place of cofactors or prosthetic groups needed to activate the enzyme. As a consequence, many critical metabolic functions are lost. These inhibitors affect virtually all enzymes, so inhibition is not selective for a specific reaction or a particular cell type. Humans and bacteria are equally susceptible to the lethal effects of these agents. Any attempt to use heavy metal inhibitors systemically as chemotherapeutic agents would eliminate the host along with the pathogen. Such nonselective enzyme inhibition accounts for the tragic consequences of lead poisoning in children who eat lead-containing paint as it peels from walls. Similar poisoning follows ingestion of food that contains high concentrations of mercury. In Japan mercury poisoning has been associat-

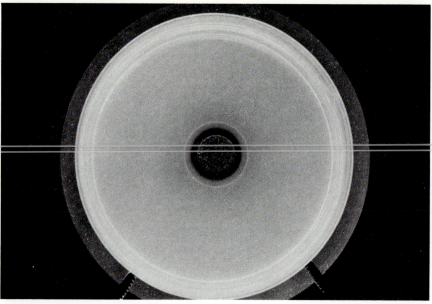

FIGURE 10-17

Inhibitory effects of silver nitrate on *Neisseria*. Area of no growth identifies site where 1% $AgNO_3$ was applied to agar surface seeded with the bacteria. Until recently silver nitrate was routinely instilled into the eyes of newborns to prevent severe eye infections that can develop if the mother has gonorrhea at the time of delivery. (Newborns now receive ocular antibiotics.) Silver inhibits bacterial growth by preventing the binding of essential cofactors to enzymes.

ed with consumption of fish caught in mercury-contaminated waters. Some metals, however, such as mercuric chloride and silver nitrate, are used as topical agents to prevent or treat infections on the surface of the body (Fig. 10-17).

OVERVIEW

Organisms are a living collection of precisely structured molecules. The structure, as well as the chemical properties of these molecules, determines an organism's physical characteristics and activities. Many of these compounds are macromolecules, large polymers used for storage and expression of genetic information, for cell structure, for energy storage, and for catalyzing metabolic reactions. The macromolecules are proteins, polysaccharides, lipids, and nucleic acids.

One group of macromolecules, the proteins, are the primary constituents of enzymes. Metabolic reactions are catalyzed by enzymes, and any factor that affects the activity of an enzyme affects the corresponding metabolic reaction. A cell can therefore regulate its own metabolism by controlling the concentration and activities of specific enzymes. Some chemotherapeutic agents inhibit microbial growth by blocking the activity of critical enzymes.

Some enzymes produced by pathogens severely damage the human body and produce symptoms of disease. Laboratory detection of extracellular enzymes produced by pathogens provides information helpful both in identifying the pathogen and in determining its virulence. All biochemical tests used for identifying microorganisms are methods of detecting which enzymes a microorganism produces.

KEY WORDS

metabolism	deoxyribonucleic acid (DNA)
macromolecule	ribonucleic acid (RNA)
monomer	nucleotide
polymer	deoxyribose
protein	hydrolysis
hydrophobic	enzyme
hydrophilic	active site
polysaccharide	coenzyme
carbohydrate	denaturation
lipid	competitive inhibition
phospholipid	

REVIEW QUESTIONS

1. List four macromolecules, their functions, and their monomers.

2. What is the significance of the precise amino acid sequence in protein?

3. How does the structure of phospholipids contribute to their role in membrane structure?

4. List three ways in which RNA differs from DNA.

5. Why are enzymes so important to living systems?

6. Discuss how each of the following affects enzyme activity:
   (a) Temperature
   (b) Competitive inhibitors
   (c) Heavy metals

Metabolic Dynamics of Life

11

All living organisms without exception need a constant supply of energy and organic building materials to maintain their life processes. These resources are needed for **biosynthesis**—the construction of molecular components in the growing cell and the replacement of these compounds as they deteriorate. Complex macromolecules in the cell perform specific biological functions, create the physical structures of the cell, and store energy for later use. As discussed in the previous chapter, the cell is a highly organized collection of molecules produced by metabolically reorganizing nutrients into cellular material. A constant source of energy is required to fuel biosynthesis and growth. Metabolism is therefore a dynamic balance between those reactions that provide the cell with energy or building materials and those that utilize them. These reactions are the keys to understanding the metabolic dynamics of life.

WHY STUDY MICROBIAL METABOLISM?

The biochemical activities of all organisms are similar in many respects. Humans and microorganisms, for example, degrade glucose to the same three-carbon compound (pyruvic acid). The role of vitamins was discovered as a result of scientific interest in the nutritional requirements of microorganisms. The study of microbial metabolism contributes enormously to our understanding of human metabolism, and clinical medicine has derived many practical benefits from that study. A few of them are listed here.

Enhanced Control of Infectious Disease Most chemotherapeutic agents operate by inhibiting specific metabolic reactions in microorganisms without interfering with human metabolism. In addition, many chemical and physical agents that reduce microbial contamination exert their antimicrobial effects by disrupting metabolic functions.

Isolation of Pathogens from Infected Persons Selective and enriched media (see Chap. 6) establish conditions favorable for the metabolic activities of the suspected pathogens, but not for those of contaminants that could otherwise overgrow the pathogen and obscure the diagnosis.

Diagnosis of Infectious Disease Identification of pathogens isolated from infected persons is largely dependent on determining the microbe's biochemical characteristics, that is, its ability to metabolize sugars and other compounds.

Understanding Protective Roles of Normal Flora The metabolic by-products of many microorganisms indigenous to humans can inhibit or kill pathogens that might otherwise cause serious disease.

Understanding Mechanisms of Infectious Disease Some pathogens survive the defenses and injure the tissues of an infected host because they produce metabolic by-products, such as toxins, that trigger adverse chemical changes in the host.

METABOLIC PATHWAYS

The chemical processes associated with life usually occur in series of reactions. Each reaction slightly modifies the previous molecule until the entire sequence is completed, or until one of the intermediate compounds is redirected into a different series of reactions. Each series of chemical changes is called a **metabolic pathway**. The chemical processes in the cell are usually accomplished by pathways rather than by single-step reactions. For example, cells degrade glucose to carbon dioxide by a series of 19 reactions. You can accomplish the same process in a single reaction by simply igniting glucose and letting it burn, but metabolic pathways allow cells to degrade molecules gradually, releasing energy in easily managed bits rather than in a single explosive reaction.

Each complete pathway ultimately produces one or more end products, temporarily generating intermediate compounds in the process. In the following theoretical pathway,

$$[A] \xrightarrow{\text{reaction 1}} [B] \xrightarrow{\text{reaction 2}} [C] \xrightarrow{\text{reaction 3}} [D] \xrightarrow{\text{reaction 4}} [E]$$

A is the *substrate* (the starting compound) and E is the final *product*; B, C, and D are called the *metabolic intermediates.* Metabolic intermediates generated by pathways may be diverted from one pathway into another sequence of reactions. In this way, hundreds of different metabolic pathways in a cell are coordinated to work together. Some of the intermediates of glucose degradation, for example, are also the substrates for the fat-producing pathway. Thus when glucose is in abundant supply, these common intermediates are diverted for storage as fat rather than completely degraded to carbon dioxide. Linking pathways in this fashion provides a mechanism for diverting surplus intermediate compounds to biosynthetic pathways. In this way, cells synthesize the new materials needed for cell growth.

ENERGY TRANSFERS

Metabolism, the total organized chemical activities of the cell, consists of two general categories of reactions—catabolic and anabolic. **Catabolism** is the degradation of complex molecules to simpler molecules. These processes generally release energy and are called *exergonic* (energy-yielding). Chemical energy that holds the atoms of the complex molecule together is released when chemical bonds are broken. **Anabolism** is the biosynthesis of complex molecules from simpler compounds. Molecular construction requires energy; the reactions are therefore said to be *endergonic* (energy-consuming). The energy supplied to these reactions is absorbed into the chemical bonds created during the formation of complex molecules.

Oxidation and Reduction

Anabolism and catabolism cannot be understood in terms of molecular complexity or energy transfers alone, since in all metabolic reactions electrons are transferred as well. These electron transfers determine whether a reaction is oxidative or reductive. **Oxidation** results in the loss of electrons, **reduction** in a gain in electrons. Whenever a molecule is oxidized, its lost electrons are captured by another molecule, which is reduced in the process. *Every oxidative reaction, therefore, is coupled with a concurrent reductive reaction.*

Catabolism of a complex molecule releases energy and electrons. Biosynthesis, on the other hand, not only consumes energy, but also requires a source of electrons. It can therefore be stated as a general rule that *highly reduced compounds are more energy-rich than highly oxidized compounds.* For example, the six carbon atoms of glucose are in a highly reduced state, and considerable energy is contained in this molecule. When an organism catabolizes glucose to six molecules of CO_2 (carbon dioxide), it transforms the carbon to a more oxidized state. This reaction is summarized in the following equation:

$$6O_2 + C_6H_{12}O_6 \longrightarrow 6CO_2 + 6H_2O + \text{ENERGY}$$

GLUCOSE
Highly reduced,
energy-rich

CARBON DIOXIDE
Highly oxidized,
energy-poor

During this reaction, 12 electrons (carried by the 12 hydrogen atoms) are transferred from glucose to oxygen, forming six molecules of water. Compared to glucose, CO_2 has little available energy, since much of the chemical energy in glucose is released during oxidation. The same is true whenever we burn wood (cellulose). The highly reduced carbons in cellulose rapidly oxidize (burn), losing their electrons and reducing oxygen to water. In addition to water, the process liberates CO_2 and generates heat. Many cellulose-digesting microbes accomplish the same task but release the energy in a controlled fashion using metabolic pathways.

Table 11-1 compares some general properties of catabolic and anabolic reactions.

ATP—The Energy Carrier

Cells derive energy by hydrolyzing complex molecules, such as starch, glycogen, and lipids, to monomers that are then oxidized by a series of reactions that sequentially release safe amounts of energy. This energy is

TABLE 11–1

Some General Characteristics of Catabolic and Anabolic Processes

	Catabolic	Anabolic
Energy exchange	Exergonic	Endergonic
Type of reaction	Oxidative	Reductive
Nature of conversion	Complex → simple (degradative)	Simple → complex (biosynthetic)
Most energy found in	Substrates	Products

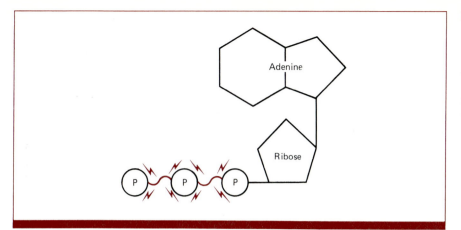

FIGURE 11-1
Skeletal representation of adenosine triphosphate (ATP) showing the two high-energy phosphate bonds ($\sim$). Adenosine diphosphate (ADP) has one less P$\sim$P bond, but is otherwise identical to ATP.

lost unless converted to a usable form that is readily available to the cell whenever it is needed for the life processes. Energy from oxidation reactions is transferred to the chemical bonds of high-energy transfer compounds, the mosticommon of which is **adenosine triphosphate (ATP)** (Fig. 11-1).

Adenosine triphosphate contains three phosphate groups attached to each other. A great amount of energy is required to form chemical bonds between pairs of phosphates. The energy required to couple adenosine diphosphate (ADP) and inorganic phosphate (abbreviated P$_i$) to form a molecule of ATP is stored in the new phosphate bond. The process is called **phosphorylation**. When the high energy phosphate bond is broken (hydrolyzed) in the reverse reaction (ATP $\rightarrow$ ADP + P$_i$ + energy), the stored energy is released. ATP formation absorbs and stores energy released from exergonic catabolic processes; ATP hydrolysis provides energy for endergonic processes such as biosynthesis of new cellular structures (Fig. 11-2). Part of the energy released in the oxidation of glucose, therefore, is immediately captured as chemical energy in ATP. The rest is lost as heat.

The cell's immediately available energy resources are represented by its wealth of ATP. Whenever the cell needs to spend energy for endergonic processes, it must have enough ATP to cover the energy cost. We can think of ATP as the cell's energy carrier and ADP as the cell's energy acceptor.

Electron Carriers

Energy transfers are usually oxidation-reduction reactions and require the exchange of electrons as well as of energy. Often the electrons are temporarily transferred to one of three coenzymes that function as electron carrier molecules. These three coenzymes are nicotinamide adenine dinucleotide (**NAD**), nicotinamide adenine dinucleotide phosphate (**NADP**), and flavin adenine dinucleotide (**FAD**). Each of these compounds can be reduced by a pair of electrons. These electrons are usually released in the form of hydrogen atoms. (An atom of hydrogen is the combination of a

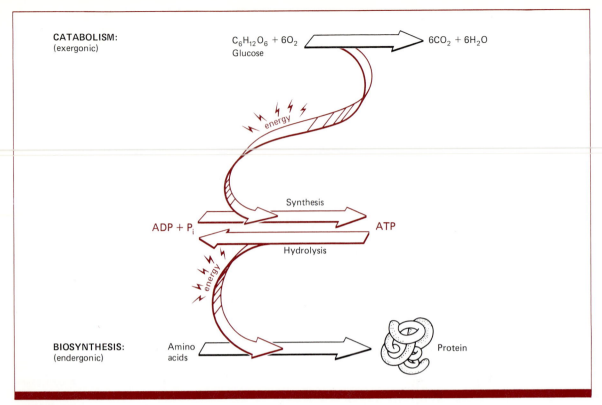

CATABOLISM:
(exergonic)

$C_6H_{12}O_6 + 6O_2$
Glucose

$6CO_2 + 6H_2O$

energy

Synthesis

$ADP + P_i$

ATP

Hydrolysis

energy

BIOSYNTHESIS:
(endergonic)

Amino
acids

Protein

FIGURE 11-2

Catabolic and biosynthetic reactions are coupled by ATP, the common denominator in energy production and utilization. Exergonic reactions generate energy for producing ATP; endergonic reactions utilize energy released by ATP hydrolysis.

proton and an electron.) Oxidized NAD and NADP are positively charged molecules (NAD^+ and $NADP^+$). Each is reduced by accepting one pair of electrons from two hydrogen atoms. Only one of the protons from the hydrogens is accepted, however, producing a neutral (noncharged) molecule containing one added hydrogen. The other proton from the pair of hydrogens is released into solution as an hydrogen ion (H^+). The reduced form of these coenzymes is represented as **NADH + H$^+$** and **NADPH + H$^+$**. During the reverse process, the coenzyme releases its two electrons and a proton. The unpaired electron reunites with a hydrogen ion from the solution. These two reversible reactions are written:

$$NAD^+ + 2H \rightleftarrows NADH + H^+$$

$$NADP^+ + 2H \rightleftarrows NADPH + H^+$$

The reduced form of FAD, on the other hand, contains both of the protons as well as both electrons. It is represented as **FADH$_2$**. Thus, in their reduced states, coenzymes are sources of electrons (and hydrogen).

The function of these coenzymes is twofold:

1. Electrons released by oxidation of organic substrateslare carried by NAD^+ and FAD to special sites where their potential energy can be converted to ATP. (This will be discussed later in this chapter.)

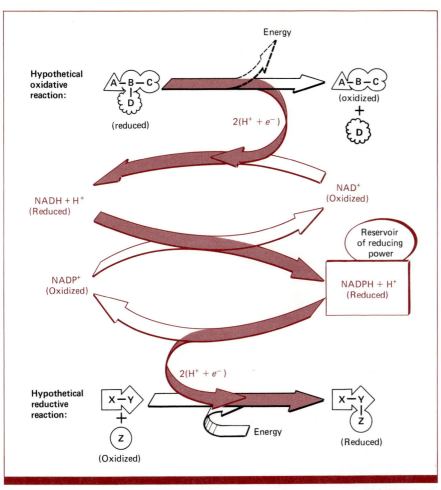

FIGURE 11-3

Formation and utilization of a pool of reducing power (NADPH+H$^+$). (The shaded portions of the arrows follow the path of the electrons.) NAD$^+$ directly accepts the electrons released from oxidation of substrates and transfers them to NADP$^+$. In this way these coenzymes carry electrons from oxidative reactions to reductive reactions. Notice that two protons (H$^+$) accompany each pair of electrons (e$^-$). One of these protons joins with the coenzyme. The other is released into solution.

2. Electrons released by the oxidation of organic substrates are transferred from NADH + H$^+$ to NADP$^+$ (Fig. 11-3). The reduced NADP (NADPH + H$^+$) may then transfer these electrons to compounds being synthesized.

Thus the oxidative events of catabolism supply energy (ATP) for cell functions and also generate the cell's *reducing power*, the supply of electrons available for biosynthesis and other reduction processes.* The pool of NADH + H$^+$ or NADPH + H$^+$ in a cell represents its reducing power.

---

*"Reducing" refers here to the opposite of oxidation, *not* to molecular degradation.

PRODUCING USABLE ENERGY

Depending on the organism, energy for ATP formation is derived either from photosynthesis or from the catabolic oxidation of highly reduced compounds.

Photosynthesis

The chemical energy present in glucose and other reduced "food" molecules originates from the sun and is trapped by the photosynthetic machinery of phototrophs (Fig. 11-4). **Photosynthesis** is a process by which many molecules of CO_2 are reduced to carbohydrates, using water to provide the electrons (and hydrogen) and chlorophyll to harness the

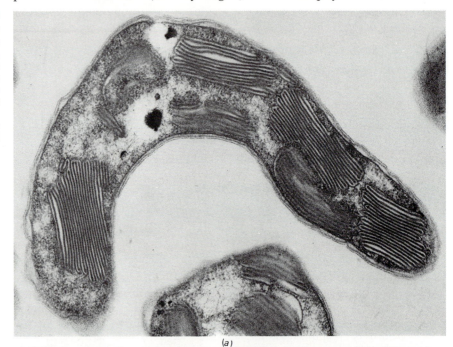

(a)

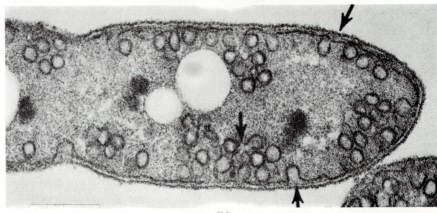

(b)

energy of light to form chemical bonds. The overall process is summarized in the following equation:

Chlorophyll absorbs light energy and releases *excited electrons* that temporarily contain this energy. These excited electrons are used to produce NADPH + H$^+$ and ATP, the reducing power and energy needed for converting CO_2 to glucose.

Photosynthesis occurs in four steps (the flow of energy is shown in color):

1.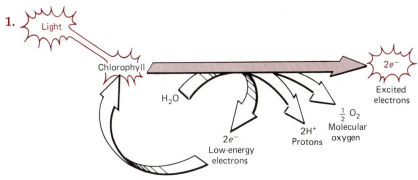

Light energy is absorbed by chlorophyll and used to excite two electrons to high-energy states. The loss of these two electrons from chlorophyll leaves a "hole." This electron hole is filled by electrons from *photolysis* (the use of light energy to split water). Removal of these two electrons from water produces two protons and an atom of oxygen gas.

2.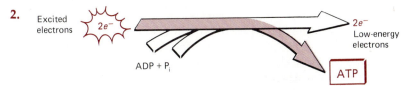

The energy in the excited electrons is converted to ATP. This process is called *photophosphorylation*. Notice that a pair of unexcited electrons remains.

3.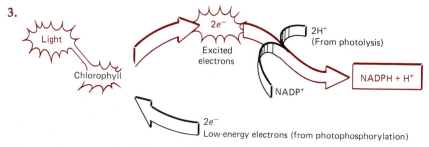

Photoexcitement of a different chlorophyll produces another pair of excited electrons. These electrons join with the protons released by splitting water (step 1) to reduce an NADP⁺ to NADPH + H⁺, generating a pool of reducing power. The electron "hole" in this chlorophyll is filled by the unexcited electrons from step 2.

4. NADPH + H⁺ (reducing power) and ATP (energy) are used to synthesize glucose from CO_2.

The oxygen generated by photolysis is eventually liberated as free molecular oxygen, the waste product of photosynthesis. Thus phototrophs supply not only food for all consumers on earth, but virtually all the molecular oxygen essential for the survival of aerobic organisms, including humans.

Photosynthesis is similar in eucaryotes and cyanobacteria. Photosynthesis among bacteria, however, uses different types of light-absorbing pigments, called *bacteriochlorophylls*. In addition, bacterial photosynthesis does not produce free oxygen because a compound other than water is the source of hydrogen atoms used for reducing CO_2 to glucose. For example, some bacteria use hydrogen sulfide (H_2S) as the hydrogen donor. This leaves elemental sulfur (S) as a waste product rather than oxygen:

$$12H_2S + 6CO_2 + \text{sunlight} \xrightarrow[\text{chlorophyll}]{\text{bacterial}} C_6H_{12}O_6 + 12S + 6H_2O$$

Other photosynthetic bacteria reduce CO_2 by using molecular hydrogen (H_2), thiosulfate, or some organic compound such as succinate or malate.

Catabolism

Nonphotosynthetic organisms are chemotrophs; they derive their energy by oxidizing reduced molecules. A few bacteria, the *chemoautotrophs*, use the energy released when they oxidize inorganic compounds and obtain their carbon as do plants and other photoautotrophs—they reduce CO_2 to organic compounds. Some chemoautotrophs, the nitrifying bacteria, are important participants in the earth's nitrogen cycle. Others oxidize hydrogen sulfide to elemental sulfur, sometimes producing huge geological deposits of this element. Table 11-2 provides some examples of inorganic chemotrophic processes.

Most nonphotosynthetic organisms, however, are *chemoheterotrophs*, obtaining both energy and carbon by oxidizing organic "food" molecules, such as glucose. Glucose is most commonly catabolized by a process called glycolysis.

Glycolysis   Some of the chemical energy in glucose can be liberated by

TABLE 11–2
Examples of Reactions by
which Chemoautotrophic
Bacteria Derive Energy
from Inorganic Compounds

Energy Source	Reaction	Organism
Hydrogen gas	$2H_2 + O_2 \rightarrow 2H_2O$	*Hydrogenomonas*
Ammonia	$2NH_3 + 3O_2 \rightarrow 2HNO_2 + 2H_2O$	*Nitrosomonas*
Nitrite	$2HNO_2 + O_2 \rightarrow 2HNO_3$	*Nitrobacter*
Hydrogen sulfide	$2H_2S + O_2 \rightarrow 2H_2O + 2S$	*Thiobacillus*

glycolysis, splitting this six-carbon sugar into two molecules of **pyruvic acid**, each containing three carbon atoms. Glycolysis is accompanied by the production of ATP and NADH + H⁺. The pathway consists of 10 metabolic reactions catalyzed by specific enzymes. Each step of the pathway redistri-

FIGURE 11-5

Abbreviated version of glycolysis. Energy and electron exchanges appear in the right margin, next to the corresponding metabolic reaction.

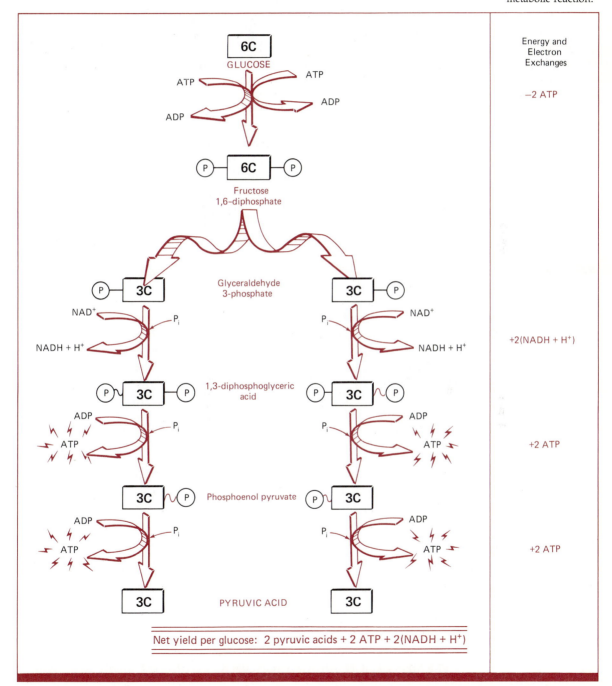

Net yield per glucose: 2 pyruvic acids + 2 ATP + 2(NADH + H⁺)

butes the energy contained in the original glucose molecule until much of it is concentrated in four high-energy phosphate bonds. This energy can then be used to form four ATP molecules. An abbreviated version of glycolysis is illustrated in Figure 11-5.

The oxidation of glucose is initiated by activating the sugar with energy obtained from two ATP molecules, a process analogous to priming the pump. This makes the stable glucose molecule more reactive. Because of this energy input, the net yield from glycolysis is two molecules of ATP per molecule of glucose, two pairs of electrons (carried by two NADH + H$^+$), and two molecules of pyruvic acid.

Production of reduced coenzyme NADH + H$^+$ during glycolysis depletes the supply of the electron acceptor NAD$^+$. NAD$^+$ is essential for oxidizing the intermediates of glycolysis, and in its absence the pathway cannot proceed. The supply of NAD$^+$ is replenished by transferring the electrons in NADH + H$^+$ to another molecule. The compound that ultimately acquires these electrons from NADH + H$^+$ determines (1) the metabolic end products of glucose oxidation and (2) whether an organism is fermentative or respiratory. If an organic compound is the final electron acceptor, the metabolic process is called **fermentation**. If an inorganic molecule, such as molecular oxygen, is the final electron acceptor, the process is termed **respiration**.

Fermentation   Pyruvic acid or its derivatives is the final electron acceptor in most fermentations. Depending on the enzymes present, these compounds may be reduced by NADH + H$^+$ to form ethyl alcohol, lactic acid, acetic acid, or a number of other metabolic by-products (Fig. 11-6). Regardless of the end product, all these processes are fermentations because an organic compound is the final electron ecceptor. Fermentation is an anaerobic process, requiring no molecular oxygen; for many obligate anaerobes this is the only method of obtaining energy.

In most cases the only usable energy released from fermentation is generated during glycolysis. For example, the familiar alcoholic fermentation characteristic of brewer's yeast, *Saccaromyces cerevisiae*, converts pyruvic acid, the end product of glycolysis, to CO$_2$ plus a two-carbon compound, acetaldehyde, which is subsequently reduced by NADH + H$^+$ to form ethyl alcohol. No additional energy is produced by these reactions. The release of carbon dioxide in the first step accounts for the familiar bubbling of alcoholic fermentation as well as the natural carbonation of beer and champagne.

USING THE BY-PRODUCTS OF FERMENTATION   The end products of microbial fermentation have provided many medically and economically valuable products. In addition to producing various alcohols, industrial fermentations have been used to produce acetone, lactic acid, and formic acid. The final product depends on the microorganism employed.

FERMENTATION AS A DIAGNOSTIC TOOL   The ability to ferment a

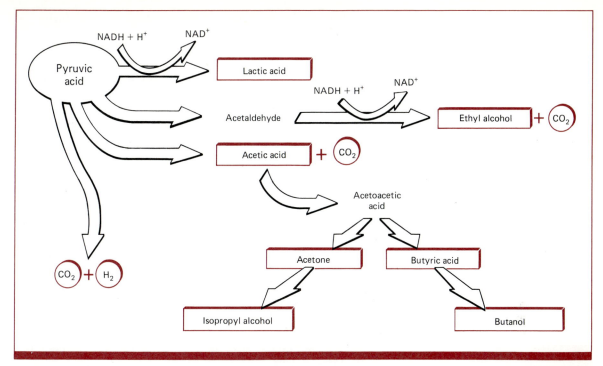

FIGURE 11-6

Some end products of
pyruvic acid metabolism.
The compound formed is a
stable property of the
organism and may be used
to help characterize or
identify the organism.

particular sugar or other substrate depends on the genetic capacity of the
organism to produce the necessary enzymes. Like other genetically determined characteristics (appearance, staining properties, and antigens),
fermentation reactions provide valuable clues for identifying microorganisms in the laboratory and are therefore important in diagnosis of
infectious disease. A fermentation test is pictured in the next group of color
photographs.

PROTECTION OF HUMANS BY BACTERIAL FERMENTATION   The
by-products of fermentation in normal bacterial flora may provide us some
protection against infectious disease. For example, the lactic acid produced
by the lactobacilli in the adult vagina opposes infection by lowering the pH
below that which can be tolerated by many vaginal pathogens. A similar
effect helps protect human skin from infection.

In culture fermentation, by-products inhibit bacterial growth
when critical concentrations of wastes have accumulated. Buffers are
included in culture media to neutralize the acidic by-products of some
microbes, thereby enhancing yields from laboratory cultivation of the
organisms.

Respiration   The incomplete oxidation of glucose by glycolysis makes it
a relatively inefficient process. At the end of glycolysis, approximately 95
percent of the chemical energy in glucose remains untapped in the

chemical bonds of pyruvic acid. Fermentation provides no mechanism for cells to directly harvest this energy. (Ethyl alcohol retains so much chemical energy it is used as an automobile fuel.) Respiration is an oxidative process that, unlike fermentation, efficiently transfers the energy remaining in pyruvic acid to the high-energy electrons of reduced NADH + H$^+$ and then into ATP.

The pair of electrons carried by NADH + H$^+$ contains considerable energy. During respiration this energy is harvested by passing the electrons through the **electron transport system** (also called the *respiratory chain*), which is a series of electron carriers. Most of these carriers are **cytochromes**, iron-containing molecules that accept high-energy electrons from the preceding member in the sequence and donate them to the next cytochrome of the respiratory chain (Fig. 11-7a). Each member of the chain is more attractive to electrons than the previous molecule. With each successive transfer, the electrons lose energy. It is believed that at three points in the chain enough energy is released to propel protons across a cell membrane, causing them to accumulate on one side. The proton accumulation represents a reservoir of energy, much like water that has accumulated behind a dam. The protons eventually release this potential energy when they flow back across the membrane through "proton pores." These regions also house enzymes for catalyzing ATP production. The energy released by this proton flow is used by the enzyme for phosphorylating ADP to ATP.

Each electron pair donated by NADH + H$^+$ contains enough energy to generate three ATP molecules when processed by the electron transport system. If the electron pair is introduced into the system as FADH$_2$ rather than NADH + H$^+$, only two ATP molecules are produced (analogous to skipping the first waterwheel of the cascade depicted in Figure 11-7b).

The respiratory chain needs a terminal electron acceptor. Only when a cytochrome transfers its electrons to the next member of the chain can it

FIGURE 11-7

(a) The flow of electrons through the respiratory chain (electron transport system) results in sequential oxidation-reduction reactions. Each cytochrome is reduced by electrons transferred from the previous cytochrome and is oxidized by losing electrons to the next member. The electrons travel through the electron transport system, releasing energy as they move toward a lower, more stable energy state. (The accompanying protons are released into solution by coenzyme Q. The final oxidation rejoins a pair of protons from solution with oxygen and the two electrons to form water.) (b) The energetics of the electron transport system may be compared to water cascading over a series of energy-generating waterwheels. The water (electrons) flows into the system in a high-energy state and is carried by buckets (NAD$^+$) to a series of waterwheels (cytochromes), which are turned by the force of the falling water. The dynamo attached to each wheel generates energy for storage in a battery (ATP). This energy is readily accessible whenever needed. Water (electrons) leaves the system in a low-energy state.

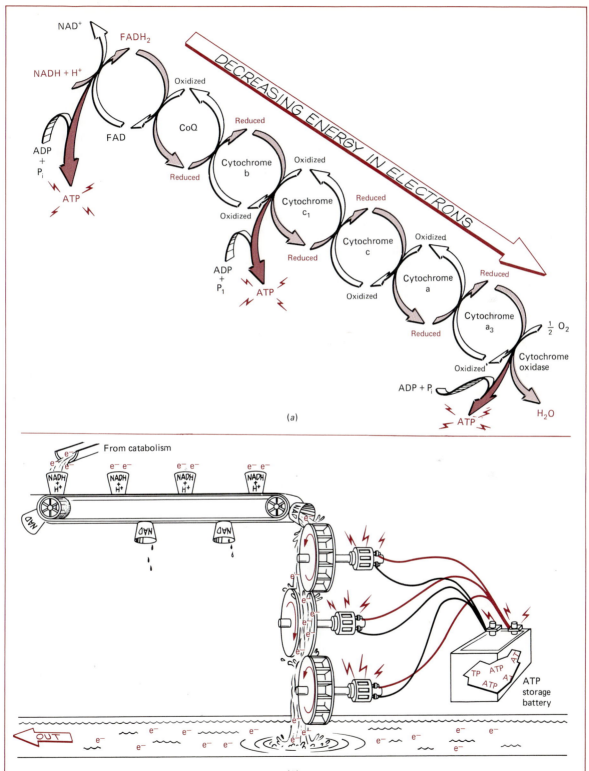

(a)

(b)

accept another electron pair. The final cytochrome, however, has no subsequent member in the chain to accept its electrons. These electrons must move out of the chain before this molecule can continue accepting electrons. In aerobic respiration, molecular oxygen, with its great affinity for electrons, is the terminal acceptor. One atom of oxygen (one-half of an oxygen molecule) accepts one pair of electrons (and hydrogen ions) to form H_2O. Thus in the presence of oxygen the chain remains open for passage of the next pair of electrons. This is the reason that oxygen is essential to the respiration of all aerobic organisms, from bacteria to humans. In the absence of molecular oxygen, most aerobic organisms die of energy starvation. The whole respiratory chain plugs up with electrons, the flow stops, and no ATP is produced. The respiratory poison cyanide has a similar effect (see color box).

Each member of the respiratory chain is physically separated from all but the immediately adjacent electron donors and acceptors. This prevents the cytochromes at the end of the chain from capturing the electrons directly from NADH + H$^+$ and releasing all the energy in a single uncontrollable event (analogous to the water in Figure 11-7b missing the waterwheels and falling directly back into the stream). Membrane compartments physically confine cytochromes to positions that assure the transport of electrons in the proper sequence. NADH + H$^+$ never contacts any chain member other than FAD, which, in turn, has access only to the next member in the sequence. In eucaryotes, cytochromes are compartmentalized in the membranes of mitochondria. In procaryotes, which have no mitochondria, cytochromes are physically segregated by their attachment to the solid surfaces of the cell membrane and the mesosome.

AEROBIC OXIDATION OF GLUCOSE The end products of glycolysis—two molecules of pyruvic acid—are the same for fermentative and respiratory organisms. Aerobic organisms, however, can harvest additional ATP from the two NADH + H$^+$ pairs produced during glycolysis. These two electron pairs can be cashed in for six ATP molecules at the electron

ORGANISMS THAT SURVIVE CYANIDE

■ The sodium cyanide pellet sits harmlessly on its platform. As long as it remains there it can kill no one. Once it falls into a container of hydrochloric acid, however, a lethal cloud of cyanide gas fills the room. The gas chamber begins its grim task.

Cyanide, a respiratory poison, binds irreversibly to the iron in a key cytochrome, preventing it from transporting electrons. The effect is the same as removing all available molecular oxygen. The electron transport system, and therefore respiration, is blocked. Any organism that relies on respiration will be unable to obtain sufficient energy from glucose and will quickly die. Fermentative organisms are not dependent on cytochrome function to obtain their energy. These microbes, therefore, would survive even if their environment were saturated with deadly cyanide.

transport system. Whereas fermentative organisms derive two ATP from each glucose oxidized by glycolysis, aerobes obtain eight ATP, two from glycolysis plus six from the electrons oxidized by the cytochromes. In addition, respiratory organisms harvest the energy remaining in pyruvic acid by catabolizing it to CO_2, its most oxidized state.

The following discussion describes how the complete oxidation of glucose by respiration releases 19 times more usable energy (ATP) than does fermentation.

AEROBIC OXIDATION OF PYRUVIC ACID   Pyruvic acid (the end product of glycolysis) is oxidized by *decarboxylation,* removing one of the carbon atoms in the form of CO_2 (Fig. 11-8). Pyruvic acid decarboxylation also produces a two-carbon fragment, called an *acetyl* group, that is temporarily hooked to the carrier molecule coenzyme A (CoA). The whole complex is called *acetyl coenzyme A.* The reaction also yields a molecule of NADH + H^+, which can generate three ATP molecules by passage of its electron pair through the electron transport system. Decarboxylation of the two molecules of pyruvic acid generated from each glucose molecule therefore yields two acetyl CoA and six ATP.

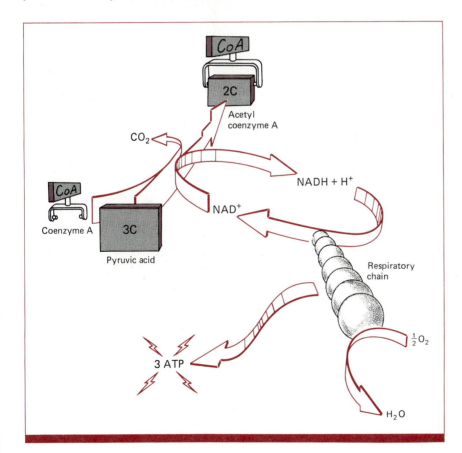

FIGURE 11-8

Decarboxylation (removal of CO_2) of pyruvic acid to form the two-carbon molecule, acetyl CoA (*CoA* = coenzyme A). This oxidation of pyruvic acid is coupled with reduction of NAD^+ to NADH+H^+, which may generate three ATP molecules when its electrons are processed by the electron transport system. Decarboxylation of the two molecules of pyruvic acid from glucose yields a total of six ATP molecules.

KREBS CYCLE Pyruvic acid decarboxylation is a metabolic bridge that transfers the carbon of glucose from the glycolytic pathway to the **Krebs cycle** (also called the **tricarboxylic acid cycle**), the pathway that completes the oxidation of glucose (Fig. 11-9). Acetyl CoA enters the Krebs cycle by being enzymatically coupled to a four-carbon molecule (oxaloacetate). The resulting six-carbon compound is citric acid. (For this reason the Krebs cycle is sometimes called the "citric acid cycle.") During the successive steps, the two carbon atoms from acetyl CoA are oxidized to two molecules

FIGURE 11-9

Simplified version of the Krebs cycle. When coupled with the electron transport system (ETS), this cyclic pathway completes the total oxidation of glucose. Each NADH+H⁺ may be sent through the ETS to yield three ATP molecules. The FADH₂ yields two ATP molecules when oxidized by the respiratory chain. Both carbon atoms in acetyl CoA (from decarboxylation of pyruvic acid) are oxidized to CO_2. Water is the end product of respiration. Net energy yield is 12 ATP molecules per acetyl CoA or 24 ATP molecules per glucose oxidized. (See text for full explanation.)

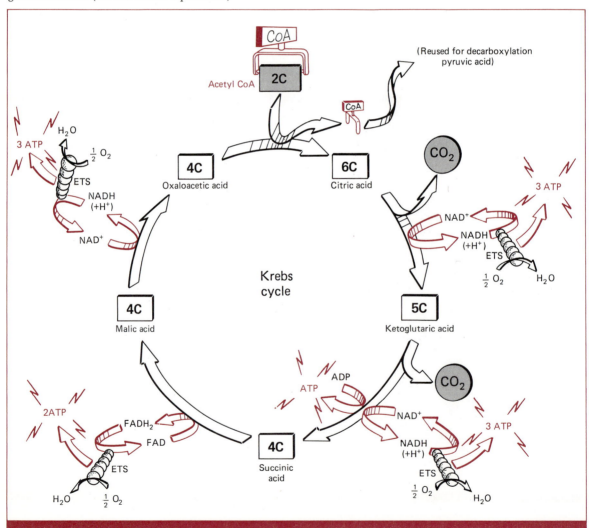

of CO_2, leaving the four-carbon oxaloacetate to accept another acetyl CoA. In the course of the cycle, some of the energy released from the oxidation of citric acid is used directly to produce one ATP molecule. Most of the energy, however, is transferred by four electron pairs to three molecules of NAD^+ (forming $NADH + H^+$) and one molecule of FAD (forming $FADH_2$). The energy from these electrons is then used to generate ATP at the electron transport system. In this way one acetyl CoA yields 12 ATP molecules when oxidized by the Krebs cycle. Since two acetyl CoAs are produced for each glucose oxidized, the final energy yield from the Krebs cycle is 24 ATP molecules. Adding this to the 14 ATP molecules derived from the previous oxidations, we find that aerobic oxidation of each

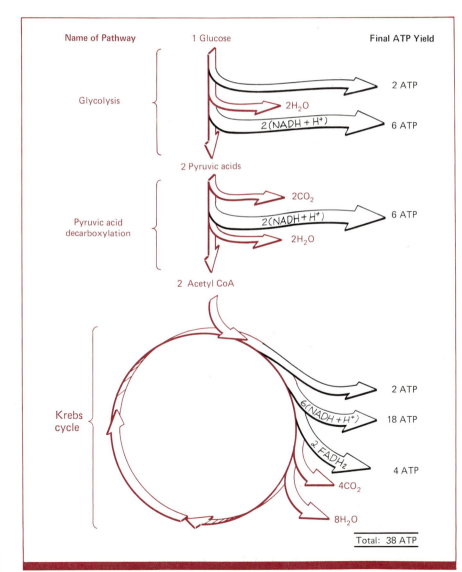

FIGURE 11-10

The complete respiratory sequence of glucose oxidation
$(C_6H_{12}O_6 + 6O_2 \rightarrow 6CO_2 + 6H_2O + 38ATP)$.

glucose molecule yields enough energy to produce 38 molecules of ATP (compared to two ATP produced by anaerobic glycolysis). Thus, more than 90 percent of the ATP synthesized from glucose oxidation depends on the availability of oxygen as the final electron acceptor. The complete sequence of aerobic oxidation of glucose to CO_2 and water is condensed in Figure 11-10.

Facultative anaerobes preferentially utilize the energy-efficient aerobic pathways when molecular oxygen is available, a phenomenon known as the **Pasteur effect**. Some industrial fermentations are carefully controlled to exclude oxygen from the culture, since in the presence of molecular oxygen the useful by-products of fermentation either are not produced or are further oxidized to undesirable compounds.

ANAEROBIC RESPIRATION Some anaerobic organisms possess respiratory chains that convert the energy of electrons into ATP in the absence of molecular oxygen. The terminal electron acceptor in these **anaerobic respirations** is an inorganic molecule other than oxygen, such as a sulfate, a nitrate, or carbon dioxide. Anaerobic respiration is less efficient than aerobic respiration; the respiratory chains are shorter and they yield fewer ATP molecules per electron pair processed. In addition, the Krebs cycle in many anaerobic organisms is incomplete, producing fewer NADH + H⁺.

The presence of molecular oxygen is actually toxic to most obligate anaerobes. Although their oxygen intolerance is not fully understood, it may be because they lack a mechanism for destroying the highly poisonous superoxide radicals (O_2^-) that are produced during oxygen-related metabolism. Aerobic organisms protect themselves by using the enzyme superoxide dismutase to combine O_2^- immediately with hydrogen ions to form hydrogen peroxide (H_2O_2). The H_2O_2 is further catalyzed to harmless H_2O and O_2 by the enzymes catalase or peroxidase. Most obligate anaerobes are unable to produce one or more of these enzymes (see color box).

The four methods by which organisms obtain chemical energy are summarized in Table 11-3.

ENERGY UTILIZATION—ANABOLISM

Much of the energy generated by catabolic reactions is used for driving anabolic reactions that assimilate simple compounds into the monomers needed for assembling

TABLE 11–3
Four Metabolic Strategies for the Production of Energy

Strategy	Type of Metabolism	Final e⁻ Acceptor	Energy Source
Photosynthesis	Anabolic	Chlorophyll and NADP⁺	Light
Fermentation	Catabolic	Organic compound	Chemical
Aerobic respiration	Catabolic	O_2	Chemical
Anaerobic respiration	Catabolic	Inorganic compound other than O_2	Chemical

AN ANCIENT GLOBAL CATASTROPHE

■ Since the appearance of life, the world has not known a more catastrophic occurrence. It resulted in the extinction of thousands of species; creatures that had prevailed for millions of years disappeared forever, banished by a new type of organism. Not even the eventual reign of humans would change the face of the planet as completely as did the rise of these terrible microbes—the cyanobacteria.

These blue-green procaryotic algae produced one of the most universally poisonous substances the world has ever known. In most places the only survivors were a few tiny organisms with the "evolutionary wisdom" to have developed some way of protecting themselves from the toxin. The cyanobacteria produced so much of this deadly substance that it saturated the oceans and the atmosphere, killing most of the world's organisms, and irreversibly altering the history of life on earth.

Eventually new forms of life emerged, organisms that not only withstood the poisonous substance, but actually became dependent on it, using it to increase their metabolic efficiency. These evolutionary innovators were, along with cyanobacteria, the first users of that deadly poison, oxygen.

Molecular oxygen, O_2, is produced as a by-product of photosynthesis. Before the earth became populated with her early photosynthetic creatures, the atmosphere was completely devoid of this substance. There was, therefore, no selective pressure to favor the development of aerobic organisms. The typical organism of that day had no defenses against oxygen's tendency to form superoxide radicals that combine with and oxidize essential cellular biochemicals. Such uncontrolled oxidation of cytoplasm is always fatal. A few organisms, however, possessed enzymes that defused oxygen's powerful destructive tendencies before it could damage their cytoplasm. Certainly, the cyanobacteria had these enzymes or they would have destroyed themselves with their own wastes. But some nonphotosynthetic microbes possessed these enzymes too, and many of these organisms eventually developed an electron transport system— a mechanism for using oxygen to tap additional energy from their food. This revolutionary development led to a totally new line of creatures, the aerobes, a line that ultimately produced humans.

Although the modern world is shared by oxygen producers and oxygen consumers, many places are still as devoid of oxygen as was earth 2 billion years ago before the emergence of photosynthesis. Within these anaerobic environments reside microbes that, like their primitive ancestors, would have little chance of survival in an oxygen-saturated environment.

proteins, nucleic acids, polysaccharides, and lipids. Since biosynthesis of these four essential types of macromolecules is an endergonic reductive process, it requires both energy (ATP) and electrons (reducing power, usually in the form of NADPH + H$^+$). Phototrophs obtain reducing power from photosynthesis. Chemotrophs utilize the electrons released as NADH + H$^+$ during catabolic reactions. Before they can be used for biosynthesis, however, these electrons must be transferred to NADP$^+$. The following equation illustrates this transfer:

$$\text{NADH} + \text{H}^+ + \text{NADP}^+ \rightleftarrows \text{NAD}^+ + \text{NADPH} + \text{H}^+$$

The direction of this reaction is determined by an organism's energy needs. When energy is abundant, the production of NADPH + H$^+$ is favored. (Biosynthesis of macromolecules is not only essential to growth, but is also an excellent method of storing energy.) Catabolic and biosynthetic reactions are therefore coupled not only by ATP (an energy carrier), but by electron carriers as well.

Monomers are manufactured by modifying intermediate compounds formed during catabolism (Fig. 11-11). These intermediates provide the cell

FIGURE 11-11

Biosynthesis of macromolecules from the intermediates of glucose catabolism. Many of these pathways are reversed to degrade macromolecules back to their precursors. Depending on the cell's needs, one type of macromolecule may be degraded to provide either energy or the raw materials to produce another type of polymer in greater demand at that moment.

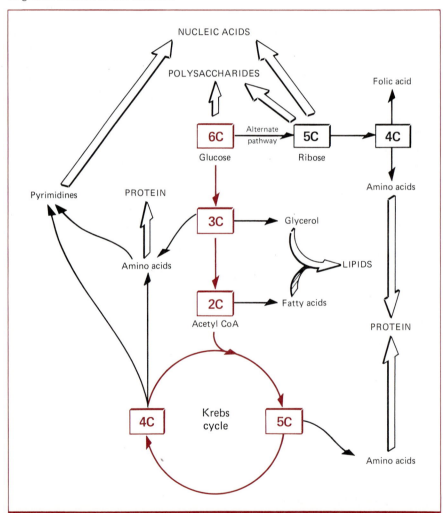

with a method of directing metabolic compounds to the pathways that best satisfy the cell's changing needs. For example, when ample glucose (and therefore energy) is available to the cell, the cytoplasm is rich in all the intermediates of glucose catabolism, such as pyruvic acid, acetyl CoA, and the Krebs cycle compounds. These catabolic intermediates are redirected to the biosynthesis of protein, lipid, polysaccharide, or nucleic acid.

Macromolecules are often broken down by reversing their biosynthetic pathways. Virtually any type of molecule can be oxidized to a form that can enter the Krebs cycle and be used for the production of ATP. Protein, for example, is metabolized to individual amino acids, many of which are then oxidized to acetyl CoA, ketoglutaric acid, or succinic acid. Each of these can then be directed into the Krebs cycle for energy production. Conversely, when energy is abundant, these intermediates can be converted to amino acids for protein synthesis.

Lipid Metabolism

Energy is most efficiently stored in the chemical bonds of lipids. The fatty acids in lipids are constructed by sequentially attaching the two-carbon acetyl group of acetyl CoA to a growing chain. Fatty acids are attached to glycerol (also an intermediate of glycolysis) to produce fats and oils. A cell is rich in lipid precursors when high concentrations of glucose are available (Fig. 11-11). When glucose availability exceeds the energy and carbon demands of the cell, some cells store the surplus by diverting glycerol and acetyl CoA into lipid synthesis. The cell can later tap this stored energy by degrading lipid to glycerol and acetyl CoA, which may be completely oxidized to CO_2 and water (or redirected to other biosynthetic pathways).

Polysaccharide (Carbohydrate) Metabolism

Sugar polymers may be synthesized as a means of energy storage (starch and glycogen are examples) or to form essential structures such as the polysaccharide backbone of peptidoglycan in procaryotic cell walls. As with lipids, several polysaccharides are more abundant in some cells during time of energy surplus.

Protein Metabolism

The amino acids that make up protein are often produced by adding an amino group (NH_2) to an organic precursor molecule. This process is called *amination*. Pyruvic acid and acetyl CoA and some of the intermediates of the Krebs cycle may be aminated to form different amino acids. Proteins, therefore, may be synthesized from the intermediates of glycolysis, pyruvic acid decarboxylation, and the Krebs cycle. In addition, amino acids can be oxidized back to these metabolic intermediates by *deamination*, the removal of NH_2. In this way cells can completely oxidize proteins and amino acids to CO_2 and water whenever ATP and electrons are needed.

Nucleic Acid Metabolism

The five-carbon sugars in nucleotides are derived by removing a carbon atom from glucose by an alternative metabolic pathway. The purines (adenine and guanine) and pyrimidines (thymine, cytosine, and uracil) are synthesized by modifying and joining together catabolic intermediates to form the ring-shaped molecules characteristic of these bases. The bases are coupled to sugar and phosphate to form the nucleotide. Nucleic acid synthesis occurs by coupling the nucleotides in a linear arrangement. As with protein, the monomers of nucleic acids are assembled in a precise predetermined sequence. Catabolism of nucleic acids yields by-products that may be completely oxidized to CO_2 and water.

REGULATION OF METABOLISM

Cells adapt to constantly changing needs by diverting metabolic intermediates from one pathway to another. How can a cell evaluate its needs and correctly "decide" whether to utilize glucose and its catabolic intermediates for energy production or for biosynthesis? It does so by regulating the concentration and activity of the enzymes that catalyze metabolic reactions. The activity of a pathway may be enhanced by producing more of the enzymes that catalyze key metabolic reactions. Conversely, a pathway may be turned off by inhibiting enzyme activity or synthesis. In such cases, enzyme activity and enzyme production are controlled by the concentration of the substrates or end products of a reaction, since these concentrations represent the cell's need for the corresponding set of reactions. Such enzyme regulation is accomplished by several strategies: feedback inhibition, induction, and repression.

Feedback Inhibition

One way cells regulate a specific metabolic pathway is by temporarily inactivating an enzyme in the pathway when the end product is plentiful. High concentrations of end product, therefore, inhibit the pathway. This process is called **feedback inhibition** since the enzyme's activity is negatively affected by the end product. As the following hypothetical pathways illustrate, the substrate (A) is used in the pathway that has the smallest amount of end product (assuming the cell requires equal amounts of D and Z).

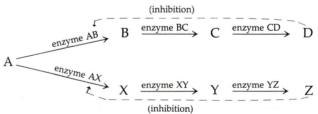

In high concentrations, one of the end products (either D or Z) binds to the first enzyme unique to its biosynthetic pathway, inhibiting its activity. Substrate A would therefore be directed into the other noninhibited pathway, allowing it to be used for synthesis of the less abundant compound. If both end products (D and Z) are abundant, both pathways are shut down, thereby conserving energy. Many biosynthetic reactions are regulated in this fashion. The compound that inhibits the pathway is called the **effector**.

Enzymes that are sensitive to feedback inhibition are **allosteric enzymes**. These enzymes possess an additional site (called an allosteric site) to which the effector molecule in the pathway can bind. Binding of this molecule to the allosteric site changes the configuration of the active site so it no longer reacts with the substrate. In this way, the effector temporarily inactivates the enzyme (Fig. 11-12). The probability of an effector binding to the allosteric site increases as the concentration of the effector increases. Thus, the greater the concentration of the end product of a pathway, the more the pathway will be inhibited. The inhibition is reversible, and the enzyme's activity is restored when the end product becomes scarce and dissociates from the allosteric site. The biosynthetic pathway operates only when its product(s) is needed.

Control of Enzyme Synthesis
Allosteric inactivation of enzymes provides immediate control but wastes energy because it provides no mechanism for halting the synthesis of

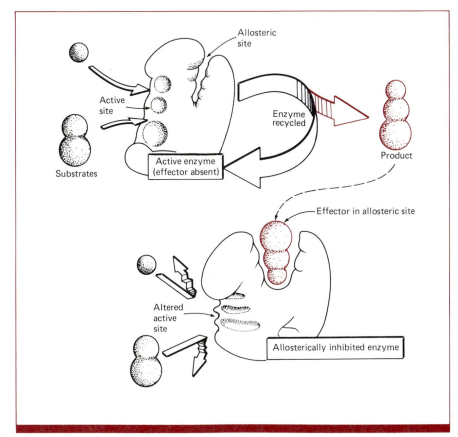

FIGURE 11-12
Inhibition of an allosteric enzyme by an effector (in this case, the end product of the metabolic pathway). Inhibition is dose-dependent—high concentrations of effector increase the likelihood of inhibition.

unneeded enzymes. In many metabolic pathways, however, elevated concentrations of the pathway's product block enzyme synthesis. Such a system is referred to as **repressible** since enzyme synthesis is repressed by high product concentrations. Many catabolic systems, on the other hand, are **inducible**. Enzyme synthesis occurs only when the corresponding substrate is available and increases according to the concentration of the substrate. *Escherichia coli*, for example, does not produce the enzymes necessary for utilizing lactose when this sugar is unavailable. The presence of lactose, however, induces the synthesis of enzymes that catabolize lactose; induction is sustained until the supply of lactose is exhausted. Thus, many organisms with the genetic potential for utilizing several sugars will only produce enzymes for metabolizing those that are available to the cell. The mechanism by which enzyme synthesis is regulated is discussed in Chapter 12.

OVERVIEW

The capacity for life is dependent on a cell's ability to extract energy and building materials from the environment and convert them into usable forms. Energy is metabolically transformed into chemical-bond energy of ATP or used for the construction of other organic molecules. All the events in the cell depend on such chemical changes, the sum total of which is called metabolism.

Photosynthetic organisms acquire their energy by photophosphorylation, transforming light energy into chemical energy in ATP. Chemotrophic organisms obtain energy by oxidizing the chemical bonds of food molecules, either by fermentation or by respiration. The electron transport system of respiratory organisms provides a means to use the energy in the electrons to form ATP, greatly increasing the efficiency of the catabolic processes. The electrons are carried to the electron transport system by the coenzymes NAD^+ and FAD.

The intermediates of catabolism provide the raw materials for biosynthetic processes necessary for the continued existence and growth of the organism. Most of the monomers of protein, lipid, polysaccharide, and nucleic acid are manufactured from intermediate compounds generated from glycolysis, pyruvic acid decarboxylation, or the Krebs cycle. The intermediates may be diverted away from catabolism toward the synthesis of needed compounds.

In this way organisms are able to respond to changes in both their chemical environment and internal metabolic needs. Therefore, metabolism is carefully controlled by a cell so that necessary reactions are turned on and unnecessary reactions are inhibited. This is accomplished by allosteric inhibition of enzyme activity and by the processes of induction and repression of enzyme synthesis. These mechanisms for metabolic control are usually sensitive to the changing concentrations of the substrate or product of the pathway being regulated.

KEY WORDS

biosynthesis

metabolic pathway

catabolism

anabolism

oxidation

reduction

adenosine triphosphate (ATP)

phosphorylation

NAD

NADP

FAD

NADH + H$^+$

NADPH + H$^+$

FADH$_2$

photosynthesis

glycolysis

pyruvic acid

fermentation

respiration

electron transport system

cytochrome

Krebs cycle

tricarboxylic acid (TCA) cycle

Pasteur effect

anaerobic respiration

feedback inhibition

effector

allosteric enzyme

repressible system

inducible system

REVIEW QUESTIONS

1. Why are metabolic pathways more advantageous to a cell than a single complete oxidation reaction?

2. Explain how ATP and electron-carrying coenzymes couple oxidative and biosynthetic reactions.

3. Describe the production of ATP and reducing power by photo-trophs.

4. Differentiate between fermentation, aerobic respiration, and anaerobic respiration.

5. How are fermentation by-products used as diagnostic tools in the identification of microorganisms?

6. Explain how aerobic organisms obtain eight ATP molecules from glucose catabolism whereas fermentative organisms produce only two ATP molecules.

7. Why must cytochromes be bound to membranes in order to function correctly?

8. Why may an organism get fat when it consumes large quantities of sugar?

9. Differentiate between feedback inhibition, enzyme induction, and enzyme repression.

10. How does an effector control the activity of an allosteric enzyme?

11. Explain why it is incorrect to write $NADH + H^+$ as $NADH_2$.

Microbial Genetics

12

Throughout history, people have wondered, "What is the secret to life?" From a biological point of view this question has been answered, at least partially, within the past 30 years. The amazing "secret" was discovered by studying the genetics of microorganisms.

The term **genetics** is derived from the word "generation" and means the study of heredity and variation among generations of organisms. An organism's characteristics are acquired from its parents by means of the fundamental units of heredity, the **genes**. These genes are linearly arranged along strands of genetic material, the chromosomes, which possess all the information necessary for reproducing offspring that, in procaryotic organisms, are exact copies of the parental organism.

Bacteria contribute enormously to our understanding of the genetics of all organisms. Many of the mechanisms discovered in relatively simple bacterial systems are very similar to corresponding mechanisms in humans. Bacteria are useful scientific models for studying genetics for several reasons:

They can be propagated so rapidly that many generations can be studied in short periods of time.

Large populations of essentially identical bacteria can be cultured from a single parental cell (essential for genetic homogeneity).

Compared to eucaryotes, bacteria are genetically simple organisms. *Escherichia coli*, for example, possesses a single chromosome that contains about 5000 genes. Human cells, on the other hand, with their 46 chromosomes and 1 million genes, are much more complex and difficult to characterize genetically.

Genetic material is readily transferred from one bacterial cell to another, so we can experimentally investigate the mechanisms of gene function.

Bacteria require much less laboratory space than plants and animals.

GENOTYPE AND PHENOTYPE

The entire complement of genes possessed by an organism is called its **genotype**, the organism's genetic potential. Not all the genes, however, are expressed as observable characteristics at any one time. Many genes for producing enzymes necessary for digestion and utilization of nutrients, for example, are "turned on" only in the presence of these nutrients. The characteristics expressed at any given time compose the organism's **phenotype**.

Temperature, pH, age, and humidity are additional influences that may dictate which genes are expressed and which are turned off. For example, *Serratia marcescens* forms red colonies at 24°C and white colonies at 37°C (see color plate 16). The genotype is the same in both cultures, but the higher temperature inhibits pigment production. Thus, the phenotype has changed dramatically. Such cultural characteristics are important in laboratory identification of *Serratia marcescens,* an important opportunistic pathogen responsible for many nosocomial (hospital-acquired) infections (see Chap. 24).

In contrast to alterations in phenotypes, changes in genotype are infrequent and are called **mutations**. The new genotype is stable and is inherited by all the descendants of the *mutant* (the organism containing the mutation).

DNA—THE GENETIC MATERIAL

Except for RNA viruses, the genetic material of all genes and chromosomes is deoxyribonucleic acid (DNA) (see Nucleic Acids, Chap. 10). DNA stores specific genetic information that ultimately determines all the characteristics of an organism. Biological differences between organisms are due primarily to differences in the information encoded in their chromosomal DNA.

The biological tasks of DNA are threefold:

1. **Storage of Genetic Information** DNA is the cell's blueprint and contains all the information needed to build and maintain the unique organism.

2. **Inheritance** Genetic information is precisely transmitted to all the organism's descendants.

3. **Expression of the Genetic Message** Information stored in DNA is translated into protein molecules that direct cellular activities and determine cellular characteristics.

The manner in which genetic information is stored, inherited, and expressed is basically the same in all organisms. The scientific detective work involved in explaining how these three tasks are accomplished is an exciting chapter in the quest to discover life's secrets.

Task 1—Information Storage

The unique structure of DNA accounts for its ability to store genetic information. DNA is a linear molecule made up of repeating units of the four deoxyribonucleotides that contain the bases adenine (A), thymine (T), guanine (G), and cytosine (C) (Fig. 12-1; see also Chap. 10). In the double-stranded molecule, each adenine pairs with thymine and each guanine pairs with cytosine. The two strands, therefore, are *complementary* —the nucleotide sequence in one strand will by reflected by a complementary sequence in the opposite strand.

The Genetic Code The linear sequence of nucleotide bases can be arranged in an infinite number of orders. This sequence carries a message written in **genetic code**. Just as written information is stored and transmitted by virtue of the linear sequence of letters to form meaningful words and

FIGURE 12-1

The four nucleotides in deoxyribonucleic acid (DNA) are chemically bonded in a linear sequence; only a single strand of the normally double-stranded molecule is shown here.

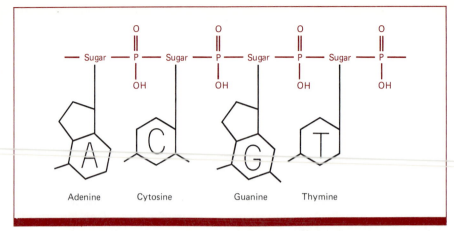

Adenine   Cytosine   Guanine   Thymine

sentences, the genetic information necessary for determining every characteristic of an organism is encoded in the order in which the nucleotides are arranged along the DNA. In the code we call written language, whenever the letters of a word are changed, the meaning of the word itself changes. Similarly, the meaning of the genetic message changes if the sequence of "letters" (nucleotides) is changed. *Mutation*, therefore, may be defined more specifically as a permanent change in the nucleotide sequence in DNA. This may change the meaning of the stored information, thereby changing the characteristic of the organism that was specified by the DNA "word" (gene). Different nucleotide sequences result in organisms with different characteristics, just as two sentences make completely different statements by virtue of their unique sequence of letters.

Task 2—Information Transmission to Offspring (Heredity)

A cell ensures that all progeny cells get the necessary genetic information by producing exact copies of its chromosomal DNA and giving one copy to each of the daughter cells. This duplication process begins with the enzyme-catalyzed separation of the two complementary strands of the DNA molecule (Fig. 12-2). As the strands separate, segments of unpaired, single-stranded DNA are produced. All cells maintain a reservoir of free nucleotides that are not tied up in DNA. In the presence of polymerizing enzymes, these free nucleotides combine with complementary unpaired nucleotides in a DNA molecule. Nearly as rapidly as the DNA strands separate, these free nucleotides associate and bind with the unpaired regions, building a new complementary strand on each side. Because of the specificity in base pairing, two identical DNA molecules are produced, each containing one strand from the parent molecule and one newly synthesized complementary strand. Each progeny cell acquires one of these DNA double strands, which contains precisely the same genetic message that was stored in the original DNA molecule. This is the mechanism of heredity.

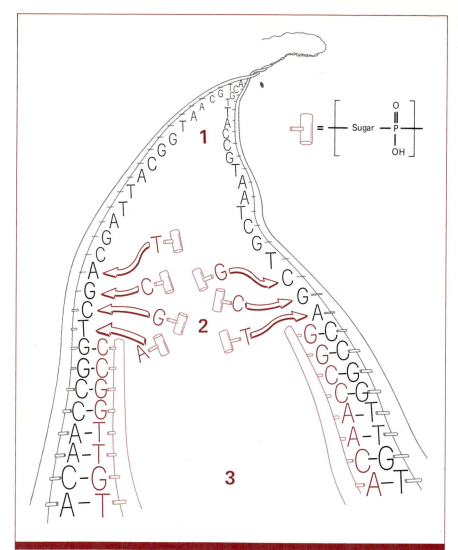

FIGURE 12-2
Mechanism of chromosome
duplication by DNA
replication. The newly
synthesized strand is
shown in color. (1) Strands
separate; (2) free
nucleotides are
enzymatically assembled
along complementary
nucleotides of the
single-stranded sections;
and (3) the two finished
strands are identical in
sequence to the original
DNA molecule.

261

DNA—THE GENETIC
MATERIAL

Task 3—Translation of Genetic Information into Expressed Cellular Characteristics

Cells read their genetic messages and follow the specific commands encoded in the DNA. Each gene on a chromosome is a "written" message that directs the formation of a single protein. This is called the "one gene–one protein" explanation of genetic control. Thus the types of genes present in the DNA determine the types of structural proteins and enzymes made by the cell. The enzymes, in turn, catalyze the cell's metabolic reactions, building a unique organism with characteristic properties. *The ultimate control DNA exerts over the cell is in dictating which enzymes and structural proteins are synthesized by the cell.*

The sequence of nucleotides in the DNA of a gene determines the function of a protein by determining its amino acid content and sequence. A change in the order of nucleotides in a gene may change the order of amino acids in the corresponding protein. Such changes may alter or destroy the activity of the protein, thereby changing the cellular characteristic produced by that protein. •

Protein Synthesis How does the nucleotide sequence of a gene determine the amino acid sequence of a protein? Proteins are synthesized at "workbenches" called ribosomes, distributed throughout the cytoplasm. The genetic message, therefore, must get from the DNA to the ribosomes to direct the synthesis of proteins. Carrying the message to the ribosomes is the role of **messenger RNA (mRNA)**.

Ribonucleic acid (RNA) is very similar to DNA in that it is a linear polymer of nucleotide sequences. The primary differences between DNA and RNA are (1) the nucleotide sugar of RNA is ribose; (2) unlike DNA, RNA usually forms single strands; and (3) in RNA uracil replaces thymine. Uracil pairs with adenine exactly as thymine does.

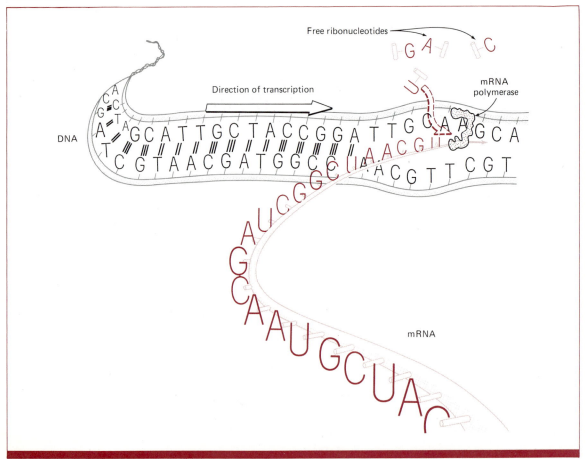

The first step in transmitting the genetic message from DNA to the ribosome is to encode the precise information into a molecule of mRNA. Cells do this by assembling a linear molecule of mRNA along a temporarily single-stranded portion of the DNA molecule (Fig. 12-3). The process resembles DNA replication, except that free RNA nucleotides are specifically base-paired with the corresponding DNA nucleotides. A chain of mRNA is synthesized, using one DNA strand as a template to determine its nucleotide sequence. The new mRNA is complementary to the DNA strand along which it was synthesized; therefore, the mRNA contains the same genetic information as DNA by virtue of its nucleotide sequence, which was determined by the sequence in DNA. The process of synthesizing these mRNA molecules is called **transcription** because DNA's genetic message is transcribed (copied) from the DNA template into a molecule of mRNA. Transcription requires free RNA nucleotides, a DNA template, and specific enzymes such as **RNA polymerase**, which catalyzes the assembly

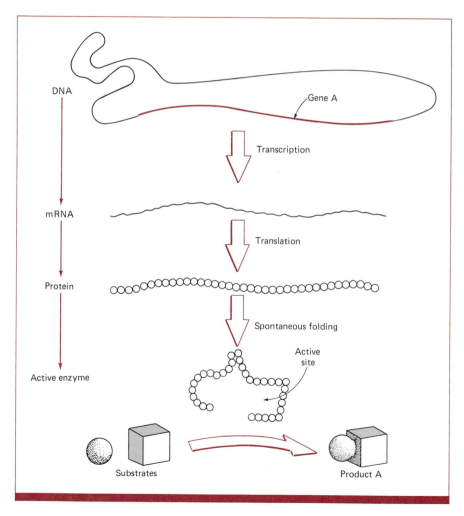

FIGURE 12-4

Flow of genetic information: The nucleotide sequence of gene A is encoded into a molecule of mRNA which directs the synthesis of a specific protein. The protein gains its unique activity by folding into its functional three-dimensional shape. In this case the protein is an enzyme that catalyzes the production of a particular characteristic (called product A here). Notice that DNA, mRNA, and protein are all linear polymers.

of ribonucleotide subunits into a polymer. This enzyme is also called transcriptase.

Just as you may transcribe mental information into a written message and send it to a person who will translate your instructions into your intended action, the RNA polymerase transcribes genetic information from DNA into a message (mRNA) which is sent to the ribosomes for "translation" into a particular protein (Fig. 12-4).

TRANSLATING THE MESSAGE The genetic message encoded in mRNA is translated by reading the code in groups of three nucleotides (triplets). These triplet sequences in mRNA are called **codons**. Each codon specifies a particular amino acid to be inserted in the protein chain. For example, anytime the codon CUC appears in mRNA, the amino acid leucine will be inserted at that point in the protein as it is synthesized. The RNA triplet UAU always specifies the amino acid tyrosine. CGG always directs the insertion of the amino acid arginine, and so on until the entire message is read and the protein is completed. Since the 20 common amino acids may be arranged in a virtually infinite number of orders, the types of proteins and enzymes possible are unlimited. The manner in which the nucleotide sequence in DNA ultimately determines the amino acid order of protein is illustrated in Figure 12-5.

The process of protein synthesis is called **translation** and is a complicated process requiring several enzymes, amino acids, RNAs, ribosomes, and high-energy transfer compounds such as ATP. Before an amino acid can be inserted into the protein, it must be attached to **transfer RNA (tRNA)**. These are molecules that carry amino acids to the surface of ribosomes. Transfer RNAs also provide the means by which each amino

FIGURE 12-5

The sequence of "triplets" in DNA determines the sequence of codons in RNA and, therefore, the sequence of amino acids in the enzyme.

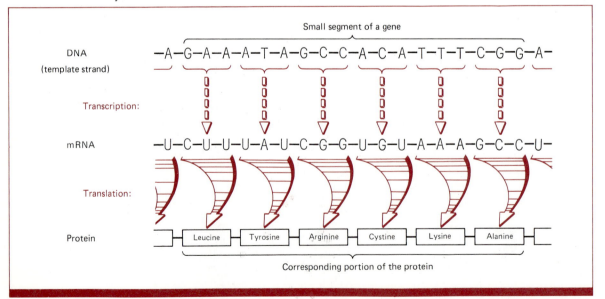

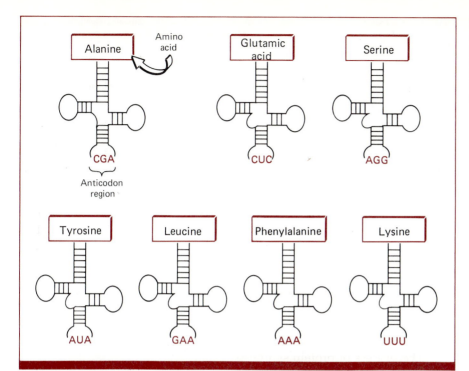

FIGURE 12-6

A few typical transfer RNA (tRNA) molecules, each carrying the correct amino acid that corresponds to the appropriate triplet sequence in mRNA.

acid can recognize its corresponding codon on the mRNA, so it may be inserted into the protein whenever the appropriate codon occurs. The cell maintains a pool of different tRNAs specific for each of the different amino acids. Each tRNA has a unique region of three nucleotides, called an **anticodon**, which is capable of binding to the complementary base triplet (codon) on the mRNA. In Figure 12-6 the tRNA with the anticodon CUC carries only the amino acid glutamic acid and recognizes only one codon (GAG) on mRNA. Thus, the codon GAG always specifies the insertion of glutamic acid into the growing protein chain. Each tRNA, therefore, recognizes one triplet on mRNA and inserts the correct amino acid when that triplet appears in the mRNA sequence.

Translating the genetic code is made possible by the fact that each of the 20 amino acids has at least one unique codon (Table 12-1). Since the four nucleotides can be arranged in 64 possible triplet sequences, most amino acids have more than one codon, any one of which will specify its addition to the growing protein chain. There are three codons (UGA, UAG, and UAA), however, for which no amino acid is specified. These codons are known as *nonsense codons* and are used by cells to terminate the synthesis of a complete protein.

A simplified version of the events of translation is represented in Figure 12-7. (Many of the enzymes and cofactors have been omitted.) Each step in the figure is numbered to correspond to the following description:

Step 1 The mRNA attaches to the ribosome with the first codon (AUG) in

TABLE 12-1
The Genetic Code

Amino Acid	Codons Specific For Each Amino Acid
Alanine	GCU, GCC, GCA, GCG
Asparagine	AAU, AAC
Aspartic acid	GAU, GAC
Arginine	CGU, CGC, CGG, CGA, AGA, AGG
Cysteine	UGU, UGC
Glutamic acid	GAA, GAG
Glutamine	CAA, CAG
Glycine	GGU, GGC, GGA, GGG
Histidine	CAU, CAC
Isoleucine	AUU, AUC, AUA
Leucine	UUA, UUG, CUU, CUC, CUA, CUG
Lysine	AAA, AAG
Methionine	AUG
Phenylaline	UUU, UUC
Proline	CCU, CCC, CCA, CCG
Serine	UCU, UCC, UCA, UCG, AGU, AGC
Threonine	ACU, ACC, ACA, ACG
Tryptophan	UGG
Tryosine	UAU, UAC
Valine	GUU, GUC, GUA, GUG
No corresponding amino acid	UGA, UAG, UAA (the nonsense codons)

position to accept the tRNA displaying the complementary anticodon (UAC). The mRNA shown is only a portion of the entire molecule, which would likely be at least 600 nucleotides in length. Notice the free tRNAs available in the cytoplasm. Each tRNA is carrying its specific amino acid.

Step 2   The tRNA with anticodon UAC is base-paired with the first codon (AUG) on the mRNA. The amino acid on the opposite end of the tRNA is aligned on the ribosome. The first codon on mRNA is usually AUG, the *initiator codon*. This triplet assures that translation begins with the correct nucleotide. If the message were initiated at the second nucleotide, the remaining triplets would be incorrectly read. UGC would replace AUG as the first codon, the remaining triplets would be read out of synchrony, and the resulting protein would be a useless product of biological gibberish. The synchrony in which the triplets are read is called the *reading frame*.

Step 3   The second tRNA (in this case, the anticodon is GAG, the amino acid is leucine) base-pairs with the CUC codon. The amino acids line up next to each other on the ribosome.

Step 4   The two amino acids are now enzymatically joined together, forming a peptide bond (see Chap. 10). The first amino acid (methionine) dissociates from its tRNA. The growing protein is now two amino acids long.

Step 5   The entire tRNA, mRNA, and amino acid complex is shifted over three nucleotides, so a new mRNA codon is brought into the location for binding its specific tRNA. The first tRNA is released from the ribosome complex.

Step 6   This time the codon is ACU, which binds with the anticodon UGA

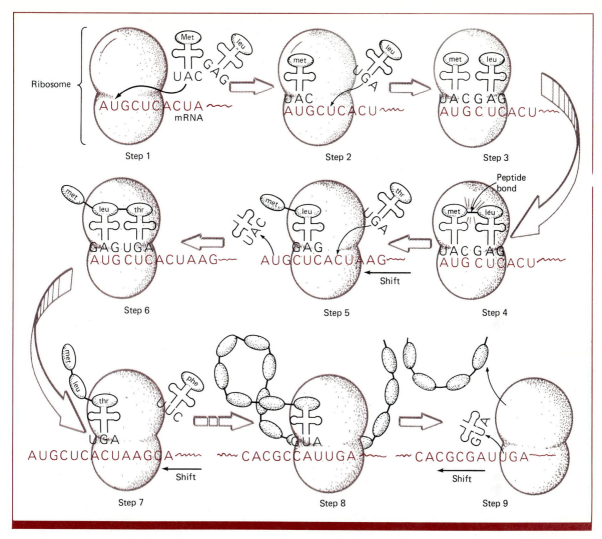

FIGURE 12-7

Simplified representation of translation as it occurs in the cytoplasm. (See text for description of steps.)

of the tRNA carrying the amino acid threonine. Another peptide bond is formed, and the growing protein acquires another amino acid.

Step 7 Once again, as in step 5, the complex shifts to accommodate another tRNA. The process continues in this fashion, the chain of amino acids growing in the proper sequence until the entire protein is completely synthesized.

Step 8 The process nears its completion with *protein termination.* There are three protein-chain-terminating codons that specify no amino acid. These are the nonsense codons UGA, UAA, and UAG. Since no tRNA binds with a nonsense codon, no amino acid is available for peptide bond formation. The amino acid inserted just before the nonsense codon thus becomes the terminal amino acid on the growing protein chain.

Step 9 The final step in the process is the release of the completed protein. When the entire complex is shifted, the finished protein is released from the ribosome.

The specific sequence of amino acids composing the protein polymer determines the protein's characteristic three-dimensional shape. This shape is essential to the function of the protein.

Protein synthesis occurs very rapidly in cells. Ribosomes attach to mRNA and begin translation while the mRNA is still being transcribed from the DNA. Soon after a ribosome has translated the initial codons of the mRNA, the initiation site is available for the attachment of another ribosome. Thus, the efficiency of protein synthesis is enhanced because many ribosomes can translate a single message at the same time, generating a chain of ribosomes held together by mRNA. This complex is called a **polysome**. Since each ribosome on the polysome is synthesizing its own protein strand, a single molecule of mRNA may be simultaneously generating hundreds of identical proteins (Fig. 12-8).

We have seen how the genetic information stored in the specific nucleotide sequence of a chromosome determines the characteristics of an organism by directing the synthesis of proteins which have unique amino acid sequences. The order of the amino acids ultimately determines the enzymatic activity of a protein, which in turn directs specific metabolic reactions that build the organism. The organism inherits this DNA

FIGURE 12-8

Electron micrograph of protein synthesis in progress. The darkly stained ribosomes attach to the mRNA as soon as transcription begins, accounting for the different lengths of polysomes in this photograph. Although not detectable in the photograph, a protein chain is growing from each ribosome on the complex.

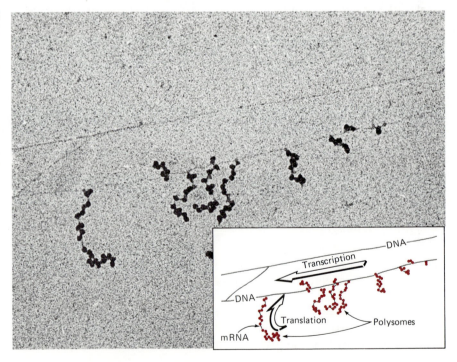

REVERSING THE FLOW OF GENETIC INFORMATION

■ The central dogma of molecular biology states that genetic information flows from DNA to RNA to protein by the processes of transcription and translation. Some activities of the oncogenic RNA viruses, however, have challenged the inflexibility of this rule. Chromosomes of cells infected with these viruses were found to have DNA sequences complementary to the single-stranded RNA genome of the virus. In other words, a double-stranded DNA copy of the viral genome apparently integrates into the host cell's chromosome. This led to speculation that there may be another kind of polymerase, one that uses virus RNA as a template to create a complementary strand of DNA. The search for the new enzyme culminated in 1970 with the discovery of **reverse transcriptase** (RNA-dependent DNA polymerase) in the virions of these oncogenic viruses. During replication of the retroviruses (*retro*, backward), as these oncogenic RNA viruses are now called, the enzyme promotes the assembly of a single-stranded DNA molecule along the RNA viral genome. A second strand of DNA is then synthesized, this one complementary to the first. The result is a double-stranded DNA molecule that may integrate into the host chromosome.

This unconventional flow of genetic information, however, did not convince many biologists that a similar reversal of translation would ever be discovered. Reversing transcription from one type of nucleic acid to another was understandable; converting amino acid sequences to nucleotide sequences was quite another matter. Translation is much more complicated than transcription. The discovery of a new type of infectious agent, however, has some investigators believing that reverse translation may soon become biological fact. These agents, called **prions**, appear to be composed entirely of protein. These particles apparently replicate and show genetic inheritance in spite of their lack of nucleic acid. Some scientists suggest that an RNA intermediate is synthesized from the protein, its nucleotide sequence dictated by the amino acids in the protein. This strand of nucleic acid is then conventionally translated into hundreds of infectious prions. (Most investigators currently believe it is more likely that prions may contain undiscovered nucleic acid or that the prion is a product of a normally inactive host gene that is induced by exposure to the prion. Thus the prion may be a self-induced protein.)

If the prion is really serving as a template for RNA production, it may be possible that cellular proteins can actually change a cell's stable genetic configuration through a combination of reverse translation and reverse transcription. Such a finding would test the adaptability of our contemporary view of genetics and the molecular basis of life.

sequence from parental cells and transmits the same genetic information to its progeny. All the life processes are ultimately determined by the sequence of nucleotides in DNA.

REGULATION OF GENE EXPRESSION

The expression of some genes cannot be regulated; these genes are constantly transcribed and translated throughout the life of the cell. Such nonregulated genetic function is referred to as **constitutive enzyme synthesis**. At any given time, however, a cell will rarely express its entire genetic potential.

Resources are conserved by turning off genes for unneeded functions. Many catabolic enzymes, for example, are not synthesized unless the substrate is available. The presence of the substrate turns on transcription of the appropriate gene, the enzyme is produced, and the substrate is utilized. In the absence of the substrate, transcription is turned off. Residual mRNA deteriorates within 2 minutes; after that the unneeded enzym s are no longer synthesized. This is control of gene expression by the process of enzyme (or gene) **induction**.

Inducible genes are turned on by the *inducer*, which is usually the substrate of the corresponding catabolic pathway. Enzymes of biosynthetic pathways, on the other hand, are produced only when the end product of the pathway is needed. When this product is abundant, production of the biosynthetic enzymes is turned off by the process of enzyme (or gene) **repression**. The product of the biosynthetic pathway must be present for repression of enzyme synthesis.

Induction and repression are methods of regulating enzyme synthesis by controlling transcription of mRNA from the regulated genes in response to varying concentrations of substrates and products. The mechanism that provides a bacterial cell with this ability is called the operon.

The **operon** is a segment on the bacterial chromosome that consists of the following elements (depicted in Fig. 12-9):

FIGURE 12-9
The clustered genetic elements of the classical bacterial operon. Genes 1, 2, and 3 synthesize enzymes in a single metabolic pathway. The promoter and operator segments are located on the end of the operon where transcription is initiated. Transcription of the adjacent genes is from left to right. The R gene (regulator gene) produces a repressor protein that can bind to the operator region and turn off transcription.

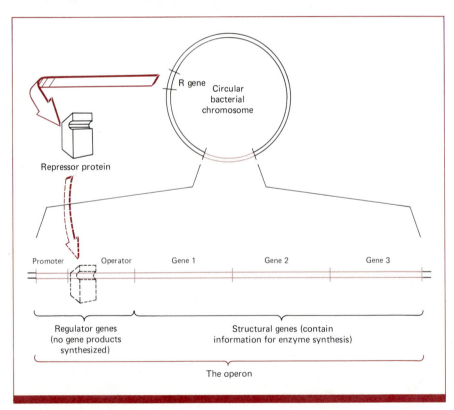

Repressor protein

R gene

Circular bacterial chromosome

Promoter Operator Gene 1 Gene 2 Gene 3

Regulator genes (no gene products synthesized)

Structural genes (contain information for enzyme synthesis)

The operon

1. Genes for producing the enzymes of a single metabolic pathway. The genes are physically adjacent to one another and are regulated as a unit.

2. A regulator region consisting of a promoter and an operator. The **promoter** is the site where mRNA polymerase (transcriptase) binds to the DNA and initiates transcription of the genes for enzyme production. The **operator** region lies adjacent to the promoter and is the binding site for a specific *repressor protein*. When the operator site is occupied by repressor, mRNA polymerase cannot bind to the promoter on mRNA, and transcription, therefore, is blocked. This turns off the synthesis of those enzymes translated from this mRNA.

3. A **repressor gene** (R gene) located in another portion of the bacterial chromosome. This gene codes for the production of the repressor protein that attaches to the operator region. The repressor protein has the additional ability to specifically bind with a small effector molecule that determines whether an operon will be turned on or turned off.

Inducible Operons
For many catabolic pathways, the substrate must be present before the pathway's enzymes are produced. When present, the substrate induces the transcription of the operon by inactivating the repressor and preventing it from binding to the operator (Fig. 12-10). The operon is repressed when no substrate (inducer) is available to prevent the repressor from binding to the operator. In this state, transcription, and therefore enzyme synthesis, is blocked.

Repressible Operons
Repressible operons differ from inducible operons in several ways:

1. They produce and regulate the enzymes of biosynthetic pathways instead of catabolic ones.

2. The effector is the end product of the pathway.

3. The repressor protein is inactive in the *absence* of the effector. The effector, therefore, functions as a **co**repressor that must be present before the operon can be turned off.

The enzymes of repressible pathways are thus produced whenever the pathway's final product (P), which is also the corepressor, is in low concentration (and therefore needed by the cell) (Fig. 12-11a). Under such conditions, the repressor protein cannot bind to the operator and the operon is said to be derepressed. When the product is in ample supply, it activates the repressor, which binds to the operator and blocks transcription of the genes. This is a state of enzyme (or gene) repression (Fig. 12-11b).

The properties of inducible and repressible operons are compared in Table 12-2.

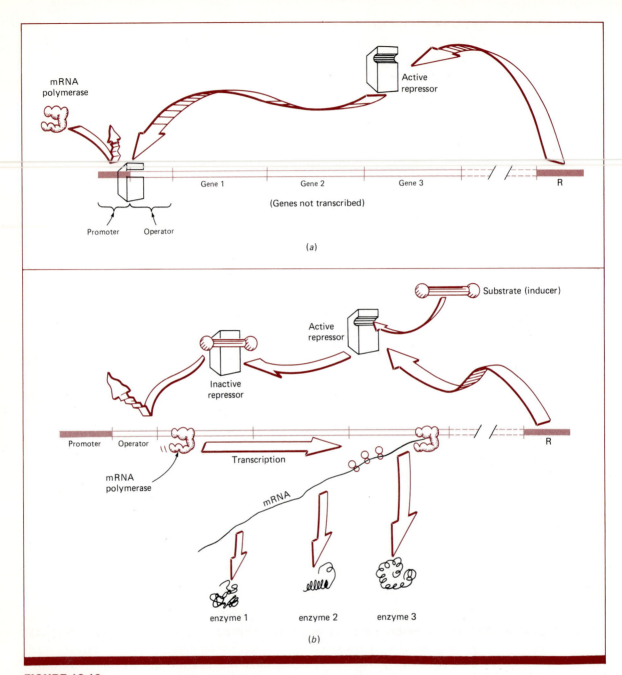

FIGURE 12-10

Regulation of an inducible operon: (*a*) In the absence of the inducer (here, the substrate S) the repressor protein binds to the operator and blocks transcription of genes by preventing binding of MRNA polymerase. The operon is said to be repressed. (*b*) When the inducer is present, it binds to the repressor and prevents its attachment to operator. Transcription of genes occurs unimpeded. The operon is induced, enzymes are produced, and the substrate is utilized.

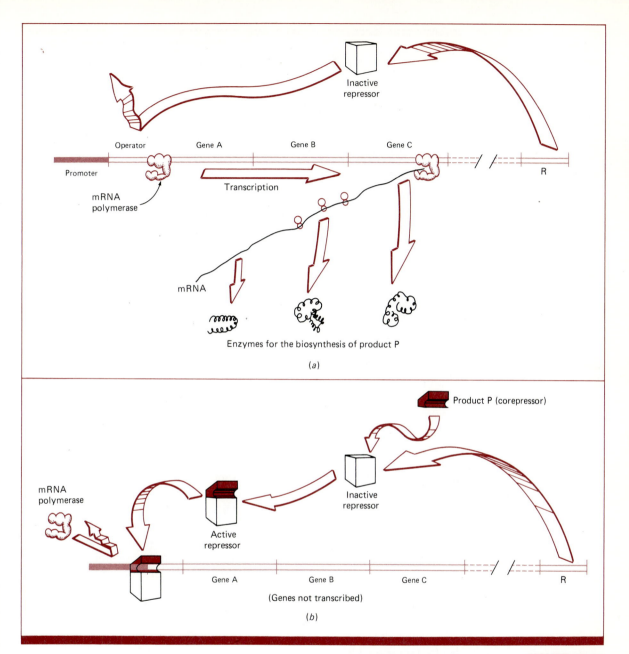

FIGURE 12-11

Regulation of a repressible operon: (*a*) In the absence of the corepressor (the product P) the inactive repressor protein fails to bind to the operator. Transcription proceeds, and the operon is derepressed. (*b*) When the corepressor is present, it binds to and activates the repressor protein. The repressor-corepressor complex attaches to the operon and blocks transcription. The operon is repressed.

	Inducible	Repressible
Type of pathway regulated	Catabolic	Anabolic (biosynthetic)
Type of effector	Inducer	Corepressor
Relationship of effector to pathway	Substrate	End product
In absence of effector operon is	Repressed	Derepressed
In presence of effector operon is	Induced	Repressed
Native repressor is	Active	Inactive
Repressor-effector complex is	Inactive	Active

Control of enzyme synthesis by induction or repression supplements feedback inhibition of enzyme activity (see Feedback Inhibition, Chap. 11) to regulate cellular metabolism. Unlike feedback inhibition, the operon prevents the synthesis of unneeded enzymes, thereby conserving valuable resources.

MUTATIONS

Unlike the temporary phenotypic changes of induction and repression, mutations (alterations in genotype) are infrequent and stable. Mutations are due to changes in nucleotide sequence. These include replacement of a single nucleotide by a different one (a *point mutation*); loss or addition of one or two nucleotides, shifting the reading frame so that the message is transcribed out of synchrony (a *frame-shift mutation*); and loss of a large segment of a gene. Cells with altered genotypes are called **mutants**. The original type of organism with the nonmutant genotype is called the **wild type**.

Mutations occur in all species, from viruses to humans. They commonly result in the production of a nonfunctional gene product, an inactive protein. If the protein is critical to the survival of the organism, the mutation is lethal. Occasionally, however, a mutation increases an organism's ability to survive. An important example is the mutation that permits a microbe to survive the effects of an antibiotic to which it was formerly susceptible. Antibiotic-resistant mutants are often responsible for infectious diseases that do not respond to the usual chemotherapy.

Mutagens

The probability of a mutation occurring spontaneously in a single bacterial gene is, on the average, about 1 in 1 million. This low frequency, however, is often increased by exposing an organism to agents that affect DNA. Any physical or chemical agent that increases the rate of mutation is called a **mutagen**.

Because mutations are often lethal, some mutagens are utilized as germicidal agents. *Ultraviolet* (uv) *light*, for example, damages the DNA of microorganisms. However, because of its poor penetrating power, ultraviolet light will not pass through glass, plastic, and some liquids. Its

BACTERIAL ALLIES IN THE WAR AGAINST CANCER

■ Bacteria have been recruited to "sniff out" substances that are possible carcinogens (cancer causers). About 90 percent of the known carcinogens are also mutagens. A potentially carcinogenic substance could therefore be detected by determining its ability to cause mutations. Bacteria provide an exceptionally sensitive, rapid, and inexpensive system for detecting the mutagenic properties of such substances. By exposing bacteria to a suspected carcinogen and subsequently determining the mutation rate for a particular trait, laboratories can screen hundreds of compounds for mutagenic activity within a relatively short time. The procedure, called the **Ames test**, uses bacteria that, because of a mutation, can only grow on minimal media that have been supplemented with a particular nutrient. (Mutants that acquire additional nutritional requirements are called **auxotrophs**.)

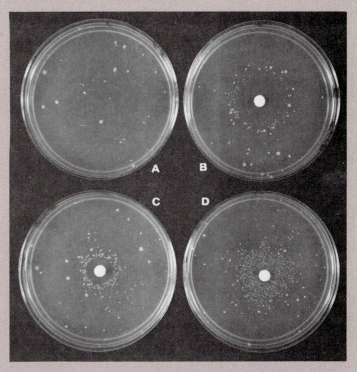

A standardized number of these test bacteria are seeded onto the surface of a solid medium that lacks the nutrient required for their growth. A mutation in the correct gene could reverse the original mutation, restoring the test bacteria's ability to synthesize the needed nutrient and proliferate on an unsupplemented minimal medium. Therefore, the number of colonies that develop represents the number of bacterial cells that have such a mutation. This is the control plate and indicates the rate of spontaneous mutation, since no agent was included that might affect the mutation rate.

The seeded test plate is identical to the control plate except for a small amount of the test substance spotted in the center of the agar. If the number of colonies that arise around the test substance is greater than the colony number on the control plate, the suspected material is shown to be a mutagen. In the photographs above, the substances in the centers of plates B, C, and D are all demonstrable mutagens, one of which is a commonly employed food additive. Plate A contains the spontaneous mutants. The Ames test has identified mutagenic potential in hair dyes, cigarette smoke, cured meats, and other products to which many persons are frequently exposed.

germicidal activity is limited to exposed surfaces or to liquids that are vigorously agitated.

Ultraviolet light with wavelengths around 260 nm also induces mutations in the superficial cells of humans. Sunlight contains mutagenic ultraviolet light, and overexposure to it is associated with an increased incidence of skin cancer. In addition, looking directly into ultraviolet light may severely damage the retina of the eye and cause blindness. Ultraviolet light with wavelengths of 280 nm, however, is neither germicidal nor harmful to humans. This "black light" merely induces fluorescence and does not cause cellular or ocular damage.

Other forms of radiation (x-rays and gamma, beta, and alpha rays) also produce mutations. In addition, people are frequently exposed to many chemical mutagens, such as those in tobacco smoke. Bacteria can be used as indicators to detect compounds with mutagenic properties (see color box).

GENETIC TRANSFER IN BACTERIA

Procaryotic organisms are capable of donating genetic information (DNA) to recipient cells, which may acquire new characteristics as a result. Some of these genetic transfers have a profound impact on the outcome of infectious disease in humans.

Genes are transferred among bacteria in one direction, from donor to recipient. In most cases, only part of the DNA is transferred. Once this DNA fragment is in the recipient, it may align with the corresponding segment on the existing chromosome. The DNA may then recombine by breakage of the host chromosome and reunion of the free ends with the newly received DNA fragment (Fig. 12-12). **Recombination** therefore stably incorporates the new genes into the recipient's chromosome. The newly integrated fragment is the same length as the DNA that is replaced. The excised fragment is destroyed by the enzyme DNase.

Although the two DNA segments are homologous—that is, they possess genes for the same functions—the recipient cell may receive a different form of a gene, thereby acquiring a new property from the donor cell. For example, a mutant, nonfunctional gene may be replaced by a homologous wild-type gene from the donor. If the altered characteristic is observable it may be used to indicate whether genetic transfer has occurred. The appearance of one of the donor's traits in a recipient cell suggests the transfer of genetic information.

The three mechanisms for gene transfer in bacteria are transformation, transduction, and conjugation.

Transformation

Destruction of a cell does not necessarily destroy its genetic material. When bacteria lyse, they often release their DNA into the surrounding medium. This DNA retains its ability to direct the synthesis of specific proteins if introduced into another viable cell. Recipient cells absorb onto their cell surface pieces of the released DNA, which are then transported across the

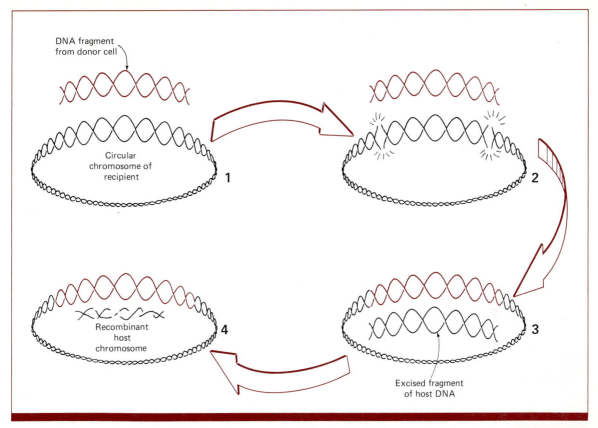

DNA fragment
from donor cell

Circular
chromosome of
recipient

1

2

Recombinant
host
chromosome

4

3

Excised fragment
of host DNA

FIGURE 12-12

Integration of newly received DNA into the existing
chromosome of the recipient cell by process of breakage
and reunion. (1) Donor fragment aligns with homologous
portion of recipient's chromosome. (2) Recipient
chromosome is cleaved by DNAase, and the strands are
broken. (3) The free ends of the recipient's chromosome
are attached to the free ends of the donor's DNA
fragment, closing the DNA circle. (4) DNAase destroys
the replaced chromosomal fragment.

cell membrane into the cytoplasm and incorporated by recombination into
the cell's chromosome. The recipient cell often acquires new characteristics
as a result. This transfer of genetic information by free, extracellular DNA
is called **transformation**.

The phenomenon of transformation was demonstrated in 1944 in the
famous experiment that proved DNA to be the genetic material (Fig. 12-13).
The experiment showed that pure DNA isolated from a pathogenic
encapsulated strain of *Streptococcus pneumoniae* (the pneumococcus) provided nonencapsulated strains of this organism with the genetic ability to
form capsules and, therefore, the ability to cause pneumonia.

Transformation may be an important natural mechanism of genetic
transfer among bacteria. There is evidence that in infected animals some

277

FIGURE 12-13

The transformation experiment that proved DNA to be the genetic material. The ability to synthesize a capsule, which is necessary for virulence, is transferred to an avirulent (nonencapsulated) strain of pneumococcus by cell-free DNA extracted from the encapsulated strain.

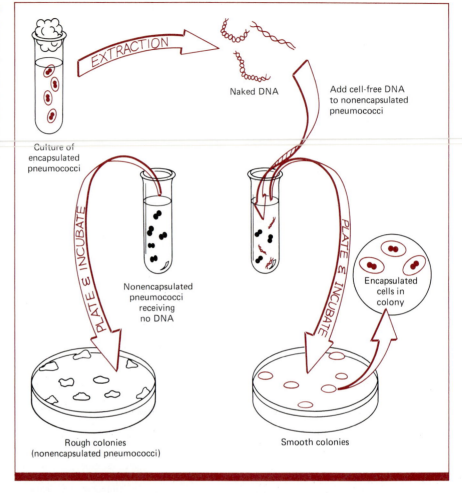

Naked DNA

Add cell-free DNA to nonencapsulated pneumococci

Culture of encapsulated pneumococci

EXTRACTION

PLATE & INCUBATE

PLATE & INCUBATE

Nonencapsulated pneumococci receiving no DNA

Encapsulated cells in colony

Rough colonies (nonencapsulated pneumococci)

Smooth colonies

strains of bacteria may transform harmless normal flora organisms into dangerous pathogens. The significance of transformation in nature, however, is still somewhat obscure.

Transduction

Bacteriophage-mediated gene transfer is called **transduction**. Bacterial genes can become accidentally enclosed in a bacteriophage capsid during replication of the virus in a host cell (Fig. 12-14). This happens because, during viral infection, the host chromosome may be degraded into small pieces, and occasionally a fragment similar in size to the replicating viral chromosomes becomes encapsidated. Transducing particles, viral capsids that contain bacterial genes, are liberated with normal progeny viruses during lysis of the host cell.

When a transducing particle subsequently infects a susceptible host cell, the genes from the previous bacterial host are introduced into this new cell. Since the capsid lacks viral genes, no bacteriophage synthesis occurs.

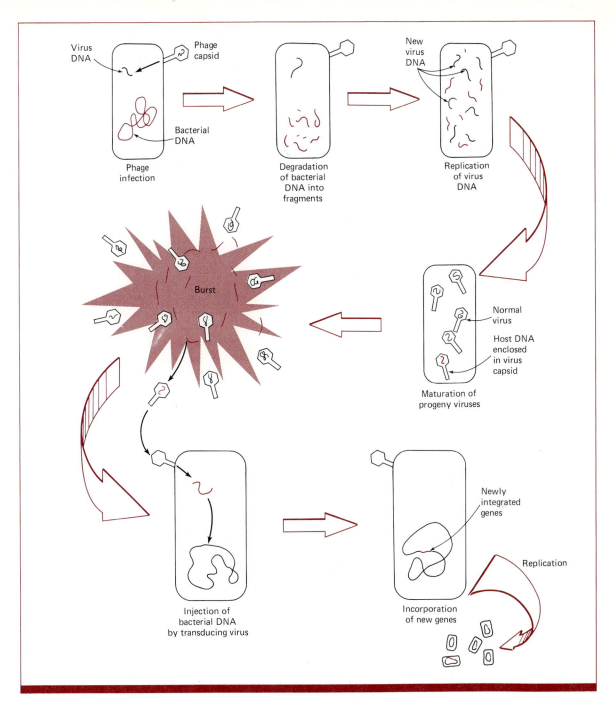

FIGURE 12-14

Events in generalized transduction. The transducing particle has enclosed a fragment of bacterial DNA (color fragments) by mistake. When it infects another cell, the injected DNA may confer new genetic characteristics to the recipient.

The newly introduced bacterial genes may recombine with the homologous segment of the bacterial chromosome.

Another type of transduction occurs when DNA of a temperate bacteriophage excises from the host chromosome incorrectly, picking up a few adjacent bacterial genes and leaving some of the prophage DNA. This phage-host DNA hybrid replicates and becomes encapsidated, so all the progeny virions in the cell are transducing particles, each of which can infect a new host cell and integrate into the chromosome. The recipient cell acquires the characteristics specified by the newly obtained bacterial genes. This type of gene transfer is called *specialized transduction* because only those genes that lie adjacent to the phage integration site can be transferred.

Conjugation

Some gram-negative bacteria have the ability to attach to other gram-negative cells and transfer genetic material. This process is called **conjugation**. Externally, the donor differs from the recipient by the presence of at least one sex pilus extending from its surface. The sex pilus connects the two cells during conjugation (Fig. 12-15). A conjugation tube develops through which DNA is transferred.

Although conjugation superficially resembles some aspects of sexual reproduction in eucaryotes, it has several important differences. It is not a method of reproduction (no fertilization occurs), but rather is a means of genetic transfer. Genetically, the donor may differ from the recipient only by the presence of a small, circular nonchromosomal piece of DNA called an **F factor** (fertility factor). Genetic information on the F factor provides a

FIGURE 12-15
Conjugation between bacteria. One donor cell (right) in contact with two recipient bacteria, each connected by two F pili. The nodules along each pilus are bacteriophages that have been used here to enhance the visibility of the pili.

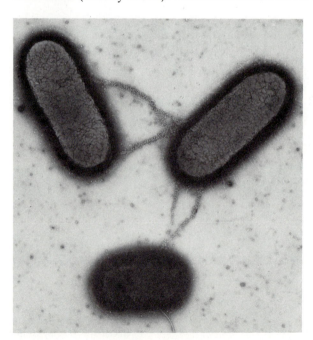

bacterial cell with everything needed to be a donor, including the capacity to synthesize the sex pilus.

The F factor may exist as a free element in the cytoplasm, replicating independently of the bacterial chromosome, or it may integrate into the chromosome and be replicated whenever the cell produces a copy of its own chromosome. Conjugation may, therefore, have one of two outcomes depending on whether the F factor is free or integrated.

1. **F+ Donors** When the F factor is not integrated, recipient (F−) cells are readily converted to donor (F+) cells (Fig. 12-16a). The donor does not sacrifice its F factor, but replicates it and gives away the copy. In this way, a single F+ cell introduced into a culture of F− recipients can lead to the rapid conversion of all the cells to F+. F+ donors do not transfer any chromosomal genes, only the F factor.

FIGURE 12-16

(a) Conjugation between F+ and F− bacterial cells, producing two F+ cells. No genes of the bacterial chromosome are transferred; thus no genetic recombination occurs. (b) Hfr donor transfers DNA to recipient. The new genes recombine with homologous portions of the recipient's chromosome. Since the F factor is rarely transferred, the recipient is not converted to a donor.

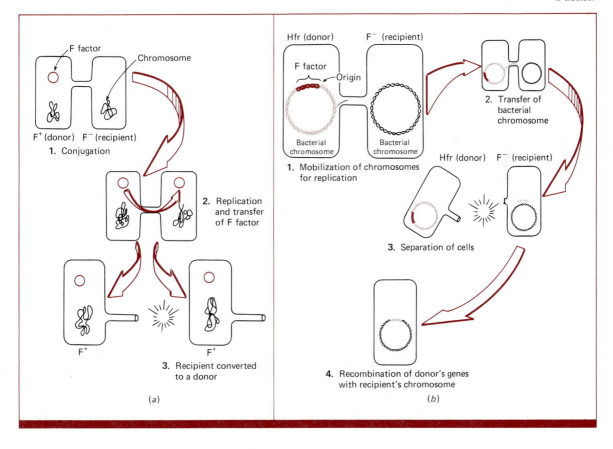

2. Hfr Donors If the F factor integrates into the bacterial chromosome, the resulting donor is called an Hfr cell. The term Hfr denotes **high** frequency of **r**ecombination, referring to the fact that, unlike F+ donors, Hfr cells commonly donate copies of genes on the bacterial chromosome (Fig. 12-16b). Chromosomal replication begins at the *origin* (the site adjacent to the integrated F factor). The replicated genes pass through the conjugation tube until the cells separate. Conjugation rarely lasts the 90 minutes required to transfer an entire copy of the donor's chromosome; thus, because the F factor is always the last thing to be transferred, recipients are rarely converted to donor cells. In the recipient, the newly acquired bacterial genes recombine with the chromosome. As a result, the recipient may receive new genetic traits from the donor.

Integrated F factors occasionally excise from the chromosome and the Hfr cell reverts to an F+ cell. Sometimes an error is made during the excision process and the bacterial genes adjacent to the integrated F factor are incorporated into the free F factor. These bacterial genes are replicated and transferred with the F factor. Bacteria containing such hybrid F factors are called *F-prime (F') cells.*

Although bacterial conjugation is usually associated with gram-negative bacteria, a novel mechanism for gene transfer among streptococci and staphylococci probably requires physical contact between these gram-positive cells. The resulting conjugal union is not mediated by pili.

Plasmids

The F factor is only one of several types of small, circular pieces of nonchromosomal DNA that commonly reside in the cytoplasm of many

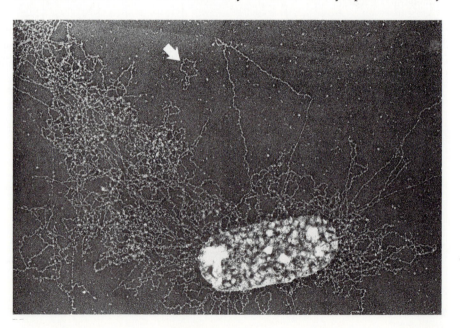

bacteria. All extrachromosomal pieces of DNA in procaryotes are called **plasmids** (Fig. 12-17). Most known plasmids confer new genetic traits to the host bacterium.

The F factor is a plasmid that integrates into the bacterial chromosome. Most plasmids, however, remain autonomous in the cytoplasm. They may be transferred by conjugation, transduction, or transformation. Most plasmids that are transferred by conjugation contain two genetic regions. One segment codes for the sex pilus and other characteristics of donor bacteria; the rest of the plasmid contains genes that provide additional traits to the plasmid-containing cell. For example, one type of plasmid, the **R factor** (resistance factor), contains genes that convert drug-sensitive bacteria to antibiotic-resistant cells (see color box).

Plasmid-associated genetic traits are usually not essential for the survival of the cell, although in certain environments they can provide some selective advantages over cells that lack the traits. The trait depends on the type of plasmid. For example, plasmids may confer on their host

THE R FACTOR—A THREATENING PLASMID

■ One of the most medically alarming aspects of bacterial conjugation was first reported in Japan in the 1950s. In several hospitals, the number of cases of certain infectious diseases (especially dysentery) that were incurable by available antibiotics increased dramatically. The pathogenic bacteria responsible for the diseases had suddenly acquired a resistance to a multitude of antibiotics to which they previously had been susceptible. Researchers discovered that a plasmid was responsible for transferring antibiotic resistance from resistant bacteria to sensitive bacteria by the process of conjugation. They called this plasmid an R factor (resistant factor). R factors were found to be similar to F factors in the manner in which they rapidly spread through the bacterial population, converting recipients to donors. In addition, R factors confer resistance to many commonly employed chemotherapeutic agents, including tetracyclines, chloramphenicol, streptomycin, sulfanilamide, kanamycin, neomycin, and penicillin. Bacteria that harbor an R factor are also more resistant to many commonly used disinfectants that contain nickel, cobalt, and mercury.

The problem of multiple drug resistance among bacteria is worsening because of the excessive and indiscriminate use of antibiotics. This practice favors the exclusive survival of bacteria resistant to these selective pressures. The problem is further compounded by the fact that transfer of R factors is not limited to passage between members of the same bacterial genus. If R factors are harbored in gram-negative members of the normal human flora (*Escherichia coli*, for example) multiple antibiotic resistance may be rapidly transferred to the dangerous gram-negative pathogenic bacteria that cause typhoid fever and dysentery, as well as to *Pseudomonas, Serratia*, and *Proteus*, bacteria that annually cause thousands of nosocomial infections.

R factors have also been found in *Staphylococcus* (another common cause of nosocomial infections) and a few other gram-positive bacteria. R factors in gram-positive cells may be transmitted by transformation, although conjugation between gram-positive bacteria has also been reported.

bacteria the ability to produce toxins, degrade oil, produce some antibiotics, synthesize enzymes for cheese production, metabolize camphor, and cause tumors in plants.

Genetic Engineering

Plasmids are important tools for researchers in the field of **recombinant DNA technology** (also called genetic engineering). Genetic manipulation can introduce foreign genes into bacterial cells. These foreign genes may come from other species, even from eucaryotic cells (since the genetic code is universal for all cells). Substances such as insulin, which were formerly

FIGURE 12-18

Cleavage of DNA by restriction endonuclease, producing two fragments with single-stranded sticky ends.

284

MICROBIAL GENETICS

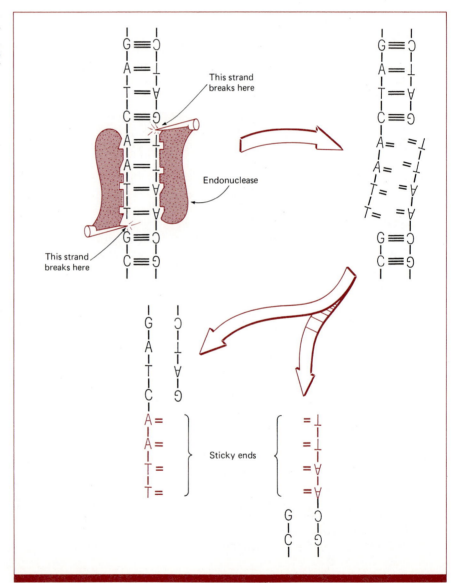

available only by extraction from animal cells (often an expensive, low-yield process), may be cheaply produced by genetically engineered bacteria that have been given the appropriate eucaryotic gene. These cells can then be grown in standard laboratory media, abundantly producing the desired product. Recombinant DNA technology has already provided medicine and science with inexpensive sources of several hormones and the antiviral substance interferon, which also shows promise as a treatment for cancer.

The transfer of DNA from eucaryotic cells to procaryotic cells does not occur naturally. Scientists use plasmids as carriers of the foreign genes so that they will be recognized, absorbed, and maintained by recipient bacteria. The foreign DNA is physically integrated into the plasmid by the use of **restriction endonucleases**, enzymes normally produced by bacteria to protect themselves from viral and other foreign DNA. Restriction endonucleases recognize short sequences of DNA bases and break each of the double DNA strands whenevor these sequences occur. The endonuclease in Figure 12-18, for example, breaks the strand after AATT. In this instance, the complementary sequences in the two strands are identical in

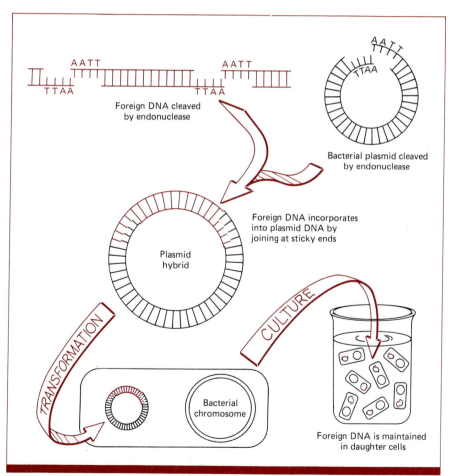

Foreign DNA cleaved by endonuclease

Bacterial plasmid cleaved by endonuclease

Foreign DNA incorporates into plasmid DNA by joining at sticky ends

Plasmid hybrid

TRANSFORMATION

CULTURE

Bacterial chromosome

Foreign DNA is maintained in daughter cells

FIGURE 12-19
Formation and cloning of recombinant DNA plasmid. The relative size of the plasmid has been enlarged here for diagrammatic purposes.

opposite directions. Thus a four nucleotide overlap occurs between the breaks, producing short, single-stranded regions. Because these single-stranded regions have a tendency to rejoin with any single-stranded region that has a complementary sequence, they are referred to as *sticky ends*. All DNA fragments broken by the same endonuclease can join with each other by complementary sticky ends, regardless of the source of the DNA. Thus, human DNA can be incorporated into a bacterial plasmid by cleaving both with the same type of endonuclease and allowing them to rejoin by recombination (Fig. 12-19).

Once the foreign DNA is integrated into the plasmid, the hybrid plasmid is transferred to recipient bacteria by transformation. Bacteria that acquire the plasmid transcribe and translate the foreign genes into expressed traits. The newly acquired DNA is also transmitted to all progeny cells, producing a clone of the foreign genes.

Recombinant DNA research has provided scientists with the technology to produce new gene combinations. The commercial potential of recombinant organisms was launched in 1980 when the first patent on a life form was granted to the scientists who created it.

OVERVIEW

The characteristics of an organism are determined by genetic information encoded in its DNA. Genetic instructions are inscribed in the sequence of nucleotides of the double-stranded helix. Precise replication of DNA assures that, except for mutations, progeny bacteria inherit the same genetic instructions possessed by their parent cells. Mutations occur when the genetic code is altered.

Genetic instructions in DNA direct the synthesis of proteins. In this way, encoded information becomes expressed as cellular characteristics. These proteins can be structural components of the cell or enzymes that control the cell's metabolism. Every physical or chemical property of an organism corresponds to a gene (or genes) that directs its production. The information in DNA is transcribed into a complementary molecule of mRNA. At the ribosomes the mRNA serves as a template that dictates the order in which amino acids assemble into a growing protein chain. This amino acid sequence determines the activities of the protein.

Cells regulate gene expression in response to their changing needs or environmental conditions. The bacterial operon provides a mechanism by which whole blocks of metabolically related genetic functions may be induced or repressed. Constitutive enzymes on the other hand, are constantly synthesized. Unlike induction and repression, mutations are permanent changes in the genetic code, often producing new or altered properties in the mutant. The frequency of mutation may be accelerated by mutagens. Since these agents may cause cancer, determining the mutagenic properties of a substance helps identify potential carcinogens.

Bacteria can transfer genetic information from a donor to a recipient.

The transferred genes can be stably incorporated into the recipient's chromosome by recombination. The DNA may be transferred in bacteriophage capsids (transduction), by physical contact between cells (conjugation), or as free nucleic acid liberated into the surrounding medium and subsequently absorbed by recipient cells (transformation). The transferred genetic material is a copy of either a portion of the donor's chromosome or a plasmid. Plasmids may provide bacteria with the ability to conjugate, to survive treatment with a multitude of antibiotics, or to produce toxins that can injure or kill people. Plasmids are also the vehicles used by genetic engineers in the field of recombinant DNA technology.

KEY WORDS

genetics	operon
gene	promoter
genotype	operator
phenotype	repressor
mutation	mutant
genetic code	wild type
messenger RNA (mRNA)	mutagen
transcription	Ames test
RNA polymerase	auxotroph
codon	recombination
translation	transformation
transfer RNA (tRNA)	transduction
anticodon	conjugation
polysome	F factor
reverse transcriptase	plasmid
prion	R factor
constitutive enzyme synthesis	recombinant DNA technology
induction	restriction endonuclease
repression	

REVIEW QUESTIONS

1. Differentiate between (a) genotype and phenotype, (b) transcription and translation, (c) induction and repression, and (d) transduction and transformation.

2. How does the nucleotide sequence in DNA determine the amino acid sequence in a protein?

3. Explain the role of tRNA in protein synthesis.

4. What would be the effect of a mutation that creates a nonsense codon in the middle of a gene?

5. Compare conjugation when the donor is an Hfr cell versus an F+ cell.

6. What is the function of the following parts of the bacterial operon?
   (a) Operator
   (b) Promoter
   (c) Effector
   (d) Repressor protein

7. How are restriction endonucleases used to splice genes for genetic engineering?

Control of Microorganisms

13

Microbes present a constant threat to health, jeopardize our food supplies, and destroy many useful materials. Fortunately, through physical or chemical methods, microorganisms can be reduced in number or completely eliminated from an environment or an infected person. These protective measures are particularly important in the hospital, where high concentrations of potentially pathogenic organisms are found. Microbial controls are also instrumental in the safe preparation of food, pharmaceutical agents, cosmetics, and other products for use in or on the human body.

ANTIMICROBIAL EFFECTS

The control of microorganisms often depends on establishing conditions that cannot be tolerated by microbes. Antimicrobial conditions are createe by microbicidal or microbistatic agents (Fig. 13-1). **Microbicidal** (*-cide*, kill) agents kill microorganisms and therefore have an irreversible and permanent effect. **Microbistatic** (*static*, standstill) agents inhibit microbial growth and multiplication, thereby preventing an increase in the number of microorganisms. Microbistatic agents do not kill or eliminate microorganisms. The microbe persists and can resume growth once the agent is removed. Therefore, microbicidal agents are generally preferred over microbistatic ones. **Germicidal** is another general term which refers to the destruction of microorganisms. Agents that specifically kill (or inhibit) bacteria, fungi, or viruses are referred to as bactericidal (or bacteristatic), fungicidal (or fungistatic), or virucidal (or virustatic). Sporicidal agents kill bacterial endospores, the most resistant forms of microbes.

FIGURE 13-1

Effects of microbistatic and microbicidal agents. Organisms may be inhibited or killed by physical or chemical antimicrobial agents.

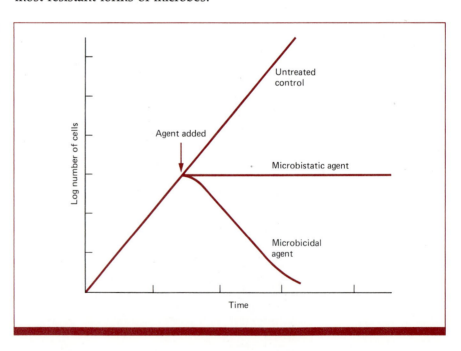

Antimicrobial agents perform one or more of the following processes.

■ **Sterilization** eliminates all forms of life, including vegetative cells, spores, and viruses. During sterilization, organisms are either killed or are physically removed from the objects being treated. Sterility is an absolute term—there is no such thing as "almost sterile." An object or environment is not sterile as long as it contains even a single viable microbe.

■ **Disinfection** eliminates the vegetative forms of most potentially hazardous and pathogenic organisms, but does not assure the elimination of all microbes. Bacterial spores, tubercle bacilli, and many viruses are particularly resistant to common disinfectants. Like sterilizing agents, disinfectants are used only on inanimate objects and never on body surfaces. Disinfection is generally employed if sterilization is either impossible or unnecessary. The purpose of disinfection is to minimize the risk of infection or product spoilage by reducing the number of microbes, especially pathogens in the inanimate environment.

■ **Antisepsis** is the inhibition or destruction of microorganisms on the surface of living tissue. Antiseptics must be milder than disinfectants to avoid harming the tissues on which they are used.

■ **Chemotherapy** is treatment of infectious disease by introducing into the human body antibiotics or other chemicals that inhibit or kill microorganisms. An antibiotic is a chemical synthesized by microorganisms, usually a bacterium or fungus, which at very low concentrations (micrograms per milliliter) inhibits or kills other microbes. Many drugs similar to antibiotics are chemically synthesized in the laboratory rather than harvested from microbes. The term "antibiotic" has come to include both the natural and synthetic antimicrobial agents. Other substances produced by microorganisms—for example, organic acids, alcohos, and enzymes—also have antimicrobial activity but only if they are used at much higher concentrations. Since chemotherapeutic agents are used inside the human body, they must exhibit selective toxicity for the target microorganisms, with little or no toxicity for human tissues. The unique characteristics of these antimicrobial agents is further discussed in Chapter 14.

■ **Preservation** prevents microbial proliferation in prepared products either by the addition of a chemical preservative, by storage in cold or frozen states, or by dehydration. These techniques retard spoilage of and growth of pathogens in foods, pharmaceutical preparations, and biological products for use in or on the human body.

The most appropriate antimicrobial strategy is one that limits the microbial population without damaging the person or object being treated. When possible, the best way to do this is to avoid microbial contamination. **Aseptic techniques** are precautions that help prevent contamination of culture materials (see Chap. 6), equipment, personnel, or environment. Because aseptic procedures prevent accidental introductions of microorganisms, they constitute a major element in combating disease and spoilage.

Factors Affecting Antimicrobial Activity

Some antimicrobial agents are microbicidal under one set of conditions and microbistatic under others. They may lose effectiveness as concentrations decrease or as conditions become suboptimal. Factors that influence the activity of antimicrobial agents are (1) the susceptibility of the microorganism, (2) the concentration or dose of the agent, (3) the length of exposure, (4) the number of microorganisms, and (5) environmental conditions.

Microbial Susceptibility   Microbes vary in their response to different antimicrobial agents. Vegetative bacteria, fungi, and enveloped viruses are usually most susceptible to destruction. Vegetative cells of the mycobacteria that cause tuberculosis and leprosy, however, are covered by a waxy coating that protects them from many antimicrobial chemicals. In addition, the hepatitis B virus and some fungal spores are resistant to most disinfectants and are persistent problems in hospitals. *Bacillus* and *Clostridium* (endospore-forming bacteria) are especially difficult to eliminate.

Concentration or Dose of the Agent   Diluting microbicidal chemicals usually weakens their antimicrobial activity. At lower concentrations they become microbistatic or lose antimicrobial activity altogether. The antimicrobial effects of temperature or radiation also depend on intensity of the exposure. Low doses may inhibit growth, whereas high doses may result in sterilization. With a few important exceptions, the more concentrated or intense the exposure to any germicidal agent, the more likely that target organisms will be destroyed.

Length of Exposure   Microorganisms die when physical or chemical conditions irreversibly damage essential cell components. All organisms present, however, do not die rapidly and simultaneously when a critical exposure is achieved, because microbial death is a function of time—the longer microbes are exposed to potentially lethal conditions, the more microbes will be killed (Fig. 13-2). For many germicides, if exposure time is long enough, the probability of even a single cell surviving becomes so low that sterilization is practically assured. In contrast, microbistatic agents are effective only as long as they are present and must be used during the entire time that inhibition is to be maintained.

Number of Microorganisms   Antimicrobial effectiveness also depends on the initial concentration of the microbial population. As the number of microbial contaminants increases, either the exposure period to or concentration of the agent must increase to achieve acceptable levels of decontamination. Dust-covered objects, for example, are usually heavily contaminated (each gram of dust contains about 1 million organisms). The

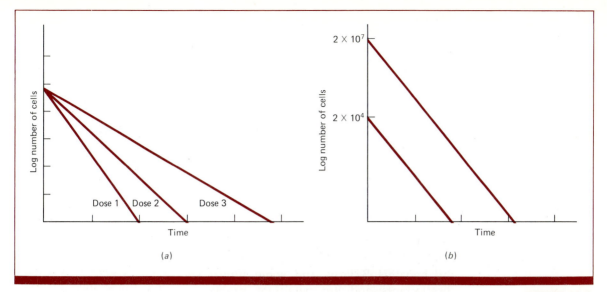

Cell death following exposure to a microbicidal agent. (*a*) The time required to kill identical concentrations of cells depends on the dose of the germicidal agent. In this diagram, dose 1 is the most concentrated and dose 3 is the least concentrated. (*b*) When the dose is constant, the time required to kill increases as the starting number of cells increases.

number of organisms in fecal matter and pus is even greater. Removing microbes by washing objects in a detergent and rinsing with water dramatically reduces microbial contamination and increases the likelihood that subsequent antimicrobial treatment will be adequate.

Although sterilization kills all microorganisms, it does not necessarily eliminate all harmful microbial effects. Endotoxins and other bacterial products often remain on objects and are consequently introduced into the body. Washing and rinsing remove microorganisms and reduce endotoxin contamination.

Environmental Conditions   Temperature, pH, and moisture affect the efficiency of most antimicrobial agents. In addition, some chemical agents are absorbed by organic materials (blood, mucus, feces, and tissue) that severely reduce antimicrobial effectiveness. These agents therefore cannot be used on the skin. To prevent interference by organic debris, objects can be rinsed prior to disinfection. Some antimicrobial agents are impeded by soaps and detergents that remain as thin films on skin or object surfaces. This difficulty can be avoided by thorough rinsing prior to disinfection or antisepsis.

PHYSICAL AGENTS FOR CONTROLLING MICROBES

The most common physical methods of control utilize moist heat, dry heat, radiation, or filtration.

Moist Heat

Although any organism can be killed by excessive heat, the lethal temperature depends on the heat resistance of the organism and the amount of water in the environment. Moist heat, especially steam, effectively kills cells by coagulating their proteins (critical enzymes, for example). In the absence of water, heat does not coagulate protein. Dry heat kills cells by poorly understood procssses that require much higher temperatures than moist heat. Three of our most common antimicrobial processes rely on moist heat: pasteurization, boiling, and autoclaving (saturated steam under pressure). Of these only autoclaving assures sterilization.

Pasteurization   In order to prevent undesirable microbes from spoiling wine, Louis Pasteur gently heated the juice to kill contaminants before inoculating with yeast to start fermentation. This process became known as *pasteurization*. Similar processes are used today by dairies, breweries, and other food industries to prevent disease transmission and spoilage. The selected combination of temperature and duration of heating is one that kills the most heat-resistant pathogens commonly transmitted by that medium without damaging product quality. The target pathogen for pasteurization of milk, for example, is *Coxiella burnetti*, the rickettsia that causes Q fever. It is destroyed by 30 minutes of heating at 62.8°C (or 15 seconds at 71.6°C). Although this treatment does not kill all microbes, it does eliminate the common pathogens. For example, the bacteria that cause tuberculosis and brucellosis, two milk-borne diseases, are less heat-resistant than *Coxiella burnetti* and are readily destroyed by pasteurization. Pasteurization has reduced the incidence of milk-borne disease, although occasional outbreaks still occur among consumers of raw (unpasteurized) cow's milk.

Boiling   Heating increases the temperature of water until it reaches 100°C, the temperature of boiling at normal atmospheric pressure at sea level. Further heating fails to raise the temperature of water because the additional heat energy escapes the liquid in steam.

Although most vegetative cells are eliminated by 10 minutes of boiling, some endospores, notably those that cause botulism, survive 5 hours of continuous boiling. This process is therefore considered a procedure for disinfection rather than for sterilization. It is used when small instruments are to be disinfected or when there is no access to equipment needed for sterilization (for example, in emergency deliveries of babies at home). Surgical instruments, however, should always be sterilized and not merely disinfected.

Although the higher temperature of boiling water kills more effectively than pasteurization, it is not routinely used in food industries because such high temperatures often damage the product.

Autoclaving (Use of Steam under Pressure)   The most common instrument for sterilizing heat-stable materials is the **autoclave**. This

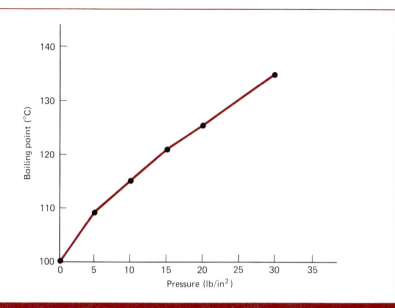

FIGURE 13-3

The boiling point of water as a function of pressure. Each point on the line indicates the pressure needed to change the boiling point of water to the temperature indicated by the same point.

CAN YOU CAN OR CAN'T YOU CAN?

■ In the early nineteenth century, the French military suffered major troop losses because of famine. To prevent these non-combat losses, French officials conducted a contest to develop a successful method for preserving foods. The winner, Nicholas Appert, heated food in sealed glass jars, thereby inventing the process of canning. Heating kills spoilage microbes and sealing keeps out contaminants. Appert, however, had no knowledge of microbes or why his process worked. He believed that heat destroyed "ferments," nonliving factors in air that spoiled food.

Today, the canning method of food preservation is routinely used in industry and in the home. Two processes are generally employed, the boiling-water bath and the high temperature-pressure method. Boiling easily eliminates the major spoilage bacteria, yeasts, and molds. Some foods contain only these contaminants and are sterilized after a few minutes at 100°C. Fruits, for example, are high in acid (pH below 5), a condition that prevents the survival of heat resistant pathogens, notably *Clostridium botulinum*. Some nonacid foods can also be safely preserved by boiling and sealing. Such foods as jams, syrups, and sweetened condensed milk contain high sugar concentrations that inhibit most spoilage microbes and pathogenic bacteria, including endospore formers.

Most low-acid foods require sterilization and must be exposed to temperatures that exceed 100°C. The only practical canning methods that sterilize these foods without destroying them depend on pressure cookers and autoclaves. "Pressure canning" (actually, high-temperature canning) is essential for ensuring the safety of meats and vegetables. Botulism outbreaks can usually be traced to nonacid products that have been boiled or otherwise processed using inadequate temperatures.

instrument sterilizes with saturated steam under pressure. The pressure increases the boiling point of water, thereby increasing the temperature to which water can be heated (Fig. 13-3). Pressure cookers, for example, build pressures within the closed vessel so heated water vaporizes to steam at temperatures above 100°C (see color box). The autoclave, like a pressure cooker, uses increased pressures to raise the temperature required to produce steam. Cells are destroyed by the higher temperatures and not by the effects of pressure. The dynamics of the autoclave are explained in Figure 13-4.

Standard autoclaves are usually operated at 15 lb/in² above atmospheric pressure and at a temperature of 121°C. Most spores are killed after 15 minutes at these high temperatures (Fig. 13-5). Other autoclaves use even higher temperatures (132 to 136°C) and pressures (27 to 33 lb/in²) but for shorter periods of time (3 to 10 minutes). This is an advantage in emergencies and for sterilizing rubber and other materials that may deteriorate with prolonged heat exposures.

FIGURE 13-4

Steam penetration through an autoclave chamber. Steam entering the top of the chamber expels cooler, heavier air out the exit at the bottom. The exit is sealed automatically when the chamber is filled with steam at the appropriate temperatures. Alternatively, in some autoclaves the air is removed with a vacuum pump. Most exhaust systems have a valve that automatically closes when pure steam begins to escape.

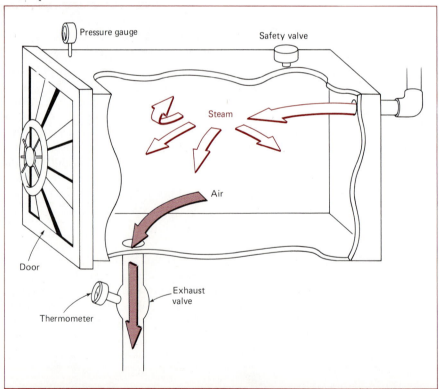

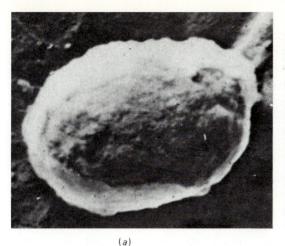

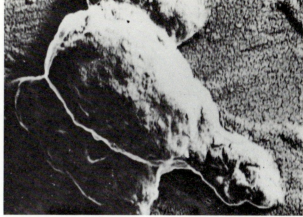

(a)   (b)

FIGURE 13-5

Electron micrograph of (*a*) intact bacterial endospore and (*b*) disrupted endospore after autoclaving.

297

PHYSICAL AGENTS FOR CONTROLLING MICROBES

The autoclave uses saturated steam to kill everything in its chamber. *Saturated steam* is water vapor at the temperature at which it is produced. At this temperature steam readily condenses on the surface of cooler objects, and in doing so, tremendous amounts of heat energy are transferred from the steam to the object. *Saturated steam heats an object about 2500 times more efficiently than does hot air at the same temperature.* In addition, dry articles (dressing, linens, etc.) acquire the moisture needed for sterilization at these temperatures. As steam condenses, it creates a vacuum that draws more steam into the autoclave. This process continues until the entire load has been penetrated by heat and thus brought to the temperature of the chamber. Longer periods of time are needed for heat to completely penetrate large volumes and bulky items. The chamber temperature is then maintained for the time needed to ensure sterilization.

After liquids have been autoclaved, pressure should be returned to normal by exhausting the steam very slowly from the chamber. Otherwise, the temperature of the liquid will exceed the boiling point at the reduced pressure. Slow exhausting prevents boiling over of loosely capped liquids, or explosion of liquids packed in airtight containers. After dry objects have been autoclaved, the steam may be rapidly exhausted. Most modern autoclaves are equipped with separate cycle controls for liquids and dry materials so that the exhaust process can be tailored to the articles being autoclaved.

All items should be packaged to prevent recontamination when removed from the autoclave. Paper, cotton, and cloth packaging become wet during autoclaving and are permeable to contaminants unless allowed to dry thoroughly before removal from the autoclave. Some autoclaves create a vacuum at the end of the cycle to draw off residual moisture.

Materials not suitable for autoclaving include powders, heat-sensitive compounds, and water-insoluble substances (such as mineral oil) that steam cannot penetrate. Microbes suspended in oil are exposed only to dry heat during autoclaving.

Dry Heat

Dry heat is generally used in three ways: flaming, incineration, and baking. A flame is commonly used to sterilize loops and needles so that microorganisms may be transferred without contamination. The mouths of test tubes and containers are also routinely flamed, although the heat is not sufficient to sterilize these surfaces. Scalpels and other sharp metal instruments are damaged by constant flaming and are usually packaged and autoclaved. Incineration is useful for destroying heavily contaminated materials or tissues in the hospital and cultures of pathogenic microbes discarded in the laboratory. If the entire article is completely burned and there is no unburned material expelled in the exhaust, the elimination of the pathogenic agents is assured.

Baking in hot-air ovens requires prolonged exposure to temperatures between 150 and 180°C. The length of exposure depends on how readily heat can penetrate the material, since all parts to be sterilized must reach critical temperatures. Higher temperatures require shorter periods of exposure (Table 13-1). Temperatures below 150°C require too much time to be practical. Temperatures above 180°C often damage materials. Hot-air ovens are used for sterilizing materials such as glassware or metal instruments that can tolerate prolonged heat exposure and powders, oils and waxes that are either destroyed or not effectively sterilized by the moist heat of the autoclave. Water-based liquids, however, can be heated only to 100°C and cannot be sterilized in hot-air ovens. Similarly, articles made of rubber, plastic, or fabric may be destroyed by intense heat and should never be exposed to baking.

Radiation

Gamma rays, x-rays, cosmic rays, ultraviolet light, even visible light are all forms of radiation. When these rays strike an organism, energy may be absorbed by the cells, often causing cell damage or death. Radiation with the shortest wavelengths has the greatest energy and is therefore the most lethal. Ionizing radiation and ultraviolet radiation are two types of radiation used in microbial control.

Ionizing Radiation   Some rays possess so much energy that they cause biologically active molecules to lose electrons. This results in production of ionized molecules that no longer perform critical cellular functions. Such high-energy **ionizing radiation** is an effective sterilizing agent. High doses of radiation kill everything they strike and are a commonly employed

TABLE 13-1

Conditions for Dry Heat
Sterilization

Operating Temperature		
°C	°F	Sterilization Time, hours
121	250	>6
150–160	302–320	>3
160–170	320–338	2–3
170–180	338–356	1–2

alternative to the autoclave for sterilizing plastic petri dishes and other heat-sensitive materials.

Two types of ionizing radiation are commonly used for sterilization: *gamma rays*, emitted by radioactive elements (cobalt 60, for example), and *electron accelerators*. The penetrating power of gamma radiation makes it useful for sterilizing large loads or bulk items. Exacting safety precautions are required to shield people from the radiation, since it is equally detrimental to human cells. In contrast to gamma rays, high-energy electrons penetrate less efficiently and are consequently less hazardous. They are used to sterilize smaller, individually wrapped articles.

Ultraviolet Radiation   Cellular DNA absorbs the energy of radiation at wavelengths between 250 and 260 nm and forms aberrant chemical bonds between adjacent thymine nucleotide bases. These *thymine dimers* distort the DNA strands and impair replication and transcription. This interferes with the expression of genes. Thymine dimers are lethal when they occur in genes for essential functions or when DNA replication is blocked.

The germicidal effects of ultraviolet rays are dose dependent—longer exposures and greater doses (higher wattage or less distance from the source of radiation) increase the number of vegetative cells killed. Some bacterial endospores are protected against ultraviolet irradiation by substances in the spore coat that absorb light waves. Ultraviolet light is not, therefore, a sterilizing agent but is used as a disinfecting agent. The major limitation of ultraviolet light, however, is its poor penetrating power. Although these rays pass readily through dust-free air and clear water, they fail to penetrate ordinary glass and many plastics, turbid solutions, thin films of grease, and milk. Effectiveness also decreases as the distance from the radiation source increases. In addition, ultraviolet light can severely damage the retina of someone who looks directly at the bulb, and prolonged skin exposure contributes to the development of skin cancer.

Filtration

Liquids and gases can be sterilized by passing them through filters. Microorganisms are retained by filters that have pores smaller than the size of the microbes (Fig. 13-6). The filter acts as a strainer, a microbial sieve. Standard bacteriological membrane filters are composed of nitrocellulose and have pore diameters of 0.45 μm, small enough to prevent passage of most bacteria. Since the elastic mycoplasmas and any small viruses pass through pores of this size, standard bacteriological filters do not actually assure sterilization. This is an important limitation to the practical applications of filtration, and many filtered solutions should be tested for the presence of viruses before they are used for treating patients. On the other hand, filters can be used to separate viruses from their host cells—for example, in the purification of live viral vaccines. Filtration is used for preparing heat-labile culture media components, pharmaceuticals, and biological solutions.

Other types of filters are made of porcelain, glass, or fibrous materials

FIGURE 13-6

Removal of microorganisms
by filtration. Too large to
pass through the pores,
bacteria collect on the
surface of the membrane
filter (pore size here is 0.22
μm diameter).

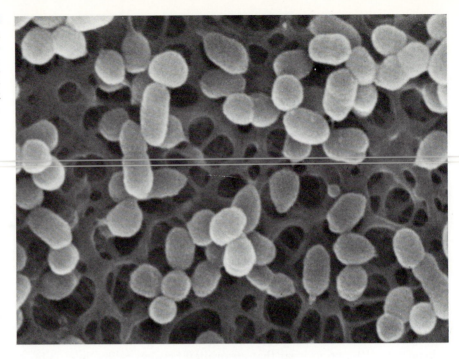

CONTROL OF
MICROORGANISMS

(cotton, asbestos, and paper) that do not contain pores of uniform size, but remove cells by the tortuous path they must follow through the filter. If the filter is thick enough, particles are absorbed, intercepted, or settled out on the filter materials. This is why cotton can be used as stoppers for microbial cultures. Fibrous filters must be kept dry because liquids transport microorganisms across the open spaces in the filter. Fibrous filters are, therefore, only used for filtering gases. Some common applications of fibrous filters in air filtration include

Cotton or gauze masks that filter organisms released in expired air, trapping many of the microbes shed by the wearer (Masks must be changed when they become damp from exhalation.)

Cotton plugs in flasks, test tubes, pipettes, and air lines used for bubbling sterile air through liquid media in which aerobic microorganisms are being cultured

Filters used to prepare mixtures of gases for respiratory therapy

Filters in ventilation systems that provide sterile air to operating rooms

The *laminar flow hood*, used for reducing the danger of infection while manipulating infectious microorganisms and for preventing contamination of sterile materials (Fig. 13-7)

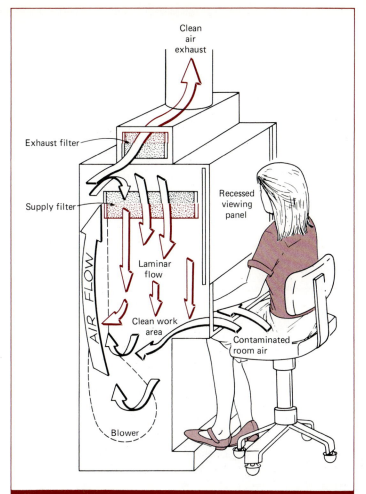

FIGURE 13-7

The laminar flow hood. Black arrows indicate contaminated air; colored arrows indicate filtered air. Room air is filtered before entering the working chamber and moves in a single direction. Contaminated air from the chamber is filtered again before being either recirculated in the system or exhausted to the outside.

The properties of common physical methods of microbial control are summarized in Table 13-2.

The Mechanical Scrub

The most common source of infections acquired in the hospital is the unwashed hands of medical personnel. Conscientious hand washing by all persons in contact with patients is perhaps the most important precaution in preventing nosocomial diseases. Careful and adequate hand-washing procedures should be employed by physicians, nurses, therapists, and anyone whose hands are exposed to pathogenic organisms and susceptible patients. Hands should always be washed *before and after contact with each patient* and after exposure to secretions and excretions that may be sources of infectious agents.

Hand washing reduces the number of transient organisms on the skin

TABLE 13-2

Physical Methods of
Microbial Control

Procedure	Usual Conditions	Primarily Recommended For	Major Limitations
Moist Heat			
Pasteurization	62.8°C for 30 min or 71.6°C for 15 s	Milk, wine, and other beverages	Not completely bactericidal
Boiling	100°C for 10 min or more	Noncritical disinfection of instruments (or when emergencies preclude use of sterile instruments)	Not sporicidal or virucidal
Autoclaving	15 lb/in² and 121°C for 15 to 30 min	Heat-stable solutions and equipment	Restricted to materials that can withstand heat and be penetrated by moisture
Dry Heat (baking)			
	170–180°C for 1 to 2 h	Glassware; some metal instruments; powders, oils, and waxes; materials that are impervious to or damaged by steam	Restricted to materials that can withstand higher temperatures
Radiation			
Ionizing	2.5 Mrads	Heat-sensitive plastic materials; pharmaceuticals	Expensive to operate; requires elaborate safety precautions
Ultraviolet	260 nm for prolonged periods	Air and surfaces	Poor penetrability
Filtration			
	Membrane or fibrous filters, pore size as appropriate	Heat-sensitive liquids; air	Viruses and mycoplasma not fully eliminated

surface. Although the hands cannot be sterilized, most transient organisms can be removed by 30 seconds of proper scrubbing with soap and water. Resident microbes found in sweat ducts and hair follicles of the skin, however, cannot be readily dislodged. These microbes are a threat to patients with reduced defenses, so scrubbing must often be supplemented by wearing sterile gloves and gowns. Hand washing is also an important precaution to be exercised by food preparers and handlers to prevent outbreaks of food-borne illness.

CHEMICAL AGENTS FOR CONTROLLING MICROBES

Chemicals That Sterilize

Only three chemicals are recommended for use as sterilizing agents: ethylene oxide, formaldehyde, and glutaraldehyde. When used properly, these agents dependably kill all microbes, including bacterial endospores

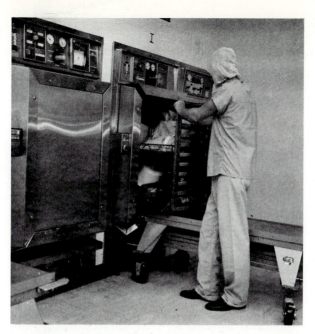

FIGURE 13-8
Ethylene oxide treatment of
materials in the hospital.
This chamber is specifically
designed for ethylene oxide
sterilization and is not
interchangeable with the
autoclave it resembles.

and the hepatitis B virus. All three chemicals kill organisms by damaging their proteins and nucleic acids. Like filtration and radiation, these chemicals do not rely on heat forkilling and are often referred to as agents of "cold sterilization".

Ethylene oxide is a gas used in an apparatus resembling an autoclave (Fig. 13-8). Air is evacuated from the chamber and replaced with the germicidal gas. Since ethylene oxide is toxic for humans, all traces of the gas must be eliminated after sterilization, a process which takes from 24 hours to 1 week at room temperature or 4 to 8 hours in heated aerators. Other disadvantages associated with ethylene oxide sterilization include its highly combustible nature and its inability to penetrate many materials. Only materials that are permeable to the gas can be sterilized by this method. Metallic foils and impermeable plastics cannot be used for wrapping. The major advantage of ethylene oxide sterilization is the low operating temperature, which makes it suitable for treating many heat-sensitive products.

Glutaraldehyde and *formaldehyde* solutions are often used for sterilizing small heat-sensitive instruments by soaking them for at least 10 to 12 hours. Like ethylene oxide, these chemicals are toxic irritants and must be removed before the instruments can be safely used. This is achieved by thoroughly rinsing with sterile water. Objects rinsed in unsterile water or handled with poor aseptic technique during rinsing are no longer sterile.

Chemical Disinfectants and Antiseptics

Depending on the spectrum of organisms that they kill, antimicrobial chemicals are classified as high-, intermediate-, and low-level germicides (Table 13-3). *High-level germicides*—ethylene oxide, glutaraldehyde, and

TABLE 13-3

Levels of Germicidal Action: The Ability of Antimicrobial Chemicals to Kill Various Forms of Microorganisms

| Chemical Agent | Germicidal* against | | | | | |
| | Bacteria | | | Fungi | Viruses | |
	Vegetative	Tubercle Bacillus	Spores		Lipid, Medium Size	Nonlipid, Small
High-Level Bactericide Ethylene oxide; glutaraldehyde; formaldehyde	+	+	+	+	+	+
Intermediate-Level Bactericide Iodine plus alcohol; chlorines; ethyl alcohol; phenolics	+	+	−	+	+	+
Low-Level Bactericide Iodophors; quaternary ammoniums; mercurials; chlorhexidine	+	−	−	+	+	−

+ = effective; − = ineffective.
Source: Adapted from R. Bartlett, "Control of Hospital-Associated Infections," in *Manual of Clinical Microbiology*, American Society of Microbiology, 1974.

formaldehyde—have the capacity to kill all microbes and spores. These germicides are used for chemical sterilization. Under less stringent conditions these chemicals can also act as *intermediate-level germicides*, readily killing vegetative cells and most viruses but not most endospores. For example, a 2% solution of glutaraldehyde kills most organisms in 10 minutes, but destruction of endospores may require 10 hours of exposure. Whenever spore formers are not a concern, intermediate-level disinfection is appropriate. The antimicrobial activity of *low-level germicides* is limited to a few types of bacteria, fungi, and enveloped viruses. Most antiseptic agents are low-level germicides.

Phenolics   The original antiseptic used by Lister to prevent surgical sepsis was **phenol** (also known as carbolic acid). Today various phenol derivatives with less irritating or corrosive properties are used in place of phenol. One group of derivatives, the *cresols*, is the active ingredient in commercial disinfectants such as Lysol. These agents rapidly kill most vegetative cells, including the mycobacterium that causes tuberculosis, and are stable in the presence of soaps, detergents, and organic matter. In addition, cresols require little water for antibacterial effectiveness and therefore remain active as water evaporates from surfaces during the drying process.

Another group of phenol derivatives are the *bisphenols*. (Hexachlorophene is the most familiar example.) This mild bacteriostatic agent controls the growth of gram-positive organisms, particularly *Staphylococcus aureus* on the skin. Bathing infants with soaps containing 3% hexachlorophene dramatically reduces the incidence of infection in hospital nurseries. This practice was discontinued following the discovery that hexachlorophene is absorbed through the skin and causes neurologic damage in infants. It is still used for hand washing by nursery personnel in an attempt to limit disease transmission. Pregnant women are advised not to use hexachlorophene because its absorption through the mother's skin may damage the fetus.

Surfactants Surfactants are *surface active agents* that interfere with the normal interaction between a cell's surface and its aqueous environment. Surfactants are either soaps or detergents with a cleansing action that helps physically remove microbes. In addition, some surfactants kill cells by disrupting their membranes and are therefore disinfectants. For example, **quaternary ammonium compounds** (commonly called "quats") are cationic (positively charged) surfactants that absorb to the negatively charged surface of bacteria, altering cell membrane permeability and killing the cell. Quaternary ammonium compounds are most effective against grampositive bacteria. *Pseudomonas, Proteus,* and most other important opportunistic gram-negative pathogens are poorly controlled by these agents. Bacterial endospores and tubercle bacilli are also not affected. Quats are inactivated by soaps left on skin after washing and by the cellulose fibers of gauze wound dressings, and they are of limited use as antiseptics.

Alcohols Ethanol and isopropanol are used to reduce the number of microbes on skin sites designated for hypodermic injections or for the withdrawal of blood. They are also used for disinfecting thermometers and some other small instruments. They can be used alone or in combination with other antimicrobial agents, usually iodine, formaldehyde, or quaternary ammonium compounds. Alcohol diluted with water to concentrations of 50 to 90% kills cells by coagulating essential proteins. Alcohol is not used at full strength because water is necessary to prevent dehydration of the cell (protein coagulation is impeded in the absence of water). The alcohol-water mixture rapidly inactivates most vegetative bacteria and many viruses, but endospores are unaffected. Alcohol is also a solvent that helps dislodge microbes embedded in fats and oils on the skin. Because they evaporate quickly from surfaces, alcohols are useful only for short-term disinfection or antisepsis.

Halogens Some of the oldest and still most useful germicidal agents are compounds containing the halogens iodine or chlorine. Free *chlorine* is released from compounds used to treat water for drinking, swimming pools, whirlpool baths, and hot tubs. It effectively eliminates such common waterborne pathogens as *Escherichia coli, Salmonella typhi,* and *Entamoeba histolytica*. Continued disinfection, however, requires maintaining adequate chlorine levels. High temperatures and rapid circulation increase the rate of chlorine evaporation, quickly reducing concentrations to ineffective levels. These circumstances are associated with outbreaks of skin infections caused by *Pseudomonas aeruginosa* in swimming pools, hot tubs, and whirlpools. Automatic chlorinators monitor and maintain free-chlorine levels at adequate concentrations. Some municipalities use chlorine to treat wastewater and sewage, although much higher concentrations are required because chlorine is inactivated by organic materials. The wisdom of this practice is questionable (see Chap. 25).

Hypochlorites are chlorine-containing inorganic compounds such as bleach. They are commonly used as disinfectants, especially by the food and dairy industry, to decontaminate equipment. Strong hypochlorite solutions are effective against hepatitis B virus, a particularly difficult pathogen to eliminate (Table 13-4). In addition, it is the only chemical that can consistently inactivate the slow virus that causes Creutzfeld-Jakob disease (see Viral Diseases, Chap. 9). These compounds are not used as antiseptics because they irritate skin and tissues. However, several organic chlorine compounds, such as chloramine, are nonirritants and are used for treating skin and wounds, where they release hypochlorous acid, the active ingredient in hypochlorites.

Another commonly employed halogen is *iodine*. In dilute solutions of water or alcohol (tincture of iodine) it rapidly inactivates microbes by irreversibly combining with proteins. Iodine is useful for disinfecting thermometers, for reducing microbes on skin sites selected for surgery or needle puncture, and for treating cuts and wounds. Unfortunately, iodine stains skin and fabric and stimulates nerve endings in skin. (Who can forget the sensation of iodine on a scraped knee?) Pain can be avoided by using **iodophors,** complexes of iodine and surfactants that slowly release the iodine as a potent nonirritating antiseptic.

Chlorhexidine This popular antiseptic is active against both gram-positive and gram-negative bacteria and some fungi, but is ineffective against mycobacteria, bacterial spores, and viruses. Chlorhexidine is nontoxic and retains antimicrobial activity for several hours, even in the presence of soaps and organic matter. Unlike hexachlorophene, it is not absorbed through the skin, so there is no danger to the fetus of a mother using chlorhexidine. Chlorhexidine is often a component of antiseptic lotions used in surgical scrubs.

Heavy Metals Ions of heavy metals readily bind with and inactivate proteins, even in very low concentrations. Their effect is not selective for

TABLE 13-4

Recommended Procedures
for Inactivation of
Hepatitis B Virus

Agent	Time
Heat	
Boiling water (100°C)	10 min
Autoclave (121°C)	15 min
Dry heat (160°C)	2 h
Chemical	
Sodium hypochlorite (0.5–1%)	30 min
Formalin (40% in water)	12 h
Formalin (20% in 70% alcohol)	18 h
Glutaraldehyde (2%)	10 h
Ethylene oxide	Variable

Source: Adapted from *Morbidity and Mortality Weekly Report,* supplement to vol. 25, no. 3, May 7, 1976.

TABLE 13-5

Common Uses for
Chemical Disinfectants
and Antiseptics

Use	Agent(s) Employed
Handscrub	
Routine washing	Soap, detergent, or iodophor, water, and "elbow grease"
Preoperative scrub; high-risk situations	Iodophor, chlorhexidine, hexachlorophene
Skin Preparation	
Routine injections	Alcohol (often used with an iodophor)
Surgical	Iodophor, chlorhexidine, hexachlorophene
Instrument Decontamination*	
Lensed instruments for internal examinations of the body	Ethylene oxide, glutaraldehyde
Thermometers	Alcohol and iodine
Other small medical instruments	Alcohols, quaternary ammoniums, iodophors, phenolics, chlorines, formaldehyde, or glutaraldehyde soak
Environmental Control	
Linens and clothing	Ethylene oxide or chlorines
Floors, walls, and other surfaces	Phenolics, chlorines, iodophors, quaternary ammoniums
Utensils	Chlorines, quaternary ammoniums
Water	Chlorines
Preservatives in Pharmaceutical Preparations	Phenol, alcohol, quaternary ammoniums, chlorhexidine, mercurials

*Any instrument that will be introduced into the tissues or bloodstream *must* be sterilized. The preferred method for heat-stable items is autoclaving.

microbes so, as antiseptics, they must be used in dilute concentrations and only topically. The most common of these antiseptics are the *mercurials* (compounds containing mercury), mercurochrome and merthiolate, and the silver-containing agents, silver nitrate and silver chloride. Silver nitrate solutions are used to irrigate infected urinary bladders and to prevent eye infections. Until recently replaced by antibiotics, silver nitrate drops were added to the eyes of all newborn children in the United States to prevent *Neisseria gonorrhoeae* infections acquired during the birth process.

The common uses of chemical antimicrobial agents are summarized in Table 13-5.

EVALUATION TESTS FOR STERILITY

Although autoclaving, baking, ionizing radiation, and some chemical treatments theoretically kill all life forms, they fail to sterilize when improperly used. Undetected failures can have serious, even fatal, consequences (for example, injecting contaminated solutions directly into people). All sterilization procedures must be carefully and consistently monitored to detect failures and assure sterility. Several approaches can be adopted to evaluate the sterility of treated items:

■ Recording devices on sterilizing equipment can be used to measure operating conditions—temperature, moisture, pressure, gas content, and length of exposure to the sterilizing condition. The readings indicate any deviations from standard conditions that may impede sterilization. Even ideal readings, however, do not guarantee that the entire load has been subjected to identical sterilizing conditions.

■ Since it is not possible to individually monitor every article for contamination, items can be randomly selected for evaluation. This approach, however, depends on accurately predicting which media and culture conditions will most likely detect any pathogens present. It also depends on luck, since only a few items are sampled.

■ Chemical indicators can be placed throughout the load to provide evidence of local sterilizing conditions. Tapes and strips impregnated with sensitive chemicals change color after the conditions necessary for sterilization are achieved. Some indicators are sensitive to temperature, others to steam, gas, or radiation (Fig. 13-9a and b).

■ The most accurate test to determine whether the load is sterile uses *biological indicators,* most commonly **spore strips**. These are pieces of filter paper impregnated with highly resistant bacterial spores (Fig. 13-9c). They are placed in areas where sterilizing conditions are most difficult to achieve—the lower front of the autoclave where air pockets may reside, the coolest areas in a hot air oven, or the interior of bulky items. When the sterilization cycle is completed, the strips are aseptically removed, placed in broth, and incubated. The absence of growth indicates that sterilization was achieved. The choice of bacterial spores depends on the sterilization process being monitored. The most resistant organism is selected since its destruction would indicate that more sensitive microbes have also been killed. Spores of *Bacillus subtilis* are most resistant to ethylene oxide and dry-heat processes, whereas spores of *Bacillus stereothermophilus* are preferred for monitoring steam sterilization. *Bacillus pumilus* endospores are most resistant to gamma radiation.

FIGURE 13-9
Indicators of sterility: (a) Chemical indicators monitor conditions (time, temperature, humidity, and gas concentration) in the sterilizer. Indicator strips are placed in strategic areas throughout the load. (b) Chemical indicators can be used to label each item. (c) Biological indicators contain organisms that possess extreme resistance to the sterilizing agent.

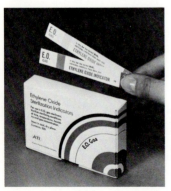

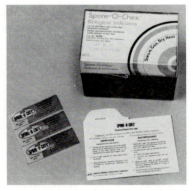

(a)   (b)   (c)

Despite adequate monitoring, failures in sterilization still occur, principally because of two human errors—failure to choose the appropriate procedure and, too often, failure to even place the item in the sterilizer.

Sterility does not always guarantee safety. Material contaminated with endotoxin retains the ability to injure when introduced into the body. Most articles to be used internally are rinsed prior to sterilization and samples of the sterile objects are evaluated for endotoxic contamination. (See color box, Chap. 4.)

EVALUATION OF DISINFECTANT AND ANTISEPTIC ACTIVITY

The danger of ineffective disinfection or antisepsis is reduced by evaluating the antimicrobial effectiveness of the procedure. Disinfectants are currently evaluated by laboratory tests or by "in-use" tests.

Laboratory Tests

The activity of antimicrobial chemicals may be determined in the laboratory under defined conditions and against specific test organisms. The results indicate the possible usefulness of an agent but cannot assure that the chemical will be active under actual conditions of use.

Microbial activity of disinfectants is often compared to that of phenol. The results are expressed as a ratio called the **phenol coefficient (P.C.)**.

$$\text{P.C.} = \frac{\text{Antimicrobial activity of test agent}}{\text{Antimicrobial activity of phenol}}$$

Chemicals that are more effective than phenol have coefficients greater than 1. The P.C. of a bactericidal agent is meaningful only if the agent's mode of action is similar to that of phenol.

Another laboratory test, the **use-dilution method** evaluates the ability of a germicide to kill a standard number of microbes on an object's surface. Stainless steel cylinders soaked in a test culture are usually used as the contaminated carrier surfaces. The contaminated cylinders are soaked in different strengths of the disinfectant, after which they are subcultured in broth. For each disinfectant, the highest dilution (weakest solution) that kills all cells on the test cylinder is the ideal concentration for disinfection.

In-Use Tests

Laboratory tests often fail to reveal antimicrobial effectiveness in real-life situations with all their undetermined variables. In-use tests are designed to evaluate disinfectants under actual conditions of use. The objective is to obtain accurate counts of the microbial contaminants that remain on articles after treatment with a particular antimicrobial agent.

Following routine treatment with a test disinfectant, one of three procedures is usually used to determine the number of surviving microbes:

■ **Direct Sampling of Surfaces with Agar Plates** RODAC (replicate

FIGURE 13-10

The medium in the RODAC plate makes direct contact with surfaces to be sampled. Like other media used to test the effectiveness of antimicrobial chemicals, it contains agents that inactivate disinfectants or antiseptics that may remain on the object tested.

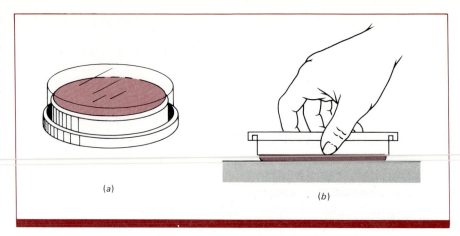

(a)   (b)

organisms *d*irect *a*gar *c*ontact) plates are designed so that smooth surfaces can be sampled after they have been cleaned with the test disinfectant. The agar in these plates forms a convex (rounded) surface that extends beyond the lid of the lower plate and therefore makes direct contact with flat materials against which it is pressed (Fig. 13-10).

CONTROL OF
MICROORGANISMS

■ **Swabbing** Inaccessible areas of uneven porous surfaces must be sampled with moist swabs, which are then streaked on agar plates. The medium is incubated, and the resultant colonies indicate the number of organisms that survived disinfection.

■ **Rinsing** Following disinfection, small objects may be rinsed with or immersed in standard volumes of sterile water. The number of viable microbes eluted from the object is determined by plate count of the rinse water.

Like disinfectants, antiseptics are also evaluated by in-use tests. Since contaminated hands are such an important source of disease, many of the techniques are primarily concerned with determining the number of organisms on the hands following scrubbing. This may be done by:

1. Swabbing the treated skin surface and streaking agar plates with the swabs

2. Directly pressing the skin against the solid surface of a growth medium (Fig. 13-11)

3. Rinsing the hands with sterile water, plating the water on solid medium, and counting the colonies that develop after incubation

4. Pressing a strip of sterile tape against the skin surface and placing tape on a solid medium

5. Swabbing the interior of surgical gloves after use to determine the

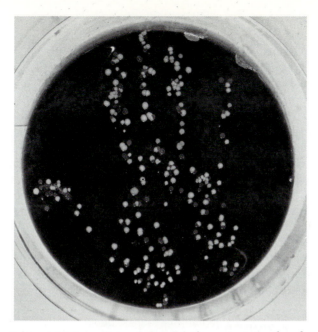

FIGURE 13-11
Hand impression made on sterile blood agar in a pie plate and subsequently incubated.

number of organisms that could have been released if a puncture developed.

All of these procedures have their flaws. The final evaluation of antimicrobial agents is the reduction of detrimental activities of microbes, particularly a decreased rate of spoilage and infectious disease.

OVERVIEW

Antimicrobial agents are employed to destroy potentially harmful microbes or inhibit their growth. Microbicidal agents kill microbes by causing irreparable damage to essential components. Microbistatic agents only inhibit the growth of microorganisms and do so only while in contact with the microbe. Many physical and chemical agents are used to eliminate microbial contaminants from inanimate articles. Some chemicals are nontoxic enough for use on the surface of the human body.

Sterilization, the complete elimination of all forms of life, is achieved by moist heat (autoclaving), dry heat, radiation, filtration, and three germicidal chemicals—ethylene oxide, glutaraldehyde, and formaldehyde. The selection of a sterilization procedure depends on the ability of the object to withstand the treatment. Success of the treatment depends on length of exposure and concentration of the agent, as well as on the nature and number of the contaminating microbes. A greater number of microbes requires more exposure time or increased doses of the antimicrobial agent to achieve sterility. The success of sterilization procedures may be evaluated with chemical or biological indicators. Chemical indicators determine if the conditions in the apparatus were adequate for sterilization. Biological indicators determine whether the sterilization process has actually killed an especially resistant test organism, usually bacterial endospores.

Unlike sterilization, disinfection does not destroy all microbes, but eliminates potentially harmful contaminants. Boiling, pasteurization, and most antimicrobial chemicals are disinfectants. Bacterial endospores are particularly resistant to these agents. Low-level disinfectants can be used as antiseptics on body surfaces. These chemicals include alcohols, quaternary ammonium compounds, iodine and iodophors, mercurials, silver-containing compounds, and chlorhexidine. They are most effective if the microbial populations on the body surface are first reduced by mechanical methods such as washing. This preparatory step also removes organic debris that interferes with the antimicrobial activity of many disinfectants. Disinfectants are evaluated in the laboratory either by comparing their activity to phenol (phenol coefficient method) or by determining the highest dilution effective against a test organism (use-dilution method). In-use tests evaluate both disinfectants and antiseptics under actual conditions of use.

KEY WORDS

microbicidal agent

microbistatic agent

germicidal agent

sterilization

disinfection

antisepsis

chemotherapy

preservation

aseptic technique

pasteurization

autoclave

ionizing radiation

ultraviolet radiation

filtration

ethylene oxide

phenolic

quaternary ammonium compound

iodophor

chlorhexidine

spore strip

phenol coefficient

use-dilution method

REVIEW QUESTIONS

1. List five factors that influence the activity of antimicrobial agents.

2. Why is it advisable to wash highly contaminated objects with soap and water before treating with antimicrobial agents?

3. Differentiate between the following: sterilization, pasteurization, antisepsis, and disinfection.

4. What are the advantages and limitations of the following?
  (a) Boiling   (b) Autoclaving
  (c) Irradiation   (d) Dry heat
  (e) Ethylene oxide   (f) Filtration

5. List four procedures used for monitoring sterilization processes.

6. List five chemical antiseptics and the conditions recommended for their use.

7. What are the advantages of in-use tests over laboratory tests of antiseptics and disinfectants?

Antibiotics and Chemotherapy

14

The search for agents to cure infectious disease began long before people were aware of the existence of microbes. Some of these early attempts were even successful. Eating the bark of the cinchona tree, for example, prevented and frequently cured malaria. Only in modern times was quinine shown to be the active ingredient in the bark. Such successes were rare, however, and millions of lives were lost to diseases that today are readily cured with **chemotherapeutic agents**, chemicals administered to people for treating disease.

Some of the most effective chemotherapeutic agents are **antibiotics**, chemicals produced by microorganisms that, in very low concentrations, selectively kill or inhibit the growth of other microbes. Antibiotics that are nontoxic to human cells can usually be safely introduced into an infected person to combat pathogens.

The antibiotic era was ushered in with the accidental discovery of penicillin in 1929 (see color box). Most of the other medically important antibiotics in use today were discovered between 1939 and 1963 (Table 14-1). The most effective agents were isolated from the molds *Penicillium* and *Cephalosporium* and from members of the bacterial genera *Streptomyces* and *Bacillus*. All these organisms are common contaminants in the soil and air.

TABLE 14-1

Some Major Chemotherapeutic Agents

Antibiotic	Year of Discovery	Source	Spectrum of Activity
Penicillin G	1928	*Penicillium*	Gram-positive bacteria and Neisseriae
Sulfa drugs*	1935	Chemical synthesis	Gram-positive and gram-negative bacteria
Griseofulvin	1939	*Penicillium*	Fungi
Streptomycin	1943	*Streptomyces*	Gram-negative bacteria; mycobacteria
Bacitracin	1945	*Bacillus*	Gram-positive bacteria
Chloramphenicol	1947	*Streptomyces*	Gram-positive and gram-negative bacteria
Polymyxin	1947	*Bacillus*	Gram-negative bacteria
Tetracycline	1948	*Streptomyces*	Gram-positive and gram-negative bacteria
Cephalosporin	1948	*Cephalosporium*	Gram-positive bacteria
Neomycin	1949	*Streptomyces*	Gram-negative bacteria
Nystatin	1950	*Streptomyces*	Fungi
Erythromycin	1952	*Streptomyces*	Gram-positive bacteria
Cycloserine	1954	*Streptomyces*	Gram-positive bacteria
Amphotericin B	1956	*Streptomyces*	Fungi
Vancomycin	1956	*Streptomyces*	Gram-positive bacteria
Metronidazole*	1957	Chemical synthesis	Protozoa; anaerobic bacteria
Kanamycin	1957	*Streptomyces*	Gram-negative bacteria
Rifamycin	1957	*Streptomyces*	Mycobacteria
Gentamicin	1963	*Micromonospora*	Gram-negative bacteria

*Because these drugs are not produced by microbes they are not considered antibiotics. But like antibiotics, their inhibiting effects are selective against microorganisms, and they are consequently effective chemotherapeutic agents.

SELECTIVE TOXICITY

Antimicrobial agents are chemotherapeutically valuable only if they are **selectively toxic**, that is, produce no serious undesirable side effects in the person being treated. The safest drugs are those that, like penicillins, cephalosporins, and sulfa drugs, interfere with metabolic processes unique to procaryotes. Penicillins and cephalosporins, for example, inhibit the synthesis of the peptidoglycan required for a functional bacterial cell wall. These targets are not shared by eucaryotic cells and provide the basis for the drugs' selective toxicity.

Drugs that are toxic for eucaryotic pathogens (fungi or protozoa) are generally less selective. The host and the pathogen share so many cellular similarities that an agent affecting the microbe will likely have an adverse effect on human cells. Many drugs that effectively eliminate eucaryotic pathogens have such toxic side effects they are used only when the danger of the patient dying from the disease exceeds the danger of toxic reactions to the drug.

Drugs that are selectively toxic against viruses are even more scarce. Because viruses become integral parts of their host cells, it is usually impossible to interfere with virus processes without also damaging the host. Consequently, most viral infections are incurable by antimicrobial drugs. The natural defenses of the host, however, limit the course and consequences of most viral infections.

ANTIMICROBIAL RESISTANCE

The discovery of antibiotics fostered hope that infectious disease would soon be eradicated. Dangerous staphylococcal infections, for example, seemed universally curable by penicillin. However, within a year after the commercial introduction of penicillin for wide-scale use against staphylococcal infections, most of the target pathogens were resistant to the antibiotic. Pencillin resistance occurred in many other pathogens, as well. In fact, resistant pathogens appear soon after the introduction of any antibiotic. The more an antibiotic is used, the more likely it is that drug-resistant organisms will emerge.

Antibiotics do not create resistant cells or cause resistant mutations. They only selectively favor the survival and proliferation of drug-resistant strains which are usually an insignificant subpopulation within the vast majority of sensitive cells. Prolonged exposure to antibiotics, however, eliminates most sensitive cells, allowing drug-resistant organisms to become the dominant microbes (Fig. 14-1). Therefore, overuse of an antibiotic decreases its effectiveness. Fortunately, in the United States the restricted access to antibiotics limits the numbers of antibiotic-resistant pathogens in the general community. Because of widespread use of antibiotics in hospitals, however, most pathogens isolated from hospitalized patients are resistant to one or more antibiotics.

FIGURE 14-1

The emergence of antibiotic resistance by mutation and selection. A population of antibiotic-sensitive bacteria often contains a small number of antibiotic-resistant cells that appear by random mutation. In the presence of antibiotics, most sensitive cells are killed or inhibited; a few resistant cells are unaffected and continue to grow. Prolonged exposure to the drug prevents sensitive cells from repopulating the area, allowing resistant microbes to predominate. Infection by antibiotic-resistant cells can no longer be controlled by the drug. In the absence of the antibiotic, the survival of drug-sensitive bacteria is favored over drug-resistant cells.

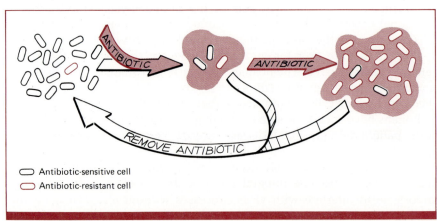

Antibiotic-sensitive cell
Antibiotic-resistant cell

Genetics of Antibiotic Resistance

Antibiotic resistance is a genetically determined trait acquired either by mutation in the bacterial chromosome or by direct transfer of R-factor plasmids from antibiotic-resistant strains to sensitive recipients (see Chap. 12).

Random chromosomal mutation produces in any bacterial population a few cells that are resistant to antibiotics. These mutants become significant only when antibiotic exposure favors their survival over sensitive strains. Transfer of plasmids by conjugation or transformation, however, rapidly transmits antibiotic resistance among normal flora. The overuse of antibiotics creates an environment that continually favors resistant bacteria and, in this way, establishes a reservoir of R factors in the normal flora. Since plasmids can be transferred among different species, pathogens may acquire R factors from normal flora bacteria (Fig. 14-2). For example, *Escherichia coli*, a harmless intestinal resident, may donate R factors to the pathogens that cause dysentery or typhoid fever or to opportunists such as *Pseudomonas*. Such "plasmid promiscuity" allows for extensive spread of antibiotic resistance throughout a heterogeneous population of bacteria. Since R factors usually provide resistance to a number of drugs, prolonged

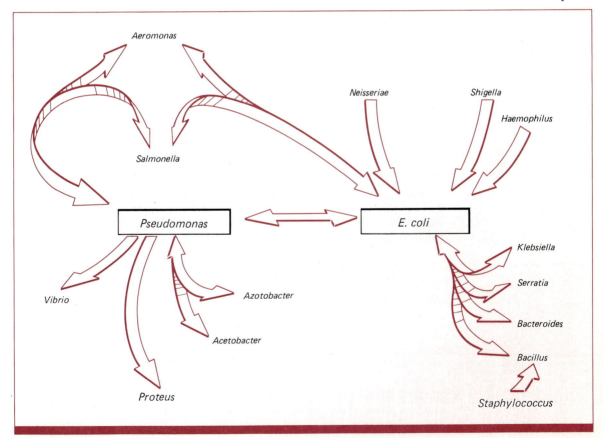

exposure to even a single antibiotic may favor the proliferation of bacteria resistant to several drugs.

Antibiotics are extensively used as additives in feeds for pigs, cattle, chickens, and other farm animals to encourage increased meat production. Unfortunately, this practice selects for the survival of microorganisms that may contain R factors. Federal statutes prevent the sale of meat that contains residual antibiotics. However, antibiotic use in livestock feeds poses another serious danger that is not controlled by these statutes. The microorganisms shed by the living animal and during slaughter contain R factors that may transfer antibiotic resistance to human pathogens and normal flora.

CHOOSING THE BEST CHEMOTHERAPEUTIC AGENT

No single antimicrobial drug is safe in all patients or effective against every infectious disease. The following factors influence the therapeutic value of antimicrobial drugs and must be considered if the patient is to receive the most effective chemotherapy.

The susceptibility of the pathogen to the chemotherapeutic agent

The drug's spectrum of activity

Possible adverse reactions to the drug

The site of infection and the drug's ability to reach those tissues

Metabolism of the drug in the body

Duration of treatment

Interaction with other drugs the patient may be taking

Susceptibility of the Pathogen

For an antimicrobial drug to be of value it must be able to kill or inhibit the pathogen within the body. Isolation and identification of the pathogen may suggest an appropriate drug; however, the emergence of drug-resistant microbes makes it impossible to predict effectiveness based solely on the microbe's identity. The isolated pathogen should be tested for susceptibility to various antimicrobial chemicals to assure that it can be killed or inhibited by the agent selected. The methods of testing for antimicrobial susceptibility and drug resistance will be discussed later in this chapter.

Spectrum of Activity

Antimicrobial agents are either narrow-spectrum or broad-spectrum drugs according to the range of microorganisms against which they are *usually* effective. **Broad-spectrum** drugs affect a wide number of microorganisms, whereas **narrow-spectrum** agents are more limited in the types of cells affected. Some narrow-spectrum drugs, for example, inhibit only gram-positive bacteria. Usually the ideal agent has the narrowest spectrum that is effective against the identified pathogen, since broad-spectrum agents disrupt the normal microbial flora that contribute to the ecological balance and health of the host. However broad-spectrum drugs such as tetracyclines, chloramphenicol, and sulfa drugs are often used to treat mixed infections caused by several pathogens or for emergency situations in which there is no time to wait for laboratory results. In most cases such "shotgun therapy" is unwarranted.

Adverse Drug Reactions

Chemotherapeutic agents may have mild-to-fatal side effects. These may be general symptoms of chills, fever, headache, nausea, or rash. More severe toxic reactions may damage the liver, kidney, or nervous system. For example, drugs that accumulate in the kidney cause renal damage and should not be used in older persons or people with previous renal disease. Antimicrobial agents that pass through the placenta and cause fetal damage should not be used in a pregnant woman even though these drugs are harmless to the mother. Similarly, antibiotics that are secreted in breast milk should not be used by mothers who are breast-feeding.

Perhaps the most common adverse effects of antibiotic therapy are the secondary infections that develop because the normal flora were disrupted by broad-spectrum antibiotics. These **superinfections** are caused by fungi or bacteria whose growth is usually controlled by competition from the normal flora. If these opportunistic pathogens are resistant to the antimicrobial agent used, they may rapidly replace the disrupted flora and cause diarrhea, vaginitis, severe inflammation of the colon, or pneumonia. Sometimes superinfections can be controlled by withdrawing the antibiotic and allowing the normal flora to repopulate. Many times, however, another antibiotic is used to eliminate the opportunistic organisms. Development of opportunistic superinfections can usually be avoided by using an effective drug with the narrowest spectrum of activity.

Site of Infection and Drug Distribution within the Body

Antimicrobial agents have no effect unless they reach the site of infection in concentrations high enough to incapacitate the pathogen. Antibiotics that cannot cross the barrier between the blood and the central nervous system, for example, are useless for treating meningitis unless injected directly into the cerebrospinal fluid. Abscesses, walled-off localized accumulations of microorganisms, are protected against chemicals that fail to penetrate the abscess walls. Similarly, the gall bladder protects the typhoid bacillus from the effects of antimicrobial agents, often creating "healthy" carriers that continually shed typhoid bacteria in their feces in spite of antibiotic therapy. Intracellular pathogens are protected from penicillin, streptomycin, and other antibiotics that penetrate poorly into human cells. Tetracyclines easily enter host cells and are therefore effective against rickettsial and chlamydial diseases. Sulfonamides are ineffective against pathogens in necrotic (dead) tissue, which contains compounds that compete with the drug.

The amount of antimicrobial agent found at the site of infection also depends on the route of administration. Localized infections are best treated with drugs that accumulate at the site of infection. Gastrointestinal infections, for example, respond best when treated by agents that are taken orally, are poorly absorbed from the intestine, and therefore remain at the site of infection. Local infections of the urinary bladder are best treated with agents that concentrate in the urine even though serum drug levels may be too low to be antimicrobial.

For effective treatment of many diseases the antibiotic must attain elevated concentrations in the patient's blood. Intravenous injections of penicillin, for example, ensure high concentrations throughout the bloodstream within 30 minutes. The intramuscular route is not as rapid, and somewhat lower concentrations are achieved. Oral administration is even slower and yields a substantially lower blood concentration (Fig. 14-3).

Drugs that are too toxic for internal use may be valuable as topical agents if the infection is localized on the body's surface.

Metabolism of the Drug

Many drugs are metabolized by the body. In some cases these changes increase their antimicrobial effectiveness. Prontosil, for example, is inactive until converted by the body to sulfonilamide (see color box). Metabolic changes, however, usually diminish antimicrobial effectiveness. For example, many chemotherapeutic agents are destroyed by the low pH of the stomach. Others are bound and inactivated by serum proteins. Therefore, the amount of effective antibiotic in the body may be considerably lower than the amount administered.

322

ANTIBIOTICS AND
CHEMOTHERAPY

FIGURE 14-3

The influence of route of administration on blood levels of penicillin G.

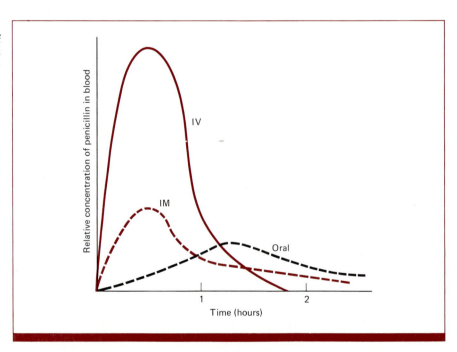

PRONTOSIL, A WONDROUS CURE IN CAMOUFLAGE

■ Theoretically it shouldn't have worked. Yet this miraculous new drug, the first of its kind, saved thousands of lives in the 1930s. The wonder drug, called prontosil, was the product of the German scientist, Domagk. He had synthesized this simple sulfur-containing dye with hopes that its tendency to stain bacteria but not human cells would translate into selective toxicity when given to persons with bacterial infections. From the start prontosil showed evidence of doing just that. The drug cured mice of the streptococcal infections Domagk had experimentally given them. Without protosil, they died. People suffering from similar infections were also cured with prontosil. The drug was accepted for widespread use, and soon the number of women dying from uterine infections following childbirth was reduced by more than half. Its ability to cure infectious diseases was unquestionable.

Attempts to determine how the drug worked, however, baffled scientists, who found that prontosil does not kill or even inhibit bacteria growing on artificial media. Yet the drug readily cures many diseases caused by these prontosil-"resistant" pathogens. This apparent paradox was resolved by the discovery that prontosil is metabolized in the body to sulfonilamide, a compound that inhibits bacterial growth both in the body and on artificial media. Sulfonilamide specifically interferes with a metabolic reaction essential for bacterial growth but not needed by eucaryotic cells, thereby accounting for its selective toxicity. Domagk was fortunate to have used animals for studying prontosil; had he used bacterial inhibition tests he may never have discovered it. Sulfonilamide was the first of the sulfa drugs, which even today, are among our most valuable antimicrobial chemicals for treating infectious disease.

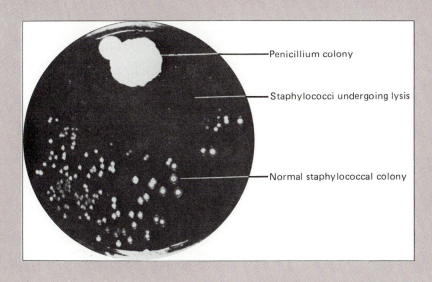

Penicillium colony

Staphylococci undergoing lysis

Normal staphylococcal colony

Duration of Treatment

Most drugs are metabolized or excreted before the infectious agents are eliminated. These drugs must be periodically readministered to maintain therapeutic levels. The schedule of administration is critical, and even a

323

delay of one-half hour may reduce the effectiveness of therapy. This problem is partially solved by including another chemical that prolongs the drug's effect. For example, procaine delays absorption of penicillin from an intramuscular injection site so there are therapeutic concentrations in the bloodstream for longer durations. Probenecid administered with penicillin delays excretion of the drug and is routinely included in the single large dose of penicillin used for treating gonorrhea.

Symptoms often disappear before all pathogens are eliminated, but the cells that remain can resume growth and cause a recurrence unless therapeutic concentrations of the chemotherapeutic drug are maintained until all pathogens are eliminated from the infected patient. Unfortunately many persons stop taking their medication as soon as symptoms subside; consequently, their diseases recur.

Drug Interactions

Antimicrobial agents are sometimes administered in combination to increase the spectrum of activity against mixed infection or when treatment of undiagnosed disease is urgent. Drug combinations are best employed when they result in *synergism,* an enhancement of activity greater than the sum of the two agents when used alone. For example, carbenicillin, (a penicillin derivative) weakens the bacterial wall structure and enhances the penetration of gentamicin into the interior of the cell, where this second drug impairs protein synthesis. Synergistic combinations are sometimes effective against infections that normally do not respond to either drug alone. Synergistic combinations allow some drugs to be used in lower concentrations, thereby reducing negative side effects. In some cases, combination therapy reduces the likelihood that drug resistance will develop; mutants resistant to one antibiotic are eliminated by the other drug. On the other hand, combination therapy favors the emergence of multiple-resistant strains that contain R factors.

Some antimicrobial agents *antagonize* each other. Bacteriostatic agents such as tetracycline, for example, should not be used with penicillin, which kills only actively growing cells. Other drugs may inactivate or precipitate one another and cannot be administered simultaneously in a single solution. Antibiotics may react with other types of drugs or even with food, sometimes producing toxic reactions in the patient. For example, metronidazole (Flagyl), an antibiotic commonly used, for treating *Trichomonas* vaginal infections, interferes with the metabolism of ethyl alcohol. Toxic levels of ethyl alcohol may accumulate if alcoholic beverages are consumed during the course of metronidazole therapy.

ANTIMICROBIAL MECHANISMS

Microbistatic chemotherapeutic agents impede microbial growth until the host defense mechanisms can eventually destroy the pathogens. Microbicidal agents are usually able to kill cells with no assistance from the host defenses. These agents function by selectively disrupting (1) cell wall synthesis, (2) cell membrane function,

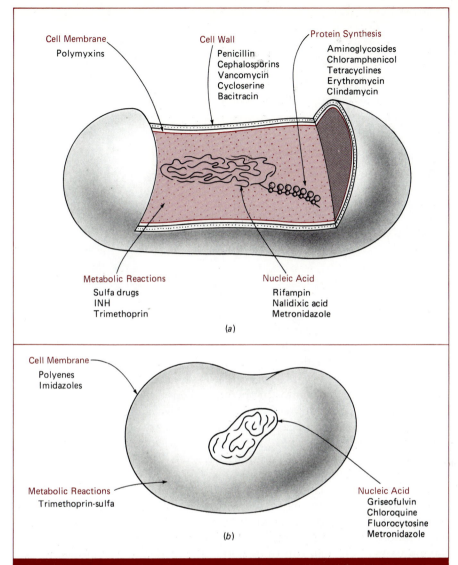

FIGURE 14-4
Microbial targets of the commonly used chemotherapeutic agents: (*a*) Antibacterial drugs and (*b*) drugs used against eucaryotic pathogens.

Cell Membrane
Polymyxins

Cell Wall
Penicillin
Cephalosporins
Vancomycin
Cycloserine
Bacitracin

Protein Synthesis
Aminoglycosides
Chloramphenicol
Tetracyclines
Erythromycin
Clindamycin

Metabolic Reactions
Sulfa drugs
INH
Trimethoprin

Nucleic Acid
Rifampin
Nalidixic acid
Metronidazole

(*a*)

Cell Membrane
Polyenes
Imidazoles

Metabolic Reactions
Trimethoprin-sulfa

Nucleic Acid
Griseofulvin
Chloroquine
Fluorocytosine
Metronidazole

(*b*)

325

ANTIMICROBIAL
MECHANISMS

(3) protein synthesis, (4) nucleic acid synthesis, or (5) other specific metabolic reactions. The major drugs and their microbial targets are identified in Figure 14-4.

Target: Bacterial Cell Walls

Agents that prevent the synthesis of peptidoglycan produce osmotically fragile bacterial cells that lyse unless kept in a medium that prevents the influx of water (Fig. 14-5). Bacteria with defective cell walls survive poorly in the human body. Antibiotics that inhibit cell wall synthesis include bacitracin, vancomycin, cycloserine, penicillins, and cephalosporins. Be-

FIGURE 14-5

Neisseria gonorrhoeae after treatment with penicillin. (a) Cells lyse and all that remain are membrane fragments; (b) in medium that prevents the influx of water the cells are distorted but do not burst.

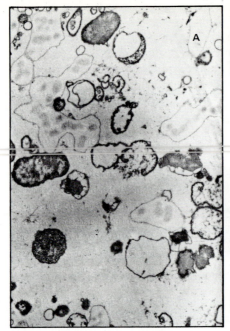
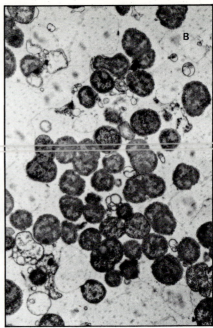

ANTIBIOTICS AND CHEMOTHERAPY

cause none of these agents destroys existing peptidoglycan, they are effective only against actively growing bacteria. Gram-positive bacteria are most sensitive to many of these agents because the outer membrane of most gram-negative cell walls tends to prevent penetration of the antibiotic to its site of action. Eucaryotic cells lack peptidoglycan and are completely resistant to penicillins and cephalosporins, which are among the least toxic of all antibiotics. Cycloserine, bacitracin, and vancomycin, however, have toxic effects unrelated to their antimicrobial activity that reduce their value as chemotherapeutic agents.

Penicillins Penicillin may have several antibacterial mechanisms. Its effect on the cell is determined by *penicillin-binding proteins* in the cell membrane. Depending on the type of binding protein that couples with the antibiotic, penicillin can cause inhibition of peptidoglycan synthesis, activation of autolytic enzymes in the cell, cellular filamentation, or other abnormal configurations that interfere with growth. In some species, penicillin tends to cause bacteriostatic changes rather than cell lysis.

Penicillin G, originally discovered by Fleming, is still one of the most useful antimicrobial drugs available. The antibiotic is a metabolic by-product of the blue-green mold *Penicillium chrysogenum*. (Other types of natural penicillins are produced by altering the nutrients in the growth medium.) Various chemical groups can be attached to the penicillin core molecule, producing penicillin derivatives with additional useful properties. These antibiotics are referred to as *semisynthetic derivatives* because part of their structure is the result of fungal metabolism and part is the result of chemical manipulation.

These derivatives differ from penicillin G in three ways:

Acid Resistance Penicillin G, which is destroyed by stomach acids, must be administered by injection. In contrast, penicillin V, oxacillin, and ampicillin are acid-resistant and are administered orally.

Penicillinase Resistance Some bacteria produce **penicillinases,** enzymes that inactivate the antibiotic by hydrolyzing it to an ineffective form. They are therefore resistant to the antibiotic and may even protect penicillin-sensitive organisms that are present in mixed infections. The penicillin derivatives methicillin, oxacillin, and nafcillin have been altered to a form resistant to penicillinase hydrolysis. These derivatives are much less active than penicillin G against penicillin-sensitive microbes and are recommended only for treating diseases caused by penicillinase-producing bacteria.

Broad-Spectrum Activity Some penicillin derivatives are effective against gram-negative bacteria, most of which are not killed by penicillin G. Ampicillin, amoxicillin, and carbenicillin readily penetrate outer membranes of gram-negative cells.

Table 14-2 summarizes the properties of some penicillins that are currently employed in therapy.

Although penicillins are among the least toxic antibiotics, approximately one in twenty patients experiences some allergic reaction to penicillin treatment. Usually the symptoms are mild rashes, but the occasional severe reaction can kill a person within 15 minutes of penicillin injection. Patients with a history of penicillin allergy should be treated with alternative drugs.

TABLE 14-2

Properties of Natural and Semisynthetic Penicillins

Name of Drug	Important Properties
Penicillin G	Low toxicity*; acid-sensitive; pencillinase-sensitive; inactive against most gram-negative bacteria; poorly absorbed from gastrointestinal tract; rapidly absorbed and excreted following injection
Penicillin G with procaine	Prolonged duration of therapeutically effective concentrations
Penicillin V (phenethicillin)	Acid-resistant
Methicillin, nafcillin	Penicillinase-resistant; less effective than penicillin G against nonpenicillinase producers
Oxacillin, cloxacillin, icloxacillin	Acid and penicillinase-resistant; less effective than penicillin G against nonpencillinase producers
Ampicillin, amoxacillin	Acid-resistant; active against some gram-negative bacteria
Carbenicillin	Active against gram-negative bacteria, including *Proteus* and *Pseudomonas*

*All penicillins may produce adverse reactions in allergic persons.

FIGURE 14-6

General structure of
natural penicillin G and
cephalosporin C. A
number of derivatives with
different properties can be
created by modifying the R
group, the chemical region
shown in color.

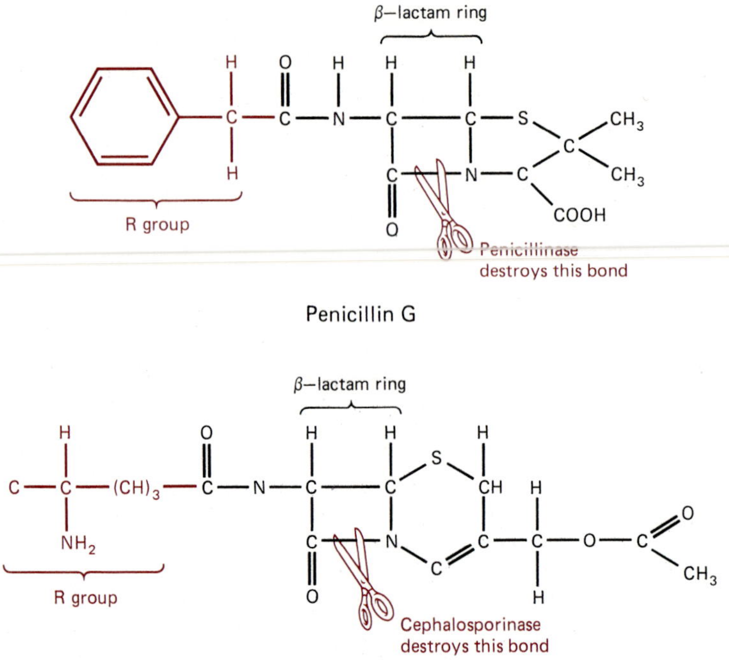

Penicillin G

Cephalosporin C

Cephalosporins Cephalosporins are similar to penicillins in structure and activity but are produced by molds in the genus *Cephalosporium*. Like penicillin, semisynthetic derivatives of the natural product, cephalosporin C, are used in various clinical situations. The molecular structure of the cephalosporins is similar to that of penicillin (Fig. 14-6). Cephalosporins are not destroyed by penicillinases and can be prescribed for treating infections caused by penicillinase-producing bacteria. They are, however, inactivated by cephalosporinases, enzymes produced by some resistant bacteria. Because of the chemical similarities between the antibiotics, persons allergic to penicillins are usually not treated with cephalosporins.

Target: Cell Membranes

Some antibiotics kill cells by interfering with the normal permeability of the cell membrane. The most important inhibitors of membrane function are the *polyene* antibiotics, compounds that react with sterols of eucaryotic membranes. The most therepeutically valuable of the polyenes are those that bind with ergosterol, the sterol in fungal membranes. Human cells contain cholesterol instead of ergosterol and are therefore not as susceptible as fungi to these agents. The sterols in human cells, however, are somewhat affected by the polyenes. Toxicity for the patient, therefore, should be considered when using these antibiotics.

Polyenes are among the few antimicrobials available for the treatment of cryptococcosis, coccidiodomycosis, aspergillosis, mycormycosis, candidiasis, and other potentially fatal systemic fungal infections. The drug most often used is amphotericin B, an antibiotic whose therapeutic and toxic doses are so close that full therapeutic doses are used only in life-threatening situations.

The tendency of amphotericin B to increase the porosity of the cell membrane, however, is used to facilitate the entry of other drugs into the fungal cell. Lower, less toxic doses of amphotericin B are often administered in synergistic combination with otherwise ineffective drugs that alone cannot reach their intracellular sites of action.

Another polyene, nystatin, is too toxic for systemic therapy and is primarily used as a topical agent to eliminate *Candida* infections from oral or vaginal mucous membranes. Systemic use of nystatin would eliminate the patient along with the pathogen.

Imidazoles are another group of agents that interfere with the cell membrane of fungi. Unlike the polyenes, these inhibit the *synthesis* of ergosterol. Imidazoles appear to be effective broad-spectrum antifungal drugs with few or no serious side effects. The most promising drug for treating systemic mycoses is ketoconazole, which has been used successfully against several pathogenic molds and yeasts (Fig. 14-7). Ketoconazole is also effective against the dermatophytes.

Another group of membrane inhibitors, the *polymyxins,* bind to the phospholipids in bacterial membranes and alter their permeability. This causes leakage of small molecules and cell death. Polymyxins are especially valuable as topical agents for controlling gram-negative bacterial infections in burn patients. Their toxic side effects, however, prevent their systemic use.

Target: Protein Synthesis

Some antibacterial chemicals can selectively inhibit bacterial protein synthesis by disabling procaryotic ribosomes (70 S) while causing little deleterious effect on eucaryotic ribosomes (80 S). The selective toxicity of

FIGURE 14-7
The effect of ketoconazole on a chronic yeast infection: (*a*) Before treatment and (*b*) after treatment.

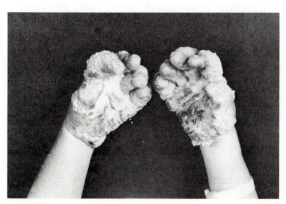

(a)

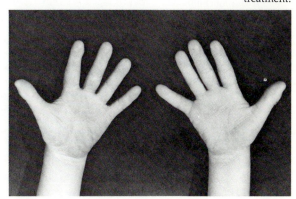

(b)

these agents is often enhanced by their ability to pass through bacterial membranes more readily than through eucaryotic membranes.

Aminoglycosides   The *aminoglycosides* are a group of structurally similar antibiotics that attach to the bacterial ribosome and interfere with accurate translation of the genetic code. Unfortunately they also affect the small ribosomes in mitochondria (see Chap. 2), which perhaps accounts for their severe side effects. Prolonged use or high dose of any aminoglycoside can cause deafness. Aminoglycosides are therefore used systemically only against serious gram-negative infections. They are poorly absorbed from the gastrointestinal tract and are usually administered by injection. The aminoglycosides include streptomycin, gentamicin, neomycin, tobramycin, and kanamycin (and its semisynthetic derivative, amikacin). With the exception of gentamicin, all aminoglycosides are produced by *Streptomyces* species. (Gentamicin is spelled "micin" rather than "mycin" to denote that its origin is *Micromonospora,* not *Streptomyces.*)

Tetracyclines   The *tetracyclines* are a group of broad-spectrum antibiotics that prevent the binding of transfer RNA to 70 S ribosomes. Since 80 S ribosomes are not affected, the inhibition is specific for procaryotes.

Tetracyclines are acid-stable and readily absorbed from the gastrointestinal tract, so they are administered orally. Milk products and iron-containing foods reduce antimicrobial effectiveness and should be avoided when medication is taken. Because tetracyclines readily penetrate cell membranes, they are the drug of choice for treating intracellular bacterial infections caused by chlamydias, rickettsias, and *Brucella.*

The antimicrobial spectrum of tetracyclines is so broad that their extended use disrupts the normal flora and encourages secondary infections by tetracycline-resistant staphylococci or the yeast *Candida albicans.* In addition, these antibiotics discolor developing teeth (Fig. 14-8) and may

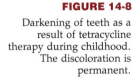

330

ANTIBIOTICS AND
CHEMOTHERAPY

FIGURE 14-8

Darkening of teeth as a
result of tetracycline
therapy during childhood.
The discoloration is
permanent.

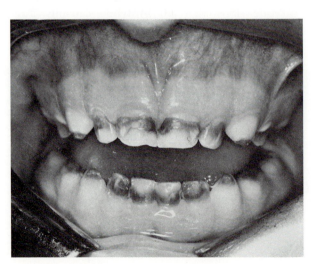

retard normal growth of bones. Tetracyclines should therefore not be prescribed for pregnant women (they cross the placenta) or for growing children.

Erythromycin, Clindamycin, and Chloramphenicol Several ribosome inhibitors prevent peptide bond formation during bacterial protein synthesis. Because they do no structural damage to ribosomes, they are all bacteriostatic agents. *Erythromycin* is primarily active against gram-positive bacteria and commonly used to treat infections in persons allergic to penicillins. *Clindamycin* is used for treating abscesses and systemic infections caused by gram-negative anaerobic bacteria. Its use, however, may upset the normal anaerobic intestinal flora, thereby promoting the growth of *Clostridium difficile,* an intestinal opportunist that causes severe, sometimes fatal, colitis (inflammation of the bowel). *Chloramphenicol* is an inexpensive broad-spectrum antibiotic that readily diffuses into spinal fluid, the gall bladder, and other body sites that are inaccessible to many other chemotherapeutic agents. Unfortunately, in some persons chloramphenicol has caused fatal aplastic anemia, the complete loss of the bone marrow's ability to produce red blood cells. In the United States, chloramphenicol is used only for treating typhoid fever and other life-threatening infectious diseases that fail to respond to safer antibiotics. Because it can be inexpensively synthesized, however, chloramphenicol is one of the most commonly used antibiotics in many areas of the world.

Target: Nucleic Acids

Rifampin is one of the few agents in this category selective enough for safely treating infectious diseases. This drug inhibits transcription of mRNA from DNA by binding to and inactivating bacterial mRNA polymerase. Rifampin is especially effective against *Mycobacterium* species and is therefore an important drug for treating tuberculosis and leprosy. Since rifampin-resistant strains emerge rapidly and chemotherapy continues for long periods of time, it is usually used in combination with other antimycobacterial agents.

Nalidixic acid degrades the replicating bacterial chromosome. High concentrations are excreted in the urine, making nalidixic acid useful for treating urinary tract infections, especially those caused by gram-negative organisms. In the serum, however, therapeutic levels are never achieved.

Flucytosine indirectly inhibits nucleic acid replication in fungi by preventing the synthesis of some nucleotides. It does not have a similar effect on humans because only fungi contain the enzyme that activates flucytosine. It is often used in conjunction with amphotericin B for treating systemic *Candida* and *Cryptococcus* infections. *Griseofulvin* also interferes with DNA replication in some fungi. It is used for treating dermatophyte infections of skin, nails, and scalp. The pathogens are not killed by griseofulvin, but are ultimately shed with sloughed skin or nails, a process requiring months of antibiotic therapy.

Metronidazole is most commonly employed to treat vaginal infections caused by the protozoan *Trichomonas vaginalis,* but is effective against other pathogenic protozoa as well. Sensitive organisms metabolize metronidazole to a compound that binds to DNA and rapidly kills the cell. This synthetic drug is also one of the most active agents against anaerobic bacteria. Almost no aerobic bacteria are affected.

Target: Bacterial Metabolism

Several important chemotherapeutic agents called **antimetabolites** competitively inhibit bacterial metabolic reactions. The antimetabolites are usually substrate analogs, compounds with structures that closely resemble the substrate of an enzyme and therefore compete for the enzyme's active site. (See Chemical Inhibitors of Enzyme Activity, Chap. 10.) If the concentration of the inhibitor is high enough, it successfully competes with the substrate and prevents its conversion to products. Inhibitors that interrupt reactions essential to microbial growth but not needed for human metabolism may be used for chemotherapy.

Sulfa drugs, for example, are analogs of para-aminobenzoic acid (PABA) (Fig. 14-9a). Normally PABA is enzymatically converted to folic acid, a coenzyme essential for growth. Sulfa drugs react with the enzyme but are not converted to products. They tie up the active site and prevent the production of folic acid (Fig. 14-9b). Bacteria that must synthesize their own folic acid (they cannot absorb folic acid from the medium) are inhibited by

FIGURE 14-9

(*a*) Note the similarity in structure between PABA and sulfa drugs. They are structural analogs of one another. (*b*) Inhibition of folic acid synthesis by sulfa drugs.

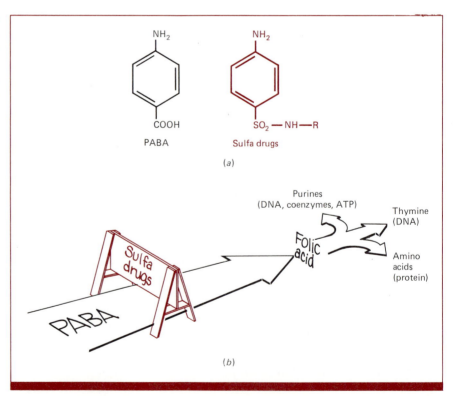

the sulfa drugs. Human cells obtain all their folic acid from external sources and are therefore not affected by inhibitors of folic acid synthesis. Consequently the sulfa drugs are selectively toxic for bacteria. Because inhibitor binding is reversible, sulfa drugs are bacteriostatic agents.

A variety of folic acid inhibitors have additional advantageous properties. Sulfamethoxazole is more soluble in urine than other sulfa drugs and therefore more effective against urinary tract infections. Increased solubility also reduces the harmful precipitates that may otherwise be deposited in the kidney. *Trimethoprin* blocks a later step in the folic acid pathway. Trimethoprin and sulfa drugs interact synergistically and are often prescribed together. Unlike sulfa, trimethoprin interferes with a step that is critical to human metabolism as well as to bacterial metabolism. Fortunately, the affinity of trimethoprin for the critical enzyme is 50,000 times greater in bacteria than in human cells. At therapeutically effective doses, therefore, toxicity to the patient is inconsequential.

Another antimetabolite, *isoniazid* (INH) is thought to competitively inhibit NAD synthesis in mycobacteria and is used for treating tuberculosis. Unlike the sulfa drugs, INH is bactericidal. Since INH resistance is common among mycobacteria, this antimetabolite is usually used in combination with rifampin and other antibiotics.

ANTIVIRAL AGENTS

Despite years of intensive investigations, currently only a few chemotherapeutic chemicals are licensed for the treatment of viral disease. None of these agents cures the disease for which it is used. The drug *amantadine* blocks the penetration of type A influenza virus into cells, but has no effect on the virus once it has entered and is replicating. Amantadine may prevent influenza infection and can shorten the duration of symptoms by 24 hours if administered within 2 days of disease onset. It is most often employed to protect older people and others at risk of contracting fatal cases when epidemic outbreaks are expected.

Acyclovir is a specific inhibitor of herpesvirus DNA synthesis. The drug is altered by a viral protein and is activated only in virus-infected cells. It is used topically on oral or genital herpesvirus lesions to shorten the course of the initial episode. The drug does not prevent recurrences once the virus is established in the nervous system. When administered systemically, it may prevent overwhelming herpesvirus infections in transplant recipients and other immunocompromised patients and is also effective against herpes zoster (shingles) in normal hosts. Acyclovir has few side effects and may be the first in a new generation of antiviral agents for widespread treatment of herpesvirus infections.

Idoxuridine was the first antiviral agent licensed in the United States. For more than 20 years it has been used topically for treating herpesvirus infections of the eye, but is too toxic for systemic therapy. Idoxuridine is an inhibitor of DNA replication and both cell DNA and viral DNA may be affected. Cells of the cornea are not affected because they are not actively dividing. *Ara A* is also used for ocular herpes, but because of its toxicity, it is used internally only for treating patients with potentially fatal herpesvirus diseases such as encephalitis.

The safety and antiviral effectiveness of several unlicensed compounds are being investigated, and they may soon be commercially available. Interferon is promising as an effective therapeutic agent against many viral diseases. Bromovinyl deoxyuridine (BVDU) is several times more effective against herpes zoster than acyclovir.

MECHANISMS OF ANTIMICROBIAL RESISTANCE

Resistance to antimicrobial agents may be natural or acquired. Some microorganisms are naturally resistant because they lack the target that the antibiotic affects or because the drug cannot permeate to its site of action. Fungi and protozoa, for example, possess no peptidoglycan and are naturally resistant to penicillin and other inhibitors of bacterial cell wall synthesis. Sensitive microbes, on the other hand, may acquire resistance by gaining the ability to

Inactivate or destroy the antibiotic

Alter their own membranes so they are no longer permeable to the agent

Alter the target site so it is no longer affected by the drug, or

Develop a mechanism to bypass the target metabolic reaction.

Inactivation of Antibiotics
Many microorganisms produce extracellular enzymes that destroy an antibiotic's activity. Penicillinases, for example, are produced by many bacterial species, including *Staphylococcus*, *Bacillus*, *Pseudomonas*, *Proteus*, *Mycobacteria*, *Yersinia*, *Salmonella*, and *Shigella*. Other bacteria may produce enzymes that chemically modify antibiotics to forms that are poorly absorbed by the microbe.

Decreased Antibiotic Uptake
Most chemotherapeutic agents must be able to penetrate the cell wall and cell membrane in order to achieve effective concentrations at an internal target site. A modification in the cell membrane may reduce its permeability to the drug, thereby increasing the microbe's resistance. Altered cell membrane permeability, however, does not confer resistance against penicillin and cephalosporins since these antibiotics block extracellular assembly of peptidoglycan.

Alteration of the Target Site
Microorganisms commonly acquire resistance when a structure or enzyme that is normally impaired by an antibiotic is modified and is no longer

recognized by the drug. Bacteria resistant to streptomycin, for example, produce modified ribosomes to which the antibiotic cannot bind. Protein synthesis continues unimpaired in these bacteria even in high concentrations of streptomycin.

Bypassing the Target Metabolic Reaction

Some bacteria acquire resistance to antimetabolites by bypassing the metabolic step inhibited by the drug. For example, bacteria that acquire the ability to absorb folic acid no longer depend on the biosynthetic step blocked by sulfa drugs. The drug continues to interfere with the reaction, but the formerly sensitive cell is unaffected because it has an alternate source of this essential coenzyme.

Cross Resistance

Sometimes an organism develops **cross resistance,** or insensitivity to several related antibiotics. For example, a single cellular modification may provide resistance to all the tetracyclines. Penicillinase-producing bacteria inactivate several types of penicillins. For this reason, when an organism is resistant to an antibiotic, the alternative agent is usually selected from an unrelated group of antimicrobial compounds.

PRACTICAL USE OF CHEMOTHERAPEUTIC AGENTS

Antimicrobial drugs are used therapeutically or prophylactically. **Chemotherapy** is antimicrobial treatment of an existing disease. Administering drugs to prevent infectious disease is called **chemoprophylaxis** (*prophylactic,* preventative).

Chemoprophylaxis

Prophylactic antibiotic therapy is usually not justified because of potential side effects and because, in most cases, there is no verification that it prevents disease. Chemoprophylaxis has been proven effective only in the following situations.

1. Antimicrobial drugs administered before and during surgery on heavily contaminated body sites such as the gastrointestinal and genitourinary tracts help reduce the likelihood of postsurgical infection. The bowel cannot be sterilized, however, and reliance on chemoprophylaxis instead of good aseptic technique, increases postsurgical infection rates.

2. Persons whose heart valves have been damaged by rheumatic heart disease receive chemoprophylaxis before even minor surgeries. During dental manipulation (including teeth cleaning), for example, microorganisms indigenous to the mouth may enter the bloodstream, colonize damaged areas of the heart, and cause subacute bacterial endocarditis, a serious infection of the heart valves.

3. Individuals with depressed immunological responses are extremely susceptible to opportunistic pathogens. Chemoprophylaxis provides some protection to these vulnerable persons, although no antibiotic combination provides complete protection.

4. Healthy people travelling to countries where malaria is prevalent are advised to take chloroquine to prevent this disease, since exposure to the pathogen is virtually inevitable. Because chloroquine may cause nausea, some persons refuse chemoprophylaxis and contract malaria as a result. Similarly, during epidemics of influenza, persons most likely to die if they contract the disease may be given amantadine.

Since most infections cannot be prevented by chemoprophylaxis, it is dangerous to apply this technique indiscriminately. Chemoprophylaxis exposes patients to the danger of allergic or toxic drug reactions and disrupts the protective normal microbial flora, thereby encouraging secondary infections. Chemoprophylaxis also encourages the proliferation of dangerous antibiotic-resistant pathogens and the R factors they may contain.

Chemotherapy

Ideally, before any antibiotic is administered, a clinical specimen is collected from the infected patient, the infectious agent is isolated and identified, and the drugs to which the pathogen is susceptible are determined. When a disease is serious or especially painful, the patient's symptoms may guide the selection of an antibiotic for use while waiting for laboratory diagnosis. Because these guesses are sometimes wrong, specimens for laboratory determination should be collected *before* administering the antibiotic. If the patient fails to improve during the initial therapeutic regimen the pathogen's drug sensitivities identified by laboratory results will suggest the best antibiotic.

Antibiotic Susceptibility Tests Once isolated, the pathogen's drug sensitivities are determined by observing its ability to grow in the presence of the drug, using one of the following methods.

THE DISC DIFFUSION METHOD Drug sensitivity is typically determined by a disc diffusion method. Inoculum size, type of medium, incubation conditions, and disc potency are carefully standardized in a procedure called the **Kirby-Bauer technique** (Fig. 14-10). The diameter of the growth inhibition zone indicates whether the pathogen is resistant or sensitive to the drug in the disc. (The organism is considered "sensitive" only if inhibited by a concentration of the drug that can be achieved at the site of infection.) This procedure reveals the pathogen's susceptibility to many different drugs in a single, easily performed procedure. The technique is unusable, however, for bacteria that are slow growers, obligate anaerobes, or too fastidious to grow on the standard assay medium. It also

FIGURE 14-10

Kirby-Bauer method of determining an organism's antibiotic sensitivities. Antibiotic-impregnated discs are placed on the surface of a solid medium that has been seeded with the isolated pathogen. After 24 hours of incubation, each antibiotic has diffused into the agar. Antibiotics that inhibit microbial growth produce a clear zone around the disc in which no organisms grow. The diameter of a zone indicates the amount of drug that is required to inhibit the organism in the infected person. If that amount cannot be attained in the body, the microbe is considered resistant.

is only semiquantitative, since zone sizes may be affected by even minor variations in procedure or preparation of material.

MINIMUM INHIBITORY CONCENTRATION A more accurate method of determining a pathogen's drug sensitivity is to measure the **minimum inhibitory concentration (MIC)**, the smallest amount of the drug that inhibits the multiplication of the pathogen. If therapy is to be effective, at least this concentration of the drug must be maintained at the site of infection until all of the pathogens are eliminated. MIC is usually determined by a broth dilution method either in test tubes or in panels of small wells (Fig. 14-11). A standard inoculum of the pathogen is incubated in a series of tubes (or wells) containing decreasing concentrations of the antibiotics being tested. If the drug inhibits the microbe at the concentration in the tube, no growth appears; the organism grows only in those concentrations below that required for inhibition. Therefore, the highest dilution (the lowest concentration) showing no visible growth is the MIC. Cells from the tubes showing no growth can be subcultured in media lacking antibiotics to determine if the inhibition is reversible or permanent. In this way the *minimum bactericidal concentration (MBC)* is determined. The

337

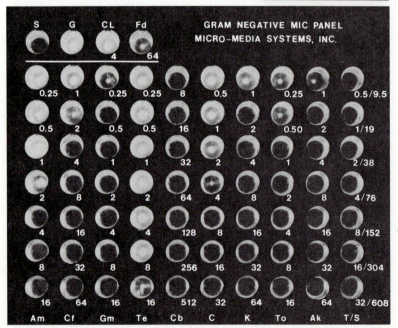

MIC
test

Concentration
of antibiotic
(μg/ml)

0   0.5   1   2   4   8   16

MBC
test

(a)

GRAM NEGATIVE MIC PANEL
MICRO-MEDIA SYSTEMS, INC.

S	G	CL	Fd						
		4	64						
0.25	1	0.25	0.25	8	0.5	1	0.25	1	0.5/9.5
0.5	2	0.5	0.5	16	1	2	0.50	2	1/19
1	4	1	1	32	2	4	1	4	2/38
2	8	2	2	64	4	8	2	8	4/76
4	16	4	4	128	8	16	4	16	8/152
8	32	8	8	256	16	32	8	32	16/304
16	64	16	16	512	32	64	16	64	32/608
Am	Cf	Gm	Te	Cb	C	K	To	Ak	T/S

(b)

FIGURE 14-11
Measuring antimicrobial effectiveness:
(*a*) The MIC is the lowest concentration in a serial dilution
of the antimicrobial agent that inhibits the growth of the
test organism. To determine minimum bactericidal
concentration (MBC) cells from tubes showing no growth
in MIC tests are plated onto a solid medium lacking
antibiotics. Organisms that have been killed fail to grow.
The lowest antibiotic concentration that kills the test
organism is the MBC. The MIC for this drug is 2 μg/ml,
whereas its MBC is 8 μg/ml. (*b*) The microdilution test for
determining the MICs of several antibiotics. Each vertical
row contains a dilution series of a different antibiotic
identified by initials at the bottom of the panel. Growth of
the organism is evident by turbidity in the wells.

MBC is nearly always higher than the MIC since it usually requires more
antibiotic to kill an organism than to merely inhibit its growth.

AUTOMATED METHODS Automated instruments rapidly determine
drug sensitivities by measuring turbidity with a spectrophotometer.
Growth inhibition is evidenced by a lack of turbidity in the culture. The
pathogen is first incubated in the presence of different antibiotics. No
growth occurs in those tubes containing drugs that inhibit the microbe,
and no change in optical density is observed. If the agents are ineffective,
the microbes will continue to multiply and optical density will increase.
Spectrophotometers detect slight changes in turbidity, so susceptibilities
may be determined in 4 hours rather than the day or more required by the
Kirby-Bauer technique.

THE ANTIBIOGRAM In practice an isolated pathogen is tested against a
few well-chosen antimicrobial agents, generating a sensitivity pattern
called an **antibiogram.** Since pathogens generally react similarly to closely
related antibiotics, an antibiogram predicts susceptibilities to more drugs
than actually used in the test. Thus only one member of each antibiotic
group need be tested. (Most aminoglycosides, however, do not display
cross resistance and must be tested individually.) Disc dispensers conveniently apply up to 12 discs on one plate for Kirby-Bauer susceptibility
testing.

Some common bacterial diseases and the antibiotics usually used to
treat them are listed in Table 14-3. These recommendations should be
supported with antibiogram data on the pathogen.

The number of drugs that can be employed against eucaryotic pathogens is quite limited (Table 14-4), and antibiotic susceptibility testing is
rarely employed. Fortunately, except for malaria, the development of
resistance to these drugs has not been a serious problem.

Monitoring Drug Levels Knowing the concentration of antibiotics in
body fluids can help one to ascertain whether therapeutic levels have
actually been reached at the site of infection. This is particularly important
when chemotherapy fails to promote patient recovery, even when the

Pathogen	Disease	Drug of Choice
Gram-Positive Cocci		
Staphylococcus aureus		
Nonpenicillinase producer	Abscesses; endocarditis; pneumonia	Penicillin
Penicillinase producer	osteomyelitis; septicemia	Penicillinase-resistant penicillin
		Penicillin
Streptococcus pyogenes	Scarlet fever; puerperal fever; erysipelas	Penicillin
Streptococcus viridans	Endocarditis	Penicillin
Streptococcus, enterococcus group	Endocarditis	Ampicillin or penicillin with aminoglycoside
	Urinary tract infection	Ampicillin or penicillin
Streptococcus pneumoniae	Pneumonia	Penicillin
Gram-Negative Cocci		
Neisseria gonorrhoeae	Gonorrhea	Tetracycline or penicillin
Neisseria meningitidis	Meningitis	Penicillin
Gram-Positive Bacilli		
Bacillus anthracis	Anthrax	Penicillin
Clostridium tetani	Tetanus	Penicillin
Clostridium perfringens	Gas gangrene	Penicillin
Clostridium difficile	Pseudomembranous colitis	Vancomycin
Corynebacterium diphtheriae	Diphtheria	Erythromycin
Gram-Negative Bacilli		
Bacteroides fragilis	Wound infections; bacteremia	Clindamycin
Bordetella pertussis	Whooping cough	Erythromycin
Brucella	Brucellosis	Tetracycline
Escherichia coli	Urinary tract infection	Sulfonamides
	Other infections	Gentamicin
Francisella tularensis	Tularemia	Streptomycin
Haemophilus influenzae	Meningitis	Chloramphenicol plus ampicillin
	Pneumonia	Ampicillin
Klebsiella pneumoniae	Urinary tract infection	Sulfonamide
	Other infections	Gentamicin
Legionella	Legionnaires' disease	Erythromycin
Leptospira buccalis	Vincent's infection	Penicillin
Proteus mirabilis	Urinary tract infection	Sulfonamide
	Other infections	Ampicillin
Pseudomonas aerugenosa	Urinary tract infection	Carbenicillin
	Other infections	Gentamicin with carbenicillin
Salmonella typhi	Typhoid fever	Chloramphenicol
Salmonella enteriditis	Gastrointestinal infections	Ampicillin†
Shigella	Bacillary dysentery	Trimethoprin-sulfamethoxazole
Vibrio cholerae	Cholera	Tetracycline†
Yersinia pestis	Plague	Streptomycin
Acid-Fast Bacilli		
Mycobacterium tuberculosis	Tuberculosis	INH with rifampin
Mycobacterium leprae	Leprosy	Dapsone with rifampin
Actinomycetes		
Actinomyces israelii	Actinomycosis	Penicillin
Nocardia sp.	Nocardiosis	Trisulfapyrimidine

TABLE 14-3 (continued)

Pathogen	Disease	Drug of Choice
Chlamydia		
Chlamydia psittaci	Pneumonia	Tetracycline
Chlamydia trachomatis	Trachoma	Tetracycline
	Pneumonia	Erythromycin
	Urethritis	Tetracycline
	Lymphogranuloma venereum	Tetracycline
Mycoplasma		
Mycoplasma pneumoniae	Pneumonia	Erythromycin
Rickettsia		
All species	Rocky Mountain spotted fever; typhus; Q fever	Tetracycline
Spirochetes		
Treponema pallidum	Syphilis	Penicillin
Treponema pertinue	Yaws	Penicillin

*In each case, in vitro susceptibility tests should confirm the pathogen's sensitivity to the drug selected.
†Antibiotics are used in severe cases only.

TABLE 14-4

Drugs Recommended Against Common Eucaryote Pathogens

Pathogen	Disease	Drug of Choice
Fungi		
Aspergillus spp.	Aspergillosis	Amphotericin B
Candida albicans	Disseminated candidiasis	Amphotericin B
	Oral or vaginal candidiasis	Nystatin
	Chronic mucocutaneous candidiasis	Ketoconazole
Coccidioides immitis	Coccidioidomycosis	Amphotericin B
Cryptococcus neoformans	Cryptococcosis	Amphotericin B with flucytosine
Dermatophytes	Tineas (ringworm)	Clotrimazole (topical); miconazole (topical)
Histoplasma capsulatum	Histoplasmosis	Amphotericin B
Algae		
Prototheca	Protothecosis	Amphotericin B
Protozoa		
Entamoeba histolytica	Amebiasis	Metronidazole with or without diiodohydroxyquin
Giardia lamblia	Giardiasis	Quinacrine
Leishmania sp.	Leishmaniases	Stibogluconate*
Plasmodium sp.	Malaria	Chloroquine
Pneumocystis carinii	Pneumocystic pneumonia	Trimethoprin-sulfamethoxazole
Toxoplasma gondii	Toxoplasmosis	Pyrimethamine plus trisulfapyrimidines
Trichomonas vaginalis	Trichomoniasis	Metronidazole
Trypanosome sp.	Chagas's disease	Bayer 2502*
	Sleeping sickness	Suramin*

*These drugs are available from the Centers for Disease Control.

pathogen was shown to be sensitive to the agent used. The drug should have a serum level that is between 2 and 8 times the MIC of the organism in order to ensure that the tissue concentrations, which are lower than those in the serum, will reach the MIC. Drug concentrations are also monitored to prevent harmful side effects. Patients receiving amphotericin B or aminoglycoside therapy, for example, may develop impaired kidney or liver function if the drug exceeds safe levels.

Monitoring antibiotic concentration in tissue fluids is usually performed by placing a sterile paper disc saturated with the patient's serum (or body fluid) on solid media seeded with the test organism and measuring the zone of inhibition around the disc (Fig. 14-12a). This diameter is compared with zones around discs that contain known amounts of the drug so that the zone size can be translated into actual drug concentrations (Fig. 14-12b).

ANTIMICROBIAL THERAPY: THE FUTURE

The most difficult problem associated with antimicrobial chemotherapy is the emergence of antibiotic-resistant microorganisms. The development of new antibiotics

FIGURE 14-12

Measuring drug concentrations in the patient's body fluids: (a) Paper discs containing known concentrations of antibiotic and one unknown serum sample are placed on a seeded agar plate and incubated until zones of inhibited growth can be measured. The results of the test in this diagram indicate a zone size of 15 mm for the serum sample. (b) The unknown concentration in the patient's serum can be determined from the data generated by the control information. In this case, the patient's serum contains 5 µg/ml of active antibiotic.

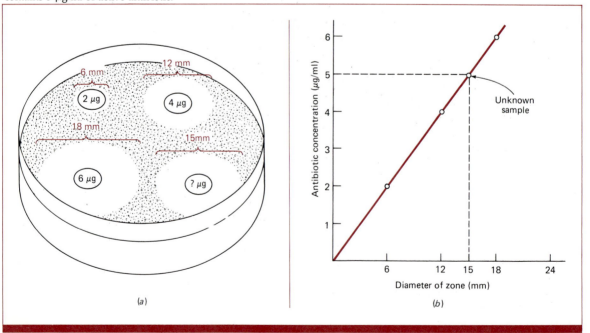

that are effective against the resistant organisms provides only temporary solutions. Bacteria resistant to the new drugs quickly appear and may transfer this resistance on plasmids. The rational approach to this problem is to greatly reduce the indiscriminate use of antibiotics. This is the only way to discourage the proliferation of antibiotic-resistant microbes and safeguard the effectiveness of existing antimicrobial drugs. The success of this strategy depends on adoption of the following measures:

Reduce the need for antimicrobial chemotherapy by preventing the transmission of disease (see Chaps. 16, 18, and 24).

Develop antibiotic susceptibility tests that provide accurate results more rapidly so that use of ineffective or broad-spectrum drugs will be reduced.

Rely on microbiological laboratory information to reveal which drugs are effective against the isolated pathogen.

Do not treat viral disease with antibacterial chemotherapeutic drugs.

Limit the prophylactic use of antibiotics to those infrequent situations for which its effectiveness has been demonstrated.

Reserve certain antibiotics for use *only* against infections that fail to respond to the recommended drugs. If the medical community would uniformly agree to restrict usage of these antibiotics, most microbes would rarely be exposed to the reserve drugs, thereby discouraging the proliferation of resistant pathogens. When needed, these antibiotics would likely be effective.

The overuse of antibiotics has contributed to a profoundly dangerous medical threat to modern health care. Within some hospitals as many as 50 percent of the patients receiving antibiotics have no laboratory cultures. Up to 70 percent of patients receiving antibiotics do not even show evidence of infections. Many hospitals require that all major prescriptions be reviewed and approved by personnel attuned to the problems of antibiotic abuse and aware of the alternatives for treatment. Automatic stop orders ensure that no antimicrobial drug is administered for more than 48 hours without a physician review. If its continued use is truly required, the drug must be reordered.

The public should also be informed of the limitations and hazards of antibiotics. Physicians too often succumb to pressure from patients who expect "wonder drugs" to cure even the mildest of ailments. Public awareness may relieve some of this pressure to overprescribe.

If current guidelines governing judicious use of antimicrobial agents were observed, the problem of drug resistance could be controlled. The future effectiveness of these important weapons against infectious disease depends on adoption of a conscientious and responsible approach to antimicrobial chemotherapy by members of the medical community.

Antibiotics and chemotherapeutic drugs that are selectively toxic to microbes are important agents for treating and preventing infectious disease. They may be used before infection occurs (chemoprophylaxis) or for eliminating an invading pathogen from the body (chemotherapy). In most cases, chemoprophylaxis is unwarranted.

Chemotherapeutic drugs are microbicidal or microbistatic. Microbistatic drugs inhibit microbial growth until host defenses ultimately eliminate the pathogens.

Chemotherapeutic drugs must be selectively toxic for microbes before they can be safely used. Among the most effective and least toxic antibacterial drugs are penicillins and cephalosporins, agents that kill bacteria by inhibiting cell wall synthesis. Sulfa drugs are also highly specific for bacteria since they interfere with bacterial metabolic processes not needed by eucaryotic organisms. Drugs that inhibit protein synthesis (aminoglycosides, tetracyclines, erythromycin, chloramphenicol, and clindamycin) or cell membrane function (polymyxins) are somewhat less selective in action and are therefore more likely to cause side effects. Antibiotics such as the polyenes which are directed against eucaryotic microbes are even less selective. Most can cause serious toxic reactions in patients receiving normal therapeutic doses. Few effective antiviral drugs are available.

Microorganisms may acquire drug resistance by mutation or by receiving R factors from a resistant donor. The resistant cell may inactivate the drug or prevent it from reaching its site of action. Alternatively, the target site may be altered so it no longer reacts with the drug. Regardless of the mechanism, drug-resistant microbes usually become the dominant organisms in environments that contain the corresponding drug.

The uncontrolled use of antibiotics encourages the proliferation of drug resistance among pathogens and normal flora. Transfer of R factors to other bacteria increases the probability of therapeutic failure. In addition, many of these drugs may cause side effects—fever, rash, superinfections, and even death. These problems can be minimized by limiting chemoprophylaxis to situations of proven effectiveness, isolating pathogens and identifying their drug sensitivities, and determining the patient's medical history to prevent allergic and toxic reactions.

It is usually impossible to predict a pathogen's drug sensitivity, so laboratory tests are used to determine which drugs are most effective. Therapeutic effectiveness of several antibiotics can be evaluated by disc diffusion tests or by determining MICs. Similar methods are used to test a patient's body fluids to determine levels of active antibiotic at the site of infection. Automated techniques identify a pathogen's antibiotic sensitivities within 4 to 6 hours of isolation.

In addition to the pathogen's drug sensitivities, other factors to consider when choosing the best agent for antimicrobial chemotherapy include the drug's access to the site of infection, how it interacts with other medications the patient is taking, and the duration of treatment.

KEY WORDS

chemotherapeutic agent

antibiotic

selective toxicity

broad-spectrum drug

narrow-spectrum drug

superinfection

penicillinase

antimetabolite

cross resistance

chemotherapy

chemoprophylaxis

Kirby-Bauer technique

minimum inhibitory concentration (MIC)

antibiogram

REVIEW QUESTIONS

1. List six factors that should be considered when selecting an antibiotic for therapy.

2. Differentiate between:
(a) Broad-spectrum and narrow-spectrum antibiotics
(b) Natural and semisynthetic penicillins
(c) Chemotherapy and chemoprophylaxis
(d) MIC and MBC

3. Why are so few drugs effective in treating viral infections?

4. Describe the mechanism of selective toxicity of (a) penicillins, (b) sulfa drugs, (c) cephalosporins, (d) aminoglycosides, and (e) tetracyclines.

5. Identify the group of microorganisms susceptible to polyenes and explain why the drugs cause serious side effects.

6. Discuss the disadvantages of the following: (a) tetracyclines, (b) penicillins, (c) gentamicin, and (d) chloramphenicol.

7. How do drug-sensitive strains acquire the genetic information that transforms them into drug-resistant strains?

8. What factors contribute to the proliferation of drug-resistant microbes?

9. List three mechanisms of drug resistance in bacteria.

10. Distinguish between the Kirby-Bauer test and MIC determinations.

11. What is an antibiogram?

12. List five methods for discouraging the development of antibiotic-resistant microbes.

Host-Parasite Interactions

15

People are populated by an extraordinary number of microorganisms. We become colonized by microbes at the moment of birth and remain so throughout life. There are more microorganisms on one human body than there are people on earth. In the colon alone, the bacteria number in the billions. Many of these microbes have established a **commensal relationship** with the host; that is, their presence neither overtly harms nor benefits the inhabited person. Other microbes actually benefit their hosts by providing nutrients, aiding in food digestion, or preventing the establishment of more dangerous pathogenic microbes. Even these beneficial organisms, however, can invade and kill an unprotected host. Fortunately, microorganisms are usually confined to safe locations on the body's surfaces by an elaborate defense system, so that, although colonization is common, disease is relatively infrequent.

Even so, infectious disease strikes all too often. In this chapter we will describe those attributes of microbes and their hosts that influence the outcome of the host-parasite interaction. (The mechanisms of tissue injury will be discussed in Chap. 17.)

COLONIZATION VERSUS DISEASE

With the exception of diseases caused by the inoculation of microbes directly into the bloodstream, infectious disease is always initiated by **colonization**, the establishment of microbes on the skin or mucous membranes. Microbial colonization of a body site may result in one of several outcomes (Fig. 15-1). If the microbes are eliminated without affecting the host, colonization is said to be *transient*. If colonization is *stable*, the established microbes proliferate on the body surfaces, as illustrated by our normal indigenous flora. This can be considered one form of **infection**, growth of microbes in the body or on its surfaces. Often, infection has no detrimental effects on the host. **Disease** follows infection only if the presence of the microbe or its growth by-products directly injures the infected person or elicits a host reaction that damages tissues. Infection is therefore distinct from disease. The two terms are nonetheless often used interchangeably.

A microbe's **pathogenicity** is its ability to cause disease. The degree of pathogenicity is referred to as **virulence** and reflects two properties: (1) *infectivity* (how easily the microbe survives the normal host defenses and establishes infection) and (2) *severity* of the damage it causes the infected host. Infectivity is quantified by determining a pathogen's **infectious dose**, the number of microorganisms required to cause observable infectious disease. The greater the pathogen's infectivity, the fewer microbes will be required to cause disease symptoms. The result is expressed as ID_{50}, the number of microbes required to cause disease in 50 percent of the laboratory animals experimentally infected with the pathogen.

A pathogen's infectivity and its ability to damage the host are independent properties; its virulence may be due to either its infectivity or

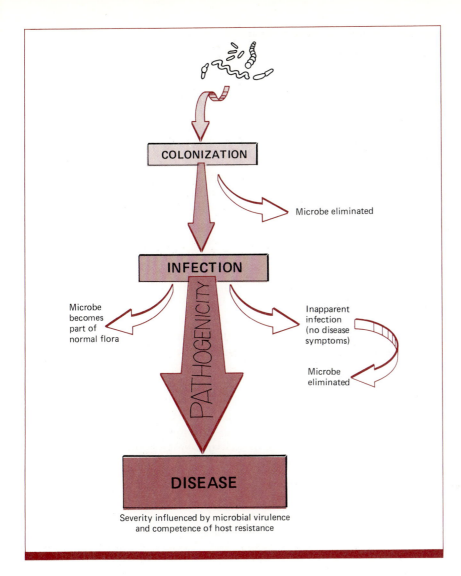

FIGURE 15-1

Some possible outcomes
following exposure of a
potential host to
microorganisms. (See text
for description of events.)

COLONIZATION

Microbe eliminated

INFECTION

Microbe
becomes
part of
normal flora

PATHOGENICITY

Inapparent
infection
(no disease
symptoms)

Microbe
eliminated

DISEASE

Severity influenced by microbial virulence
and competence of host resistance

its severity, or both. For example, the low infectivity of *Mycobacterium tuberculosis* contrasts sharply with the severity of disease it causes. Most persons exposed to this pathogen eliminate the microbe before it becomes established, yet the few persons who develop tuberculosis acquire serious, often life-threatening disease. Cold viruses and agents of most common childhood diseases have high infectivity but cause little damage. Most persons exposed to these pathogens become infected, develop a mild illness, and recover without suffering severe injury. The most dangerous pathogens are high in both infectivity and severity, a combination that secured a place in history for the agents of plague and smallpox.

Many pathogens cause disease in otherwise healthy persons. Some pathogens, however, cause disease only in persons exposed to an overwhelmingly large number of microbes or in persons who are incapable of resisting infection because of impaired host defenses. If a microbe causes disease *only* in persons with impaired defenses it is called an *opportunistic pathogen*. These organisms are common environmental contaminants or even members of our normal flora.

DYNAMIC BALANCE

The outcome of exposure to microorganisms depends upon a dynamic balance between the ability of the host to resist infection and the ability of the potential invader to cause disease (Fig. 15-2). Many factors influence the balance of this host-parasite relationship. Factors that help a microbe either to survive host defenses or to injure the infected host increase its ability to cause infectious disease. These microbial attributes that increase either infectivity or disease severity are collectively referred to as **virulence factors**. They tend to shift the balance in favor of the offending microorganism.

Fortunately, normal human defenses usually counteract microbial virulence factors and we stay healthy. This resistance is largely due to three lines of defense that any potential invader must penetrate before causing serious disease. These three lines of defense are (1) mechanical, chemical, and microbial defenses on the body's surfaces; (2) phagocytic cells that engulf and destroy invading organisms that penetrate the surface defenses; and (3) specific immune mechanisms that aid in the destruction of foreign microorganisms and toxins. The immune system "remembers" an invader and prepares the body to better resist subsequent encounters with the same pathogen or toxic substance.

FIGURE 15-2

Dynamic balance between host defenses and microbial virulence. Usually the balance favors the host.

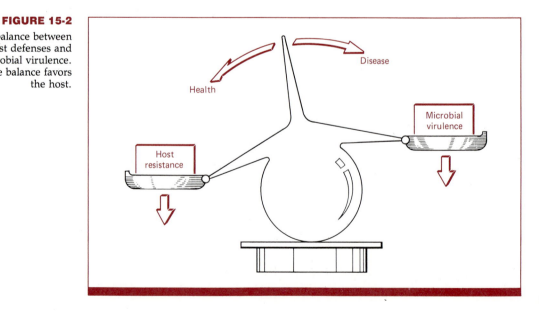

MICROBIAL VIRULENCE

The ability of a microorganism to cause infectious disease depends on its ability to establish infection, survive the host's defenses, and injure the infected host.

Virulence Factors that Encourage Establishment of Infection

Most microbes establish infection by attaching to host tissues, thereby anchoring themselves to the target tissues and reducing the likelihood of being eliminated from the host. The pili of some gram-negative bacteria attach to mucous surfaces of the intestinal, urinary, or genital tracts. Attachment protects the microbe from being flushed out by the flow of urine or the movement of the intestinal contents. Some bacteria adhere to their target host tissues by the surface of their capsules or glycocalyses. Attachment of influenza virus to the surfaces of the respiratory tract is aided by neuraminidase, an enzyme that is an integral part of the virus envelope. It dissolves neuraminic acid, an important component in the mucus that normally covers the respiratory cells and protects them from viral attachment. Neuraminidase exposes these cells to the virus.

Virulent and avirulent strains of some microbes differ solely in their ability to attach to host tissues. For example, toxin-producing strains of *Escherichia coli* cause no diarrheal disease unless they possess pili for attachment to the intestinal mucosa. Genes for pili production reside on a plasmid. Nonpiliated *Escherichia coli* can acquire the ability to attach by conjugating with a virulent donor. On the other hand, virulence disappears with the loss of the plasmid.

The infectivity of most pathogens can be neutralized by preventing attachment. Many successful vaccines stimulate the host to produce antibodies that block the binding sites on the pathogen's surface.

Virulence Factors that Promote Pathogen Survival and Host Injury

Once infection has been established, the development of disease depends on the microbe's ability to successfully evade host defenses. Several virulence factors protect the microbe from host resistance mechanisms. Many of them also increase the pathogen's ability to multiply in host tissues and spread from the original site of infection. In doing so the pathogen often damages host tissues as it gains access to nutrients and portals of exit from the host. These virulence factors are discussed according to their effects.

Toxic Factors Some pathogens damage the host and reduce host defenses by producing exotoxins or endotoxins (or both).

Exotoxins are soluble proteins produced by microorganisms and secreted into their surroundings (*exo*, external). They may be carried by the bloodstream to any part of the body. In this way, a localized infection can have fatal consequences. For example, *Clostridium tetani* infections are restricted to the site of the wound into which the organism was introduced.

The organism itself destroys local tissue only, but from this focus it produces a powerful exotoxin which is dispersed throughout the body. The toxin attacks motor nerves, triggering involuntary muscle contractions that may be strong enough to break the patient's spine. Death is usually due to paralysis of the patient's respiratory muscles.

Some exotoxins clearly benefit the microbe by facilitating invasion of tissues, protecting against host defenses, or promoting transmission to a new host. For example, **enterotoxins**, exotoxins that affect the intestinal mucosa, produce an explosive diarrhea that encourages shedding of intestinal pathogens and increases the likelihood of their transmission to new hosts. Similarly, pertussis toxin causes violent coughing ("whooping cough") that showers the air with the respiratory pathogens.

Exotoxins are the important virulence factors in the pathogenesis of diphtheria, scarlet fever, botulism, tetanus, gas gangrene, bacterial dysentery, cholera, and whooping cough (pertussis). The major symptoms of all these diseases are caused primarily by exotoxin; nontoxin-producing strains are incapable of causing these diseases. The ability to neutralize a pathogen's exotoxin often confers immunity. The effects of some exotoxins are described in Table 15-1.

Endotoxin is the lipid A component of lipopolysaccharide found in the

TABLE 15-1

Diseases Caused by Exotoxin-Producing Pathogens

Disease	Organism	Effect of Toxin
Diphtheria	*Corynebacterium diphtheriae*	Damages heart, nerves, and liver
Tetanus	*Clostridium tetani*	Alters nerve function; paralyzes muscles in a state of contraction
Botulism	*Clostridium botulinum*	Blocks nerve impulses; paralyzes muscles in a state of relaxation
Scarlet fever	*Streptococcus pyogenes*	Causes rash by injuring capillaries
Toxic shock syndrome	*Staphylococcus aureus*	Causes rash, fever, and shock
Pertussis (whooping cough)	*Bordetella pertussis*	Causes necrosis of epithelial lining of upper respiratory tract
Gas gangrene	*Clostridium perfringens*	Causes necrosis of affected tissue
Dysentery	*Shigella dysenteriae*	Causes neurological impairment
E. coli gastroenteritis	Some strains of *Escherichia coli*	Triggers outpouring of water into colon, causing diarrhea
Food poisoning	*Staphylococcus aureus*	Stimulates vomit reflex center in the central nervous system
Food poisoning	*Clostridium perfringens*	Stimulates vomit reflex center
Cholera	*Vibrio cholera*	Triggers outpouring of water into colon, causing severe diarrhea

gram-negative cell wall. It is released from bacteria during cell disintegration and triggers white blood cells to discharge chemicals into the blood and surrounding tissues. These released substances induce fever and pain at the site of infection. Rash may develop as the result of capillary hemorrhage. In large doses endotoxin prevents normal capillary constriction, resulting in a severe drop in blood pressure, sometimes leading to fatal endotoxic shock.

Antiphagocytic Factors   Capsules are the most important virulence factors that protect many bacteria from being engulfed and destroyed by the host's phagocytic cells. Capsules cover surface components to which phagocytes attach, an essential step for engulfment and destruction of the target cell. Some capsules allow attachment of the phagocyte but prevent the binding of a chemical that promotes ingestion of the microbe. The antiphagocytic activity of capsules is responsible for the infectivity of many important pathogens. For example, encapsulated strains of *Streptococcus pneumoniae* avoid destruction by phagocytosis and cause dangerous lower respiratory disease. Nonencapsulated strains of the same organism are nonpathogenic. Other pathogens that can produce antiphagocytic capsules include *Staphylococcus aureus*, *Neisseria meningitidis*, *Klebsiella pneumoniae*, *Cryptococcus neoformans*, and *Salmonella typhi*.

   Some bacteria produce soluble antiphagocytic substances. *Leukocidin*, for example, is one of several virulence factors that poison phagocytic leukocytes (white blood cells). *Coagulase* is an enzyme that triggers the clotting of plasma around the site of infection. In this way some strains of *Staphylococcus aureus* may protect themselves from phagocytic cells, which cannot penetrate the wall of clotted fibrin. Some pathogens may elaborate antichemotactic factors that reduce chemotaxis, the movement of phagocytes toward the site of infection.

   Perhaps the most ingenious methods of confronting the phagocytic defenses are those used by intracellular parasites that survive after being engulfed by phagocytes. These pathogens may even use the phagocytes to their own advantage (see Chap. 17).

Spreading Factors   Pathogenic bacteria may discharge enzymes that dissolve host tissue and destroy physical barriers that would limit the spread of the microbes throughout the infected body. These spreading factors may cause alarming injury to the patient. The tissue destruction associated with gas gangrene, for example, is largely due to the production of *collagenase* by the pathogen. This enzyme destroys protein of skin, bone, and cartilage, literally liquefying tissues in the process. The pathogen also produces *lecithinase* (also called alpha-toxin), an enzyme that depolymerizes host cell membranes, destroying additional physical barriers. Another spreading factor, *hyaluronidase*, dissolves hyaluronic acid, the unifying matrix of connective tissue. Hyaluronidase is believed to facilitate the spread of *Streptococcus pyogenes* and *Staphylococcus aureus* from their initial sites of infection. *S. pyogenes* also produces *fibrinolysin*, an enzyme

TABLE 15-2

Some Factors That Increase
Microbial Virulence

Virulence Factor	Increases Microbial Ability to	Representative Microbes	Mechanism of Action
Pili	Establish infection	*Neisseria, gonorrheae; Escherichia coli*	Facilitate attachment to target tissue
Capsule	Establish infection Resist host defenses	*Cryptococcus neoformans; Streptococcus pneumoniae; Klebsiella pneumoniae*	Facilitates attachment Resists phagocytosis
Neuraminidase	Establish infection	Influenza virus	Facilitates attachment
Exotoxins	Damage the host (advantage to pathogen often obscure)	*Corynebacterium diphtheriae; Clostridium tetani; Staphylococcus aureus*	Interfere with key physiological processes
Endotoxins	Injure host tissue and survive host defenses	Most gram-negative pathogens	Release endogenous pyrogens (induce fever); cause hemorrhage and rash; block capillary contraction, resulting in circulatory collapse, shock, and death
Leukocidin	Survive host defenses	*Staphylococcus aureus*	Kills phagocytic leukocytes
Coagulase	Survive host defenses	*Staphylococcus aureus*	Walls off site of infection in a protective fibrin clot
Collagenase	Spread from initial infection site	*Clostridium perfringens*	Dissolves protein of bone, skin, and cartilage
Lecithinase	Spread from initial infection site	*Clostridium perfringens*	Destroys host cell membranes
Hyaluronidase	Spread from initial infection site	*Streptococcus pyogenes*	Dissolves hyaluronic acid, the ground substance of connective tissue
Fibrinolysin	Spread from initial infection site	*Streptococcus pyogenes*	Dissolves fibrin clots

that dissolves fibrin clots which tend to wall off the infection site and prevent the dissemination of the pathogen.

A number of additional enzymes elaborated by pathogenic bacteria may increase microbial virulence, but their actual role in pathogenesis is obscure. For example, *hemolysins* are extracellular products that destroy red blood cells. How the lysis of red blood cells contributes to a microorganism's ability to invade and damage the host is poorly understood. The

TABLE 15-3

The Three Lines of
Host Defense Against
Foreign Invaders

Host Defense	Makes Use of
Surface Defenses (nonspecific)	
Mechanical	Skin; mucosa; cilia; cough, sneeze, and epiglottis reflexes; hair
Chemical	Low pH; fatty acids; lysozyme
Microbial	Bacterial competition
Phagocytic Defenses (nonspecific)	Phagocytic cells; inflammatory response
Acquired Immunity (specific for each disease)	Humoral immunity (B-cell immunity); cell-mediated immunity (T-cell immunity)

contribution of hemolysins to virulence may be their additional ability to destroy phagocytic cells.

Some virulence factors and their role in causing disease are summarized in Table 15-2.

HOST RESISTANCE

Three lines of host resistance usually confine microbial growth to the surface of the body where most microbes are harmless. Any factor that interferes with these defenses, however, can promote serious infectious disease. The efficiency of a person's surface defenses, phagocytic defenses, and immune system plays a critical role in preventing transient colonization from developing into severe disease (Table 15-3).

Surface Defenses

Infectious disease is initiated by the microbe's contact with the surfaces of the body. The host uses mechanical, chemical, and microbial mechanisms to combat these constant challenges.

Mechanical Defenses   The body is surrounded by its most essential line of defense, the skin and the mucous membranes that line the respiratory tract, alimentary tract, genitourinary tract, and conjunctiva (Fig. 15-3). These barriers keep the vast majority of microorganisms out of the body's more vulnerable interior. Cuts, puncture wounds, and other trauma breach the barriers and may result in minor local infections or serious systemic disease. Weakened skin or mucous membranes are more susceptible to these traumatic injuries. Malnutrition may encourage the development of infectious disease by weakening the integument (skin). Similarly, respiratory mucous membranes can be weakened by the noxious effects of tobacco smoke, air pollutants, and several gaseous anesthetics.

Dust and airborne particles often carry microbes that cause respiratory infections. The airways of the respiratory tract are protected by special mucous membranes called the *ciliated mucosa*. These cells secrete a layer of mucus that provides an additional mechanical safeguard. The sticky mucus efficiently traps particles that get into the respiratory tract. The mucus-

FIGURE 15-3

The skin and mucosa, shown here in color, are effective barriers between the environment and the internal regions of the body (shaded).

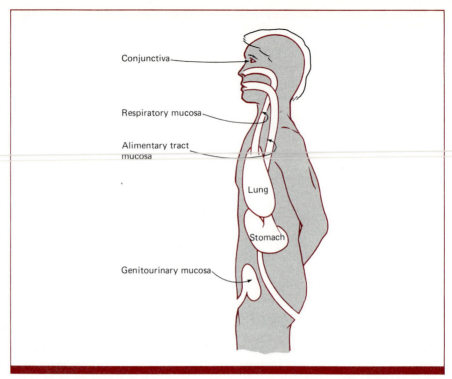

Conjunctiva

Respiratory mucosa

Alimentary tract mucosa

Lung

Stomach

Genitourinary mucosa

secreting cells possess cilia that move the mucus blanket about 1 inch every minute, sweeping trapped microbes harmlessly into the throat where they are swallowed. The mucus also helps prevent viral infections by blocking attachment of viruses to their cellular adsorption sites.

The respiratory mucosa may temporarily dry out during sudden reductions in humidity. Moving from a warm, moist environment to a cold, dry climate, for example, often encourages infectious respiratory disease because the invading pathogens slide through the dry airways without being trapped in mucus.

Airborne particles are also moved out of the respiratory tract by two mechanical reflexes, the cough and the sneeze, that forcefully expel particles. Additional mechanical protection for the lungs is provided by the *epiglottis*, a flap of tissue that, during swallowing, covers the opening leading to the lower respiratory tract (trachea, bronchi, and lungs). The epiglottis directs swallowed substances into the esophagus and stomach, preventing solids and liquids from entering the vulnerable lower airways. Chronic alcoholism suppresses all three of these physical responses, allowing pathogen-laden particles and liquids deep entry into the lungs. Because of these impairments to the mechanical defenses of the body, many alcoholics develop pneumonia. Pneumonia may also follow near-drowning accidents because tremendous numbers of bacteria are washed into the lungs by the inhalation of water.

Other mechanisms provide additional mechanical protection. Inhaled air is tumbled as it is drawn through the nasal hairs. This turbulence increases the probability that particles will be trapped in the nasal mucus. Body hair also reduces the number of microbes settling onto the skin. The flow of urine and tears flushes microbes from the urinary tract and eyes. The sloughing of the dead surface layer of our skin effectively discharges many microbes that have attached to these epithelial cells. A similar mechanism protects the gastrointestinal tract, which sloughs about 1 pound of its own cells each 2 days. (The entire intestinal epithelium is replaced every 36 hours.)

Chemical Defenses Mucous secretions, tears, sweat, and saliva all contain *lysozyme*, an enzyme that destroys the cell walls of many gram-positive bacteria (Fig. 15-4). Fatty acids in sweat and earwax have bacteriostatic and fungistatic properties. The stomach contains a high concentration of hydrochloric acid, which rapidly kills microbes. The high acidity of the adult vagina protects its membranous surfaces against colonization by many types of pathogens.

Microbial Defenses The body's resident bacterial flora provide another important line of defense. The *resident flora* continually compete with potentially pathogenic microbes for the limited nutrients and space on the body's epithelial surfaces. Because the normal flora are already established, these beneficial microorganisms usually win the competition and prevent either initial infection or unrestricted growth of pathogenic bacteria and fungi. Resident bacteria also have the ability to alter their local

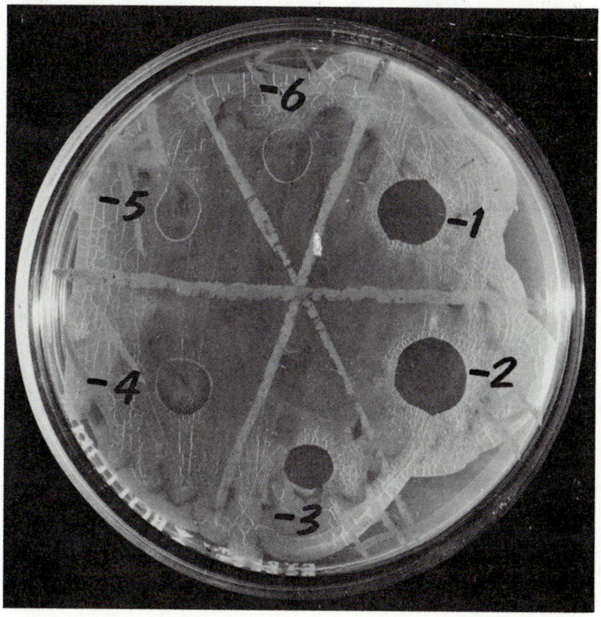

FIGURE 15-4
Destruction of a gram-positive bacterium by lysozyme in tears. To demonstrate this effect, various dilutions of tears were placed on an agar plate seeded with *Micrococcus*. Growth inhibition is observable where higher concentrations were applied to the plate.

environment so that it is generally unfavorable for the growth of most pathogens. The lactobacilli that normally inhabit the adult human vagina, for example, produce lactic acid as a by-product of growth. This lowers the pH of the vagina to between 4.0 and 4.5, intolerably acidic for most pathogens. Metabolic by-products of skin flora create acidic conditions on the epidermal surfaces (a pH of 3.0 to 5.0) that oppose the growth of many potential pathogens.

Phagocytic Defenses

When infectious agents penetrate the surface defenses of the body, they are usually devoured by **leukocytes** (white blood cells) or other phagocytic cells. These cells engulf and destroy foreign particles by *phagocytosis* (Fig. 15-5a). Most phagocytic cells in the circulation contain large numbers of lysosomes (also called *granules*) in their cytoplasm. Lysosomes contain peptidoglycanases, lipases, and proteases, plus chemicals that oxidize and destroy engulfed cells. Following engulfment of an invading microbe into an intracytoplasmic vacuole, lysosomes migrate to the vacuole membrane

FIGURE 15-5

(a) A macrophage engulfing a yeast cell. *(b)* Phagocytic destruction of engulfed particles: Following engulfment, lysosomes (shown in color) discharge their antimicrobial contents into the phagocytic vacuole by membrane fusion. Because the lysosomes (granules) disappear, the process is called degranulation.

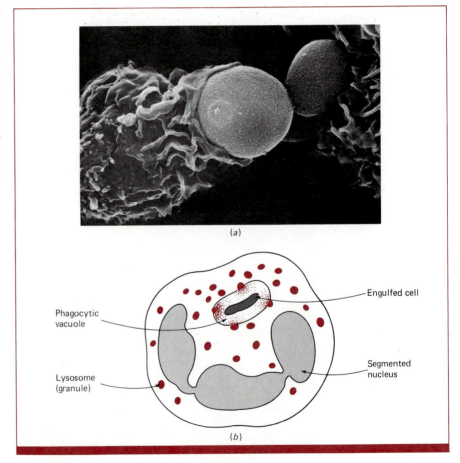

(a)

Phagocytic vacuole

Engulfed cell

Lysosome (granule)

Segmented nucleus

(b)

and discharge their contents (Fig. 15-5b). The phenomenon, called *degranulation*, occurs by fusion of lysosomal and vacuole membranes and accounts for the disappearance of lysosomes soon after engulfment. The discharged chemicals kill and digest the engulfed particle in the vacuole. Human tissue can also be damaged by the lysosomal contents. The vacuole membrane, however, usually prevents tissue injury by housing the discharged chemicals, preventing their sudden release into the surrounding tissue.

Phagocytic Cells Lysosomes are characteristic of a group of white blood cells called *granulocytes*. (The cells found in human blood are depicted in Fig. 15-6.) Of the three types of granulocytes found in normal blood, **neutrophils** are most abundant. (See color box.) These cells are also

FIGURE 15-6

The mature blood components found in the fluid of human blood: Erythrocytes (red blood cells), thrombocytes (platelets), and several types of white blood cells. The granulocytes are the neutrophils, eosinophils, and basophils. The agranulocytes are the lymphocytes and monocytes (which differentiate into actively phagocytic macrophages, not pictured here because of variation in appearance among different types). Except for erythrocytes, which carry oxygen, all these cells directly contribute to the body's defense. Thrombocytes aid in blood clot formation, which, in addition to preventing blood loss, helps restrict the spread of microbes. Neutrophils and macrophages are actively phagocytic. Lymphocytes are the body's immune cells, producing antibiodies and other factors that contribute to specific acquired immunity.

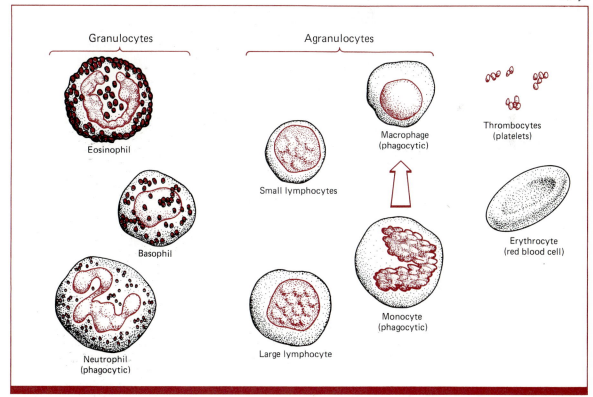

Granulocytes

Agranulocytes

Eosinophil

Basophil

Neutrophil (phagocytic)

Small lymphocytes

Large lymphocyte

Macrophage (phagocytic)

Monocyte (phagocytic)

Thrombocytes (platelets)

Erythrocyte (red blood cell)

DO THEY HAVE TO TAKE MY BLOOD?

■ Blood samples provide valuable clues for diagnosing illness. The differential blood count is perhaps the most routine of these tests. The patient's blood is stained with Wright's stain, a special dye that allows exact identification of the various granulocytes, lymphocytes, monocytes, and thrombocytes. The stained smears are microscopically examined to determine the numbers of each cell type per cubic millimeter of blood. The concentrations of these cells often suggest the nature of the patient's illness.

About 4 billion neutrophils are produced every hour in the bone marrow of the normal healthy person. Most of these are retained in the bone marrow, which houses a reserve of about 3000 billion neutrophils in addition to the 35 million or so in circulation. During bacterial infection these cells are quickly released from the marrow, sometimes increasing the number of circulating neutrophils 30-fold. An increase in circulating leukocytes is called *leukocytosis* and is indicative of active infection. Some severe diseases, such as typhoid fever and influenza, cause *leukopenia* (a lower than normal number of leukocytes). This occurs when phagocytes are depleted by an overwhelming number of microbes. Increased numbers of lymphocytes (lymphocytosis) are characteristic of whooping cough and infectious mononucleosis. In the latter disease many of the cells are abnormal in appearance. Monocytosis (an increased number of monocytes) is typically observed in malaria. Allergic conditions stimulate an increase in eosinophils.

Leukemia is a disease in which there is an uncontrolled proliferation of leukocytes and their immature precursors before they have acquired their protective activity. In acute leukemia the predominant leukocytes in the blood are immature, leading to an impaired defense against infection. The disease is usually fatal within 6 months. In chronic leukemia, most of the white blood cells are mature, but they are so numerous that red blood cells don't survive. Thus, in chronic leukemia, leukocytosis is usually accompanied by a progressive anemia (reduction in red blood cells).

called polymorphonuclear leukocytes (PMNs) because of their irregular multilobed nuclei. Neutrophils are actively phagocytic and are usually the first protective cells to arrive at the site of trauma or infection. They do their work rapidly and are fairly short-lived. Two other types of granulocytes, eosinophils and basophils, are not phagocytic, but contribute to defense by producing substances that help regulate the host response to injury or infection. (See Inflammation below.)

The other group of phagocytic cells is made up of *agranulocytes,* so called because they have fewer dark-staining lysosomes. Nonetheless, lysosomes are present in numbers great enough to effect destruction of engulfed particles during phagocytosis. Agranulocytes include the *lymphocytes* (immune cells discussed in Chap. 16) and *monocytes*. Monocytes have a single large oval or horseshoe-shaped nucleus. Circulating monocytes migrate into tissue, where they enlarge and differentiate into actively phagocytic **macrophages**. Macrophages are most commonly found attached to tissues of the liver, spleen, lymph nodes, and bone marrow and along the blood and lymph vessels. They engulf foreign particles and debris from blood and lymph as it flows through these regions. These *fixed macrophages* therefore function as a filter to clean debris from the blood and trap potential pathogens. *Wandering macrophages* migrate to the lungs, spleen, and other sites where microbes are likely to be encountered. (The lung macrophages are called "dust cells.") Macrophages also travel to areas of trauma or infection to participate in the body's overall protective response. The monocytes and macrophages are the long-lived members of the body's *mononuclear phagocytic system*, a functional network of cells once called the reticuloendothelial system.

Inflammation Phagocytosis is an essential component of **inflammation** —the body's attempt to localize and destroy infectious microorganisms and repair damaged tissues. The events of inflammation, described in Figure 15-7, characteristically elicit four symptoms: swelling, pain, redness, and heat. These symptoms are partially the result of increased vascular permeability, which allows plasma to escape the local vessels and accumulate at the site of injury, and of chemotaxis, the migration of white blood cells toward the source of injury or infection. Inflammation protects the host by promoting phagocytosis of microbes, by localizing infection with walls of clotted plasma, and by producing an *exudate* (pus), which, in some cases, allows direct drainage of microbes and tissue out of the host's body (as from a pimple, for example). The symptoms of inflammation also alert us to the presence of infection, increasing the likelihood of early diagnosis and treatment.

The pain associated with inflammation often encourages the use of corticosteroids or other anti-inflammatory agents to reduce discomfort and suffering. Although this is occasionally warranted (for treating chronic inflammation of a noninfectious origin, for example), the body's overall resistance to infectious disease is compromised by such an approach (see color box). Thus, while cortisone improves the way an infected person feels by relieving the symptoms of inflammation, the infection may progressively worsen. Chronic emotional stress can have a similar anti-inflammatory effect due to the long-term release of corticosteroids from the adrenal glands. The effects of diabetic coma are also anti-inflammatory, so uncontrolled diabetes mellitus reduces resistance to infection. Overexposure to x-rays can depress bone marrow production of phagocytic leukocytes and allow the normal nasopharyngeal microbes to cause fatal opportunistic infections of the lower respiratory tract, bloodstream, or central nervous system.

Immunity

The body's third line of defense is the immune response. Immunity differs from other types of resistance in that it is acquired during a person's lifetime following encounters with microbes or foreign substances. Furthermore, the resultant immunity specifically protects against the single

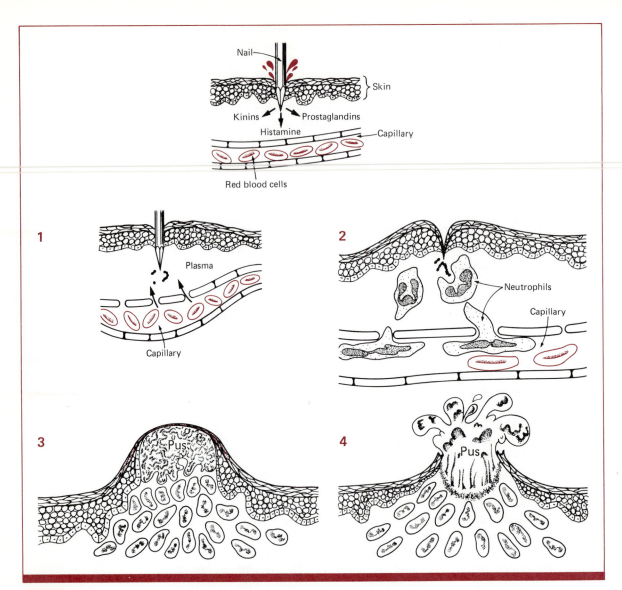

FIGURE 15-7

The inflammatory response following tissue injury and release of chemical mediators from disrupted host cells. (1) Chemical mediators cause dilation and increased permeability of blood vessels in area of injury. This allows more blood into the injured area and promotes the escape of fluid from vessels into the surrounding tissues. The area swells and gets hotter (because blood carries heat). The swollen tissues hurt because of pressure on sensory nerve endings and aggravation by the chemical mediators. Redness is due to the additional blood in the inflamed tissues. (2) Chemical mediators also attract phagocytic cells, which escape the capillary by diapedesis, squeezing between the cells of the intact capillary wall. Chemotactic chemicals continue to attract the phagocytic cells until they reach the site of injury, where they engulf and destroy microorganisms and liberate enzymes that digest dead or injured host tissue. (3) Continued influx of phagocytes (and water) produces an exudate, an aggregate of leukocytes, digested host tissue, microbial debris, and serum. (4) Pus may eventually drain from the abscess.

■ The use of cortisone (corticosteroids) to suppress inflammation can have severe consequences. Cortisone has been shown to affect mobilization of leukocytes and phagocytosis in the following ways:

■ Cortisone delays the attachment of leukocytes to the capillary wall, thereby interfering with diapedesis (the movement of leukocytes from inside blood vessels to the surrounding tissue). Effect: fewer phagocytes at the site of infection.

■ Cortisone reduces chemotaxis and migration of leukocytes. Effect: fewer phagocytes at the site of infection.

■ Cortisone discourages phagocytic engulfment. Effect: phagocytes at the infected site fail to attack pathogens.

■ Cortisone stabilizes lysosomes so they fail to fuse with the vacuole membrane. Efficiency of degranulation is reduced. Effect: engulfed pathogens are not destroyed.

The overall effect of cortisone therapy is the reduction of the body's resistance to infectious disease.

---

type of organism or substance that induced the response. For example, immunity to diphtheria does not confer immunity to unrelated diseases such as measles. Immunity will be discussed in the next chapter.

Additional Resistance Factors

Our body fluids contain nonspecific or semispecific substances that contribute to our resistance to potential pathogens. *Interferons* are a group of soluble proteins produced by host cells during infection by viruses and some other microbes as well. Uninfected cells in the area exposed to viral-induced interferon become resistant to viral infection (see Chap. 9). Interferons also modulate the activity of our phagocytic and immunological defenses. *Beta-lysin* is a soluble blood chemical that attacks bacterial membranes.

Transferrins are iron-binding proteins that reduce the free iron levels in the body. The presence of free iron is necessary for the growth of microbes. Iron is also required by humans, but we can extract it from iron-binding proteins as we need it, whereas bacteria cannot. Reducing free iron levels in the blood therefore helps protect against microbial proliferation. The strategy is enhanced by mononuclear phagocytes that remove even more free iron from the blood during infection.

Endogenous pyrogens are fever-inducing proteins housed in vacuoles of phagocytic cells where they have no effect on the host. When stimulated by the attachment of some foreign substances or microbes, these cells discharge their endogenous pyrogens. These substances trigger an elevated body temperature. Although fever can itself damage human cells, it is believed to provide some protection by elevating temperatures above that which is optimal for microbial growth and by speeding up the mobilization and protective efficiency of the body's defenses.

Complement is a complex group of serum proteins that act in concert with the immune system to facilitate bacterial lysis. The complement system will be discussed in greater detail in Chapter 16.

OVERVIEW

The dynamic balance between microbial virulence and host defenses determines whether host-parasite interactions result in elimination of microbial contaminants, colonization, infection, or disease. The balance is shifted to favor the microbe whenever host resistance is reduced or microbial virulence is sufficiently great enough to overwhelm host defenses. Several virulence factors produced by microbes increase their infectivity or the severity of the diseases they cause. Many pathogens possess surface components that attach and anchor the microbe to host cells, making their removal by mechanical host defenses more difficult. Toxic factors kill protective host cells or trigger coughing, diarrhea, and other activities that promote spread of the pathogens to new susceptible hosts. Antiphagocytic factors aid the establishment of infection by discouraging engulfment by leukocytes. Spreading factors dissolve fibrin clots and host tissue that are barriers to microbial dispersal throughout the infected person.

Humans have three lines of defense against microbial invasion. Microbes are usually restricted to the body's surface by our primary lines of defense. These include our skin, mucous membranes, and ciliated mucosa; our cough, sneeze, and epiglottis reflexes; antimicrobial chemicals in sweat, saliva, tears, and mucus; the acidity of the stomach, skin, and vagina; and our normal microbial flora. Microbes that penetrate into our bodies are usually arrested by our second line of defense, the phagocytes. Phagocytosis is coordinated with a number of tissue changes that efficiently bring the phagocytic cells into the area of injury and infection. This explains the protective value of inflammation. The inflammatory response is initiated when chemical mediators are released from damaged tissue. These mediators increase vasodilation and chemotactically summon neutrophils to the area. Macrophages arrive later to ingest microbes and dead neutrophils.

Other nonspecific substances that enhance resistance include interferons, beta-lysins, endogenous pyrogens, transferrins, and complement.

The third line of defense is acquired immunity. It is acquired by exposure to a pathogen or foreign substance and specifically protects against later challenges by the same foreign agent.

KEY WORDS

commensal relationship

colonization

infection

disease

pathogenicity

virulence

infectious dose

virulence factor

exotoxin

enterotoxin

endotoxin

leukocyte

neutrophil

macrophage

inflammation

endogenous pyrogens

REVIEW QUESTIONS

1. Differentiate between colonization, infection, and disease.

2. Differentiate between exotoxins and endotoxins. List five diseases that are caused by exotoxin-producing pathogens.

3. Describe three biological effects of endotoxin.

4. Briefly describe how each of the following factors contributes to an organism's virulence:
    (a) capsules
    (b) leukocidin
    (c) coagulase
    (d) collagenase
    (e) pili

5. Describe three mechanical defenses and how each protects the host from invading organisms.

6. How do some of the body's chemical secretions protect against disease?

7. Briefly describe the events of the inflammatory response.

8. How do neutrophils, fixed macrophages, and wandering macrophages differ in function?

Acquired Immunity

16

Some microbes and toxic substances inevitably penetrate the surface defenses and escape immediate destruction by phagocytosis. These foreign intruders are usually rendered harmless by the body's third line of defense, **acquired immunity**. Immunity is a specific type of resistance that (1) is acquired during a person's lifetime as a result of exposure to specific foreign substances, (2) usually protects against the single type of pathogen or toxin that induced the response, and (3) commonly provides long-term protection against the agent that stimulated the immunity.

Immunity explains why persons who recover from diseases such as measles enjoy lifelong protection against future encounters with the same pathogen. It also accounts for the success of *vaccination*, the use of nonpathogenic variants of dangerous microbes (or their products) to stimulate protection against the virulent form of the pathogen.

ANTIGENS

Any substance that specifically stimulates an immune response when introduced into the body is an **antigen**. Several characteristics are known about their nature.

■   They are usually composed of protein or polysaccharide, or contain these macromolecules as a major constituent.

■   Some are soluble, others are particulate. Soluble bacterial exotoxins are antigenic. The structural components of the cell that secreted the toxin are particulate antigens. A bacterium may have several types of particulate antigens (Fig. 16-1).

■   In general, larger molecules stimulate a more intense immune response than smaller molecules.

■   Antigens must be recognized by the host as *foreign* before they stimulate an immune response. This is perhaps their most important characteristic, since the system must selectively eliminate foreign antigens without damaging the host's tissues in the process. A substance that is antigenic (recognized as foreign) in one host may be completely tolerated (recognized as "self") in another individual. Kidney transplants, for example, are rejected because the proteins of the transplanted organ, which were recognized as "self" by the donor, are foreign to the recipient and attacked by the immune system. Similarly the red-blood-cell antigens of different persons may be distinct enough to trigger an immune reaction if transfusion mismatches occur. Thus, type B blood cells are antigenic in a recipient with type A blood, yet are immunologically tolerated by type B persons. The mechanism that prevents us from rejecting our own tissue is called **immunological tolerance**. In general those constituents that were present in our bodies during fetal development are tolerated as "self." Most other potentially antigenic substances introduced into the body after birth are recognized as foreign and stimulate an immune response.

ACQUIRED IMMUNITY

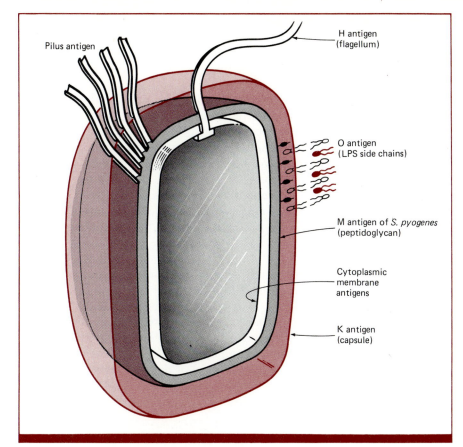

Pilus antigen

H antigen
(flagellum)

O antigen
(LPS side chains)

M antigen of *S. pyogenes*
(peptidoglycan)

Cytoplasmic
membrane
antigens

K antigen
(capsule)

FIGURE 16-1

Many parts of bacterial
cells are distinctly
antigenic. These antigens,
which have been
consolidated here into a
single diagram, are often
identified by assigning
each a letter. The
specificities of antigens can
be used to distinguish
between strains within a
single species. Two
different strains of the
food-borne pathogen
Salmonella typhimurium, for
example, will have
different O antigens.

■ An antigen contains chemically distinct sites, called **antigenic
determinants**, that define its specificity. These are the molecular regions
against which the immune response is directed. Small molecules that are
nonantigenic may sometimes become antigenic determinants when coupled to large carrier molecules. These small compounds are called **haptens**.
Haptens react with their specific antibodies (or immune cells), but unless
attached to larger carrier molecules, they are too small to stimulate
antibody production.

Antigens may stimulate humoral immunity or cell-mediated immunity, the two branches of our immune system.

HUMORAL IMMUNITY

Humoral immunity is due to the production
of a special class of protein molecules, called antibodies, that are soluble in
the body fluids (the "humors"). An **antibody** is produced by a vertebrate
host in response to the introduction of antigens into the body and is

capable of specifically binding with the antigen that stimulated its formation. The reaction often protects the host against many detrimental effects of these intruding substances.

Antibodies

Antibodies are also called **immunoglobulins** (Ig) because they mediate immune reactions and belong to a class of proteins called globulins. Antibodies are *monospecific* molecules; they combine only with the single *type* of antigenic determinant that stimulated their formation. Most antibodies are also *bivalent*. They possess two identical reactive sites and can couple with two identical antigenic determinants. An antibody does not react with dissimilar antigenic determinants, even if both types of determinants are on the same antigen molecule (Fig. 16-2). The reaction between antibody and antigen results in the formation of an *antigen-antibody complex*.

Antibody Structure Although some variations exist among their structures, most antibodies consist of four protein chains linked together in what is usually illustrated as a Y-shaped structure (Fig. 16-3). The two shorter chains, called *light* (L) *chains*, are covalently linked to the branches of the longer *heavy* (H) *chains*. Each chain has variable and constant regions. The specificity of the antibody's combining sites is determined by the amino acid sequence in the variable regions of both the H and L chains. The amino acid sequence in the constant region determines other charac-

FIGURE 16-2
Antigens and antibodies can combine with each other if the reaction site of the antibody is complementary to an antigenic determinant on the antigen. Each antibody pictured here has two identical reaction sites and can therefore combine with two identical antigenic determinants. A single antigenic molecule (or cell), on the other hand, may contain multiple determinants and may react with hundreds of antibodies of many different specificities.

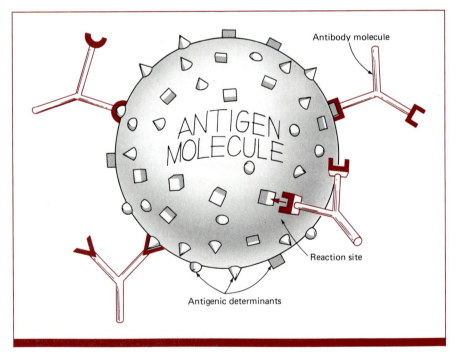

Antibody molecule

Reaction site

Antigenic determinants

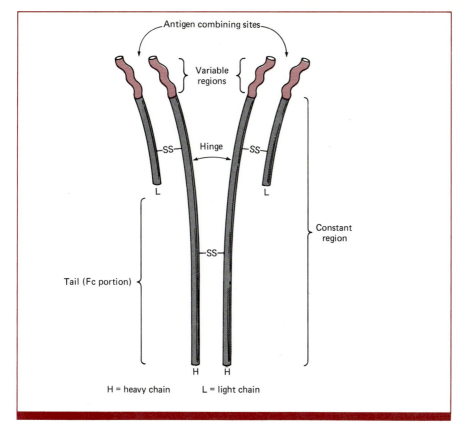

FIGURE 16-3
Structure of the most
typical antibody, called
immunoglobulin G (IgG).
The amino acid sequence
of the variable region
(shown in color)
determines the antigen
specificity of the antibody.
The rest of the molecule is
identical for all IgG
produced by each person.
The angle of the hinge
varies according to antigen
size. Large antigens may
bend the hinge to an angle
of 180°.

Antigen combining sites

Variable regions

—SS— Hinge —SS—

L   L

—SS—

Constant region

Tail (Fc portion)

H   H

H = heavy chain   L = light chain

teristic properties of an antibody, such as its ability to cross certain tissue
barriers. These constant region properties define the five major classes of
immunoglobulins:

1. Immunoglobulin G (IgG) is the most common class of antibodies,
composing 80 percent of those found in the blood. IgG is typically
Y-shaped, bivalent, and crosses the placenta, so a mother's IgG antibodies
help protect her developing fetus. When coupled with antigen, it fixes a
group of serum proteins called complement. (Complement fixation will be
discussed later in this chapter.)

2. Immunoglobulin M (IgM) antibodies consist of five Y-shaped subunits
linked together by their "tails," called the *Fc region*. These are the first
antibodies to appear after initial exposure to an antigen. They fix complement, but do not cross the placenta.

3. Immunoglobulin A (IgA) is a class of antibodies found in two forms. In
the serum-IgA structure resembles that of IgG. IgA's other form, called
secretory IgA, is the principal antibody found in saliva, mucus, and other
external secretions. Secretory IgA is a dimer composed of coupled IgA

TABLE 16-1
Some Properties of the Five Immunoglobulin Classes

Immuno-globulin Class	Structure	Molecular Weight, daltons*	Percent in Blood	Location	Crosses Placenta	Fixes Complement
IgG		150,000	75–80	Blood and tissue fluids	Yes	Yes
IgM		900,000	6–7	Blood and tissue fluids	No	Yes
IgA		170,000†	15–21	Saliva, mucus, and secretions	No	No
IgE		200,000	<1	Skin, respiratory tract, and tissue fluids	No	No
IgD		180,000	<1	Serum	No	No

*One dalton equals the weight of an atom of hydrogen.
†This is the weight of *each* immunoglobulin in the *secretory IgA dimer* illustrated here.

antibodies. It is readily secreted across mucous membranes, providing local protection in areas such as the alimentary and genitourinary tracts.

4. Immunoglobulin E (IgE) attaches by its Fc region to certain host cells, leaving its antigen-combining sites available for binding with antigen. IgE mediates one group of allergic reactions, the best known of which are hay fever and asthma. Its protective function is unknown.

5. Immunoglobulin D (IgD) is the least understood type of antibody. It is believed to be necessary for the differentiation of immune cells.

Some properties of all five immunoglobulin classes are presented in Table 16-1.

Mechanisms of Antibody Action   Formation of the antigen-antibody complex is in itself usually not enough to protect against the potentially adverse effects of a pathogen or a harmful substance. The reaction must alter the antigen so that it is inactivated, killed, prevented from spreading throughout the body, or rendered more susceptible to other defense mechanisms. Antibodies protect hosts in the following ways:

■ **Neutralizing antibodies** react with either viruses or toxins and block their harmful effects. *Antitoxins* are formed in response to exposure to toxic antigens. Many people who survive botulism, tetanus, diphtheria, or poisonous snakebites have developed antitoxins that neutralize the corresponding toxin in their blood. *Viral neutralizing antibodies* prevent viral disease by occupying the attachment sites on the virus, thereby neutralizing infectivity.

■ **Agglutinins** are antibodies that react with particulate antigens (such as whole bacteria) and cause agglutination, or clumping (Fig. 16-4). Agglutin-

FIGURE 16-4

(a) Agglutination. The antibody reacts with its particulate antigen (bacterial cell), causing it to agglutinate (upper frame). A nonagglutinated control is shown in the lower frame. (b) Agglutination is due to the bivalent nature of antibodies. Each antibody can attach to identical antigenic determinants on two separate antigen particles (cells). Other particles are bound to these in a similar fashion. The process continues, creating a network of clumped cells.

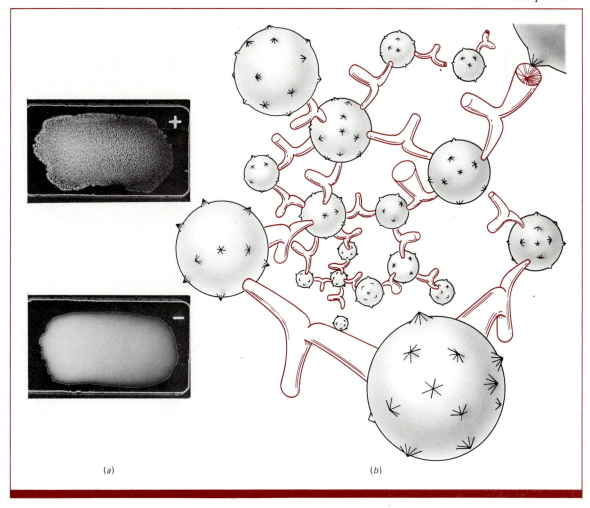

(a)   (b)

FIGURE 16-5

Immunoprecipitation of a soluble antigen and specific antibody in agar. The soluble antigen (center well) and various preparations of antibody (peripheral wells) diffuse through agar and, if they react, form visible lines of precipitation. Such reactions can be used to specifically detect or identify chemical substances or antibodies in a patient's serum.

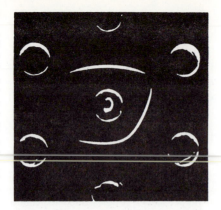

374

ACQUIRED IMMUNITY

ation opposes the spread of pathogens through the body. In addition, each phagocyte can engulf a large number of agglutinated bacteria in a single "bite." Many more phagocytes would be required to engulf the same number of unclumped bacteria.

■ **Precipitins** are immunoglobulins that react with soluble antigens and convert them to a solid precipitate (Fig. 16-5). Most soluble molecules, such as extracellular virulence factors secreted by infecting bacteria, are inactive when precipitated. In addition, precipitated molecules settle out as solids that are more easily phagocytized.

■ **Opsonins** are antibodies that promote phagocytosis of the antigens with which they react. Once the antibody binds with its antigen, its Fc region couples to a receptor site on the surface of the phagocyte, forming an antibody bridge between phagocyte and antigen. The importance of opsonins is illustrated by their role in prevention or recovery from pneumonia caused by heavily encapsulated *Streptococcus pneumoniae*. In the absence of antibodies, the antiphagocytic capsule prevents engulfment of the offending bacteria. Anticapsular antibodies, however, neutralize the antiphagocytic properties of the capsule, and the pathogen becomes easily engulfed and destroyed by the macrophages in the lung.

■ **Complement-fixing antibodies** lyse cells (especially gram-negative bacteria) in the presence of a series of blood proteins called *complement* (C'). Complement-fixing antibodies combine with their antigenic target cells before the inactive complement components sequentially assemble into their active forms on the antigen-antibody (Ag-Ab) complex. Complement proteins alter the cell membrane so that it leaks cytoplasm and disrupts the cell. These membrane lesions are microscopically evident as thousands of tiny holes (Fig. 16-6). The lytic action of complement is less effective against gram-positive bacteria because of the additional protection afforded by their thicker cell walls.

The protective action of complement is not limited to cell lysis. Soluble chemotactic fragments are released during assembly of complement on the antibody–target cell complex. These free components attract phagocytes to

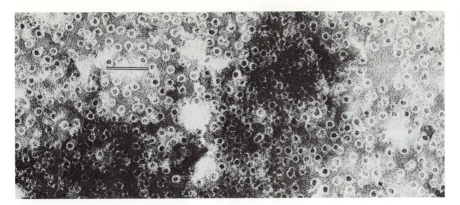

FIGURE 16-6
Complement-mediated
damage is evident as holes
in the cell membrane.

the area of reaction. In addition, one complement component binds to the bacterial cell wall and to surface receptors of phagocytic leukocytes, thereby promoting engulfment of the target bacteria. Unlike cell lysis, the chemotactic and opsonizing actions are effective against both gram-positive and gram-negative bacteria.

Complement works in conjunction with antibodies to provide three types of protection: lysis, chemotaxis, and opsonization (Fig. 16-7). The whole process is called *complement fixation* because soluble complement is irreversibly "fixed" to the Ag-Ab complex and is therefore removed from serum.

FIGURE 16-7
Three protective activities
of complement fixation.
(See text for full
explanation.)

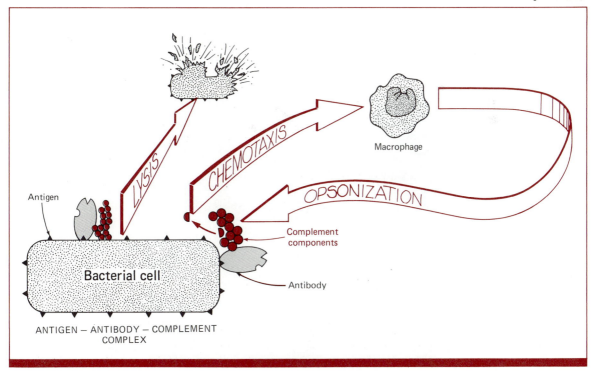

A few complement components can combine with some microbes in the absence of antibodies. This *alternate complement pathway* releases chemotactic and opsonizing factors that promote phagocytosis of the microbe, but the target cell is not lysed.

Persons devoid of serum complement are susceptible to a number of recurrent diseases even if their ability to produce antibodies and immune cells is unimpaired.

Most immunoglobulins possess several protective capabilities. For example, an antibody that agglutinates *Klebsiella pneumoniae* may also opsoninize this bacterium, lyse the pathogen by complement fixation, and precipitate a soluble antigen released from this microbe. In addition to protecting the immune host, these antibody activities can be used in the laboratory for detecting and identifying microbes and evaluating host immune responses.

Antibody-Producing Cells   Special nonphagocytic white blood cells, called **lymphocytes**, are responsible for immune responses. Humoral immunity is associated with lymphocytes that have passed through a special lymphoid organ that programs these cells for participation in antibody production. This organ was first discovered in chickens, where it is called the bursa of Fabricius. The bursa-programmed cells are thus called **B lymphocytes**, or simply B cells (Fig. 16-8). The bursa equivalent in humans appears to be lymphatic tissue in the gut, bone marrow, or perhaps the tonsils. The B lymphocyte population in each person can recognize approximately 10^8 distinct antigens; therefore virtually any antigen can stimulate a humoral immune response. Each B cell has receptors on its surface that recognize a specific antigen. The B cell is programmed to respond when these surface receptors react with the

FIGURE 16-8
B lymphocytes as viewed with the scanning electron microscope.

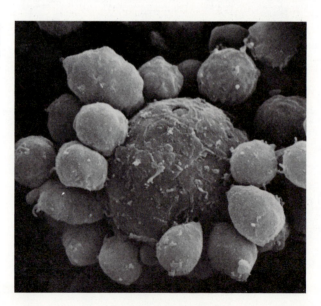

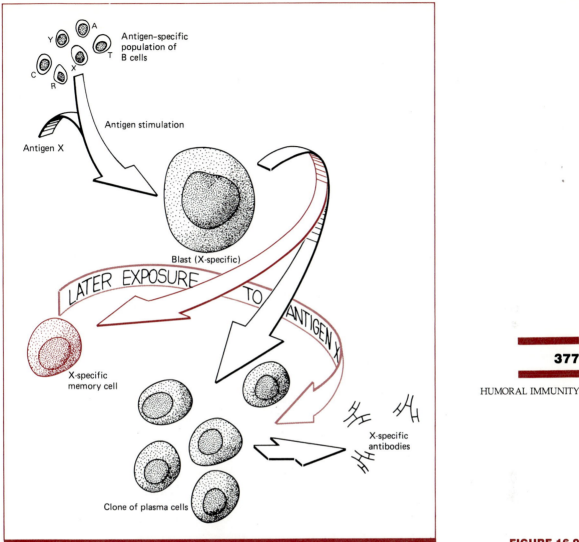

FIGURE 16-9

Clonal selection theory of antibody production. An antigen-specific B cell is selected from other B cells when stimulated by the corresponding antigen (represented here by hypothetical antigen X). If antigen persists, a pool of antibody-producing plasma cells and a few memory cells eventually develop. Later exposure to the same antigen rapidly activates memory cells (shown in color) to differentiate into plasma cells, proliferate, and produce antibodies.

specific antigen. Initial exposure to the antigen triggers the B cell to proliferate, forming a large clone of cells (Fig. 16-9). Continual stimulation increases the number of these B lymphocytes, which differentiate into smaller antibody-producing **plasma cells**. Each clone of plasma cells manufactures antibodies that specifically react with the antigenic determinant that stimulated the initial proliferation. In addition to plasma cells, some lymphocytes differentiate into **memory cells** that produce no antibody. When later exposed to the same antigen, however, these memory cells rapidly differentiate into antibody-producing plasma cells. This may occur even years after initial exposure to the antigen.

Dynamics of Antibody Production   Antibody production occurs in two distinct stages. The **primary response** follows exposure to an antigen that the host has never before encountered. During this time, antigen-specific B lymphocytes proliferate, and the host becomes *sensitized* to that antigen so that each subsequent exposure to the same antigen stimulates a **secondary response**. These two stages are graphically illustrated in Figure 16-10, which shows antibody production following exposure to an antigen that, prior to time 0, had never been introduced to this host.

PRIMARY RESPONSE   Initial exposure to the antigen is followed by a latent period of 3 to 30 days. During this time no antibody is detectable in the blood. The length of the latent period depends on several variables, including the nature and concentration of the antigen, its route of administration, and the immunological competence of the host. After the latent period, the antibody *titer* (concentration) increases for a short period of time, then diminishes and eventually disappears. This initial production of antibody is called the primary response.

FIGURE 16-10

Dynamics of antibody production. During the primary response the host becomes sensitized to the antigen so that antibody production is accelerated and intensified during the secondary response. Each subsequent exposure boosts the magnitude of the anamnestic response until the maximum titer is achieved. Both primary and secondary responses may be induced by a single antigenic exposure if the antigen persists in the body throughout the duration of the primary response.

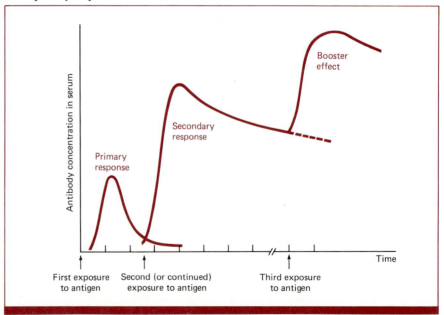

SECONDARY RESPONSE   If the same antigen is reintroduced into this individual or if the antigen persists through the duration of the primary response, it stimulates a secondary response. The secondary response is distinguished from the primary phase by several factors:

1.   A lower dose of antigen is needed to trigger antibody production.

2.   There is a much shorter latent phase.

3.   Much higher antibody concentrations are produced.

4.   There is a much longer duration of peak antibody titers.

The secondary response therefore provides greater immunological protection at a much faster rate than the primary response. In addition, each subsequent exposure acts as a *booster* to stimulate higher antibody concentrations until maximum titers have been achieved. This may require three to five booster exposures.

The secondary response is a product of immunological recall provided by memory B cells generated during the primary response. Antigenic exposure quickly activates these memory cells even years later when antibody is no longer detectable. Because of this recall phenomenon, the secondary response is also referred to as the **anamnestic** (without amnesia) **response**, or simply, the memory response.

The anamnestic response accounts for the long-lasting immunity that follows recovery from many infectious diseases, even if the infection was *asymptomatic* (without obvious symptoms). People are susceptible during their initial exposures to most pathogens because several days are required to develop protective levels of antibody. If we survive this period of delay (the primary response), the secondary immune response eventually arrests the disease. Thus a single infection may provide us with long-lasting immunity to subsequent disease because additional exposures elicit rapid secondary responses that eliminate the pathogen before it can cause disease again. The strategy behind vaccination is to induce immunological memory specific for a pathogen by administering an antigenically identical yet harmless variant of either the microbe or its principal virulence factor. In this way we acquire immunity without running the risk of developing disease during the vulnerable primary response. Periodically administered boosters enhance protection and maintain protective antibody titers.

CELL-MEDIATED IMMUNITY (CMI)

The second general type of immunity is mediated by special cells called **T lymphocytes** (Fig. 16-11). T cells are programmed for participation in **cell-mediated immunity**, (**CMI**) by passage through the *thymus gland*. T cells produce no antibodies, but provide protection in the manner described below and shown in Figure 16-12.

Receptors on the surface of T cells recognize and bind with specific antigens. Antigen attachment activates the T cells and triggers them to

FIGURE 16-11

T lymphocytes as viewed
with the scanning electron
microscope.

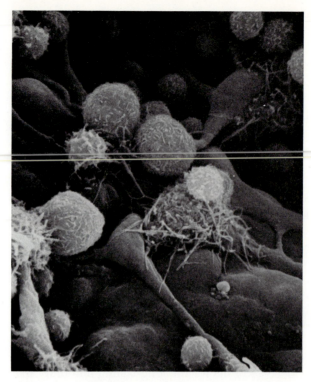

380

ACQUIRED IMMUNITY

proliferate, forming an expanded clone of immunocompetent lymphocytes
in the area of the body where the antigen is concentrated. Among these
rapidly proliferating cells are three subpopulations of lymphocytes, functionally defined as (1) *effector* cells, (2) *regulator* cells, and (3) *memory* cells.

Effector Cells

Effector cells are the lymphocytes that provide the actual protection. For
example, **T killer lymphocytes** (cytotoxic T cells) physically attach to their
target cells and destroy them by membrane disruption and lysis. Such
antigen-specific cytotoxicity enables our immune system to selectively
eliminate some tumor cells and viral infected host cells that acquire new
antigenic determinants which are recognized as foreign. **Delayed hypersensitivity lymphocytes** (DHS cells) are effector cells that bind to their
target antigen and release soluble substances called **lymphokines**. Although DHS cells cannot directly damage their cellular targets, lymphokines hasten the destruction of the foreign cells. Several lymphokines have
been described; most of them attract macrophages and increase the
phagocytic efficiency of these cells. The following list describes several
lymphokines.

Chemotactic factors attract macrophages to areas of lymphocyte concentration (and therefore to the area of antigen concentration).

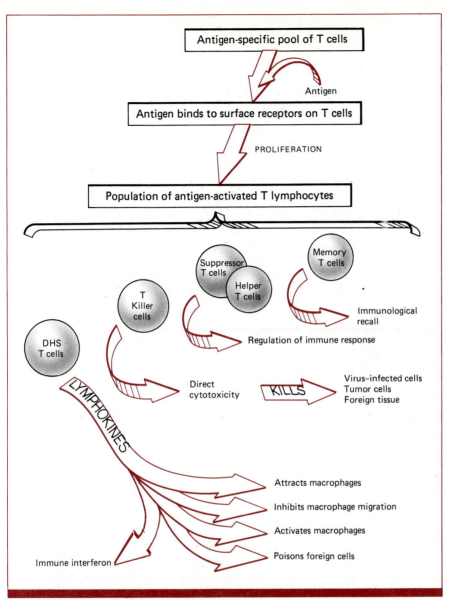

FIGURE 16-12
The development of
cell-mediated immunity
following antigenic
stimulation. (See text for
description of events.)

Antigen-specific pool of T cells

Antigen

Antigen binds to surface receptors on T cells

PROLIFERATION

Population of antigen-activated T lymphocytes

Suppressor T cells

Helper T cells

Memory T cells

T Killer cells

Immunological recall

DHS T cells

Regulation of immune response

LYMPHOKINES

Direct cytotoxicity

KILLS

Virus-infected cells
Tumor cells
Foreign tissue

Attracts macrophages

Inhibits macrophage migration

Activates macrophages

Poisons foreign cells

Immune interferon

Migration inhibition factor (MIF) reduces the mobility of macrophages so that they accumulate around the activated T cells. MIF prevents macrophages from moving away once they reach the site of antigen stimulation.

Macrophage activating factor (MAF) greatly enhances phagocytosis of any foreign particles in the area of antigen stimulation. MAF creates a population of "angry macrophages" with larger, more numerous lysosomal granules. The result is enhanced intracellular killing capacity following engulfment of any cell.

Specific macrophage arming factor (SMAF) enhances the ability of macrophage to kill the specific antigenic target cell.

Lymphotoxin nonspecifically destroys potentially pathogenic cells. Unfortunately host cells may also be damaged by lymphotoxin, which may account for some tissue damage associated with delayed-type hypersensitivity (discussed later in this chapter).

Immune interferon produced by antigen-activated lymphocytes aids the arrest of infections by increasing the cytotoxicity of T cells and by triggering the production of antiviral proteins in neighboring cells (see Chap. 9).

The overall effect of lymphokines is enhanced activity of our nonspecific protective mechanisms, especially the phagocytic macrophage system. In some instances, as with SMAF, these defenses are directed against the specific antigen that induced CMI. Antigenic challenge, however, boosts the CMI defenses in a nonspecific fashion so that an unrelated pathogen may be simultaneously eliminated by the stimulated system.

Regulator Cells

Regulator cells are T lymphocytes that help govern the intensity of the immune response and that provide one of the avenues of cooperation between CMI and humoral immunity. **Helper T cells**, for example, cooperate with B cells to initiate an antibody response. This may provide a fail-safe system against accidentally triggering antibody production, since both T and B cells must recognize an antigen as foreign before an immune response is launched. As a result, the probability of accidentally attacking "self" tissues is reduced. **Suppressor T cells** affect B cells in the opposite fashion, decreasing the production of antibody. This helps avoid excessive immune responses that might damage the host.

Memory Cells

A third pool of antigen-stimulated T lymphocytes are **sensitized memory cells**. These quiescent cells remain in the host long after elimination of the challenging antigen. They retain the potential for rapid activation whenever the host is exposed to the corresponding antigen, at which time they proliferate into a pool of effector lymphocytes. This CMI recall response is similar to the humoral anamnestic response and decreases the duration of the latent period (the time between antigen stimulation and development of protective immunity) from about 20 days to 48 hours.

CMI Antigens

Cell-mediated immunity is usually directed against whole cells—protozoa, fungi, some bacteria (especially those that are intracellular parasites), and viral-infected host cells. Resistance and recovery from infectious diseases caused by these agents depend on the T cell response. Although these antigens may also induce antibody production, immunoglobulins fail to protect against many of these diseases. CMI also functions as an immune surveillance system which detects and destroys tumor cells that occasionally develop from normal host cells. The foreign tissues of transplanted human organs or skin grafted from other persons are rejected and destroyed by T cells that become sensitized to these tissue antigens. Corticosteroids and other drugs that suppress CMI in the recipient can be used to prolong graft survival. Such immunosuppression, however, enhances the danger of fatal infectious disease caused by viruses, fungi, protozoa, and certain bacteria. (See Other Immunological Disorders later in this chapter.)

NATURAL KILLER CELLS

Natural killer (NK) cells are neither macrophages nor T or B lymphocytes, yet they lyse a variety of virus-infected cells and tumors. Great numbers of natural killer cells are found in persons infected with mumps and herpesviruses. Natural-killer-cell activity appears to be increased by interferon released from virus-infected cells. The antitumor activity of natural killer cells may be important in defending against cancer. This activity is enhanced by the lymphokine, immune interferon. This substance, as well as a variety of immunopotentiators, is being investigated as a possible anticancer therapy (see color box).

ACTIVE AND PASSIVE IMMUNITY

Immunity, whether humoral or cell-mediated, is acquired either actively or passively, depending on whether the immune person actively produces the protective antibody (T cells) or receives presynthesized antibody (sensitized lymphocytes) from another individual. Both active and passive immunity can be acquired naturally or artificially (see Table 16-2).

Active Immunity

Active immunity is the production of antibodies or specialized lymphocytes by the host as a result of exposure to a foreign antigen. It is characterized by (1) a latent period of at least 2 weeks between initial exposure to the antigen and development of protective immunity, and (2) an extended duration of immunity that often lasts for years.

Natural active immunity develops following exposure to an antigen by natural infection. Immunity is directed against the specific infectious agent

Active Immunity: Immunized individual produces antibody.
Natural immunity follows natural infection.
Artificial immunity follows vaccination (immunization).

Passive Immunity: Immunized individual receives antibody produced by another individual.
Natural immunity is due to transfer of maternal antibodies to fetus or suckling newborns.
Artificial immunity is induced by injecting presynthesized antibody into recipient.

TABLE 16-2
Four Categories of
Acquired Humoral
Immunity

IMMUNOPOTENTIATORS: PRIMING THE IMMUNOLOGICAL PUMP

■ The search for safer, more effective cancer treatments is complicated by the similarity between tumor cells and normal human cells. Most agents that can kill tumors are similarly lethal for normal cells. Therefore, most chemotherapeutic agents for treating cancer produce toxic side effects. This disadvantage could be avoided by using agents that have no lethal effects at all, but rather stimulate cell-mediated immunity to naturally eradicate the tumor. This approach is called **immunopotentiation**. Furthermore, while conventional cytotoxic agents for cancer therapy compromise the patient's resistance to pathogens, immunopotentiators would oppose opportunistic infections. Is it possible to find such remarkable agents?

Several microorganisms or their products are reported to favorably affect the treatment of cancer by immunopotentiation. The most extensively tested of these agents is Bacille Calmette-Guerin (BCG), the vaccine against tuberculosis. BCG is an attenuated strain of the tubercle bacillus that stimulates cell-mediated immunity. The effects of T cell enhancement are nonspecific; not only is tuberculosis prevented but, in many cases, malignancies are also eradicated. The effect is most pronounced against skin cancer. Pertussis vaccine and nonviable preparations of *Corynebacterium parvum* also promise effectiveness in launching an immune-based attack against cancer.

In addition to stimulating CMI, these agents may also enhance antibody production.

Many immunopotentiators are *adjuvants*, substances that enhance the immune response when injected with an antigen. The enhancement is often specific for antigens administered with the adjuvant. For example, killed *Bordetella pertussis* enhances immunity against diphtheria and tetanus when administered with the corresponding toxoids. The DPT—diphtheria, pertussis (whooping cough), tetanus—vaccine thus provides better immunity than would the toxoids given separately.

Several chemical agents show promise as immunopotentiating agents against virus infections. Isoprinosine is reported to shorten the mean duration of Type 2 herpes from 10 to 4 days while reducing the incidence of recurrences. Although it has no direct antiviral activity, isoprinosine apparently enhances the efficiency of the immune system to eradicate the viruses and resolve the lesions. Some scientists feel that interferons are the most exciting of the immunopotentiators. In addition to stimulating antiviral proteins, interferons increase the effectiveness of the immune system. Other specific lymphokines have also been shown to be effective immunopotentiators. The search for additional nonspecific immunopotentiators promises safer, more effective ways to manage malignancy and many infectious diseases.

as well as its toxic by-products. Immunological recall in both cell-mediated and humoral responses maintains protection for months and often for years. The protective efficiency of natural active immunity is demonstrated by the infrequency of second attacks of smallpox, chickenpox, mumps, and measles. Even in the absence of overt symptoms, these infections may stimulate lifelong immunity.

Artificial active immunity is similar to natural active immunity except for the nature of the antigen and the method of introduction into the host. Instead of natural infection by a potentially virulent microbe, a vaccine is intentionally introduced into the body by a clinical procedure such as injection. The vaccine is antigenically similar to a pathogen (or to its toxic by-products) but has been treated so that it can be administered to people with little danger of disease. The vaccine sensitizes the immune system to the corresponding pathogen, inducing immunity without the danger of infectious disease developing during the latent period. In immunized persons, natural exposure to the corresponding virulent pathogen triggers a protective anamnestic response.

Preparation of Vaccines   An effective vaccine retains the corresponding pathogen's antigens but none of the pathogen's ability to damage the host. This is accomplished in a number of ways.

KILLED ORGANISMS   A pure culture of the pathogen is exposed to an agent that will kill the microbe without altering the surface antigens. Injection of these killed organisms into a host provides safe initial exposure to the pathogen and subsequent resistance to disease. Typhoid fever vaccine, for example, is a formalin-killed preparation of the pathogen *Salmonella typhi*. Vaccines containing killed organisms (or inactivated viruses) have been used to prevent polio (Salk vaccine), cholera, typhoid fever, influenza, whooping cough, rabies, and epidemic typhus. Most of these vaccines require several booster shots before protection is adequate.

ATTENUATED ORGANISMS (LIVE VACCINES)   The most effective vaccines are those that are capable of actually multiplying in the host, mimicking the early stages of natural infection. Many pathogens may be **attenuated**, that is, alive but weakened in their capacity to cause severe disease. Vaccines containing attenuated organisms are especially effective because they can be introduced by natural routes. The attenuated microbes harmlessly propagate in the body to provide a greater and more prolonged antigenic stimulation. Furthermore, the attenuated pathogens may be shed by the immunized individuals and vaccinate other susceptible persons.

The attenuated Sabin polio vaccine, for example, is generally superior to the Salk vaccine (an inactivated poliovirus preparation) because it is administered orally and harmlessly replicates in the intestine, similar to the early stages of poliomyelitis. Unlike virulent poliovirus, the attenuated variant causes no paralysis. The Sabin vaccine stimulates the formation of secretory IgA antibodies which neutralize viruses in the intestine before they invade the bloodstream. The inactivated Salk vaccine, on the other hand, must be injected into the muscle, inducing little protection in the intestine. In addition, "live" Sabin vaccine virus is shed in the feces of vaccinated persons. Since poliomyelitis is naturally transmitted by the oral-fecal route, people may be inadvertently vaccinated by ingesting attenuated viruses shed by vaccinated individuals. In addition to protecting against poliomyelitis, attenuated vaccines protect against influenza, measles, mumps, and many other diseases.

Unlike killed vaccines, however, attenuated microbes have been known to regain virulence and have caused serious disease in vaccinated persons. Only those attenuated organisms unlikely to revert can be safely used in vaccines.

PURIFIED ANTIGEN FRACTIONS Some vaccines contain only the antigens against which the protective immune reactions are directed. For example, purified extracts of the capsular antigens from virulent *Streptococcus pneumoniae* stimulate the host to produce opsonizing antibodies that effectively protect most immunized persons.* Antigenic fractions are used in vaccines against cholera, plague, hepatitis B, and some types of pneumonia and meningitis.

*This vaccine, called pneumovax, contains capsular antigens of the 14 most prevalent virulent strains of *Streptococcus pneumoniae*. About 80 percent of all cases of pneumococcal pneumonia are caused by one of these 14 strains.

FIGURE 16-13

Vaccines against toxins are prepared by chemical modification of active exotoxin. The resulting harmless toxoid still retains its antigenic determinants, so antibodies induced by immunization with toxoid neutralize the active exotoxin.

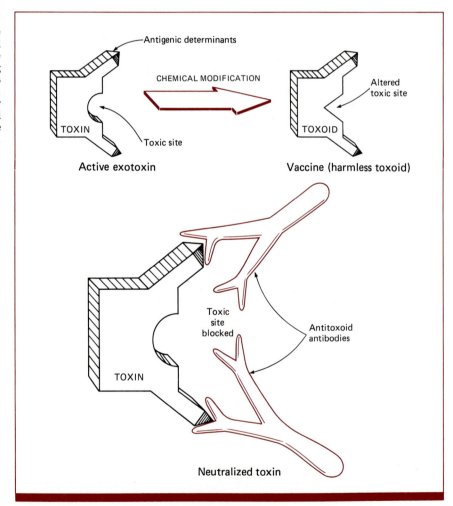

386

TOXOIDS The dangerous symptoms of tetanus, diphtheria, and a few other diseases are due almost entirely to the effects of soluble exotoxins secreted by the pathogen. (See Toxic Factors, Chap. 15.) Active immunity against these diseases can be induced by injecting a **toxoid**, an exotoxin that remains antigenically unaltered but has been chemically treated to destroy its poisonous properties (Fig. 16-13). An injection of toxoid stimulates production of antibodies (antitoxins), thereby inducing immunity against the corresponding disease.

SYNTHETIC PEPTIDES Through the use of sophisticated laboratory technology the amino acid sequence of purified protein antigens can be precisely determined. Once the sequence is known, it is possible to synthesize peptides of approximately 20 amino acids that may represent antigenic determinants against which protective immunity is directed. Such peptides have been prepared from hepatitis B virus surface antigen, *Streptococcus pyogenes* M protein, diphtheria toxin, and a number of other proteins. The advantage of synthetic vaccines is that they ensure exclusion of contaminating materials that might harm the host. The widespread use of synthetic vaccines awaits the development of high-yield systems for their production.

The various types of vaccines used to prevent infectious diseases among humans are listed in Table 16-3. Unfortunately many diseases cannot be controlled by vaccination. (See color box.)

Passive Immunity

Antibodies produced by active immunization can be transferred to a nonimmune recipient. The antibody recipient is then said to be *passively*

TABLE 16-3

Vaccines Used in Humans
to Prevent Infectious
Disease

Type of Vaccine	Disease	Etiological Agent
Attenuated (infectious)	Poliomyelitis*	Poliovirus
	Influenza	Influenza virus
	Measles	Measles virus
	Mumps	Mumps virus
	Rubella	Rubella virus
	Tuberculosis	*Mycobacterium*
	Yellow fever	Yellow fever virus
Killed or inactivated	Whooping cough†	*Bordetella pertussis*
	Typhoid fever	*Salmonella typhi*
	Epidemic typhus	*Rickettsia prowazekii*
	Rabies	Rabies virus
	Influenza	Influenza virus
	Poliomyelitis (Salk)	Poliovirus
Antigen fractions	Cholera	*Vibrio cholerae*
	Plague	*Yersinia pestis*
	Meningitis	*Neisseria meningitidis*
	Pneumonia	*Streptococcus pneumoniae*
	Hepatitis	Hepatitis B virus
Toxoid	Tetanus†	*Clostridium tetani*
	Diphtheria†	*Corynebacterium diphtheriae*

*For Sabin vaccine.
†Routinely administered during the first year of life as the combined vaccine, DPT (diphtheria, pertussis, tetanus).

WHY NOT VACCINATE AGAINST ALL INFECTIOUS DISEASES?

■ Some early immunologists dreamed of developing a broad complement of vaccines that would create a population of people resistant to all infectious diseases. Yet today we still cannot immunize against many important diseases; gonorrhea, syphilis, and even the common cold continue to strike millions of persons each year. The lack of suitable vaccines against many diseases is due partially to the nature of these pathogens and partially to characteristics of the immune system. The following lists a few of the barriers to immunization:

■ Some pathogens, such as *Neisseria gonorrhoeae*, are so nonimmunogenic that they fail to trigger protective levels of immunity. Extensive efforts to produce immunogenic vaccines against gonorrhea have, as of 1984, yet to yield a promising product.

■ Infections localized on the body's skin or mucous membranes are not affected by circulating antibodies and are generally controlled by secretory IgA. Unfortunately, immunological recall is less efficient with these secretory immunoglobulins than with circulating antibodies. Thus uncomplicated gonorrhea and other diseases restricted to the urogenital mucosa stimulate poor natural immunity during infection and are less likely to be controlled by vaccines. In addition, *Neisseria gonorrhoeae* produces a protease that protects the pathogen by enzymatically hydrolyzing IgA.

■ Some pathogens suppress immunological response during natural infection. Unfortunately vaccines prepared against these diseases (malaria, syphilis, and leprosy) may similarly depress immunity rather than enhance it.

■ Some viruses become sequestered within host cells where they are protected from the body's immune elements. Herpesviruses, for example, persist in nerve cells following disappearance of the epithelial lesions. Lesions may recur when the virus migrates directly along infected neurons

immunized. Although these antibodies provide immediate protection, they are eventually depleted and are not replaced by the body. Passive immunization, therefore, provides only temporary protection, usually lasting no more than a few weeks. Antibodies can be transferred by natural or artificial routes.

Natural passive immunity helps protect newborns from infectious disease. The fetus, which is incapable of producing antibodies, becomes passively immunized by acquiring maternal antibodies, antigen-specific immunoglobulins produced by the mother and transferred across the placenta to the fetus. Maternal antibodies persist for many weeks following birth and help protect the baby until he or she can actively produce antibodies. (Immunological competence begins during the third to sixth month after birth and may not be adequate for 2 or 3 years.) Additional protection is provided by maternal antibodies that are passively transferred in breast milk. Breast milk also contains 25 percent of the mother's daily production of monocytes. Such protection cannot be acquired from a commercial infant's formula or cow's milk.

to skin, even in the presence of specific circulating antibodies.

- Vaccine production usually requires laboratory cultivation of the pathogen to provide a source of antigen. Unfortunately some pathogens, notably *Treponema pallidum* (syphilis), cannot be readily grown in vitro.

- The common cold viruses, although extremely antigenic, are so numerous it would be virtually impossible to develop vaccines against them all. More than 100 antigenically distinct cold viruses would have to be included in a complete vaccine. Such a vaccine would not only be impractical but would be of little value because poor immunity is induced by simultaneous administration of so many antigens.

- Some pathogens periodically alter their surface antigens. The new antigens are not recognized by immune cells sensitized to the previous determinants. Influenza is the most extensively documented example of a pathogen's tendency toward antigenic shifts. Although influenza vaccines are available, they are type-specific. A new vaccine must be developed whenever a new, antigenically distinct strain appears.

- Vaccines are most successful against diseases controlled by humoral immunity. Few vaccines are available against diseases that are controlled primarily by CMI. For example, we cannot immunize effectively against any diseases caused by protozoa or fungi.

- Some pathogens immunologically camouflage themselves by coating themselves with host blood proteins. In this way the surface antigens of plasmodia (malaria) and trypanosomes (encephalitis) are hidden from immune effectors. Vaccination against these diseases may trigger the formation of antibodies and lymphocytes that can't locate their camouflaged antigens.

Artificial passive immunity is a valuable therapeutic tool when immediate protection is required, with no time to wait for active immunity to develop. Antibodies produced by one individual are injected into a recipient to provide a temporary supply of protective circulating antibodies. Antitoxin immunoglobulins, for example, are produced by actively immunizing a horse with tetanus toxoid and concentrating the gamma globulin fraction from the animal's blood collected several weeks later. Patients in danger of developing tetanus are injected with this antitoxin preparation, which rapidly neutralizes the potentially lethal exotoxin. These persons could otherwise die of the disease before producing their own antibodies. Protective antibodies can be prepared against several poisons that are lethal unless quickly neutralized in the body. *Antiserum* (blood serum* that contains antibodies of known specificity) is available for use against botulism, tetanus, and the poisonous bites of spiders and snakes.

---

*Serum is the liquid that remains when blood is allowed to clot and the clot is removed. It retains most of the proteins and other soluble substances found in blood.

Antibodies for passive immunization are also administered as *hyperimmune serum* obtained from persons who have been actively immunized with an appropriate vaccine. Hyperimmune human sera are available for use against rabies, measles, pertussis, and tetanus.

Unfortunately, artificial passive immunization may cause **serum sickness**, a side effect that develops in response to massive amounts of foreign protein found in these sera. The danger of serum sickness increases with each successive exposure. For example, persons receiving horse serum for the second time are already immunologically sensitized to its antigens. They quickly produce antibodies against the foreign horse-globulin proteins, generating enormous numbers of antigen-antibody complexes that are deposited in capillary walls and in the kidney, often causing severe, sometimes fatal disease. Serum sickness may also follow passive immunization with human serum, but the danger is diminished because the proteins are less foreign. Passive immunization with purified antigen-specific antibody preparations further reduces the danger of serum sickness (see color box). Maintenance of active immunity by routine vaccination against tetanus and diphtheria eliminates the need for the potentially dangerous injections of foreign sera to combat these diseases.

Unlike humoral immunity, CMI cannot be passively transferred by serum. Immune lymphocytes from a sensitized donor, however, can provide temporary cell-mediated immunity. Furthermore, a soluble component, called *transfer factor*, extracted from immune lymphocytes, can specifically sensitize a nonimmune recipient's T cells. Transfer factor has been successfully used to control severe *Candida albicans* infections in persons with apparently deficient CMI against this opportunistic yeast.

THE REMARKABLE POTENTIAL OF MONOCLONAL ANTIBODY

Obtaining a preparation of antibody specific for a single antigen has traditionally required immunizing laboratory animals with the desired antigen and later collecting the immune serum. Unfortunately this serum contains antibodies specific for many other antigens. These undesired antibodies are removed by precipitation with their antigens. The remaining immunoglobulin then has to be purified. Because this procedure is very expensive and yields little antibody, crude preparations of antibody are usually used.

An ingenious new technique, however, has made this approach to antibody production obsolete. Pure monospecific antibodies can now be produced in cell culture rather than in animals. Because all the cells in a culture are descendants of a single parental cell, they all produce identical products. A culture grown from one isolated B cell produces immunoglobulin specific for a single antigen. Immunoglobulin manufactured by cloned cell cultures is called **monoclonal antibody**.

Unfortunately, normal B cells cannot be continuously cultured in vitro. However, lymphocytes from persons with myeloma (a bone marrow tumor) grow readily in vitro and produce large amounts of nonspecific immunoglobulin. A myeloma cell and an antigen-specific B lymphocyte can be fused

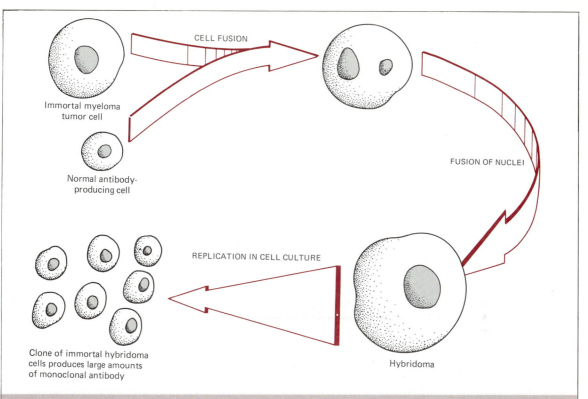

CELL FUSION

Immortal myeloma
tumor cell

Normal antibody-
producing cell

FUSION OF NUCLEI

REPLICATION IN CELL CULTURE

Clone of immortal hybridoma
cells produces large amounts
of monoclonal antibody

Hybridoma

to form a single *hybridoma cell* which can be continuously cultured and is committed to the production of the single antibody type specified by the B cell (see figure). The resulting pure antigen-specific monoclonal antibody is easily harvested from the culture fluid, enormously increasing the yield of monospecific antibody against virtually any antigen desired.

The medical applications of monoclonal antibodies are remarkable. Early diagnosis of cancer is facilitated by using monoclonal antibody that is specific for tumor antigens to detect cancer in patients before symptoms develop. The same antitumor antibodies may also improve the effectiveness of cancer treatment; if chemically attached to cytotoxic chemotherapeutic agents, these antibodies, which are specific for the tumor and fail to bind to normal tissues, can deliver the cytotoxic agent directly to the malignant cells in a concentrated form, enhancing tumor killing and reducing toxicity to nor-

mal host cells. This approach is 100 times more lethal to tumors than the use of either the drugs or the antitumor antibodies alone.

The hybridoma technique promises new approaches to treating otherwise poorly controlled diseases. For example, monoclonal antibodies against autoantibodies (those directed against "self") may be valuable in treating autoimmune diseases, thereby reducing the need to use dangerous immunosuppressive agents. Influenza, rabies, and malaria may all be treated by passively immunizing with pure antibody against the pathogens. Indeed the effectiveness and safety of passive immunization in general is enhanced by the elimination of unwanted antibodies specific for other antigens. This greatly reduces the amount of foreign protein introduced into the body, resulting in fewer immune complexes and thereby minimizing the danger of serum sickness.

HYPERSENSITIVITY

Sensitizing our immune cells to specific antigens provides us with immunological memory and protection. Sometimes, however, immune sensitization to antigens causes immunological overreactions that damage our tissues. This is the condition we call **hypersensitivity**, or, more commonly, **allergy**. One of ever six Americans suffers from allergy.

As with immunity, there are two basic types of hypersensitivity: (1) immediate-type hypersensitivity and (2) delayed-type hypersensitivity. **Immediate-type allergy** is dependent on the B cell immune system and is mediated by antibodies. Sensitized individuals often respond within minutes after exposure to an *allergen* (an antigen that elicits an allergic response). In contrast, **delayed-type hypersensitivity** is associated with sensitized T lymphocytes. Sensitized people react to allergen stimulation 24 to 48 hours after exposure (thus the name "delayed" hypersensitivity). Table 16-4 compares the important properties of delayed-type hypersensitivity with those of the most common form of immediate-type allergy.

Immediate-Type Hypersensitivity

The most prevalent form of immediate-type allergy is due to the production of a unique class of antibodies called **immunoglobulin E (IgE)**.* This

*IgE-mediated allergy is also called type I hypersensitivity.

TABLE 16-4
Comparison of IgE-Mediated Immediate-Type Hypersensitivity* with Delayed-Type Hypersensitivity

	Immediate	Delayed
Mediated by	Humoral antibody (IgE)	T lymphocytes
Symptoms elicited by release of soluble factors:	Histamine; serotonin; slow-reacting substance of anaphylaxis; other pharmacologically active mediators	Lymphokines; lysosome contents released into surrounding tissue
Time (after initial allergen exposure) required for sensitization:	Days	Months
Reaction time in sensitized person:	Immediately after allergen exposure	24–48 h after allergen exposure
Symptoms relieved by:	Anthistamines; epinephrine; sometimes corticosteroids	Corticosteroid
Allergy can be passively transferred to recipient by:	Serum	Lymphocytes or transfer factor
Desensitization achieved by:	Small doses of allergen (induces blocking antibody or suppressor T cells)	No known mechanism
Symptoms:	Acute; reaction is short-lived, although damage may be severe and permanent	Chronic; effect may persist

*This type of allergy represents one of three major types of immediate hypersensitivities.

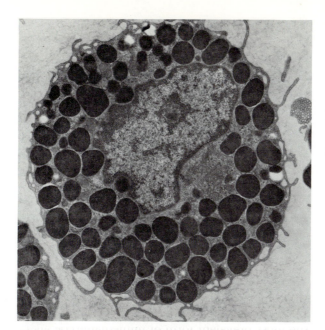

FIGURE 16-14

Electron micrograph of a mast cell shows large granules that contain histamine and other mediators of immediate hypersensitivity. Mast cells are located in tissue. Basophils play a role similar to mast cells in allergy, but circulate in the blood.

special antibody (also called *reagin*) is similar to IgG in structure and antigen reactive sites. It is formed in response to antigen (allergen) stimulation and specifically reacts with the eliciting allergen. The unique property of IgE, however, resides in the antibody's Fc region, the end opposite the antigen-binding sites. Once formed, this unique immuno-globulin attaches by its Fc region to host *mast cells* and basophils, cells with large numbers of granules (Fig. 16-14). IgE-coated mast cells are primed for allergic response. (Each sensitized cell may contain 500,000 IgE molecules on its surface.) Within the granules of the mast cell are chemicals such as *histamine* that trigger rapid changes in capillaries and smooth muscles. These chemicals are normally kept in cellular compartments and very little reaches the surrounding tissues. However, when IgE on the primed mast cells reacts with the corresponding allergen, the sensitized cells immediate-ly degranulate, releasing these chemicals into the surrounding tissues where they trigger the general symptoms of allergy: itching, edema (accumulation of fluid in tissues), smooth muscle spasms, and vascular dilation. The sequence of events in immediate-type allergy is depicted in Figure 16-15.

The location of these symptoms depends on the distribution of the mast cells (they are predominantly found in the gastrointestinal and respiratory tracts, connective tissues, and skin) and the route of exposure to the eliciting allergen. Hay fever (allergic rhinitis), for example, is an upper respiratory response to inhaled allergens, often airborne pollen or fungal spores. The response evokes watery discharges from the eyes and nose because of histamine-induced capillary leakage in these areas. Asth-ma attacks are due to release of mediators from mast cells in the bronchioles, where the mediators trigger smooth muscle spasms that

FIGURE 16-15

Mechanism of immediate-type hypersensitivity. Such allergy only occurs after immunoglobulin E attaches by its Fc region to mast cells or basophils. IgE-coated mast cells are primed for allergic response. Subsequent reaction with specific allergen bridges two adjacent IgEs, causing membrane distortion and release of the chemical mediators that trigger allergy symptoms.

394

ACQUIRED IMMUNITY

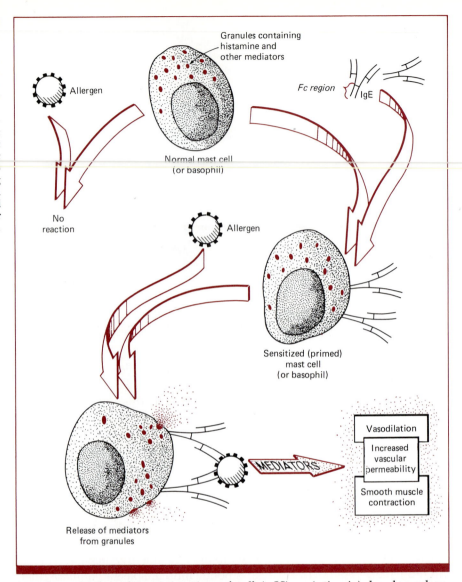

interfere with breathing (sometimes fatally). Hives (uticaria) develop when subdermal mast cells degranulate, causing vascular fluid to leak into skin tissue and form itchy red swellings. Degranulation of intestinal mast cells elicits some of the symptoms of food allergy, although the allergen may be absorbed from the intestine and trigger hives and other generalized allergic symptoms. Most IgE-mediated allergies can be symptomatically treated by using histamine antagonists such as the antihistamines in commercial allergy medications.* The best prevention is to avoid contact with the allergen if possible.

---

*Antihistamines are not used to treat asthma, the major symptoms of which are due to release of mediators other than histamine from mast cells.

Systemic Anaphylaxis   One form of hypersensitivity can kill a person within 10 minutes of allergen exposure. Systemic **anaphylaxis** occurs when many body sites containing sensitized mast cells and basophils simultaneously release large quantities of chemical mediators into body fluids. The bronchioles contract and trap inhaled air in the lungs. The capillaries dilate and blood pressure plunges, causing shock. The patient may quickly die of asphyxiation or circulatory failure. Administration of adrenalin (epinephrine) early in the anaphylactic reaction may save a life by counteracting the effects of the chemical mediators of anaphylaxis. Antihistamines are of limited value in systemic anaphylaxis because many of the symptoms are caused by additional substances released along with histamine from degranulating mast cells. Anaphylactic shock is most commonly associated with systemic injections of drugs (notably penicillin), horse serum used for passive immunization, insect venom, and even allergens used for treating allergy. (See Allergy Desensitization below.) A patient's history of allergy should always be considered prior to administering antibiotics, horse serum, or any foreign protein.

Sensitization and Shock   Allergen exposure elicits an allergic response only in persons sensitized to the antigen. Sensitization occurs when allergen-specific IgE has been produced as a result of previous exposure to the antigen and the allergen-specific IgE has attached to host mast cells. Since this sensitized state requires previous exposure to the allergen, the first encounter with any antigen fails to induce allergic symptoms. This initial exposure is called the *sensitizing dose* and is analogous to a primary immune response. Once sensitized to a specific allergen, subsequent exposures (called *shocking doses*) trigger IgE-primed mast cells to degranulate, inducing the physiological symptoms of hypersensitivity. Subsequent allergen exposures act as boosters of the anamnestic responses, thereby intensifying the extent of the allergic reaction. Consequently, people who show only mild allergic reactions to penicillin or insect venom may, with subsequent exposures, develop life-threatening anaphylactic reactions.

Allergy Desensitization   The symptoms of IgE-mediated allergy can be averted by preventing allergens from reacting with IgE-coated cells. This may be accomplished by *allergy desensitization.* Allergic individuals can be desensitized by continual injection of controlled concentrations of the specific allergen(s) to which the person is allergic. These continual low doses of antigens are believed to stimulate the production of other antibodies, called *blocking antibodies,* that compete with IgE for the allergen. Blocking antibodies are IgG and therefore do not fix to mast cells. They are believed to effectively saturate the system and remove the allergen before IgE can react with it, thereby preventing the development of allergic symptoms. Desensitization may also be assisted by the proliferation of suppressor T cells that inhibit formation of allergen-specific antibody.

Results of desensitization, however, are still inconsistent. The technique is only effective in a portion of the patients treated this way.

Delayed-Type Hypersensitivity

Delayed-type allergy* is an exaggerated CMI response that damages host tissue. Hypersensitive persons respond to certain pathogens and chemical allergens with a cell-mediated reaction that in itself accounts for many of the symptoms of disease.

Tuberculosis is the classical example of a disease that induces delayed-type hypersensitivity. In an attempt to eradicate the infection, host T lymphocytes move to the region of infection and release lymphokines, attracting macrophages to the area, where they are "aroused" by MAF. The intracellular mycobacteria are eliminated by killing the infected host cell. Host protection thus requires some tissue damage. The pathogen, however, is sometimes not destroyed and may continue to replicate within the phagocytes, providing continual and prolonged antigenic stimulation. In addition, sensitized lymphocytes react with the allergen and produce a lymphokine that transforms more lymphocytes into tuberculin-sensitized cells, greatly exaggerating the response. The continual infiltration of sensitized T cells and macrophages into the area and the subsequent release of lysosomal enzymes damage the surrounding tissue. This tissue necrosis in turn further intensifies chronic inflammation, producing a self-perpetuating prolonged allergic disorder. Symptoms of tuberculosis, therefore, are caused more by the human "protective" response to infection than by the primary virulence of the bacteria. Immunity is a biological double-edged sword. CMI may kill the pathogen, but delayed hypersensitivity may kill the infected person.

Delayed hypersensitivity can occur anywhere in the body where T cell–activating allergens localize. In sensitized persons a 24- to 48-hour delay in onset of allergic symptoms follows exposure to the allergen. This is why 2 days are required for positive tuberculosis skin tests to develop. These skin tests are performed by injecting purified tuberculin extract containing the allergen to which tuberculin-sensitized lymphocytes react. Localized delayed hypersensitivity (redness and swelling) at the site of injection indicates a positive test (Fig. 16-16). A positive tuberculin skin test is due to the presence of T lymphocytes sensitized to the pathogen, suggesting previous exposure to the tubercle bacillus (but not necessarily presence of active disease). A negative skin test suggests little or no contact with the pathogen. Tuberculosis skin tests, most commonly the Mantoux test or Tine test, are routinely administered as a screening method to detect possible cases of tuberculosis among college students, health care practitioners, and other large populations. Skin test findings are often the first diagnostic sign suggesting a person has tuberculosis.

Other allergic reactions that are classed as delayed-type hypersensitivity are (1) *contact dermatitis*, allergic skin reactions to poison ivy plants,

---

*Delayed-type hypersensitivity is sometimes called Type IV hypersensitivity.

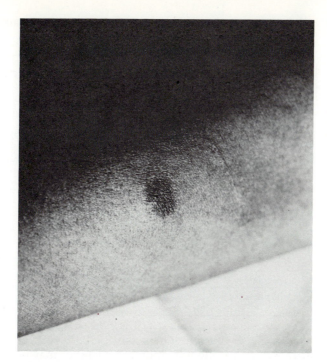

FIGURE 16-16

A positive tuberculosis skin test. Injection of purified protein from the tubercle bacillus triggers a delayed-type allergic response, indicating prior exposure (and sensitization) to the organism.

397

OTHER IMMUNOLOGICAL
DISORDERS

cosmetics, soaps, drugs, and certain metals; and (2) allergies to infections caused by some fungi, parasites, viruses, and a few bacteria. The common denominator of all these delayed hypersensitive reactions is the participation of allergen-sensitized T lymphocytes. Antibody and histamine play no role in this type of allergy.

OTHER IMMUNOLOGICAL DISORDERS

Autoimmune Disease

The immune system may partially lose its ability to distinguish some "self components" from foreign antigens. In other words, the body no longer tolerates certain indigenous antigens and begins to reject some of its own tissues as foreign. The result is **autoimmune disease**, immunity to one's self. Usually this break in immunological tolerance is antigen- (or tissue-) specific, and the disease is localized, not an overall rejection of all the tissues of a host. Occasionally, however, as with systemic lupus erythematosis (SLE), the tissue destruction is systemic and often fatal.

Autoimmunity is poorly understood. None of the causes of autoimmune diseases (a few of which are listed in Table 16-5) have been clearly defined. Perhaps infection by a microbe with antigens similar to human tissue triggers formation of antibodies that cross-react and attack the tissue antigen as well as the microorganism. Or perhaps a tissue antigen hidden from the immune system during fetal development failed to be processed as "self." Later physical trauma could release some of the antigen into the circulation. This exposes the immune system to the tissue antigen and may trigger autoantibody formation. At least five such theories have been

TABLE 16-5

Some Autoimmune
Diseases in Humans

Disease	Target Tissue	Clinical Effect
Hashimoto's hypothyroiditis	Thyroid	Goiter
Graves's disease	Thyroid	Hyperthyroidism
Addison's disease	Adrenal gland	Adrenal insufficiency
Autoimmune aspermatogenesis	Spermatozoa	Infertility
Autoimmune hemolytic anemia	Red blood cells	Anemia
Myasthenia gravis	Motor nerve receptors on muscle cells	Neuromuscular dysfunction
Rheumatoid arthritis	Synovial membranes	Inflammation of joints and connective tissue
Allergic encephalitis	Brain	Inflammation of the brain
Systemic lupus erythematosis	Antigens in cell nucleus (DNA)	Lesions on heart, vessels, and kidney
Goodpasture's syndrome	Glomerular basement membrane	Kidney failure

advanced to explain why immune tolerance may be lost. Although none of these theories can explain all cases of autoimmune disease, some facts are well established. The incidence of autoimmunity increases with age. More than half the persons older than 70 have autoantibodies (although not all have overt disease). Females are more likely to develop autoimmune disease than males. And oddly, left-handed persons are more frequently autoimmune victims than are right-handed individuals.

Controlling autoimmune diseases is difficult. The target of some autoantibodies can be surgically removed and functionally replaced. The function of a removed thyroid gland, for example, can be provided by orally administering thyroid hormone. Most organs damaged by autoimmunity, however, must be physically replaced with a donor's organ. The recipient's immune system usually attacks and damages the transplant even more efficiently than it did the original organ. Such graft rejection and the symptoms of autoimmunity itself can be diminished by intentionally suppressing the immune response, but such immunosuppressed persons are vulnerable to fatal opportunistic infections. Treatment of autoimmune disease will surely be improved by determining why autoantibodies are produced and by developing methods of either reestablishing immunological tolerance or suppressing antigen-specific immunity without compromising immunity in general.

Immune Complex Disease

*Immune complex diseases** result from the deposit of antigen-antibody

---

*Immune complex disease is sometimes called Type III hypersensitivity.

complexes on capillary or renal membranes. These immune complexes usually fix complement and damage membranes of the host by the same mechanism that lyses gram-negative bacteria. Passive immunization may trigger serum sickness because of the overwhelming number of immune complexes that are formed. These immune complexes are too numerous for disposal by phagocytes and are deposited in kidneys, joints, and blood vessels where they cause inflammation and injury. A similar condition, localized in the kidney, may follow infection by *Streptococcus pyogenes* and some persistent viruses. Acute glomerulonephritis is a kidney disorder due to deposition of immune complexes on renal membranes, which are subsequently damaged when complement fixes to these complexes.

Immunodeficiency

Immunodeficiency is a deficit in either T-cell or B-cell immunity. Such deficiencies can be either natural or induced (Table 16-6). Naturally occurring absence of B-cell immunity is called *agammaglobulinemia,* or *hypogammaglobulinemia* if the amount of antibodies is abnormally low. Persons with compromised antibody production are especially prone to opportunistic infections by pyogenic (pus-producing) bacteria, especially *Staphylococcus aureus* and *Streptococcus pyogenes.* Agammaglobulinemia is usually a genetically acquired disorder or is caused by tumors of the lymphoid organs. Antibody production may also be depressed by dietary protein deficiencies, since amino acids are the building blocks of immuno-

TABLE 16-6

Some Natural Immunodeficiencies

Deficiency	Caused by	Effect on Immunity		Increases Susceptibility To
		Humoral	**Cell-Mediated**	
Agammaglobuli-nemia	Genetically acquired B-cell deficiency	Depressed	Normal	Pyogenic bacteria*
Hypogammagobul-inemia	Dietary protein deficiency	Depressed	Normal	Pyogenic bacteria*
Complement deficiency	Lack of C'$_3$†	Normal	Normal	Pyogenic bacteria*
T-cell deficiency	Lack of thymus development	Somewhat depressed	Severely depressed	Candida, certain viruses, tumors
T-cell deficiency	Certain infections	Usually normal	Depressed	Secondary bacterial infections
Severe combined deficiency	Lack of stem cell	Depressed	Depressed	All the above
T-cell or B-cell deficiency	Lymphoid malignancies	Depressed	Depressed	All the above
Acquired immune deficiency syndrome (AIDS)	Unknown	Normal	Depressed	Pneumocystis infections; Kaposi's sarcoma

*Pyogenic (pus-forming) bacteria include *Staphylococcus aureus, Streptococcus pyogenes, Streptococcus pneumoniae, Neisseria meningitidis,* and *Haemophilis influenzae.*
†C'$_3$ is one of the components of the compliment system.

globulins. Complement deficiencies also compromise immunity because the protective action of many antibodies depends on the lytic and inflammatory action of these proteins.

T-cell immunodeficiencies occur in persons who fail to develop a functional thymus, so lymphocytes cannot be processed to T cells. These persons are very susceptible to recurrent diseases caused by fungi, viruses, and intracellular bacteria. Deficits in CMI also increase the probability of malignant tumors as much as 1000-fold. In addition, impaired CMI compromises antibody production, since helper T cells are needed to induce humoral immunity against some infectious diseases. A few pathogens depress T-cell immunity during natural infections; immunosuppression is a common feature of syphilis, leprosy, and malaria. Cell-mediated immunity is also crippled by tumors of the lymphoid organs (for example, leukemia and Hodgkin's disease) or by the inability of lymphocyte precursor cells to develop into immunocompetent T cells.

One especially alarming CMI deficiency is *AIDS (acquired immune deficiency syndrome)*, an apparently transmissible form of immune paralysis. Persons with AIDS are especially susceptible to opportunistic pulmonary infections by *Pneumocystis carinii* (see Chap. 20) and to Kaposi's sarcoma, a malignancy of the connective tissue. Nearly half the victims of AIDS die within 1 year of diagnosis. This disease is primarily transmitted by sexual contact; it is discussed in more detail in Chapter 22.

Cell-mediated immunity may be intentionally suppressed in transplant recipients to prolong graft survival or in persons with severe delayed-type hypersensitivity. Immunosuppression may be induced by drugs that depress lymphocyte proliferation, by irradiation with x-rays, or by administration of *antilymphocyte serum (ALS)*. Lymphocyte-specific antibodies of ALS passively immunize recipients against their own lymphocytes, thereby preventing graft rejection and other T-cell-mediated activities. All such immunosuppressed persons are susceptible to opportunistic infectious disease, and, unless their immune status is allowed to return to normal, they die of infectious disease, even during treatment with broad-spectrum antibiotics.

SEROLOGY AND SKIN TESTING: USING THE IMMUNE RESPONSE FOR DIAGNOSIS

Diagnostic techniques that depend on specific reactions between antigens and antibodies can be performed in vitro. These are called **serological tests**. Their diagnostic reliability is due to two important properties of the immune response:

1. Immunity is acquired only as the result of exposure to an antigen.

2. The reaction of antibodies (or T lymphocytes) with the corresponding antigen is highly specific.

We can therefore obtain valuable diagnostic information by demonstrating the presence of an elevated blood concentration of antibody that specifically reacts with a laboratory preparation of the suspected microorganism, its extracted antigens, or its extracellular products. Elevated antibody titers indicate that the person has been exposed to the pathogen and has responded immunologically. An extremely high antibody titer suggests an active role of the corresponding microbe in causing the patient's disease.

Because several days are required to respond immunologically to a pathogen, a patient's specific antibody concentration will be higher late in the course of a disease than in the acute phase. Thus, an antibody titer that, in a late serum sample, is at least four times greater than that of serum collected several days earlier implicates the corresponding pathogen as the etiological agent. For diseases of short duration, this test is of limited therapeutic value since the patient's disease will have run its course by the time the second serum sample is collected. Many serious diseases, however, progress more slowly.

All of these tests require a source of antigen. For microbes that are difficult to culture, however, laboratory preparations of their antigens are not readily available. In many cases, fortunately, another substance is substituted for the natural antigen. This is possible because of coincidental similarities in antigenic specificities. These antibodies may cross-react with *heterophil antigens*. These are antigens of unrelated plants or animals. For example, antibodies generated against the virus that causes infectious mononucleosis cross-react with an antigen on the surface of sheep erythrocytes. Detecting elevated titers of antibody that agglutinates sheep red blood cells is therefore diagnostic for infectious mononucleosis.

In addition to demonstrating antibodies in the patient's body fluids, laboratory preparations of antibodies with known specificity can be used to rapidly identify unknown microorganisms or their by-products of growth.

Serology

Although most serological reactions depend upon the same theoretical principle, the methods of detecting the specific antibody-antigen reactions differ. Agglutination, precipitation, complement fixation, viral neutralization, and toxin neutralization are observable activities that can be used to detect these positive reactions. For example, antibodies in serum from a patient in the later stages of typhoid fever will visibly agglutinate standard laboratory preparations of the etiological agent *Salmonella typhi*. Antibodies against soluble antigens extracted from this pathogen can be identified by precipitation. The serum of people suffering from some viral diseases contains neutralizing antibodies that, when mixed with a standard preparation of infectious viruses, will block the virus's ability to infect otherwise susceptible tissue culture cells. Some antibodies fix complement (C') when they react with the antigen. The amount of C' removed when a patient's serum is mixed with the antigens of a suspected pathogen is an indication of positive serological reaction (Fig. 16-17). Some commonly employed serological tests are described in Table 16-7.

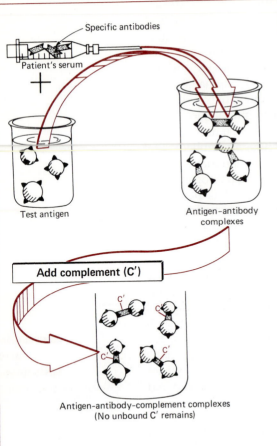

Specific antibodies

Patient's serum

+

Test antigen

Antigen–antibody complexes

Add complement (C′)

Antigen–antibody–complement complexes
(No unbound C′ remains)

Add indicator system

Sheep red blood cells (RBC)

Anti-sheep RBC antibody

Sheep RBC-antibody complex

No hemolysis

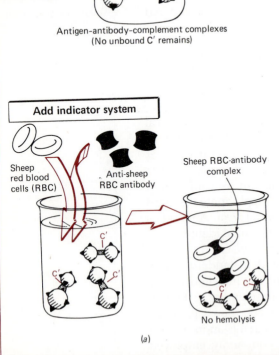

(a)

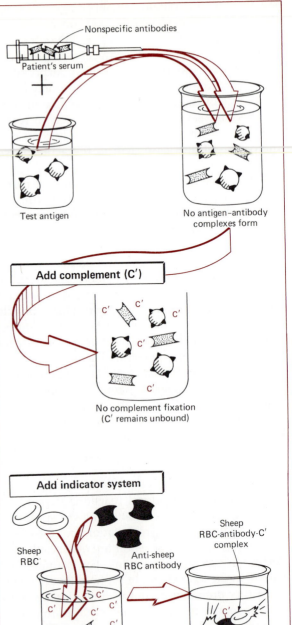

Nonspecific antibodies

Patient's serum

+

Test antigen

No antigen–antibody complexes form

Add complement (C′)

No complement fixation
(C′ remains unbound)

Add indicator system

Sheep RBC

Anti-sheep RBC antibody

Sheep RBC-antibody-C′ complex

Hemolysis

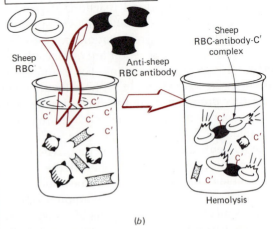

(b)

FIGURE 16-17

(a) Positive complement fixation test. In the test system, the patient's serum contains complement-fixing antibodies specific for the test antigen. The resulting antigen-antibody complexes fix and remove complement. When the indicator system is added, no complement remains to lyse the antibody-coated sheep red blood cells. Thus, a positive complement fixation test shows no lysis of indicator cells. **(b)** Negative complement fixation test. The patient's serum contains no complement-fixing antibodies specific for the test antigen. No antigen-antibody complexes form, and no complement is removed from the test system. When the indicator system is added, complement is available to combine with and lyse antibody-coated sheep red blood cells. Thus, a negative complement fixation test shows lysis of indicator cells.

TABLE 16-7

Some Common Types of Serological Tests for Disease Diagnosis*

Type of Test	Representative Disease	Usually Identifies	Visible Evidence of Positive Reaction
Agglutination	Typhoid fever	Pathogen isolated from patient	Clumping occurs.
Immunoprecipitation	Diphtheria	Toxin produced by isolated bacteria	Line forms in area of agar where antigen and antibody diffuse into each other.
Viral neutralization	Rabies	Virus or specific antibody	Neutralized viruses fail to proliferate in experimentally infected cell culture.
Complement fixation	Coccidioidomycosis	Antibody in patient's serum	There is no hemolysis of indicator RBCs.
Fluorescent antibody technique	Syphilis	Antibody in patient's serum	Antigen glows when observed under fluorescence microscope.
Toxin neutralization	Botulism	Toxin in food, patient's serum, or feces	Experimental animal shows no sign of toxicosis when administered antitoxin with test material.
Radioimmune assay	Hepatitis B	Viral antigens in patient's serum	Very little radioactive antigen-antibody complexes are evident in precipitate.
ELISA	Viral gastroenteritis	Virus in patient's feces	Visible products develop when substrate is added to mixture of antigen-antibody complex and antihuman IgG conjugated with enzyme.

*In addition to their contribution to disease diagnosis, most of these tests are valuable research tools for biological and biochemical studies.

Some antibodies are labeled to facilitate detection of antigen-antibody reactions. Antibodies can be tagged with fluorescent dyes that glow when exposed to ultraviolet light. These immunoglobulins coat their cellular antigens so that the antibody-coated microbes glow when observed with the fluorescent microscope (Fig. 16-18). Viral-infected human cells can also be rapidly detected by mixing them with fluorescent-labeled antibody specific for the virus.

FIGURE 16-18

Positive fluorescent
treponema antibody (FTA)
test for syphilis.

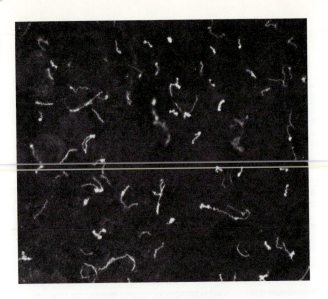

404

The *enzyme linked immunosorbant assay (ELISA)* is an immunological test that uses an enzyme-mediated reaction as an indicator of antigen-antibody reaction. The procedure is described in Figure 16-19. ELISA is an extremely sensitive technique for rapidly detecting specific antibodies or antigens. Virtually any enzyme that converts its substrate to a detectable product can be used as the indicator.

Standard antigen preparations labeled with radioactive iodine (^{125}I) provide another highly sensitive method for detecting minute amounts of antigen. The technique, called *radioimmunoassay (RIA)* employs the following procedure:

1. A known antibody preparation is mixed with the sample for assay. Any specific antigen in this sample will complex with the antibody.

2. ^{125}I-labeled antigen is then added to this mixture and the radioactive antigen reacts with any remaining unbound antibody. The number of radioactive antigen-antibody complexes are thus inversely proportional to the amount of antigen in the unknown sample.

3. All antigen-antibody complexes are precipitated and separated from solution, and the radioactivity of the precipitate is determined. (Unreacted ^{125}I antigen remains in solution and is discarded.) The radioactivity of the precipitate is inversely proportional to the amount of antigen in the test specimen. If this sample had large amounts of antigen, it would react with most of the antibody, leaving little or no immunoglobulin to react with ^{125}I antigen; very few radioactive antigen-antibody complexes could form and the precipitate would have little radioactivity. Conversely, if the original sample had small amounts of antigen, the binding sites of the antibodies would still be available to react with the ^{125}I antigen, yielding a highly radioactive precipitate.

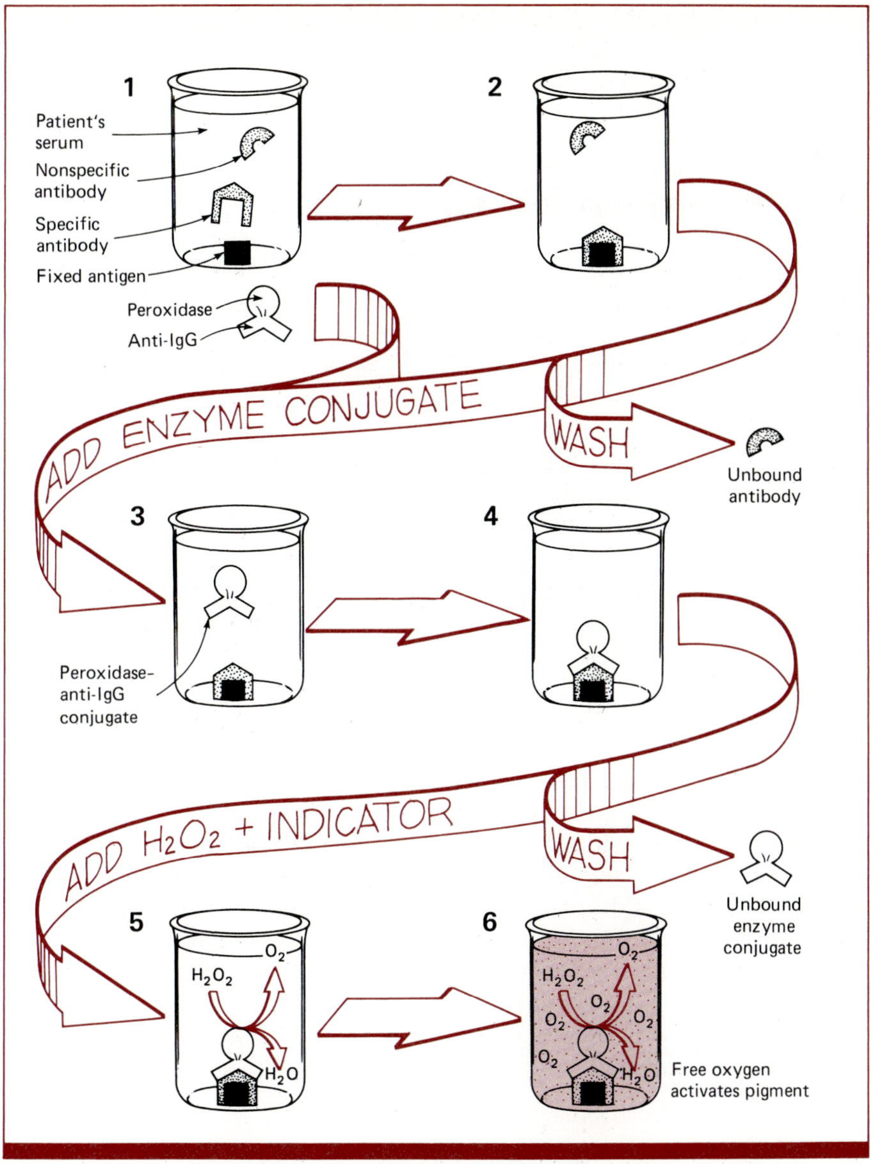

FIGURE 16-19

Detecting specific antibody in a patient's serum by ELISA. (1) Antigen is fixed to solid surface, and then the patient's serum is added. (2) If specific antibody is present it will bind to antigen and be fixed to the container. Nonspecific antibody will fail to bind. The serum and all unbound antibodies are washed away before next step. (3) Enzyme conjugate is added. The conjugate is prepared by chemically linking the indicator enzyme (in this case, peroxidase) with antihuman IgG antibody (prepared by injecting human IgG into another animal). (4) The anti-IgG of the enzyme conjugate will bind to the IgG of the fixed antigen-antibody complex. This in turn fixes the peroxidase to the container. Unbound enzyme conjugate is washed away before next step. (5) The enzyme's substrate, H_2O_2 (peroxide) and a colorless indicator chemical are added. (6) Peroxidase converts H_2O_2 to H_2O and O_2. In the presence of O_2 the indicator changes color. This is a positive reaction. Negative sera contain no specific antibody and consequently fail to provide immunoglobulin bridges that link the enzyme conjugate to the fixed antigen. All peroxidase is therefore removed during the second wash (after step 4), and the substrate is *not* converted to products. Negative sera thus show no color.

RIA and other techniques capable of detecting and identifying biological molecules in minute amounts are indispensable tools of research, industry, medicine, and forensic laboratories.

The serology of specific diseases will be considered in the later chapters.

Limitations of Serological Tests Several limitations of these diagnostic tests should be recognized. Specific antibody may persist in a person's blood for years after infection, so positive serology on a single serum sample may indicate previous infection rather than recent or active disease. Antibodies are usually undetectable for several days following onset of initial infection, so serological tests early in the disease may produce false negative results. Early blood samples should nonetheless be collected in order to compare with the antibody concentrations found in blood samples collected late in the disease (during convalescence). *A four fold rise in specific antibody concentration between the acute and convalescent stages of the disease verifies the etiological role of the corresponding microorganism.* Although elevated levels of specific antibodies may be diagnostic, they do not necessarily imply immunity since antibodies are nonprotective in some diseases.

Skin Tests

Infection by a microorganism that triggers ÇMI can be detected by demonstrating sensitized T lymphocytes in the infected person. Subdermal injection of an antigenic extract of the suspected microorganism causes a local delayed-type hypersensitive reaction at the site of injection. Because lymphocytes are sensitized by antigen exposure (usually during natural infection), a positive skin test suggests current or previous infection by the microbe. The test is diagnostically useful for initial preliminary screening to identify possible cases of tuberculosis, histoplasmosis, brucellosis, and a number of other diseases that stimulate T-cell reactions. Since hypersensitivity exists for life, a person who is positive will remain that way (unless immunosuppressed). Active cases can be differentiated from past infections by additional diagnostic tests, such as isolation and identification of the pathogen.

OVERVIEW

Immunity, the body's third line of defense against disease, is acquired during a person's lifetime as the result of exposure to antigens. The resulting immunity specifically protects against pathogens that possess the provoking antigen. Protection may be due to the production of soluble antibody molecules (humoral immunity) or the activation of T lymphocytes (cell-mediated immunity). Antibodies are subdivided into five immunoglobulin classes: IgG, IgM, IgA, IgE, and IgD. Antibodies specifically react with the antigen that stimulated their formation, often changing it in a manner that aids elimination of the antigen from the host. Such changes include agglutination, precipitation, opsonization, virus or toxin neutralization, and complement fixation. A single antibody molecule may demonstrate several of these activities.

Antibodies are synthesized by plasma cells derived from B lymphocytes programmed to recognize specific antigens. Initial exposure to an antigen triggers the events of the primary response. Antigen-specific B lymphocytes proliferate into clones of plasma cells and memory cells. Immunological recall is due to these memory cells. Subsequent antigenic challenges trigger memory cells to rapidly differentiate into antibody-producing plasma cells and proliferate, accounting for the events characteristic of the secondary (or anamnestic) immune response. Immunological memory is responsible for acquired immunity following disease or vaccination. Humoral immunity protects against some bacteria and viruses, toxins, and other extracellular products of pathogens.

Cell-mediated immunity also shows immunological recall, but no antibodies are produced. Instead, T lymphocytes are activated by antigen exposure, generating three cellular subpopulations. Effector T cells may kill their target cell by direct contact or they may produce lymphokines that attract and activate phagocytes, inhibit their migration so they accumulate in the region of antigen stimulation, trigger the production of antiviral proteins, or poison target cells. Lymphokines may also convert nonspecific T lymphocytes to antigen-sensitive activated effector cells thereby enhancing the response. Regulator T cells suppress B-cell activity to protect against immune-mediated tissue damage and are required as helpers to elicit a humoral immune response. Memory cells store the information necessary to elicit a rapid recall response upon subsequent antigenic exposure. CMI is usually directed against intracellular parasites, protozoa, fungi, virus-infected host cells, and tumors.

Immunity may be actively or passively acquired. Natural active immunity often follows infectious disease, whereas vaccination stimulates artificial active immunity. Antibodies are passively acquired transplacentally or in breast milk (natural passive immunity) or transferred from donor to recipient by hypodermic injection (artificial passive immunity). Active immunity provides long-lasting protection after an initial latent period of several days to weeks. Passive immunity confers immediate protection that disappears within a few weeks.

Immune-mediated host damage can be caused by hypersensitivity (allergy), autoimmune disease, or immune complex disease. Hypersensitivity is the immediate type if mediated by immunoglobulins and the delayed type if mediated by sensitized T lymphocytes. Autoimmunity follows partial loss of immunological tolerance and results in overt disease if the autoantibodies attack and injure host tissues. Immune complex disease results from deposits of antigen-antibody complexes in the membranes of capillaries and kidneys, which are then obstructed and damaged by complement fixation. The symptoms of all these diseases may be diminished by immunosuppressive agents (such as cortisone) but at a cost of increased susceptibility to opportunistic infections. Persons with natural immunodeficiencies are also prone to opportunistic infections, which are often fatal if the deficiency completely paralyzes B-cell or T-cell-mediated immunity.

Immunological techniques aid in disease diagnosis. Serology is used to demonstrate elevated titers of pathogen-specific antibodies in the serum of infected patients. A fourfold rise in antibody titer between the acute and convalescent stages of disease indicates the etiological role of the corresponding microbe. Antigen-specific T-cell immunity can be determined by skin testing. Persons who have been exposed to a pathogen usually show a positive skin test against its extracted antigens.

KEY WORDS

acquired immunity

antigen

immunological tolerance

antigenic determinant

hapten

humoral immunity

antibody

immunoglobulin

neutralizing antibody

agglutinin

precipitin

opsonin

complement-fixing antibody

lymphocyte

B lymphocyte

plasma cell

memory cell

primary response

secondary response

anamnestic response

cell-mediated immunity

T lymphocyte

T killer cell

delayed hypersensitivity (DHS)

lymphocyte

lymphokine

helper T cell

suppressor T cell

sensitized memory cell

natural killer (NK) cell

immunopotentiation

attenuation

toxoid

serum sickness

monoclonal antibody

allergy

immediate-type hypersensitivity

delayed-type hypersensitivity

immunoglobulin E (IgE)

anaphylaxis

autoimmunity

serological test

REVIEW QUESTIONS

1. How do haptens differ from antigens?

2. Why is IgG said to be a monospecific bivalent molecule? Could the same terms be applied to the other four classes of antibodies?

3. State six ways in which antibodies protect against pathogens or their by-products.

4. What is the role of memory cells in the immune response?

5. How do the primary, secondary, and booster responses differ?

6. What is the major role of each of the following subpopulations of T cells?
    (a) T killer cells
    (b) DHS cells
    (c) Regulator cells

7. Describe the action of four lymphokines.

8. Distinguish between (a) active and passive immunity, (b) natural and artificial immunity, and (c) immediate- and delayed-type hypersensitivity.

9. Describe the advantages of attenuated vaccines as compared with killed vaccines.

10. What role does IgE play in allergic reaction?

11. Describe three immunologically mediated disorders.

12. Explain how an antigen-antibody response aids in the diagnosis of disease. Give five examples.

General Concepts of Pathogenesis

17

The infectious disease process is considerably more complicated than can be explained by the simplistic belief that a single pathogen always causes a particular disease and that exposure to the pathogen always has the same outcome. *Staphylococcus aureus*, for example, can produce a broad spectrum of disease, from simple furuncles (boils) to fatal pneumonia, toxic shock, and meningitis. *Streptococcus pyogenes* may cause mild pharyngitis (sore throat) in some persons and rheumatic heart disease or severe kidney injury in others. On the other hand, a single disease syndrome, such as pneumonia, can be attributable to any one of a number of bacteria, fungi, viruses, or protozoa. In this chapter we will discuss the factors that determine which of many possible outcomes develop following our exposure to potential pathogens and the mechanisms by which the growth of microbes in our bodies can endanger our lives.

OUTCOMES FOLLOWING EXPOSURE TO PATHOGENS

In the absence of clinical treatment, several possible consequences follow exposure of a susceptible person to pathogenic microorganisms:

1. No infection—Disease fails to develop and no organisms are shed.

2. Inapparent (subclinical) infection—No clinical symptoms develop; a carrier state may be established and may persist. (Carriers are apparently healthy persons who shed pathogens and can therefore transmit disease.)

3. Mild acute symptomatic disease—Symptoms rapidly develop and quickly disappear.

4. Severe acute disease—Debilitating or fatal symptoms quickly develop.

5. Chronic infectious disease—Symptoms persist for months or years.

6. Latent infection—Pathogen remains dormant in the host following apparent recovery; subsequent recurrences are common.

Many explanations account for this array of results (see color box). For example, the more microorganisms introduced into the body, the more likely the host is to develop overt infectious disease. In addition, the nature of the pathogens, how they are introduced into the host, and the host response to the pathogens determine **pathogenesis** (sequence of events and mechanism of tissue injury) and **clinical manifestations** (symptoms of disease). Table 17-1 provides a glossary of terms used in discussing infectious disease.

WHY DOESN'T EVERYONE GET SICK?

■ "I never get sick!" To a person who picks up every bug going around, such a claim seems remarkable. Why don't we all catch cold during the cold season? Aren't we all exposed to the same microbes?

The following are some variables that at least partially explain this diversity in disease resistance:

Race Some infectious diseases are more prevalent and severe in some races than others. Tuberculosis is more virulent for American Indians than for people of European descent. Europeans are more susceptible to syphilis than are Chinese, who have been exposed to the disease for thousands of years longer, time enough for natural selection to create a syphilis-resistant population. Similarly, blacks are more resistant to malaria than whites, and whites are more resistant to coccidioidomycosis than blacks.

Age Very young or very old persons are most susceptible to infectious disease. The immune systems of infants and very young children are not fully mature. The efficiency of the immune system deteriorates with age. Many persons who die of "old age" actually succumb to infectious disease.

Sex Anatomical, hormonal, and biochemical differences between men and women help explain why some diseases are more prevalent in one sex than the other. For example, women are much more likely to contract urinary tract disease because of their shorter urethras.

Occupation Infectious disease is a persistent occupational hazard of certain professions. Such hazards, however, are usually created by increased exposure to pathogens rather than by diminished host resistance. Veterinarians and animal control officers, for example, have a higher incidence of rabies. Meat packers are more likely to acquire Q fever, brucellosis, and other cattle-borne diseases. Because health care professionals are frequently exposed to persons shedding virulent human pathogens, the hospital is one of the most hazardous places to work. This risk is reduced to acceptable levels only by diligent adherence to medical asepsis (see Chap. 24, "Nosocomial Infections").

Stress Prolonged physical or emotional crisis, fatigue, or depression alter hormonal balance and reduce resistance to disease. Stress-compromised persons often suffer outbreaks of oral or genital herpes lesions or become susceptible to more severe diseases. Trench mouth (acute necrotizing ulcerative gingivitis) is a stress-associated disease caused by normally minor members of the indigenous flora.

Underlying Disease Some diseases compromise a person's resistance to microbial invasion. Persons with severe diabetes mellitus, for example, are more susceptible to vaginitis, urinary tract infections, and blood infections. Rheumatic fever damages the heart valves and makes them susceptible to colonization by normal flora microbes that get into the bloodstream during tooth extraction or even routine dental drilling. Minor dental procedures can result in fatal heart infections (subacute bacterial endocarditis) in these persons. People with acute leukemia or other cancers that prevent the formation of white blood cells often die of infectious disease.

Climate The seasonal incidence of influenza and colds testifies to the role of climate and seasonal changes in causing epidemics of respiratory diseases. Cold, dry winter conditions can dehydrate the mucous membranes and paralyze the ciliated mucosa. In hot, humid climates, the accumulation of moisture on the body encourages the growth of skin pathogens.

Nutrition Malnourishment contributes to

the incidence and severity of several diseases. Cholera is usually fatal only in malnourished persons. Measles is 100 to 400 times more prevalent among undernourished African children than it is in the United States and is often fatal or permanently debilitating among the impoverished.

Hygiene and Habitat Because of overcrowding and low community hygiene, high-density urban areas have a much higher incidence of disease. As with occupation-related disease, this problem is one of increased exposure rather than of decreased resistance to disease.

TABLE 17-1

Terminology Commonly
Used in Discussing
Infectious Disease

Colonization	Establishment of microorganisms on a body surface (See Chap. 15.)
Infection	Microbial proliferation in host tissue (See Chap. 15.)
Incubation period	Time interval between exposure to a pathogen and appearance of disease symptoms
Prodromal period	Earliest phase of a developing condition
Inapparent (subclinical) infection	Infection that causes no clinically apparent symptoms
Carrier	Person with inapparent infection who is shedding pathogens and can therefore transmit disease
Latent infection	State in an infection during which no symptoms are manifest; often recognized only after the later emergence of overt illness
Local infection	Infection limited to a single body site
Focal infection	Localized infection from which microbes spread to distant body sites
Systemic infection	Actively proliferating microorganisms disseminated throughout the body (not merely on epithelial surfaces)
Septicemia	Presence of actively proliferating microorganisms in the blood; also called "blood poisoning"
Infectious disease	Injury of host tissue due to infection (See Chap. 15.)
Pathogenesis	Sequence of events during development of disease and mechanism(s) by which tissues are injured
Clinical manifestations	Observable symptoms of disease
Acute disease	Disease of rapid onset and short duration
Chronic disease	Slowly progressing disease of long and often indeterminable duration
Communicable disease	Disease that may be transmitted from infected hosts to uninfected individuals; (also called "contagious disease")
Endogenous disease	Opportunistic disease caused by normally harmless microorganisms that inhabit the body; often occurs in hosts with compromised defenses or following transfer of normal flora to vulnerable body sites
Exogenous disease	Disease acquired by exposure to pathogens from sources external to the body
Convalescence	Recovery from disease
Relapse	Recurrence of disease after apparent recovery

The Nature of the Pathogens

Extracellular versus Intracellular Parasites
Extracellular parasites usually resist phagocytic engulfment and multiply in body fluids and tissue spaces. **Intracellular parasites**, on the other hand, reproduce inside host cells. For example, some intracellular pathogens survive destruction within phagocytes, which then become the microbe's host cells. Intracellular and extracellular parasites characteristically elicit different symptoms. *Streptococcus pneumoniae*, an encapsulated extracellular bacterium, produces a rapidly progressing acute pneumonia that can quickly kill a person. The patient who survives this initial episode rapidly recovers because of the production of opsonizing antibodies that neutralize the capsule's antiphagocytic properties. Intracellular parasites, on the other hand, usually produce chronic diseases with less rapid onsets, much longer recovery times (even with chemotherapy), and a greater likelihood for clinical relapse. *Mycobacterium tuberculosis* exemplifies a typical intracellular parasite (Fig. 17-1).

These characteristics of intracellular infections are largely due to the protection afforded the parasites by their host cells. Humoral antibodies have little access to intracellular pathogens and fail to contribute to their eradication and to patient recovery. Intracellular pathogens are attacked by CMI, which eliminates the microbes by killing the infected host cells. Some of these pathogens, however, may escape destruction by CMI and cause recurrences months or years after recovery. In addition, intracellular infections are poorly controlled by chemotherapeutic agents that either fail to penetrate the infected host cells or are ineffective in its cytoplasm. In

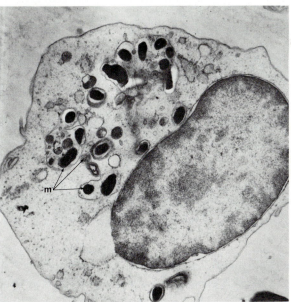

FIGURE 17-1

Mycobacterium tuberculosis within an alveolar macrophage. The microbes *(m)* survive phagocytosis and multiply in the phagocytic cell.

Property	Extracellular Parasite	Intracellular Parasite
Mechanism of resistance to phagocytes	Resists engulfment	Resists intracellular destruction
Effect of antimicrobial therapy	Usually effective	May be ineffective because of intracellular protection of pathogen
General nature of disease	Acute	Chronic
Immunity	Usually protective against subsequent attacks	Often fails to eradicate the organism; recurrences common
Immune response	B-cell mediated	T-cell mediated
Common clinical manifestations	Host damage due to microbial by-products	Host damage due to hypersensitivity, chronic inflammation, or other host tissue reactions

addition to protecting intracellular pathogens, infected host cells (especially the mobile phagocytes) may spread the infection throughout the body.

The distinctions between extracellular and intracellular parasites are summarized in Table 17-2.

Unlike typical intracellular parasites, most viral pathogens elicit rapidly progressing acute diseases that usually subside spontaneously several days following onset of symptoms. This pattern is typical of colds, influenza, and the common viral diseases of childhood. In the course of infection, most viruses antigenically alter their host cells, which become recognized as foreign and are eliminated by CMI. Viruses also stimulate interferon production which aids in recovery. After escaping host cells, viruses may be neutralized by antibodies when humoral immunity is sufficiently stimulated. Other viruses, however, such as the agents of herpes and measles, establish latent or slow virus infections that may persist in the host long after the acute phase has resolved because they reside in tissues protected from the immune system.

Microbial Interactions   Although many diseases result from the proliferation of a single pathogen, some disease conditions are caused by the presence of several pathogens. *Coinfections,* or mixed infections, develop from the introduction of several different pathogens into the body, usually by traumatic injury. Gas gangrene, an extreme in mixed infection, can be caused by as many as 10 coinfecting microbial species introduced during surgical or accidental trauma. Each type of microbe could cause disease alone or in combination with the other coinfectants. A chemotherapeutic agent that is effective against all participating pathogens should be selected, since a single species that escapes treatment may continue to cause disease.

Obligate coinfections occur whenever two or more organisms elicit a particular set of symptoms only in combination with one another. Singly, in pure culture, they are incapable of causing the disease. For example, amebic dysentery is thought to be caused by a dual infection with *Entamoeba histolytica* and intestinal bacteria. The latter establish conditions necessary for the growth of the protozoa, which in turn invade and destroy cells of the large intestine.

Secondary infections (also called *superinfections*) often occur following reduction of host resistance by a primary infection. Some pathogens paralyze the ciliated mucosa or depress antibody production, thereby increasing the likelihood of secondary infection. In some cases, the secondary infection is more dangerous than the primary one. For example, serious pulmonary disease caused by opportunistic bacteria, most commonly *Streptococcus pneumoniae* and *Haemophilus influenzae,* often follows measles, influenza, or whooping cough. Chemoprophylaxis used in patients suffering from minor viral illnesses often defeats its own purposes by favoring the development of secondary infections by antibiotic-resistant pathogens.

Microbial interactions that are antagonistic may favor the host. The ability of the normal human flora to prevent the uncontrolled proliferation of potential pathogens is perhaps the most important example of host protection due to microbial antagonism.

The Nature of the Host: Predisposing Factors

Any influence compromising the defenses described in Chapters 15 and 16 may alter the outcome of a person's exposure to pathogens by predisposing the individual to the development of infectious disease. Predisposing factors can be categorized as intrinsic or extrinsic. *Intrinsic predisposing factors* are due to naturally occurring malfunctions of the body's defenses, including diabetes, circulatory disturbances, bone marrow failure, disorders of the mononuclear phagocytic system, or any of the other factors listed in Table 17-3. *Extrinsic predisposing factors,* on the other hand,

TABLE 17-3

Some Intrinsic Factors That Predispose Humans to Develop Infectious Disease

Predisposing Factor	Effect
Hormonal imbalance	Suppresses inflammation and antibody production
Diabetes	Impairs granulocyte function; may prevent protective cells in circulation from reaching local tissues
Extreme youth	Precedes maturation of immune system
Old age	Brings immunological senescence
Allergy	May impede access of immune effectors to local tissue
Pregnancy	Rapid hormonal fluctuations may reduce inflammation and antibody production; anatomical changes may predispose for infections of the genitourinary tract
Immune deficiency diseases	Defects in B or T cells lead to partial or complete loss of immunity
Rheumatoid arthritis	Reduces chemotaxis of phagocytes, promoting infections of skin and lower respiratory tract
Cystic fibrosis	Depresses phagocyte function; causes poor clearance of mucus from lungs
Renal failure	Depresses CMI; impairs chemotaxis

Predisposing Factor	Effect
Alcoholism	Reduces efficiency of mechanical respiratory defenses; impairs leukocyte function
Smoking	Impairs alveolar phagocytosis; impairs escalator activity of lower respiratory tract, reducing cilia activity by triggering hypersecretion of mucus
Malnutrition	May reduce inflammation, CMI, complement and antibody production; impairs mechanical barriers
Chronic stress	Increases adrenal activity which may reduce inflammation and antibody formation
Viral disease	May suppress immune response; reduces efficiency of ciliated mucosa
Trauma	Bypasses primary lines of defense, providing access to deeper tissues
Burns	An especially dangerous traumatic injury because normal skin structure is destroyed, not merely disrupted; also impairs phagocytes
Antimicrobial therapy	Alters normal flora; selects for antibiotic-resistant microbes; may encourage growth of opportunistic fungi
Anti-inflammatory therapy with corticosteroids	Depresses phagocytic defenses; depresses interferon function; depresses antibody production by suppressing lymphocyte and monocyte proliferation; promotes diabetic state
Anticancer therapy (ionizing radiation, cytotoxic chemotherapy)	Depresses antibody formation; decreases bone marrow production of leukocytes; depresses phagocyte formation; may compromise mechanical defenses
Immunosuppression to accommodate transplants	Depresses T-cell response
Surgical procedure	Provides portal of entry to vulnerable internal tissues; introduces foreign bodies that may be contaminated

are created by external influences—through trauma or medical procedures for treating other clinical disorders, for example, or through unhealthful personal qualities, notably habitual alcohol consumption, smoking, fatigue, prolonged emotional stress, or malnutrition (Table 17-4).

Repeated infections by organisms of low virulence often indicate serious underlying immunological deficiencies. For example, children who suffer repeated episodes of opportunistic infections by *Staphylococcus aureus* or by such common environmental saprophytes as *Serratia* or *Pseudomonas* may have leukemia, which impairs the function of phagocytes, or chronic granulomatous disease (CGD). Children with CGD have defective phagocytes that readily engulf invading microbes but lack one of the critical lysosomal chemicals needed to kill the engulfed bacteria. Phagocyte defects, leukemia, and T-cell deficiencies predispose for systemic candidiasis and other fatal opportunistic infections (Fig. 17-2).

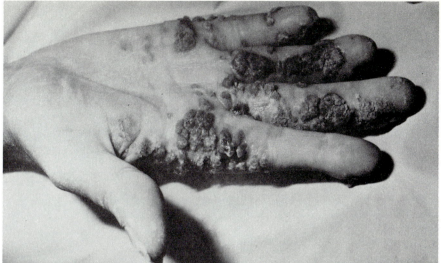

PATHOGENESIS OF INFECTIOUS DISEASE

The actual mechanisms by which a pathogen accomplishes its detrimental effects are thoroughly understood in only a few of the less complex diseases, in which symptoms are caused almost entirely by the effects of a single exotoxin (diphtheria and tetanus, for example). *Staphylococcus aureus,* on the other hand, is known to produce at least 23 factors that may contribute to its ability to invade virtually every human organ and tissue. No single factor, however, is absolutely necessary for this bacterium to elicit disease; strains that lack any one of them are still pathogenic. Unfortunately, the pathogenesis of many infectious diseases is similarly unclear.

Sequence of Events During Infection

Sites of Infection   The affinity most pathogens exhibit for a limited number of preferential tissues is called **tissue tropism** (also called organotropism). Tissue tropism not only determines the location of the *primary site* of colonization (the first habitat infected, sometimes with little or no damage to the area), but also the predilection for specific *secondary sites* of infection once the initial lines of defense have been breached. Tissues that lack surface receptors complementary to the microbe's attachment sites usually remain uninfected, although they may still be damaged by extracellular toxins or cytocidal substances released from either the microbe or the neighboring injured tissues. To initiate infection, therefore, pathogens must break into the body through a portal of entry that allows them to reach tissues displaying the appropriate surface receptors (Fig. 17-3).

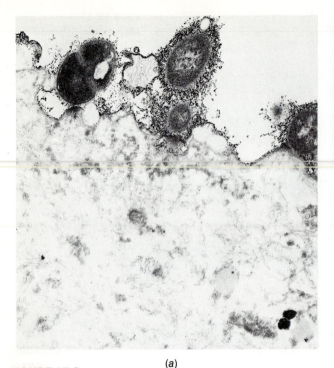

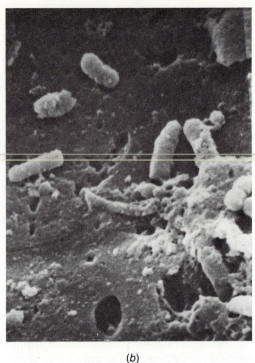

(a)

(b)

FIGURE 17-3

Bacterial adherence:
(a) Establishment of initial
infection by *Neisseria
gonorrhocae*, which adheres
to epithelium of genital
mucosa via pili ("holdfasts").
(b) *Escherichia coli* attachment
to intestinal villi in normal
nonpathogenic role with
healthy host.

Tissue tropism is also influenced by nutritional and environmental conditions at various body sites. Obligate anaerobes are restricted to areas of low oxygen tension—for example, the intestinal tract or regions of a deep wound. Some organisms cause only skin lesions because they cannot tolerate temperatures above 33 to 34°C. The dermatophytes' ability to use keratin, a protein found in skin, hair, and nail, encourages their growth on these areas of the body.

Tissue tropism is partially determined by inhibitory substances that protect body sites from microorganisms. Gram-negative bacteria, for example, infrequently infect normal skin because of their sensitivity to the high salt and fatty acid content of the epidermis. Neither salt nor fatty acids, on the other hand, inhibit the gram-positive *Staphylococcus aureus*, which causes a variety of skin lesions from boils to impetigo.

Spread of Infection Following infection, disease develops by at least one of the following patterns (Fig. 17-4):

1. The organism multiplies locally at the primary site of infection and fails to penetrate the regional epithelium. Because the etiological agent does not invade surrounding tissues, these organisms are referred to as **noninvasive**. The proliferating pathogen may either elicit local symptoms or, if it is **toxigenic**, may produce an exotoxin that diffuses into the bloodstream and affects distant tissues.

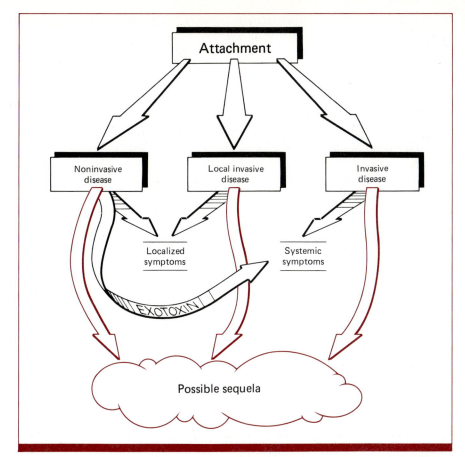

FIGURE 17-4
The possible sequence of events that elicit disease following initial attachment to the primary site. (See text for explanation.)

2. The organism penetrates, multiplies in, and destroys the regional epithelium or other surrounding tissue at the primary site of infection, causing **localized invasive diseases**. The characteristic intestinal ulceration of dysentery, for example, is due to invasion of the intestinal epithelium by the pathogen.

3. The pathogen penetrates into deeper tissues and subsequently disseminates to secondary sites throughout the body. These **invasive** pathogens usually multiply at both the primary site and the invaded secondary tissue.

Invasive diseases have much higher mortality rates than localized infections. Invasion of secondary tissues and organs commonly follows **septicemia**, the condition in which microorganisms multiply in the bloodstream. Since all living tissues in the human body are supplied with blood, *hematogenous* spread (*hemato*, blood) may lead to invasion of virtually any organ system of the body. For example, strep throat (pharyngitis caused by *Streptococcus pyogenes*) often leads to severe disease when the pathogen invades the bloodstream and settles in a number of secondary sites. If the lungs are invaded, pneumonia may develop. Skin lesions are common indicators of septicemia. Arthritis signals bone and joint involvement. Visceral complications may injure the heart, kidney, spleen, liver, or any of the internal organs. Meningitis or encephalitis may occur when the streptococci extend from the circulatory system to the central nervous system across a functional boundary called the blood-brain barrier. Pathogens can also spread by the flow of the lymphatic fluid. Since the blood and lymphatic systems flow into one another at the thoracic duct, the lymphatic routes have similar dynamics to those described for hematogenous dissemination.

Pathogens may be carried by direct extension to tissues adjacent to the primary site. For example, pneumonia may develop following transfer of upper respiratory microbes directly into the lower respiratory tract. Pathogens causing pharyngitis may extend to the adjacent auditory canals and cause otitis media (middle-ear infection). From there the infection may further extend to the mastoid sinuses and to the brain, causing encephalitis or meningitis.

Some pathogens cause severe diseases only following invasion of secondary tissue. For example, the primary sites for poliovirus infection are the throat and intestine. In approximately 1 percent of infected persons, however, the virus spreads hematogenously from the alimentary tract to motor neurons, (those that stimulate movement). The resulting paralysis characteristic of poliomyelitis is entirely due to secondary invasion rather than to primary infection.

Sequelae Some diseases develop only after the initial infection is resolved. These disorders, called **sequelae**, may be recurrences of a latent infection characterized by symptoms that are different from—and seemingly unrelated to—the primary infection. Zoster (shingles), for example, occurs only in persons who have had chickenpox. Unlike the disseminated skin rash of the initial disease, the painful lesions of shingles are restricted to regional skin areas and are due to reactivation of dormant viruses "hidden" in the host's nerve cells. Some evidence suggests that cervical carcinoma and Burkitt's lymphoma may be sequelae of herpesvirus infection.

Other sequelae seem to be immunologically mediated, a detrimental consequence of eradicating the initial disease. Following recovery from minor *Streptococcus pyogenes* infections, some people suffer acute glomerulonephritis (kidney inflammation) or rheumatic heart disease (destruction of the heart valves). Both of these poststreptococcal sequelae may be due to humoral antibodies specifically directed against the streptococci. These antibodies seem to produce either an autoimmune disorder (rheumatic heart disease) or an immune complex disease (glomerulonephritis) soon after the streptococci are eliminated.

A partial list of important infectious disease sequelae is provided in Table 17-5.

TABLE 17-5

Sequelae of Some Common Infectious Diseases

Sequel Disorder	Target Tissue	Primary Disease	Etiological Agent	Interval between Primary Disease and Sequela	Proposed Mechanism
Zoster (shingles)	Skin	Chickenpox	Herpes zoster	Years	Recrudescence of old infection
Subacute sclerosing panencephalitis	Central nervous system	Measles	Measles virus	Years	Slow virus infection
Rheumatic heart disease	Heart valves	Streptococcal pharyngitis	*Streptococcus pyogenes* (beta-hemolytic streptococcus)	18 days	Autoimmunity
Acute glomerulonephritis	Kidney	Streptococcal infection of throat or skin	*Streptococcus pyogenes* (beta-hemolytic streptococcus)	10 days	Immune complex disease
A few cancers	Lymphatic tissue, cervix, pharynx	Herpes infections, infectious mononucleosis, etc.	Herpesvirus, Epstein-Barr virus	Unknown	Latent infection

Mechanisms of Tissue Injury

The human body develops characteristic symptoms in response to infectious disease. Initial indications of infection are usually generalized symptoms: malaise (overall discomfort), headache, listlessness, loss of appetite, weakness, and body aches. As the infection progresses, symptoms become more dramatic. Fever develops, sometimes accompanied by confusion, dehydration, weight loss, skin rash, changes in pulse and blood pressure, and other signs of tissue destruction or metabolic imbalance. Persons with severe diseases show potentially fatal signs of septic shock, pulmonary distress, coronary dysfunctions, brain or neurological impairment, and disorders of other critical organ systems. Why does the infected person suffer such damage simply because microbes are replicating in the body?

Microbial Factors that Elicit Disease Symptoms The simplest (but only partially correct) way to explain the tissue damage that accompanies infection is to claim that *cytotoxicity* (death of host cells) accounts for all the symptoms of infectious disease. Cytotoxicity can be caused by pathogens growing intracellularly or on surface membranes, or it can be caused by direct contact with toxins and destructive enzymes (Table 17-6). Although these microbial factors are responsible for many disease symptoms, this oversimplified explanation creates more questions than it can possibly answer. For example, what accounts for the severe, often fatal disease symptoms evoked by those pathogenic microorganisms that produce no identifiable toxins or factors that directly harm host cells?

Factor	Effect Leading to Clinical Symptoms
Whole virus	Lyses and destroys host cells, (possibly releasing substances such as histamine that elicit additional disease symptoms)
Diphtheria toxin	Kills cells by inhibiting protein synthesis, primarily in tissue of heart, liver, and kidneys
Tetanus and botulinum toxins	Influence transmission of nervous impulses at synapse
Erythrogenic toxin	Causes rash by capillary necrosis
Lecithinase	Destroys lecithin in host cell membranes
Other cytolytic enzymes	Decompose human tissue by enzymatic digestion
Hemolysins	Causes anemia due to lysis of red blood cells
Peptidoglycan	Is cytotoxic for host cells
Coagulase	Stimulates intravascular clotting of blood
Enterotoxin	Affects gastrointestinal mucosa, causing diarrhea
Endotoxin	Triggers release of endogenous pyrogens; intravascular clotting and vasomotor disturbances that elicit fatal shock; myocardial depression; local hemorrhage
Edema-producing substances	Cause fluid accumulation in lungs of pneumonia patients

Damage of Host Origin

Our arsenals for fighting off bacteria are so powerful, and involve so many different defense mechanisms, that we are in more danger from them than from the invaders. We live in a midst of explosive devises; we are mined.

Lives of a Cell, Lewis Thomas

Many normally protective responses of the human body can severely harm the host during or after infection. Some host-mediated responses that have been implicated in the pathogenesis of several important diseases are summarized in Table 17-7 and are discussed below.

INFLAMMATION The normally healthy inflammatory response may become a chronic nuisance when combating pathogens that survive and proliferate inside phagocytes. The defenses remain mobilized in a futile effort to eradicate the intruders; this "inflammatory impotence" may continue for months or even years. Chronic inflammation eventually causes necrosis of regional tissues because of extracellular accumulation of destructive lysosomal contents released from phagocytic vacuoles. The area becomes avascularized, denying the infected site life-sustaining nutrients and oxygen. Whenever this phenomenon occurs in the respiratory tract, the characteristic swelling and exudate formation may cause airway obstruction and asphyxiation (especially if the patient is a child). This threat often necessitates the use of anti-inflammatory agents to reduce

Activity	Effects Leading to Clinical Symptoms
Chronic inflammation	Influx of inflammatory cells; release of endogenous substances (such as histamine and lysosomal enzymes) and subsequent tissue reactions
Killing of infected host cells (autocytolysis)	Elimination of crucial host cells if recognized by the immune system as being infected by viruses
Release of endogenous cellular substances	Increased fever; necrosis; smooth muscle contraction; respiratory obstruction
Autoimmunity	Destruction of "self" tissue by autoantibodies
Immune complex formation	Circulatory and renal obstruction; complement activation at site of aggregation, with subsequent tissue destruction
Hypersensitivity (allergy)	Release of endogenous substances and associated effects (see above); chronic inflammation; tissue necrosis

the immediate danger of closing the small passageways in the lower respiratory tracts of children. While these agents improve symptoms and make the patient feel better, they depress host resistance mechanisms and contribute to progressive infection. Ironically, they mask evidence of these deteriorating conditions.

The clotting of blood around local areas of inflammation tends to imprison pathogens in the initial infection site, preventing more dangerous invasive disease. When the bloodstream is showered with large numbers of microbes, however, the clotting response may be triggered throughout the circulatory system, causing *disseminated intravascular coagulation*. This fatal complication represents another example of how our normally protective responses may contribute to disease symptoms when overexpressed.

AUTOCYTOLYSIS Occasionally the host must destroy its own cells (*autocytolysis*) to protect itself from microorganisms propagating inside the target cell. Viral-infected cells are selectively destroyed by CMI, thereby eliminating the source of progeny pathogens. This protective process in itself may kill so much human tissue the host suffers overt symptoms of disease.

RELEASE OF ENDOGENOUS SUBSTANCES Tissue destruction may result from the release of endogenous substances either from infected host cells or from host cells exposed to endotoxin. These physiologically active substances are ordinarily sequestered inside host cell lysosomes or other membrane-bound vacuoles. Disruption of these compartments releases their contents into the surrounding tissue and bloodstream where they elicit localized or systemic reactions. These endogenous substances induce fever, stimulate contraction of smooth muscles in respiratory and vascular systems, necrotize tissue, and trigger systemic reactions, often fever and shock.

AUTOIMMUNITY The infected host may also damage itself by produc-

FIGURE 17-5

Fluorescence micrograph of renal membrane from patient with acute glomerulonephritis. The antigen-antibody complexes in the kidney can be detected with fluorescent antibody that specifically reacts with human antibody in the complex.

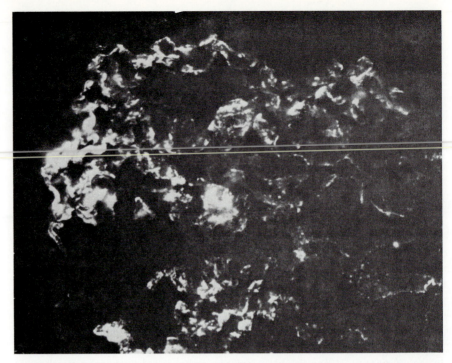

GENERAL CONCEPTS OF
PATHOGENESIS

ing autoantibodies against its own tissues, as may be the case with poststreptococcal rheumatic heart disease. Similar antigens are shared by the human heart valves and the surface of some strains of *Streptococcus pyogenes*. After eliminating the offending pathogen, some of the remaining antibodies may attack cross-reacting cardiac tissue and cause rheumatic heart disease.

IMMUNE COMPLEX DISEASE Occasionally the body accumulates huge aggregates of antigen-antibody complexes, which are deposited in the capillary bed of the vascular system or on the filtering membranes of the kidneys, causing acute glomerulonephritis (Fig. 17-5). The fixation of complement by these antigen-antibody aggregates contributes to the destruction of renal and vascular membranes. The resulting immune complex diseases often follow infections by beta-hemolytic streptococci (*Streptococcus pyogenes*) and some viral diseases.

HYPERSENSITIVITY Some infectious diseases trigger delayed-type hypersensitivity. This response causes most of the symptoms of tuberculosis, leprosy, most fungal diseases, and a few viral and protozoan infections. Although hypersensitivity may actually eliminate the pathogen, in some persons the response produces chronic inflammation, necrosis, pulmonary damage, and enhanced pathogen dissemination due to drainage of abscesses containing virulent microbes into surrounding tissue spaces. Immune-mediated host injury is discussed in more detail in Chapter 16.

It may appear ironic that our defense mechanisms may be recruited as participants in our own destruction. Such tricks of nature, however, are exceptional. Without our defense mechanisms, there would be nothing to defend.

OVERVIEW

The pathogenesis of infectious disease is a function of the etiological agent's ability to directly injure tissue and to induce detrimental host responses. Microbial properties that elicit direct tissue injury include toxins, enzymes, and viral lysis of host cells. Diseases due to these factors are usually acute, and, except for viruses, the corresponding pathogens are generally extracellular. Conversely, pathogens that injury the host solely by triggering adverse tissue responses are usually intracellular, and the corresponding diseases are generally chronic. Adverse host responses include chronic inflammation, hypersensitivity, release of endogenous substances, autocytolysis, and autoimmunity. Inadequate host responses may also predispose the host to develop opportunistic disease. Such predisposing factors may be intrinsic or extrinsic.

Pathogenesis is largely determined by the microbe's invasiveness and toxigenicity. Invasive pathogens may disseminate throughout the body. Exotoxin-producing microbes, on the other hand, may cause systemic disease without invading tissues. Many pathogens are both invasive and toxigenic.

The pathogen's tissue tropism determines which body sites will most likely be infected and depends on differences in nutritional and environmental conditions of various tissues. Concentrations of protein and keratin, as well as pH, osmotic pressure, and oxygen tension, influence whether a pathogen may grow at a particular body site. Selective inhibition by fatty acids or substances present in local tissue prevents the growth of many bacteria that may otherwise have an affinity for the protected body site. Tissue tropism is also determined by the presence of complementary receptor sites that allow adherence and colonization by the pathogen. A combination of these factors determines the primary and secondary sites of infection.

Three types of infectious diseases are described on the basis of the pathogen's ability to spread throughout the body: noninvasive, localized invasive, and invasive. Invasive organisms may spread to other tissues by the bloodstream (hematogenous route), by lymphatic channels, or by direct extension to adjacent body areas.

Sequelae occasionally develop following recovery from some infectious diseases. These disorders bear no symptomatic resemblance to the initial disease. The sequela may be a reactivation of latent infection or a detrimental consequence of the immune response against the pathogen.

KEY WORDS

pathogenesis

clinical manifestation

extracellular parasite

intracellular parasite

tissue tropism

noninvasive pathogen

toxigenic pathogen

localized invasive disease

invasive pathogen

septicemia

sequela

REVIEW QUESTIONS

1. Distinguish between the major properties of diseases caused by extracellular and intracellular pathogens.

2. What factors determine a pathogen's tissue tropism?

3. List five extrinsic and five intrinsic factors that predispose the host to develop infectious disease and discuss what measures, if any, can reduce the likelihood of infection.

4. Describe three types of interactions between microbes that influence their ability to cause disease.

5. How does the immunological response sometimes contribute to the pathogenesis of disease?

GENERAL CONCEPTS OF
PATHOGENESIS

Principles of Epidemiology

18

Our history has been and continues to be shaped by **epidemics**, sudden outbreaks of infectious disease in a community. Many of the same factors that determine the occurrence of disease in an individual also influence the spread of infectious disease throughout populations. Epidemics occur when a virulent pathogen is introduced into a population of susceptible people. The affected population may be as small as a single family unit or as large as the global community. Environmental conditions that bring organisms and hosts together encourage the spread of epidemic disease. These environmental conditions include any physical, chemical, biological, or social factors that are essential to the survival and transmission of the infectious agent. Only when these three major elements—an infectious agent, a susceptible host, and proper environmental conditions—work in concert does infectious disease emerge and persist in a community (Fig. 18-1). The spread of disease can be prevented by interrupting the chain of events leading to infection. This is the ultimate goal of **epidemiology**, the study of disease distribution in populations and of the factors that influence this distribution.

ETIOLOGIC AGENTS AND THEIR RESERVOIRS

Diseases of microbial origin may be caused by pathogens or their toxic products. The place (or places) where the pathogen normally lives and multiplies is its **reservoir.** The reservoir may be a plant, an animal, or a human, or it may be an inanimate environment such as water or soil. Most diseases originate from a single reservoir; a few diseases, however, have several reservoirs. A list of the reservoirs of some important pathogens is presented in Table 18-1.

FIGURE 18-1

Elements essential for the occurrence of disease within the population.

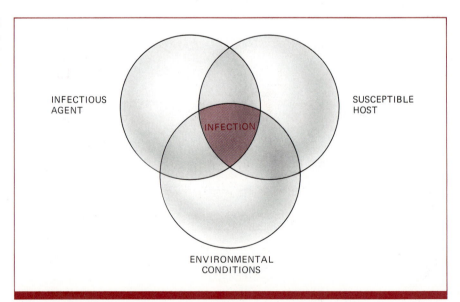

INFECTIOUS
AGENT

SUSCEPTIBLE
HOST

INFECTION

ENVIRONMENTAL
CONDITIONS

TABLE 18-1

Reservoirs of Infection

Principal Reservoir	Type of Pathogen	Some Representative Diseases
Human	Bacteria	Cholera, diphtheria, gonorrhea, leprosy, pneumonia, shigellosis, staphylococcal and streptococcal infections, syphilis, tuberculosis, typhoid, whooping cough
	Viruses	Chickenpox, hepatitis, herpes simplex, influenza, measles, mumps, polio, rubella, smallpox
	Protozoa	Amebiasis, malaria, trichomoniasis
	Fungi	Candidiasis, some ringworms
Animal	Bacteria	Anthrax, brucellosis, plague
	Chlamydias	Psittacosis
	Rickettsias	Q fever, Rocky Mountain spotted fever
	Viruses	Encephalitis, rabies, yellow fever
	Protoza	Toxoplasmosis
	Fungi	Some ringworms
Environment	Bacteria	Botulism
	Protozoa	Amebic meningoencephalitis
	Fungi	Blastomycosis, coccidioidomycosis, histoplasmosis, sporotrichosis
	Algae	Prototothecosis
Humans and one or more other reservoirs	Bacteria	Salmonellosis (animal), tetanus (soil)
	Protozoa	Leishmaniasis (animal), trypanosomiasis (animal)
	Fungi	Mucormycosis (soil)

The Human Reservoir

People are the sole reservoir for the pathogens of most human infectious diseases. Some of these agents are so fragile that they cannot live outside the body for longer than a few minutes; humans are obligate hosts for these organisms.

Infected individuals with overt symptoms of disease are obvious reservoirs that can transmit the pathogen to other susceptible hosts. Often a more important reservoir is the **carrier**, a person who sheds infectious agents but shows no clinical symptoms of disease. Carriers are particularly important contributors to the spread of many diseases when infection is not suspected. In these cases the carrier seeks no medical attention and takes no precautions against transmission. For example, 40 to 80 percent of people with gonorrhea are asymptomatic. Many are unaware that they have the disease, remain untreated, and continue to infect other people.

For many diseases, temporary carrier states occur during the *incubation period*—the time between exposure to the agent and onset of clinical symptoms. A susceptible person exposed to mumps, for example, begins shedding infectious viruses about 48 hours before the first symptoms appear. The pathogens of some diseases continue to be shed during the *convalescent period* that follows disappearance of symptoms.

Carriers of certain pathogens may shed organisms for long periods of time. Chronic carriers of *Salmonella typhi*, for example, if left untreated may shed the bacterium for their entire lives. Such carriers are often responsible for outbreaks of typhoid fever. Carriers of hepatitis B maintain the virus in their bloodstreams, and the disease can be transmitted from them by transfusion or by articles contaminated with their blood.

People are also reservoirs for the myriad of organisms that compose the normal flora. These organisms may become the agents of opportunistic infections in hosts with compromised resistance. Opportunistic infections also may occur in persons with normal defenses following *autoinoculation*, the transfer of microbes indigenous to one area of the body to another area where they may cause disease. Thus a person may be an endogenous reservoir of his or her own infectious disease.

The Animal Reservoir

Humans are innately resistant to many pathogens that affect other animals. Just as some diseases occur only in humans, some are restricted from us. Thus we can freely interact with many of our pets with little danger of infection by their pathogens. Some diseases, however, are transmitted from vertebrate animals to humans. These diseases are known as **zoonoses.** The reservoir of plague, for example, is usually the rodent population. Rabies is carried by many domestic and wild mammals. The human is usually an *incidental host,* not critical to the natural life cycle of the zoonotic pathogen.

Although a few diseases (especially rabies) are equally severe in both animals and humans, most zoonotic pathogens cause more severe symptoms in people. In addition, some pathogens elicit different types of symptoms in the different hosts. *Chlamydia psittaci,* for example, causes severe pulmonary infection in people and gastrointestinal infections in birds.

Environmental Reservoirs

Soil may be transiently contaminated with pathogens released from animal or human gastrointestinal tracts or from carcasses. Soil also serves as permanent habitat for a few virulent human pathogens. These dangerous microbes include *Clostridium tetani, Clostridium botulinum, Clostridium perfringens,* and the fungi responsible for the systemic mycoses. The fungi in particular have rigid requirements that restrict their geographic distribution to regions where the soil contains the nutritional and environmental factors conducive to growth. Most of the microbes found in soil are harmless saprophytes. Many of these, however, are opportunistic pathogens of people.

Water is an important reservoir of human disease, especially when contaminated by raw sewage. In many places such contaminated water is used for drinking, cooking, washing, or recreation. A single outbreak of waterborne disease may affect thousands of people. Feces from infected persons are the most common source of water contamination.

MECHANISMS OF TRANSMISSION

To cause disease, a pathogen must be able to survive the transfer from its reservoir to a susceptible host.

The reservoir for the majority of human diseases is the human body. The pathogens usually escape the infected person in respiratory droplets, feces, skin exudates (pus), or blood removed by a biting arthropod or hypodermic syringe, and are transmitted by direct or indirect routes.

Direct Transmission

Direct transmission is the immediate transfer of the infectious agent from the reservoir to a new host with no intervening intermediary. For diseases with human reservoirs, direct transmission may be vertical or horizontal. **Vertical transmission** is the spread of disease from parent to offspring by an infected sperm or egg, by passage of pathogens across the placenta during fetal development, or during the birth process. Gonorrhea, syphilis, genital herpes, and rubella can all be vertically transmitted from an infected mother to her child, causing blindness, congenital malformations, or death. **Horizontal transmission** is the spread of disease from person to person within a group. Direct horizontal transmission may occur by contact with contaminated respiratory droplets that are briefly suspended in the air (Fig. 18-2) or by physical contact with infected persons. Pathogens that can only be transferred by direct human-to-human contact, such as the agents of gonorrhea and syphilis, are generally fragile microbes, incapable of surviving more than a few minutes outside their natural reservoir.

Some horizontally transmitted diseases are acquired by direct contact with infected animals. Rabies, for example, develops in persons bitten by an infected animal (the reservoir). Anthrax is an occupationally related disease of veterinarians, agricultural workers, and textile workers who handle infected animals or their hides. (Tourists have also acquired anthrax from souvenirs adorned with contaminated goat or sheep hair—an example of indirect transmission.)

FIGURE 18-2
Microbe-laden droplets
produced during
(a) sneezing and (b) talking.

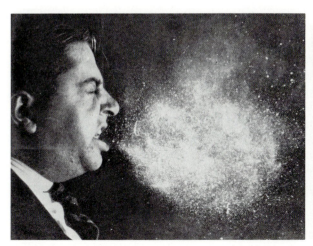

(a)   (b)

Indirect Transmission

Indirect transmission occurs when the infectious agent is transferred by an intermediary—either a vehicle, a vector, or a contaminated (rather than infected) person.

Vehicles   Vehicles are nonliving materials capable of transmitting infectious agents. If the vehicle is a contaminated inanimate object, it is called a **fomite**. Bedding, clothing, cooking and eating utensils, surgical instruments, and furniture are common fomites. Other vehicles are contaminated air, food, and water. Contaminated drugs, blood, or materials used in intravenous therapy occasionally transmit diseases within the hospital. Vehicles usually become contaminated by contact with either the reservoir or with respiratory droplets, pus, urine, or feces shed from infected persons.

Some vehicles support the growth of pathogenic microbes; others merely maintain microbial viability. For example, dry bedclothes may harbor dangerous microbes but provide little to promote their growth. Tuna salad, on the other hand, may be a microbial gourmet delight. Vehicles that allow microbes to proliferate enhance the likelihood of transmitting disease by increasing the numbers of microbes to which a person may be exposed. Vehicles that fail to support microbial growth are nonetheless dangerous if the concentration of the infectious agent initially shed onto the vehicle is sufficient to establish infection in a host.

Air is one of the most common vehicles of disease transmission. Droplets are released from an infected host during coughing, sneezing, or even talking. These droplets settle out of the air quickly because of their large size. As the moisture evaporates, however, the droplets form smaller, lighter particles called *droplet nuclei*. These particles still contain all the microbes in the original droplet and are capable of remaining suspended in air indefinitely. Furthermore, they are small enough to penetrate the primary respiratory defenses and become entrenched in the lower respiratory tract. Droplet nuclei are therefore important vehicles in the transmission of respiratory diseases. A single droplet nucleus reaching the alveoli may contain enough organisms to cause serious pulmonary disease.

Airborne pathogens may also be transmitted by dust particles that are often produced by agitation of soil, contaminated laundry, or surfaces on which microorganisms have settled. Fungal spores are often transmitted by the dust generated when contaminated soils are agitated. For example, the number of cases of coccidioidomycosis (valley fever) increases when wind activity is greatest. People who work in the soil of the San Joaquin Valley of California are especially endangered. In this area, outbreaks of coccidioidomycosis are reported among members of archeologic digs, construction workers, military personnel on maneuvers, and Scout explorers.

Vectors   Living organisms that are intermediaries in disease transmission are called **vectors**. The most common vectors are fleas, mosquitos,

TABLE 18-2
Some Important Vectors
and the Diseases They
Transmit

Vector	Type of Pathogen	Reservoir	Representative Disease
Flea	Bacterium	Wild rodents	Plague
	Rickettsia	Rats	Endemic typhus
Tick	Bacterium	Wild animals, ticks	Tularemia
	Rickettsia	Wild rodents	Rocky Mountain spotted fever
Fly	Protozoan	Humans, animals	Sleeping sickness (African trypanosomiasis); leishmaniasis
	Chlamydia	Humans	Trachoma
Mosquito	Virus	Humans, animals	Yellow fever; dengue; eastern equine, western equine, Japanese, and St. Louis encephalitis
	Protozoan	Humans	Malaria
Louse	Rickettsia	Humans	Epidemic typhus
Mite	Rickettsia	Mites	Scrub typhus
Bugs	Protozoan	Humans, animals	Chagas's disease (American trypanosomiasis)

flies, ticks, lice, and true bugs (Table 18-2). The arthropods are either mechanical or biological vectors.

Mechanical vectors pick up microorganisms on their feet or other body parts. Flies often give microbes a free ride from feces to foods. The mechanical vector is simply a living vehicle and is not necessary for the multiplication or development of the microorganism. Humans may also be mechanical vectors. In fact, the most common vector in the hospital environment is the person whose hands are constantly in contact with pathogens and patients. This is why proper attention to hand washing reduces the incidence of hospital-acquired infections.

Biological vectors are infected and not merely contaminated. Pathogens multiply within these vectors and usually cannot be transmitted to humans directly from an infected vertebrate host. Biological vectors may introduce the microbe by biting the host (Fig. 18-3) or by depositing contaminated

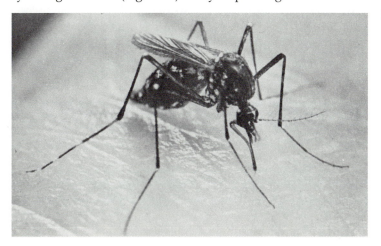

FIGURE 18-3
Aëdes aegypti, the mosquito vector of yellow fever.

feces or other excretions on the skin. Any breaks in the skin surface allow microorganisms to penetrate, especially when the person scratches the contaminated bite site.

Portal of Entry

To cause disease, an organism that has escaped from its reservoir must enter a new host through a **portal of entry**, a site that provides access to tissues where environmental and nutritional conditions are conducive to establishing infection and where local defense mechanisms fail to subdue the pathogen. These portals of entry are the respiratory tract, alimentary tract, genitourinary tract, unbroken skin (or mucous membrane), and traumatic openings in the epithelium caused by insect bites, surgical wounds, burns, and punctures. The properties of the microbe determine which portal, if any, can be used to gain access to susceptible host tissues (see Microbial Virulence, Chap. 15).

Most pathogens escape the infected host through the same portal used to enter the body. Nonetheless, additional *portals of exit* may develop. For example, syphilis is usually acquired through the genitourinary tract, but

---

UNUSUAL SOURCES FOR DISEASE

■ The incidence of infectious disease can be reduced by identifying reservoirs of infection and common routes of transmission, then avoiding or controlling these sources. Occasionally, however, disease is acquired by unconventional mechanisms that are difficult or impossible to anticipate. For example, in 1979 a woman died of rabieslike symptoms several weeks after receiving a corneal transplant. She had no history of animal bites. A subsequent autopsy, however, revealed that the donor of the cornea had undiagnosed rabies when he died. Several other fatal cases of transplantation rabies have been documented.

Another fatal neurological disease was acquired through electrodes inserted into a patient's brain to test its function. This disease, called Creutzfeld-Jakob disease (CJD), causes a progressive deterioration of the brain. The electrodes were contaminated when they were used in the brain of another CJD victim 6 months earlier. Although the electrodes were immersed in formalin, a disinfectant, for a half year, the slow virus that causes CJD remained infectious and claimed another victim.

An outbreak of typhoid fever in New York City was ultimately traced to contaminated apples sold by a single fruit vendor. Epidemiologists were surprised, however, to find that the vendor was not a carrier of typhoid. Further investigation revealed that the apples were stored directly below a leaky sewage pipe coming from the apartment above, which was occupied by a typhoid carrier.

Anthrax, a disease usually acquired by exposure to bacterial endospores on the hides of infected animals, is rare in the United States because of strict control programs. The disease is usually associated with persons who handle livestock or their hides. Several years ago, however, an outbreak of anthrax occurred among individuals who had no apparent exposure to the usual animal sources of anthrax. Further investigation revealed that all infected persons possessed or used shaving brushes made of animal hair. The brushes were imported and were indeed contaminated with endospores of *Bacillus anthracis*.

the pathogen invades the bloodstream and also causes skin lesions that shed *Treponema pallidum*.

A major goal of microbiology, epidemiology, and medicine is to control diseases by preventing the entry of pathogens into the body. (A few examples of unusual mechanisms of disease transmission are discussed in the color box.) In later chapters (Chaps. 20 to 23), specific diseases will be discussed according to their portals of entry. This approach emphasizes techniques that can be used to successfully interfere with the transfer of pathogens from reservoir to susceptible human hosts.

Several mechanisms of transmission are summarized in Table 18-3.

THE SUSCEPTIBLE POPULATION

Some diseases sweep rapidly through a community soon after the pathogen is introduced into the population. At another time or in another population, the same pathogen may cause no noticeable increase in disease incidence. Whether a pathogen causes a few isolated cases of disease or a full-scale epidemic is largely determined by differences in host susceptibility. These differences are primarily influenced by the status of our natural and acquired defense mechanisms. The magnitude of any epidemic is determined by the immune status of the persons in the affected population. An outbreak of disease may immunize so many persons in a community that subsequent epidemics by the same pathogen are unlikely. Such a reduction in both the size of the reservoir and the number of susceptible persons in the community is called *herd immunity*, and is the goal of widespread artificial

TABLE 18-3
Mechanisms of Disease
Transmission

Mechanism of Transmission	Representative Diseases	Portal of Entry
Direct Transmission		
Direct contact	Gonorrhea, syphilis	Genitourinary tract
	Staphylococcal infections	Skin
	Rabies	Traumatic break in skin
Direct spray of respiratory droplets	Measles, influenza	Respiratory tract
Indirect Transmission		
Vehicle-borne		
Food or water	Botulism, staphylococcal food poisoning, cholera, salmonellosis, shigellosis, typhoid	Alimentary tract
Fomites	Hepatitis B	Needle wound
	Tetanus, gas gangrene	Traumatic break in skin
Vector-borne	Malaria, plague, many viral and rickettsial infections	Vector bite
Airborne		
Droplet nuclei	Chickenpox, tuberculosis	Respiratory tract
Dust	Histoplasmosis, coccidioidomycosis	Respiratory tract

immunization programs. Other factors that decrease the resistance of a population and increase the incidence and severity of disease include widespread malnourishment and air that contains unhealthful pollutants.

METHODS OF EPIDEMIOLOGY

Epidemiologists are detectives who try to identify etiological agents and discover the events that contribute to diseases. They also determine the effectiveness of preventive measures that interfere with the chain of infection. Epidemiologists collect information that reveals patterns of disease; they determine the frequency of illness, the time and place each person acquired the disease, and the characteristics of those who become ill and those who escape illness. These determinations help epidemiologists discover factors that contribute to the spread of disease.

Quantifying Disease Occurrence

Epidemiologists continually monitor disease **morbidity** (the number of cases of each disease) and **mortality** (the number of people who die from the disease). Morbidity and mortality rates measure the amount of disease or death in relation to the population size. These values are used to compare the occurrence of disease and death in different populations. If differences exist between populations, epidemiologists try to discover the conditions responsible for low disease rates and attempt to establish those conditions in other populations.

Determining morbidity and mortality requires collecting accurate data. To assist in this endeavor, some diseases have been declared **notifiable diseases,** those that by law must be reported to local or state public health officials whenever cases are diagnosed (Table 18-4). Most diseases, however, are underreported because not all affected persons seek medical help. Reported information may reveal sporadic outbreaks and unusual increases in disease incidence and can be used to avert an epidemic by implementing control measures.

Identifying Patterns of Disease

Frequency Patterns Morbidity and mortality of some diseases fluctuate in cyclic variations that are often related to seasonal changes. Seasonal outbreaks of many vector-borne diseases are related to the size of insect populations necessary for disease transmission. Respiratory infections generally increase in frequency during cold winter months when people tend to crowd together indoors, encouraging the transmission of these infections. In the past, polio was more prevalent during summer when swimming is popular. Unchlorinated water was often contaminated with feces from polio-infected persons. A major environmental reservoir of the Legionnaires' bacillus is air conditioning units, which perhaps explains why most cases of Legionnaires' disease occur between July and October.

Many diseases are **endemic** in a community, that is, they occur with

TABLE 18-4
Notifiable Diseases

Disease	Cases Reported in U.S. in 1981
Bacterial	
Gonorrhea	990,864
Salmonellosis	39,990
Syphilis	31,266
Tuberculosis	27,373
Shigellosis	19,859
Meningococcal infections	3,525
Pertussis	1,248
Tick-borne typhus	1,192
Typhoid fever	584
Legionellosis	408
Tularemia	288
Rheumatic fever	264
Leprosy	256
Brucellosis	185
Psittacosis	124
Botulism	103
Leptospirosis	82
Tetanus	72
Flea-borne typhus	61
Cholera	19
Plague	12
Diphtheria	5
Anthrax	0
Viral	
Chickenpox	200,766
Hepatitis	57,929
Infectious mononucleosis*	11,719
Rabies (animal)	7,118
Mumps	4,941
Measles	3,124
Rubella	2,077
Encephalitis	322
Congenital rubella	19
Polio	6
Rabies (human)	2
Protozoan	
Giardiasis*	121,284
Amebiasis	6,632
Malaria	1,388
Fungal	
Coccidioidomycosis*	833
Histoplasmosis*	513
Other	
Aseptic meningitis	9,548
Trichinosis	206

*Reporting of these diseases is optional

constant frequency in the population. They persist because of the absence of effective controls against their spread and the presence of susceptible persons in the population. For example, cholera is endemic in areas where malnutrition creates a susceptible population and poor sanitation practices enhance transmission. Coccidioidomycosis is endemic in the southwest United States where environmental conditions are suitable for proliferation

FIGURE 18-4

Mortality curve for pneumonia-influenza showing the occurrence of an epidemic during the midwinter months, 1980–1981.

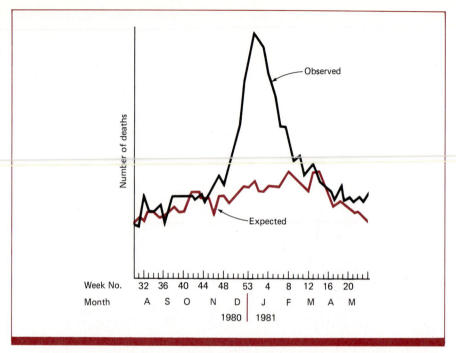

of the fungus in the soil. Histoplasmosis, on the other hand, is endemic to the Mississippi and Missouri River valleys. Some diseases are endemic in hospitals, where many routine procedures may transmit opportunistic infections among highly susceptible hospitalized patients.

A significant increase in the usual number of cases of a disease is classified as an epidemic (Fig. 18-4). The number of cases that constitute an epidemic depends on the normal frequency of the disease within the affected population. In the United States a few cases of cholera or any other rarely occurring disease would be considered an epidemic. In contrast, an increase by a few cases of tuberculosis, an endemic disease, would be epidemiologically insignificant. Occasionally epidemics spread to more than one continent or around the world. Such global outbreaks are called **pandemics**. The most commonly occurring pandemic disease is influenza.

Epidemic diseases frequently become endemic. Organisms that are introduced into a community for the first time in a population of susceptible hosts may cause rapidly spreading epidemics (see color box). After recovery, most persons have active immunity, and the epidemic runs its course and subsides. Some pathogens, however, may remain and multiply either in humans after recovery from disease, in animals, in soil, or in water. These microbes may attack susceptible persons who move into or are born into the community or persons with impaired defenses. The number of cases will usually become constant, and the disease will become endemic.

A TRAGIC GREETING FROM THE OLD WORLD

■ ". . . Lahina was about to be visited by a pestilence known as the scourge of the Pacific. On earlier trips this dreadful plague had wiped out more than half the (Hawaiian) population, and now it stood poised in a whaler (ship), prepared to strike once more with demonic force. It was the worst disease of the Pacific: measles.

"Men from the infected whaler had moved freely through the community, and on the next morning Dr. Whipple looked out his door and saw a native man, naked, digging himself a shallow grave beside the ocean, where cool water could seep in and fill the sandy rectangle. Rushing to the reef, Whipple called, 'Kekuana, what are you doing?' And the Hawaiian, shivering fearfully, replied 'I am burning to death and the water will cool me . . . you do not know how terrible the burning (fever) is,' and he sank himself in the salt water and within the day he died.

"Now all along the beach Hawaiians, spotted with measles, dug themselves holes in the cool wet sand . . . and died. Throughout Lahina, one Hawaiian in three perished."*

Such accounts were all too common among the native inhabitants of the western hemisphere and South Pacific. These native populations had never in their history been exposed to the pathogens common in Europe. They were, therefore, extraordinarily susceptible to diseases that were relatively harmless to Europeans, who had developed resistance following centuries of exposure. In the Hawaiian Islands, more than 50 percent of the native population died after European visitors and settlers first exposed them to such diseases as measles, influenza, and syphilis.

Similarly, European visitors to the new world were acutely susceptible to the pathogens endemic to these regions. Immigrant settlers in Central and South America, for example, suffered and died of yellow fever. Fortunately, the settlers who fled the disease could not carry yellow fever home with them because the mosquito vector necessary for transmitting the pathogen fails to survive in temperate European and North American climates.

*Excerpts compiled from James Michener, *Hawaii*, Random House, Inc., New York, 1959.

Distribution Patterns Every geographic area is associated with characteristic endemic diseases. Some diseases are geographically restricted because only a few locations can support growth and survival of the pathogen. Vector-borne diseases occur only in regions where the transmitting arthropod is prevalent. Thus, African sleeping sickness is limited to those areas where the tsetse fly is found.

The distribution of cases often may help reveal the source of an epidemic. The map that John Snow made during the London cholera epidemic of 1854 shows one of the most famous examples of this approach (Fig. 18-5). Snow observed that almost all the deaths seemed to occur within the vicinity of the Broad Street pump, one of the local sources of water. He had the handle of the pump removed to prevent further consumption of contaminated water. In a more recent example, the 1976 outbreak of Legionnaires' disease was traced to the common location—a

FIGURE 18-5

Snow's map of the distribution of deaths from cholera during an epidemic in London in 1854. Nearly all the deaths (indicated in color) occurred within a short distance of the Broad Street pump.

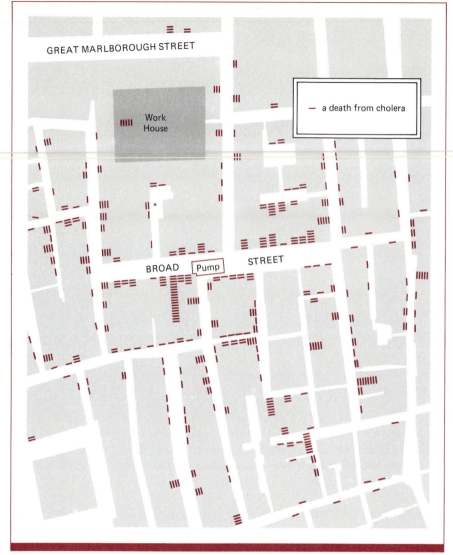

GREAT MARLBOROUGH STREET

Work House

— a death from cholera

BROAD | Pump | STREET

442

PRINCIPLES OF
EPIDEMIOLOGY

hotel in Philadelphia that was the site for an American Legion convention. Nearly all infected persons in the outbreak had attended the convention.

Studying disease distribution therefore helps epidemiologists pinpoint the sources of infection. This information is most valuable when the outbreak is a *common-source epidemic,* one that emanates from a single origin. These diseases are either noncommunicable or only poorly spread by person-to-person contact. Examples include food poisoning or any disease for which soil is the sole reservoir of the infectious form of the pathogen. Cases occur only as long as susceptible persons are exposed to the reservoir or contaminated vehicle. Most epidemics, however, are *propagated epidemics* in which any infected host can transmit the disease to other susceptible individuals. The number of cases increases as more infected hosts are produced and as long as they are in contact with a sufficiently susceptible population.

Epidemiological Typing

To verify the existence of an epidemic it is necessary to establish that all cases are caused by the same etiological agent. This generally requires isolation of the pathogen from the majority of infected persons. Identification of the pathogen's genus and species may not be sufficient to distinguish between cases caused by different strains of the same organism. For example, an outbreak of influenza may be due to a virus that has caused a previous epidemic, leaving a relatively immune population. Alternatively, the outbreak may signal the appearance of a new strain and indicate an impending epidemic. Several techniques for epidemiological typing help establish a microbial "fingerprint" to identify strains within a single species.

■ *Biotyping* is based on differences in biochemical properties and is used primarily to distinguish between members of the Enterobacteriaceae.

■ *Antibiograms* distinguish between similar organisms by revealing differing patterns of susceptibility to several antibiotics.

■ *Serotyping* ascertains differences in microbial surface antigens. This technique has the broadest range of application and is used for typing virtually all microbes.

■ *Bacteriocin typing* is generally available only in research or reference laboratories, but is an important tool in differentiating strains of *Escherichia coli, Pseudomonas aeruginosa,* and *Serratia marcescens,* as well as a few other bacteria. **Bacteriocins** are substances (usually proteins) produced by bacteria that inhibit or kill closely related bacteria. Strains of bacteria may be differentiated either by the specificity of the bacteriocins they produce or by the spectrum of bacteriocins to which they are sensitive.

■ *Bacteriophage typing* distinguishes between strains by revealing differences in susceptibility to infection by a number of phages. It is most useful for typing bacteria such as *Staphylococcus* and *Salmonella* which can be lysed by many different types of viruses (Fig. 18-6). Phage typing is often instrumental in locating and controlling sources of staphylococcal infections among hospitalized persons; it is used to identify healthy carriers of bacteria with the same phage type as those causing the nosocomial diseases.

■ *Nucleic acid restriction analysis* is one of the most sensitive methods for distinguishing between subgroups of similar viruses. Nucleic acid isolated from virions is exposed to various restriction endonucleases. Variations in nucleic acid sequences are revealed by detecting differences in the products formed by enzymatic digestion. The technique is mainly employed by

FIGURE 18-6
Phage typing of
Staphylococcus aureus.
Different strains can be
distinguished by
differences in phage
sensitivities, as indicated
by areas of lysis that
appear on a plate seeded
with the bacteria being
tested.

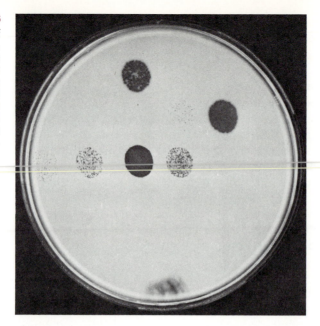

research epidemiologists and is rarely used for routine typing of viral pathogens.

BREAKING THE CHAIN

The responsibility for monitoring infectious disease rates and for planning programs to reduce their numbers falls on members of an intricate series of agencies. In the United States, the main agency is the Center for Disease Control (CDC) in Atlanta, Georgia. The CDC gathers and analyzes data on all notifiable diseases, as well as on any other diseases considered of particular concern to our health and welfare. Information collected by the CDC is published in a weekly journal, *Morbidity and Mortality Weekly Report (MMWR)*.

In the hospital, control of infectious diseases rests with an infection-control team headed by the hospital epidemiologist. The team gathers and evaluates information on all cases of infectious disease in the hospital. Their primary goal is to develop and implement programs to reduce the rate of nosocomial infections (see Chap. 24). All nosocomial infections, as well as any epidemics and unusual cases of diseases, are reported. This information is eventually centralized by the CDC which uses it to identify and control disease sources. For example, in 1970, inadvertently contaminated fluids used in intravenous therapy were distributed to many hospitals across the country. Within these hospitals, diseases appeared in patients who received the fluid during treatment. Information accumulated by the CDC alerted epidemiologists that a national epidemic was in progress. Further investigation revealed the source of the nosocomial outbreak; the contaminated fluid was recalled and additional disasters averted.

The World Health Organization (WHO) in Geneva, Switzerland, is concerned with diseases of global interest. This agency gathers information on the occurrence of these diseases and issues guidelines for the containment of their spread. The popularity of international travel increases the danger of transmitting diseases from endemic areas to susceptible populations. WHO helps establish and enforce measures that reduce this danger.

Control Measures

Control measures are usually aimed at the weakest link in the chain of events leading to infection. These measures vary according to the disease and usually include one or more of the following:

Control of Animal Reservoirs   Animal populations can be monitored for the presence of infection; infected animals may be treated or destroyed. For example, in the western United States wild rodents (and their fleas) are checked for signs of *Yersinia pestis*, the plague bacillus. Antirodent measures help reduce the chance of human infection.

Control of Vectors   Many diseases are controlled by insecticides or by eliminating arthropod breeding grounds (for example, by draining swamps). In the United States, controlling vector populations has reduced the incidence of yellow fever, malaria, typhus, Rocky Mountain spotted fever, and plague. In endemic areas where vector control has not been accomplished, wearing protective clothing and sleeping in enclosed quarters and under mosquito nets can help prevent vector-borne diseases.

Reduction of Vehicle Contamination   Good sanitation policies, improved personal hygiene habits, proper sterilization of equipment, and adherence to aseptic techniques can help reduce the number of organisms found on fomites, fingers, and environmental sources. This is especially important in preventing nosocomial diseases and food-borne infections. Proper hand washing is particularly crucial to disease control. Foods and water should be handled in a manner that prevents contamination and treated so that microorganisms cannot multiply. Chlorinating drinking water prevents proliferation of microbes and destroys most microbial contaminants. Public health inspection guidelines should be followed to minimize microbial contamination.

Treatment of Human Reservoirs   Antimicrobial agents, if available, are administered to infected persons to destroy the agent and reduce the time during which the patient sheds the pathogen. Some human carriers of typhoid, hepatitis B, and malaria, however, cannot be cured by chemotherapy. These individuals should be notified that they are carriers and restricted from occupations that would encourage the spread of their infections to susceptible people.

Interruption of Transmission   Transmission of disease within the hospital can be minimized by using physical barriers such as protective masks, clothing, and gloves when treating an infected patient, when handling infected materials, or when shedding potential pathogens. Venereal dis-

A CHEERFUL RHYME FROM A MOURNFUL EVENT

Ring-a-ring of roses,
A pocketful of posies.
Atishoo! Atishoo!
We all fall down.

■ It is ironic that this poem has become a popular children's nursery rhyme, considering the somber origin of these four bitter lines. The poem commemorates the heroic death of 226 residents of Eyam, a small village in England, during an epidemic of bubonic plague over 300 years ago. The rhyme's roses refer to the rosy hemorrhages that developed on the chest of plague victims. The posies were superstitiously believed to provide some protection from affliction. The ring refers to the unusual way the villagers of Eyam sacrificed themselves to save countless others.

The plague epidemic of 1665 had subsided, leaving 68,500 people dead in London alone. The following year, however, some of the 259 villagers in tiny Eyam were seized by raging fevers, terrible back pains, excruciating bubos (swollen lymph nodes for which bubonic plague is named), deliri-

um, and death. The plague had again erupted. In terror, the townspeople prepared to evacuate Eyam to escape the source of the pestilence. The town clergyman, however, insightfully realized that such an exodus would spread the disease to neighboring communities and eventually throughout the country. He appealed to the villagers to quarantine themselves in order to contain the disease and they agreed. A circle around Eyam 1 mile in diameter was marked off with stones. Food and other needed goods were gratefully supplied by the surrounding communities, which remained healthy because of Eyam's sacrifice. Each week the outsiders would leave supplies on the perimeter and anxiously flee.

A widespread wave of devastation may have been avoided because of the self-imposed quarantine ring around Eyam. The plague ravaged the village, however, killing nearly 90 percent of Eyam's population and leaving only 33 survivors. Although the clergyman survived, his wife was not so lucky; she was one of the last to die.

ease transmission can be interrupted either by abstinence or by the use of a condom. In fact, the condom was first developed as a protective device against venereal disease and not as a contraceptive (hence the common name "prophylactic"—disease-preventing). Any break in these barriers completely negates their protective function.

Isolation and quarantine procedures are used to prevent contact between susceptible persons and those known or suspected to be infectious. In hospitals, patients with highly infectious disease are usually isolated from susceptible persons. A comparable procedure outside the hospital, called *quarantine*, is used to reduce exposure to persons with cholera, plague, or other dangerous communicable diseases (see color box). During illness voluntary quarantine by staying home from school or work helps protect susceptible people. Many diseases, however, are most infectious during the incubation period, before symptoms appear. Since isolation usually begins after clinical disease appears, quarantine would be too late to prevent transmission. Persons exposed to someone known to have a dangerous communicable disease should immediately be quaran-

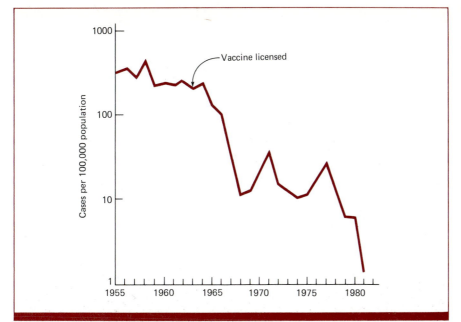

tined until it can be determined that they have not contracted the disease and are not shedding the pathogen.

Immunization of Susceptible Persons Active immunization helps to reduce the susceptible population by lowering the number of persons who develop disease and disseminate pathogens. Disease fails to spread throughout an immunized population. For example, DPT, the triple vaccine against diphtheria, pertussis (whooping cough), and tetanus, has virtually eliminated these diseases as public health problems. The incidence of polio, mumps, rubella, and measles has also been dramatically reduced by vaccines (Fig. 18-7).

Notification of Health Authorities Reporting diseases helps identify epidemics so that control measures can be implemented.

Education of the Public The public should be informed of the dangers of uncontrolled disease and of the measures they can take to prevent it. This is a major strategy in rabies control in England (Fig. 18-8). National cooperation has helped prevent the entry of virus-infected people and animals, the sole sources of rabies virus. As a result, the British Isles have been free of rabies for several years, while the European continent has seen a significant increase of the disease.

Eradication of a Disease
In the mid-1960s, the World Health Organization launched a program to eradicate smallpox. This accomplishment required the total elimination of the causative organism from all natural sources. The conquest of smallpox

FIGURE 18-8

One element of rabies
control in England is
informing the public how
they can prevent the
introduction of the disease
into the country.
Restriction measures are
rigidly enforced. Here a
notice aboard a foreign
ship warns that the ship's
dog must not be taken
ashore.

448

PRINCIPLES OF
EPIDEMIOLOGY

was our first (and, to date, only) absolute success against infectious disease. In October 1977, the last case of naturally acquired smallpox was reported (Fig. 18-9). The only smallpox viruses that remain on earth are believed to be those in a small number of laboratory collections. Unfortunately, most diseases are not susceptible to such global eradication. Smallpox was the ideal target for this program because of its unique epidemiological characteristics, which are outlined here.

■   Humans are the only known reservoir of smallpox virus. No animal, insect, or environmental reservoirs provide known "hiding places" that could sequester or harbor the virus. Once the pathogen is eliminated from the human population, it is eliminated from the world.

■   Smallpox is caused by a virus with a single unchanging antigenic type that induces solid immunity. Thus, persons immune to one smallpox virus are immune to all smallpox viruses. The susceptible population is readily reduced by vaccination, and individuals who have recovered from disease are also immune. Because it is impossible to immunize everyone, however, vaccination of large populations was in itself incapable of totally eliminating the disease.

■   Smallpox rarely causes inapparent cases that result in carrier states. Thus, a careful surveillance system identified every case virtually anywhere in the world. These individuals were isolated or quarantined to contain further spread. Because no viruses are shed during the asymptomatic incubation period, communicability begins only after clinical symptoms appear. Therefore, all persons in contact with known smallpox cases were treated by active and passive immunization, and quarantined until it was assured that they were not communicable. This combination of surveillance and containment was successful because, through international cooperation, all cases were quickly reported. In endemic areas,

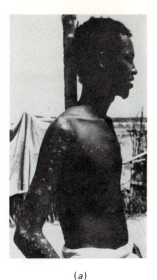 

(a)   (b)

FIGURE 18-9
(a) The world's last
recorded case of endemic
smallpox. Onset of rash
was on October 26, 1977.
(b) With the recovery of
this patient, smallpox was
eradicated from the earth.

monetary rewards were offered as incentive for citizens to report cases of smallpox to the authorities. Local workers and volunteers went to public markets and schools displaying pictures of typical cases and asking if any sick people had been seen. All such reports were quickly followed up, and smallpox was eventually eradicated.

Because there is no reservoir of smallpox left on earth, the United States no longer recommends vaccinating anyone against smallpox. America has dropped its requirement for smallpox vaccination of persons entering its borders, regardless of the traveler's origin. WHO, however, still offers a $1000 reward for any confirmed reports of person-to-person transmission of the disease. If some unknown reservoir of infection still remains, this strict surveillance should be an effective mechanism to detect it and prevent the return of the disease.

Unfortunately, most common human diseases lack at least one of the epidemiological features that made smallpox a target for eradication. The major exception appears to be measles. Within the United States, health authorities have selected the measles virus as a target for eradication, using many of the techniques successful against smallpox.

OVERVIEW

Epidemiology is the study of factors influencing the morbidity and mortality of diseases in populations. The persistence of disease depends on the continued presence of infectious agents, susceptible hosts, and conditions suitable for pathogen survival and transmission. The primary reservoirs of human infectious diseases are infected persons, either carriers or persons with overt disease. Animals, soil, and water are also reservoirs of infection.

Pathogens are transmitted from reservoirs to appropriate portals of entry in a susceptible individual by direct or indirect mechanisms. For diseases with human reservoirs, direct transmission may be vertical (from infected parent to offspring either by an infected sperm or egg, during fetal development, or during the birth process) or horizontal (by physical contact with infected persons or their infectious respiratory droplets). Indirect transmission is mediated by vehicles or vectors. Many infectious diseases are geographically limited by the distribution of the reservoir or vector.

Some diseases are endemic within a population; that is, they are characterized by fairly constant morbidity and mortality rates. Others appear as epidemics, often with seasonal patterns. A few diseases, such as influenza, can become pandemics.

Government agencies and infection-control committees in hospitals monitor diseases in order to identify outbreaks and prevent their spread. The methods used to break the chain of infection depend on the disease. They include controlling animal reservoirs, arthropod vectors, and human carriers; preventing microbial contamination of vehicles by adhering to public health and personal hygiene measures; reducing the susceptibility of the human population by immunization; and interrupting direct transmission through isolation and treatment of infected people. These methods, coupled with public awareness and cooperation, help achieve the goal of epidemiology—the eradication or containment of disease.

KEY WORDS

epidemic	fomite
epidemiology	vector
reservoir	portal of entry
carrier	morbidity
zoonosis	mortality
direct transmission	notifiable disease
vertical transmission	endemic disease
horizontal transmission	pandemic
indirect transmission	bacteriocin
vehicle	

REVIEW QUESTIONS

1. Differentiate between
   (a) Biological vector and mechanical vector
   (b) Common-source epidemic and propagated epidemic
   (c) Morbidity and mortality
   (d) Vertical and horizontal transmission

2. Explain the role of carriers in transmission of disease and disease control.

3. Describe each of the following and indicate whether it transmits disease directly or indirectly:
   (a) Respiratory droplet   (b) Droplet nucleus
   (c) Dust   (d) Arthropod
   (e) Contaminated needle   (f) Herpes blister

4. Why do some diseases display seasonal variations in the frequency of occurrence?

5. Briefly describe four methods of epidemiological typing.

6. How do some epidemic diseases become endemic?

7. List two reasons why a disease may be restricted in its geographic distribution.

8. Discuss how each of the following contributes to control of infectious disease:
   (a) Vaccination   (b) Antibiotic therapy
   (c) Quarantine   (d) Insect control
   (e) Hand washing

Clinical Specimens for Microbial and Serological Analysis

19

PRIOR TO COLLECTION
Selection of the Correct Specimen
Collection at the Proper Time

SPECIAL CONSIDERATIONS
Safety Precautions
Contaminated Specimens
Failure to Isolate the Etiological Agent
Inaccurate Concentrations

COLLECTING THE SPECIMEN
Throat-Nasopharyngeal Specimens
Specimens from Sputum
Eye and Ear Specimens
Stool Specimens and Rectal Swabs
Genital Tract Specimens
Urine Collection
Specimens from Skin Lesions
Specimens from Wounds
Specimens from Biopsy Material
Cerebrospinal Fluid (CSF) Specimens
Blood Specimens

LABORATORY EXAMINATION OF SPECIMENS

INTERPRETATION OF LABORATORY RESULTS

OVERVIEW

The human body responds with a number of different signs when infected with pathogenic microbes. Overt symptoms provide a clinical picture of the disease progress. More subtle indicators of infection are provided by laboratory tests that reveal the responses of the immune system, the quantitative and qualitative changes in tissue and blood cells, and the presence of pathogenic or toxic agents. Locating these subtle clues requires the analysis of **clinical specimens**, fluids or tissues removed from the patient's body and sent to the laboratory. Reading this "body language" is a fundamental task of the diagnostic team.

The diagnostic team often consists of a nurse, who collects most of the specimens for laboratory analysis; the laboratory staff, who examine the specimen for diagnostic clues; and the physician, who assimilates all the information into a diagnostically useful whole. Improper collection of clinical specimens often results in inaccurate laboratory findings that can seriously cripple the diagnostic function of the physician.

Rapid laboratory identification of infectious patients minimizes the danger of epidemics in both the hospital and the community. Microbiological lab findings may also alert the attending physician to immunological deficiencies or other serious underlying disorders, information that can be used to save the patient's life.

PRIOR TO COLLECTION

Before clinical material is gathered, one should decide which type of specimen to collect, the anatomical location from which it should be collected, the optimal time to obtain material that will yield the most valuable information, and the proper collection technique.

Selection of the Correct Specimen

The types of symptoms and their locations suggest what specimens should be sent to the laboratory. The specimens may be exudates from infected wounds or skin lesions, swabs from an inflamed throat, sputum from the patient with lower respiratory symptoms, feces from the infected gastrointestinal tract, or urine and purulent discharges from genitourinary tracts. Sometimes specimens can only be obtained by **biopsy**, removal of tissues by surgical procedures. Central nervous system disorders often necessitate puncture between the lumbar vertebrae to collect cerebrospinal fluid. Blood samples for culture are drawn whenever symptoms, especially rash or fever, suggest the presence of microorganisms in the patient's circulatory system. Blood is also collected for several other reasons, most commonly to determine the relative number of white blood cells. This information is useful for a variety of diagnostic purposes (see Phagocytic Cells, Chap. 15).

Collection at the Proper Time

Materials are best collected at the time when the pathogen is most likely to be at the selected site. This is usually during the acute phase of disease, as early as possible after onset of clinical illness. Specimens collected during convalescence are less likely to contain viable pathogens. Convalescent blood samples are often collected for comparison with blood specimens taken during acute phases of disease to determine if antibody concentrations have increased in the course of the patient's disease. These paired serum samples often provide solid evidence that the corresponding microorganism is the etiologic agent (see Chap. 16).

Specimens should be collected before the patient receives antimicrobial therapy because antibiotics in the patient's tissues and fluids may also be present in the clinical specimen. Since many microorganisms fail to proliferate in the presence of antibiotics, the diagnosis may be missed, especially if the lab is not informed of precollection antimicrobial therapy. Although the lab personnel can sometimes augment chances of recovering the pathogen by treating the specimen with specific inhibitors of antimicrobial agents, it is better if antibiotic therapy can be delayed at least until the initial diagnostic specimens have been obtained.

Except for emergencies, collection should also be timed so that it corresponds with the normal working hours of the laboratory staff. Delays in examining the culturing specimens may alter the numbers and kinds of microorganisms recovered. Accurate results depend on immediate processing of the specimen.

SPECIAL CONSIDERATIONS

Safety Precautions

Collection procedures pose a potential threat to the patient, to persons executing the procedure, to transporters, and to laboratory workers. Familiarity with and adherence to prescribed safety precautions reduce this danger. Invasive procedures are especially dangerous to the patient. **Invasive procedures** are techniques that introduce an instrument into the body, either through a mucosal orifice or by disrupting the body's surface, with a needle or scalpel, for example. Blood sampling by *venipuncture* (inserting a needle into a vein) may introduce surface contaminants into the bloodstream. Infection may also follow the use of contaminated instruments. Microbes may be harbored on hands or clothing of medical personnel or on unsterile instruments and equipment. The danger to the patient is reduced by using strict medical asepsis and sterile instruments.

Clinical specimens may contain pathogenic microorganisms that are hazards to clinical and laboratory personnel (see color box). The risk of subsequent infection is minimized by adopting certain protective measures:

1. Collection containers that completely enclose the specimen should be used; these discourage spillage and external contamination of the container with infectious material.

2. Any surface accidentally contaminated by potentially infectious clinical

THE HAZARDS OF LABORATORY MICROBIOLOGY

■ Each year people contract infectious disease because of accidental exposure to pathogens in laboratories. Over 4000 cases of laboratory-associated infections have been recorded, 164 of which were fatal. Over half these cases were caused by the ten agents listed in the table below. Some of these organisms are no longer the primary concern of modern laboratories because of their relative scarcity. For example, 97 percent of the accidental brucellosis and typhoid cases listed here occurred before 1955. In contrast, over 40 percent of the laboratory-associated cases of hepatitis have occurred since 1955.

The Centers for Disease Control have defined which microbes currently represent the greatest hazard in the laboratory and have classified etiologic agents into four categories. Class 1 contains agents that pose little risk of serious disease. These include *Staphylococcus epidermidis* and many other members of the normal flora. Microbes in class 4 are the most dangerous and require the highest degree of containment. The plague bacillus is a class 4 bacterium.

The greatest number of laboratory infections has occurred among persons engaged in research activity. Fewer than 33 percent have been reported for those in diagnostic laboratories. This perhaps reflects the unanticipated hazards of handling newly discovered, poorly understood, or previously unencountered microbes. In a dramatic incident that occurred in Marburg, Germany, in 1967, 31 persons handling the tissues of African green monkeys were infected by a previously unknown virus in the tissue. Six of these victims died. Researchers are microbiology's "test pilots," and many unexpected hazards will likely surface during their preliminary investigations. All persons who handle pathogenic microorganisms are at risk of infection.

Most laboratory-associated infections are acquired by contact with infectious aerosols. Random air sampling demonstrates that common laboratory manipulations release microorganisms into the atmosphere. The photos show how infectious aerosols are created by blowing out a pipette or by simply opening a stopper. Aerosols probably account for the many laboratory-associated infections that occur in the absence of an identifiable accident. The centrifuge is also an important source of infection, and has accounted for at least four fatalities. Infected experimental animals may discharge contaminated respiratory droplets into the air. In addition, bites and

Ten Most Frequently Reported Laboratory-Associated Infections*

Infection	No. of Cases	No. of Deaths
Brucellosis	426	5
Q fever	280	1
Hepatitis	268	3
Typhoid fever	258	20
Tularemia	225	2
Tuberculosis	194	4
Dermatomycosis	162	0
Venezuelan equine encephalitis	146	1
Psittacosis	116	10
Coccidioidomycosis	93	2
Total	2168	48

*Not included are 113 cases of hemorrhagic fever contracted from wild rodents in one laboratory in Russia in 1962.
Source: Adapted from R. M. Pike, *Arch. Pathol. Lab. Med.,* **102** (1978).

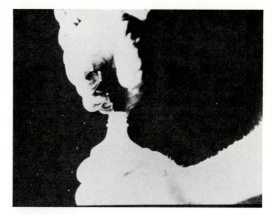

scratches from these animals occasionally cause infection. Mouth pipetting is one of the most hazardous of all laboratory manipulations, resulting in numerous cases of typhoid fever, tularemia, scarlet fever, hepatitis, and influenza. Accidental inoculation with needles and syringes also accounts for many laboratory-associated infections, as do spills of infectious materials.

Laboratory-associated infections can be reduced by establishing and enforcing rules for safe laboratory practices and by using protective equipment. Biological hoods are designed to contain infectious aerosols at the source. Sealable centrifuge tubes of unbreakable plastic minimize the possibility of leaks or breakage during high-speed spins. Hand-operated suction devices eliminate the need for mouth pipetting. Experimental animals can be properly housed to control airborne infection. Properly designed laboratories provide adequate space, restricted access to areas where highly infectious agents are being utilized, and ventilation systems that encourage airflow into areas where infectious aerosols may be generated. Such ventilation systems, however, protect only persons outside the laboratory. In the final analysis, biological safety depends on the attitude and conduct of the individual worker.

material should be immediately disinfected with a conveniently located effective germicide.

3. After use, all objects that touch the specimen should be placed in germicide or in a closable biohazard bag for subsequent decontamination.

4. The person transporting the specimen to the lab should be informed of the infectious potential of the consignment, as well as of proper disinfection procedures in the event of accidental spillage.

5. If the specimen is believed to contain dangerous pathogens, such as hepatitis virus, warnings should be clearly and conspicuously written on the requisition slip.

Contaminated Specimens

Microbial contaminants confuse diagnostic interpretation of results and may prevent detection of the etiological agent. Specimens may be contaminated by microbes from noninfected areas of the patient's body, from the individual executing the procedure, or from nonsterile equipment or containers. Sometimes contamination cannot be avoided. Upper respiratory secretions, for example, invariably contain microorganisms normally found in the nose, throat, or mouth. Other sources of contaminants are unsterile containers and instruments. Discard any "sterile" instrument that has accidentally touched an unsterile object. Never use "sterile" equipment stored in open or moist paper bags. Only dry wrapping can maintain sterility of its contents, and only then when completely sealed.

Failure to Isolate the Etiological Agent

Failure to recover an etiological agent from a specimen may be due to improper collecting or handling. All the following can cause the loss of the etiological agent and subsequent misdiagnosis.

Problem Obligate anaerobes, responsible for such diseases as tetanus, botulism, gas gangrene, and abscesses, fail to grow in the presence of molecular oxygen.

Precaution: Collecting duplicate samples provides one specimen for aerobic and a second for anaerobic processing. Since even transient exposure to normal air can kill strict anaerobes, collection instruments are often bathed in carbon dioxide, thereby eliminating molecular oxygen. Many containers for anaerobic culture are prefilled with carbon dioxide, making them safe for oxygen-sensitive pathogens. Special anaerobic transport media can also be employed.

Problem Many specimens are refrigerated if they cannot be processed immediately. The delicate *Neisseria* species that cause meningitis and gonorrhea, however, are rapidly destroyed by lower temperatures.

Precaution: Genital specimens and cerebrospinal fluid should be immediately examined and cultured, and should *never be refrigerated* if they are to be tested for the presence of *Neisseria*. In addition, they should never be inoculated onto media that is still cold from storage in the refrigerator.

Problem Many microorganisms are sensitive to the drying that occurs when specimens are not kept moist during transport to the lab.

Precaution: Organisms collected on cotton or synthetic fiber swabs should be either placed in a sterile liquid transport medium to prevent desiccation or immediately inoculated onto the appropriate culture medium.

Problem Nonspecific antimicrobial factors sometimes kill pathogens before they can be isolated and identified (Table 19-1). Human serum, for example, may contain active phagocytic cells or substances that kill microorganisms. Some types of swabs are attached to their applicator stick with an adhesive that is inhibitory to bacteria. Swabs attached to metal

CLINICAL SPECIMENS
FOR MICROBIAL AND
SEROLOGICAL ANALYSIS

Likely Location	Antimicrobial Factor	Protective Precaution
Serum	Phagocytes, chemicals in serum	Dilute in nonnutritive transport media.
Swabs	Metallic ions of wire, adhesive used to attach cotton, bactericidal substances in cotton	Replace cotton with synthetic fiber (e.g., calcium alginate); use swabs only when there is no suitable alternative.
Containers and instruments	Soaps, detergents	Rinse thoroughly after washing (prior to sterilization).
Any specimen	Chemotherapeutic agent administered prior to collection	Delay chemotherapy until after specimen collection; otherwise use media containing enzymes and agents that destroy or inhibit the antimicrobial agents used in therapy.

wires may contain metallic ions that kill microorganisms. Cotton itself contains bactericidal substances. Containers or instruments that have been inadequately rinsed after washing harbor residual soaps and detergents that quickly destroy any susceptible microorganisms in the specimen.

Precaution: Immediately inoculate the specimen or use nonnutritive transport media that dilute or inactivate antimicrobial substances in serum and other specimens. Use swabs made with synthetic fibers, calcium alginate, for example, instead of those containing cotton. It is advisable to use alternatives to swabs, such as syringe aspiration, whenever possible. All glassware should be well rinsed *(before sterilization)* to remove residual soaps.

Problem Antibiotics in clinical specimens usually obscure laboratory results, *even if the drug is having no therapeutic effect in the patient's body.*

Precaution: As mentioned earlier, antimicrobial therapy should be delayed until after the appropriate specimens have been obtained. Any antimicrobial drugs that may be present in clinical material should be identified on the culture requisition slip so the lab can take measures to neutralize their inhibitory effects. Specimens containing penicillin, for example, should be cultured on media containing penicillinase. This problem has been somewhat alleviated by the advent of a method for removing many types of antibiotics from blood by simply mixing the specimen with resin beads to which these antibiotics attach. The beads containing the drugs are then easily separated from the specimen. These counteractive measures can only be initiated if the lab is informed that the patient received antibiotics prior to specimen collection.

Problem Etiological agents in specimens cultured on improper media either fail to grow or are overgrown with the normal contaminants from the region of the body selected. This may occur when incomplete or inaccurate reports furnished with the specimen misdirect the laboratory microbiologists in their subsequent efforts to recover the pathogen.

Precaution: Specimens should be labeled with the description of the clinical material, suspected diagnosis, date, and time of collection. If the specimen is to be inoculated onto media at the patient's bedside, consultation with fully qualified microbiologists in the laboratory will aid in determining the proper medium. Identification of viruses or rickettsias may require shipping frozen samples to special laboratories.

Inaccurate Concentrations

Microbial concentrations in clinical specimens often help determine whether a microbe is causing disease or is simply a passenger in the specimen.

Problem Delays in processing collected specimens may lead to changes in microbial concentrations. Prolonged storage at room temperature encourages proliferation of microorganisms in urine, sputum, cerebrospinal fluid, and blood. Fragile pathogens may die during storage.

Precaution: When possible, inoculate clinical specimens immediately after collection. Avoid delays in delivering uninoculated specimens to the laboratory. If delay of more than 30 minutes is unavoidable, *some* specimens should be refrigerated at 4°C until they can be processed.

COLLECTING THE SPECIMEN

Using standardized procedures for obtaining each type of specimen minimizes variability and increases the reliability of the lab results. Table 19-2 summarizes the guidelines for each procedure discussed below and lists both the potential pathogens and the indigenous organisms most likely to be recovered from each type of specimen. Figure 19-1 illustrates some of the materials and instruments commonly used in collecting specimens.

Throat-Nasopharyngeal Specimens

Throat or nasopharyngeal specimens aid in the diagnosis of infectious disease of the upper respiratory tract. They are usually obtained on a sterile swab tipped with a cottonlike material such as calcium alginate wool. Because many pathogens will not survive dehydration, drying of the specimen must be prevented by immediate inoculation of culture media or by placing the swab in a transport medium and taking it immediately to the laboratory.

Obtaining the optimum throat culture requires care in avoiding the slightest contact with the teeth, gums, cheeks, tongue, and other regions heavily populated with normal flora. Such precaution will prevent the throat culture from becoming a mouth culture, which is of no diagnostic value.

Nasopharyngeal cultures are obtained by using a sterile flexible swab (usually a cotton-tipped wire) that may be passed up behind the uvula or down through the nasal passages and gently rotated.

Specimens from Sputum

Although the lower respiratory tract is usually sterile, its secretions, called

Specimen	Amount Needed	Container or Medium	Normal Flora	Common Pathogens
Throat-nasopharynx	As much as possible (from inflamed tonsils or throat, *not* tongue, teeth or cheeks)	Swab	Streptococci, staphylococci, *Neisseria, Haemophilus,* diphtheroids	Viruses, beta-hemolytic streptococci, *Corynebacterium diphtheriae, Streptococcus pneumoniae, Staphylococcus aureus, Candida albicans, Neisseria meningitidis, Haemophilus influenzae*
Sputum	2–3 cc (of sputum, *not* saliva)	Sterile cup (transport immediately)	*Staphylococcus epidermidis, Neisseria* sp., streptococci, diphtheroids, *Micrococcus*	Any organism in pure or predominant culture
Material from eye	As much as possible	Sterile swab	Small numbers of *Staphylococcus epidermidis* and anaerobic diphtheroids	*Pseudomonas aeruginosa, Haemophilus aegypticus, Moraxella* sp., *Staphylococcus aureus, Streptococcus pneumoniae,* fungi, viruses
Material from ear	As much as possible	Sterile swab or aspirate in syringe	Mixed skin flora	*Pseudomonas aeruginosa, Staphylococcus aureus, Haemophilus influenzae*
Stool	About a gram	Clean, unsterile container; swab when *Shigella* or *Neisseria* is suspected	*Bacteroides* sp., *Fusobacterium* sp., and other anaerobes; coliforms; *Staphylococcus* sp., *Streptococcus* sp., *Lactobacillus* sp.; yeasts	*Salmonella* sp., *Shigella* sp., *Clostridium perfringens, Vibrio* sp., enteropathogenic *Escherichia coli,* enteroviruses, many others
Cervical-vaginal	As much as possible	Sterile swab	Variable mixture of *Lactobacillus, Bacteroides, Clostridium,* diphtheroids, and yeasts	*Neisseris gonorrhoeae, Candida albicans, Trichomonas vaginalis,* chlamydias
Male urethral discharge	As much as possible	Sterile swab	*Staphylococcus epidermidis* and other members of normal skin or fecal flora	*Neisseria gonorrhoeae, Chlamydia trachomatis*
Urine	1 ml	Syringe or sterile cup (transport immediately)	None in the bladder (may contain genital contaminants)	Any organisms greater than 100,000/ml (excluding lactobacilli and diphtheroids)
Skin lesions	As much as possible	Sterile swab, aspirate, or biopsy material	Diphtheroids, *Propionobacterium acnes; Staphylococcus epidermidis*	Beta-hemolytic streptococci, *Staphylococcus aureus,* dermatophytes
Wound	As much as possible (after cleaning wound to remove skin flora)	Syringe or swab	None (may have contamination from skin flora)	Any organisms (*Staphylococcus aureus,* beta-hemolytic streptococci, gram-negatives, yeast, anaerobes)
Cerebrospinal fluid	1 ml	Sterile tube	None	Virtually any organism present
Blood	10 ml from two sites	Medium provided by laboratory	None	Virtually any organism present

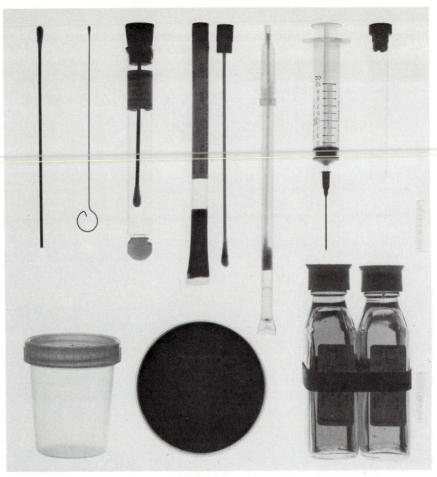

FIGURE 19-1

Some commonly used materials for collecting specimens: Top, from left: Cotton swab; flexible nasopharyngeal and urethral swab; swab and tube containing carbon dioxide for anaerobic specimens; swab and tube containing special medium and carbon dioxide for *Neisseria* sp.; culturette swab with medium for transporting anaerobic specimens; sterile syringe for collecting blood and, with needle removed, aspirating wound drainage; sterile stoppered test tube for collecting ceresbrospinal fluid, Bottom, from left: sterile plastic urine cup; plate media; duplicate culture bottles for aerobic and anaerobic cultivation of blood.

sputum, may become contaminated by oral microflora during the process of expectoration. Sputum differs from saliva, the watery secretions of the mouth that contain oral microorganisms. When these organisms contaminate the sputum specimen, they may provide false clues that can mislead the diagnostician.

Sputum may be collected by one of the three methods, expectoration, nasotracheal or orotracheal suction, or transtracheal aspiration. Ideally, expectorated sputum is collected in the morning immediately after the patient awakens. This is the time when the most sputum is available and the patient is strongest for coughing it up. The procedure should be carefully explained to the patient and the importance of elevating the material from deep in the bronchii should be emphasized. Oral flora contamination can be reduced by washing the mouth before coughing. If coughing is difficult, the process can be assisted by *nebulization*, exposure to a mist of hypertonic saline solution to loosen and liquefy bronchial secretions while stimulating a cough reflex. Collecting sputum in a large-mouthed sterile cup with a tight-fitting lid prevents spillage and contamination of the specimen.

Patients incapable of voluntary coughing and expectoration may require *tracheal suction* of sputum with a device that is passed through the mouth or nose into the lower respiratory tract. This technique does *not* reduce the problem of contamination by the normal oral or nasal flora. *Transtracheal aspiration*, on the other hand, eliminates the presence of upper respiratory organisms in the specimen. In this procedure, a sterile large-bore needle is passed through the decontaminated skin of the neck and into the trachea. Sputum is then aspirated into a syringe.

Ear and Eye Specimens

Although the human eye possesses a characteristic microflora, bacterial numbers are relatively small because of the continual flushing activity of tears and the action of antibacterial substances such as lysozyme. These natural inhibitors may also destroy bacteria in a specimen that is not immediately processed. Specimens are obtained by retracting the eyelid and stroking the infected regions with a sterile swab. Caution should be exercised to avoid touching the sensitive iris. The swab may then be placed in a sterile test tube containing just enough transport medium to keep the material moist. The specimen is immediately inoculated onto growth medium if *Neisseria gonorrhoeae* infection of the eye is suspected.

Prior to collecting a specimen from the ear, the external surfaces should be decontaminated with alcohol or an iodophore. The clinical material can be removed by sterile swab, or pus can be aspirated into a sterile syringe.

Stool Specimens and Rectal Swabs

Most pathogens that cause gastrointestinal illness reside in the lumen as well as in or on the walls of the intestine. Since these organisms are shed in the feces, it is not necessary to obtain scrapings of intestinal walls to recover them in culture. Like sputum, the optimal fecal sample is available in the morning, the first stool of the day. These specimens are most valuable when obtained early in the course of the disease. The material can be collected by scraping a small amount of feces from a fresh stool left in a bedpan or in an infant's diaper. Care should be exercised to avoid contamination of feces with urine. Unlike other specimens, stool specimens are collected in unsterile containers.

Many intestinal pathogens are sensitive to desiccation and die unless immediately inoculated or placed in a transport medium. Intestinal protozoa alter their characteristic forms in specimens that are not examined soon after collection.

A few pathogens are found in rectal ulcers and are not readily isolated from feces. In these cases, specimens taken on sterile swabs may prove more beneficial than a stool specimen. The intestinal lesions should be visually located, using a proctoscope, and directly swabbed.

Isolation of pathogenic viruses from feces usually requires shipping the specimen to the nearest viral laboratory. The feces should be frozen quickly at −40°C until processing can occur.

Genital Tract Specimens

Venereal diseases and genital infections are the most commonly reported infectious diseases in the United States; cervical-vaginal swabs, urethral exudates, and fluid from genital lesions are among the most common types of specimens collected. Unless protected, many venereal pathogens quickly die when removed from their genital sanctuaries.

Specimens from women with vaginitis are best obtained by swabbing the vaginal mucosa at the place where the discharge is most plentiful. The close proximity of vagina and rectum necessitates care to avoid contamination with fecal organisms. The swabs should then be placed in tubes of sterile saline. If trichomoniasis is suspected, the specimen should be microscopically examined within 15 minutes to locate the characteristic trichomonads, which quickly lose motility and become increasingly difficult to identify. These specimens are also inoculated into appropriate media to help detect asymptomatic trichomonas carriers.

Although blood is the predominant specimen employed in the diagnosis of primary syphilis, specimens for microscopic detection of *Treponema pallidum* may be recovered from fluid expressed from the characteristic genital lesions (chancres).

Urine Collection

Bladder urine is normally sterile and contains microorganisms only when bladder, ureters, or kidneys are infected. Even healthy persons, however, void urine that becomes heavily contaminated with the microbial residents of the urethra and external genitalia. Since these resident organisms may also cause urinary tract infection, quantitative determinations are necessary to distinguish whether these microbes are causing disease or are simply contaminants in urine specimens. For example, *Escherichia coli*, a common contaminant of urine voided from all persons, causes 80 percent of all uncomplicated urinary tract infections. Because contamination of properly collected urine is inevitable, fewer than 10,000 bacteria per milliliter are considered normal. When the concentration of enteric gram-negative bacilli exceeds 100,000 cells per milliliter, they are declared the etiologic agent. If the count is between 10,000 and 100,000 per milliliter, new specimens are collected and processed. These numbers apply only to specimens collected by the midstream-catch technique (described below).

Any procedures that alter the microbial proportions in the sample invalidate the laboratory results and confuse the diagnosis. Since urine itself is an excellent culture medium, specimens allowed to stand at room temperature for more than 30 minutes become overgrown and may lead to misdiagnosis (see color box). If delay is unavoidable specimens should be stored at 4°C, and even then for no longer than 6 hours before cultivation.

To reduce the extent of contamination during collection, voided urine should be gathered by the "clean-catch" or **midstream-catch** technique. This requires thorough cleansing of the external genitalia. Uncircumcised males should retract the foreskin and females should clean from front to back to avoid transferring rectal flora to the vagina. The patient begins voiding and, without stopping the stream, collects the midstream portion in a sterile cup until the container is no more than half full.

Alternatives to the clean-catch method are occasionally employed to reduce contamination. The once-common practice of inserting a catheter up the urethra and into the bladder does reduce microbial contamination of

WHY MRS. BAXTER'S NUMBER WAS UP

■ Because of her pain medication, Mrs. Baxter, the patient in room 415, found it impossible to void urine. The doctor gave her one last opportunity to try to voluntarily empty her bladder. She failed. A urinary catheter was inserted and withdrawn again as soon as the bladder was empty. No further instrumentation was needed. Several days later, however, Mrs. Baxter was preparing to check out of the hospital when she began to develop symptoms of cystitis, a urinary bladder infection. The doctor ordered a urine sample for quantitative bacteriological analysis.

Mrs. Baxter's urine was collected at 9:15 A.M. by the midstream-catch technique and transported to the laboratory. Mrs. Baxter's room was on the fourth floor and the lab was on the ground floor, and additional specimens were collected from other patients en route. By 12:45 all specimens were delivered together to the laboratory. Quantitative microbiology revealed Mrs. Baxter's urine to contain well in excess of 1 million *Escherichia coli* per milliliter. Because of their great numbers, *E. coli* was presumed to be the etiological agent, and ampicillin therapy was immediately begun. Antibiograms confirmed the ampicillin sensitivity of Mrs.

Baxter's *E. coli*. Nonetheless, the treatment failed. Several days later she still had her painful bladder infection.

Mrs. Baxter was misdiagnosed for two reasons: her specimen was not promptly transported to the lab, and urine is an excellent culture medium for many microorganisms, including *Escherichia coli*. This bacterium can grow at the rate of three generations per hour. Mrs. Baxter's bladder urine contained no *E. coli*. During voiding, her urine was contaminated, as could be expected, with the many microbes on her external genitals. The fresh specimen harbored around 10,000 bacteria per milliliter, a high normal count following clean-catch collection. Had the laboratory processed the urine immediately, *E. coli* would have been dismissed as a probable cause of Mrs. Baxter's infection. During 3 and one-half hours of transport however, these bacteria doubled in number about 10 times, overgrowing the real pathogen. The milliliter concentration of *E. coli* had grown from 10,000 to over a million. If Mrs. Baxter's urine had been processed immediately, or refrigerated during the delay, she would have been spared the discomfort and danger of misdiagnosis.

the sample, but is discouraged because of the danger of urinary tract infections due to this procedure. Urine specimens from patients with existing catheters, however, may be collected with a sterile syringe from either the port of the catheter or through the tubing. Both sites need decontamination prior to insertion of the needle. Because the genital flora are avoided by collecting from existing catheters, any gram-negative enteric bacterium that exceeds concentrations of 10,000/ml in catheter urine is considered the etiological agent of infectious urinary tract disease. Specimens should *never be obtained from the urine "incubating" in the catheterized person's collecting bag.*

Specimens from Skin Lesions

The skin is normally colonized by resident bacteria, mostly gram-positive cocci and diphtheroids. Skin can be partially decontaminated by proper cleansing and antiseptics.

Draining rashes or purulent (pus-producing) lesions are swabbed after cleaning with alcohol to remove the pus. Swabs that accidentally contact the surrounding skin and its indigenous flora are discarded. Closed lesions are not exposed to surface flora, and their exudates are excellent specimens when aspirated directly into sterile syringes.

Dermatophytes from cutaneous lesions cannot be obtained on a swab. These fungi are usually collected by obtaining a portion of the infected tissue or aspirating any exudate that may be present.

Specimens from Wounds

Uninfected wounds are free of established microorganisms. Many transient residents of the skin, however, may cause serious infections in wounds. Contaminants introduced into the specimen during collection may therefore confuse the diagnosis.

Infected superficial wounds usually contain aerobic organisms. Deep wounds or abscesses often support the growth of obligate anaerobes, and specimens should be cultured anaerobically, especially if the area is characterized by a foul odor or by necrotic or gangrenous tissue. The same is true of any infection of a deep sinus, the pleural cavity, or joints.

Specimens from Biopsy Material

Biopsy specimens are commonly fixed in formalin before they are microscopically examined. Since formalin kills the microorganisms in the tissue, fixed biopsy specimens are useless for culture. The fresh specimen should be divided in half and the unfixed portion sent to the laboratory for microbiological analysis.

Biopsy material may have to be aseptically ground prior to cultivation to release pathogens hidden in the tissue. Because aerosols are inevitably created by this vigorous procedure, tissues should be ground under a contagion hood. Biopsy material should be cultured both aerobically and anaerobically.

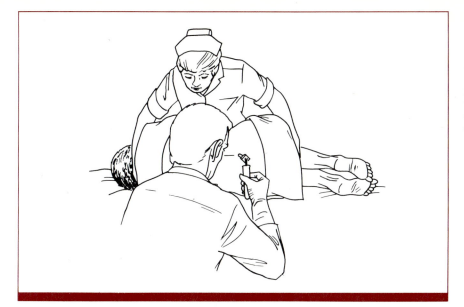

FIGURE 19-2

Collecting cerebrospinal fluid. Position patient with his knees and head flexed at an acute angle to maximize separation of interspinous spaces. Stabilize patient by placing one hand behind head and one behind knees.

Cerebrospinal Fluid (CSF) Specimens

Infectious disease of the central nervous system often causes rapidly progressing irreversible damage or death. Because of the urgency of this situation, delays in microscopic and cultural analysis of CSF should be avoided so that the most effective antimicrobial agents can be determined as quickly as possible.

Normal CSF is sterile and not likely to be contaminated by surface organisms if the skin is decontaminated with an iodophor and if the fluid is properly collected (Fig. 19-2). If Gram-stained smears of the fluid are microscopically examined immediately after collection, a complete morphological description of any microbes in the CSF smears can be returned within minutes to the physician. This information often suggests which antibiotics will most likely inhibit the microscopically observed organisms.

Blood Specimens

Septicemia, bacterial endocarditis, and most systemic infectious diseases can be diagnosed by isolating and identifying the pathogens from the blood. The value of the specimen, however, can be destroyed by inadequate preparation of the skin at the venipuncture site, since many organisms residing on the skin are opportunistic pathogens. Introduction of contaminants into blood cultures produces a microbiological "smoke screen" that diverts diagnostic attention away from the actual pathogens.

Venipuncture sites are most commonly decontaminated by using a combination of alcohol (or acetone) and iodine. Each of these antiseptics should be applied with a circular motion in an outward spiral to move residual microorganisms away from the proposed site of puncture. After

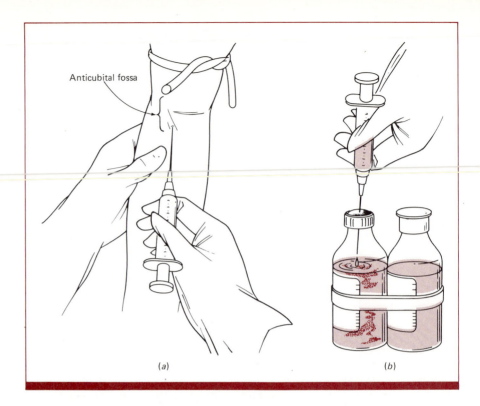

FIGURE 19-3

One method of collecting a blood specimen for culture. *(a)* With a tourniquet in place, prepare the antecubital fossa and aseptically withdraw 10 ml of blood into syringe or vacutainer, a sterile tube containing a vacuum to automatically draw in blood. *(b)* If the specimen is collected in a syringe, carefully change the needle, and inject 5 ml of blood into each of two different bottles, one containing aerobic culture medium, the other anaerobic.

applying a tourniquet, the specimen is drawn according to the procedure depicted in Figure 19-3. Careful preparation of the skin also protects the patient from being inoculated with surface microorganisms.

Enough blood should be withdrawn to inoculate both aerobic and anaerobic culture media. Before inoculating each medium, a fresh sterile needle should replace the one used to penetrate the skin, since it is likely contaminated with residual skin flora. Additional blood samples may be collected for quantitative plate counts, microscopic examination for pathogens, and differential characterization of stained blood cells. Blood specimens are also collected for immunological analysis to detect microbe-specific antibodies (see Chap. 16).

LABORATORY EXAMINATION OF SPECIMENS

Four procedures in the microbiology laboratory are critical to the diagnosis of infectious disease:

Microscopy Preliminary microscopic examination of stained smears or wet mounts of clinical material can provide early indications of the types of pathogens that may be causing the patient's disease. Gram's stains of sputum, CSF, or blood can provide evidence to guide the physician in early selection of antimicrobial agents in urgent situations that require antibiotic therapy soon after the specimen is collected. Gram's stains of bacteria in CSF, for example, can indicate bacterial etiology of suspected meningitis or encephalitis and suggest which antibacterial agents would be most effective. Acid-fast smears can alert the staff to a potential case of tuberculosis. Direct darkfield examination of fluid from suspected syphilis chancres can often reveal the presence of motile spirochetes.

Culture   Diagnosis may be confirmed by isolating, culturing, and identifying suspected pathogens. Quantitative cultures help determine the causative role of the isolated species. Some specimens may yield positive cultural data only after substantial delay. Blood cultures, for example, should be maintained for at least 3 weeks before being discarded as negative.

Antibiograms   The determination of the antibiotic susceptibilities of the isolated microorganisms is as important as species identification of pathogens. The resultant antibiograms are used as guidelines to the selection of effective antibiotic therapy (see Chap. 14).

Serology   Serological tests on patients' sera provide diagnostic information revealing the presence of antibodies that react with the antigens of specific pathogens. This information offers indirect indications of the etiology of disease, often before the pathogen can be isolated and identified. In cases where cultivation of fastidious or delicate pathogens is difficult or impossible (as with the agent of syphilis, for example), serological findings become increasingly important. Serological tests are also used to identify antigens of unknown specificity.

In addition to these procedures valuable epidemiological information gathered from clinical specimens may alert clinicians to impending hospital or community epidemics. This information can be used to facilitate diagnosis, decrease the danger of exposure to the pathogen, and guide the prophylactic administration of antimicrobial or immunological therapy to appropriate individuals. Epidemiological analysis of isolated pathogens often requires epidemiological typing to identify different subspecies of the same isolated pathogens (see Chap. 18).

INTERPRETATION OF LABORATORY RESULTS

Isolation of a microorganism in culture is not indisputable evidence that it is causing a patient's disease. The recovered microorganisms, even a potential pathogen, may be involved as (1) a member of the patient's normal flora; (2) a transient resident contaminating the surface of the patient's body; (3) an environmental contaminant introduced during the process of specimen

FIGURE 19-4

(a) Squamous epithelial cells in sputum indicative of oropharyngeal contamination. (b) Sputum containing abundant neutrophils. Absence of squamous cells indicates proper collection.

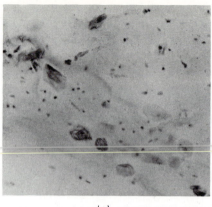 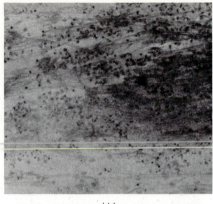

(a)   (b)

collection; or (4) the primary pathogen causing the infectious disease. Treatment will probably fail if any organism in the first three categories is incorrectly identified as the agent responsible for disease.

Several clues help determine the relationship of an isolated organism to the disease of the patient. Specimens from normally sterile body sites that yield a heavy mixture of several different types of bacteria are probably contaminated. Contamination should be suspected if a common member of the patient's microbial flora is isolated when there is no evidence to warrant suspicion of infection by opportunistic pathogens. In a properly collected sputum sample from a patient with pulmonary disease, polymorphonuclear leukocytes should be abundant. If this specimen contains squamous epithelial cells instead of leukocytes, plus a mixture of bacteria normally found in the mouth, the "sputum" should be discarded as saliva, and another specimen should be collected (Fig. 19-4).

OVERVIEW

Many of the changes that occur in an infected person provide clues that aid in disease diagnosis. These signs are often uncovered by laboratory personnel who harvest microbiological and serological information from the patient's tissues and body fluids. These clinical specimens may yield the pathogen itself or reflect immunological and cellular changes that help identify the pathogen. The laboratory's success, however, depends on the quality of specimen collection and handling, usually the responsibility of clinical personnel. Everyone who collects or handles clinical specimens should exercise all safety precautions for protecting people and the specimen itself.

Some specimens warrant special considerations. Material suspected of containing pathogenic viruses or rickettsias should be immediately frozen and transported to specially equipped laboratories. Specimens likely to contain cold-sensitive organisms, such as *Neisseria,* should never be refrigerated. Blood and material from abscesses and deep wounds, should be cultured both aerobically and anaerobically. All specimens should be processed as soon as possible to prevent loss of sensitive pathogens or overgrowth with normal flora microbes.

Respiratory tract specimens are obtained by swabbing the nasopharynx or by collecting sputum. Sterile swabs are also used to obtain specimens from surface lesions in eyes and ears and on skin and mucous membranes. Material from closed lesions, cerebrospinal fluid, or blood is withdrawn with a sterile needle and syringe. Urine and feces are collected directly. Biopsy specimens are obtained by surgically excising tissue.

In the laboratory, etiological agents are isolated and identified, antibiotic susceptibilities of the isolated pathogen are ascertained, and the patient's immunological and cellular responses are determined.

KEY WORDS

clinical specimen

biopsy

invasive procedure

sputum

midstream-catch technique

REVIEW QUESTIONS

1. What type(s) of specimen should be collected for:
   (a) Sore throat
   (b) Lower respiratory tract infection
   (c) Gastrointestinal tract infection
   (d) Urinary tract infection
   (e) Central nervous system infection

2. Describe any precautions needed to protect the value of the following specimens.
   (a) Sputum
   (b) Male urethral discharge
   (c) Urine
   (d) Blood
   (e) Biopsy material

3. What precautions should be taken to protect the health of (a) the patient and (b) the laboratory personnel during the collection and processing of clinical specimens?

4. How can a delay in culturing a specimen affect the accuracy of laboratory results?

5. Why shouldn't a CSF specimen be refrigerated?

6. Discuss three observations that suggest an organism is a contaminant in the specimen rather than the etiological agent.

CLINICAL SPECIMENS
FOR MICROBIAL AND
SEROLOGICAL ANALYSIS

Diseases Acquired through the Respiratory Tract

20

ANATOMY AND DEFENSES OF THE RESPIRATORY TRACT

PREDISPOSING FACTORS

EPIDEMIOLOGY AND DYNAMICS OF TRANSMISSION
Human Reservoirs
Environmental Reservoirs
Animal Reservoirs

RESPIRATORY-ACQUIRED DISEASES
Upper Respiratory Diseases
Common Cold (Viral) / Streptococcal Pharyngitis/Tonsillitis / Nonbacterial Sore Throat / Epiglottitis (Viral) / Adenoid Viral Infections / Diphtheria (Bacterial) / Systemic Infections / Infections of Accessory Respiratory Structures
Lower Respiratory Diseases
Pneumonia / Influenza / Pertussis (Bacterial) / Tuberculosis / Systemic Mycoses

PREVENTION AND CONTROL
Interfering with Disease Transmission
Reducing Population Susceptibility
Reducing the Reservoirs of Infection

OVERVIEW

Throughout history people have feared the air, believing that most human diseases were caused by breathing invisible "malignant vapours." Malaria, for example, literally means "bad air," because it was thought to be due to some insidious characteristic of the night air. Other diseases such as influenza were attributed to the adverse "influence" of breathing winter air. No air was above suspicion.

Although proponents of these myths appear superstitious, they were, for the most part, correct in their naive explanations. Air does in fact represent a significant morbid potential, for most infectious diseases are acquired by inhaling airborne pathogens. Furthermore, since cold, dry air may decrease the efficiency of respiratory defenses, night and winter air can contribute to the acquisition of disease.

Respiratory-acquired diseases are the most common illnesses for which people consult their physician. Pathogens of tuberculosis, meningitis, and many other serious diseases gain entrance to the body through the respiratory tract. Nosocomial diseases are commonly transmitted by this route, often following introduction of pathogens during inhalation therapy by tracheal intubation. Children still suffer from outbreaks of childhood diseases, especially chickenpox. Pandemics of influenza continue to strike millions of people. And modern medicine has yet to devise a cure for the common cold.

ANATOMY AND DEFENSES OF THE RESPIRATORY TRACT

The respiratory tract is divided into two systems, the upper respiratory tract (URT) and the lower respiratory tract (LRT) (Fig. 20-1). The opening between the two systems is called the glottis. The eyes and middle ears are physically connected to the upper respiratory tract by the nasolacrimal ducts and auditory canals and are considered accessory upper respiratory structures. Pathogens can enter the respiratory system from eyes or ears.

The warm, moist surfaces of the human respiratory tract provide ideal conditions for the growth of pathogens. Each person's daily intake of 11,500 liters of air brings numerous airborne pathogens into the lower airways and lungs. The average person inhales at least eight microorganisms each minute, about 10,000 per day. The huge surface area inside the lungs (1000 square feet, about 20 times the surface area of the skin) would be readily colonized and infected if it were unprotected. Fortunately, mechanical defenses of the respiratory tract cleanse the air and prevent deep microbial penetration. Chemical and cellular defenses further diminish the likelihood of infectious diseases (Table 20-1).

The upper respiratory system is abundantly populated with microorganisms (Table 20-2). Most are harmless members of the normal flora that contribute to the respiratory defenses by bacterial competition. Some of these organisms, however, are opportunistic pathogens that can cause disease if predisposing factors compromise resistance or if they become

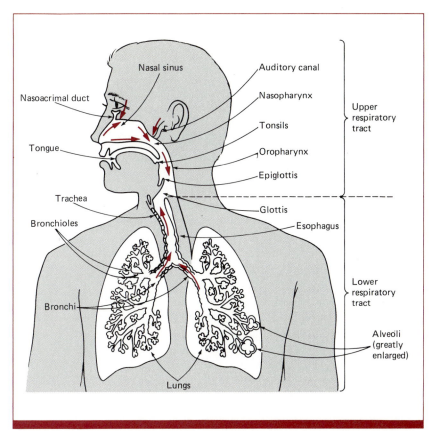

Nasal sinus

Auditory canal

Nasolacrimal duct

Nasopharynx

Tonsils

Tongue

Oropharynx

Epiglottis

Trachea

Glottis

Bronchioles

Esophagus

Bronchi

Alveoli (greatly enlarged)

Lungs

FIGURE 20-1

Anatomy of the respiratory tract (the airways are uncolored). The opening between URT and LRT is called the glottis. During swallowing it is covered by the epiglottis to prevent food from entering the LRT. Ciliated mucosa traps particles in a blanket of mobile mucus and directs them to the oropharynx. The trapped particles are disposed of by swallowing.

TABLE 20-1

Respiratory Defenses

Type	Responsible Factors*	Protective Action
Mechanical	Epiglottis	Prevents particles from entering LRT during swallowing
	Ciliated mucosa	Traps particles in mucus, which is propelled out of respiratory tract
	Nasal hairs	Tumbles inhaled air, increasing chances of trapping particles in mucus
	Neurological sensitivity of nasal mucosa	Discharges particles from URT through sneezing
	Neurological sensitivity of trachea	Discharges particles from LRT through coughing
Chemical	Interferon	Protects uninfected host cells against viral infection
	Lysozyme	Dissolves peptidoglycan in cell walls of many bacteria
Microbial	Normal flora	Compete with pathogens of URT (no normal flora in LRT)
Cellular	Alveolar macrophages (dust cells)	Phagocytize microbes that reach the alveoli
Immuno-logical	Secretory immuno-globulins	Neutralize pathogens at respiratory surfaces
	T lymphocytes	Enhance killing of pathogens

*See Chapters 15 and 16 for a discussion of each factor.

Anatomical Site	Most Common Organisms	Microscopic Characteristics
External ear (URT)	Mycobacterium sp.	Gram-positive acid-fast rods
	Corynebacterium sp. (diphtheroids)	Gram-positive pleomorphic rods
	Staphylococcus aureus	Gram-positive cocci, clusters
	Staphylococcus epidermidis	Gram-positive cocci, clusters
Eye (conjunctiva) (URT)	Branhamella catarrhalis	Gram-negative pleomorphic rods
	Haemophilus sp.	Gram-negative pleomorphic rods
	Corynebacterium sp. (diphtheroids)	Gram-positive pleomorphic rods
	Neisseria sp.	Gram-negative diplococci
	Staphylococcus epidermidis	Gram-positive cocci, clusters
	Viridans streptococci	Gram-positive cocci, chains
Nose and throat (URT)	Staphylococcus aureus	Gram-positive cocci, clusters
	Staphylococcus epidermidis	Gram-positive cocci, clusters
	Viridans streptococci	Gram-positive cocci, chains
	Streptococcus pneumoniae	Gram-positive diplococci
	Branhamella catarrhalis	Gram-negative diplococci
	Corynebacterium sp. (diphtheroids)	Gram-positive pleomorphic rods
	Haemophilus sp.	Gram-negative pleomorphic rods
	Actinomyces sp.	Gram-positive rods
	Bacteroides sp.	Gram-negative pleomorphic rods
Trachea, bronchi, bronchioles, and lungs (LRT)	No flora indigenous to this region	

entrenched in the lower airways. Fortunately, the upper respiratory defenses are usually efficient enough to prevent the introduction of pathogens into the lower tract, where infectious diseases commonly follow life-threatening courses. The lungs have no ciliated epithelium and must be defended to prevent pneumonia and other dangerous pulmonary diseases. The occasional microorganisms that reach the alveoli, the terminal portions of the respiratory tract where gas exchange occurs, are normally disposed of by **alveolar macrophages** (Fig. 20-2). Additional protection is provided by interferon, humoral antibodies, and T lymphocytes. The combined efficiency of respiratory defenses is evidenced by the sterile conditions that exist in healthy human lungs. In persons with compromised defenses, however, these fortifications may be incapable of preventing establishment of pathogens in the lower respiratory tract.

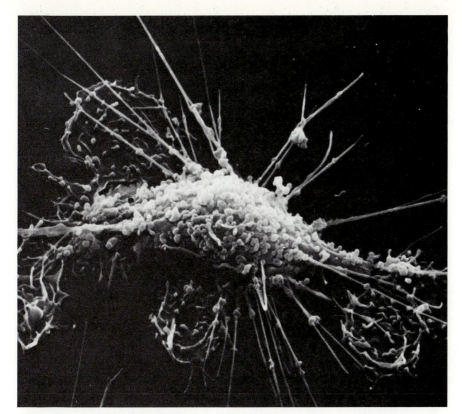

PREDISPOSING FACTORS

Many factors may predispose a person to develop respiratory disease (Table 20-3). These factors—smoking for example—injure or decrease the efficiency of the host respiratory defenses or bypass them altogether (see color box). Respiratory-acquired disease is also encouraged by factors that compromise resistance to infectious diseases in general: extremes in age, chronic stress, hormonal imbalances (such as corticosteroid excesses or deficiencies), diabetes mellitus, immunological deficiencies, disorders of cellular or phagocytic mechanisms, immunosuppressive therapy to accommodate a transplanted organ, anticancer chemotherapy, or sever malnutrition. Persons with these high-risk factors should be closely attended until corrective measures reverse their predisposition.

Not all cases of respiratory-acquired disease result from predisposing factors. Healthy individuals occasionally contract mild or even life-threatening diseases if the inoculum is large enough or if the pathogen is sufficiently virulent to overwhelm host defenses. *Yersinia pestis* (pneumonic plague) and measles virus are two examples of virulent pathogens that can cause disease in an otherwise healthy person.

Factor	Impaired Defense	Site Affected	Corrective Measure
Tobacco smoking	Ciliated mucosa, phagocytosis	LRT*	Stop smoking.
Alcoholism	Cough reflex, phago-cytosis, ciliated mucosa.	LRT	Reduce alcohol consumption.
Decreased humidity	Ciliated mucosa	URT†	Humidify the air.
Chill	Ciliated mucosa	URT	Avoid chilling.
Viral infection	Ciliated mucosa, phagocytosis	URT and LRT	Carefully monitor symptoms; use anti-biotics if culture confirms secondary bacterial infection.
Antibiotic therapy	Upper respiratory flora disrupted and bacterial competition reduced	URT	Use narrow-spectrum antibiotics; rely on antibiograms for antibiotic selection.
Certain airborne pollutants such as asbestos and silicon	Phagocytosis	LRT	Change environment.
Allergic rhinitis (hay fever)	Ciliated mucosa	URT	Avoid allergens; de-sensitize to allergins.
Invasive medical procedures, e.g., respiratory therapy and tracheotomy	All physical defenses bypassed	LRT	Use sterile equipment and aseptic procedures.
General anesthesia	Ciliated mucosa	URT	Monitor carefully.
Obstructive pulmonary diseases	Airways	LRT	Correct antecedent condition if possible.
Tonsillar inflammation	Drainage of auditory canal	Middle ear	Remove chronically inflamed tonsils.
Aspiration of foreign material	Ciliated mucosa; large inoculum introduced into LRT	LRT	Avoid factors that promote aspiration.
Acquired immune-deficiency syndrome (AIDS)	T lymphocytes	LRT	Avoid potential sources when possible.

*Lower respiratory tract.
†Upper respiratory tract.

EPIDEMIOLOGY AND DYNAMICS OF TRANSMISSION

Most airborne human pathogens are acquired from infected people or from the inanimate environment. Infected animals are the reservoirs of only a few respiratory-acquired diseases of humans (Fig. 20-3).

DYING FOR A CIGARETTE

■ Cigarette smoking has been unquestionably linked to fatal respiratory disease. Seventy percent of all persons who die of respiratory disease are smokers. Although many of these people die of lung cancer or emphysema, infectious pulmonary disease (especially pneumonia) commonly develops because of the chronic changes that accompany long-term inhalation of smoke. The following cigarette-induced physiological changes contribute to high mortality among smokers:

Chronic Cough Smoking irritates the membranes of the larynx and trachea, triggering nonspecific vigorous coughing. The violent inhalations that accompany coughing spasms encourage aspiration of microbe-laden droplets and particles, which inoculate the lower airways with opportunistic pathogens.

Increased Mucus Production Inhaled smoke stimulates respiratory mucous glands to overproduce mucus. So much mucus accumulates along the respiratory epithelium, it can no longer be elevated by the action of cilia. Instead of being carried out of the respiratory tract, trapped particles descend by gravity deep into the lower airways and lung.

Saturation of Dust Cells Smoke is an airborne suspension of carbon particles. Each lungful of inhaled smoke coats the alveoli with these particles, which are phagocytized by protective alveolar macrophages (dust cells). Unfortunately, there are a finite number of dust cells available, and these can be rendered ineffective by diverting their activities to the task of engulfing an overwhelming number of smoke particles. A few opportunistic pathogens inevitably escape destruction by this saturated line of defense.

Paralysis of Ciliated Mucosa Nicotine has been associated with temporary loss of ciliary function along the respiratory tract. The mobility of the mucous blanket is eventually recovered in most persons after they quit smoking.

Human Reservoirs

Person-to-person transmission by contaminated respiratory droplets or droplet nuclei is the most common mechanism for transmitting infectious disease. Activities that promote the production of airborne **respiratory droplets** include coughing, sneezing, talking, singing, and spitting. Kissing may also transmit pathogens into the respiratory tract, although no airborne vehicles are involved. Disease transmission primarily occurs indoors, where crowding and close human interaction promote contact with airborne respiratory secretions. Schools, military bases, hospitals, and other environments and institutions with high indoor population densities have the highest incidences of these diseases.

The most efficient vehicles of transmission are tiny **droplet nuclei** formed from the evaporation of small respiratory droplets before they settle out. Droplet nuclei are more dangerous than the respiratory droplets from which they are formed because desiccation often prolongs microbial viability. Many airborne pathogens survive for longer periods of time between hosts if suspended in dry droplet nuclei rather than in moist

FIGURE 20-3

Transmission of respiratory
acquired diseases: From
human reservoirs by
respiratory discharges and
lesion exudates; from
endogenous human
reservoirs by
autoinoculation; from
animal reservoirs by
aerosols of dander, dried
feces, or respiratory
discharges; from
environmental reservoirs
by dust or equipment used
in invasive medical
procedures.

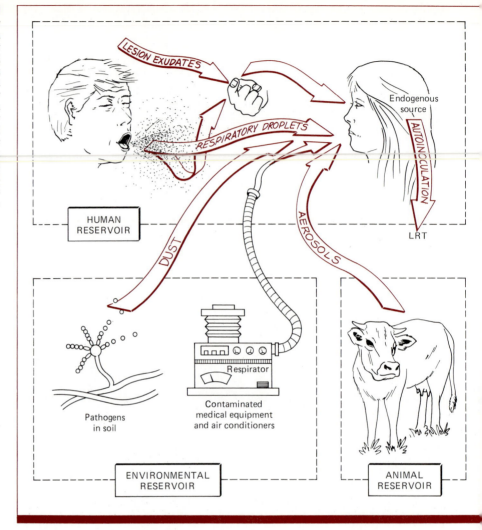

respiratory droplets. In addition, droplet nuclei are the ideal size (5 μm or less) for deep penetration into the respiratory tract. Larger particles are more readily impeded by the mechanical defenses of the upper respiratory tract. The lower humidity and temperatures of winter favor rapid evaporation of respiration secretions into droplet nuclei. Winter conditions also reduce efficiency of respiratory defenses because of drying of the mucous membranes and because of increased crowding of people indoors. The seasonal nature of acute respiratory disease may be explained by the combination of these factors.

Respiratory secretions discharged by a single cough or sneeze from an infected person can introduce up to 1000 infectious doses of pathogens into the air and onto table surfaces, eating utensils, bed linen, clothing, or fingers. Direct contact with fingers and fomites contaminated by these secretions may be a significant factor in disease transmission. Microbes from these objects often enter the respiratory tract through nasolacrimal ducts of the eyes. The traditional oronasal mask offers no protection against respiratory pathogens entering through this route.

Exudates of skin and mucous membrane lesions are another important source of respiratory pathogens. Smallpox viruses, for example, are shed from lesions, even though the portal of entry is the respiratory tract. Some pathogens invade lower airways following autoinoculation. The indigenous microbes of the upper respiratory flora may be transferred to the trachea and bronchi by invasive medical techniques, usually intubation or bronchioscopy (examining the pulmonary area with an optical device passed into the lower tract airways).

Environmental Reservoirs

Many microorganisms that live on inanimate organic matter in the environment are capable of causing human disease if inhaled. Unlike disease acquired from the human reservoir, these infectious disorders are rarely contagious. The pathogens are generally carried on dust, and the incidence of these infections is often related to the amount of dust in the air. This is especially true in the hospital, where many environmental airborne pathogens are spread by dust.

Contaminated water is another important environmental reservoir of respiratory-acquired infections that endangers hospital patients and laboratory personnel. Contaminated humidifiers, nebulizers, and air conditioning units may create infectious aerosols, as can respirators, with moist chambers that provide ideal growth conditions for many types of bacteria.

Animal Reservoirs

Occupational exposure to animals increases the incidence of respiratory zoonoses. Slaughterhouse employees, for example, are at a higher risk of contracting Q fever or pulmonary anthrax by breathing the dander of infected cattle. Some diseases are acquired by inhaling dust contaminated with dried feces from infected animals. Caves with large populations of bats, for example, are hazardous to spelunkers (cave explorers), who may acquire rabies by inhaling dried feces of rabid bats. The resulting cases are invariably fatal.

RESPIRATORY-ACQUIRED DISEASES

Pathogens entering the body through the respiratory tract generally cause localized infectious diseases. These noninvasive infections of the upper respiratory tract are the most frequently occurring human illnesses. They are usually self-limiting diseases caused by viruses and require only symptomatic treatment to relieve discomfort. Some localized infections, however, may injure secondary, nonrespiratory body sites by the action of hematogenously spread exotoxins or by immunologically mediated sequelae (see Chap. 17). Other pathogens may directly extend to the lungs or central nervous system or be spread by hematogenous or lymphatic routes.

Upper Respiratory Diseases

Ninety percent of infectious diseases of the upper respiratory tract are caused by viruses and are unresponsive to antibiotics. A few upper respiratory infections, however, are bacterial in nature and can be effectively treated with appropriate antibiotics. Proper treatment of upper respiratory diseases depends on determining whether the etiological agent is a virus, a bacterium, a fungus, or a protozoan. A summary of infectious diseases introduced through the upper respiratory tract is provided in Tables 20-4 and 20-5.

Common Cold (Viral)   The common cold (also called *coryza*) is probably the most frequent infectious human disease. Except for malaise (discomfort or uneasiness), the symptoms of a cold are entirely upper respiratory. The person suffering from a cold experiences watery eyes, nasal discharge, sneezing, swollen nasal membranes, and occasionally pharyngitis. Cold symptoms do *not* include coughing, which is indicative of lower tract involvement, nor systemic symptoms such as fever.

Most colds are caused by a group of RNA viruses called *rhinoviruses* (*rhino*, nose). These agents infect and replicate in the respiratory mucosa and trigger the symptoms associated with the cold syndrome. Because they replicate at 33°C, they are usually restricted to the cool surfaces of the upper respiratory tract. The combination of nasal discharge and sneezing facilitates person-to-person spread of the disease by showering the environment with virus-laden respiratory droplets. People are most infectious during the first 2 days of illness, the time when symptoms are most acute.

Colds are usually self-limiting infections that resolve in several days following onset of symptoms. Recovery is probably due to a production of interferon and specific secretory IgA immunoglobulin. Immunity may last for 2 years, but only protects against the same antigenic type of virus that caused the infection. More than 120 antigenically distinct cold viruses have been identified. A person would have to catch at least 120 colds, each with a different etiology, to acquire temporary immunity to all these viruses. Even then, one would still be susceptible to occasional colds caused by coronaviruses, enteroviruses, and other nonrhinoviruses.

Because of their viral etiology, colds cannot be effectively treated with antibiotics. Infrequently, secondary bacterial infection follows a cold because of decreased host resistance induced by the viral respiratory infection. This infrequent occurrence, however, does not justify the prophylactic use of antibiotics in persons with simple colds.

Streptococcal Pharyngitis/Tonsillitis (Bacterial)   One of the most frequently mismanaged of all infectious diseases is acute pharyngitis or tonsillitis, the sore throat. Although the overwhelming majority of sore throats are caused by viruses, streptococcal sore throat ("strep throat") is the second most commonly *reported* infectious disease in the United States.

Strep throat is caused by a bacterium, *Streptococcus pyogenes*, also called group A beta-hemolytic streptococcus (see color box). Symptoms of infection include fever, reddening and swelling of the mucous membranes

TABLE 20-4

Summary of Bacterial Diseases That Follow Introduction of Pathogen into the Upper Respiratory Tract and Accessory Structures

Disease	Pathogen(s)	Incubation Period	Primary Attack Site	Secondary Attack Site	Reservoir	Complications or Sequelae	Diagnostic Specimens	Diagnostic Procedures	Treatment
Strep throat	*Streptococcus pyogenes* (group A, beta-hemolytic streptococci)	1–3 days	Pharynx, tonsils	Heart valves kidney, skin, joints	Infected people	Scarlet fever, rheumatic heart disease, glomerulonephritis	Throat swab, blood	Smear, isolation of pathogen, serology	Penicillin
Diphtheria	*Corynebacterium diphtheriae*	2–5 days	Pharynx	Toxin affects heart, kidneys, liver, nerves	Infected people	—	Throat swabs	Smear, isolation of pathogen	Antitoxin, penicillin, erythromycin, tetracyclines
Otitis media	*Streptococcus pneumoniae*, *Haemophilus influenzae*, *Streptococcus pyogenes*, *Staphylococcus aureus*	Variable	Middle ear (may extend from sore throat)	Central nervous system	Infected people	Meningitis	Discharge from infected site, nasopharyngeal swabs	Smear, isolation of pathogen	Penicillin, erythromycin
Myringitis	*Mycoplasma pneumoniae*	—	Tympanic membrane	—	Infected people	Rare hearing loss due to secondary bacterial infection	Swab of discharge from site	Isolation of pathogen, serology	Unnecessary
Sinusitis	Several bacteria, especially streptococci, *Haemophilus*, and *Bacteroides*	Variable	Nasal sinuses	Eyes, facial bones	Infected people	Orbital abscess, osteomyelitis	Swab of discharge from site	Smear, isolation of pathogen	Chemotherapy, depends on pathogen
Conjunctivitis	*Staphylococcus aureus*, *Haemophilus aegypticus*	1–3 days	Conjunctiva	—	Infected people	—	Swab of infected conjunctiva	Smear, isolation of pathogen	Topical tetracyclines
Meningitis	*Neisseria meningitidis*, *Haemophilus influenzae*	3–4 days	Throat or middle ear	Meninges	Infected people	Neurological impairment, death	Spinal fluid	Smear, isolation of pathogen from spinal fluid or blood	Penicillin, chemoprophylaxis for contacts
Trachoma	*Chlamydia trachomatis*	5–12 days	Conjunctiva	Cornea	Infected people	Secondary bacterial infection, blindness	Scrapings of conjunctiva	Direct microscopy, isolation of pathogen	Tetracycline, sulfonamide
Inclusion conjunctivitis	*Chlamydia trachomatis*	5–14 days	Conjunctiva	—	Infected people, contaminated swimming water	—	Scrapings of conjunctiva	Direct microscopy, isolation of pathogen	Tetracycline, sulfonamide

TABLE 20-5

Summary of **Nonbacterial** Diseases that Follow Introduction of Pathogen Through the Upper Respiratory Tract

Disease	Pathogen(s)	Incubation Period	Primary Attack Site	Secondary Attack Site	Reservoir	Complications or Sequelae	Diagnostic Specimens	Diagnositic Procedures	Treatment
Common cold	Rhinoviruses, adenoviruses, some enteroviruses	1–3 days	Membranes of nose and throat	—	Infected people	Secondary bacterial infections	Nose or throat swabs, blood	Isolation of virus, serology	Symptomatic treatment, no specific antiviral treatment
Chickenpox	Herpesvirus varicella	2–3 weeks	Membranes of the URT	Skin and nerves	Infected people	Zoster (shingles), Reye's syndrome	Throat washings, blood	Isolation of virus, serology	No specific antiviral treatment
Measles (rubeola)	Measles virus	11 days	Membranes of the URT	Lymphoid tissue, skin	Infected people	Subacute sclerosing panencephalitis	Nasopharyngeal washings, blood	Isolation of virus, serology	No specific antiviral treatment
Mumps	Mumps virus	18–21 days	Membranes of the URT	Parotid glands, testes, ovaries, pancreas, thyroid	Infected people	Sterility, meningoencephalitis	Blood	Isolation of virus, serology	No specific antiviral treatment
Rubella (German measles)	Rubella virus	14–18 days	Membranes of the URT	Skin and lymphatics	Infected people	Damage to developing fetus	Blood, throat washings, urine	Serology, isolation of virus	No specific antiviral treatment
Smallpox	Variola virus	12–16 days	Membranes of the URT	Skin, liver, spleen, lungs	Infected people	—	Exudate from skin lesions, blood	Isolation of virus, serology, microscopic exam for Guarnieri bodies	No specific antiviral treatment
Infectious mononucleosis	Epstein-Barr virus	2–6 weeks	Membranes of the URT	Lymphoid tissue, spleen, liver	Infected people	Ruptured spleen, jaundice; Burkitt's lymphoma	Blood	Serology	No specific antiviral treatment
Amebic meningo-encephalitis	Naegleria sp.	3–7 days	Probably nasal mucosa	Brain and meninges	Water and soil	—	Spinal fluid	Direct microscopic exam	Amphotericin B with miconazole and rifampin

BRINGING ORDER TO THE CHAOTIC STREPTOCOCCI

■ Many species of *Streptococcus* inhabit the bodies of humans and a variety of other animals. Some of these, such as the viridans streptococci, are usually harmless members of the normal flora. Others, especially *Streptococcus pyogenes*, are dangerous pathogens. A system of distinguishing between the many groups of streptococci is vital to clinical practitioners and epidemiologists.

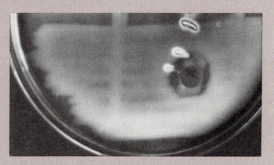

The streptococci are divided into three groups according to their lytic effects on red blood cells in blood agar. This effect, called hemolysis (Chap. 6), is due to the production of soluble extracellular substances, hemolysins, that diffuse into the agar and lyse the blood cells. Hemolysis is seen in two characteristic patterns. *Alpha hemolysis* appears as a greenish brown zone around the colony. This color change is due to the reduction of normally red hemoglobin to an unknown product. *Beta hemolysis*, on the other hand, is the complete destruction of red blood cells around the colony, producing a characteristic clear zone with no remaining color. Beta hemolysis is best observed when the blood agar culture is incubated anaerobically, because one of the chemical hemolysins functions poorly in the presence of oxygen. This hemolysin, called streptolysin O, in concert with a second hemolysin, streptolysin S, produces beta hemolysis. Most normal flora (viridans) streptococci are alpha hemolytic or produce no hemolysis at all. The dangerous pathogen, *Streptococcus pyogenes*, is beta hemolytic.

Hemolysis alone, however, must be supplemented with additional information since not all beta-hemolytic streptococci are *S. pyogenes*. Furthermore, some alpha-hemolytic streptococci (*S. pneumoniae*, for example) are virulent pathogens and must be distinguished from the normal flora. An additional system of classification was devised by Rebecca Lancefield. The Lancefield scheme divides the streptococci into immunological groups according to different antigenic specificities of the C carbohydrate, a major surface antigen. Thirteen major groups of streptococci—groups A through O (there are no groups I or J)—have been described. Although *S. pneumoniae* and the viridans streptococci possess no group carbohydrate, the Lancefield system is valuable for identifying other important pathogens, including *Streptococcus pyogenes*, often referred to as the group A beta-hemolytic streptococci.

Many clinical laboratories are not equipped to determine antigenic groups and rely on *bacitracin sensitivity* to distinguish *S. pyogenes* from other beta-hemolytic streptococci. A paper disc impregnated with the antibiotic bacitracin releases the antibiotic into the surrounding medium when placed on blood agar seeded with the organism. Group A streptococci are sensitive to low doses of bacitracin that fail to inhibit other streptococcal groups. Thus beta-hemolytic streptococci that fail to grow around the bacitracin discs on blood agar are identified as *S. pyogenes* (see photo).

Diagnosis of group A beta-hemolytic streptococcal infection may also be facilitated by detecting antibodies specific for the hemolysins. Streptolysin O is antigenic, and persons with *S. pyogenes* infections develop peak antibody titers against the hemolysin about 2 weeks after initial infection. These antistreptolysin O (ASO) titers are useful in diagnosing poststreptococcal sequelae (rheumatic fever and glomerulonephritis) that develop after the initial streptococcal infection is eradicated.

(sometimes with abscess formation), and enlargement of cervical lymph nodes. Symptoms are generally more severe in children, but usually disappear spontaneously after about a week. Occasionally, however, patients with strep throat develop dangerous complications. These complications are diagrammed in Figure 20-4. Prevention of rheumatic heart disease is the major consideration in proper management of streptococcal sore throat. Effective antibiotics, usually penicillin, administered within the first 7 days of symptoms, prevent such dangerous sequelae. The goal of penicillin therapy is *not* to reduce the symptoms of pharyngitis, which will spontaneously subside with or without antibiotic therapy. Antimicrobial management of streptococcal sore throat is aimed entirely at preventing poststreptococcal sequelae—rheumatic fever and glomerulonephritis.

Rheumatic fever is potentially fatal, and the heart valves of survivors may be permanently damaged. These persons are extremely susceptible to relapses of rheumatic fever with each subsequent *Streptococcus pyogenes* infection. They are also highly susceptible to colonization of the heart valves by oral flora microbes (especially alpha-hemolytic streptococci) that reach the damaged valves following oral trauma—tooth extraction, for example. Persons with diagnosed rheumatic heart disease are usually treated for years with large doses of penicillin to prevent recurrences of streptococcal sore throats that could trigger a fatal attack of the sequela. The gravity of these potential complications emphasizes the need for rapid and accurate identification of *Streptococcus pyogenes* from infected throats.

Nonbacterial Sore Throat   Most sore throats are caused by viruses and are usually characterized by mild, localized symptoms, general discomfort, constitutional symptoms, and few serious complications. Many of the viruses infecting the upper respiratory tract, however, can partially paralyze the mucociliary defenses throughout the lower respiratory tract and reduce bronchiolar phagocytic action. Acute pneumonia, therefore, may evolve from a mild viral infection of the upper respiratory tract. In addition, viruses may spread from an infected throat to the epiglottis, nasal sinuses, and accessory respiratory structures, the eyes and middle ears.

Epiglottitis (Viral)   Acute epiglottitis is characterized by painful swelling and reddening of the epiglottis. This disease is especially dangerous to children, whose small airways are easily closed by a swollen epiglottis. Without immediate measures to provide air to the lungs, a child may asphyxiate. Epiglottitis is usually associated with a *croup syndrome* (acute laryngotracheobronchitis), that is, inflammation of the larynx, trachea, and bronchi. Croup is commonly caused by a group of RNA viruses called parainfluenza viruses. These agents are also associated with severe pneumonia or bronchitis of children. Parainfluenza infection of adults, however, usually produces mild coldlike symptoms, perhaps because the partial immunity acquired by previous infections limits severity.

Adenoid Viral Infections   Adenoids are cervical lymphoid tissue that,

DISEASES ACQUIRED
THROUGH THE
RESPIRATORY TRACT

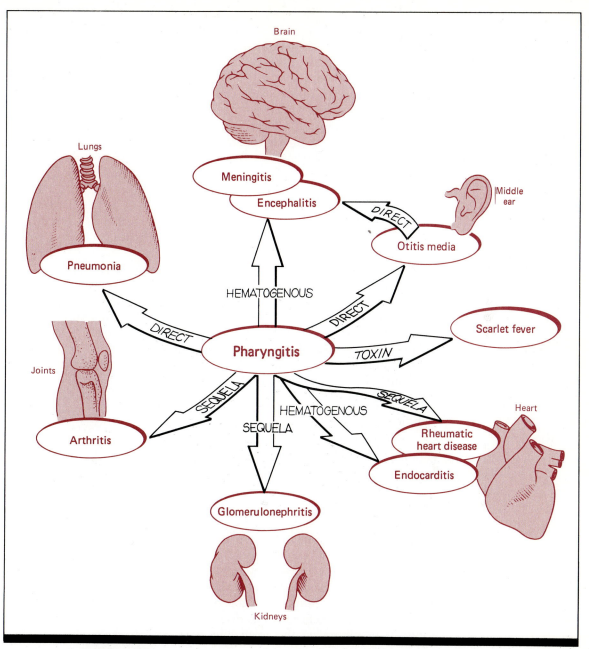

FIGURE 20-4

Possible complications following sore throat (pharyngitis/tonsillitis) caused by group A beta-hemolytic streptococci *(Streptococcus pyogenes)*. (1) Scarlet fever—systemic disease due to the effects of a soluble exotoxin produced by some strains of *Streptococcus pyogenes*. (2) Rheumatic fever—poststreptococcal sequela that usually damages heart valves and joints (arthritis) (3) Glomerulonephritis—poststreptococcal sequela that impairs kidney function. (4) Endocarditis—inflammation of the endocardium, the membranous lining of the heart. (5) Meningitis—inflammation of the meninges, the membranous covering of the brain. (6) Encephalitis—inflammation of the brain. (7) Otitis media—inflammation of the middle ear. (8) Pneumonia—inflammation of the bronchial or alveolar spaces or supporting interestitial tissue.

like the tonsils, may become enlarged as the result of viral infection. Such adenoidal tissue may remain chronically inflamed for years and predispose for secondary infection of adjacent anatomical sites. Removal of chronically inflamed tonsils and adenoids may reduce the likelihood that these complications will develop, but is a controversial approach. Because of their protective function, tonsils and adenoids are removed only when there is no alternative therapy.

The agents most commonly isolated from inflamed tonsils are the *adenoviruses*, a group of double-stranded DNA viruses that have been implicated in several upper respiratory and ocular diseases, as well as in severe pneumonia in children. Adenovirus infection of the pharynx resembles the symptoms of infectious mononucleosis and strep throat. When blood tests rule out infectious mononucleosis (see discussion later in this chapter) and throat cultures eliminate the possibility of streptococcal infection, the symptoms are likely due to adenoviruses.

Diphtheria (Bacterial) Because of successful immunization programs, diphtheria is not the uncontrolled killer it once was. Each year, however, diphtheria continues to claim some lives in the United States and Britain, and it is a more serious problem in developing countries. The disease is a localized throat infection caused by toxigenic strains of *Corynebacterium diphtheriae*. These strains produce a soluble extracellular toxin (diphtheria toxin). The infected pharynx, tonsils, and nasal region often appear to be covered with a "pseudomembrane" composed of a network of fibrin, bacteria, and white blood cells (Fig. 20-5). These local effects, however, are usually not severe enough to endanger the patient, although sometimes airway obstruction may develop. The systemic tissue injury characteristic of diphtheria is caused solely by the exotoxin. The targets most profoundly affected by the toxin are the nerves, heart, liver, and kidneys. Subsequent damage to these tissues, if untreated, may be fatal. Toxin neutralization by antibodies (antitoxin) promotes recovery and prevents future cases of diphtheria as long as antibodies or immunological memory persist.

Corynebacterium diphtheriae is a gram-positive pleomorphic rod that is transmitted by respiratory droplets from its human reservoir of infection. The few cases that occur in the United States are usually among children who failed to receive the trivalent DPT vaccine, which stimulates active immunity against diphtheria exotoxin (D), pertussis (P), and tetanus exotoxin (T). Treatment of active disease is by passive immunization with antitoxin to neutralize the toxin and with antibiotics to eliminate the infectious agent.

Systemic Infections The spread of pathogens to secondary tissues following initial upper respiratory infection is typical of chickenpox, measles, mumps, rubella, smallpox, and infectious mononucleosis. These virus diseases are highly communicable and are usually acquired during childhood or young adulthood. Except for smallpox, they are typically

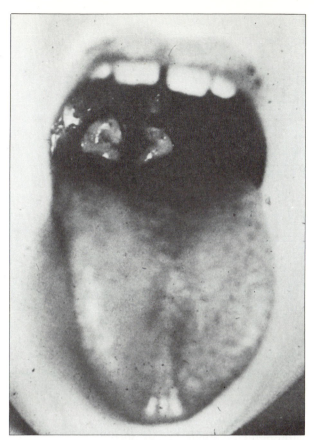

FIGURE 20-5
Grayish pseudomembrane
covering the oropharynx,
tonsils, and nasopharynx
of patient with diphtheria.

self-limited diseases and require no therapy. Occasional complications, however, may have serious or fatal outcomes.

CHICKENPOX (VIRAL) The *varicella virus* is a herpesvirus that usually causes a mild disease in children; the infection is characterized by fever and a rash. The virus is carried by the bloodstream from the infected respiratory tract to the skin, where superficial lesions appear. These lesions evolve into fluid-filled blisters which rupture, form crusts, and eventually heal without scarring. Although most cases resolve uneventfully, chickenpox can be fatal to immunologically compromised persons and babies of mothers who develop the disease shortly before delivery. Chickenpox is more severe in adults than in children. As many as one in five adult victims die from the disease. Fortunately, the infection is usually contracted by young persons. Because there is only a single antigenic type of varicella virus, immunity against chickenpox is acquired following the initial case.

Natural immunity appears to prevent subsequent episodes of chickenpox but may fail to eradicate the virus. The virus presumably persists in nerves and, years after recovery from chickenpox, may reactivate, migrate down the nerve axons and cause painful lesions at the sites where sensory nerves join the skin. This disease is called *zoster*, or shingles (Fig. 20-6).

FIGURE 20-6

Zoster (shingles) lesions
erupt at the junctures of
skin and sensory neurons.
The disease may be quite
painful.

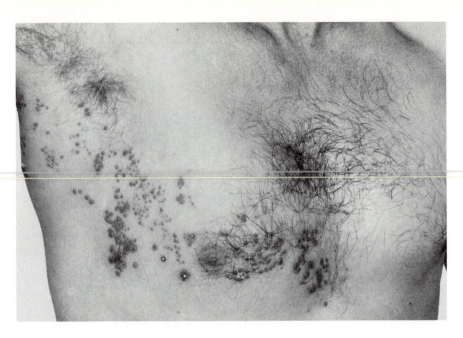

DISEASES ACQUIRED
THROUGH THE
RESPIRATORY TRACT

Although caused by the same virus, chickenpox and zoster are completely
different diseases. Zoster, which is considerably more painful than
chickenpox, only occurs in persons who have had chickenpox earlier in
life. Children exposed to viruses shed by persons with zoster develop
typical chickenpox, not zoster.

Scrapings from varicella lesions can be stained and examined microscopically. The virus is too small to be seen in the light microscope, but its
effects can often be observed. For example, varicella infections cause host
cells to fuse with each other, producing large multinucleated cells. Sometimes developing viruses aggregate in observable intranuclear inclusion
bodies. Viral-specific antigen can be rapidly detected in host tissue by the
fluorescent antibody technique. Usually, however, varicella infections are
diagnosed on the basis of their unique clinical picture.

Because chickenpox is usually neither fatal nor debilitating when
contracted in childhood, early infection is often encouraged to avoid the
severe cases of adult chickenpox. No therapy effectively shortens the
course or severity of chickenpox, but zoster can be successfully treated
with acyclovir (see Chap. 14). The use of aspirin and other salicylate
analgesics contributes to the development of a severe complication, called
Reye's syndrome (see color box). Protective immunoglobulin is recommended for immunocompromised persons who have been exposed to
active cases of either chickenpox or zoster.

MEASLES (VIRAL) Measles (rubeola) is one of the most contagious of all
infectious diseases. It is usually transmitted by respiratory droplets. Like

A MODERN MEDICAL MYSTERY

■ In 1963 an Australian pathologist, R. D. K. Reye, described an unusual disease that rapidly develops in some children. During apparent recovery from an acute viral illness, either chickenpox or influenza, the child abruptly begins vomiting and cannot stop. This is followed by disorientation, lethargy, combativeness, or other changes in mental states. Within a few minutes or hours the child may lapse into a deep coma. During this time, the brain is swelling and the liver degenerating. Three days from the start of vomiting, the child may die. The disease, known as **Reye's syndrome,** has occurred in at least 21 countries, with the highest incidence reported in the United States, Canada, and Thailand. As many as 548 cases of Reye's syndrome are reported each year in the United States. Over one-third of the afflicted children may die of the disease, and many others suffer permanent brain damage.

After 2 decades of research, the disease remains a mystery. We have yet to discover what causes the syndrome, the mechanisms responsible for the pathological changes, or why some children develop the disease while most do not. However, researchers have established a successful diagnostic and treatment approach. Diagnosis is based on the characteristic clinical picture, knowledge of previous viral infection, laboratory tests indicating liver dysfunction, and a case history that rules out other illnesses with similar symptoms. Although there is no known cure for Reye's syndrome, supportive treatment in a hospital intensive care unit often reverses the course of the disease. The objective of therapy is to control intracranial swelling. With early diagnosis and aggressive treatment, most Reye's syndrome patients recover completely.

Recent observations suggest a possible link between Reye's syndrome and the use of aspirin during the preceding viral illness. Although a cause-and-effect relationship between salicylates (aspirin) and Reye's syndrome has yet to be established, the Centers for Disease Control (CDC) have recommended abolishing the practice of giving aspirin to children with chickenpox or any influenza-like illness.

chickenpox, measles invades the bloodstream from its primary site of infection, the URT, and is carried to all parts of the body. Hematogenous dissemination is characterized by the appearance of a skin rash over the entire body. Additional symptoms of uncomplicated measles include high fever, cough, delirium, photophobia (oversensitivity to light), and conjunctivitis. Most persons with measles fully recover and acquire solid immunity that prevents subsequent attacks of the disease.

The severity of measles is often underestimated. Serious complications may be due to secondary infection by bacteria, primarily *Haemophilus influenzae*, *Streptococcus pyogenes*, and *Streptococcus pneumoniae*. The virus may also invade the central nervous system and infect the brain, causing fatal encephalitis. Even after recovery from uncomplicated measles, some persons may develop a rare degenerative disease of the central nervous system called *subacute sclerosing panencephalitis (SSPE)*. The factors that predispose for the developments of this fatal sequela are still poorly understood.

The etiological agent of measles is an RNA-containing *paramyxovirus*. It

FIGURE 20-7

Measles-infected cell
identified by specific
fluorescent antibody.

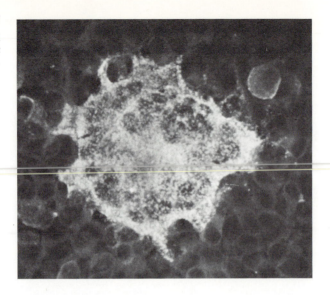

can be isolated from blood or the nasopharynx of infected persons, grown in cell culture, and identified (Fig. 20-7). Diagnosis, however, is usually based on clinical symptoms. The characteristic sign of measles is the appearance of Koplic's spots, bluish-yellow spots on the mucous membranes of the mouth. Koplic's spots appear about 2 or 3 days before the rash. Diagnosis may be confirmed by detecting high serum titers of antibodies that neutralize measles virus or fix complement when mixed with measles virus.

Chemotherapy fails to alter the course of measles, but prophylactic administration of pooled serum globulin to persons exposed to someone with measles can temporarily protect the recipient. Measles is most successfully controlled by a combination of mass immunization and containment. Since the first attenuated measles vaccine was introduced in 1963, the number of annual cases reported in the United States has dropped from about 400,000 to 1400. In other words, the incidence of measles in 1983 was less than 1 percent of that in 1963. A corresponding decrease in SSPE has been observed.

MUMPS (VIRAL) Like measles, mumps is caused by a single antigenic type of paramyxovirus and is spread by respiratory droplets and saliva. The illness has many clinical forms, but it most commonly causes fever and painfully swollen parotid glands. These glands are located between the ear and the angle of the jaw. The virus also invades the bloodstream and may elicit symptoms in the ovaries, testes, thyroid gland, pancreas, or central nervous system. (Testicular infection after puberty can lead to permanent sterility.) These symptoms may appear in either the presence or the absence of parotid gland swelling. About a third of all mumps infections are completely asymptomatic. In all cases, recovery from infection provides immunity to all forms of mumps.

The varied clinical appearances of mumps often prevent diagnosis based solely on symptoms. The virus can be propagated by injecting the patient's urine, saliva, blood, or CNS fluid into chick embryos. Serology can identify complement-fixing antibodies specific for mumps virus in the patient's serum. There is no effective treatment for mumps, but protection can be provided by immunization with an attenuated vaccine. The vaccine is especially useful for preventing sterility in postpubescent males who have never had mumps.

RUBELLA (VIRAL) Rubella (German measles) is a mild disease from which people generally recover uneventfully within 3 or 4 days. The causative agent of rubella is an RNA virus that is different from the measles (rubeola) virus. Disease symptoms include fever, mild inflammation of the respiratory membranes, some enlargement of cervical lymph nodes, and occasionally a rash. Joint pain commonly occurs in adult women but disappears upon recovery. A single infection induces immunity.

Rubella is transmitted by respiratory secretions but is poorly communicable; thus many persons don't acquire the disease until adulthood. When a woman contracts rubella during her first trimester (3 months) of pregnancy, the virus may cross the placenta and infect her developing fetus. Unfortunately, fetal tissues that develop from viral-damaged embryonic cells may be seriously deformed or their growth totally arrested. Miscarriages are common among rubella-infected fetuses, and surviving babies may be born with heart defects, impaired vision and hearing, mental retardation, and other congenital abnormalities. Many of these babies die within 1 year following birth.

Immunization with attenuated rubella virus protects a woman and her fetus if the vaccine is administered while the woman is not pregnant. Unfortunately, the living viruses in the vaccine theoretically may cause many of the same congenital defects as rubella if given during pregnancy. Therefore, rubella vaccine should never be administered to pregnant women. The recommended age for immunization is just prior to puberty. Immunity against rubella can be ascertained by hemagglutination inhibition, a test which quantitatively determines the amount of antiviral antibody in a person's blood. This information reveals the risk of fetal rubella syndrome developing if the mother has been exposed to the virus.

SMALLPOX (VIRAL) Smallpox (*variola*) is a highly contagious disease that has killed millions of people throughout history. Some of the most alarming epidemics occurred in North America after colonization by Europeans. Ninety percent of the people living along the Massachusetts coast were killed by smallpox during a 2-year epidemic in the seventeenth century.

The smallpox virus enters through the respiratory tract and hematogenously spreads to the rest of the body. Persons with smallpox initially suffer fever, chills, headache, back pain, and total fatigue. Rash appears on the face and spreads to extremities, shoulders, and chest, but rarely to the abdomen. The rash consists of fluid-filled blisters that rupture and form crusts in later stages of the disease. These lesions are called pocks. Survivors bear permanent scars after the lesions have healed. The virus also propagates extensively in the liver, spleen, and lungs. Recovery from smallpox provides lifelong immunity to the single antigenic type of variola virus.

Although the appearance of the rash is suggestive of smallpox, diagnosis should be confirmed by isolation of the virus in chick embryos or by detection of variola-specific complement-fixing antibodies in the patient's serum. Infected cells from lesions may also be stained and examined with light microscopy for the presence of characteristic cytoplasmic inclusions, called *Guanieri bodies*. It is hoped that such a diagnosis will never be made again, since the virus is now believed to be virtually extinct following a successful global eradication program (see Chap. 18).

Smallpox vaccination was one of the essential elements in eradicating this virus. The vaccine is another virus, called *vaccinia*, that is antigenically similar to variola virus. Inoculated into the skin, it causes a single sore that heals in 2 weeks, leaving the vaccinated person immune to smallpox. Because no reservoirs of smallpox remain, vaccination is no longer recommended for the general population. In fact, the danger of serious vaccine-related complications greatly outweighs the danger of acquiring smallpox. The vaccine may cause encephalitis or spread to other parts of the body, particularly in persons with eczema. Such complications are often fatal.

The last case of naturally transmitted smallpox occurred in Africa in 1977. In the United States, smallpox has been unknown for over 30 years. Nonetheless, virulent smallpox virus is stored in several labs around the world, and laboratory accidents could conceivably reintroduce the virus into its natural human reservoir. Such an accident occurred in a Birmingham, England, laboratory in 1978. Fortunately, the disease was confined to two laboratory workers. To reduce such risks, only a reference stock is preserved in the United States.

INFECTIOUS MONONUCLEOSIS (VIRAL) Infectious mononucleosis (IM) is a disease primarily of young adults that invades the lymphatic system and occasionally the liver and spleen. The etiological agent, the *Epstein-Barr (EB) virus,* usually causes fever, chills, headache, fatigue, and a painful sore throat. A few patients may also develop hepatitis, meningitis, myocarditis, or paralysis. Although the disease is rarely fatal, a few IM patients die from a ruptured spleen. Most cases are self-limiting and run their course in a few days to several weeks. During this time, the patient has an abnormally high number of mononuclear white blood cells, hence the name "mononucleosis." Infected patients also produce an unusual heterophil antibody that agglutinates sheep red blood cells. This antibody can be detected by a diagnostic blood test. Infectious mononucleosis is believed to be introduced through the URT by close oral contact. Consequently, it has been dubbed the "kissing disease." In spite of this name, the disease is not highly communicable. A single infected person rarely spreads the disease to other members of the family.

The EB virus is also linked to *Burkitt's lymphoma,* a malignant tumor of the lymphatic tissues. This tumor occurs primarily among young black Africans.

BACTERIAL MENINGITIS   After causing infections of the nasopharynx, some pathogens invade the central nervous system and cause serious infections of the meninges, the membranous lining that covers the brain. Epidemics of meningitis among children between 6 months and 2 years of age are most often caused by a gram-negative bacillus, *Haemophilus influenzae* (Fig. 20-8*a*). Frequent epidemics among adults, often military recruits, are usually caused by a gram-negative diplococcus, *Neisseria meningitidis* (Fig. 20-8*b*). Chemotherapy is effective only if the antibiotic can penetrate the "blood-brain barrier" and reach the site of infection. Penicillin is the drug of choice for *Neisseria meningitidis* infections, and ampicillin is used for *Haemophilus influenzae* meningitis. Ampicillin-resistant strains are treated with chloramphenicol. Untreated meningitis has a high mortality rate and may kill a person within 24 hours after onset of symptoms. It is one of the most serious of secondary developments following upper respiratory infection.

FIGURE 20-8

Laboratory identification of two important causes of bacterial meningitis: (*a*) Satellite phenomenon. *Haemophilus influenzae* fails to grow on blood agar because the medium lacks two essential nutrients (called X and V factors). *Staphylococcus aureus,* which readily grows on blood agar, produces enough X and V factors to supplement the adjacent medium. Thus, small colonies of *Haemophilus influenzae* develop in areas of *Staphylococcus aureus* growth. This satellite phenomenon may aid in the laboratory identification of *Haemophilus influenzae.* (*b*) Oxidase test. *Neisseria* can be distinguished from many other genera of bacteria by their ability to oxidize dimethyl and tetramethyl-phenylene diamine hydrochloride (oxidase reagent). When flooded with oxidase reagent, *Neisseria* colonies quickly change color, eventually appearing black. All the members of this genus are oxidase-positive, so the test cannot be used to distinguish between two *Neisseria* species such as *N. meningitidis* and *N. gonorrhoeae.*

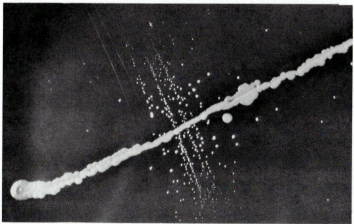

(*a*)

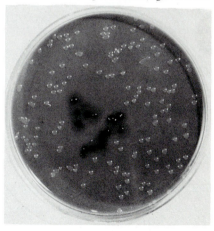

(*b*)

AMEBIC MENINGOENCEPHALITIS Several small outbreaks of fatal central nervous system infections have been caused by free-living amoebas that reside in natural bodies of fresh water. Most of these protozoa belong to the genus *Naegleria*. They apparently infect the nasal mucosa of swimmers in contaminated waters, invade the brain and the meninges, and cause meningoencephalitis. Most persons exposed to water containing these parasites fail to develop symptoms. Factors that increase susceptibility to these amebic infections have yet to be identified.

Infections of Accessory Respiratory Structures Prolonged tonsillar swelling often obstructs the auditory canal and prevents the protective draining of this system. The result may be infection and inflammation of the middle ear, a condition known as *otitis media*. Accumulation of inflammatory products in the restricted region increases the pressure on sensitive auditory mucosa and distends the tympanic membrane (eardrum). The painful pressure is often relieved by cutting the tympanic membrane before it ruptures, thereby allowing drainage. Bacterial pathogens are responsible for more than half of these cases. *Streptococcus pneumoniae*, *Haemophilus influenzae*, and *Streptococcus pyogenes* are the most common bacterial offenders. The remaining cases of otitis media are caused by viruses and are untreatable by antibiotics. Infection of the tympanic membrane itself, a condition called *myringitis*, is often caused by *Mycoplasma*, which is resistant to penicillins and cephalosporins.

Inflammation of the conjunctiva, mucous membranes that line the eyes and lids, is called *conjunctivitis* (pinkeye). Pathogens may travel to the eye from an infected nasal cavity through the nasolacrimal duct or may enter the eye directly on respiratory secretions from an infected person. *Staphylococcus aureus* and *Haemophilus aegypticus* (Koch-Weeks bacillus) are common causes of conjunctivitis.

Some pathogens that cause conjunctivitis can blind the infected person. The most common example of this is *Chlamydia trachomatis*, the cause of **trachoma.** Worldwide more people are blinded by trachoma than by any other cause. After proliferating in the spithelial mucosa of the upper eyelid, the pathogen extends to the cornea, resulting in corneal scarring and eye deformation. Secondary bacterial infection of the cornea is common and further damages vision. The disease is transmitted from infected persons by contaminated fingers, towels, or other objects carrying the pathogen. Flies have also been implicated as important mechanical vectors. The disease usually occurs among crowded populations living in unsanitary conditions. Its control depends on eliminating these conditions and rapidly treating trachoma patients with tetracycline, erythromycin, or sulfonamide.

Another serotype of *Chlamydia trachomatis* causes a less severe eye disease, called **inclusion conjunctivitis,** an infection of the conjunctiva that does not produce blindness. This disease is often contracted by swimming in unchlorinated water contaminated by chlamydias released from the genitals of infected persons and is often called "swimming pool conjunctivitis." Newborns may suffer infant conjunctivitis after exposure to the pathogen in the mother's birth canal. The disease quickly disappears following treatment with tetracycline, erythromycin, or sulfonamide.

Sinusitis, inflammation of the nasal sinus membranes, is an occasional complication of upper respiratory infection. Sinusitis is often caused by the same pathogens that cause otitis media, specifically *Streptococcus pneumoniae*, *Haemophilus influenzae*, and *Streptococcus pyogenes*, as well as many viruses. Strict anaerobes such as *Bacteroides* may also cause sinusitis.

Lower Respiratory Diseases

Unlike diseases of the upper respiratory tract, infections of the lungs and lower airways are commonly fatal if untreated. Fortunately, more than 50 percent of all lower respiratory tract infections are caused by bacteria and respond to the therapeutic effects of appropriate antibiotics. Unless the disease requires emergency treatment, selection of a chemotherapeutic agent should be determined by laboratory analysis of properly collected clinical specimens such as sputum and blood. A summary of infectious diseases that originate in the LRT is provided in Tables 20-6 and 20-7.

Pneumonia   Inflammation of the lungs, **pneumonia,** can be caused by a variety of infectious agents, as well as by chemical irritants and allergy (Fig. 20-9). Often the airways of the LRT are also inflamed, as in bronchopneumonia, for example. Inflammation of the lung's alveolar spaces is called lobar pneumonia, whereas inflammation of the interstitial cells that support the alveoli is called interstitial cell pneumonia. The type of pneumonia depends on the nature of the causative agent. If the site of initial infection is the lung or associated airways, the disease is primary pneumonia. Infections that spread from other regions of the body to the lungs are called secondary pneumonias. The introduction of pathogens into the lower airways by aspiration of foreign material such as food or gastric contents commonly results in aspiration pneumonia. It may be most useful, however, to consider these diseases by etiological agent.

PNEUMOCOCCAL PNEUMONIA (BACTERIAL)   The most common etiological agent of bacterial pneumonia is a gram-positive diplococcus, *Streptococcus pneumoniae* (also called the pneumococcus). The organism's capsule is its major virulence factor, protecting the pathogen from engulfment by alveolar macrophages. Neutralization of this antiphagocytic property by capsule-specific antibodies eliminates the organism's virulence, is responsible for natural recovery from the infection, and prevents subsequent disease with that serotype. Bacterial variants that lack capsules are nonpathogenic and may exist as normal flora in the pharyngeal regions of healthy humans.

Pneumococcal pneumonia is responsible for more mortalities than any other infectious disease of the lower respiratory tract. The disease compromises pulmonary function by impairment of gas exchange, thereby increasing CO_2 concentration in the blood. This triggers increased cardiac

TABLE 20-6

Summary of **Bacterial** Diseases that Follow Introduction of Pathogen into the Lower Respiratory Tract

Disease	Pathogen(s)	Incubation Period	Primary Attack Site	Secondary Attack	Reservoir	Complications or Sequelae	Diagnostic Specimens	Diagnositic Procedures	Treatment
Pneumonia	*Streptococcus pneumoniae*	1–3 days	Lungs	Heart	Infected people	Endocarditis, pericarditis	Sputum, blood	Smears	Penicillin
	Gram-negative bacilli	1–3 days	Lungs, bronchi,	—	Infected people	—	Sputum, blood	Smears, isolation of pathogen	As indicated by antibiograms
Primary atypical pneumonia	*Mycoplasma pneumoniae*	7–14 days	Lungs	—	Infected people	—	Sputum, blood	Smears, isolation of pathogen, serology	Tetracycline
Legionnaire's disease	*Legionella pneumophila*	5–6 days	Lungs	—	Environment	—	Sputum, lung biopsy, serum	Smears, isolation of pathogen, injection into guinea pigs, serology	Tetracycline, erythromycin
Psittacosis	*Chlamydia psittaci*	4–15 days	Lungs	Blood vessels	Infected birds	Vascular damage	Sputum, blood	Isolation of organism, serology	Tetracycline
Q fever	*Coxiella burnetii*	2–4 weeks	Lungs	Heart, meninges	Infected cattle	Endocarditis, meningitis	Sputum, blood, spinal fluid, urine	Isolation of rickettsia in tissue culture, serology	Tetracycline
Tuberculosis	*Mycobacterium tuberculosis, M. bovis*	4–6 weeks	Lungs	Gastrointestinal tract, bones, or any organ of the body	Infected people and cattle	Miliary tuberculosis	Sputum, exudates from local lesions, gastric washings, urine	Acid-fast smears, isolation of pathogen, skin test	Isoniazid with PAS, ethambutol, or rifampin
Pertussis	*Bordetella pertussis*	2–5 days	URT	Larynx, trachea, bronchi and bronchioles	Infected people.	—	Nasopharyngeal	Smears; isolation of pathogen	Erythromycin, ampicillin (unreliable)

TABLE 20-7

Summary of **Nonbacterial** Diseases That Follow Introduction of Pathogen into Lower Tissue Respiratory Tract

Disease	Pathogen(s)	Incubation Period	Primary Attack Site	Secondary Attack	Reservoir	Complications or Sequelae	Diagnostic Specimens	Diagnostic Procedures	Treatment
Influenza	Influenza virus	1–3 days	Membranes of URT and bronchi, bronchioles	Lymphoid tissue	Infected people	Primary or secondary pneumonia, Reye's syndrome	Blood, throat	Serology, isolation of virus	No specific treatment for active disease (amantadine prophylactically)
Coccidioidomycosis	Coccidioides immitis	10–21 days	Lungs	Any organ of the body	Soil	Disseminated progressive disease	Sputum, lung biopsy, pus from lesions, spinal fluid	Direct microscopic exam, isolation of fungus, skin test	Amphotericin B, ketoconizole
Histoplasmosis	Histoplasma capsulatum	5–18 days	Lungs	Any organ, especially of respiratory tract, lymph nodes, spleen, liver	Soil	Disseminated progressive disease	Sputum, biopsy of infected tissue, spinal fluid	Direct microscopic exam, isolation of fungus	Amphotericin B
Cryptococcosis	Cryptococcus neoformans	Unknown	Lungs	Most commonly meninges or brain	Soil	Meningitis, encephalitis	Sputum, spinal fluid	Direct microscopic exam, isolation of fungus	Amphotericin B, flucytosine
Blastomycosis	Blastomyces dermatitidis	Several weeks (poorly defined)	Lungs	Bones, subcutaneous tissue, central nervous system, viscera	Soil	Disseminated disease	Sputum, pus, biopsy of lesions, blood	Direct microscopic exam, isolation of fungus, serology	Amphotericin B
Paracoccidioidomycosis	Paracoccidioides brasiliensis	Several weeks (poorly defined)	Lungs (or URT)	Skin, lymphatics, internal organs	Soil	Disseminated disease	Sputum	Direct microscopic exam, isolation of fungus, serology	Trimethoprinsulfamethoxazole
Pneumocystis pneumonia	Pneumocystis carinii	1–2 months	Lungs	—	Infected persons (?)	—	Sputum	Direct microscopic exam	Trimethoprinsulfamethoxazole

FIGURE 20-9

Bronchial pneumonial infiltrate. Instead of containing air, as seen in the microscopic section of a normal lung (left photo), infiltrated alveoli (right photo) are filled with an inflammatory exudate which contains macrophages and neutrophils.

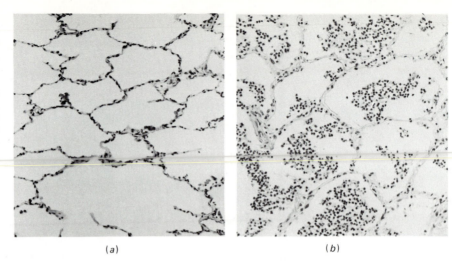

(a)   (b)

activity to accommodate the ventilatory insufficiency. Such cardiopulmonary hyperactivity contributes to the mortality of these respiratory diseases. In addition, the pathogen may spread from the LRT and cause septicemia, endocarditis, or meningitis. Pneumococcal pneumonia is especially prevalent among alcoholics, usually following aspiration of gastric contents vomited while unconscious.

Sputum specimens from persons with suspected pneumonia should be stained and examined for the presence of encapsulated gram-positive diplococci (Fig. 20-10a). Sputum should also be cultured on blood agar. The pneumococci and many normal flora streptococci are alpha hemolytic. Pneumococci, however, are sensitive to the inhibitory effects of the chemical *optochin*, while the indigenous streptococci are optochin-resistant. Thus, paper discs that have been impregnated with optochin inhibit the growth of pneumococci on agar plates (Fig. 20-10b). The pathogen can also be identified by the *quellung reaction*. Capsule-specific antibody reacts with encapsulated pneumococci, causing the appearance of capsular swelling when observed with a light microscope. The quellung test, however, is used primarily for epidemiological typing of pneumo--cocci.

Antibiotics, especially penicillin, are usually effective if given early during pneumococcal infections. Penicillin-resistant strains can be treated with erythromycin or other effective drugs. Even with antibiotic therapy, 20,000 to 50,000 people in the United States may die each year from the disease. Most of these are persons under 2 or over 50 years old. Prevention of pneumococcal pneumonia is complicated by the wide distribution of the pathogen. Healthy carriers that harbor and shed virulent *Streptococcus pneumoniae* continue to be an important source of the pathogens. Development of a universally effective vaccine has been hampered by the existence of approximately 100 types of pneumococci, each with different capsule antigens. Most cases, however, are caused by one of 14 types. A vaccine has been developed which contains purified capsular material from all 14

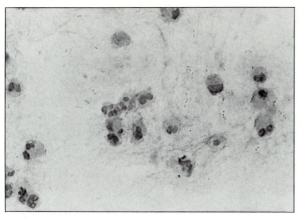

(a)

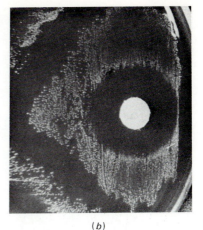

(b)

of these pneumococci. This combined vaccine (called *pneumovax*) has the potential for reducing pneumococcal disease by 60 to 65 percent.

GRAM-NEGATIVE BACTERIAL PNEUMONIA Serious pulmonary disease can also be caused by *Klebsiella pneumoniae*, a heavily encapsulated gram-negative bacillus. As with *Streptococcus pneumoniae*, the capsule is antiphagocytic. *Klebsiella pneumoniae* is a common resident of the normal gastrointestinal flora and URT of healthy humans. Most cases of *Klebsiella* pneumonia occur in persons with compromised respiratory defenses; alcoholics are especially vulnerable.

Other gram-negative bacteria that cause human pneumonia, generally among hospitalized patients, include *Escherichia coli, Pseudomonas aeruginosa, Serratia marcescens*, and *Proteus* sp. *Haemophilus influenzae* causes secondary pneumonia in persons with influenza. Pneumonia is also part of the complex clinical illnesses caused by such virulent gram-negative pathogens as *Yersinia pestis* (plague) and *Francisella tularensis* (tularemia). Antibiotic resistance among these gram-negative bacteria is common, and the choice of therapy should be confirmed by laboratory antibiograms on the isolated pathogen. The drug resistance problem is compounded by the ability of these pathogens to transfer plasmids (R factors) that confer multiple antibiotic resistance to recipient bacteria (see Chap. 12). In the hospital *Klebsiella* is believed to be one of the major sources of R factors transferable to other pathogens.

MYCOPLASMA PNEUMONIA Primary atypical pneumonia, or "walking pneumonia," is a pulmonary infection caused by *Mycoplasma pneumoniae*, a bacterium that lacks a cell wall. This organism damages the ciliated mucosa (Fig. 20-11) and causes a rapidly progressing pneumonia that quickly runs its course, usually with no dangerous complications. Acute symptoms may be temporarily debilitating, but irreversible tissue damage or fatalities are rare. Most cases are subclinical. Even inapparent infections induce permanent immunity.

FIGURE 20-11
Ciliated mucosa of the
respiratory tract.
(a) Normal and *(b)*
following infection with
mycoplasma.

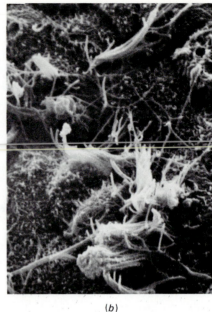

(a)   (b)

Patients with *Mycoplasma* pneumonia develop cold agglutinins in their blood. These diagnostically valuable antibodies agglutinate type O red blood cells at 4°C but not at 37°C. Since mycoplasmas lack cell walls, penicillin and cephalosporins are of no therapeutic value. The organism is sensitive to erythromycin, streptomycin, and tetracycline.

LEGIONNAIRES' DISEASE (BACTERIAL)   In 1976 a severe pneumonia swept through a group of persons attending an American Legion convention in a Philadelphia hotel. Of the 186 persons contracting the disease, 29 died. When all the attempts to isolate and identify the pathogen failed, the malady (which acquired the name Legionnaires' disease) became a national curiosity and a source of potential panic. It forced the closing of one of the nation's largest hotels and triggered speculation of sabotage. The etiological agent was eventually proved to be a previously undescribed strain of bacterium now called *Legionella pneumophila.* It was growing in the hotel's air conditioning system and was subsequently distributed throughout the building.

Since then the organism has been implicated in several sporadic outbreaks, often associated with the presence of *Legionella* in the water-cooling towers of air conditioning units. It is now known that the disease occurs worldwide. Deaths from Legionnaires' disease, however, have been reduced by accurate diagnostic procedures and by the discovery that tetracyclines and erythromycin are effective chemotherapeutic agents.

Legionella is a small, gram-negative bacterium (Fig. 20-12) that readily grows when plated on modified Mueller-Hinton agar and incubated in 5% CO_2. Laboratory identification of the pathogen may be supplemented by

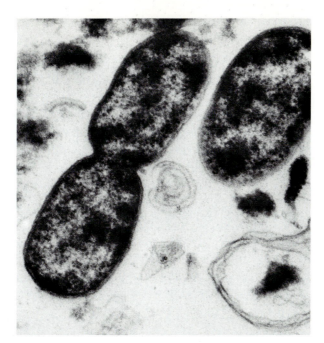

FIGURE 20-12
Microscopic appearance of *Legionella pneumophila*. In spite of its worldwide distribution and ease of cultivation in the laboratory, it remained undiscovered until the 1976 Philadelphia outbreak.

detecting a fourfold rise in blood titers of *Legionella*-specific antibody during the course of disease. The pathogen may also be demonstrated in lung biopsy tissue by staining with fluorescent antibody.

There is no human reservoir of infection and the disease is noncommunicable. Prevention depends on detecting the organism in environmental sources that would promote distribution of airborne pathogens. Such sources should be routinely disinfected with antimicrobial chemicals.

PSITTACOSIS (CHLAMYDIAL) Pulmonary infections of *Chlamydia psittaci* represent an occupational hazard for bird handlers, veterinarians, and their allied professionals. The organism produces an inapparent infection of birds of the psittacine group (parrots and parakeets) from which the disease derived its name. Actually any type of wild or domestic bird may be the source of human infection, so the disease is perhaps better termed *ornithosis*. Regardless of the type of bird infected, its feces contain the infectious agent. When inhaled by people, aerosols of dried feces deposit the organism in the lungs where they replicate and cause pneumonia. Unless treated, psitticosis pneumonia can kill one of every three persons who develop the advanced symptoms.

The etiological agent belongs to a family of small obligate intracellular bacteria called chlamydias (see Chap. 5). Although once believed to be viruses, they are actually procaryotes and are susceptible to antibacterial antibiotics that can reach the chlamydias inside the host cell. Treatment with tetracyclines usually effects complete recovery.

Q FEVER (RICKETTSIAL) *Coxiella burnetii*, a rickettsia that causes pneumonia, is acquired by inhaling aerosols from infected animals, usually cattle. It was initially called "query fever" because the identity of the pathogen was questionable. The name was subsequently shortened to Q fever. Although Q fever pneumonia is rarely life-threatening, the organism can spread to the meninges or the heart valves and cause fatal meningitis or endocarditis.

The pathogen is spread through its natural animal reservoir by infected ticks, although ticks have never been shown to transmit the disease to humans. Livestock animals often show no symptoms when infected, yet shed large numbers of rickettsias in their dander, feces, urine, and milk. Although inhalation of dust contaminated by these sources is the most common mode of transmission, the disease may also be contracted following ingestion of unpasturized milk from infected cows. Human-to-human transmission is unusual.

Q fever is an occupationally related risk for persons who work with livestock or their by-products and for laboratory workers who handle the organism. A vaccine is available to protect these high-risk individuals. Persons with active Q fever are successfully treated with tetracyclines.

PNEUMOCYSTIS PNEUMONIA (PROTOZOAN) *Pneumocystis carinii* is the only protozoan that causes primary pulmonary disease. This opportunistic pathogen is widely distributed in nature. It causes a highly contagious interstitial plasma-cell pneumonia that is especially severe when it strikes infants or immunocompromised patients. Pneumocystis is the most frequent cause of death among children with acute lymphoblastic leukemia and among persons with acquired immunodeficiency syndrome (AIDS) (see Chap. 22). In spite of the ubiquity of pneumocystis pneumonia, little is known about its mode of transmission, which is probably by inhalation. Diagnosis depends on microscopically detecting the ovoid or crescent-shaped microbes clustered in packets of eight cells within a membrane. These clusters may be found free or within phagocytes. Trimethoprin-sulfamethoxazole in combination with supportive measures decreases the mortality of pneumocystis pneumonia.

VIRAL PNEUMONIA Pneumonia caused by viruses is less commonly diagnosed than bacterial infections because virological facilities are unavailable in most clinical laboratories. Generally viruses are assumed to be the etiological agent when no pathogens can be isolated from blood or sputum. The most common agents of viral pneumonia are parainfluenza viruses, adenoviruses, and respiratory syncytial viruses. Most cases occur in small children and infants or older persons with chronic pulmonary disease. Viral pneumonia is also an occasional complication in such LRT diseases as influenza. Antibiotics are useless in treating viral pneumonias.

Influenza Influenza is a viral disease of the lower respiratory tract caused by an orthomyxovirus, an enveloped RNA virus covered with "spikes." These surface spikes are the enzymes *hemagglutinin* (also called the H antigen) and *neuraminidase* (N antigen). Neuraminidase facilitates viral infection by hydrolyzing mucoproteins in respiratory secretions, thereby destroying the effectiveness of the mucociliary defenses. The H antigen is the virus's attachment site. Neutralization of H antigen by antibodies blocks viral infectivity, whereas N-specific antibodies cannot. In addition to H and N antigens, the internal nucleoproteins are antigenic, and their specificities form the basis for dividing the virus into three types—A, B, and C. In humans, type A influenza causes more severe cases than type B. Type C rarely causes clinical disease in people.

Influenza viruses usually affect the bronchioles and bronchi rather than the alveoli and interstitium. Commonly called "flu," the disease ranges in severity from asymptomatic to mild upper respiratory infections to pneumonia and death. (Gastrointestinal disease is *not* associated with influenza. The clinical syndrome incorrectly called the "stomach flu" is caused by a different group of etiological agents discussed in Chap. 21).

The symptoms of influenza are due to viral infection and subsequent destruction of the cells lining the URT, trachea, and bronchi. The symptoms begin suddenly and include malaise, fever, chills, headache, muscle aches, and a feeling of weariness. Acute illness rapidly progresses to a period of prostration lasting 3 to 5 days, characterized by sore throat, nonproductive ("dry") cough, and some nasal obstruction. Uncomplicated influenza is self-limiting, and most people fully recover. Nonetheless, many people die each year of disease complications. Fatal cases are usually associated with secondary bacterial pneumonia, the most common cause of death in the devastating 1918 epidemic (see color box). The most predominant secondary invaders are *Staphylococcus aureus*, pneumococci, *Haemophilus influenzae*, and *Streptococcus pyogenes*. Less frequently, influenza viruses attack the alveolar mucosa and cause viral pneumonia. Between 1968 and 1981, more than 200,000 people in the United States died from some complication. Most of these people were elderly or suffered from heart or kidney disease, diabetes mellitus, severe anemia, or impaired immunological defenses at the time they contracted influenza. Because of the increased risk of severe or fatal complications, persons having any of these predisposing factors should receive annual vaccination against the prevalent strains of the influenza virus. The disease can also be prevented by administering amantadine, a chemoprophylactic agent that prevents intracellular development of influenza A virus (but not influenza B virus). Neither amantadine nor any other chemotherapeutic agent can alter the course of influenza once symptoms appear.

Although influenza outbreaks occur every year, major pandemics sweep the world approximately once every 11 years. These periodic episodes are due to the emergence of new viral strains with different surface antigens. Recovery from one antigenic variety of the flu induces immunity that prevents subsequent infection by the same strain. Periodically, however, new strains emerge with unique antigenic determinants for which previously immune persons have no immunological memory. These new viral strains therefore escape recognition by the memory cells of the immune systems and establish a new pandemic. The viral changes are due in part to **antigenic drift,** alterations of surface determinants by spontane-

THE THREAT OF REPEATING HISTORY

■ The most dramatic influenza pandemic in history occurred in 1918. In less than 2 months, the population of the entire world was engulfed in a devastating wave of illness. Half a million Americans were killed during this short period of time. More than 20 million people worldwide died of primary influenza, pneumonia, or secondary bacterial infection. Mortality was not limited to high-risk individuals; many of the deaths occurred in otherwise healthy persons between 20 and 30 years of age, including athletes. Because the pandemic occurred prior to the advent of techniques for cultivating viruses, the etiological agent was never isolated. Consequently, there was no opportunity to study the virus properties and explain the increased mortality associated with this pandemic.

Immunological analyses of sera from infected patients, however, have produced an indirect picture of the antigenic makeup of this unusually severe pathogen. Any influenza virus matching this antigenic description could theoretically cause another pandemic now, since most of the world's population was born after the 1918 epidemic and therefore have never developed immunity to this strain. Epidemiologists have been alerted to any signs of such a potentially dangerous development. Such a strain was known to exist in pigs, but it had never been shown to be infectious for people. In 1976, however, this strain suddenly emerged as a human pathogen and the alarm was sounded. The "swine flu" was coming.

The alarming development occurred when hundreds of soldiers at Fort Dix, New Jersey, were diagnosed as having influenza. The virus was isolated and found to be antigenically identical to that causing the deaths in the 1918 pandemic. One of the soldiers died of his disease. Because of this danger, the most massive vaccination campaign in history was launched, with the objective of protecting the United States population from a repeat of the 1918 tragedy. Millions of persons were immunized with attenuated swine flu virus. The New Jersey virus strain, however, proved to have little affinity for humans, and the anticipated epidemic failed to develop, even among persons who were not immunized. The vaccination program apparently played no role in epidemic prevention. We still do not know if vaccination programs can prevent a major epidemic of virulent influenza.

ous mutation, producing minor changes in antigenic structure of hemagglutinin. **Antigenic shift,** on the other hand, produces major structural changes and results in immunologically distinct strains of influenza virus. Antigenic shift is believed to occur by genetic recombination between the nucleic acid of two different strains of influenza viruses during coinfection of the same cell. The resulting recombinant virus may possess immunologically unrecognizable surface determinants.

The etiological agent causing a new epidemic bears the name of the geographical location where the virus was initially isolated and identified. This system gives rise to such names as the Russian flu (1977), the Hong Kong flu (1968), and the Asian flu (1957).

506 **Pertussis (Bacterial)** Pertussis (whooping cough) is a disease of the lower respiratory tract, where it causes severe convulsive illness in children. Most cases begin with establishment of the etiological agent, *Bordetella pertussis,* in the nasal and pharyngeal mucosa, generating mild, localized symptoms. As this bacterium infects the lower respiratory tract, it triggers a mild cough that gradually evolves into violent episodes of convulsive coughing, followed by a desperate gasp for air (a "whoop"). These coughing spasms shed copious amounts of respiratory secretions containing enormous numbers of pathogens into the environment. The symptoms also include vomiting and convulsions.

Bordetella pertussis is a small, gram-negative encapsulated rod. It is strictly aerobic and can be cultured from the infected patient's respiratory secretions. The agent is identified by staining the isolates with fluorescent-labeled antibody specific for *B. pertussis.* Fluorescent antibody can also be used for rapid identification of the agent in smears of nasopharyngeal secretions.

The effectiveness of antibiotic therapy is at best unreliable; even with chemotherapy, mortality among infected infants is one of every ten. Prophylactic measures to control pertussis by vaccination, however, have been very successful. The vaccine is a suspension of killed bacteria given to infants in combination with toxoids against diphtheria and tetanus. The DPT vaccine has reduced the morbidity of pertussis from 80 cases per 100,000 population in 1950 to a current annual rate of about 2 per 100,000.

Tuberculosis   Throughout history, tuberculosis has been among the most important of human afflictions. Even today it remains unconquered and poorly understood. It is primarily a slowly progressing infectious disease of the lungs caused by the bacterium *Mycobacterium tuberculosis* (commonly called the tubercle bacillus). The lesions develop as a result of the body's cellular response to the presence of the inhaled pathogen. The bacteria survive phagocytic destruction following engulfment by the host macrophages that infiltrate the infected site. The tubercle bacillus multiplies in these host cells and elicits a cell-mediated immune response that often protects the host from severe injury and spread of the disease. Since cell-mediated immunity (CMI) destroys many healthy cells as well as target cells, some tissue damage is an unavoidable consequence of recovering from tuberculosis. In many cases CMI fails to eliminate the pathogen and is responsible for the characteristic chronic inflammatory reactions that result in tissue injury instead of protection. The factors that determine whether protection or disease will prevail have yet to be identified (Fig. 20-13).

A person's first exposure to *Mycobacterium tuberculosis* is characterized by the development of an exudative lesion. Following macrophage engulfment of the pathogens, an acute inflammatory reaction develops, sometimes leading to local tissue necrosis. These primary complexes heal and disappear in the majority of persons exposed to *Mycobacterium tuberculosis.* In some persons, however, the pathogens become surrounded and encased in a complex network of macrophages and lymphocytes that serves to restrict the entrapped organisms. This lesion is called a **tubercle.** The center of the tubercle may *caseate,* that is, become liquefied as the macrophages disintegrate, taking on the appearance of soft cheese (Fig.

FIGURE 20-13
One property of the tubercle bacillus associated with virulence is the cord factor. This factor is a waxy substance that causes virulent strains of bacilli to aggregate into serpentine cords. Injection of the cord factor alone produces many tuberculosis symptoms in laboratory animals.

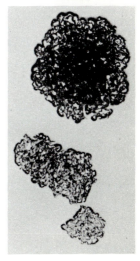

FIGURE 20-14

Appearance of caseations in lung tissue.

20-14). If no further microbial multiplication occurs, the disease may be arrested in this early state.

Active cases of primary tuberculosis occur if the growth of the bacteria in the initial focus of infection is not arrested by the host response (or by antibiotic therapy). As the mycobacteria continue to proliferate in either the initial exudative lesion or in the tubercle, they may invade the regional lymph nodes and produce new tubercles. The lymphatic system may then transport the pathogens to other parts of the body. Dissemination may also follow a hematogenous route from the lungs if the pathogen invades the pulmonary circulatory vessels. This grave development is called *miliary tuberculosis* and, although uncommon, is usually fatal.

Healed tubercles may contain viable, nonproliferating bacilli that may survive for the life of the host. These residual pathogens are believed to be responsible for many cases of active tuberculosis which develop by the process of reactivation infection. Recovery from either clinically apparent or asymptomatic infection may be followed by reactivation years later. The factors that trigger such reactivation are unknown.

Tuberculosis is usually acquired by inhaling the pathogens shed from infected persons. It may also be introduced into the body by contaminated milk from cows infected with *Mycobacterium bovis*. If inhaled, this pathogen produces a disease indistinguishable from that caused by *Mycobacterium tuberculosis*. When ingested, however, *Mycobacterium bovis* has a predilection for bone marrow rather than lung tissue. Milk-borne tuberculosis is no longer a problem in the United States and other countries where pasteurization of dairy products, which kills these and other milk-borne pathogens, is routinely employed, and where governmental inspections of herds identify and eliminate tuberculosis-infected animals.

Tuberculosis continues to elude complete control. The annual case rate in the United States is still about 24,000 (1983), and close to 10 percent of these people die from the disease. Because of the slowly progressing nature of tuberculosis, infected persons may shed pathogens for months before symptoms appear. Exposure to the organism is virtually impossible to prevent. Exposure can be reduced, however, by identifying and treating persons with early tuberculosis. Chest x-rays reveal large cavities in the lung but fail to detect the tiny initial lesions. The most effective screening device is the skin test for tuberculin hypersensitivity. When extracts of the tubercle bacillus, called **purified protein derivative (PPD),** are injected into the skin, persons with CMI against the pathogen will develop a local delayed-type hypersensitivity reaction. Positive skin tests develop early in the course of the disease, long before symptoms are apparent. A positive response indicates only that the person has been exposed to the pathogen and does not necessarily suggest the presence of active tuberculosis.

Although skin testing and chest x-rays are valuable screening devices for detecting probable cases of tuberculosis in large populations, positive diagnosis depends on identifying the etiological agent in sputum, gastric washings, urine, or biopsy material. These specimens may initially be stained and examined for the presence of acid-fast bacilli (color photo). Some nonpathogenic mycobacteria look microscopically similar, so direct microscopy provides only tentative information. The microbe can be cultured on media containing egg yolk (or oleic acid plus albumin), with penicillin or malachite green included to selectively inhibit contaminants. These inhibitors are necessary because of the pathogen's extremely long generation time of about 12 hours (36 times longer than that of *Escherichia coli*).

The tubercle bacillus is resistant to many antibacterial antibiotics and rapidly develops resistance to those agents to which it is initially susceptible. To discourage the proliferation of resistant strains, active cases of tuberculosis are treated with a combination of chemotherapeutic agents—isoniazid (INH), accompanied by para-aminosalicylic acid (PAS), ethambutol, or rifampin. Streptomycin is also effective, but long-term therapy promotes deafness. Because of the pathogen's slow growth rate and the protection afforded by the tubercle lesions, chemotherapy is reliably effective only if continually administered for about a year. The Centers for Disease Control recommend that any tuberculin-positive person receive a regimen of oral INH if (1) the conversion from negative to positive response to skin test has occurred within the previous 2 years, (2) the person is immunologically suppressed, (3) chest x-rays reveal characteristic lesions, or (4) the person is under 35 years of age. This approach protects the public from exposure to carriers, while preventing the development of active disease in the treated person. Treatment should also be administered to all cohabitants and close contacts of any person with tuberculosis. This is one of the infrequent situations that warrant the prophylactic use of antibiotics.

Many countries other than the United States have immunization programs against tuberculosis. The vaccine is a nonvirulent variant of *Mycobacterium bovis* called **BCG** (bacillus of Calmette and Guerin) that stimulates at least partial immunity against tuberculosis by inducing the

RESPIRATORY-ACQUIRED
DISEASES

proliferation of sensitized lymphocytes. The efficacy of the vaccination program is controversial, and it has not been adopted in the United States because vaccinated persons show positive skin tests, and the tuberculosis prevention system in the United States relies on the skin test for early detection of potential cases.

Systemic Mycoses Some fungal pathogens that ordinarily grow as saprophytes in the soil produce infectious spores that, when inhaled, can germinate in the human lungs and cause disease. Usually the resulting infections are mild or inapparent. In some individuals, however, these fungi can cause pulmonary necrosis, spread to the lymphatic channels and then to any tissue in the body. Advanced disseminated cases are often marked by extremely disfiguring cutaneous lesions. Death may follow destruction of any essential organ or overwhelming proliferation in the lung.

Two systemic mycoses in the United States are *coccidioidomycosis* (valley fever) and *histoplasmosis*. Both diseases share many common characteristics. For example, the etiological fungus that causes each disease is dimorphic, that is, filamentous (the mold form) in the soil and nonfilamentous in the human body (see Chap. 7). The mold which produces the infectious spores grows only in the soil, which is consequently the reservoir of infection. An infected human, on the other hand, harbors only the noninfectious, nonfilamentous form of the organism. The systemic mycoses, therefore, are noncommunicable diseases. Since they are not contracted from other human beings, their prevention depends on reducing exposure to infectious mold or changing the environment so that it no longer favors the growth of these fungi. For example, the accumulation of bird droppings enriches the soil with nutrients that encourage growth of infectious *Histoplasma*. Deserted houses, chicken coops, or any enclosed structure containing numerous bats or birds are ideal reservoirs of infection for histoplasmosis. Prevention of histoplasmosis is one of the important goals of bird-control projects such as Tennessee's effort to reduce the explosive population of starlings in the 1970s. In addition, vaccines are being tested and may soon be used to reduce susceptibility to these diseases. The agents of these diseases are restricted to characteristic geographical regions. Histoplasmosis is primarily a disease of the eastern American river valleys, where optimal growth conditions exist. Coccidioidomycosis, on the other hand, is restricted to the arid regions of the southwest United States, as well as Mexico and a small portion of South America. Most individuals residing within these endemic regions have acquired immunity to the corresponding disease, probably because of subclinical infection. Susceptible persons are usually visitors to the regions or new residents who have never been exposed to the endemic pathogens. Immunosuppression or other resistance-crippling factors also create a small population of susceptible persons. Without appropriate chemotherapy these compromised individuals may eventually die of disseminated infection.

Histoplasmosis is often mistaken for tuberculosis when preliminary diagnosis is based on chest x-rays. Differential diagnosis should be carefully established since tuberculosis (a bacterial disease) is susceptible to different antibiotics than the systemic mycoses. Diagnosis of histoplasmosis and coccidioidomycosis depends on microscopic detection of the fungi in stained tissue specimens as well as identification of the pathogens by laboratory isolation and cultivation (Fig. 20-15). In addition, these fungi induce antigen-specific delayed-type hypersensitivity in much the same way the tubercle bacillus does. Extracts of the fungi or their metabolic by-products may therefore be used for skin testing. The reliability of this technique is increasing as the antigens become refined by research. Thus, positive skin reactions to the extracted fungal antigens histoplasmin or coccidioidin provide additional data supporting the diagnosis of the corresponding systemic mycosis, although some cross-reactivity exists between the different fungi.

Several other fungi cause serious systemic mycoses when introduced into the lower respiratory tract. The yeast *Cryptococcus neoformans* usually enters the body on dust generated from soil that contains the organism. (It cannot be transmitted from an infected person.) The pathogen is especially prevalent in environments enriched with pigeon feces, such as bird roosts in buildings and barns. The disease is less frequently acquired through breaks in the skin. When exposed to *Cryptococcus neoformans,* most persons develop mild lung infections or escape infection altogether. Some persons, however, fail to eradicate the organism, which may spread to other parts of

FIGURE 20-15

Coccidioides immitis: (a) Soil phase (450×). The mold phase of this pathogen is characterized by the formation of barrel-shaped arthrospores alternating with apparently empty sections. The arthrospores are the infectious particles. *(b)* Tissue phase (1000×). In the body the fungus grows as spherules that contain fungal endospores. These can be seen in stained preparations of clinical material.

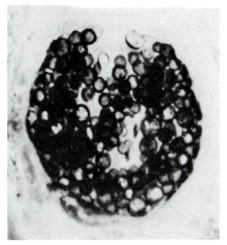

(a)   (b)

the body, especially to the central nervous system, where it causes meningitis. Cryptococcal meningitis is differentiated from other meningeal infections by microscopically demonstrating the presence of the encapsulated yeast in spinal fluid. Specimens are suspended in india ink, which accentuates the large-capsule characteristic of this pathogen. Disseminated and meningeal cryptococcosis is usually fatal unless treated, usually with amphotericin B in combination with flucytosine.

Two other systemic fungal diseases, *blastomycosis* and *paracoccidioidomycosis*, begin as infections of the respiratory tract and in a few persons spread throughout the body, causing fatal disseminated disease. Both pathogens are dimorphic fungi that produce infectious spores while growing in the soil. Blastomycosis occurs primarily in Canada and the United States, whereas paracoccidioidomycosis is endemic in South America.

Immunocompromised persons are susceptible to invasions of the respiratory tract by several opportunistic fungi. These agents are usually harmless contaminants widely distributed in most natural environments. For example, several species of *Aspergillus* (especially *A. fumagatis*) cause invasive pulmonary disease (aspergillosis) in persons with diabetes, cancer, tuberculosis, or an immunosuppressed condition. Similarly, mucormycosis occurs in compromised persons infected with molds of the genera *Mucor* and *Rhizopus*. These fungi grow extremely rapidly and can cause disseminated infections that are often diagnosed after the victim's death. Aspergillosis and mucormycoses are difficult to control because exposure to the pathogens is virtually impossible to avoid. Amphotericin B is the antibiotic most often employed, but many persons with these diseases die in spite of treatment.

PREVENTION AND CONTROL

Interfering with Disease Transmission

The incidence of respiratory-acquired infections can be decreased by controlling production of and exposure to infectious droplet nuclei. An educated public better knows how to control respiratory discharges. Persons with respiratory diseases should be instructed to stay at home. Sneezes and coughs contained in disposable handkerchiefs shower the air with fewer pathogens. Symptomatic medications can reduce the production of infectious aerosols by suppressing excessive sneezing, coughing, and production of mucous secretions.

Additional precautions in the hospital or clinic help prevent the transmission of respiratory-acquired diseases. These include, above all, proper *hand washing* before and after patient contact. In addition, patients shedding highly contagious, virulent pathogens should be isolated, and all individuals entering the room should wear masks. Other measures include wearing clean gowns to protect clothing; careful sterilization, disinfection, or disposal of all linen, utensils, and equipment (especially equipment to be introduced into the mouth or respiratory tract); careful handling and sterilization of all dishes and trays from potentially infectious patients (dietary personnel should be informed if the utensils are possibly infectious); gentle handling of laundry to minimize turbulence that disseminates airborne pathogens; and thorough disinfection of the room after the infectious patient vacates. Air filtration systems promise to be effective (although expensive) methods for removing infectious droplet nuclei from hospital air.

Reducing Population Susceptibility

The most effective method for creating a population of resistant people is active immunization with vaccines. Vaccines are successfully used against the airborne pathogens of smallpox, diphtheria, pertussis, rubella, measles, mumps, and with limited success, influenza and pneumococcal pneumonia.

A few diseases are so potentially dangerous that exposure to an infected person warrants prophylactic administration of antibiotics. For example, all close contacts of any person with confirmed tuberculosis or meningococcal meningitis should receive the appropriate regimen of antibiotics. General resistance to infectious respiratory disease may also be promoted by identifying and correcting factors, such as tobacco smoking or alcoholism, that increase susceptibility.

Reducing the Reservoirs of Infection

The infected person suffering from overt symptoms should be quickly diagnosed, treated, and isolated until pathogens are no longer being shed. Unrecognized cases of infection represent the greatest infectious hazard to others. Inapparent carriers continually distribute airborne pathogens to susceptible persons, who may acquire overt disease or themselves become carriers. Detection of carriers, as well as of people in the incubation phases of disease, is often impossible in the general community. In a more restricted population, however, such as a military base or a school, bacteriological sampling of all persons usually detects healthy carriers. Such an effort should begin at the first sign of a potential epidemic. Anyone identified as an asymptomatic carrier should receive immediate chemotherapy (if appropriate) to prevent the development of overt disease and to alleviate the carrier state if possible. All persons who have had close contact with a carrier should be notified and, if the disease is extremely dangerous, should also be cultured and treated. Epidemics may be prevented by such prophylactic measures.

Respiratory therapy equipment may be especially dangerous if saprophytes are allowed to proliferate in the moist chambers of the units. Such contaminated equipment serves as a vehicle for introducing these opportunistic pathogens deeply into the susceptible respiratory tract. All instruments and equipment that might produce potentially contaminated aerosols should be carefully monitored and maintained to prevent contamination and growth of microbes.

OVERVIEW

The respiratory tract is the major portal of entry for human pathogens. Although the anatomy and defenses of the airways are usually capable of preventing disease, many factors reduce respiratory defenses and predispose for subsequent infection. In addition, some virulent pathogens may overwhelm the intact defenses of healthy persons.

Most infectious diseases begin by microbial colonization of the upper respiratory mucosa, where they cause localized symptoms. Although infections are usually limited to these areas (as in the common cold), some pathogens spread to the bloodstream and cause disease in the skin, liver, spleen, or central nervous system—measles, chickenpox, and infectious mononucleosis are examples. Some noninvasive pathogens produce a toxic substance that circulates hematogenously and causes systemic damage, as in diphtheria and scarlet fever. Accessory respiratory structures may become infected by direct extension of pathogens from the URT, as seen in otitis media and myringitis. One of the most important URT pathogens is *Streptococcus pyogenes*, which may trigger the dangerous sequel disorders, rheumatic heart disease and acute glomerulonephritis.

The lower respiratory tract may also be infected by direct extension from the URT, often by members of upper respiratory flora. Many cases of pneumonia are caused by direct extension of these opportunistic pathogens. Influenza and pertussis are externally acquired LRT diseases that cause epidemics. Chronic lung infections may follow inhalation of *Mycobacterium tuberculosis* or spores of the fungi that cause systemic mycoses. Inhalation of dried feces from birds with psittacosis can cause pneumonia in people. Legionnaires' bacillus, often circulated by air conditioning systems, also causes pneumonia in many persons inhaling contaminated aerosols.

Immunization helps prevent many diseases that enter through the respiratory tract. People may now be vaccinated against measles, mumps, diphtheria, smallpox (no longer recommended), pertussis, pneumococcal pneumonia, influenza, and, in some countries, tuberculosis. Accurate diagnosis determines whether the etiological agent is bacterial and therefore susceptible to antibiotics. Most pathogens that become established in the upper respiratory tract, however, are viruses that produce self-limiting diseases.

KEY WORDS

alveolar macrophage

respiratory droplet

droplet nucleus

pharyngitis

Reye's syndrome

pneumonia

antigenic drift

antigenic shift

tubercle

purified protein derivative (PPD)

BCG vaccine

REVIEW QUESTIONS

1. Describe five defenses unique to the respiratory tract.

2. How do the following predispose for respiratory infection?
   (a) Smoking
   (b) Viral infection
   (c) Respiratory therapy
   (d) Alcoholism

3. Why are droplet nuclei more dangerous than respiratory droplets?

4. Describe the sequel disorders associated with recovery from each of the following: (a) measles, (b) chickenpox, and (c) streptococcal pharyngitis.

5. Why is zoster, a disease of the skin and nerves, considered a respiratory-acquired disease?

6. Why does rubella, usually a mild disease, warrant a national program of control?

7. Describe four agents of pneumonia and the treatment for each.

8. Discuss the limitations that hamper the effectiveness of or restrict the development of vaccines against each of the following: (a) influenza, (b) pneumococcal pneumonia, (c) chickenpox, and (d) the common cold.

9. Tuberculosis causes more deaths in the United States than any other infectious disease. Discuss the role of CMI in the pathogenesis of the disease.

10. Why are histoplasmosis and coccidioidomycosis noncommunicable?

Diseases Acquired through the Alimentary Tract

21

Like the respiratory tract, the alimentary tract is constantly exposed to environmental microorganisms, and food- and waterborne pathogens have historically been responsible for large outbreaks of serious diseases. In 1900, for example, in the United States diarrheal diseases were responsible for more than 150 deaths per 100,000 population. The annual incidence of reportable diseases acquired through the alimentary tract is now estimated to be approximately 7 per 100,000; these cases are seldom fatal.

The threat of infection and the seriousness of most of these diseases have been reduced primarily by advances in food preparation and storage and in personal and environmental sanitation. These procedures help reduce the contamination of ingestible items and the transmission of disease. The effectiveness of this approach is illustrated by the dramatic reduction of milkborne diseases following the introduction of pasteurization in 1908. Other foodborne diseases, unfortunately, have not been as easy to control.

In many countries poor sanitation, overcrowding, poverty, and malnutrition continue to encourage the transmission of disease by contaminated food, water, and fingers. Cholera has been endemic in India at least as long as recorded history. Infectious diarrheas continue to be a serious problem in scores of developing nations, particularly among children below the age of 5; such infections kill more children than any other infectious disease syndrome.

ANATOMY AND DEFENSES OF THE ALIMENTARY TRACT

The alimentary tract is essentially a single long tube, lined by mucous membranes, that traverses the body from the mouth to the anus (Fig. 21-1). Its primary function is the digestion of food, absorption of the resulting nutrients, and elimination of indigestible waste products. Additional accessory structures—the salivary glands, liver, gallbladder, and pancreas—secrete substances that are needed for digestion into the digestive tract. For this reason, these organs are often considered part of the alimentary system. They are also vulnerable to infection by pathogens that enter the body through the digestive tract.

Although the intestinal tract is constantly barraged with microbial invaders that enter as contaminants of foods and fingers, the interactions of several natural factors protect the organs of the system from disease (Table 21-1). Few microbes can survive the acidic conditions in the stomach which contains hydrochloric acid and has a pH of 1.5 to 2.5 (see color box). The rhythmic contraction of the intestinal wall, **peristalsis**, removes microbes that are not securely attached to the mucosal surfaces. These organisms are swept out of the gastrointestinal tract with the feces. The organisms that remain have a mechanism for firm attachment to the mucosal surfaces.

DISEASES ACQUIRED
THROUGH THE
ALIMENTARY TRACT

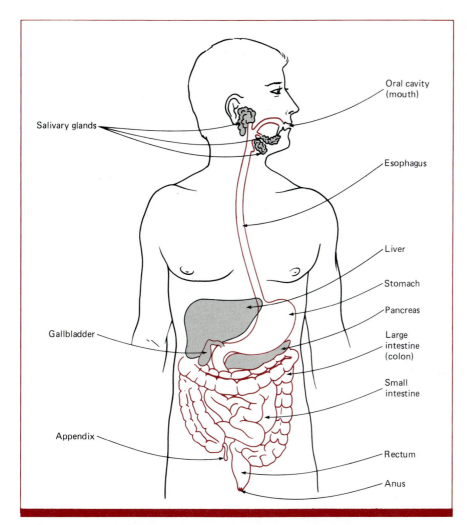

FIGURE 21-1

The human alimentary tract. The digestive canal is in color; the accessory organs are in gray.

Salivary glands

Oral cavity (mouth)

Esophagus

Liver

Stomach

Pancreas

Large intestine (colon)

Small intestine

Gallbladder

Appendix

Rectum

Anus

TABLE 21-1

Gastrointestinal
Defenses

Type	Responsible Factor	Protective Action
Mechanical	Peristalsis	Moves unattached microbes out of the gastrointestinal tract
	Mucous lining	Prevents microbial attachment to epithelium
Chemical	Stomach acid	Low pH opposes growth of many microbes
Microbial	Normal flora	Compete with pathogens; produce substances that inhibit other microbes
Cellular	Leukocytes	Phagocytize microbes
Immuno-logical	Secretory IgA T lymphocytes	Neutralizes toxins, viruses, and bacteria Enhance killing of pathogens

SURVIVING THE STOMACH

■ The stomach acts as a decontamination chamber, and its highly acid secretions are considered one of the best barriers against infection. Some factors, however, reduce the effectiveness of the stomach's assault, thereby increasing the likelihood of a microbe's successful passage to the intestines. Pathogens that enter through the GI tract survive the stomach in one of the following ways.

■ Some protozoa form protective cysts that are resistant to acid destruction. Cyst forms of *Entamoeba histolytica*, *Giardia lamblia*, and *Balantidium coli* pass through the stomach unharmed. In the intestines, the fragile trophozoites emerge and reproduce.

■ The number of organisms that survive passage through the stomach increases when large doses of a pathogen are ingested within a single inoculum. For most alimentary-acquired pathogens the ingested dose that assures infection is at least 10^8 or 10^9 microbes. Under most circumstances only grossly mishandled materials will contain such high microbial concentrations.

■ For pathogens that are highly virulent, only a few organisms may need to arrive at the intestines in order to initiate a disease process. As few as 100 *Shigella*, for example, may produce a severe dysentery.

■ A rapid transit through the stomach shortens the period of exposure to acidic secretions and allows contaminating organisms to escape destruction. In general, fluids and semisolids are transported fastest, while meals of high fat content are held in the stomach longer and increase the likelihood that ingested pathogens will be destroyed.

■ The stomach acid may be buffered by the medium in which the microbes are transported. For example, neutralization of the acid with sodium bicarbonate (antacids) can decrease the infectious dose for cholera from 10^9 to 10^5 organisms, a 99.99 percent reduction.

Some microbes that cause alimentary-acquired diseases don't have to survive the stomach's acidity. They induce disease by the effects of toxins they form while growing in the contaminated food prior to ingestion.

Host lymphocytes, macrophages, and phagocytes all participate in the local cell-mediated immune response of the digestive tract. Serum immunoglobulins are capable of entering the intestines but are probably destroyed by enzyme digestion before they can stem a pathogen's attack. On the other hand, secretory IgA produced by local lymphoid tissue plays an important role in resistance to many infections within the oropharynx and intestinal tract. Although the protective effect of IgA is usually short-lived, it is crucial for preventing the adherence of pathogens to the mucosal surface, for eliminating pathogenic bacteria, and for neutralizing viruses and toxins. Microbial competition between pathogens and the normal flora indigenous to the alimentary tract also contributes to the defenses of this system.

Normal Microbial Flora

Although the intestinal tracts of infants are devoid of microorganisms at birth, colonization begins within hours, in fact even with the first cry. Within 2 weeks a complement of indigenous microbes is established. By adulthood, impressive numbers of bacteria, yeasts, and protozoa are harbored as part of the normal flora (Table 21-2). Alpha-hemolytic streptococci reach concentrations greater than 10^9 per milliliter of saliva. The major repository of microorganisms, however, is the colon, crowded with microbes in greater numbers than are obtained in laboratory batch cultures grown under optimal conditions. (The colon contains approximately 10^{11} organisms per gram of intestinal material.) Obligate anaerobes in the colon generally outnumber other microbes 1000 to 1.

The normal flora help restrict pathogens from the alimentary tract. Some organisms produce and maintain an anaerobic environment that is

TABLE 21-2

Normal Flora in the Alimentary Tract of a Healthy Adult

Anatomical Site	Most Common Organisms	Microscopic Characteristics
Mouth		
Saliva and teeth	*Streptococcus* sp.	Gram-positive cocci, chains
	Lactobacilli	Gram-positive rods
	Veillonella sp.	Gram-negative diplococci
	Bacteroides sp.	Gram-negative pleomorphic rods
	Fusobacteria	Gram-negative rods with tapered ends
Oropharynx	*Streptococcus* sp.	Gram-positive cocci, chains
	Branhamella catarrhalis	Gram-positive diplococci
	Corynebacterium sp. (diphtheroids)	Gram-positive pleomorphic rods
	Staphylococcus sp.	Gram-positive cocci, clusters
Stomach	Usually sterile	
Small Intestines	Lactobacilli	Gram-positive rods
	Streptococcus faecalis	Gram-positive cocci, chains
	Bacteroides sp.	Gram-negative pleomorphic rods
	Candida albicans	Yeast
Large Intestines	*Bacteroides* sp.	Gram-negative pleomorphic rods
	Coliforms	Gram-negative rods
	Streptococcus faecalis	Gram-positive cocci, chains
	Clostridium sp.	Gram-positive rods
	Fusobacteria	Gram-negative rods with tapered ends
	Anaerobic Lactobacilli	Gram-positive rods
	Staphylococci	Gram-positive cocci, clusters
	Peptostreptococci	Gram-positive cocci, chains
	Entamoeba coli	Amoeboid trophozoite, cysts with eight or more nuclei
	Trichomonas hominis	Flagellated protozoa

inhospitable to aerobic pathogens. Many indigenous organisms produce antibiotics or release ammonia, organic acids, alcohols, or other fermentation by-products that interfere with the survival or multiplication of some intestinal pathogens. Indigenous organisms also occupy the mucosal surfaces and leave few vacant attachment sites for pathogens. Conventional flora form dense layers that act as remarkably effective barriers to the establishment of most pathogens.

PREDISPOSING FACTORS

Many factors may increase susceptibility to disease acquired through the alimentary tract or increase severity of these illnesses (Table 21-3). Malfunctions of the acid-producing machinery of the stomach reduce its effectiveness as a decontamination chamber. Peristalsis impairment encourages local overgrowth of microorganisms in the small intestine, where the microbial population may increase a million-fold as a result. Impaired immunological systems may fail to restrict proliferation of pathogens. For example, removal of the tonsils, a lymphoid tissue in the oropharynx, predisposes for paralytic polio. Antibiotic therapy disrupts the competitive microflora of the normal gastrointestinal tract and increases susceptibility to doses of pathogens that would be harmless to the undisturbed intestinal ecosystem. Malnutrition may deprive the immune system of the amino acids needed to make immunoglobulins.

TABLE 21-3

Predisposing Factors That Encourage Infectious Disease of the Alimentary Tract

Factor	Resultant Impairment	Precautionary Measures
Decreased production of stomach acid	Elevates pH of stomach	Minimize exposure to food-borne and water-borne pathogens.
Gastrectomy (removal of stomach)	Deprives system of acid	Minimize exposure to foodborne and water-borne pathogens.
Antacid therapy for ulcers	Neutralizes antimicrobial acidity in stomach	Use lowest effective dose
Gastrointestinal obstructions; blind loops	Create regions not evacuated by peristalsis	Use surgical procedure and/or antibiotic therapy
Tonsillectomy	Reduces production of secretory IgA	Remove tonsils only when alternative approaches fail.
Antibiotic therapy	Disrupts normal flora and subsequently reduces bacterial competition	Use narrow-spectrum antibiotics; rely on antibiograms for antibiotic selection.
Malnutrition	Reduces immunological competence	Provide an adequate diet.
Infancy	Precedes immunological incompetence; normal flora not established	Minimize exposure to feces and contaminated foods.

Because of this, proper diet can substantially reduce the morbidity and mortality associated with diarrheal diseases in impoverished nations. As with most other infections, alimentary-acquired diseases are considerably more prevalent among children than adults.

EPIDEMIOLOGY AND DYNAMICS OF TRANSMISSION

Organisms enter the alimentary tract through air, foods, unwashed hands, and countless numbers of contaminated objects which find their way into the mouth during the course of an average day. For example, as you read this sentence you may have your fingers, a pencil, or a pen placed against your lips or teeth. Since most diseases acquired through the alimentary tract have high infectious doses, direct person-to-person contact is generally not a factor in transmission. The infectious dose is usually found only on heavily contaminated vehicles. This contamination may be due to the deposit of large numbers of pathogens directly onto the vehicle. More commonly, small numbers of organisms are deposited in food or water, and the microbes proliferate in the vehicle, where they reach concentrations capable of establishing infection. When sanitary practices are completely disregarded, however, large quantities of pathogens can be transferred directly from person to person. This pattern of transmission is responsible for outbreaks in day-care centers and mental institutions. Direct person-to-person spread is also an important mechanism in the transmission of diseases that require very small infectious doses, such as bacterial dysentery. However, the general inefficiency of direct person-to-person transmission probably accounts for the lower incidence of gastrointestinal diseases compared to incidence of infections acquired through the respiratory tract.

Lesions on skin and mucous membranes of food handlers are important sources of foodborne illness. Most pathogens that enter the alimentary tract and establish infections, however, are acquired from food or water that has been contaminated with feces. These organisms are spread by untreated water supplies, flies, or fingers. Fecally contaminated fingers may go directly into the mouth or onto other objects that may be placed in the mouth. Several infectious diseases—most notably, hepatitis, shigellosis, amebiasis, and giardiasis—can be transmitted directly in feces during oral-anal intercourse or by oral-genital contacts following anal intercourse.

Some fecally transmitted pathogens are harbored in the intestinal tract of animals. Such microbes contaminate soil (and crops grown in the contaminated soil), water supplies, and foods derived from infected animals. For example, most *Salmonella* infections result from ingesting contaminated meat, poultry, eggs, or unpasteurized milk. These foods may cross-contaminate other foods that are prepared by the same handler or with the same instruments.

Although many pathogens found in soil and water are temporary inhabitants shed from human or animal colons in fecal matter, some pathogens exist independently in the environment. *Clostridium perfringens*

FIGURE 21-2

Major routes of
transmission of diseases
introduced through
alimentary tract. Although
not shown, a few diseases
may be transmitted by
transfer of feces directly to
the mouth by fingers,
fomites, or flies.

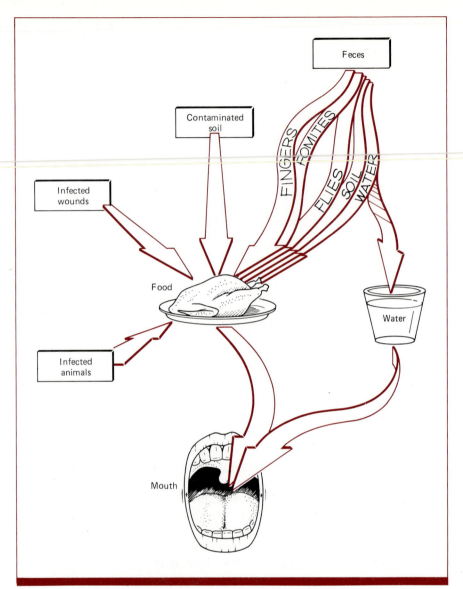

524

DISEASES ACQUIRED
THROUGH THE
ALIMENTARY TRACT

and *Clostridium botulinum*, which multiply freely in soil, can accidentally
contaminate food and cause foodborne illness. *Vibrio parahaemolyticus* and
the algal dinoflagellates, found only in marine environments, contaminate
fish and shellfish. Some fungi are plant pathogens that also cause human
disease in people who consume food prepared from the infected crops.

The major routes of transmission of diseases acquired through the
alimentary tract are illustrated in Figure 21-2.

TABLE 21-4

Factors Contributing to Foodborne Outbreaks in the United States during 1980

Factor	Number of Outbreaks Associated with Factor
Improper holding temperature	87
Inadequate cooking time	27
Contaminated equipment	16
Unsafe source*	10
Poor personal hygiene	29
Other	13

*Food contaminated prior to harvesting with agents that are not inactivated by routine preparation or cooking.

Source: Foodborne Diseases Surveillance, Annual Summary, 1980, Centers for Disease Control, February 1983.

Transmission Cycle of Foodborne Diseases

Foodborne diseases are usually associated with failures to prevent either contamination of food or proliferation of microbes in the contaminated foodstuff (Table 21-4). The following conditions are usually associated with outbreaks of these diseases:

1. A *vulnerable* food is contaminated with a foodborne pathogen. Vulnerable foods have nutrients, moisture, osmotic pressure, and pH that encourage multiplication of the pathogen. These characteristics of the food determine what types of illness it may transmit. For example, the pathogen of botulism proliferates best in canned foods with low-acid contents. Eggs or meat are suspected when the etiological agent is *Salmonella*.

2. Microbes reproduce to high numbers if the contaminated food is kept at temperatures between 20 and 35°C. In these conditions, population densities high enough to ensure the ingestion of an infectious dose may be reached in a few hours. Such conditions generally encourage toxin production as well. Although this temperature range is most dangerous, some enteric pathogens can grow at temperatures as low as 4°C. Thus, if contaminated foods are incubated for long periods, even at low temperatures, they may become effective vehicles of disease transmission.

Foodborne illness is frequently associated with ingesting unrefrigerated foods sold in vending machines, canned goods, bakery products, and catered foods that are poorly refrigerated during transport. Foods kept warm on steam tables or hot trays at temperatures below 60°C are actually being incubated, encouraging explosive microbial growth. Pathogens can reproduce when contaminated foods are cooled too slowly or refrigerated in such large volumes that the center portions reach 4°C too slowly. Thawing at room temperatures allows microbes to proliferate on the surface while the interior of an item is still frozen.

Adequate cooking discourages outbreaks of many foodborne diseases. Thus, foods that are cooked (or reheated) at temperatures above 60°C and served immediately are usually not dangerous. Higher temperatures are required to destroy heat-stable toxins or spores which may germinate before food is served.

Even foods that are properly cooked and refrigerated can be subsequently contaminated by unclean kitchen utensils (the serving spoon, for example), fingers, or microbes in the air. These foods remain at incubation temperatures during the initial serving, and the leftovers served later may be heavily contaminated.

Transmission Cycle of Waterborne Diseases

Water may be a vehicle for disease transmission if it is heavily contaminated by pathogens in feces. Contamination is usually due either to poor sanitary habits or to inadequate programs for treating public water supplies. Many outbreaks have been traced to failures in established sewage disposal systems—for example, breaks in pipes or contamination during construction or repair. Natural disasters, such as earthquakes or floods, can cause overflow of untreated sewage into freshwater supplies. The majority of waterborne pathogens cannot multiply in water, but many can survive in water and retain infectivity for long periods.

ALIMENTARY-ACQUIRED DISEASES

Intoxications are true food poisonings; they follow consumption of food containing preformed toxins produced by microbial growth prior to ingestion. Some of these chemicals, the **enterotoxins,** alter the physiology of the intestinal tract. Others are **neurotoxins** that affect nerve functions after absorption and distribution throughout the body by the circulatory system. Because microbial multiplication is not required after ingestion, incubation periods for many intoxications tend to be short, ranging between 2 and 8 hours, although some, like botulism, may require several days before symptoms appear.

Unlike intoxications, infections are due to the multiplication of viable microbes in the host. Incubation periods are generally longer for infections than intoxications; usually about 8 to 48 hours elapse before symptoms appear.

Intoxications

Botulism   The most potent toxin yet identified is produced by the gram-positive anaerobic rod *Clostridium botulinum*. As little as 1 μg of toxin ingested on a single string bean or a few peas can inhibit nerve function and produce fatal paralysis. Several grams of toxin evenly distributed throughout the world would probably be sufficient to kill every human on earth. This toxin is produced when foods contaminated with *Clostridium botulinum* endospores are kept under anaerobic conditions. Canned foods, for example, contain little or no air, a condition that encourages spore germination and bacterial multiplication, especially in low-acid foods such as meat, poultry, fish, string beans, beets, corn, and some fruits. In other words, these canned foods are ideal incubation chambers for *C. botulinum*. Botulism is rarely associated with high-acid foods because the germination of spores is inhibited at pH's below 5.3. Outbreaks following ingestion of home-canned jalapeño peppers, however, illustrate the hardy nature of the pathogen.

The disease derives its name from *botulus,* the Latin word for sausage, because the earliest recorded cases were traced to the consumption of contaminated sausage products. In Europe most cases are still due to sausages and other home-preserved meats. In the United States 75 percent of all botulism cases reported to the Centers for Disease Control (CDC) are associated with home-canned fruits and vegetables.

Most cases are due to failure to kill the heat-resistant bacterial endospores during the canning process. Reliance on boiling, for example, is dangerous. The spores of *Clostridium botulinum* survive 5 hours in boiling water. Vulnerable foods are best processed by sterilizing them (and their containers) in a pressure cooker to achieve conditions similar to those of an autoclave. Unlike the temperature-resistant endospores, the toxin is very sensitive to heat and can be destroyed by a few minutes of boiling. The number of cases associated with homemade products can be reduced by processing foods at the time, pressure, and temperature required to kill the spores or by boiling foods for 10 minutes before they are eaten to inactivate the toxin.

Botulism is occasionally associated with contaminated commercially canned products. The incidence is less frequent than with home-canned foods because of strict regulations and constant monitoring. Commercial canned foods are prepared under conditions specifically designed to prevent botulism. Surveillance by public health authorities often recognizes any failures in this system in time to prevent widespread distribution of contaminated foods. Suspected foods are immediately recalled, and mass media announcements inform the public of products that may contain the lethal toxin.

Unlike many other foodborne pathogens, *Clostridium botulinum* sometimes produces easily recognized signs of its presence in contaminated foods. For example, anaerobic metabolism may yield foul-smelling by-products. Carbon dioxide or hydrogen gas may accumulate in the container and cause a noticeable bulge or even an occasional explosion. The slightest evidence suggesting *C. botulinum* contamination should be reported to local health authorities immediately. Unfortunately, the development of an obnoxious odor or gas depends on the nature of the food or on the strain of the organism, and sometimes there are no telltale signs, even in heavily contaminated food. Since the toxin is so deadly, reliance on the "taste test" to determine edibility could be fatal.

Symptoms of botulism usually begin to appear between 12 and 36 hours after consumption of the toxin. These symptoms are neurological, not gastrointestinal, and include weakness, blurred or double vision, and paralysis. If untreated, respiratory or cardiac failure and death may occur within 3 to 6 days. Diagnosis is confirmed by immunological identification of toxin in serum or feces. The only successful treatment for botulism is antitoxin given early in the course of the disease to neutralize the toxin. Since there are seven distinct serotypes of toxin, each of which is neutralized by a different antitoxin, it is important to identify the specific serotype or to treat with multiple antitoxins. Since the disease is not an infection, antibiotics are of no therapeutic value. About 5 percent of botulism victims die each year in the United States.

Staphylococcus aureus Food Poisoning

The most common cause of food poisoning in the United States is consumption of an enterotoxin produced by certain strains of *Staphylococcus aureus* in vulnerable foods. The symptoms are generally self-limiting, and the disease is of shorter duration than most other food poisonings. Because of this, staphylococcal intoxications are usually not reported to health authorities. However, it is believed that many incidents labeled as "ptomaine poisoning"—upset stomach, indigestion, and "stomach flu"—may actually be due to staphyloccus enterotoxin. Thus the true incidence of this disease is not known.

The staphylococci usually enter the food from a human source, usually an infected food handler. They are most commonly shed from nasal secretions or from infected wounds, boils, and abscesses. Toxigenic *Staphylococcus aureus* have also been isolated from cattle, and the milk from these cows has occasionally been responsible for outbreaks. Protein-rich foods serve as excellent culture media for staphylococci, especially foods rich in eggs or milk—for example, bakery goods and cream-filled pastries, custards and salad dressings. Meat and poultry also support the growth of *S. aureus*. If contaminated foods are maintained at temperatures between 20°C and 35°C for several hours, the pathogens multiply and release enough toxin to elicit symptoms of intoxication. Ham and turkey left to cool at room temperatures or bakery goods stored without refrigeration are particularly vulnerable to *Staphylococcus*. Thanksgiving dinners are often the ideal meal for the staphylococcal agents, particularly when they are prepared by a cook who fails to practice measures required to safeguard against foodborne disease.

Staphylococcal toxin is referred to as an enterotoxin because the symptoms are primarily related to the gastrointestinal tract. These symptoms include severe nausea, vomiting, cramps, and occasional diarrhea. The toxin produces these symptoms by affecting the vomiting center of the brain. Generally, the time between consumption of the food and the appearance of symptoms is 2 to 4 hours. The illness rarely lasts more than 1 or 2 days and requires no treatment.

Staphylococcus food poisoning is typically identified by its characteristic symptoms, the nature of the foods involved, the shortness of the incubation period, and the isolation of the bacteria from the implicated food. It is most important to identify the source of the organism so that measures to prevent repeated outbreaks can be implemented. Staphylococci can be isolated on media containing 7.5% NaCl. Most toxigenic strains of *Staphylococcus aureus* produce coagulase. They are all beta hemolytic and ferment mannitol. Phage typing (see Chap. 18) of isolates of enterotoxin-producing *S. aureus* provides epidemiologic evidence for tracing the source of an outbreak.

Because of its short incubation period, *S. aureus* was capable of causing a rather unique outbreak on an airliner flying from Tokyo to Copenhagen in 1975. After eating contaminated ham, 277 persons in economy class developed staphylococcus food poisoning while still in flight. The organism had been shed from a lesion on the hand of a food-service employee in

Anchorage, an intermediate stop where the food was prepared. The food was held at room temperatures for 6 hours and stored for 14 hours at 10°C prior to delivery to the plane. The ham was held another 8 hours before it was served. Fortunately, the cockpit crew ate first-class meals.

Mycotoxicoses Some fungi when proliferating in suitable foods, produce toxic substances called **mycotoxins.** For example, many feed products and human foods containing peanuts are contaminated with *Aspergillus flavus,* a fungus that produces *aflatoxin.* In large doses, this toxin causes liver damage. In smaller amounts, aflatoxin has been implicated as a cause of cancer. Other mold species, primarily from the genera *Penicillium* and *Aspergillus,* produce mycotoxins while growing in stored food products. These toxins can damage the kidney, central nervous system, and gastrointestinal tract.

Molds and mold spores are widely distributed and readily contaminate foodstuffs. Mycotoxin production, however, is enhanced in specific foods—peanuts, cottonseed, wheat, soybeans and corn—especially when temperature and humidity are high. Mycotoxicoses are prevalent in tropical areas where peanuts are used as an inexpensive source of protein for malnourished children. In more temperate climates, the toxin is elaborated during storage of contaminated foods in warm, moist silos. In the United States, federal statutes require that animal feeds and many foods for human consumption be surveyed for detectable mycotoxin levels. These statutes prohibit the sale of products that surpass minimum concentrations.

Paralytic Shellfish Poisoning Several species of dinoflagellate algae elaborate a lethal neurotoxin for which no known antidote exists. Blooms of these algae, a condition known as red tide, contain huge quantities of toxigenic dinoflagellates (see Chap. 8). Humans are exposed to the toxin when they eat shellfish that have concentrated the algal product in their tissues while feeding in contaminated waters. The resulting disease in humans is paralytic shellfish poisoning (PSP). Since it cannot be effectively treated, protection against PSP depends on preventing the disease by monitoring conditions in coastal waters and restricting the consumption of products known or suspected to be contaminated. The success of these control measures is evident by the absence of any case of PSP in the United States during 1982.

The major pathogens responsible for intoxications are summarized in Table 21-5.

Infectious Diseases

Pathogens entering through the alimentary tract can cause any of four types of disease processes: (1) infections of the oral cavity; (2) noninvasive infections restricted to the lumen or mucosal surfaces of the intestines; (3) locally invasive infections of the intestinal mucosa; and (4) invasive infections that cause pathological effects at other body sites.

TABLE 21-5
Important Microbial Agents of Food Intoxications

Pathogen	Disease	Cases Reported in U.S. 1981	General Characteristics	Major Vehicle	Incubation Period	Clinical Picture	Diagnosis	Treatment
Staphylococcus aureus	Staphylococcal food poisoning	2934	Gram-positive coccus, heat-stable toxin	Meats, salads containing mayonnaise	2–4 h	Nausea, vomiting, diarrhea	Detection of enterotoxin in food; isolation of organism with same phage type from food, stools, or vomitus of victims and/or skin or nose of food handler	None
Clostridium botulinum	Botulism	22	Gram-positive rod, strict anaerobe, heat-labile toxin	Home-canned, low-acid foods	12–36 h	Speech difficulty, double vision, dry mouth, nausea, paralysis; death due to cardiac arrest or respiratory failure	Detection of toxin in serum, feces, or food; isolation of organism from food or stools	Antitoxin therapy
Aspergillus flavus	Fungal food poisoning	*	Aflatoxin	Contaminated nuts and grains	Unknown	Liver damage	Identification of toxin	None
Gonyaulax sp.	Paralytic shellfish poisoning	0	Dinoflagellate algae, heat-stable toxin	Shellfish	30 min–3 h	Numbness of lips, mouth or face; upper and lower gastrointestinal symptoms	Detection of toxin in shellfish or of toxin-producing algae in waters from which shellfish were gathered	Induce vomiting

*No information was collected on *Aspergillus* food intoxications.

Infections of the Oral Cavity

DENTAL CARIES Microbes that colonize the oral cavity must attach to the cheek, tongue, tooth, or gingival (gum) surface to resist the flushing action of saliva. Some microbes are firmly embedded in a sticky matrix of organic materials. On teeth this complex material is the carbohydrate dextran, produced by *Streptococcus mutans*, a member of the oral flora. Dextran and its embedded microorganisms are called **plaque** (Fig. 21-3). Plaque cannot be removed simply by rinsing with water. Even most toothbrushing leaves significant material between teeth and at the gingiva. The large populations of bacteria concentrated in this sticky matrix produce acidic metabolic by-products, acids that can destroy tooth enamel and cause *dental caries* (tooth decay). Plaque formation is essential to the development of dental caries because it cements microbes to the tooth surface, thereby concentrating the acids at these sites. In the absence of plaque, dental caries fail to develop, because saliva dilutes the acid and washes it from the mouth. The precursor for plaque formation is sucrose, ordinary table sugar. (Other common sugars cannot be used by *Streptococcus mutans* to form dextran.) Thus, diets high in sucrose increase the incidence of dental caries by encouraging plaque formation and by providing a sugar substrate for fermentation and acid production. Sucrose can therefore convert a "sweet tooth" to an "acid tooth," one that is prone to caries.

PERIODONTAL DISEASES Other bacteria, most notably anaerobes and spirochetes, reside in the spaces between the teeth and gums. These microbes, in conjunction with organisms in plaque, cause *periodontal disease*, a progressive infection of the soft tissues around the teeth. Tissue necrosis and tooth loss are the ultimate consequences of untreated periodontal diseases, which are the predominant cause of tooth loss in adults over 35 years of age. Both periodontal diseases and caries can best be prevented by daily flossing of teeth to disrupt bacterial colonies and prevent accumulation of plaque.

FIGURE 21-3

(*a*) A complex mixture of filamentous and nonfilamentous bacteria makes up the flora found in plaque. (*b*) Teeth that have been stained to demonstrate accumulated plaque.

(*a*)

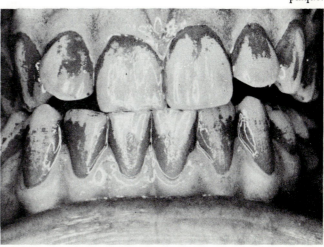

(*b*)

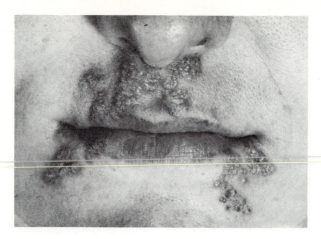

INFECTIONS OF THE ORAL MUCOSA   Type 1 herpesvirus and *Candida albicans* are the two agents most commonly responsible for infections of the oral mucosa and tongue. Herpesvirus causes *gingivostomatitis*, tiny vesicles or ulcers on the lips and in the anterior portions of the oral cavity. The lesions probably develop from an inflammatory response to virus replication in the oral mucosa (Fig. 21-4). Herpesviruses are readily spread to other persons either directly in saliva or by contaminated fomites. Serological studies indicate that the infection is quite common; most persons have herpes-specific antibodies. In fact, most children have already encountered the virus by the age of 5 years. The vast majority of infections are asymptomatic. Even overt attacks are normally self-limiting, and the lesions disappear within a few days. However, the viruses are not eliminated from the body, but remain in a latent state in local nerve cells. A variety of triggering factors, including emotional stress, sunlight, fatigue, or infectious illness, may precipitate the reappearance of lesions on the skin. Because common colds or febrile disease often trigger a recurrence, the lesions are popularly referred to as "fever blisters" or "cold sores." As with most other viral infections, oral herpes is not cured by chemotherapy, although treatment with acyclovir during the primary episode reduces the duration of symptoms and the shedding of the virus, and may prevent the establishment of latency.

Oral infections with *Candida albicans* are called *thrush*. The disease often affects infants before they develop their competitive bacterial flora, persons on antibiotic or steroid therapy, persons with diabetes, and persons with immunologically debilitating diseases. Oral candidiasis is a nonfebrile, noninvasive disease characterized by elevated white patches of yeast on the surface of the tongue and mucosa (Fig. 21-5). Topical treatment with the antibiotic nystatin or an imidazole effectively eliminates the infection, but recurrences are common as long as the underlying predisposing condition persists.

INVASION THROUGH THE ORAL MUCOSA   Microbes pour into the

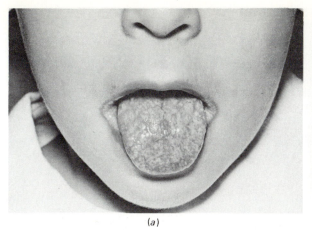

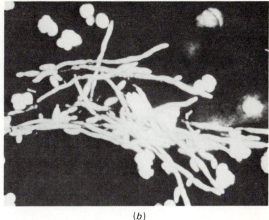

(a)

(b)

FIGURE 21-5
(a) Thrush, or oral candidiasis. (b) *Candida albicans* yeast and pseudohyphae stained with fluorescent antibody.

533

ALIMENTARY-ACQUIRED
DISEASES

bloodstream whenever the oral membranes are disrupted by disease, dental manipulations, or minor accidents, such as biting one's cheek. Usually these microbial intruders are halted by phagocytic defenses in the blood before systemic invasion can occur. The introduction of oral streptococci into the bloodstream of persons with previously damaged heart valves, however, may have serious consequences. These bacteria may colonize abnormal heart valves and cause *subacute bacterial endocarditis*. Penicillin chemoprophylaxis has proven effective in preventing the development of this potentially fatal complication. Such prophylactic therapy is recommended for all persons with defective heart valves when they receive even such routine dental treatment as teeth cleaning.

Although its major route of entry into the human body is through parenteral inoculation of skin, hepatitis B virus can invade the bloodstream through cuts and breaks in the oral mucosa and eventually infect the liver. Contaminated dental instruments are often implicated as vehicles of hepatitis infection. The role of infected dentists in direct transmission is uncertain. It appears that dentists more commonly acquire the infection from their patients than transmit it to them.

Infections of the Gastrointestinal (GI) Tract NONINVASIVE DISEASES Most pathogens acquired through the alimentary tract are noninvasive and confined to the intestines, where they cause **gastroenteritis,** a syndrome characterized by diarrhea, nausea, or vomiting. Most noninvasive GI pathogens produce enterotoxins while multiplying in the small intestines. Enterotoxins cause water to be released from the body into the intestine at such a rapid rate that the normal absorptive mechanism of the large intestine is overwhelmed. The result is the production of watery diarrhea. Because these pathogens do not invade the tissues of the intestinal mucosa, they rarely cause fever, dysentery (red blood cells in stools), or fecal leukocytosis (white blood cells in stools). These microbes do, however, attach to the surface of the intestinal wall and are therefore

TABLE 21-6
Characteristics of Major Noninvasive Pathogens of Gastroenteric Disease

Pathogen	General Characteristics	Cases Reported in U.S., 1980	Infectious Dose	Incubation Period	Major Vehicle	Clinical Picture	Diagnosis	Treatment
Vibrio cholerae	Gram-negative curved rods, highly motile, enterotoxin producer	9	10^8	1–3 days	Water	Diarrhea, dehydration	Isolation of vibrio from stool or vomitus	Fluid and electrolyte replacement
Clostridium perfringens	Gram-positive rod, strict anaerobe, enterotoxin producer	1,463	10^8	8–12 h	Meat	Diarrhea, cramps	Isolation of organism with same serotype in food and stool or isolation of $>10^6$ organisms per gram in food	None
Escherichia coli	Gram-negative rod, enterotoxin producer	500	10^8	12 h	Food, water	Diarrhea, dehydration	Isolation of organism with same serotype in food and stool	Fluid and electrolyte replacement in severe cases
Giardia lamblia	Trophozoite and cyst forms	12,947	10 cysts	1–4 weeks	Water	Diarrhea, dehydration	Identification of organism in feces or duodenal drainage by microscopic examination	Metronidazole or quinacrine
Clostridium botulinum	Gram-positive rod, strict anaerobe, neurotoxin producer	68	Unknown	Unknown	Honey	Constipation, loss of muscle function	Identification of organism or toxin in feces	Supportive
Clostridium difficile	Gram-positive rod, strict anaerobe, toxigenic		High concentration of normal flora	Variable	—	Diarrhea, intestinal necrosis, colitis, formation of pseudomembrane	Identification of organism in feces, detection of toxin in feces	Vancomycin

not readily eliminated with the feces. A summary of pathogens that cause noninvasive infectious diseases of the gastrointestinal tract is provided in Table 21-6.

Cholera   Cholera is caused by the growth of enterotoxigenic *Vibrio cholerae* in the intestines. These comma-shaped, gram-negative, polarly flagellated bacteria are usually transmitted by water that has been heavily contaminated with feces or vomitus of persons suffering from the disease. Food, fingers, and flies may infrequently serve as vehicles of cholera transmission but are not important in most outbreaks. The cycle of transmission is therefore easily controlled in areas where sanitary disposal of feces generally prevents the ingestion of fecally contaminated water. Unfortunately, sewage facilities in most of the world's cholera-endemic areas remain inadequate.

Cholera enterotoxin is produced by the pathogen as it grows in the infected gastrointestinal tract. The enterotoxin absorbs onto membranes of the intestinal epithelial cells and stimulates adenyl cyclase, a membrane-associated enzyme that catalyzes the formation of *cyclic AMP* (cAMP). Accumulation of cAMP results in massive secretion of salts and water from each affected cell (Fig. 21-6). The rapid loss of water produces a watery diarrhea that may cause the cholera patient to lose 20 liters of fluid in a single day. Such dramatic losses of water lead to severe dehydration, thickening of the blood, loss of blood volume, circulatory collapse (shock), and death if not rapidly treated. Some cholera victims die within 6 hours after the onset of symptoms. Half the persons with untreated cholera die of the disease.

Cholera can be preliminarily diagnosed by the characteristic appearance of the watery stools, which are virtually free of feces and contain mostly mucus, epithelial cells, and enormous numbers of *Vibrio cholerae*. These features give the stools a "rice water" appearance. Diagnosis is supported by direct microscopic examination of stool specimens for the presence of the pathogen. Isolation of a pathogen that agglutinates in the presence of *V. cholera*-specific antisera confirms the diagnosis.

In treating cholera, the overriding concern is the continual replacement of fluids and electrolytes until the pathogens are eventually eliminated by the host. In severe cases, intravenous administration is the most rapid method of returning the body fluids to equilibrium (Fig. 21-7). In most cases, fluid and electrolyte balance can be maintained by the oral administration of electrolytes in the presence of glucose. This mixture can even be made conveniently and inexpensively at home by combining corn syrup (glucose), table salt (sodium chloride), baking soda (sodium bicarbonate), and cream of tartar (potassium tartrate). The glucose is essential because it stimulates uptake of sodium and subsequent osmotic absorption of water. The uptake of water in the absence of glucose is insufficient to replace the fluid loss. This simple treatment reduces the mortality rate of cholera to less than 1 percent.

Outbreaks of cholera have decimated communities since ancient

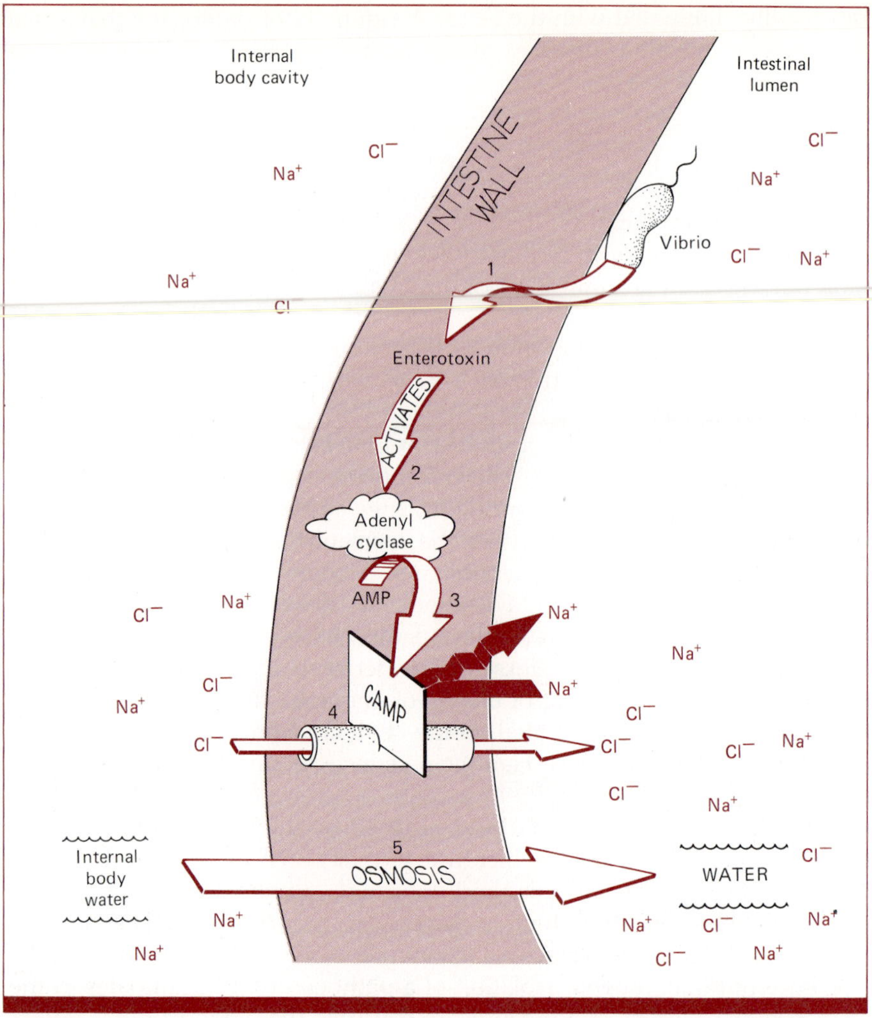

FIGURE 21-6

Mode of action of cholera toxin. After attachment to the surfaces of intestinal epithelial cells, *Vibrio cholerae* produces an enterotoxin (step 1) that activates the enzyme adenyl cyclase (step 2), which converts AMP to cyclic AMP (cAMP) (step 3). Overproduction of cAMP triggers active export of chloride ions (Cl^-) and inhibits uptake of sodium ions (Na^+) (step 4). This produces an osmotic gradient that draws huge quantities of water from the body into the lumen of the small intestine (step 5), causing a watery diarrhea and rapid dehydration of the cholera victim.

times. The disease continues to be endemic in Asia, where two-thirds of all cases in 1982 occurred. Since 1961, however, cholera has been pandemic, slowly spreading to other areas—into the middle east (1966), eastern Europe (1970), Africa (1970), western Europe (1971), and many Pacific islands (1977). Approximately 40 countries, mostly in Asia and Africa, are currently affected. Unfortunately, the number of cases reported annually has been increasing in those areas where malnutrition is common and

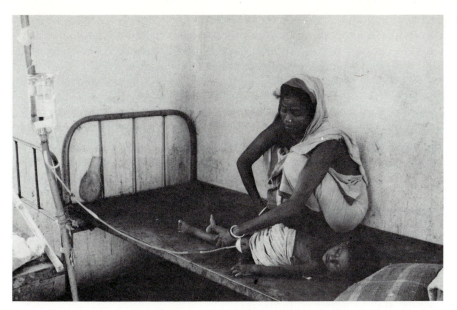

where improvements in sanitation and water purification are neglected. Cholera has been successfully controlled in the United States in spite of the persistence of toxigenic strains of the pathogen. The few sporadic outbreaks are usually caused by consumption of shellfish caught in contaminated coastal waters.

Prevention of cholera depends on rapidly treating active cases, properly disposing of sewage, and, when sewage treatment facilities are inadequate, boiling water before it is used for drinking, cooking, or washing dishes. A vaccine against cholera is available, but it is only 50 percent effective and confers immunity for only 6 months. In its current state, the vaccine is useless for preventing disease among residents of endemic areas. It may be beneficial, however, in protecting persons traveling to cholera-endemic regions.

Escherichia coli **gastroenteritis**   *E. coli* is a major cause of diarrhea and dysenterylike syndromes that have inspired such imaginative names as "Montezuma's revenge," "turista," "the trots," and "Delhi belly." As members of the normal intestinal flora, most strains of *E. coli* are avirulent and elicit no disease in the gastrointestinal tract. Some strains, however, are pathogenic and cause gastroenteritis. Although a few strains of *E. coli* (called *enteroinvasive E. coli*) can cause local invasive disease of the intestinal epithelium, most pathogenic strains are noninvasive and produce an enterotoxin while growing in the infected person's intestine. These strains are called *enterotoxigenic E. coli*. The toxin is physically and antigenically related to cholera toxin and stimulates a similar, although milder, watery diarrhea. Unlike cholera, most persons living in endemic regions are resistant to the disease because they have developed antibodies that neutralize the toxin, which plays havoc with travelers to these areas.

Endemic regions are usually characterized by poor sanitation which encourages microbial multiplication in food and water contaminated with feces. Recent evidence indicates that "traveler's diarrhea" is more often acquired from fecally contaminated raw fruits and vegetables than from water and ice. The pathogens do not appear to pose a major health threat when they are imported with a returning vacationer, since the unsanitary environment necessary for perpetuating an epidemic does not come home with the traveler. Other strains of the bacterium, called *enteropathogenic E. coli*, cause infectious diarrhea among infants in the United States by an undefined mechanism.

The large number of these bacteria required to initiate infection plays an important role in the epidemiology of *E. coli* gastroenteritis in the United States, where sanitation facilities usually protect people from exposure to infectious doses. Infected infants, however, shed such high concentrations of organisms in their feces that the disease is readily transmitted by improperly washed hands or inadequately decontaminated fomites such as scales and thermometers. Outbreaks in the United States are most commonly associated with inadequate precautions taken by hospital staff in newborn nurseries.

Enterotoxigenic *E. coli* isolated from feces or food are distinguished from normal flora inhabitants by serological tests and by demonstrating enterotoxin production. Because of the short duration of the disease, the etiology of most cases of traveler's diarrhea is not determined, and antimicrobial therapy is of little benefit. Severe cholera-like *E. coli* infections, however, require fluid replacement therapy. Epidemics of diarrhea among infants in hospitals are controlled by isolation of all babies with diarrhea and treatment with antibiotics to which the pathogen is sensitive.

The genetic information for toxin production resides on a plasmid which can apparently be transferred by conjugation in much the same way antibiotic resistance is transmitted. Fortunately, no major population shift to toxigenic strains has occurred among *E. coli* or other gram-negative enteric organisms. However, those strains that acquire the toxin plasmid are also likely to acquire multiple antibiotic resistance plasmids, making severe cases of these infections difficult to treat. Another plasmid that carries information for pilus production is also essential for pathogenicity. Without pili the organisms cannot attach to the surface of the intestinal epithelium and are flushed out of the region by the movement of the intestinal contents.

Clostridium perfringens **gastroenteritis** The versatile pathogen *Clostridium perfringens* is most often associated with gas gangrene, but it also one of the most prevalent agents of gastroenteritis. In both food and the intestine, this pathogen produces an enterotoxin that causes symptoms much like those of cholera and enterotoxigenic *Escherichia coli* infections, although the disease is usually milder. Diagnosis is rarely attempted and treatment is unnecessary because the duration of disease is usually 1 day or less.

Some experts believe that *C. perfringens* has the broadest environmental distribution of all pathogenic bacteria. Vegetative cells and endospores

DISEASES ACQUIRED
THROUGH THE
ALIMENTARY TRACT

are widely distributed in the soil and are normally found in small numbers in gastrointestinal tracts of humans and animals. The disease is acquired from food in which the organism has proliferated to significant numbers. Foods may be contaminated by animal or human feces or by water, soil, or dust containing the organism. Cooking at temperatures above 60°C destroys the vegetative forms, but spores generally survive these temperatures. Heating actually compounds the problem by driving off oxygen, producing the anaerobic conditions that encourage germination and growth of the pathogen. The organism proliferates if foods are stored at temperatures between 5 and 60°C or are slowly reheated at temperatures close to 46°C (135°F), the optimum growth temperature for *C. perfringens*. Most outbreaks are traced to cafeterias and restaurants that maintain food on steam tables at temperatures close to optimum for growth of this pathogen.

Meats and poultry prepared in large quantities by fast-food establishments or by catering services are common sources of *C. perfringens* gastroenteritis. These outbreaks, which affect large numbers of people, can be avoided if foods are prepared in small portions. Although normal cooking may never reach sporicidal temperatures, the surviving clostridia pose little threat unless allowed to proliferate in food prior to ingestion. The interior of foods stored in large volumes cools very slowly during refrigeration and becomes an ideal incubator for microbes that survive heating. Food in smaller volumes is more likely to cool thoroughly before significant bacterial growth can occur.

Although prevalence of the organism in the environment makes it impossible to avoid contamination of foods by *C. perfringens*, attention to personal hygiene and proper processing techniques can reduce the levels of contamination and therefore keep the number of organisms below the infectious dose. Meats and other high-protein foods should always be handled with the assumption that spores and vegetative cells are present. Foods should be cooked thoroughly and served immediately or cooled rapidly and kept refrigerated until served or recooked. Foods can be safely reheated or safely kept warm if temperatures are at least 60°C (140°F).

Clostridium botulinum **Infections** Although botulism is an intoxication, infection by *Clostridium botulinum* can cause *infant botulism*, a disease first reported in 1976 that affects children up to the age of 8 months. The pathogen establishes an infection in the baby's intestines, where it synthesizes the toxin that causes the neurological symptoms of botulism. The resistance of adults to botulism infections is probably due to their normal intestinal flora, which successfully competes with the pathogen. It is presumed that infants younger than 8 months have not yet established the competitive intestinal environment of adults. Epidemiologic studies have pinpointed one major source of the pathogen: honey used in infant formulas. Current guidelines recommend that honey not be fed to children in the first year of life. It is also believed that 10 percent of all cases of sudden infant death syndrome ("crib death") are due to infant botulism.

FIGURE 21-8

Many *Giardia lamblia* trophozoites adhering to the human intestinal surface.

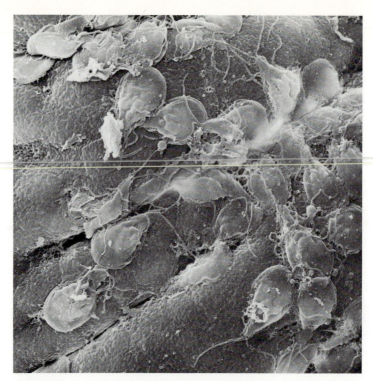

Giardiasis The flagellated protozoan *Giardia lamblia* causes a noninvasive diarrhea of the small intestine. It is the most frequent agent of waterborne diarrhea in the United States. The cysts are shed in human or animal feces, survive in water, and are not destroyed by routine chlorination procedures. However, they are usually eliminated from public water supplies by filtration. Untreated water taken directly from rivers, lakes, or streams must be boiled for 1 minute to remove the threat of infection. *Giardia* cysts are resistant to destruction by stomach acid. In the small intestines the trophozoites emerge and reproduce. The organism has sucking discs that allow it to adhere strongly to the walls of the small intestine, physically covering and mechanically interfering with the absorptive surface (Fig. 21-8).

Diagnosis is by microscopic identification of cysts or trophozoites in feces. Trophozoites are present in feces only during acute stages of diarrhea and quickly lose their characteristic motility if samples are not observed within 1 hour of collection. Cysts are stable and are observed in solid stools as well as liquid. Specimens to be examined for cysts may be held overnight. *Giardia* trophozoites that are firmly attached to the mucosa of the small intestines may be isolated by swallowing a weighted nylon string, which is carried by peristalsis to the small intestines. After 5 hours, the string is withdrawn and examined for the presence of trophozoites. The infection is treated with metronidazole.

Pseudomembranous colitis An occasional complication of the prolonged oral administration of antibiotics is *pseudomembranous colitis*, a necrotizing infection of the gastrointestinal tract caused by toxigenic *Clostridium difficile*. This organism, normally a minor member of the gastrointestinal flora, proliferates to high concentrations in the colon when other normal bacterial inhabitants are inhibited, primarily by clindamycin or ampicillin. (Oddly, *C. difficile* may also be sensitive to these antibiotics—see color box.) *Clostridium difficile* produces a necrotizing exotoxin that may cause fatal losses of massive fluid volumes. The name of the disease reflects the layer of inflammatory products that accumulates on the surface of the intestinal wall, giving the false appearance of a membrane. This clinical picture, coupled with a history of antibiotic therapy, provides a preliminary diagnosis of pseudomembranous colitis. The high mortality rate (30 to 90 percent) of untreated cases is lowered by simply discontinuing the initial antibiotic. Vancomycin is given to control the population of *C. difficile*.

PSEUDOMEMBRANOUS COLITIS: A SERIOUS COMPLICATION

■ It was a simple infection—nothing of serious consequence. Laboratory cultures indicated the causative agent, and the patient received antibiotic therapy. As the infection resolved, however, the patient was stricken with *colitis*, severe inflammation of the large intestine. An examination of the bowel revealed it to be covered by gray membranous patches, beneath which the tissue was ulcerating and dying. If the patient continued to take the antibiotic, he could soon die of classic *antibiotic-induced pseudomembranous colitis*.

Although antibiotics precipitated the situation, this person's medical dilemma was caused by *Clostridium difficile*, an anaerobic spore-forming bacillus that normally resides in small numbers in the healthy intestine. Its growth is limited by competition from the other members of the normal intestinal flora. Antibiotic therapy, especially with clindamycin, ampicillin, cephalothin, or metronidazole, which suppress the other members of the indigenous flora, allows *C. difficile* unrestricted growth. The bacterium releases a cytotoxin that induces bowel necrosis and other symptoms of pseudomembranous colitis. In many cases, colitis disappears when the principal antibiotics are withdrawn and the normal flora reestablishes itself. This process is usually helped along by treatment with vancomycin to reduce the pathogenic clostridia population in the colon.

Investigators have long assumed that *C. difficile* causes this disease because of its resistance to the eliciting antibiotics. In many cases this is accurate. But even antibiotic-sensitive strains of this opportunistic pathogen elicit colitis. In fact, all strains of *C. difficile* are sensitive to ampicillin, one of the drugs that most often induces the disease. Why aren't the clostridia suppressed with the rest of the flora?

Recent evidence suggests another pathogenic mechanism accounts for this paradox. While moderately suppressing drug-sensitive bacteria, ampicillin encourages the growth of other bacteria, those that produce beta-lactamase, an enzyme that inactivates the antibiotic. The clostridia, no longer inhibited by the destroyed drug, overgrow the bowel before the protective members of the normal flora repopulate. It may therefore be possible to prevent ampicillin-induced colitis by administering a beta-lactamase inhibitor with the antibiotic.

LOCALLY INVASIVE DISEASES Some gastrointestinal pathogens penetrate the intestinal epithelium and cause local tissue injury. Locally invasive bacteria and protozoa usually attach to and destroy the mucosal surface of the lower small intestine and colon, eliciting fever, diarrhea, and leukocytes in the feces. Often the invasion and inflammation are severe enough to bring about **dysentery,** the presence of blood in the stools, due to epithelial necrosis and ulceration. Locally invasive pathogens include bacteria in the genera *Shigella* and *Salmonella,* enteroinvasive *Escherichia coli, Vibrio parahaemolyticus, Yersinia enterocolitica, Campylobacter jejuni,* and the protozoan *Entamoeba histolytica.* In addition to gastroenteritis, some of these pathogens may invade the bloodstream and spread to other body organs. Their primary effects, however, are due to local invasion of the gastrointestinal tract.

Some characteristics of pathogens that cause locally invasive gastrointestinal infections are summarized in Table 21-7.

Shigellosis Dysentery is a clinical condition of varied etiology characterized by fever, bloody diarrhea, and fecal leukocytosis. Classical *bacillary dysentery* is associated with infections with any of four species of *Shigella* (*S. dysenteriae, S. flexneri, S. boydii,* and *S. sonnei*). The disease is also called *shigellosis. S. dysenteriae* causes a more severe disease than the other *Shigella* species. Fortunately, more than 90 percent of cases reported in the United States are caused by the less virulent *S. flexneri* and *S. sonnei*.

Unlike most enteric infections, shigellosis can be initiated with a low infectious dose (10 to 100 bacteria). Person-to-person spread is therefore not only possible but is the major mode of transmission. Poor personal hygiene practices encourage spread of the disease. Children who are either unaware of or who tend to disregard protective hygiene habits such as hand washing are implicated in most outbreaks. Over two-thirds of the reported cases and most of the deaths due to *Shigella* occur in children between 1 and 9 years of age. Nursery schools and institutions for children are the most common sites of exposure. Foods or fomites may also be contaminated with enough organisms to cause disease without proliferation in the vehicle prior to ingestion. The bacteria are extremely sensitive to dry environments, however, and will not survive on many fomites. Food may also be contaminated by flies that have fed on human feces.

Shigella infections range from asymptomatic cases to life-threatening dysentery. The organisms penetrate and multiply in the superficial epithelium of the colon, where they are believed to release endotoxin. The endotoxin triggers inflammation, which causes local damage. When absorbed into the bloodstream, the endotoxin causes the fever associated with the disease. *Shigella* also produces an exotoxin that causes diarrhea. The severity and outcome of the illness are influenced by the status of the patient, the size of the infecting dose, and the virulence of the pathogen. Like most infections, shigellosis is particularly dangerous to the compromised host, as evidenced by increased fatality rates among hospitalized persons.

TABLE 21-7
Characteristics of Locally Invasive Pathogens of the Gastrointestinal Tract

Pathogen	Cases Reported in U.S., 1980	Infectious Dose	Incubation Period	Major Vehicle	Clinical Picture	Diagnosis	Antimicrobial Treatment
Shigella sp.	19,041	10–100	12–50 h	Fingers, fomites, food	Fever, diarrhea, pus and blood in feces	Isolation from feces or ulcer	Ampicillin if severe
Salmonella enteriditis	33,715	10^5	8–40 h	Food	Diarrhea, fever, nausea; fecal leukocytes	Isolation from feces and food	Unnecessary
Escherichia coli	*	10^8	24 h	Food	Diarrhea, fever, pus and blood in feces	Isolation of same sero-type from feces and food	Unnecessary
Vibrio parahaemolyticus	12*	10^7	8–24 h	Seafood	Diarrhea, cramps, nausea, occasion-ally fever	Isolation from feces	Unnecessary
Campylobacter jejuni	162*	10^6	1–7 days	Water, food,	Diarrhea, abdomi-nal pain, fever	Isolation from feces or food	Unnecessary
Yersinia enterocolitica	*	—	3–7 days	Food, water	Diarrhea, fever, abdominal pain	Isolation from feces or food	Unnecessary
Entamoeba histolytica	6,632	10 cysts	2–4 weeks	Water, food	Diarrhea, pain, nausea	Microscopic identifica-tion of cyst or tropho-zoites in feces; ingested red blood cells in trophozoites	Metronidazole
Gastroenteritis viruses Rotavirus	—	—	16–48 h	Food, water	Diarrhea, dehy-dration	Electron microscopy or serologic tests to detect virus in feces	None
Norwalk agents	—	—	16–48 h	Food, water	Diarrhea, abdomi-nal pain		None

*These are not notifiable diseases. Some cases are reported when associated with investigated outbreaks.

A preliminary diagnosis of dysentery is provided by microscopic detection of leukocytes and red blood cells in feces. *Shigella* can be isolated from fecal specimens or from rectal swabs on selective, differential, or enrichment media that discourage overgrowth by normal flora. Treatment is usually limited to supportive therapy and the replacement of fluids and electrolytes , and the disease is allowed to run its course. In severe cases, antimicrobial drugs, usually ampicillin, tetracycline, or trimethoprin-sulfamethoxizole, are given. Recovery confers serotype-specific immunity.

Salmonellosis *Salmonella*, another genus of gram-negative bacilli, contains more than 1500 serologically distinct organisms that are capable of causing gastroenteritis in humans. These numerous serotypes have been consolidated into three species based on ecological considerations. *Salmonella typhi*, the causative agent of typhoid fever, is a single serotype highly adapted to humans. *Salmonella choleraesuis* is a single serotype commonly found among fowl, cattle, swine, and other nonhuman sources. It is rarely associated with human disease. The third species, *Salmonella enteriditis* encompasses all the other serotypes, most of which survive equally well in humans and animals. These organisms may cause an acute gastroenteritis of short duration, bacteremia, or a potentially fatal systemic infection called enteric fever.

Salmonella is the most commonly reported bacterial agent of gastroenteritis. These bacteria are shed from the gastrointestinal tracts of animals and, less frequently, humans. Salmonellosis is the most widespread of all zoonoses. The bacteria may be shed in the feces of infected animals or released during slaughter and processing of the meat. Foods obtained from animal sources (eggs, poultry, milk, and sausage) pose high risks for *Salmonella* contamination (Fig. 21-9). Special care in cooking and refrigerating these foods minimizes the potential for *Salmonella* survival and growth prior to ingestion. Pets are also a source of the bacteria and are commonly implicated in a fecal-oral spread. Pet turtles have been responsible for several outbreaks of *Salmonella* gastroenteritis, but dogs, cats, birds, and fish also harbor and shed the organism. Like humans, animals acquire *Salmonella* in the foods they eat, often from contaminated bonemeal and meat by-products. Federal regulations attempt to prevent the distribution of feeds containing *Salmonella* by prohibiting interstate transport of such materials. Unfortunately, unless entire herds or flocks are free of the organism, a few infected animals can readily transmit the pathogen during transport to the slaughterhouse, during confinement within crowded holding pens, and through contaminated machinery in the slaughterhouse. Eggs, another common source of *Salmonella*, become contaminated within the hen's oviduct or by contact with fecal material from infected hens. Organisms on the egg surface may be introduced into food products. The pathogens are usually destroyed by cooking; outbreaks are often associated with pastries, desserts, or drinks containing raw eggs. Outbreaks of salmonellosis have also been caused by marijuana that presumably had been contaminated with animal feces during growth or storage.

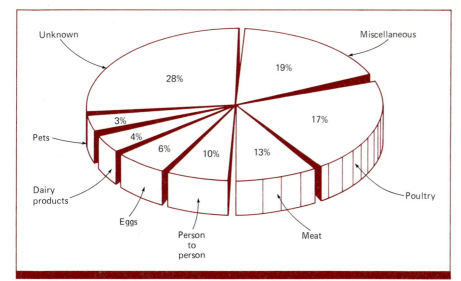

FIGURE 21-9

Mode of transmission in 500 outbreaks of human salmonellosis in the United States, 1966–1975. Miscellaneous category includes over 50 vehicles, each of which individually was responsible for less than 3 percent of outbreaks. *Samonella Surveillance Report No. 127, Annual Summary* 1976, Centers for Disease Control, November 1977.)

Because of the widespread distribution of the organisms and the numerous opportunities for spread, outbreaks of *Salmonella* gastroenteritis are common and difficult to control. Fortunately, *Salmonella* gastroenteritis is usually mild. In fact, most cases are believed to be asymptomatic. The organisms invade the mucosa of the large and small intestines causing inflammation, moderate fever, diarrhea, abdominal pain, and nausea. The symptoms usually appear within 6 to 36 hours after ingestion of the contaminated food, and most patients recover in 2 to 4 days. In compromised persons, the organisms may invade the bloodstream and spread to other body sites. The mortality rate is less than 1 percent of those infected.

Enteroinvasive *Escherichia coli* Gastroenteritis   A few strains of *Escherichia coli* can invade the epithelial cells lining the colon. These enteroinvasive *E. coli* cause fever and a dysentery syndrome that is similar to shigellosis. Clinical manifestations result from bacterial replication in the invaded tissue rather than from enterotoxin production. Enteroinvasive *E. coli* can be identified by their ability to cause conjunctivitis when inoculated in the eyes of guinea pigs. Antibiotics are used only when intestinal disease is severe.

***Vibrio parahaemolyticus* Gastroenteritis**   *Vibrio parahaemolyticus* is a halophile most commonly found in marine environments. This bacterium causes gastroenteritis following the ingestion of contaminated seafood. It has been recognized as a major cause of food poisoning outbreaks in Japan since 1956 and has been identified with increasing regularity in the United States in recent years, especially during summers, when the pathogen grows freely in coastal waters and may infect oysters, shrimp, crabs, and fish. During cold seasons, the organisms are submerged in the marine sediment. The disease is associated with eating contaminated raw or inadequately cooked seafood. Cooked foods may be recontaminated if the

545

ALIMENTARY-ACQUIRED
DISEASES

same surface or utensils are used for cooked and uncooked foods, or if the food handler fails to wash hands between manipulations. If recontaminated food is permitted to remain at temperatures between 20 and 35°C, the vibrios can rapidly multiply to infectious concentrations. The disease is rarely transmitted directly from an infected person.

Symptoms of *Vibrio parahaemolyticus* gastroenteritis are similar to those of foodborne illnesses caused by *Salmonella, Shigella,* and enteroinvasive *Escherichia coli*. Diagnosis, therefore, depends on isolating the organism from the patient's feces. Standard media used for the identification of enteric organisms will not support the growth of the marine vibrio. Special media containing increased concentrations of sodium chloride are used when *V. parahaemolyticus* is the suspected pathogen. Preventive measures depend on adequate cooking of seafood, storing seafoods at low temperatures, and avoiding the cross-contamination of cooked food by raw seafood. (Cooked and raw food should be handled in separate areas with different utensils, and hands should be washed when going between areas.)

Campylobacter **Gastroenteritis**   *Campylobacter jejuni,* a gram-negative vibrio first recognized as a human pathogen in 1947, causes as many as 11 percent of diarrheas worldwide and at least 5 percent of cases in the United States. When appropriate diagnostic techniques are employed, this pathogen is isolated more frequently than either *Salmonella* or *Shigella.*

Campylobacter jejuni is a pathogen of many animals and is transmitted to humans by ingestion of contaminated food, particularly raw milk or water. The organism can survive in fresh water for up to 5 weeks. It may also be spread to people, usually infants, by direct contact with fecal material from infected pets. Gastrointestinal symptoms with diarrhea begin 3 to 5 days after infection. It is common to find both blood and leukocytes in stools. Most infections are self-limiting; however, relapses occur in 20 percent of cases.

Healthy persons rarely harbor *Campylobacter jejuni,* and isolation of this vibrio from feces is conclusively diagnostic. *C. jejuni* is fastidious and must be grown in special media in a microaerophilic environment. Growth is optimal at 42°C. Severe infections are treated with fluid and electrolyte replacement therapy and with erythromycin.

Yersiniosis   *Yersinia enterocolitica* is a gram-negative coccobacillus that occasionally causes gastroenteritis in humans following ingestion of fecally contaminated food or water or by direct contact with contaminated feces. The disease is characterized by diarrhea and/or severe abdominal pain, usually accompanied by a fever. These symptoms often lead to a false diagnosis of appendicitis, and as many as 10 percent of persons with yersiniosis needlessly undergo appendectomies. About 20 percent of the patients infected with *Yersinia enterocolitica* develop disorders believed to be autoimmune. These include arthritis, erythema nodosum, and inflammation of the iris.

Y. enterocolitica grows best between 25 and 30°C on most media used for culturing enteric bacteria. The pathogen is more easily isolated if fecal specimens are treated with one of several enrichment techniques prior to plating. Severe infections can be treated with antibiotics.

Amebic Dysentery   Ingestion of water or food containing cysts of the protozoan *Entamoeba histolytica* may lead to a severe and sometimes fatal dysentery syndrome. Humans are the sole reservoir of this pathogen. Most infected persons, however, are asymptomatic. The ingested cysts pass through the stomach; the trophozoites emerge in the lower small intestine and multiply in the lumen of the colon. Cysts reform and are shed in feces. Approximately 450 million people are believed to be asymptomatically infected, each capable of shedding 3 million cysts daily. In some persons this carrier state continues for years.

Intestinal disease varies from mild diarrhea to severe dysentery, characterized by fever and frequent bloody, mucus-laden diarrhea. Disease develops when *Entamoeba histolytica* invades the mucosa, producing abscesses that ultimately enlarge into ulcers. The infection may spread from the colon to the liver and less often to lungs and brain. Liver abscesses sometimes occur in the absence of intestinal symptoms. Conditions that predispose to invasion are poorly understood. Infection is most prevalent and severe in areas of the world where crowding and poor sanitary conditions promote fecal-oral spread.

Diagnosis is confirmed by microscopic identification of the trophozoite in stool specimens or abscess aspirates. Trophozoites are fragile, and samples must be examined within an hour of collection or stored in the presence of a preservative. The greatest problem in diagnosis is distinguishing *Entamoeba histolytica* from similar amoebas that are part of the normal intestinal flora. Finding trophozoites containing red blood cells is diagnostic because only invasive *E. histolytica* characteristically ingest erythrocytes. Cysts can be found only in formed stools and are therefore shed only during asymptomatic or mild infections. When cysts can be found, they have one to four nuclei (unlike the normal flora *Entamoeba coli*, which have eight or more nuclei per cyst). Serologic tests are usually used to confirm diagnosis of extraintestinal infections.

Acute dysentery is usually treated with metronidazole in combination with diidohydroxyquin to ensure the destruction of amoebas in three sites—the bowel lumen, bowel wall, and invaded organs. Control of amebic dysentery depends on sanitary disposal of feces and treatment of carriers. Since persons with acute disease do not shed cysts, it is not necessary to isolate them. Travelers to endemic areas can reduce the likelihood of infection by boiling all suspected water prior to drinking and eating only cooked foods.

Viral Gastroenteritis   The source of the viruses that cause gastroenteritis is usually the infected intestines of humans, and the mode of transmission is primarily fecal-oral, either in food or water or by direct contact with contaminated feces. Gastroenteritis viruses are believed to be responsible for diarrheagenic illnesses that kill as many as 18 million people each year.

Even before their discovery, the existence of such agents was postulated in order to explain the many cases of gastroenteritis from which no bacterial agents could be cultured. In the early 1970s electron microscopic examinations of stool samples revealed the cause of many of these undiagnosed cases. The *Norwalk viruses* cause epidemic outbreaks of gastrointestinal illness in schools, families, and communities, and may be one of the causes of the erroneously named "stomach flu," a syndrome that is clearly not caused by influenza virus. Epidemic viral gastroenteritis usually is a self-limited illness that lasts 1 or 2 days. The *rotaviruses,* on the other hand, are a major, if not *the* major, cause of severe diarrhea among infants and young children. Rotavirus infections often leave children so severely dehydrated that they require hospitalization to replace the lost body fluids. The nature of the pathogenic mechanisms is not certain, but both types of viruses alter the absorbing surfaces of the small intestines, probably by multiplying in and damaging intestinal epithelial cells. Viral gastroenteritis is diagnosed by serological tests or by using an electron microscope to detect the virus in feces (Fig. 21-10). Development of vaccines has been hampered by difficulty in cultivating the viruses in the laboratory.

INVASIVE DISEASES   A few pathogens enter the gastrointestinal tract and subsequently invade tissues in other parts of the body. Although their characteristic symptomology is displayed outside the alimentary tract, all of these organisms are excreted within feces and are transmitted primarily through a fecal-oral route. Characteristics of invasive diseases are summarized in Table 21-8.

FIGURE 21-10
Rotaviruses isolated from
human feces.

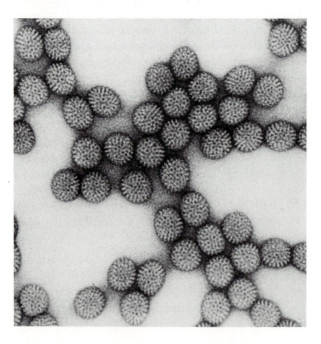

TABLE 21-8
Characteristics of Invasive Pathogens That Cause Extraintestinal Infections

Pathogen	Disease	Cases Reported in U.S., 1981	Major Vehicle	Incubation Period	Clinical Picture	Diagnosis	Antimicrobial Treatment
Salmonella typhi	Typhoid	584	Food, water	1–3 weeks	Fever, malaise, abdominal pain	Positive blood culture during first week, positive stool or urine specimen	Chloramphenicol, ampicillin
Hepatitis A virus	Hepatitis	25,802	Food, water	10–45 days	Fatigue, nausea, abdominal pain, jaundice	Clinical and epidemiological findings	None
Poliovirus	Paralytic	6	Salivary spread, food, water	7–12 days	Fever, headache, paralysis	Positive throat culture, spinal fluid, or rectal swab	None polio
Toxoplasma gondii	Toxoplasmosis	196	Raw meat, cat feces	5–21 days	Neurological and ocular damage in severe cases	Positive serological tests, isolation of pathogen from body fluids	Pyrimethamine plus sulfa drugs in severe cases

Enteric Fever A small number of *Salmonella* strains possess sufficient virulence to invade beyond the intestinal tract and produce a severe systemic infection, characterized by invasion of lymphoid tissue and prolonged fever. This clinical syndrome is called **enteric fever**; its classical form is *typhoid fever*. The pathogen, *Salmonella typhi*, multiplies in the intestinal epithelium, and the progeny are engulfed by macrophages, which fail to kill the bacteria. The parasites are transported in host macrophages to regional lymph nodes, invade the bloodstream, and may localize in the liver, spleen, gallbladder, kidneys, bone marrow, heart, lungs, and lymphoid tissue of the gastrointestinal tract. Fever slowly elevates to an average of 104°F and the skin may become dotted with small hemorrhages called rose spots. Gastrointestinal symptoms appear late in the course of the disease and are usually characterized by constipation followed by bloody diarrhea. Vomiting and abdominal tenderness may also occur. The severity of disease ranges from asymptomatic or mild to fatal infections. The mortality rate of untreated typhoid fever is 10 to 15 percent. This rate can be reduced by antibiotic therapy. Several *Salmonella enteriditis* serotypes cause a milder form of enteric fever. The disease, called *paratyphoid fever,* is similar to typhoid but is much less severe.

Enteric fever is transmitted by the fecal-oral route. The disease develops 1 to 3 weeks following the ingestion of the pathogens in food or water that has been contaminated by the feces of overtly sick victims or carriers. Poor personal hygiene habits, especially in food handlers, facilitate the direct inoculation of foods. Improper treatment of public water supplies or inadequate disposal of human excretion can cause outbreaks of typhoid in large populations. During the course of the disease, the infected gallbladder pours pathogens into the feces, seeding the environment with infectious bacteria. The gallbladder often continues to shed the organism during convalescence and sometimes remains infected after the patient recovers. Such asymptomatic persons become healthy carriers of typhoid fever.

Typhoid fever is diagnosed by identifying the pathogen isolated from blood, feces, or urine. Enrichment media containing selenite or tetrathionate favor the growth of *Salmonella* over gram-negative bacteria of the normal flora. These enriched cultures are then plated onto selective differential media to obtain isolated colonies of the pathogen. The bacteria are identified by biochemical tests and agglutination with *Salmonella*-specific antiserum. Rising antibody titers against *S. typhi* can be demonstrated by comparing serum samples drawn during acute and convalescent stages of the disease.

Control of the disease is complicated by the existence of relatively large numbers of chronic carriers. Two to five percent of those recovering from typhoid fever become permanent carriers. Controlling the disease depends on identifying these carriers and barring them from occupations entailing food handling or care of debilitated persons who may be extremely susceptible to infection. If these recommendations are followed, typhoid carriers pose no threat to the general public. The carrier state may also be eliminated by appropriate antibiotic therapy. Although chloramphenicol, despite its potentially severe side effects (including aplastic anemia), is still the most effective drug available for treating active cases of typhoid fever, ampicillin is the antibiotic of choice for treating carriers. Unlike chloramphenicol, ampicillin accumulates in high concentrations in the gallbladder. Antibiotic resistance is becoming a more frequently encountered problem. When drug therapy fails, surgical removal of the gallbladder is the only reliable way to eliminate continued shedding of pathogens (see color box).

Polio *Poliovirus* is a small RNA virus often spread by fingers, foods, flies, and fomites, the traditional vehicles of fecal contamination. This virus produces mild gastrointestinal infections in more than 90 percent of the persons it infects. It multiplies in lymphoid tissues of the tonsils and in the regional lymphoid tissue of the intestines. Local IgA usually restricts spread of the virus to other body sites. If the virus should spread to the bloodstream, IgG neutralizes it before it can infect other organs. In fewer than 1 percent of cases, however, the poliovirus infects the central nervous system. Sometimes this leads to a meningitis that lasts about 10 days and then subsides. A more severe complication evolves when the virus damages motor neurons and causes **paralytic poliomyelitis.** This form of the disease may be fatal if the muscles controlling respiration and swallowing are paralyzed. Factors that predispose for paralytic complications include tonsillectomy and exposure to an especially large dose of virulent poliovirus. Although polio is characteristically a disease of children, the danger of paralysis is greater in adults. The spread of virus through the infected body is illustrated in Figure 21-11.

Until the introduction of the inactivated poliovirus vaccine (**Salk vaccine**) in 1955, polio was a major cause of paralysis in the United States. It was not as prevalent in developing countries, where poor sanitation

FIGURE 21·11

The spread of poliovirus
through the human body.

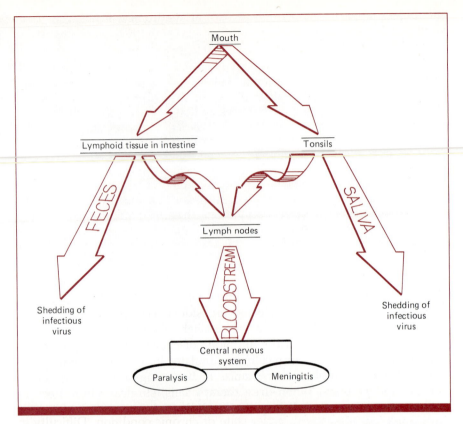

Mouth

Lymphoid tissue in intestine — Tonsils

FECES — SALIVA

Lymph nodes

BLOODSTREAM

Shedding of
infectious
virus

Shedding of
infectious
virus

Central nervous
system

Paralysis — Meningitis

resulted in early exposure to the virus, milder cases, and permanent immunity. In the United States children were more likely to escape natural exposure to the virus because of better sanitation, producing an older population of susceptible persons, with a greater risk of paralytic sequelae. The Salk vaccine was developed after advances in tissue culture techniques permitted viruses to be propagated in large quantities. In 1961 the introduction of attenuated oral vaccine (**Sabin vaccine**) further reduced the disease, so that in 1980 only eight paralytic cases were reported in the United States, and those were restricted to populations that remained unvaccinated, usually because of religious objections to medical procedures. These cases emphasize the continued existence of poliovirus in the environment and the need for rigid enforcement of immunization requirements.

Polioviruses are **enteroviruses,** a group of RNA viruses that usually enter the body via the alimentary tract (or respiratory tract) and cause asymptomatic infections or mild localized disease in the intestine. There are three categories of viruses in this group, *Coxsackieviruses, echoviruses,* and polioviruses. Although usually inapparent and undiagnosed, infections by these viruses can be systemic or localized in nonintestinal sites such as the meninges. The usual illnesses caused by Coxsackieviruses, for example, are severe sore throats with exudative lesions (called herpangia), an influenzalike respiratory disease (called summer grippe), and aseptic meningitis. Echoviruses also cause several syndromes, including a diarrhea that is particularly severe in newborns, a common coldlike illness, disseminated infections with skin rash and fever, and aseptic meningitis. Enteroviruses are primarily spread by the fecal-oral route, although respiratory secretions may also transmit these infections.

Hepatitis A   At least three distinct viruses are capable of causing **hepatitis**, a disease of the liver that results in jaundice (yellowing of the skin and other tissues). These viruses are *hepatitis A virus (HAV)*, *hepatitis B virus (HBV)*, and a yet-unidentified agent or agents referred to as *non-A non-B (NANB) hepatitis virus*. Although all three may be acquired through the gastrointestinal tract, hepatitis B and NANB hepatitis virus are primarily acquired by parenteral inoculation; their prevention and control, therefore, differ from prevention and control of hepatitis A. Because of these epidemiologic differences, heptatis B and NANB hepatitis viruses will be discussed in Chapter 23.

   Hepatitis A is transmitted primarily by the fecal-oral route, either in contaminated water and food (especially raw or poorly cooked clams and oysters from contaminated waters) or by direct person-to-person spread. The latter mechanism of transmission is especially prevalent in institutional settings. Because it is transmitted directly, hepatitis A has been popularly called "infectious hepatitis." Hepatitis A is characterized by an incubation period of 2 to 6 weeks during which the virus multiplies in the gastrointestinal tract, spreads to the blood, and invades the liver, where it may proliferate asymptomatically or cause jaundice. Patients usually recover in 6 weeks, acquiring immunological protection against hepatitis A. There is no permanent carrier state or chronic condition. Difficulty in propagating the virus has hindered studies on replication and pathogenisis, as well as the development of vaccines. Diagnosis by radioimmunoassay (RIA) and ELISA tests immunologically detect the virus in feces or blood. Although treatment of hepatitis is limited to supportive therapy, prophylactic administration of immune serum globulin provides temporary protection for persons exposed to the virus. Such passive immunization is recommended for laboratory workers and hospital personnel who will be in contact with infectious patients or clinical specimens. Gowns, gloves, and masks provide added protection.

Toxoplasmosis   Humans and many animals harbor the asexual cyst stage of the protozoan *Toxoplasma gondii* in various extraintestinal tissues, particularly in muscle and brain. A sexual form, the oocyst, occurs only in cats and is shed in the animals' feces. Infection is usually acquired by ingestion of cysts in raw or undercooked meats. Ingestion of oocysts in contaminated food or water or on fingers is another primary source of infection.

   The disease is characterized by the intracellular multiplication of the trophozoite in many extraintestinal tissues. Infections of immunologically competent persons are usually asymptomatic or mild, with fever, headache, and muscle pain. Symptoms subside spontaneously as an immune response develops. The parasites, however, remain in the tissues in a latent cyst form, and reactivation of the infection may occur if the host becomes immunologically compromised. Such endogenous reinfections usually result in severe neurological symptoms.

Toxoplasma gondii is especially dangerous when the trophozoite is transmitted from an infected mother to the fetus. Transplacental transmission occurs only when the mother has the *initial* disease during pregnancy but does not occur during reactivation infections. Fetal infection may be asymptomatic, result in fetal death, or cause congenital toxoplasmosis. Infected infants are born with neurologic and ocular damage or develop it later. Pregnant women should be tested for immunity to *T. gondii* and if not immune should avoid contact with cats and eating raw meat.

Infections are diagnosed by demonstrating rising antibody titers in the patient's serum samples. Only severe cases are treated with pyrimethamine-sulfonamides.

DIAGNOSIS

The clinical symptoms of many gastrointestinal disturbances are strikingly similar despite the differences in etiologic agents. A presumptive diagnosis is sometimes based on the length of the incubation period and on the nature of the vehicle of transmission. Foodborne illness that occurs within 1 hour after ingestion is probably of nonmicrobial origin. Onset of symptoms between 1 and 7 hours suggests staphylococcal food poisoning, especially if vomiting is the primary symptom. When the incubation period is between 8 and 14 hours, *Clostridium perfringens* is the probable pathogenic agent. If the incubation period is greater than 14 hours, an invasive pathogen is likely to be responsible.

The most objective criterion for diagnosing microbially induced disease is isolation of the pathogen or toxin from the patient's stool or vomit and, if available, from the incriminated vehicle. Unfortunately, in approximately two-thirds of reported outbreaks the responsible vehicle is either not identified or has been discarded. Isolation of pathogens from fecal specimens is especially difficult because feces are generally teeming with microorganisms of the normal flora, which may obscure detection of the pathogen. Infections caused by *Escherichia coli*, *Clostridium perfringens*, and *Clostridium difficile* are particularly difficult to diagnose since these organisms are part of the commensal population of the bowel.

Although negative stool cultures are not uncommon, pathogens may be isolated more readily if specimens are collected at the appropriate time and processed under conditions that cater to the individual needs of the suspected microbes. The most appropriate fecal specimen is collected during the acute stage of a diarrheal illness. During this time, the repeated evacuations of the bowel reduce the concentration of normal flora. Samples should be examined microscopically and inoculated onto the appropriate culture media as soon as possible after collection. This prevents the overgrowth of the commensal flora as well as the death of some of the more fragile pathogens such as *Shigella*.

In several cases, microscopic examination of stool samples yields valuable diagnostic information. Protozoan diseases, for example, are diagnosed primarily by the microscopic identification of trophozoites and cysts in feces. In addition, the presence of fecal leukocytes is presumptive evidence of local invasion of the intestinal wall by *Shigella, Salmonella,* or *Entamoeba.* The differential white cell count can be used to distinguish between these invasive pathogens. Polymorphonuclear leukocytes predominate in fecal smears from patients with *Shigella* and *Salmonella* infections, whereas mononuclear cells predominate in stool samples from patients with amebic dysentery.

Because symptoms of most GI illnesses are so similar, a variety of media that allow the multiplication of the entire spectrum of suspected agents must be used. Primary plating media for fecal samples, water, or incriminated foods should include:

Blood agar or other enriched media that will support the growth of virtually all organisms within a sample

A selective and differential medium for growing gram-negative enteric bacilli (usually MacConkey's agar or Eosine-methylene blue agar)

Enrichment broths or highly selective media that are designed for the isolation of *Salmonella* and *Shigella*

Selective media for vibrios (In the United States the most likely vibrio to be encountered is the halophile *Vibrio parahaemolyticus.*)

Anaerobic media for oxygen-free growth of clostridia

If necessary, further characterizations are performed by biochemical or serological tests. Although this diagnosis process is often completed after the patient recovers, the information may help identify the existence of an epidemic and help prevent spread of disease.

THERAPY

The majority of infections acquired through the alimentary tract are self-limited, lasting from 1 to 7 days. Most persons recover with supportive care at home and do not require the use of antibiotics. Antibiotic therapy is further contraindicated because nonbacterial pathogens are often responsible for the disease. Antimicrobial therapy is generally recommended for treating typhoid fever or severe cases of shigellosis or when the infected host is debilitated and cannot depend on the body's defenses to fend off the invading pathogen.

If repeated diarrheas result in critical fluid and electrolyte loss, prompt replacement of fluid and electrolytes is the most essential treatment. Mild cases are treated by a temporary diet of salt-rich soups and juices. Severe cases require intravenous therapy to replace lost body fluids and electrolytes.

PREVENTION AND CONTROL

Some preventive measures can be implemented by individuals; others are directed by community and public health authorities. A joint effort by food processers, consumers, and the medical community is the best insurance against widespread disease.

Individual Control Measures

Transmission of most diseases acquired through the digestive tract is enhanced by carelessness or ignorance. The following guidelines help protect individuals from ingesting pathogen-laden food or water or from oral contact with contaminated vehicles.

1. Persons directly or indirectly handling food should keep hands clean by using proper hand-washing techniques before and after food preparation and after exposure to fecal material or uncooked meat. Handling foods as little as possible discourages contamination. Gloves are recommended to prevent shedding of the resident skin flora, such as toxigenic staphylococci. Working surfaces, utensils, and processing equipment should be maintained in a sanitary state by frequent cleaning.

2. All persons should avoid working with foods when they have any infection. Mandatory screening programs covering applicants for positions as food handlers help identify carriers of typhoid fever or viral hepatitis. Even employees with minor sores can be the source of foodborne epidemics. Persons are more likely to report infections if they suffer no stigma or loss of pay when they are temporarily unable to work because of transient infections.

3. All foods to be eaten raw should be thoroughly washed and prepared in areas other than those used for the preparation of meats, poultry, seafood, and other potentially contaminated food. Fruits and vegetables may be contaminated from feces in fertilizers or from soil, as well as from individuals handling the products.

4. Cooking kills foodborne pathogens only if lethal temperatures are reached in all portions of the food.

5. Prompt and proper storage of food prevents the multiplication of microbes. Refrigeration temperatures should be below 5°C. Pathogens have little opportunity to proliferate if all parts of the food reach these bacteristatic conditions quickly.

6. Appearance, odor, and taste are not always accurate indicators of whether food is fit for consumption. Nonetheless, any food that smells foul, produces a gas, or is otherwise suspected of being contaminated should not be eaten, or even tasted.

Community Control Measures

Several governmental and private agencies have established guidelines to monitor water and food supplies prepared for large-scale consumption;

these agencies also regulate programs that discourage the spread of disease imported into the United States from other areas of the world. Public health measures include the following:

Purification and chlorination of water supplies

Pasteurization of milk and dairy products

Monitoring foods for the presence of pathogens or for an unusually high coliform count, which indicates fecal contamination and the probable presence of fecal-borne pathogens

Proper sewage disposal that eliminates pathogens and discourages vector multiplication

Establishment and supervision of sanitary guidelines for commercial manufacturing firms and eating facilities

Control of animal and human carriers of disease

Control of fly populations to discourage contamination of food and fomites by these vectors

Vaccination programs to protect against polio, cholera, and typhoid fever, the three most serious diseases acquired from food, water, and feces (Fig. 21-12)

Establishment of an information network to detect early evidence of disease outbreaks and to implement measures to control the source of the pathogenic agent. Physicians must notify the Centers for Disease Control of each case of notifiable disease diagnosed. Amebiasis,

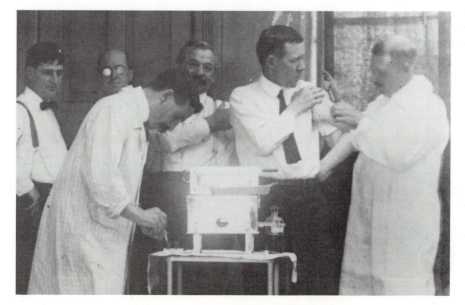

FIGURE 21-12
The administration of the
first typhoid vaccine.

botulism, cholera, hepatitis, polio, salmonellosis, shigellosis, and typhoid fever are all notifiable diseases acquired through the alimentary tract.

Health Detectives

Outbreaks of reportable diseases are nearly always detected by public health authorities. In contrast, episodes of nonnotifiable diseases of the alimentary tract are usually identified only when large numbers of individuals become ill following exposure to a common contaminated source. In either situation public health investigators attempt to identify the causative agent and its reservoir, determine vehicles and conditions contributing to transmission, and reduce further infections.

Investigative field workers interview persons exposed to the suspected source, physicians treating the victims, and, in cases of foodborne disease, personnel who prepared the suspected foods. Field workers also inspect the premises where the foods were prepared, stored, and served and collect any food samples still available and clinical specimens from food handlers. Laboratory personnel attempt to isolate a pathogen or toxin from the samples collected by field workers. The combined field and laboratory data often reveal the source of the outbreak and the vehicles. Armed with this information, authorities can usually halt the outbreak and prevent its recurrence.

Contaminated food is usually identified by the nature of the pathogen (determined from clinical findings) and the foods eaten by exposed persons. For example, any food eaten by those who become ill and not eaten by those who remained healthy is a likely candidate. In outbreaks involving large numbers of people, however, not all persons who ate the contaminated food will become ill, and some ill persons will deny eating it. These anomalies are probably due to faulty memories, variations in amounts consumed, and differing susceptibilities. Epidemiologists sort this out by compiling a table such as that presented in Table 21-9. An outbreak is most likely caused by the food that shows the highest difference in attack rates between people who did and who did not eat the food. For example, as seen in Table 21-9, in an outbreak among guests at a

TABLE 21-9

Food-Specific Attack Rates Following a Charity Luncheon

Food Served	No. of Persons Who Ate Specified Food				No. of Persons Who Did Not Eat Specified Food				Relative Risk*
	Ill	Not ill	Total	% ill	Ill	Not ill	Total	% ill	
Egg salad	54	11	65	83	6	15	21	28	3.0
Macaroni and cheese	27	8	35	77	33	17	50	66	1.2
Cottage cheese	40	18	58	69	20	7	27	74	0.9
Tuna salad	49	14	63	78	15	7	22	68	1.1
Ice cream	31	9	40	78	29	16	45	64	1.2

*Relative risk = $\dfrac{\text{Percent of ill among those who ate the food}}{\text{Percent of ill among those who did not eat the food}}$

charity luncheon it is apparent that the chance of getting ill was approximately three times greater among persons who ate the egg salad than among persons who did not. No such difference was noticed with other vulnerable foods. Thus the egg salad is implicated as the vehicle in this outbreak. Further field investigation often reveals the source of the pathogen. In this case, the egg salad may have been contaminated by an infected food handler, by raw eggs used in mayonnaise, or by contaminated water.

OVERVIEW

The alimentary tract is a portal of entry for microbes that cause diseases of the oral cavity and gastrointestinal tract and for a few pathogens that invade other areas of the body. Most healthy people are susceptible to these pathogens and will become infected if exposed to highly contaminated food or drink. Consumption of preformed microbial toxins also causes disease. Neurotoxins (botulinum toxin, dinoflagellate toxin, and some mycotoxins) affect the nervous system, whereas toxins produced by *Staphylococcus aureus* or *Clostridium perfringens* induce gastrointestinal symptoms.

Most microbes that enter the alimentary tract must attach firmly or they will be forced out by the constant flushing of its contents. Microbes in the oral cavity adhere to teeth and contribute to the development of dental caries and periodontal disease. Local lesions of the oral mucosa are caused by *Candida albicans* and herpesvirus. Some pathogens of the gastrointestinal tract survive the acid environment of the stomach and cause disease by producing enterotoxins (*Vibrio cholerae*, enterotoxigenic *Escherichia coli*, *Clostridium perfringens*, *Clostridium botulinum*, and *Clostridium difficile*). These enterotoxins trigger the release of large volumes of fluids and electrolytes into the bowel, resulting in profuse diarrheas. *Giardia lamblia*, on the other hand, physically interferes with the absorptive capacity of the intestines.

Locally invasive pathogens penetrate the epithelial surface of the intestines and usually cause dysentery (*Shigella*, *Salmonella enteriditis*, enteroinvasive *Escherichia coli*, *Vibrio parahaemolyticus*, *Campylobacter jejuni*, *Yersinia enterocolitica*, and *Entamoeba histolytica*). Gastroenteritis viruses invade these surfaces, but rarely cause dysentery.

When noninvasive or locally invasive pathogens cause the loss of large volumes of fluids and electrolytes, prompt replacement is critical to the patient's survival. Chemotherapy may not be necessary, the diseases often resolving spontaneously.

Some pathogens penetrate through the intestine wall into the bloodstream and disseminate. Among these pathogens are microbes that produce enteric fever, toxoplasmosis, polio, and hepatitis.

Most foodborne diseases can be avoided if foods are prepared with care to prevent microbial contamination and growth. Waterborne diseases are usually controlled by adequate treatment of public water supplies.

peristalsis

intoxication

enterotoxin

neurotoxin

mycotoxin

plaque

gastroenteritis

dysentery

enteric fever

REVIEW QUESTIONS

1. How do the following protect against alimentary-acquired disease?
   (a) Peristalsis
   (b) Low pH
   (c) Normal flora
   (d) Secretory IgA

2. What virulence factors are used by alimentary-acquired pathogens to overcome normal host defenses?

3. Why is direct contact not a major means of transmission for most pathogens that enter through the alimentary tract?

4. Compare the pathogenesis of *Staphylococcus* food poisoning, cholera, bacillary dysentery, and typhoid fever.

5. What is the role of sucrose in the development of dental caries and periodontal disease?

6. Identify the pathogen(s) most likely to be transmitted in each of the following vehicles, and describe how best to prevent associated diseases:
   (a) Home-canned foods
   (b) Shellfish
   (c) Foods with high egg content
   (d) Untreated mountain springwater
   (e) Honey

7. Why are most diseases of the gastrointestinal tract not treated with antibiotics?

Diseases Acquired through the Genitourinary Tract

22

The human genitourinary (GU) tract is the portal of entry for many of our most annoying and feared pathogens, some of which continue to cause epidemics in spite of modern control measures. More cases of gonorrhea are reported to public health officials in the United States than of all other infectious diseases combined. Every 12 seconds another person contracts this disease, which may be more prevalent than the common cold among sexually active persons. Another genitally acquired disease, syphilis, injures the heart and central nervous system. Moreover, since a woman's genital tract is also the birth canal, infections in a pregnant woman's vagina can be transmitted to her child during birth and cause permanent impairment or fatal diseases in the newborn child. Urinary tract infections are also recurrent problems, bringing discomfort to many people and threatening the lives of others. Most GU pathogens are sufficiently virulent to establish infection in spite of the protective mechanical, chemical, and cellular defenses of the human genitourinary tract.

ANATOMY AND DEFENSES OF THE GENITOURINARY TRACT

Urinary Tract

The anatomy of the upper urinary tract is similar for both men and women (Fig. 22-1). The kidneys remove waste materials from the blood and transfer them in the form of urine through the ureters into a "holding tank" called the *urinary bladder*. Since filtration of wastes is an essential function, people who lack kidney function survive only if another kidney is surgically transplanted or if their blood is routinely filtered by dialysis machines. Fortunately, the kidneys are located well within the body where they are isolated from pathogens in the external environment. The flow of urine flushes out microbes that gain entry into the lower urinary tract. This hydrokinetic washing is assisted by sphincter muscles at each end of the ureters that help prevent *reflux*, a backflow of urine that could carry microorganisms into the kidney. Immunological and cellular defenses also protect the kidneys from pathogens disseminated through the bloodstream.

The lower urinary tract consists of the bladder and the *urethra*, the tube through which urine flows out of the body during voiding. The external opening of the urethra is located in the anterior vulva in women and at the distal end of the penis in men. The male urethra is approximately five times longer than the female urethra. The length of the male urethra helps protect males from **cystitis** (inflammation of the bladder).

The urinary tract of males is also protected by fluid secreted by the prostate. These secretions contain spermidine, zinc, and other antibacterial chemicals. In females the acidity of vaginal fluids helps protect the region around the urethral opening from colonization by microbes that can cause urinary tract infection.

Physical and chemical properties of urine prevent the growth of some microbes. Its low pH and high concentrations of dissolved waste products

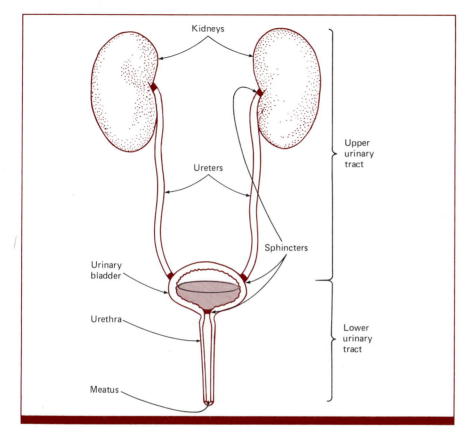

FIGURE 22·1
Anatomy of the urinary
tract. It is similar for both
sexes, except for the final
segment of the urethra,
which runs through the
penis in males and to the
external vulva in females.

Labels in figure: Kidneys, Ureters, Sphincters, Urinary bladder, Urethra, Meatus, Upper urinary tract, Lower urinary tract

are antibacterial. *Urea,* one of the major urinary constituents, inhibits the growth of some bacteria. However, urine is an excellent culture medium for the growth of other bacteria and supports the growth of a wider range of pathogens when its physical or chemical properties are even slightly altered—by hormonal imbalance, for example. Consequently, pathogens that do manage to invade the urinary tract may find the conditions ideal for growth, producing painful or even dangerous diseases.

The urinary bladder is so well protected that its contents are normally sterile. Microbes are flushed from the urethra by the flow of urine, and reflux is prevented by a set of sphincter muscles. The occasional microbes that reach the bladder will likely become trapped in a layer of sticky mucus that prevents their attachment to the underlying susceptible epithelial cells of the bladder. Normal sloughing of the cells that line the urinary tract sheds attached bacteria before they become entrenched.

Genitals

The male and female genital tracts are illustrated in Figure 22-2. Each system produces gametes for sexual reproduction. The male gametes, called *sperm,* travel from the *testes* through accessory sexual ducts to the urethra. The sperm are suspended in a liquid, called *semen,* produced by

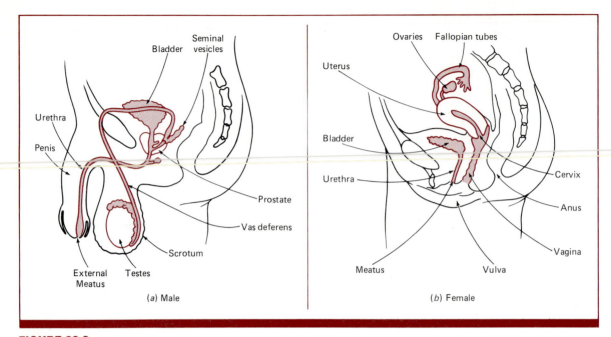

FIGURE 22-2

Anatomy of the (*a*) male and (*b*) female genital tracts and accessory structures.

the prostate and seminal vesicles. Millions of sperm are present in each ejaculate from a fertile man. This system is protected by bactericidal secretions of the seminal vesicles and prostate, the length of the male urethra, and flushing by urine and semen.

The female gametes, called *ova* or "eggs," are discharged from the *ovaries* and swept into the *fallopian tubes* where they may be fertilized by sperm. The egg enters the *uterus*, and, if fertilized, implants and develops into a fetus. Unfertilized gametes continue through the cervix into the vagina and finally out the external genitals, the *vulva*.

Because of their anatomy, females are more vulnerable than males to many GU infections. The moist, warm environment of the vagina provides a hospitable location for many microorganisms. The vaginal opening, unprotected by a sphincter or valve, facilitates the introduction of potentially pathogenic organisms into the vagina. Pathogens are often introduced by the insertion of the penis or a tampon. In addition, the close proximity of the anus encourages the accidental transfer of rectal flora to the vulva and vagina.

In spite of its vulnerability the human vagina usually remains healthy. Phagocytic inflammatory cells and secretory IgA antibody play important roles in resistance to colonization by pathogens and recovery from infections. The lysozyme content of cervical secretions contributes to the overall bactericidal activity in the vagina and uterus. The healthy state of the vagina can be largely attributed to the protective role of the normal vaginal flora that oppose colonization and proliferation of pathogens. These

TABLE 22-1

Defenses of the
Genitourinary Tract

Type	Responsible Factor	Mechanism of Protection
Mechanical	Location of kidneys	Isolates kidneys from external source of microbes
	Flow of urine	Flushes away microbes
	Sphincters	Prevent reflux (backflow)
	Length of male urethra	Puts more distance between bladder and external source of microbes
	Mucous lining in bladder	Prevents microbial attachment to bladder epithelium
	Sloughing of epithelium	Sheds attached microbes
Chemical	Low pH of urine	Opposes growth of many microbes
	Dissolved wastes in urine	Osmotic pressure opposes growth of some microbes
	Urea in urine	Is bacteriostatic for some bacteria
	Prostatic secretions	Contain bacteriostatic substances
	Cervical secretions	Contain lysozyme that helps protect vagina and uterus
Microbial	Normal vaginal flora	Lactobacilli maintain low pH inhospitable to most microbes.
Cellular	Leukocytes	Phagocytose pathogens
Immunological	Secretory IgA	Neutralizes pathogens at GU surface

bacteria convert glycogen to lactic acid, maintaining a vaginal pH lower than most pathogens can tolerate.

The defenses of the GU tract are summarized in Table 22-1.

Normal Flora of the Genitourinary Tract

The upper urinary tract and bladder are normally sterile. The microbes that inhabit the healthy GU tract, however, can contribute to host defenses or can become agents of disease. Lactobacilli are predominant in the vagina of a woman of childbearing age. Small numbers of the vaginal pathogens *Candida albicans* and *Trichomonas vaginalis* may also reside in the healthy vagina. In large concentrations, however, they elicit vaginal inflammation, or *vaginitis*.

During childbearing years the flora of women are influenced by the action of estrogen and progesterone, hormones that begin to appear shortly before puberty and subside at menopause. These hormones maintain the vaginal glycogen concentrations necessary to sustain the growth of the predominant lactobacilli in the vagina. Prior to puberty and after menopause the vaginal flora are less homogeneous, consisting primarily of coliforms, diphtheroids, and streptococci. The uterus, fallopian tubes, and ovaries of the healthy female are free of microorganisms, as are the prostate, vas deferens, and testes of the healthy male. The microbiota indigenous to the normal GU tract are described in Table 22-2.

PREDISPOSING FACTORS

Although many pathogens can overwhelm the healthy GU tract and establish infectious disease, others colonize and injure only genitourinary systems with compromised defens-

TABLE 22-2

Microorganisms Indigenous
to the Genitourinary Tract

Region	Organism
Upper urinary tract	Sterile in healthy individuals
Bladder	Sterile in healthy individuals
Urethra, male and female (outer third only)	*Streptococcus, Mycobacterium smegmatis, Bacteroides* spp., *Neisseria*, gram-negative enteric bacilli
Vagina (during childbearing years)	*Lactobacillus* (predominant), minor concentrations of gram-positive bacilli, diphtheroids, gram-negative bacilli, gram-positive cocci, *Candida albicans, Trichomonas vaginalis*
Uterus, fallopian tubes, ovaries, vas deferens, testes	None
Surface of penis	Representative of skin flora

es (Table 22-3). Any factor that obstructs the flow of urine or promotes reflux interferes with the hydrodynamic removal of pathogens. Obstructions may be due to scarring or strictures (abnormal narrowing) in the urethra or ureters, the presence of urinary calculi ("stones"), or the growth of tumors. Habitual overdistention of the bladder forces a backflow of urine into the kidneys and can lead to **pyelonephritis** (inflammation of the kidney). Defective sphincters may also encourage backflow. Urine remaining in the bladder after partial voiding may become a reservoir for organisms that cause cystitis. Incomplete emptying of the bladder may be due to interrupted flow during voiding (some persons habitually stop voiding before bladder is empty) or to structural abnormalities in the bladder that produce a "pool" of residual urine.

Females are more affected by these factors because of the short distance microbes must travel to reach the bladder (approximately 4 cm vs. 20 cm in the adult male). Any trauma to the urethral opening in women may facilitate the introduction of microbes into the bladder. For this reason, some women frequently contract bladder infections after sexual activity, a condition referred to as "honeymoon cystitis." Voiding urine immediately after intercourse reduces the likelihood of developing this infection. In men, the most common lower urinary tract infection is **urethritis** (inflammation of the urethra).

Urinary catheters are associated with cystitis in both sexes. These instruments often transfer surface microbes into the bladder in sufficient numbers to cause infectious disease. When catheterized individuals are victims of spinal cord injury or other disorders that impair nervous control of sphincters, infections of the urinary tract are almost inevitable. Recurrent or prolonged infectious disease of the lower urinary tract predisposes for pyelonephritis by providing a reservoir of pathogens that can ascend from the bladder to the kidneys.

Disruption of the normal vaginal flora by broad-spectrum antibiotic therapy or underlying diseases such as diabetes mellitus often leads to **vaginitis** (inflammation of the vagina). Vaginitis is more common during menopause, pregnancy, or while taking oral contraceptives because of the estrogen and progesterone imbalances associated with these conditions.

TABLE 22-3

Some Factors Predisposing for the Development of Genitourinary Tract Disease

Factor	Site of Disease	Preventive Measure
Overdistension of bladder	Kidney	Prompt voiding of bladder when full
Tumor of lower urinary tract	Bladder, urethra	Excision of tumor
Incomplete emptying of bladder	Bladder	Proper voiding habits, surgical correction of "sump" in bladder
Diminished neurological control of sphincters	Bladder, kidneys	Antimicrobial chemoprophylaxis
Trauma to female urethra	Bladder	Urination after intercourse
Catheterization	Bladder	Proper aseptic precautions
Recurrent lower urinary tract infection	Kidneys	Prompt diagnosis and treatment of cystitis to prevent ascent of infection to kidneys
Obstruction of ureter	Kidneys	Surgical correction, dissolution of calculi (stones)
Obstruction of urethra	Bladder	Surgical correction, dissolution of calculi (stones)
Bacterial prostitis	Urethra	Early diagnosis and treatment of prostitis
Hormonal imbalance	Vagina, bladder	Correction of imbalance if possible
Diabetes	Vagina	Insulin therapy
Pregnancy	Vagina, bladder	Hygiene, frequent voiding
Disruption of normal vaginal flora	Vagina	Avoidance of indiscriminate antibiotic therapy

Prolonged stress and all its undefined consequences increase the incidence of urogenital disease in susceptible persons.

DYNAMICS OF TRANSMISSION

Sexual Transmission

Most pathogens of the genitourinary tract are transmitted from exudates and open genital sores of infected persons. Infectious lesions may also be located on the anus, perineum, or mouth. Most of these pathogens are sensitive to drying and survive only briefly outside the host, sometimes for less than 1 minute. They are most often transmitted by direct contact during sexual activity, which protects them from exposure to the hostile conditions in the inanimate environment. Diseases that are virtually always transmitted by sexual contact have been referred to as *venereal diseases* (after Venus, the goddess of love). Venereal diseases are a category of **sexually transmitted diseases** (**STDs**) that also include infections often transmitted by nonvenereal routes but occasionally by sexual activity (Table 22-4).

TABLE 22-4
Sexually Transmitted
Diseases (STD) and Their
Estimated Annual
Incidence in the United
States

Disease	Estimated Annual Incidence*
Primarily Sexually Transmitted	
Gonorrhea	2,700,000
Syphilis	500,000
Nongonococcal urethritis (NGU) in men	2,500,000
Trichomoniasis	2,000,000–3,000,000
Genital herpes	300,000
Lymphogranuloma venereum Chancroid Granuloma inguinale	1,000
Occasionally Sexually Transmitted	
Cytomegalovirus Viral hepatitis Molluscum contagiosum Shigellosis Amebiasis Giardiasis Salmonellosis Camylobacteriosis	No data are given on relative frequency of sexual transmission, but all these diseases are common among promiscuous male homosexuals.

*Data obtained from National Institute of Allergy and Infectious Diseases

Sexually transmitted diseases are currently the most prevalent of the notifiable infectious diseases in the United States, where more than 1 million new cases of gonorrhea are reported to the Centers of Disease Control each year. Although antibiotics readily cure persons with gonorrhea, the epidemic shows few signs of subsiding. Some of the complex factors that aggravate the continual spread of STDs are listed below:

■ Asymptomatic carriers of gonorrhea, syphilis, genital herpes, and trichomoniasis serve as hidden reservoirs of infection. These persons are usually unaware they have the disease until informed by an infected sexual contact.

■ Fear of painful treatment, public disclosure, or social condemnation creates reluctance to seek medical attention and prolongs the time that the pathogen is shed.

■ Failure to notify the infected individual's sexual contacts that they have been exposed to a person with STD increases the probability that an unsuspecting individual will transmit the disease to others.

■ Failure to treat the patient's sex partner often leads to reinfection. Such failures are especially common for trichomoniasis, a painful vaginitis in women, an ordinarily asymptomatic infection in men. Untreated men often reinfect recently cured women.

■ More convenient methods of birth control have decreased the use of condoms, which are effective physical barriers against venereal disease.

■ Sexual activity with multiple partners enhances the probability of exposure to a variety of sexually acquired pathogens.

■ No protective immunity follows recovery from the major sexually transmitted diseases. Scientific efforts to develop effective vaccines have so far proved fruitless.

■ The emergence of antibiotic-resistant pathogens represents a potential problem in treating gonorrhea with penicillin. Another disease, recurrent genital herpes, is currently incurable.

Until vaccines against the major STDs are available, the best preventive measures are those enacted on an individual basis. These include avoiding sexual contact as long as genital lesions are present, getting prompt medical attention whenever one suspects possible venereal disease, and, if diagnosed as having an STD, waiting until medical confirmation of cure before resuming sexual intercourse. All sexual contacts and persons subsequently exposed should be immediately notified and treated. Even in the absence of symptoms, sexually active persons can determine through medical checkups if they are possible carriers.

Nonsexual Genitourinary Transmission

Many diseases of the GU tract are transmitted by autoinoculation of normal flora organisms, usually from the intestinal tract to the urinary tract. The most common manifestation of this is cystitis in women. Autoinoculation may occur during menstruation (saturated pads and the strings of tampons providing a fluid "bridge" between anus and vagina) or any other activity that would encourage contamination of the vulva with fecal material. Cystitis is also common following the insertion of a urinary catheter or *cystoscope* (a fiber-optics tube passed through the urethra to view the bladder's interior). Careless techniques during these procedures encourage the introduction of indigenous or transient microorganisms of the external genitals into the bladder. Contaminants on inadequately "sterilized" equipment or unwashed hands are also important sources of infection.

URINARY TRACT DISEASES

Infectious diseases of the urinary tract are categorized as either ascending or descending. If a pathogen is introduced through the urethra and extends to the bladder (and perhaps to the kidneys) the infection is **ascending**. Occasionally, pathogens spread hematogenously from another infected body site to the kidneys, and from there extend downward to the bladder. Such **descending infections** are much less common than ascending disease but are more dangerous to the patient because of the accompanying systemic involvement. They must be differentiated to determine the prognosis and appropriate treatment.

The most common urinary tract pathogen is *Escherichia coli*, which is usually mechanically introduced into the urethra (or bladder) from its indigenous site in the intestine. *Proteus, Pseudomonas, Enterobacter,* and enterococci are other intestinal bacteria that frequently cause the same condition. If these organisms bypass the mechanical defenses of the urinary tract, they attach to the bladder mucosa, usually by pili, and proliferate. In uncatheterized males the urethra is usually the target of these infections; thus males more commonly develop urethritis than cystitis. Most cases of urethritis in men, however, are believed to be sexually transmitted and will be discussed with the STDs.

Whenever an organism ascends from the ureters to the kidney, pyelonephritis or other severe renal diseases may develop. Early diagnosis and treatment of lower urinary tract infections reduce the risk of developing this potentially fatal complication.

Urinary tract infections are generally characterized by **bacteriuria**, bacteria in the urine. Quantitative microbiological assays of urine can identify the probable pathogen. This information may indicate whether the infection is ascending or descending. Intestinal bacteria are more likely to be the cause of ascending infection, especially if blood cultures are negative. Descending infections are often caused by pathogens that characteristically infect other areas of the body and spread to the kidneys by the bloodstream. In these cases, bacteremia can be detected by blood cultures.

Contaminating organisms from the distal portions of the urethra and the external genitals may confuse the diagnosis unless bacterial counts are determined using fresh midstream urine specimens collected by a "clean-catch" technique (described in Chap. 19). Any exudate discharged from the urethra should be microscopically examined and cultured *prior to* collection of urine.

Most cases of cystitis are caused by bacteria that respond to sulfonamides, ampicillin, tetracyclines, or nalidixic acid. Antibiotic resistance is common among urinary tract pathogens, however, emphasizing the need for determining the isolated pathogen's antibiotic susceptibilities. Recurrent infections suggest possible underlying anomalies which, once corrected, reduce the incidence of urinary tract infections (see Predisposing Factors).

NONVENEREAL GENITAL INFECTIONS

Candidiasis

One of the most common genital pathogens is the fungus *Candida albicans*, which causes the form of vaginitis commonly referred to as a yeast infection. The usual reservoirs of infection are the colon and the vagina itself, where the opportunistic pathogen resides as a minor component of the normal flora, held in check by the resident lactobacilli. Antibacterial antibiotics, pregnancy, oral contraceptives, or diabetes mellitus can disrupt this balance and predispose for the development of an extremely persistent and painful vaginitis. Recurrent *Candida* vaginal infections may also be associated with occasional transmission by oral-genital or anal-genital sexual activity. Although men rarely develop *Candida* genital infections, they may be a source of infection for women. Cases of candidiasis that

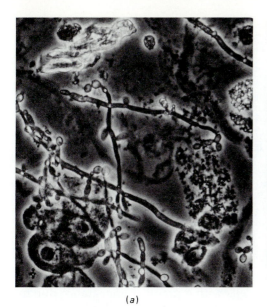
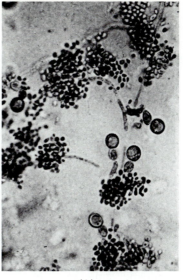

(a)   (b)

FIGURE 22-3

Candida albicans: (*a*) Vaginal exudate showing pseudohyphae and budding yeast cells. (*b*) Typical appearance when cultured on chlamydospore agar, with large chlamydospores, blastospores, and chains of pseudohyphae.

occur in the absence of identifiable predisposing factors indicate that other, poorly defined underlying conditions may also contribute to this disease.

Candidiasis is diagnosed by finding the characteristic fungus in vaginal exudates. Direct microscopic examination reveals pseudohyphae and yeast cells easily distinguished from resident bacteria (Fig. 22-3*a*). Cultivation of the fungus on chlamydospore agar, a nutritionally deficient medium, stimulates the production of blastospores (budding yeast cells), chains of pseudohyphae, and chlamydospores, the three microscopic hallmarks of *Candida albicans* (Fig. 22-3*b*).

Since *Candida* is a fungus, it fails to respond to antibacterial chemotherapy. The most effective agents against *Candida* vaginitis are nystatin, miconazole, or clotrimazole, applied topically as a cream or as a vaginal suppository. Treating the woman's sex partner may prevent reinfection. Because nystatin is not absorbed through the intestinal wall it can also be taken orally to decrease the yeast population in the colon, a probable reservoir for the vaginal pathogen. Oral nystatin reduces the problem of reinfection by autoinoculation but does not kill the yeast already in the vagina.

Nonspecific Vaginitis

One form of vaginitis is believed to be caused by a normal flora bacterium that, for unidentified reasons, proliferates beyond its normal limits. This organism, *Gardnerella vaginalis*, may be responsible for many cases of vaginitis that can't be diagnosed as candidiasis or trichomoniasis. It is not known whether the disease is sexually transmitted, but it is uncommon among women with no history of sexual intercourse. Often, male partners of women with the disease are colonized in their urethras. The thin, milky

FIGURE 22-4

Unlike the smooth normal epithelial cells on the left, the clue cell, characteristic of *Gardnerella vaginalis* infection, is coated with bacteria, giving it a granular appearance that is most apparent at the edges of the cell.

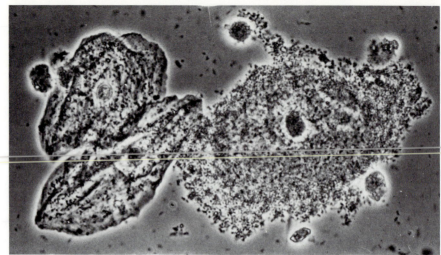

vaginal discharge of women with nonspecific vaginitis usually contains diagnostic "clue cells," epithelial cells with a granular appearance due to the attached coccobacilli (Fig. 22-4). In addition, a fishy odor is produced when the vaginal discharge is mixed with 10% KOH. Nonspecific vaginitis caused by *G. vaginalis* responds quickly to metronidazole therapy, leading some investigators to suspect that obligate anaerobes may also play a role in the disease.

Toxic Shock Syndrome

In 1980, a new and potentially fatal disease exploded into the public's awareness. The disease, called *toxic shock syndrome (TSS),* had surged in incidence from a few sporadic cases each year to nearly 1000 cases occurring in 1979 and 1980. Ninety-nine percent of these cases were in women, and in 98 percent of these women the symptoms appeared during a menstrual period. Moreover, most of these women were using tampons at the time, especially the superabsorbent products. From 1978 to 1980 the increased sales of these tampons paralleled the increased incidence of TSS. The most popular of these tampons was removed from the market in late 1980. Since then, the incidence of TSS has dropped dramatically. Toxic shock syndrome killed about 8 percent of the people affected during this period. A few cases have also occurred in males and children.

The disease strikes with a sudden onset of fever, vomiting, and diarrhea. Blood pressure plunges and the patient goes into shock. A sunburnlike rash desquamates and peels about 10 days after onset. There is usually a purulent discharge from the vagina. The disease is caused by *Staphylococcus aureus* that is either an asymptomatic resident of the genitourinary tract or is inoculated into the vagina during tampon insertion. This bacterium causes a local infection in the genitourinary tract. The pathogen elaborates an exotoxin that, in sufficient concentration, causes the symptoms of TSS, possibly by increasing a person's susceptibility to otherwise inconsequential levels of endotoxin. In addition, the toxin alters immunity and phagocytic functions, increasing the susceptibility of TSS patients to gram-negative bacterial infections. The extended time that superabsorbent tampons remain in the vagina may contribute to the development of TSS by allowing *Staphylococcus aureus* to reach the higher concentrations necessary for producing enough toxin to elicit symptoms. Women with signs of TSS should discontinue using tampons until *S. aureus* is eradicated from the vagina. The few cases not associated with menstruation were apparently caused by *S. aureus* in local lesions of the skin, lungs, or bones.

Diagnosis of TSS depends on the clinical picture (fever, hypotension, rash, and desquamation) and culturing toxigenic *S. aureus* from the vagina, cervix, urine, feces, or blood of affected patients. Supportive therapy and treatment with penicillinase-resistant antibiotics usually results in complete recovery.

SEXUALLY TRANSMITTED DISEASES

Gonorrhea

Although 1 million new cases of gonorrhea are reported to the CDC each year, most cases are not reported, and the annual morbidity of this disease is estimated to be 2.7 million cases. The etiological agent is the gram-negative diplococcus, *Neisseria gonorrhoeae*. Its primary route of transmission is through sexual contact. Infected persons may either develop symptoms characteristic of classical gonorrhea or become asymptomatic carriers. Carrier states are especially common among infected women in the early stages of disease. The large number of gonorrhea carriers in both sexes presents a major obstacle to the control of this venereal disease. Carriers may unknowingly transmit gonorrhea to their sex partners for months or years.

Overt disease in adult males differs from gonorrhea in females, although in both sexes the infection is initiated by adherence of the pathogen by its pili to the epithelial cells in the genitals (occasionally the pharynx or rectum are initial sites of infection). This occurs in the urethra of men and the vagina and cervix of women. The bacteria invade and multiply in epithelial cells, causing inflammation. Three to five days following exposure, most infected males experience symptoms characteristic of urethritis—painful urination and discharge of a thick white exudate from the penis. If untreated, the pathogen may produce scarring in the urethra (predisposing for recurrent urinary tract infections) or may infect the vas deferens. Scarring in these accessory sex ducts may result in permanent sterility, as does the testicular destruction that occasionally accompanies untreated gonorrhea in men. Gonorrhea may also extend to the prostate.

In women, gonorrhea begins as a mild cervical infection that frequently remains undetected. These patients often report to a physician only when notified by an infected sex partner. From the cervix the pathogen may enter the uterus and finally the fallopian tubes, where painful inflammation, called *salpingitis*, may develop. Such conditions may cause sterility by scarring and completely obstructing the fallopian tubes. Partial scarring leaves a small opening that may allow the passage of sperm but not the fertilized egg. The embryo cannot reach the uterus and develops in the fallopian tube instead. Tubal pregnancies, also called ectopic pregnancies, are a dangerous complication of untreated gonorrhea in women. Gonorrhea is a major cause of *pelvic inflammatory disease (PID)* in women. PID includes any infection of the uterus, fallopian tubes, or ovaries.

In both sexes *Neisseria gonorrhoeae* may invade the bloodstream and cause *disseminated gonococcal infection*. The most common targets for circulating gonococci are the joints (gonorrhea is the most common cause of arthritis in a single joint), the meninges, the endocardium, and the myocardium. The organism may also be transmitted by fingers from infected genitals to the eyes, where untreated infection can cause corneal ulceration and blindness.

Gonorrhea may be transmitted from an infected mother to her infant during delivery through the infected birth canal. Infected newborns develop a serious infection of the eyes, *ophthalmia neonatorum,* which can lead to blindness. Until 1982 this disease was controlled in the United States by instilling drops of a 1% silver nitrate solution into the eyes of all newborns. Because chlamydias, which are insensitive to silver nitrate, may also cause this disease, the CDC now recommend using erythromycin or tetracyclines for prophylaxis. For infants born to mothers with overt gonorrhea, an injection of penicillin prevents neonatal infections at other sites.

Diagnosis   Clinical specimens for diagnoses include the purulent discharge of males collected on a sterile swab from the opening or outer portion of the urethra, and swabbed material from the cervix of women. The pharynx, rectum, and eyes are sampled whenever evidence suggests they may be infected. Direct microscopic examination of Gram-stained specimens often reveals gram-negative diplococci inside pus cells (polymorphonuclear leukocytes) (Fig. 22-5). The intracellular location aids in distinguishing the gonococcus from nonpathogenic *Neisseria sicca* and *Neisseria subflava*, resident bacteria similar in appearance. Identification ultimately depends on culturing the gonococcus on Thayer-Martin agar or chocolate agar and incubating (35°C) in capneic conditions (10% CO_2) for 24 to 48 hours. *Neisseria* colonies darken when flooded with oxidase test reagent; in other words, they are oxidase-positive. *Neisseria gonorrhoeae* is differentiated from other oxidase-positive gram-negative diplococci by differential biochemical tests.

A rapid test for identifying gonorrhea in men is based on detecting endotoxin in urethral exudates using the limulus assay (see color box, Chap. 4). The test provides a presumptive diagnosis of gonorrhea within 30 minutes. Because of gram-negative bacteria in the normal flora, a positive limulus assay on cervical exudates does not necessarily indicate

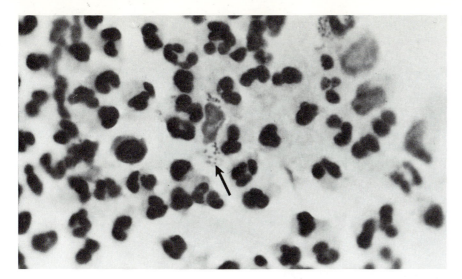

FIGURE 22-5
Gram's stain of exudate from patient with gonorrhea. The characteristic gram-negative diplococci are located inside the pus cells.

gonorrhea. A negative test, however, suggests an etiology other than *N. gonorrhoeae*.

Therapy Penicillin is the antibiotic of choice for treating gonorrhea, although the current effective dosage is about ten times greater than that of 20 years ago. Some strains of *N. gonorrhoeae* are completely resistant to penicillin. Many of these strains were originally isolated from infected military personnel returning to the United States from Viet Nam or from sexual contacts of these persons. The incidence of penicillin-resistant gonorrhea has since spread throughout the United States. Antibiotic resistance may be transmitted to the gonococcus by R factors during conjugation with other bacteria, including *Escherichia coli*. Persons with gonorrhea who fail to respond to penicillin are usually treated with spectinomycin. This antibiotic is also recommended for treating gonorrhea in patients with penicillin allergy, although it is ineffective when the disease is in the pharynx. Tetracycline is used in these cases. All sexual contacts of patients with gonorrhea should be located and treated as rapidly as possible. Since syphilis and gonorrhea may be transmitted together, persons diagnosed as having gonorrhea should be serologically tested for syphilis as well. Although penicillin also cures syphilis, spectinomycin does not. As many as 45 percent of all gonorrhea patients are coinfected with chlamydias (see Nongonococcal Urethritis). The CDC is evaluating the effectiveness and side effects of a combined antimicrobial regimen of ampicillin and tetracycline to treat these coinfections.

Syphilis

Before the antibiotic era in medicine, syphilis (historically referred to as "large pox" or simply "the pox") was one of the most dreaded diseases. Late syphilis often left its victims mentally degenerated, neurologically

FIGURE 22-6
Scanning electron
micrograph of *Treponema
pallidum*.

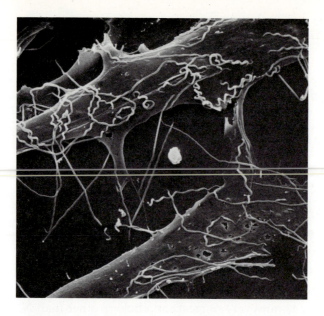

incompetent, consumed by destructive lesions, and frequently dead. Although these outcomes are far less common today because of accurate diagnosis and effective therapeutic regimens, people still die of syphilis, and babies continue to be born with tragic congenital anomalies because of exposure during gestation in an infected mother.

Syphilis is a venereal disease caused by an especially fragile spirochete, *Treponema pallidum* (Fig. 22-6). Ninety-five percent of all syphilis cases are transmitted by sexual activity. The disease can also be acquired congenitally following transplacental passage to a developing fetus or by nonsexual contact with infectious lesions—for example, between medical personnel and patients or between adults and children. Nonsexual transmission is usually associated with the pathogen's entry through breaks in the skin.

Syphilis occurs in three well-defined stages: primary, secondary, and tertiary. Usually a latent period occurs after the secondary stage. The *primary stage* is characterized by the development of a single lesion, called a **chancre**, at the site where the pathogen entered the body (Fig. 22-7a) Although usually on the genitals, chancres may also be found on the anus, mouth, hand, and breast. The lesion is hard, rubbery, and completely painless, and usually appears 7 to 21 days following infection. By this time the spirochetes have spread to the regional lymphatics and invaded the bloodstream. The chancre itself is extremely infectious. Examination of its exudate by darkfield microscopy often reveals swarms of motile, tightly coiled spiral-shaped bacteria. Since the chancre is painless, it often escapes detection when located in the vagina or on the cervix. Women may therefore transmit the disease with no knowledge that they are infected. Undetected chancres may also develop in the anal canal, a problem primarily common among male homosexuals. After several days the

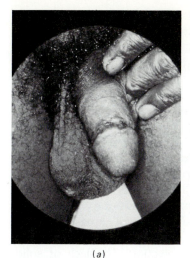

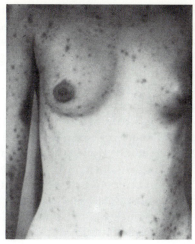

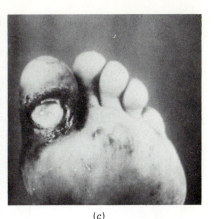

(a)

(b)

(c)

FIGURE 22-7
Characteristic
manifestations of syphilis:
(a) Chancre of primary
stage; (b) rash of secondary
stage; (c) gummas of late
(tertiary) syphilis

577

SEXUALLY TRANSMITTED
DISEASES

chancre spontaneously disappears, often fostering the belief that the disease has cured itself.

Untreated persons with syphilis enter the *secondary stage* within 3 to 4 months after initial infection. The most common manifestation is a skin rash on the trunk, extremities, and genitals (Fig. 22-7b). The rash has a characteristic predilection for the palms of the hands and soles of the feet. These disseminated lesions are painless and itchless. Additional lesions, which are highly infectious, may appear on the mucous membranes of the genitals and mouth. Secondary syphilis is therefore more communicable than the primary stage of disease with its single infectious chancre. Transplacental passage from a pregnant woman to her fetus may also occur at this stage.

Secondary syphilis is often accompanied by generalized symptoms that include low-grade fever, arthritis, aches in the shin, and less frequently, damage to the nervous system, liver, and spleen. Secondary syphilis spontaneously subsides within a few weeks to several months after onset. In untreated persons one of three developments follows the secondary stage. About one-third apparently recover from the disease, are no longer infectious, and show no further evidence of syphilis. Another third of untreated persons enter the *latent stage,* an asymptomatic phase that is recognized by the persistence of positive blood tests (see Diagnosis). Latent syphilis is not communicable by sexual contact, but may still cross the placenta and infect a developing fetus. Relapses of secondary syphilis may occur during latency. *Tertiary syphilis* develops in the other third of untreated cases following a period of latency. This may happen 3 years after initial infection or as long as 40 years later. Tertiary syphilis is also known as the destructive phase because of its devastating lesions, called **gummas** (Fig. 22-7c). These lesions are hypersensitive responses to the presence of residual spirochetes. They develop in virtually any tissue or organ system and may resemble the symptoms of many other diseases. For this reason, late syphilis is known as the "great imitator." If the gummas occur in the nervous system, neurosyphilis develops, causing insanity, deafness, or blindness. Partial paralysis affects the coordinated movements of walking, resulting in a gait sometimes referred to as the "syphilis shuffle." As with cardiovascular syphilis (gummas in the heart or blood vessels), neurosyphilis is potentially fatal.

Although 31,000 cases are reported each year in the United States, not all cases are reported and the actual number may be as high as 500,000. Today only a small percentage of these cases progress to the tertiary stage. Most cases (and their contacts) are identified and treated while in the primary or secondary stages. Campaigns to detect the disease in pregnant women help reduce the incidence of congenital syphilis. Antibiotic treatment very early in the pregnancy can avert this tragedy.

Like most venereal diseases, recovery from syphilis does not increase one's resistance to subsequent reinfection. Producing a disease-resistant population by artificial immunization has evaded medical efforts; all trial vaccines have failed to provide adequate protection. The lack of solid immunity may be due to the pathogen's ability to specifically suppress lymphocyte-mediated cellular immunity to treponemal antigens during the primary stage. Although the resultant immune paralysis does not affect humoral antibody production, even high titers of *Treponema pallidum*–specific antibodies are nonprotective against subsequent attack.

Diagnosis   The presence of a chancre, a rash characteristic of secondary syphilis, or a history of sexual contact with an infected person should always be followed by laboratory tests to confirm the diagnosis. These tests either (1) demonstrate serological reactions to the presence of the pathogen or (2) detect the motile spirochetes in the lesions.

Serological tests detect one of two types of antibodies in the serum of persons with syphilis. **Reagin** is an antibody mixture found in elevated concentrations in persons with syphilis; it is usually absent in uninfected people. Although reagin is not specific for *Treponema pallidum*, its detection provides the basis for presumptive screening tests. The most common of these tests are the **VDRL (venereal disease research laboratory)** test and the **RPR (rapid plasma reagin)** test. Both procedures rely on precipitation

FIGURE 22-8

Positive FTA-ABS test for *Treponema pallidum*–specific antibodies in a patient's serum. (1) Serum from patient suspected of having syphilis is added to known preparations of *T. pallidum* fixed to a microscope slide. (2) *T. pallidum*–specific antibodies attach to spirochetes, thereby coating the surface with IgG. Nonspecific antibodies remain unbound. (3) The slide is washed to remove unbound nonspecific antibodies. The only human IgG remaining is that attached to the treponemes. (4) Fluorescent-labeled antibody specific for human IgG is added to the slide. (5) Labeled antibody combines with the *Treponema*–specific human IgG antibody on the test bacteria, which becomes coated with fluorescent dye. The spirochetes glow when viewed with the fluorescent microscope. Sera from patients who do not have syphilis contain no *T. pallidum*–specific antibody, so fluorescent-labeled anti-IgG will not coat the test bacteria. Thus in negative tests, the spirochetes do not fluoresce. (*b*) Photomicrograph of positive FTA-ABS test.

between serum reagin and *cardiolipin*, a soluble antigen extracted from normal beef hearts. This cross-reaction is used to rapidly screen large human populations for syphilis. Sometimes reagin tests produce biologi-

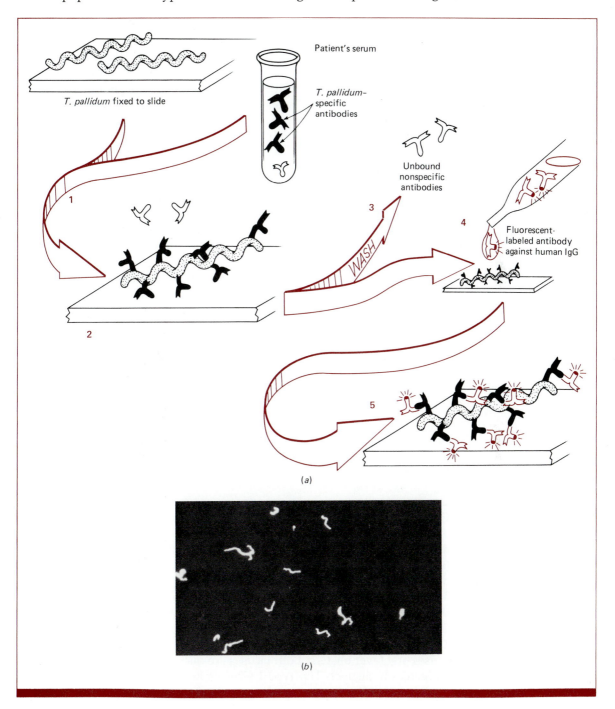

(a)

(b)

cally false positives, especially among persons with malaria, infectious mononucleosis, tuberculosis, hepatitis, leprosy, or any condition that elevates serum reagin levels. Positive reagin results are therefore always followed by tests for specific treponemal antibody.

The **specific treponemal antibody tests** are more reliable than determinations of reagin activity because they detect antibodies produced only by people with syphilis. The patient's serum is mixed with laboratory cultures of *Treponema pallidum*. Positive reactions are usually detected by indirect fluorescent microscopy or by darkfield microscopy. The **fluorescent treponema antibody (FTA-ABS) test** is called an *indirect* immunological test because the reaction between antigen and antibody cannot be directly detected. The antigen-antibody complex is indirectly detected by adding fluorescent-labeled antibody specific for human IgG (Fig. 22-8).

Living *Treponema pallidum* grown in rabbit testes are used for detecting serum antibody by the **Treponema pallidum immobilization (TPI) test**. The motility of these living treponemes is observed under darkfield microscopy. Specific antibodies in the sera of patients with syphilis, when combined with living treponemes, immediately immobilize the test bacteria. Sera from healthy persons fail to inhibit motility because they contain no treponemal-specific antibodies.

Treponema pallidum cannot be grown readily on inanimate media. Cultivation in rabbit testes is laborious and expensive, making the TPI and FTA-ABS tests uneconomical for screening large populations for syphilis. Thus, specific treponemal tests are usually reserved for use on persons who have positive serum reagin tests or who have clinical signs of syphilis.

Therapy   Syphilis is uniformly responsive to penicillin, although neurosyphilis requires especially high doses over long periods of time. Antibiotics, however, do not reverse tissue damage that occurs during the tertiary stage of the disease. The ease with which syphilis may be cured may actually contribute to its epidemic spread. Persons complacent about such an "easily cured" disease often take no precautions to avoid contracting or spreading it.

Genital Herpes

Genital herpes is a sexually transmitted disease characterized by blisters on the genitals and surrounding area. Its actual incidence rate is undetermined because physicians are not required to report cases to the Centers for Disease Control. Nonetheless, it is known that the incidence of genital herpes has dramatically escalated since the late 1960s (Fig. 22-9). More than 5 million Americans currently have the disease and at least 300,000 join their ranks each year. Genital herpes is caused by *herpes simplex virus* (HSV). The majority of genital infections are caused by the Type 2 virus, although 15 percent are caused by HSV Type 1, the virus that usually causes oral fever blisters. Both viruses are microscopically identical but are antigenically distinct. The Type 1 virus shows a predilection for the upper

RADICAL CURES FOR SYPHILIS

Before it could be cured by chemotherapy, the great pox (syphilis) was one of the world's most feared and tragic diseases. Because no effective treatment existed, one-third of all syphilis-infected adults suffered neurological deterioration and paralysis, horrible skin disfigurements, mental degeneration, insanity, circulatory disturbances, and eventually death. Mothers with syphilis commonly gave birth to fatally infected babies. Many remedies were practiced, most without positive effect. Two radical approaches, however, were marginally successful because of the acute heat-sensitivity of the pathogenic bacterium.

It had long been observed that some persons recovered from syphilis following episodes of high fever. Such observations provided the inspiration for the "hot box." Patients were placed in a heated chamber until their body temperature was elevated well above normal. After 48 hours of such treatment some patients actually recovered from their infections. Others died from the treatment. Another drastic approach was to intentionally infect a syphilitic person with malaria, a disease causing periodic episodes of very high fever. Once the induced fever killed the pathogen of syphilis, malaria was treated with quinine. Although this mode of treatment successfully cured some people of syphilis, many others had both syphilis and malaria for life.

regions of the body whereas the Type 2 virus usually strikes below the waist. Genital herpes, like oral herpes, tends to recur, often following sexual intercourse, menstruation, stress, common cold, or febrile illness. Recurrences are reactivations of latent infection and are not reinfections.

The characteristic lesion is a fluid-filled vesicle which erupts and becomes an ulcer. These open sores are teeming with virus and are very

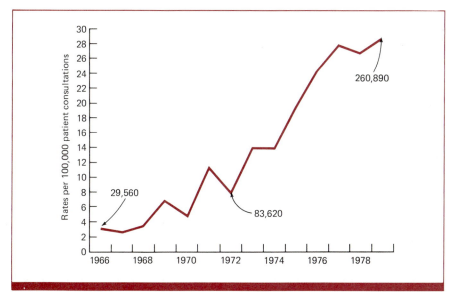

FIGURE 22-9
Estimated rate of patient consultations for genital herpes infections in the United States, 1966–1979. Because these data are limited to consultations with private physicians, the actual number of cases is much greater than indicated. (*Morbidity and Mortality Weekly Report, 31:138 (1982).*)

FIGURE 22-10

Herpes simplex virus Type 2 genital lesions, (*a*) in the male and (*b*) in the female.

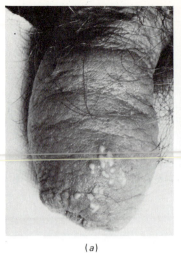 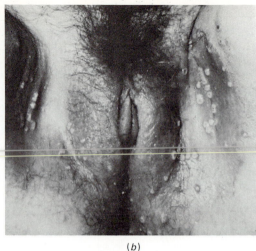

(*a*)   (*b*)

582

DISEASES ACQUIRED
THROUGH THE
GENITOURINARY TRACT

infectious. They occur anywhere on the penis in men and on the vagina, vulva, and cervix in women (Fig. 22-10). The anus, perineum, buttocks, and thighs of either sex may also be ulcerated. Evidence suggests that the disease is most communicable while active lesions are present, although virus may also be shed in urine and genital secretions.

The lesions of genital herpes are frequently painful, but they sometimes go undetected, especially in the vagina and on the cervix. Women with these painless lesions are often unaware they are infected and shedding the virus until notified of subsequent outbreaks in a sex partner. The disease may also cause psychological trauma because of embarrassment, fear of recurrence, and knowledge that no cure is available. In most victims, recurrences eventually cease, although it may take years.

One of the most serious complications of venereal herpes is infection of the newborn by its mother. Such infections are acquired during normal delivery through the birth canal of a woman with open lesions. The consequences of neonatal herpes are much more severe than those of adult herpes. Infected infants may develop blindness, meningitis, disseminated necrosis of the skin, or other potentially fatal disorders. The incidence of neonatal herpes is reduced by cesarean delivery whenever the mother has active genital sores. Although rare, a primary attack of genital herpes during early pregnancy, however, may cause abortion or premature delivery.

Another alarming consequence of genital herpes in women is an increased likelihood of developing cancer of the cervix. Although no cause-and-effect relationship has been established, statistical studies show a definite association between the two diseases. Because of this epidemiological link it is sound practice for women who have suffered genital herpes infection to be examined biannually for evidence of cervical carcinoma. When diagnosed and treated in its early stages, cervical carcinoma is readily cured.

Diagnosis and Treatment Persons with genital herpesvirus infections have an elevated titer of type-specific HSV antibody. These antibodies can be detected and quantified by determining the ability of the patient's serum to neutralize the infectivity of a stock preparation of herpesvirus. The pathogen may be isolated from lesions and propagated in tissue culture for subsequent identification. The most rapid diagnostic technique microscopically detects multinucleated giant cells and characteristic inclusion bodies in stained cells scraped from lesions.

As with any viral diseases antibiotics do not cure or even shorten the duration of infection. Although several chemotherapeutic agents have been licensed for treating herpesvirus infections, the most promising is **acyclovir**. When administered early in the course of the primary infection, acyclovir shortens the duration of discomfort and viral replication and may reduce the likelihood of recurrences. It is of no value, however, against dormant viruses or for preventing recurrences once latency is established.

Nongonococcal Urethritis

Pain during urination is one of the hallmark symptoms of gonorrhea in men. About half of all men with urethritis, however, do not have gonorrhea. They have an infection of the urethra that was previously called "nonspecific urethritis" because its etiology was not well established. Because *Neisseria gonorrhoeae* was eliminated as the possible pathogen, the disease acquired the name **nongonococcal urethritis** (**NGU**). Although not a reportable disease, NGU is the most commonly occurring STD among men in the United States, with an annual estimated incidence of 2.5 million cases. It primarily occurs in men, many of whom contract the infection during sexual activity with asymptomatically infected women (in some women cervical inflammation or PID may be evident). It is a slightly milder disease than gonorrhea, with less purulent discharge. Although the etiology remains somewhat obscure, by far the most common causes of NGU are certain serotypes of *Chlamydia trachomatis*. Many other cases of NGU are caused by *Ureaplasma urealyticum* (formerly called T-strain mycoplasma) or, less commonly, by the yeast *Candida albicans* or the protozoan *Trichomonas vaginalis*, often associated with vaginitis in the infected man's sex partner. Because none of these agents of NGU responds to penicillin, Gram-stained smears of exudates are examined to rule out gonorrhea. In most cases of NGU (except those caused by yeast and protozoa), erythromycin or tetracyclines are effective in curing the disorder. Both sex partners should be treated simultaneously to prevent reinfection of the male and to cure asymptomatic infections in the female. These asymptomatic infections progress to PID in about 1 percent of sexually active women. *Chlamydia trachomatis* causes conjunctivitis of newborns exposed to maternal vaginal infections during birth. Unlike ophthalmia neonatorum caused by *Neisseria gonorrhoeae*, this eye infection fails to respond to topical drops of silver nitrate. Because of this, the CDC recommend immediate postpartum instillation of either tetracycline or erythromycin for all newborn infants. This treatment eliminates both pathogens from the eyes.

Trichomoniasis

Vaginal infections by the flagellated protozoan *Trichomonas vaginalis* are characterized by genital irritation and an unpleasant-smelling yellowish purulent discharge. The organism normally resides harmlessly on the healthy vaginal mucosa and in the urethra and prostate of some men. *T. vaginalis* may cause disease, however, when sexually transferred from men to women. The organisms disrupt the local pH balance and elicit a local inflammatory response resulting in vaginitis. Although infected males rarely experience symptoms, occasionally they develop prostatitis or urethritis (NGU). Cystitis may occur in either sex.

Trichomoniasis is diagnosed by microscopically detecting the flagellated protozoa in purulent genital discharges (Fig. 22-11). Metronidazole (Flagyl) is the only effective chemotherapeutic agent. It should be simultaneously administered orally to both sex partners, even though the male appears uninfected; otherwise asymptomatic males reinfect cured females.

Genital Warts

For centuries, condylomata acuminata, or *genital warts,* have been a common STD in young adults. They are caused by a *papilloma virus,* similar to the one that causes common skin warts. The two viruses are immunologically distinct, however, and also differ in their rates of growth, genital warts developing much faster than skin warts. The lesions are benign tumors that appear on the external genitalia, but may also occur in the mouth, anus, urethra, and, rarely, the bladder. The duration of the disease is unpredictable. In many persons it spontaneously subsides in 6 months; in others the lesions persist for years. Eventual recovery is due to a CMI response. If CMI is depressed, however, the warts may progress to giant lesions that may destroy genital tissue, especially of males. The lesions also enlarge during pregnancy and may interfere with birth, sometimes neces-

FIGURE 22-11
Appearance of *Trichomonas vaginalis* in vaginal exudate using diagnostic light microscopy.

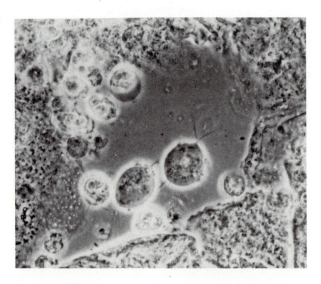

sitating delivery by cesarean section. Secondary bacterial infections complicate some cases of genital warts.

The characteristic appearance of the lesions is often enough to suggest a diagnosis of genital warts. Occasionally they may be mistaken for lesions of secondary syphilis or, in the tropics, for granuloma inguinale. They are differentiated by lack of positive tests for syphilis and absence of Donovan bodies (discussed below). As with most virus diseases, no antimicrobial treatment is very satisfactory. Most cases respond to topical applications of podophyllin, an agent that inhibits mitosis and cell proliferation. Because of podophyllin's toxicity and oncogenic potential, many physicians prefer to destroy the lesions physically by electrically burning them or freezing them with liquid nitrogen. Podophyllin treatment should be avoided during pregnancy.

Infrequently Encountered Sexually Transmitted Diseases

Three additional diseases transmitted by sexual activity are usually found in residents of tropical areas and occur infrequently in the United States, Canada, and Europe. *Chancroid* is a highly infectious venereal disease caused by a small gram-negative bacillus, *Haemophilus ducreyi*. It is often confused with syphilis because its characteristic lesions, called soft chancres or chancroids ("chancrelike"), are similar in appearance to the lesions in primary syphilis. Unlike a syphilis chancre, however, these lesions are painful and bleed easily. They appear on the genitals and surrounding areas 3 to 5 days following exposure. The lesions may extend to extragenital sites by autoinoculation with pus from the original ulcers. The organism can be isolated from lesion exudates for diagnosis. Treatment with tetracyclines and sulfonamides has been effective in the past, but antibiotic resistance has made these treatments unreliable. The CDC currently recommend treatment with erythromycin.

Lymphogranuloma venereum is caused by serotypes of *Chlamydia trachomatis* that differ from the serotypes that cause NGU. The sexually transmitted disease begins with the formation of a painless genital lesion which showers the bloodstream with pathogens 1 to 2 weeks later. The pathogen can invade the meninges, eyes, and joints and may persist in the lymphatics, causing stasis and subsequent elephantiasis (swelling due to accumulation of lymphatic fluid). The chlamydias are microscopically visible in inclusions within the cytoplasm of infected cells. Diagnosis can be confirmed by growing the isolated organism in cell culture or embryonated eggs. Treatment requires weeks, sometimes months, of tetracycline therapy, often supplemented by surgical intervention to reduce elephantiasis.

Granuloma inguinale is caused by the gram-negative bacillus *Calymmatobacterium granulomatis*. The disease is characterized by peripherally radiating lesions surrounding the single initial nodule that occurs at the site of entry. Extensive tissue destruction may occur. Although granuloma inguinale is primarily a venereal disease, it can also be transmitted by sand flies, respiratory secretions, urine, and feces. It is diagnosed by microscopi-

Disease	Pathogens	Incubation Period	Route(s) of Transmission
Cystitis	*Escherichia coli* and other gram-negative rods	1–5 days	Autoinoculation from skin or colon
Vaginitis	*Candida albicans*	Variable	Overgrowth of established yeasts; autoinoculation from colon; perhaps from colonized penis
	Gardnerella vaginalis	—	Overgrowth of normal flora
	Trichomonas vaginalis	4–20 days	Sexual contact
Toxic shock syndrome	*Staphylococcus aureus*	—	Autoinoculation or existing infection
Gonorrhea	*Neisseria gonorrhoeae*	3–5 days	Sexual contact; neonatal transmission
Syphilis	*Treponema pallidum*	10–70 days	Sexual contact; transplacental transmission
Genital herpes	Herpes simplex virus Type 2	2–12 days	Sexual contact; neonatal contact at birth
Nongonococcal urethritis	Usually *Chlamydia trachomatis; Ureaplasma urealyticum*	5–7 days or longer	Sexual contact
Chancroid	*Haemophilus ducreyi*	3–5 days	Sexual contact
Lymphogranuloma venereum	*Chlamydia trachomatis*	7–21 days	Sexual contact
Granuloma inguinale	*Calymmatobacterium granulomatis*	Unknown	Sexual contact; flies, urine, feces

cally observing the encapsulated etiological agent in stained scrapings from lesions. The bacteria often appear as rodlike forms, called *Donovan bodies*, in the cytoplasm of host cells.

Diseases Occasionally Transmitted by Sexual Contact

Some diseases that have nonsexual routes of transmission may be occa-

Complications	Clinical Specimens	Laboratory Diagnosis	Treatment
Pyelonephritis	Exudates; urine; blood	Quantitative determinations of microbes in urine	Chemotherapy determined by antibiograms; correct predisposing factors
None	Vaginal exudates	Direct microscopy; culture on chlamydospore agar	Suppositories or cream containing nystatin, miconazole, or clotrimazole
None	Vaginal exudates	Direct microscopy for "clue cells"; culture	Sulfathiozole; metronidazole
None	Vaginal exudates	Direct microscopy	Metronidazole; restore normal vaginal pH
Severe shock; death	Vaginal exudates; urine; feces; blood	Recovery of *S. aureus* in culture	Penicillinase-resistant antibiotics
Sterility; PID; tubal pregnancy; disseminated infection; opthalmia neonatorum	Exudates from penis or cervix or from newborn's eyes	Direct microscopy; culture	Penicillin; spectinomycin
Neurosyphilis; cardiovascular syphilis; congenital infection	Blood, fluid, or tissue from lesion	Serology; direct exam with darkfield or fluorescence microscopy	Penicillin; tetracycline; erythromycin
Meningitis; cervical cancer; congenital defects	Lesion exudates; blood; spinal fluid	Virus isolation; examine for inclusion bodies; serology	Acyclovir during primary stage; none thereafter
Neonatal eye infections	Exudates from penis	Gram's stain to rule out gonorrhea	Erythromycin; tetracycline
Spread to nongenital sites	Exudates from lesions	Direct microscopy; culture	Erythromycin
Systemic disease	Exudates or biopsy from lesions; blood	Direct microscopy; culture serology; skin test for hypersensitivity	Tetracycline; sulfonamide
Destruction of genital tissue; spread to other organs	Lesion biopsy	Detect Donovan bodies by direct microscopy; culture in chick embryo	Tetracycline; sulfonamide

sionally sexually communicable. Viral hepatitis, for example, is usually transmitted by oral-fecal or parenteral routes. Well-documented outbreaks of hepatitis among populations of male homosexuals, however, have revealed an epidemiological pattern characteristic of sexually transmitted disease. Other diseases that fall into the "occasional STD" category are enteric infections with *Shigella, Giardia lamblia, Entamoeba histolytica,* and

Campylobacter jejuni. Like hepatitis, their sexual transmission is most common among male homosexuals.

A summary of diseases acquired through the GU tract is provided in Table 22-5.

INTRAUTERINE INFECTIONS

Some pathogens circulating in the blood of pregnant women may cross the placenta and cause disease in the developing fetus. Many of these pathogens that cause mild disease in adults can severely debilitate or kill a fetus, especially if infection occurs early in pregnancy while differentiation of fetal tissues is most active. Injury to differentiating cells is amplified because the organs and tissues that develop from these cells will also be damaged. Thus **intrauterine infections** cause severe fetal malformations, mental retardation or other nervous system impairments, and serious defects in internal organs. Many of these infected infants who survive to term die shortly after birth.

Cytomegalovirus (CMV) causes the most common infections of the developing fetus and is an important cause of mental retardation in children. As many as 7 percent of all newborns may be subclinically infected with this herpesvirus. Overt disease, which occurs in 15 to 20 percent of these cases appears before the fourth month and may injure any organ system. CMV infections may be contracted during birth as well as by transplacental passage. The disease is also horizontally transmitted by contact with respiratory secretions, urine, or semen shed from an infected person, and can be acquired during sexual contact. Sexual transmission is especially common among male homosexuals.

As do other herpesviruses, CMV tends to establish latent infections that may cause recurrences. Most cases of overt CMV disease in adults are probably due to reactivation of these latent infections. Adult cases are most commonly associated with pregnancy, immunosuppression, and multiple blood transfusions, especially in the large volumes required for open-heart surgery. The disease may be diagnosed by immunofluorescence using labeled CMV-specific antibody.

Cytomegalovirus infections also suppress the immune system and may play an important role in the pathogenesis of *acquired immune deficiency syndrome (AIDS)* (see color box).

Congenital syphilis is a serious intrauterine infection acquired transplacentally from mothers with primary, secondary, or latent syphilis. More than half of the infected fetuses die before birth or shortly after. Many of those that survive suffer debilitating disease of the bones, nervous system, viscera, or skin, the disease sometimes appearing months or years after birth. Treatment with penicillin within the first 16 weeks of pregnancy cures the mother and prevents congenital syphilis.

Women who are infected by the protozoan *Toxoplasma gondii* during pregnancy may transmit the pathogens to their fetuses, resulting in possible mental retardation or injury to several organs later in life.

DISEASES ACQUIRED
THROUGH THE
GENITOURINARY TRACT

A FRIGHTENING NEW DISEASE

■ A mysterious medical development emerged for the first time in 1979 with the sudden appearance of a new disease, one that kills well over half the people who contract it.* The disease, called *acquired immune deficiency syndrome (AIDS)*, has become a serious public health problem and has generated some hysterical public reactions. In the first 3 years after recognition, AIDS killed more people than did Legionnaires' disease or toxic shock syndrome. In the 15 months following June 1981, 593 persons in the United States contracted the syndrome and 41 percent of them had died by the following September. The case fatality rate jumps to 60 percent for persons who have had the disease longer than a year. The incidence of AIDS is increasing at an alarming rate, having doubled every 6 months since mid-1979.* Seventy-five percent of the victims of AIDS are sexually active homosexual or bisexual men. Heterosexuals account for 13 percent, and most of them are (or were) intraveneous drug abusers, although women who are steady sex partners of men with AIDS have also contracted the disease. Other heterosexuals who have acquired AIDS include Haitians, persons with hemophilia and others receiving blood transfusions, and infants who may have contracted the disease transplacentally or at birth. Most cases of AIDS cluster in large metropolitan areas, especially on the coasts. The frequency of the disease is roughly 10 times higher in San Francisco and New York than in the rest of the country.

The disease is a serious deficiency in the cell-mediated immune system that leaves the person vulnerable to cancer or to other pathogens that are normally eliminated by CMI. Persons with AIDS are especially prone to develop otherwise rare opportunistic diseases. *Pneumocystis carinii* pneumonia (see Chap. 20) is by far the most common, although about one-third contract a formerly rare cancer called Kaposi's sarcoma. These opportunistic diseases are severe and are usually the cause of death in persons with AIDS.

Unlike CMI, humoral immunity does not appear to be directly impaired by AIDS. The defect seems to primarily depress the helper T cell subpopulation with or without significant reduction of suppressor T cells. This causes a shift in balance between helper and suppressor cells and might explain the serious suppression of CMI function. Natural killer cells, believed to be important in surveillance and destruction of cancer cells, are also diminished by AIDS.

The cause of this suddenly acquired immunosuppression is still a mystery, although it appears to be a virus that replicates in lymphoid cells. This virus is the human T-cell leukemia-lymphoma virus (HTLV-III).

Serological evidence reveals that exposure to this virus is more common in high-risk populations than in the general public. Many persons exposed to the virus develop a mild or subclinical infection and suffer no serious complications. The disease shows a pattern of horizontal transmission similar to the sexual transmission of hepatitis B, a disease that is prevalent among male homosexuals as well as drug addicts. There is no evidence that AIDS is transmitted through casual contact.

So far no explanation accounts for all the characteristic patterns of this new disease. However, in trying to understand and prevent AIDS, researchers are uncovering a wealth of clues as to the normal functions of the immune system and why they fail. The fight against AIDS may strengthen our understanding of cancer and immune deficiency diseases.

---

*Based on data available in 1984.

Toxoplasma cysts in the tissue are resistant to chemotherapy, so the disease is virtually incurable. Active cases can be arrested, however, by combination pyrimethamine-sulfonamide therapy, which kills the pathogen in the trophozoite stage.

Congenital rubella syndrome is an intrauterine infection that affects fetuses of mothers who contract rubella during the first trimester of pregnancy. Many of these fetuses die before birth. Survivors frequently suffer mental retardation and damage to the cardiovascular and nervous system. Immunization of women (prior to pregnancy) using an attenuated rubella virus vaccine has greatly reduced the incidence of congenital rubella.

Women with herpes Type 1 or Type 2 infections may transmit the virus to a developing fetus. In these cases disseminated disease usually causes spontaneous abortions or stillbirths due to liver or central nervous system damage. Whereas transplacental passage of herpes is infrequent, the virus is more often transmitted to a neonate during delivery through an infected birth canal. Neonatal herpes usually kills the infant. Other diseases acquired during birth include gonorrhea infections (especially opthalmia neonatorum), candidiasis, and neonatal group B streptococcal disease. Group B streptococcus (*Streptococcus agalactiae*) is a normal vaginal resident in 20 to 30 percent of pregnant women. Although most infants escape these infections, a few develop neonatal disease, usually within the first week of life. About half of these infected children die, usually from septicemia, pulmonary invasion, or meningitis. Half the survivors of meningitis are blind, deaf, mentally retarded, or stricken with epilepsy or cerebral palsy. Premature infants are at greatest risk. The risk of neonatal group B streptococcal disease may be reduced by administering penicillin at the onset of labor to women whose vaginas are colonized by this bacterium.

TABLE 22-6

Some Important
Intrauterine Infections of
the Fetus

Disease	Pathogen	Portal of Entry into Mother	Effect on Fetus
Cytomegalovirus infection	Cytomegalovirus (CMV)	Alimentary tract, genitourinary tract	Mental retardation; fatal liver, spleen, and neurological damage
Congenital syphilis	*Treponema pallidum*	Genitourinary tract	Abortion or postnatal death; bone deformity; liver and kidney disease; skin and mucous membrane lesions
Toxoplasmosis	*Toxoplasma gondii*	Alimentary tract	Injury to retina or central nervous system; mental retardation
Congenital rubella syndrome	Rubella virus	Respiratory tract	Mental retardation; damage to nerves and cardiovascular system
Congenital herpes infection	Herpes simplex Types 1 and 2	Alimentary tract, genitourinary tract	Viremia (usually fatal)

The most important infections of the developing fetus are presented in Table 22-6. In addition to these, the agents of mumps, chickenpox, measles, poliomyelitis, influenza, viral hepatitis, and malaria occasionally cause intrauterine infections.

OVERVIEW

The genitourinary tract is a portal of entry for pathogens of the upper and lower urinary tract and for agents of sexually transmitted diseases. Urinary tract disease often correlates with anatomical anomalies, introduction of invasive medical instruments, or other predisposing factors. Sexually transmitted diseases, on the other hand, are mainly a function of exposure to the pathogen by direct intimate contact.

Serious consequences may follow unattended cases of diseases acquired through the genitourinary tract. Lower urinary tract infections can ascend from the bladder to the kidneys, causing fatal pyelonephritis. Syphilis produces slow neurological degeneration in some of its victims as well as possibly fatal heart disease. Gonorrhea may block the reproductive ducts and result in sterility. Congenital disease and infections of the newborn present an especially dangerous threat for the developing child of a mother with active venereal disease. Whereas adult symptoms of many STDs are localized in the genitals, the fetus or child is vulnerable to deforming or fatal systemic infection. Early VDRL tests for pregnant women, cesarean delivery when the mother has an active case of genital herpes, and instillation of antibiotics in the eyes of newborns have helped to avoid these tragedies.

The epidemic proportions of the three major venereal diseases—gonorrhea, syphilis, and genital herpes—as well as of nongonococcal urethritis (NGU) emphasize the need for improved management. Control measures depend on encouraging people to submit for routine checkups, seek rapid diagnosis at the earliest indications of genitourinary tract disease, obtain prompt therapy, notify contacts, and refrain from sexual relations until the disease is cured.

KEY WORDS

cystitis	chancre
pyelonephritis	gumma
urethritis	reagin
vaginitis	VDRL test
sexually transmitted disease (STD)	RPR test
ascending infection	acyclovir
descending infection	intrauterine infection
bacteriuria	

1. Describe five characteristics of the genital and urinary tracts that oppose infection. Discuss several predisposing factors that reduce the effectiveness of these defenses.

2. Why are women more prone to genitourinary tract infections than men?

3. Why is descending urinary tract disease more dangerous than ascending urinary tract disease? How can these diseases be diagnostically differentiated?

4. Compare the pathogenesis, diagnosis, treatment, and possible complications of gonorrhea, syphilis, and genital herpes.

5. Discuss two nonsexual ways that genital infections may be acquired.

6. For each of the following diseases describe the pathogen(s) and therapy.
   (a) Toxic shock syndrome
   (b) Nongonococcal urethritis
   (c) Genital warts
   (d) Candidiasis
   (e) Trichomoniasis

7. Why are reinfections common in candidiasis and trichomoniasis? How do these repeated episodes differ from recurrences of genital herpes?

DISEASES ACQUIRED
THROUGH THE
GENITOURINARY TRACT

Diseases Acquired through the Skin

23

Although microbes constantly bombard the skin, most fail to even colonize the surface. Only a few pathogens can overcome this barrier and cause local cutaneous infection. Injury to the skin, however, creates a portal through which many otherwise harmless microbes can gain entry. Most organisms that infect minor cuts and abrasions are restricted to the site where they entered. These cutaneous infections are usually more irritating than dangerous, affecting an estimated 20 percent of the people in the United States and even a larger number of people worldwide. Some organisms are directly deposited into deeper tissues and the bloodstream by **parenteral* inoculation** through the skin, which introduces microbes through cuts, punctures, burn wounds, or bites of infected animals or arthropods. Some of the world's most severe and prevalent infectious diseases are transmitted by parenteral inoculation.

ANATOMY AND DEFENSES OF THE SKIN

The skin, the largest organ of the body, is composed of the epidermis and the dermis (Fig. 23-1). The exposed surface of the **epidermis** is the stratum corneum, several

FIGURE 23-1

Anatomy of the human skin. Microorganisms are normally found only on the stratum corneum in sweat ducts and hair follicles.

*Literally, parenteral means entering by some means other than through the alimentary canal, but it refers particularly to the introduction of material into blood vessels or subcutaneous tissue.

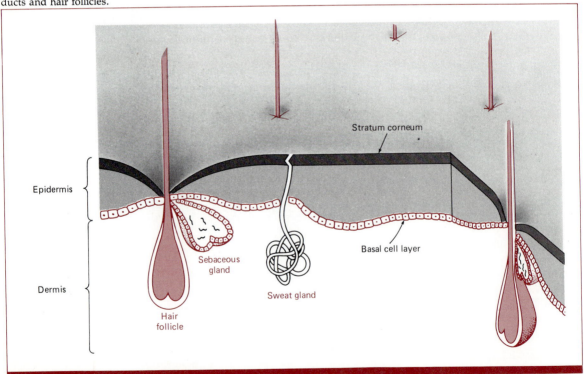

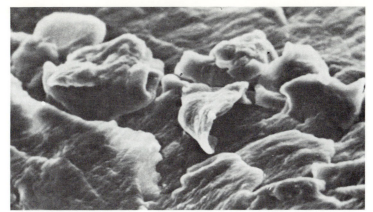

overlapping rows of flattened dead cells rich in **keratin**, a protein that is insoluble in water. The stratum corneum is impervious to most pathogens that settle on the skin. Below the keratinized cells, a living layer of basal cells continually forms new progeny that force the older cells of the basal layer toward the surface. During their 40- to 60-day journey, the cells of the basal layer produce keratin, alter their morphology, and die, eventually becoming a part of the dead keratinized layer. The dead cells of the stratum corneum are constantly shed, and the skin maintains a uniform thickness (Fig. 23-2). Microbes on the skin are usually sloughed with the dead epidermal cells.

The **dermis** lies beneath the epidermis and consists of connective and muscle tissue. The dermis contains nerve endings, blood and lymphatic vessels that supply nutrients to living epidermal cells, and protective cells capable of an inflammatory response and wound repair. *Hair follicles* are embedded in the dermis. Cells at the base of the follicle produce a keratin-rich strand of hair. Nails are also keratinized structures produced by specialized cells in the dermis. *Sebaceous glands* lie adjacent to hair follicles and empty their oily secretions, called **sebum**, into the follicle. Sebaceous glands are always associated with hair follicles, together forming a *pilosebaceous unit. Sweat glands* lie deeper within the dermal layer than hair follicles. Sweat is composed primarily of water that contains a low concentration of salts, nutrients, and nitrogenous wastes.

Cutaneous Defense Michanisms

A variety of factors join forces to toughen the skin against traumatic disruption and make it inhospitable for colonization by pathogens (Table 23-1). As long as the skin remains physically intact, fairly dry, and continues to slough its dead surface cells at its normal rate of about 1 million every 40 minutes, most microbes cannot permanently colonize, penetrate, or even survive on the skin.

Microbes deposited on the skin are presented with a disappointingly arid landscape. Desiccation is probably the single most effective element

TABLE 23-1

Skin Defenses

Type	Responsible Factor	Protective Action
Mechanical	Intact stratum corneum	Provides barrier to penetration
	Epithelial sloughing	Removes surface microbes
	Low water content	Creates poor environment for microbial proliferation
Chemical	Sebum	Contains antimicrobial fatty acids
	Sweat	Contains antimicrobial concentrations of salt and lysozyme
	Low pH	Creates poor environment for microbial proliferation
Microbial	Normal flora	Compete with pathogens
Cellular	Leukocytes	Localize infection by inflammatory response
Immunological	Immunoglobulins	Neutralize pathogens that penetrate to dermis
	T lymphocytes	Enhance killing of pathogens in dermis

working against prolonged survival of most gram-negative organisms on intact skin. Gram-positive bacteria appear to be less inhibited by the lack of water. This defense is also less effective in the armpits, groin, and other moist areas of the body. The armpit may harbor 1 million bacteria per square centimeter, approximately 100 times the population of the forearm.

Sweat and sebum contain chemicals that inhibit the growth of some organisms. Lipids and fatty acids in sebum are fungicidal and confer resistance to some dermatophytoses ("ringworm" infections). This resistance is acquired only after puberty when the activity of sebaceous glands increases. The salinity and lysosomal constituents of sweat are mildly bacteriostatic but do not outweigh the tendency of sweat to promote microbial multiplication by providing moisture. Overall, chemical factors are probably not as important in protecting the skin from infection as are mechanical defenses and the normal skin flora.

Normal Flora of the Skin

Despite antimicrobial physical and chemical mechanisms, no region of the skin is sterile. In fact, in the armpits and the groin, the microbial flora is as abundant as that of the soil. Some of the skin's microbes are transient residents; others are established members of the permanent flora (Table 23-2). The most commonly isolated microbes are staphylococci and diph-

TABLE 23-2

Normal Flora Commonly Found on Adult Skin

Microorganism	Microscopic Description
Staphylococcus epidermidis	Gram-positive coccus in clusters
Staphylococcus aureus	Gram-positive coccus in clusters
Propionobacterium acnes	Gram-positive pleomorphic rods
Corynebacterium sp.	Gram-positive pleomorphic rods
Candida sp.	Yeasts
Streptococcus pyogenes	Gram-positive coccus in chains

theroids, harmless bacteria that resemble the diphtheria bacillus and use the fatty nutrients in sebum. The normal skin microbes are scattered over the stratum corneum or firmly entrenched in pilosebaceous units where they form microcolonies of 1000 to 10,000 cells. Vigorous washing and antiseptics cannot completely remove these organisms from their protected habitat.

The resident skin flora generally interfere with the establishment of pathogenic microorganisms. This competition is so effective that a fairly stable population of microbial inhabitants is maintained on the skin. Temporary fluctuations in the members of the skin flora are due to the presence of transient organisms. These microbes fail to become firmly entrenched and are readily removed by washing, especially with an antiseptic soap. Transient microbes are poorly adapted to the skin, and many die within 2 to 3 hours; after 24 hours most contaminants disappear entirely, although they are being constantly replaced by other transient microbes from the environment.

The skin is always a potential source of pathogens. Virtually any microbe on skin may cause disease if the integrity of the skin is broken or host defenses are otherwise compromised.

Defenses Against Parenteral Infection

Microbes that penetrate the epidermis into the underlying living layers of the skin trigger an inflammatory response that subjects the intruder to macrophage and polymorphonuclear leukocyte attack. This activation of the inflammatory defenses usually restricts the microbe to a limited area within the dermis. Cellular and humoral immunity also helps prevent microbial invasion of the bloodstream and internal organs.

PREDISPOSING FACTORS

Many situations reduce the efficiency of the skin defenses and encourage infectious disease (Table 23-3). High humidity and constant sweating promote accumulation of moisture and proliferation of an assortment of pathogens on the skin. Many cutaneous diseases are most prevalent in the tropics, where these conditions are common. Waterproof materials over the skin create localized high-moisture conditions. Plastic pants used over diapers, for example, may result in concentrations of microbes that cause diaper rash. Prolonged use of plastic bandages, rubber gloves, and some shoes not only promotes growth of pathogens, but may also soften the skin until it breaks. Microbes use these portals to gain entry to the underlying tissues and bloodstream. Excess moisture frequently accumulates on feet, armpits, and groins of people with *hyperhidrosis* (excessive sweating). The condition is augmented in the folds of skin on obese persons or in areas covered by nylon underwear. Tight clothing and the friction of movement force microbes into minor skin abrasions. Dishwashers, bartenders, and other persons

Factor	Effect of Impairment	Precautionary Measures
Administration of corticosteroids	Decreases rate of surface shedding	Normal epithelial shedding returns when medication is ceased.
Increased androgens	Elevate sebum concentration	Cleansing of skin to discourage accumulation of fatty acids
Increased moisture	Creates a more hospitable environment for proliferation of pathogens	Use of absorbent powders and clothes made of fabrics that allow evaporation to keep surface dry
Obesity	Promotes accumulation of surface moisture in body folds	Loss of weight
Invasive medical procedures	Disrupt physical integrity of skin	Use of sterile equipment and aseptic technique
Tight clothing	Disrupts physical integrity of skin through continual abrasion	Wearing of looser clothing
Wounds	Expose underlying tissue to environment	Cleansing of wound; avoidance of contamination
Minor cuts and abrasions	Disrupt physical integrity of skin	Good hygiene and sanitary habits to prevent wound colonization
Burns	Expose underlying tissues, diminish immunological capacity	Avoidance of contamination; use of topical antibiotics
Underlying debilitating diseases	Cause malfunctions of the immune system and decreased phagocytic function	If possible, treatment of underlying disease
Malnutrition	Impairs immunological system	Adequate diet

whose hands are constantly immersed in water are also susceptible to opportunistic skin diseases. Keeping the body dry reduces the opportunities for infection.

Traumatic injuries produce breaks in the skin that may lead to life-threatening infectious diseases. Persons with burns, surgical incisions, or other serious wounds are highly susceptible to infection. Wounds expose underlying tissue to dangerous opportunistic pathogens. In addition, the reduced blood supply to the traumatized tissue often impairs the ability of protective immune and phagocytic cells to reach the affected area. Impaired circulation to injured tissue produces anaerobiasis and tissue death, creating an ideal environment for the growth of anaerobic bacteria, the most frequent and dangerous cause of wound infections. Debris left in the wound tends to further aggravate the tissue, prolong inflammation, and predispose to infection. Hypodermic injections may also provide a portal of entry for opportunistic pathogens unless proper medical asepsis is employed to protect patients from microbes on the skin, needle, or on the hands of medical personnel, visitors, and the patients themselves.

Corticosteroids decrease the skin's rate of epithelial shedding, often permitting cutaneous pathogens to become entrenched. Administering these drugs to persons with existing skin disease often intensifies the condition. As with most infections, those acquired on or through the skin are more prevalent and severe in persons suffering from malnutrition, diabetes, cancer, and defects of the immune system. Many persons with T-cell immunodeficiency, for example, suffer overwhelming cutaneous infections caused by the normally noninvasive yeast *Candida albicans*.

DYNAMICS OF TRANSMISSION

Most microbes that cause skin or parenteral infections are opportunistic organisms that normally reside in the environment, in animals or arthropods, or on human skin as members of the normal flora. The diseases they cause are usually "dead-end" infections, that is, they are not readily communicable from infected persons to new hosts. Of those diseases that people contract from infected animals, only a few are directly transmitted. These include rabies and ratbite fever, both acquired by the bite of infected animals. Tularemia and anthrax enter the human body through minor cuts and abrasions on skin in contact with infected animals or their dander. Tularemia and anthrax most commonly afflict hunters, animal breeders, and slaughterhouse workers.

Many diseases are transmitted from animal reservoirs to humans by arthropods. These vectors also transmit some diseases from person to person. Most arthropod-borne diseases are noncommunicable, and their continued presence depends on the existence of a population of infected arthropods. Endemic regions are therefore restricted to areas inhabited by the suitable vector. For this reason, most of these diseases cannot be exported by infected travelers to new geographical locations.

Some microbes that are normally harmless saprophytes in the environment are dangerous when they contaminate objects that cause injury to the skin. Contaminated thorns, nails, hypodermic needles, or other sharp objects simultaneously injure the skin and inoculate the wound with pathogens that thrive in the damaged tissue and cause disease. Sharp objects may also inoculate the wound they cause with organisms residing on the surface of the skin at the site of injury. Microbes remaining on the skin following improper cleansing and decontamination of puncture sites, for example, readily enter the body as the hypodermic needle is inserted. Wounds may also become contaminated after injury, especially if underlying tissue is exposed to the environment. Burn wounds are among the most susceptible to infection. Open wounds are frequently contaminated by normal flora microbes, either by autoinoculation or by another person. Wounds may also be contaminated by contact with healthy carriers of potential pathogens or with persons transiently colonized by pathogens from other sources. Usually people serve as mechanical vectors of diseases acquired through open wounds; that is, they indirectly transmit microbes from another source. Those few pathogens that infect intact skin, however, are usually directly communicable from infected people.

DISEASES OF THE SKIN

Infectious disease of the skin may occur in the epidermis (**cutaneous infections**) or in the dermis or fatty tissues (**subcutaneous infections**). Most skin rashes are not considered skin infections because they are cutaneous expressions of systemic diseases acquired by another route. Measles and chickenpox, for example, are acquired through the respiratory tract.

Infections of the epidermal layer generally develop after direct contact with the reservoir. Most pathogens require at least minor breaks in the epidermis to establish an infection. The lesions may be flat red areas (a *macule*), elevated (a *papule*), or filled with either clear fluid (a *vesicle*) or pus (a *pustule*). Lesions that have burst and necrosed are especially susceptible to secondary infections.

Dermatophytoses

These skin diseases are caused by a group of fungi called dermatophytes that have an affinity for tissue containing high concentrations of keratin, specifically, skin, hair, and nails (see Chap. 7). Dermatophytes cause a variety of infections commonly known as ringworm. The appearance of the lesions led to the erroneous belief that the infected skin harbored worms beneath its surface. These diseases are named "tinea" (a gnawing worm) followed by another word to designate the infected site. For example, tinea capitis is ringworm of the scalp, tinea corporis is ringworm of the body, tinea pedis is the disease popularly known as athlete's foot, tinea cruris is jock itch, and tinea unguium is ringworm of the nails.

The dermatophytes that cause these diseases belong to three genera of fungi, *Epidermophyton*, *Microsporum*, and *Trichophyton*, each with an affinity for particular types of keratinized structures. *Epidermophyton* infects only skin and nails, *Microsporum* invades only skin and hair, and *Trichophyton* causes disease in all three body sites. None of these organisms can proliferate in living tissue, and all are restricted to the nonliving surface of the body.

The principal dermatophyte reservoirs are infected people or animals, especially dogs, cats, and, in rural areas, cattle, horses, and wild animals. Only one dermatophyte, *Microsporum gypseum*, is found primarily in the soil. Transmission is by direct contact with infectious lesions or by contact with infected hair or skin scales that have been shed onto clothing, combs, floors, showers, towels, or other fomites. The soil and animal fungi usually elicit a more marked inflammatory response from the human host than do the organisms that have human reservoirs. The latter appear to be adapted to their host and tend to cause more chronic infections.

In most forms of **tinea capitis** the hair becomes brittle and breaks off. This form of hair loss is temporary, and the hair grows back when the disease is resolved. However, one form of tinea capitis, called **favus**, is characterized by suppurative (pus-discharging) or crusted lesions on the scalp that scar when they heal and cause permanent hair loss.

Children are most susceptible to *Microsporum* infections; adults are virtually resistant to all members of this genus. People of all ages, however, are susceptible to *Trichophyton* species.

Typical ringworm of the smooth (body) skin is characterized by a flat, spreading, circular lesion representing the continued radial growth of the fungus (Fig. 23-3). Peripheral areas of the lesion contain the living fungus and are often marked by inflammation and pustule formation resulting in a red elevated margin (resembling a worm). Alternatively, the lesion may be scaly or crusted. The center of the lesion usually heals and appears like normal skin. Lesions may merge, forming large confluent areas of infection. Infections of the foot or of other moist surfaces generally lead to extensive scaling of the skin, which may be accompanied by vesicle formation. The appearance of the lesions cannot be used to distinguish between the numerous agents of tinea corporis or tinea pedis.

Tinea unguium occurs in up to one-third of people with tinea corporis, presumably by direct extension of the dermatophyte from skin to nails. The infection can also be contracted from contaminated fomites or surfaces. Injured nails are most likely to be affected. Any dermatophyte can cause tinea unguium although *Microsporum* species are rarely responsible. The fungus may grow as patches on the nail surface, or it may invade beneath the nail plate, causing an accumulation of keratin debris. Infected nails thicken, become discolored, brittle, and cracked. These infections are usually chronic and rarely resolve without drug treatment.

The characteristic lesions on the infected scalp, skin, or nail is usually distinct enough to suggest dermatophyte infections. Ultraviolet light helps distinguish scalp lesions produced by *Trichophyton* species from those produced by *Microsporum* species. Hair invaded by *Microsporum* fluoresces under ultraviolet light. Clinical suspicion of dermatophyte infection is confirmed by direct microscopic examination of infected hair, nails, or scrapings from the active edge of a skin lesion. Prior to examination, the specimen is heated with 10% potassium hydroxide directly on a microscope slide. This procedure digests the keratin debris that might otherwise obscure the characteristic septate hyphae and spore structures of the fungus. A variety of media stimulate formation of characteristic spores when the fungus is incubated at 22 to 30°C. These spores aid in identifying

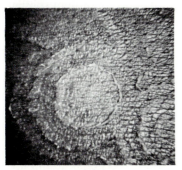

(*a*)

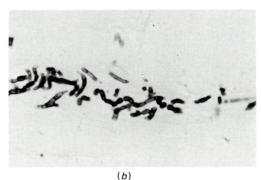

(*b*)

FIGURE 23-3
Tinea corporis: (*a*) Circular lesion on the face and, (*b*) microscopic appearance of the dermatophyte in tissue.

the dermatophyte. Dermatophytes are often slow growers and may require 10 days or longer to sporulate.

Many dermatophyte infections can be eradicated by treatment with topical antifungal agents, especially imidazoles (miconazole, ketoconazole, or clotrimazole) or tolnaftate. Griseofulvin is an antibiotic that, when taken orally, inhibits fungal growth while stimulating an increased rate of shedding of the infected keratinized layer. Fungi in skin cells are sloughed and removed during daily bathing. Several months of griseofulvin therapy may be required. Clipping hair and nails short physically removes much of the pathogen. This reduces the probability of the pathogen extending from the initially infected site. Since fungi thrive in moist, warm environments, keeping lesions dry and cool usually shortens the duration of the disease. Dusting powder absorbs excess moisture. Loose-fitting clothing made of natural porous fibers that allow evaporation helps keep lesions dry. Open sandals are recommended for persons with athlete's foot. Bandages increase local humidity, but gauze dressings protect affected skin from friction, irritation, and secondary bacterial infection. Shaving beards or other infected areas should be discontinued until lesions have healed. Simultaneous treatment of infected pets prevents reinfection. In spite of these measures, some cases of cutaneous mycoses persist.

Infected persons may avoid spreading the disease by washing themselves daily to remove loose hair and skin before it can be shed to other persons; by decontaminating combs, clothing, and other items that contact the lesions; and by avoiding activities that may transmit the disease to others. Sharing towels or clothing encourages transmission of ringworm infections. Floors in dressing rooms, locker rooms, and public showers frequently harbor skin and hair shed from infected persons and should be frequently disinfected with an antifungal agent.

Candidiasis

Candida albicans is a yeast that, in addition to its many systemic manifestations, may cause cutaneous mycoses, often on body sites where excessive moisture accumulates. *Candida* infections frequently occur in skin folds of obese individuals, under pendulous breasts, in armpits, in the crotch, and on nails and hands that spend a lot of time immersed in water or covered by nonporous gloves. The yeast multiplies in the cutaneous layers and causes local skin lesions. Persons with impaired immunological or cellular defenses may develop lesions that cover the entire body surface and the mucous membranes, a disease called *chronic mucocutaneous candidiasis* (Fig. 23-4a).

The organism can be isolated from the lesion, grown on Sabouraud's dextrose agar, and identified by biochemical tests. *Candida albicans* is unique among yeasts because it possesses the ability to form *pseudohyphae* (chains of elongated yeast cells), blastospores, and *chlamydospores* when grown on cornmeal agar. *Candida albicans* is also the only yeast that forms germ tubes after 2 hours in serum at 37°C (Fig. 23-4b).

Cutaneous *Candida* infections are usually treated topically with nysta-

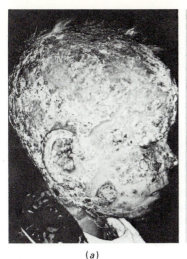 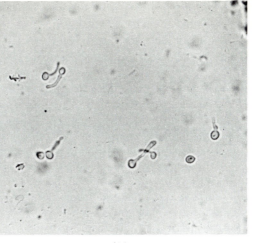

(a)   (b)

FIGURE 23-4

(a) Chronic mucocutaneous candidiasis. A hereditary immunodeficiency in the T-cell response predisposed this person for *Candida* infection of the skin and mucous membranes. (b) Characteristic microscopic appearance of germ tubes formed by *Candida albicans* in serum.

tin, miconazole, halprogen, or clotrimazole. Oral ketoconazole also helps eliminate these infections. As with other opportunistic infections, controlling the conditions that promote proliferation of the pathogen is the best way to prevent *Candida* infections. Unfortunately, many of these predisposing factors cannot be easily corrected or controlled.

Pyogenic Skin Infections

Some skin infections are characterized by formation of pus as a product of local inflammation. These **pyogenic** (pus-producing) pathogens are bacteria, most commonly *Staphylococcus aureus* or *Streptococcus pyogenes*. These two bacteria are particularly dangerous for neonates and surgical patients. In hospitals *Staphylococcus aureus* is the most frequent cause of skin lesions on newborns. Preventing infections is difficult because healthy carriers of *Staphylococcus* and *Streptococcus* are undetected sources of infection. An estimated 50 percent of all healthy persons harbor one of these bacterial species in their upper respiratory tract. Exposure to these potential pathogens is probably universal. These diseases are also acquired by autoinoculation, making their control even more difficult. Widespread outbreaks usually do not occur among persons with normal defenses. But when these defenses fail, *Staphylococcus aureus* and *Streptococcus pyogenes* may cause a number of cutaneous infections:

■ **Folliculitis** is a mild infection of a hair follicle by staphylococci. Bacteria grow in an obstructed hair follicle or sebaceous gland and elicit an inflammatory response, producing an abscess called a *furuncle* (boil). Folliculitis in the follicle of an eyelash is referred to as a *sty*. Abscesses of the skin of the neck and back tend to spread through the underlying fatty and connective tissues to adjacent follicles, causing a *carbuncle,* an extensive lesion that may develop into serious, deep-seated infections requiring surgical drainage and antibiotic therapy.

■ **Impetigo** is a skin infection that most often occurs in young children, especially on the skin of the face and legs. The disease often results from a mixed infection of *Streptococcus pyogenes* and *Staphylococcus aureus,* although either species alone can be the cause. The lesions begin as pustules that rupture and form scabs or crusts (Fig. 23-5). Bacteria in the discharged pus spread the disease to adjacent areas on the skin and to other persons. The infection is usually restricted to the skin, but impetigo in infants may seed the bloodstream with bacteria and lead to systemic disease, kidney injury, and death. Impetigo is a highly communicable disease often causing epidemics in families, schools, and nurseries. One strain of *Staphylococcus aureus* is also responsible for another severe form of impetigo called toxic epidermal necrolysis or *staphylococcal scalded skin syndrome (SSSS)*. This strain produces *exfoliatin*, a toxin that causes the epidermis to separate from the dermis, resulting in blisters, reddening, and extensive peeling of the top layers of skin. The disease may be fatal if accompanied by bacteremia. The toxin primarily affects the skin of newborns; older children and adults are usually resistant to SSSS.

■ **Erysipelas** is an inflammation that advances through the local lymphatic system. It is caused by beta-hemolytic streptococci that probably enter the dermis through minute fissures in the skin. Painful red lesions usually spread over the face and legs and may cause bacteremia in debilitated patients.

FIGURE 23-5
Impetigo caused by staphylococci and streptococci.

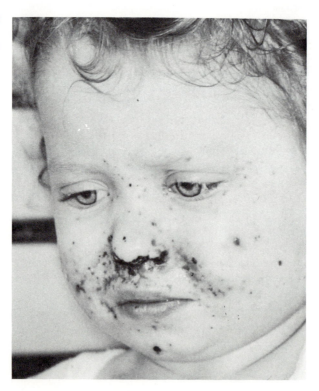

Pus discharged from any of these purulent skin infections is easily obtained for microscopic examination and culture. Pus in a closed lesion is collected by aspirating with a needle and syringe. Blood should also be cultured when fever accompanies cutaneous infections. During outbreaks, nasal cultures of suspected carriers help identify the source of the pathogen. Gram's stains of pus reveal cocci with typical chain (streptococci) or cluster (staphylococci) arrangements. Laboratory identification, however, is confirmed by biochemical and serological tests that differentiate the pathogens from members of the commensal flora similar in appearance, especially *Staphylococcus epidermidis* or nonpathogenic streptococci. The most commonly used tests for identifying *Staphylococcus aureus* determine its ability to produce coagulase, an enzyme that coagulates blood plasma, and to produce characteristic colonies on mannitol salts agar. Phage typing of *S. aureus* is often performed to determine the source of infection. *Streptococcus pyogenes* is identified by production of beta hemolysis on blood agar and sensitivity to low potency (2 μg/ml) bacitracin. (See color box, Chap. 20.)

Localized cutaneous infections require no systemic antibiotic therapy. Lesions are cleaned and topically treated with antiseptic or antibiotic ointments. In cases of impetigo the crusts are removed prior to treatment. The once popular technique of bathing infants in hexachlorophene reduced the incidence of infections, but it has been discontinued since the discovery that the antiseptic is toxic to the central nervous system if absorbed through the infant's skin.

Because many strains of *Staphylococcus aureus* produce penicillinase, antibiotic susceptibility tests are important to determine the best treatment for serious infections. Penicillin-resistant staphylococci can be eliminated by nafcillin, methicillin, or erythromycin. Streptococci are rarely resistant to penicillin, which remains the drug of choice in the absence of coinfections by penicillinase-producing staphylococci.

Acne

Acne is a localized inflammation of hair follicles that mainly affects adolescents. The disease occurs in two stages. During the first stage, excessive sebaceous secretions accumulate in hair follicles that have been blocked by keratinized cells. These plugged follicles are called *comedones* (blackheads). Contrary to popular belief, the dark color in blackheads is not dirt, but is accumulated skin pigment of keratinized cells. In the second stage, excess sebum is converted to fatty acids by the enzyme lipase, produced by *Propionobacterium acnes*, a normally harmless inhabitant of the follicular canal. The fatty acids irritate the skin and cause the inflammation characteristic of acne. The inflamed follicles develop into pustules which, in severe cases, burst and release bacteria into the surrounding tissue. The lesions may become severe enough to cause permanent pits and scars. Low doses of tetracycline and erythromycin reduce the bacterial population and therefore the production of free fatty acids. This treatment is discontinued when antibiotic-resistant *P. acnes* become predominant. Penicillin is ineffective because it is not secreted into sebaceous glands. Because of side effects to the fetus, tetracyclines are not prescribed for the control of acne in pregnant women. In 1982, *retinoic acid*, a synthetic derivative of vitamin A, was licensed for treating serious acne that failed to respond to antibiotic therapy. This compound effects a marked improvement or recovery in most persons with acne by suppressing sebum production. The side effects of dryness, hair loss, and mild muscle pain limit its use.

Cutaneous Virus Infections

Since viruses require living cells in which to reproduce, they do not infect the stratum corneum. Most virus infections that produce cutaneous lesions do so only as a secondary manifestation of disease that begins in the respiratory or gastrointestinal tract. Only three viruses enter through the epidermis and cause localized lesions in living skin cells.

Wart viruses stimulate excessive multiplication of infected epidermal cells, producing benign tumors called **papillomas** (warts). Warts may regress spontaneously or persist for many years. They can be physically destroyed by cryocautery (freezing) or electrosurgery. The herpes simplex viruses may cause vesicles on skin surrounding the mouth, genitals, or anus (see Chaps. 21 and 22). Some of the poxviruses (molluscum contagiosum, orf, cowpox, milker's nodule) cause pustules at the sites where they enter the skin. These viruses are spread by direct contact with infectious lesions. Humans are the sole reservoir of the poxvirus that causes *molluscum contagiosum*. Lesions may develop on any part of the body, but in adults are most frequent in the genital area. *Orf, cowpox*, and *milker's nodule* are poxvirus diseases transmitted to humans by contact with infected animals. These poxvirus infections usually resolve spontaneously within a few months.

Characteristics of skin infections are summarized in Table 23-4.

PARENTERALLY ACQUIRED DISEASES

Another arena for microbial attack is located in or beneath the dermal layer following parenteral inoculation of pathogens through the skin. Parenterally acquired pathogens may cause localized infections at their inoculation sites, or they may invade surrounding tissue, enter the lymph or blood systems, and cause systemic disease.

Diseases Acquired Through Animal Bites

Rabies Each year more than 30,000 people in the United States suspect they may have been bitten by rabies-infected animals. Yet no more than four confirmed cases per year occurred between 1960 and 1982 (see color box). Before the development of effective vaccination programs little could be done to save the lives of persons bitten by rabid animals. Even today, once symptoms appear, the disease follows an invariably fatal course (only

TABLE 23-4
Some Characteristics of Infectious Diseases of the Skin

Disease	Pathogen	Reservoir	Diagnostic Procedure	Antimicrobial Treatment
Dermato-phytoses	*Microsporum, Trichophyton, Epidermophyton*	Humans, animals, or soil	Microscopic examination of lesion or culture	Topical fungicides, oral griseofulvin
Candidiasis	*Candida albicans*	Humans	Microscopic examination of lesion or culture	Nystatin, imidazoles
Folliculitis	*Staphylococcus aureus*	Humans	Clinical picture	None
Impetigo	*Staphylococcus aureus*	Humans	Isolation and identification of pathogen from lesion	Erythromycin
	Streptococcus pyogenes	Humans	Isolation and identification of pathogen from lesion	Penicillin
Erysipelas	*Streptococcus pyogenes*	Humans	Isolation and identification of pathogen from lesion	Penicillin
Acne	*Propionobacterium acnes*	Humans	Clinical picture	Tetracycline
Molluscum contagiosum	Poxvirus	Humans	Microscopic identification of viral inclusion bodies in cells from lesion	None

three persons are believed to have survived rabies). Its unrelenting progression of neurological symptoms is characterized by increased salivation, irrational behavior, convulsions, and hydrophobia (fear of water). The latter symptom develops as a result of painful spasms in the throat when the victim attempts to swallow. Death usually occurs within a week after onset of symptoms. Many victims have ended their own lives in the face of inevitable madness.

THE COST OF ONE RABID DOG*

■   On May 10, 1980, a dog in Yuba County, California, was placed under observation after it bit three persons in a parking lot in the Olivehurst area. Because the dog appeared ill, it was killed, and tissues were tested and found positive for rabies on May 12. The subsequent investigation by Sutter-Yuba County Health Department personnel eventually resulted in the identification of 70 persons, who received antirabies prophylaxis because of known or probable exposure to the dog. Because investigators found that only 20 percent of the dogs and cats in the area had up-to-date vaccination, special clinics were held in which 2000 dogs were vaccinated; over 300 unclaimed dogs and cats were destroyed.

No persons or other animals were known to develop rabies as a result of this episode. However, the costs generated by this single rabid dog were estimated as: $92,650 for human antirabies treatment, $4190 for animal vaccination and veterinary services, and $8950 for health-department and animal-control programs. The total cost of the episode was $105,790, or over $1500 per person treated, not including lost work time, patient travel time, and costs of the 6 months quarantine imposed on animals exposed to the rabid dog.

*Morbidity and Mortality Weekly Report, **30**:527 (1981).

The rabies virus is readily transmitted in populations of domestic and wild animals. Skunks are the major reservoir in the United States, but dogs, cats, cattle, raccoons, bats, and foxes (the major reservoir in Europe) are also potential sources of rabies. In all animals except bats, the virus causes a fatal infection. Prior to the appearance of symptoms, however, the virus is released into salivary secretions. As the infection progresses, the animal becomes agressive and tends to bite, thereby promoting the spread of disease. Vicious attacks are not necessary for disease transmission, however, because even licking an abrasion on the skin or mucous membrane may be sufficient to inoculate a lethal dose of virus.

The virus particles replicate in muscle tissue at the site of inoculation, invade the regional nerve endings, and eventually reach the brain. From the central nervous system the virus migrates through nerves to other parts of the body, including the salivary glands. The incubation period varies between 2 to 8 weeks, depending on the site of inoculation and the size of the infecting dose. Symptoms develop more rapidly when the site of virus entry is close to the brain.

Transmission may infrequently occur by routes other than direct inoculation from an infected animal. Rabies may follow inhalation of dust heavily contaminated with rabies virus. Cave explorers run a high risk of contracting airborne rabies by inhaling dried feces of rabid bats. The disease has also been transmitted by surgical transplant of corneas obtained from persons with undiagnosed rabies at the time of death.

The appearance of typical neurological symptoms and the history of an animal bite suggest a presumptive diagnosis of rabies. The virus may sometimes be isolated by inoculating mice with saliva, but attempts are often unsuccessful. Rabies-specific fluorescent antibody detects virus in frozen specimens of skin, corneal impressions, or mucosal scrapings. When there is no evidence of animal exposure or when symptoms are atypical, the disease is usually diagnosed following autopsy by the microscopic observation of cytoplasmic inclusion bodies, called *negri bodies,* in the victim's brain cells (Fig. 23-6).

Anyone exposed to rabies should be treated to prevent the disease, since a fatal outcome is practically guaranteed once symptoms develop. Immediate action is critical if clinical disease is to be avoided. The inoculation site should be thoroughly cleansed with soap and water, detergent, alcohol, iodine, or quaternary ammonium compounds in order to physically remove and inactivate any virus remaining in the wound. Passive immunization with hyperimmune antirabies serum provides immediate protection, especially if injected directly around the wound as well as intramuscularly. In addition, a series of daily injections of rabies vaccine builds active immunity before the virus attacks the nerve cells. The original rabies vaccine developed by Louis Pasteur was obtained from homogenized brain tissue of experimentally infected rabbits. The virus was inactivated and administered in 14 to 21 painful inoculations under the abdominal skin. A newer, more potent vaccine uses viruses propagated in tissue culture, chemically inactivated, and injected intramuscularly in only

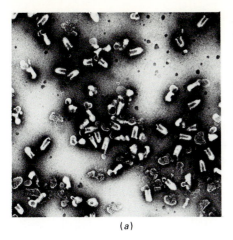

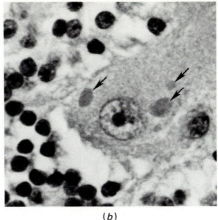

FIGURE 23-6

(a) Electron micrograph reveals the characteristic bullet shape of the rabies virus. (b) Light microscope reveals negri bodies (arrows) in brain cells from a dog with rabies.

(a)   (b)

four to six shots. Because of the small risk of encephalitis following vaccination, only those persons in danger of developing rabies receive the vaccine. This includes anyone bitten by an animal that is not confirmed free of rabies—that is, by an animal not available for examination and of a species known to carry rabies. Treatment is discontinued if the attacking animal is caught and proved to be free of disease.

The following guidelines are recommended to help reduce the danger of human rabies by reducing the number of rabies-infected animals.

■ Routinely vaccinating pets protects them from acquiring and transmitting the disease.

■ Confining pets to the home reduces the likelihood of exposure to rabid animals.

■ Reporting to local authorities any animal suspected of having rabies better enables the authorities to capture, examine, and, if necessary, destroy these sources.

■ Isolating persons with suspected rabies helps protect others from their infectious secretions. The disease may be transmitted from people by materials soiled with salivary secretions, tears, and urine. The virus is not present in the blood.

■ Monitoring the population of wild animals alerts public health officials to naturally occurring *epizootics* (epidemics among animals) and to the increased danger of disease spreading to people.

■ International travel restrictions requiring animals to be quarantined or vaccinated before they are admitted to areas that are free of rabies helps prevent the emergence of the disease in these regions. Approximately 30 countries have maintained a rabies-free animal population by rigid enforcement of such regulations.

In addition to these measures, persons who work at occupations that increase the risk of contracting rabies should be actively immunized with the vaccine. Persons at risk include veterinarians, wildlife conservation personnel, and laboratory or kennel workers who may handle rabid animals.

Ratbite Fever   This disease is uncommon in the western hemisphere. The rare cases that occur in the United States are usually caused by the bite of a rat infected with *Steptobacillus moniliformis*. A similar disease in the far east is due to *Spirillum minor*. Both these organisms may be members of the normal rat flora. The bacteria invade the bloodstream, producing a disease characterized by chills and fever and a disseminated rash. Approximately 10 percent of untreated cases die from the disease. *Streptobacillus moniliformis* can be isolated on artificial media and identified for diagnosis. *Spirillum minor* may be grown in animals inoculated with materials from lesions, blood, or lymph of an infected person. Chemotherapy with penicillins or tetracyclines is usually effective. Prevention is best accomplished by controlling the rat population (see Plague).

Table 23-5 presents the major features of rabies and ratbite fever.

Arthropod-Borne Diseases

Plague   The terrifying black death that decimated Europe in the fourteenth century killed an estimated one-quarter of the population. **Plague** is caused by a gram-negative bacillus, *Yersinia pestis*. The microbe is a natural pathogen of various rodents and is transmitted among these animals by fleas. Infected fleas also transmit the disease to humans, especially when the population of infected rats in urban areas becomes dangerously large. In the course of feeding on the blood of infected rodents, fleas ingest large numbers of *Y. pestis*. The bacteria multiply in the flea's intestinal tract and obstruct the passage of food through the GI tract. During the next blood meal, this obstruction encourages regurgitation of bacilli-laden blood into the bite wound. Its hunger unsatisfied, the infected flea will usually attack and inoculate several hosts before it dies. As rats become infected and die,

TABLE 23-5

Some Characteristics of Diseases Transmitted by Animal Bites

Disease	Pathogen	Incubation Period	Major Reservoir	Diagnostic Specimen	Diagnostic Procedure	Treatment
Rabies	Rabies virus	2–8 weeks	Canines, skunks, bats	Saliva, tears, CSF, urine, corneal smears	Animal inoculation; immunofluorescence	Immune globulin and vaccine
Ratbite fever	*Spirillum minor; Streptobacillus moniliformis*	3–10 days	Rats	Lymph, blood, pus	Isolation and identification of organism; inoculation into animals	Penicillin

their fleas seek alternative hosts such as humans. Thus, attempts to control disease transmission by destroying rats after the disease is already established in the rodent population often promotes human disease by forcing the fleas to feed on people or other available hosts. It is therefore important to control rat populations before they become significant reservoirs in a community. In the United States, the success of urban rat-control programs has virtually eliminated plague in the cities. But the pathogen is still harbored in the wild rodents of Arizona, New Mexico, and California. When contracted from wild rodents, the disease is called **sylvatic plague**. Occasional outbreaks of sylvatic plague still occur among persons living in or visiting these endemic regions. The threat of human plague is greater in areas of the world where urban rats are likely to encounter wild rodents, the principal reservoir.

In classical **bubonic plague**, the pathogen multiplies in the infected person's lymph nodes, which develop into swollen, inflamed, and painful masses called **bubos** (Fig. 23-7a). Organisms can disseminate from the lymphatics through the circulation and establish localized infections in the liver, spleen, lungs, meninges, and skin. Hemorrhage and cyanosis of the skin are responsible for the black lesions that give the disease its popular name (black death).

Bubonic plague cannot be transmitted from human to human by fleas. However, when organisms spread to the lungs, the disease may be transmitted directly from human to human by respiratory droplets from the infected person. This dangerous form of the disease, called **pneumonic plague**, is extraordinarily contagious. Inhalation of a single bacterial cell can initiate infection. Pneumonic plague can cause explosive epidemics of deadly disease that is as easily transmitted as influenza or the common cold. Unlike colds, however, the mortality rate for untreated cases of

FIGURE 23-7

Characteristics of bubonic plague: (a) Bubos are enlarged lymph nodes caused by an inflammatory response to the presence of *Yersinia pestis*. (b) *Yersinia pestis* in smear prepared with pus from a bubo shows characteristic bipolar staining.

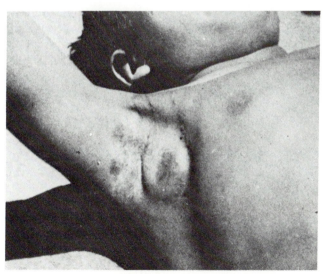

(a)

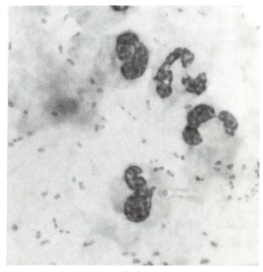

(b)

pneumonic plague approaches 100 percent. (Mortality rates for untreated cases of bubonic plague are about 50 to 75 percent.) Pneumonic plague is usually limited to the lungs, and there is no formation of bubos.

Yersinia pestis is isolated on blood agar from bubo biopsies, blood, sputum, or spinal fluid. This bacterium is gram-negative and appears as a bipolar bacillus that resembles a safety pin (Fig. 23-7*b*). Agglutination with plague antisera, fluorescent antibody tests, or bacteriophage typing definitively identifies the pathogen. Anyone handling clinical specimens or cultures containing plague bacillus should exercise precautions to prevent aerosolization or accidental inoculation of this dangerous pathogen.

Plague usually can be cured with antibiotics if they are administered early. Streptomycin is the drug of choice, but tetracycline and chloramphenicol are also effective within 8 to 24 hours after onset of symptoms. Penicillin is totally ineffective against *Y. pestis.*

Control of the disease ultimately depends on reducing encounters with infected rats and their fleas. Wild rodent populations are impossible to eliminate, but people are at risk only if they enter endemic areas or if the pathogen is transferred to rats in urban centers. The danger of urban plague is reduced by several measures. Surveillance of wild rodent populations identifies sylvatic plague areas so people can be warned to avoid these regions. When these areas are close to urban centers, control of rodents and fleas reduces the danger of epidemic plague. In the cities this is achieved by poisoning and trapping rats and by proper garbage disposal to starve them. Fleas are destroyed by treating rat runs and harborages with insecticide and by spraying houses in threatened areas. Insecticide powder should be applied directly onto persons at high risk of exposure. Ships and aircraft leaving areas endemic for plague should be free of rodents. Quarantine of cargo and fumigation of holds or warehouses may be necessary to ensure safety.

Isolation of persons with pneumonic plague helps prevent airborne epidemics. Anyone in close contact with a plague victim should be protected by tetracycline chemoprophylaxis. An inactivated plague vaccine confers temporary immunity, but is recommended for use only when the risk is so great that precautionary measures may be insufficient to prevent infection.

Relapsing Fever This disease is caused by the spirochete *Borrelia recurrentis* and is transmitted from rodents to humans by ticks and from human to human by body lice. Relapsing fever is prevalent in areas of Africa, Asia, and South America, although tick-borne relapsing fever occasionally occurs in the mountain regions of the western United States. The primary symptom is a fever that disappears and recurs over a period of weeks. The victim may suffer as many as 10 relapses. The mortality rate in untreated cases is 2 to 10 percent. These relapses are due to the pathogen's ability to repeatedly change the antigenic structure of its surface as the patient acquires immunity against the existing antigenic determinants. The disease is diagnosed by microscopic detection of the spirochete in blood. Tetracyclines are the preferred antibiotics. Preventive measures are similar to those discussed for tick- and louse-borne typhus (discussed below).

Rickettsial Diseases Ticks, lice, fleas, and mites are vectors of a variety of diseases caused by rickettsias (Table 23-6). With the exception of epidemic typhus, all these diseases have nonhuman reservoirs, and people are incidental hosts. Because they all require vectors for transmission, none of these diseases is directly communicable from person to person.

Rickettsias have had an almost unparalleled impact on world history, decimating armies and causing widespread famine. Epidemic typhus is believed to have contributed to Napoleon's defeat by Russia, and it significantly increased the mortality rates among soldiers in World Wars I and II.

Spotted fevers are a group of clinically similar diseases, worldwide in distribution, with the specific etiologic agent varying with geographic region. **Rocky mountain spotted fever** is the most common and severe rickettsial disease in the United States, where it is harbored in dogs and

TABLE 23-6
Arthropod-borne Rickettsial Diseases

Disease	Etiologic Agent	Vector	Reservoir	Geographic Distribution	Positive Weil-Felix Agglutination*	Treatment
Spotted fevers						
Rocky Mountain spotted fever	*Rickettsia rickettsii*	Tick	Wild rodents, dogs, ticks	Western hemisphere	OX-19 and OX-2	Tetracycline, chloramphenicol
Boutonneuse fever	*Rickettsia conorii*	Tick	Wild rodents, dogs	Africa, Europe, middle east, India	OX-19 and OX-2	Tetracycline, chloramphenicol
Rickettsial-pox	*Rickettsia akari*	Mite	Mites, mice	United States, Russia, Africa, Korea	None	Tetracycline, chloramphenicol
Typhus						
Epidemic typhus	*Rickettsia prowazekii*	Body louse	Humans	Worldwide	OX-19	Tetracycline, chloramphenicol
Brill-Zinsser disease	*Rickettsia prowazekii*	[Recurrence of epidemic typhus]		North America, Europe	Usually none	Tetracycline, chloramphenicol
Endemic typhus	*Rickettsia typhi*	Flea	Rats	Worldwide	OX-19	Tetracycline, chloramphenicol
Scrub typhus	*Rickettsia tsutsuga-mushi*	Mite	Mites, rats	Southeast Asia, southwest Pacific	OX-K	Tetracycline, chloramphenicol

*Strains of Proteus OX agglutinated with sera from persons with the disease.

FIGURE 23-8

Rocky Mountain spotted
fever. The rash usually
appears first on the
extremities.

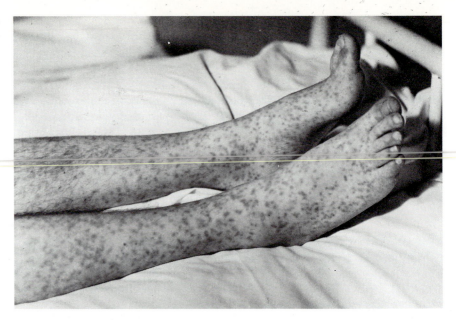

wild rodents and transmitted by ticks. The etiological agent, *Rickettsia rickettsii*, multiplies in the tick's salivary gland and readily enters the bloodstream of a mammal bitten by the infected arthropod. Ticks may also acquire the rickettsia by transovarian transmission from their infected mother. In humans, the disease begins with a high fever (105°), headache, and muscle pain, followed by a rash that first appears on the ankles and wrists, then spreads over most of the body (Fig. 23-8). Rickettsias multiply in the endothelial cells lining blood vessels, causing capillary hemorrhage responsible for the purple-black rash. If untreated, fatal disseminated intravascular coagulation frequently occurs. Currently, most cases of Rocky Mountain spotted fever occur in the eastern United States, although a few years ago it was a serious problem in the Rocky Mountain states. Several other tick-borne spotted fevers occur in different parts of the world. These include boutonneuse fever (Africa and India), Queensland tick typhus (Australia), and north Asian tick fever. All these diseases are caused by species antigenically similar to *Rickettsia rickettsii*. **Rickettsialpox** is a mild spotted fever last reported in the United States in 1969. It is transmitted from infected mice to humans by mites.

Epidemic typhus is the prototype disease in the typhus group. Unlike other rickettsial diseases, however, the human is the sole reservoir of infection. Epidemic typhus is transmitted from human to human by body lice. The louse becomes infected when it feeds on a victim of the disease. During subsequent meals it releases its rickettsial inhabitants by defecating onto human skin. People inadvertently force the rickettsias into the bite wound by rubbing or scratching the area. The clinical symptoms of epidemic typhus are similar to those of Rocky Mountain spotted fever—high temperature, rash, headache, and muscle pains. The spleen, liver, myocardium, and nervous system may be injured in severe cases. Most persons who recover from epidemic typhus acquire permanent immunity. In some people, however, epidemic typhus may recur years later in a milder form known as **Brill-Zinsser disease**, which is caused by reactivation of latent infection in the lymph nodes. The pathogen responsible for epidemic typhus is *Rickettsia prowazekii* named after Howard Ricketts and S. von Prowazek, who both died as a result of accidental infections while researching the disease. The last outbreak in the United States occurred in 1921.

Endemic typhus (murine or flea-borne typhus) is generally a milder disease than epidemic typhus. The pathogen, *Rickettsia typhii*, is harbored in infected rats, where it causes an inapparent, long-lasting infection. The organism is transmitted among rats or to humans by fleas. Once infected, the flea remains an active vector for life, which may be as long as a year. Fewer than 50 cases are usually reported in the United States each year.

Scrub typhus, transmitted by the bite of a mite infected with *Rickettsia tsutsugamushi* (*tsutsugamushi*, dangerous bug) is endemic in central, eastern, and southeastern Asia. These regions are usually characterized by scrub overgrowth on terrain that serves as a natural habitat for the wild rodents that are hosts to mites. Scrub typhus was almost unknown to the western world until World War II when allied troops entered areas endemic for the infected mites. Similar outbreaks occurred among soldiers in Viet Nam.

Most clinical laboratories are not equipped with the facilities or personnel necessary for isolating rickettsias. Preliminary diagnosis depends on the **Weil-Felix reaction**, a serological test that takes advantage of an immunological coincidence—most rickettsial infections stimulate formation of cross-reacting antibodies that agglutinate variants of *Proteus* OX, a strain of *Proteus vulgaris* (Table 23-6) (see color box). Blood specimens may be mailed to special laboratories where diagnosis can be confirmed by growing the pathogens in host cells. The intracellular parasites are usually identified by fluorescent antibody techniques (Fig. 23-9).

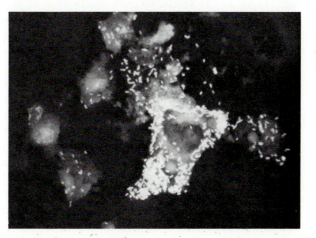

FIGURE 23-9
Rickettsia rickettsii. The intracellular location of the bacteria is revealed by fluorescent antibody.

616

DISEASES ACQUIRED
THROUGH THE SKIN

Rickettsial diseases are treated with tetracyclines and chloramphenicol, antibiotics that reach the pathogen's intracellular locations. Control measures are directed at preventing vector bites and reducing reservoirs of infection. These measures include eliminating rodent harborages and reducing the vector population by a combination of good sanitation and insecticides.

Virus Diseases   More than 350 arthropod-borne viruses (arboviruses) are believed to cause disease in vertebrates, including humans. Most of these cause zoonoses, with humans as incidental hosts. They are transmitted from animals to humans by the bite of a vector, usually a mosquito or tick (Table 23-7). These viruses cause one of three clinical syndromes: (1) benign fever often accompanied by a rash but with a short course and few complications; (2) hemorrhagic fever characterized by a purplish rash due to rupture of capillaries in the skin; or (3) disease of the central nervous system, usually encephalitis.

The benign viral fevers usually last a week or less and resolve without serious consequences, although hemorrhagic fever and encephalitis may be a rare complication. These benign fevers primarily occur in rural or jungle areas where the vectors and natural reservoirs are prevalent. Most arbovirus-induced benign fevers are restricted to tropical areas and include **dengue fever**, bunyamwera, chikungunya, O'nyong-nyong, Rift Valley, West Nile, and sandfly fevers. In the United States, Colorado tick fever is a similar but rarely reported disease. Although clinically similar, each of these geographically restricted diseases is caused by a different virus. Because infection normally confers permanent immunity, children are more likely to contract these diseases than adults.

TABLE 23·7

Important Arthropod-borne
(Arbo) Viral Diseases

Disease	Vector	Reservoir	Mortality Rate, %	Geographic Distribution
Dengue fever	Mosquitos	Humans	Low	Tropics and sub-tropics
Yellow fever	Mosquitos	Monkeys	5–50	Tropics
Dengue hemor-rhagic fever	Mosquitos	Humans	5–10	Southeast Asia
St. Louis en-cephalitis	Mosquitos	Birds	5–10	Southern, central, and western United States
Eastern equine encephalitis	Mosquitos	Birds, horses	50–70	Eastern United States
Western equine encephalitis	Mosquitos	Birds, horses	3	Western United States
Colorado tick fever	Ticks	Small mam-mals	Low	Rocky Mountain states
California en-cephalitis	Mosquitos	Rabbits, ro-dents, squir-rels	Low	North, central, and southeast United States
Venezuelan equine encepha-litis	Mosquitos	Rodents	Low	Northern South America, Central America, and the United States

The best known arbovirus hemorrhagic fevers are **yellow fever** and **dengue hemorrhagic fever**. Each disease ranges in severity from asymptomatic to fatal. Hemorrhagic fevers are characterized by bleeding of the gums and gastrointestinal tract, purpura (hemorrhage into the skin), and thrombocytopenia (decreased number of platelets in blood).

Yellow fever virus is restricted primarily to the jungles of Central and South America and Africa. In the jungle, yellow fever virus is naturally propagated in monkeys and transmitted between monkeys by mosquitos. Infected mosquitos also spread the disease to people in the endemic regions. This is the cycle for *jungle* yellow fever. If the disease spreads to urban centers, it is transmitted among infected persons by domestic *Aëdes* mosquitos. Thus, *urban* yellow fever is epidemic, and humans rather than monkeys are the reservoir.

Because of vector control measures and an effective vaccine, urban yellow fever no longer decimates large human populations (see color box). *Aëdes aegypti* mosquitos still exist in many areas, including the United States, where, fortunately, there are no reservoirs of yellow fever. In many of the world's tropical regions, however, the disease remains endemic.

Dengue hemorrhagic fever is an important cause of disease and death, primarily among children in tropical Asia. Persons with antibody to dengue virus may develop this presumably immune-mediated disease. Thus only individuals who have recovered from dengue fever or infants

THE PANAMA CANAL: WHY WASN'T IT BUILT SOONER?

Although the economic and political advantages were immeasurable, the French had given up in defeat. Panama had simply claimed too many lives. Too many of the engineers sent there had developed fever and muscle pains, begun vomiting, and died within a week. There was no treatment and no known means of prevention. The waterway connecting the oceans would have to wait until yellow fever, known in many areas as the "black vomit," could be conquered.

In the late 1890s the United States began where the French had quit. A yellow fever commission, led by Major Walter Reed, was dispatched to Havana, where the disease had killed more American soldiers than had the Spanish-American war. The commission's major task—determine how yellow fever was transmitted. Most researchers believed yellow fever to be a "filth disease," similar to typhoid fever, spread by human feces. Improving sanitary conditions, however, failed to reduce its incidence in Cuba. Clean Havana was as disease-ridden as was filthy Havana. The disease spread in a curious pattern, sometimes striking one side of a street but not the other. It was more prevalent in low wetlands than nearby highlands and was usually spread in the direction of the prevailing wind. The disease flourished in hot weather and disappeared with cold and frost.

These observations led Reed to investigate an unpopular theory proposed 20 years earlier by the ridiculed "mosquito doctor," Carlos Finlay. Finlay believed the disease was acquired by inoculation with a "living miniature hypodermic needle"—an infected mosquito. But because he was unable to demonstrate mosquito-borne transfer of the disease, his theory was not considered seriously by his contemporaries. Reed's group discovered the reason for Finlay's failures. Finlay had mosquitos bite yellow fever patients and then let these freshly charged insects bite healthy persons. Unfortunately for his theory, these persons remained healthy. With carefully timed experiments and a handful of courageous volunteers, Reed showed that mosquitos become infected only when they draw blood from a person during the first 4 days of the fever. In addition, the infected mosquito's bite can transmit the disease only after the pathogen has incubated in the insect for 7 to 10 days.

Reed also ruled out other possible modes of transmission. Healthy volunteers were confined for 3 weeks in special mosquito-proof houses where they slept on bed sheets and blankets taken from yellow fever wards and wore pajamas removed from the bodies of yellow fever victims. All these volunteers remained healthy. Reed wrote "the bubble of the belief that clothing can transmit yellow fever was pricked by the first touch of human experimentation." To show that these men were not immune to the disease, two of them were intentionally exposed to infected mosquitos and both contracted the disease.

With experiments like these, Walter Reed conclusively proved that the mosquito *Aëdes aegypti* was the vector of yellow fever. This discovery provided the first successful tactic for controlling the disease. Mosquito-reduction programs allowed construction of the Panama Canal to begin in 1904, 4 years after Walter Reed's yellow fever team arrived in Havana.

with maternal antibody are susceptible. The initial symptoms resemble benign dengue fever and appear about 5 days after inoculation of dengue virus by the aëdes mosquito. As the fever subsides, however, the patient may develop hemorrhages and, in severe cases, vascular defects that cause fluid loss. Unless the fluid is replaced, death from shock may follow within hours.

Arbovirus encephalitis is more commonly observed in the United States than other arthropod-borne diseases. The neurotropic arboviruses all cause similar diseases, with symptoms ranging from mild fever to convulsions, paralysis, coma, and death. Disease severity varies among the different agents, but they all may cause irreversible neurological damage that leaves survivors permanently retarded or paralyzed. The most common neurotropic arboviral disease in the United States is St. Louis encephalitis (SLE), which causes the majority of deaths. Eastern equine encephalitis and western equine encephalitis cause fatal infections in horses as well as humans. In spite of its name, California encephalitis is most prevalent in the Mississippi and Ohio River valleys. Venezuelan equine encephalitis was originally a disease of Central and South America but it spread to the southern United States. Most outbreaks of arbovirus encephalitis occur in summer and fall and are conspicuously absent during winter months when mosquitos are dormant. Mosquitos are believed to acquire the infection from sources other than infected people, because the low-level viremia that occurs in humans is not a rich supply of viruses. Birds, horses, rodents, squirrels, and rabbits develop high-grade viremias, and their blood provides mosquitos a stock of readily accessible virus. Early detection of these viruses in their natural animal reservoirs may alert health officials to an impending epidemic before a single person contracts the disease. Sentinel chickens have been used to monitor the danger of St. Louis encephalitis in Florida since 1977. These domestic birds are periodically examined for evidence of infection by SLE virus. Mosquito-control programs and public announcements to avoid the vector have prevented many cases of arbovirus encephalitis.

Arboviruses stimulate formation of virus-specific antibodies. These diseases are most often diagnosed by detecting rising antibody titers. The viruses that cause benign and hemorrhagic fevers can be isolated from the blood of infected persons by propagation in cell culture. Recovery of viruses from the blood of patients with arbovirus encephalitis is difficult because viremia is mild and short-lived.

Although no specific antimicrobial agent is available for treating patients with arbovirus infections, supportive therapy to reduce symptoms can minimize disease severity. Replacement of intravascular fluids caused by hemorrhagic fevers helps prevent fatal shock.

Although vaccines are available against yellow fever, Japanese B encephalitis, and the equine encephalites, vector control is the most important preventive measure. Eliminating mosquito breeding grounds gives long-term protection, while spraying human habitats with insecticide reduces spread during epidemic outbreaks. Personal protection is also afforded by the use of repellents, protective clothing, and mosquito nets.

Protozoan Infections   The major arthropod-borne diseases caused by protozoa are summarized in Table 23-8.

TABLE 23-8

Arthropod-borne Protozoan Diseases

Disease	Pathogen	Vector	Reservoir	Incubation Period	Clinical Specimen	Diagnostic Feature	Treatment
Cutaneous leishmaniasis	*Leishmania tropica*	Sand fly	Rodents, canines	Days to months	Tissue	Ovoid, nonflagellated trophozoite	Antimonials
Mucocutaneous leishmaniasis	*L. braziliensis*	Sand fly	Rodents, canines	Months	Tissue	Ovoid, nonflagellated tophozoite	Antimonials
Visceral leishmaniasis	*L. donovani*	Sand fly	Humans, rodents, canines	2–4 months	Tissue	Ovoid, nonflagellated trophozoite	Antimonials
African trypanosomiasis	*Trypanosoma rhodesiense*	Tsetse fly	Wild game, cattle	2–3 weeks	Blood lymph, CSF	Flagellated trophozoite	Suramin, melarsoprol
	T. gambiense	Tsetse fly	Humans	Months to years	Blood, lymph, CSF	Flagellated trophozoite	Pentamidine, melarsoprol
American trypanosomiasis	*T. cruzi*	Reduviid bug	Humans, many animals	5–14 days	Blood Tissue	Flagellated trophozoite Nonflagellated trophozoite	Bayer 2502
Malaria	*Plasmodium vivax*	Anopheles mosquito	Humans	14 days	Blood	Enlarged erythrocytes with stippled cytoplasm and amoeboid trophozoites	Chloroquine
	P. malariae	Anopheles mosquito	Humans	30 days	Blood	Normal-sized erythrocytes; band-formed trophozoites	Chloroquine
	P. falciparum	Anopheles mosquito	Humans	12 days	Blood	Normal-sized erythrocytes, ring-shaped trophozoites	Chloroquine, quinine for resistant strains
	P. ovale	Anopheles mosquito	Humans	14 days	Blood	Enlarged erythrocytes	Chloroquine

LEISHMANIASIS The flagellated protozoa that cause **leishmaniasis** are usually transmitted from infected humans, canines, or rodents through the bite of the sand fly. The protozoa are lodged in and block the insect's proboscis, making feeding impossible. During its numerous attempts to feed, the fly inoculates the parasites into the host. They are then engulfed by macrophages, lose their flagella, and multiply in the macrophages by a series of binary fissions, ultimately rupturing the parasitized cell and releasing progeny which continue to invade new white blood cells (Fig. 23-10). The infected white cells may remain localized in the skin at the site of inoculation or may carry the pathogen to mucosal surfaces of the nose and mouth or to various internal organs (spleen, lymph nodes, or liver) where the protozoa can multiply and be released into the bloodstream. Thus, three forms of leishmaniasis occur: cutaneous (oriental sore); mucocutaneous (espundia or American leishmaniasis); and visceral (kala-azar). The species of *Leishmania* determines which of these diseases will develop. Geographic distribution is dictated by the occurrence of the corresponding vector.

Cutaneous leishmaniasis is usually manifested as an ulcerated lesion that heals spontaneously. Infections that spread to mucous membranes of the nasopharynx, however, may be fatal. When the blood is infected, the disease is characterized by fever and enlarged lymphoid organs. The victim becomes weak and will usually die if no treatment is provided.

Leishmaniasis is diagnosed by microscopic observation of the nonflagellated form of the protozoa in appropriate specimens (lesions, bone marrow, spleen, liver, lymph nodes, or blood) depending on the form of the disease. Sodium antimony gluconate, available only from the CDC, is

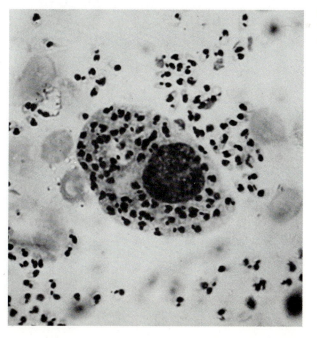

FIGURE 23-10
Leishmania reproducing within a macrophage.

recommended for treating the infection. Other antimonials and amphotericin B are sometimes effective. Control measures rely on interrupting the transmission cycle. Controlling the canine population, for example, decreases the number of human cases when dogs are the major reservoir. Insecticide spraying programs, eliminating fly-breeding habitats, avoiding sand-fly-infested areas, and wearing protective clothing all help control disease spread.

AFRICAN TRYPANOSOMIASIS Like *Leishmania*, trypanosomes are flagellated protozoa that invade the bloodstream after the bite of an arthropod vector. These parasites are pleomorphic, varying from long and slender to short and stumpy during their life cycle. Trypanosomes live in blood plasma and do not infect blood cells. Three species cause human disease, and many others infect and kill domestic animal species, thereby destroying food sources in affected areas. One such parasite prevented the successful introduction of horses for transportation in Africa.

African trypanosomiasis (African sleeping sickness), caused by *Trypanosoma gambiense* or *Trypanosoma rhodesiense,* is transmitted by the bite of the tsetse fly (Fig. 23-11). Since tsetse flies are found only in Africa and southern Arabia, the disease is restricted to these regions.

Fever, headache, and enlarged lymph nodes are the earliest signs of disease. When the organisms spread to the central nervous system, victims exhibit lethargy, severe weight loss due to anorexia, mental deterioration,

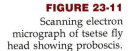

622

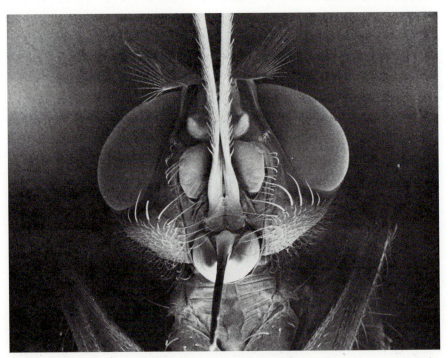

FIGURE 23-11
Scanning electron micrograph of tsetse fly head showing proboscis.

and somnolence (the daytime sleeping from which the disease derives its name). If untreated, most cases are fatal.

Trypanosoma gambiense has a human reservoir, whereas *Trypanosoma rhodesiense* lives in wild animals. The gambian disease runs a prolonged course. This increases the period of shedding, allowing the parasite ample time to infect new hosts before killing the infected person. *T. rhodesiense* is more virulent for humans and usually results in death in less than 1 year of infection.

Diagnosis is confirmed by observing trypanosomes in lymph, blood or, in later stages, cerebrospinal fluid. Suramin is effective for treating early *T. rhodesiense* infections, whereas pentamidine is used for early stages of *T. gambiense* infections. When either protozoan has spread to the central nervous system, melarsoprol is used. In the United States, these drugs are available only from the CDC. Prophylactic administration of pentamidine confers protection against *T. gambiense* for 3 to 6 months. However, because this drug is potentially toxic and because drug-resistant strains are becoming prevalent, pentamidine is used only by persons at high risk of being exposed to the pathogen.

AMERICAN TRYPANOSOMIASIS **American trypanosomiasis**, also known as **Chagas's disease**, is caused by *Trypanosoma cruzi*, a protozoan found among domestic and wild animals from South America to Mexico and in many southern states of the United States. The vectors are bloodsucking insects called reduviid bugs. These bugs commonly seek a blood meal by biting the lips, accounting for their nickname, the kissing bug. When an infected bug feeds, it deposits contaminated feces on the skin surface. The protozoa usually enter the body by being rubbed into the bite wound, where they multiply and frequently cause a swollen skin lesion called a *chagoma*. As the infection progresses, fever develops, local lymph glands enlarge, and the parasites disseminate to the bloodstream and tissues. Acute disease is most common in children and may be fatal if severe myocarditis or meningoencephalitis develop. Most infections, however, are chronic and are not fatal. Some chronic infections are asymptomatic; the others are usually characterized by various degrees of myocardial damage.

Chagas's disease is diagnosed by demonstrating the pathogen's presence in blood, either by direct microscopic examination, by culturing the organisms, by inoculation into mice, or by allowing an uninfected vector to feed on the patient and after a few weeks examining the bug's feces for trypanosomes. There is no treatment that is both safe and effective against American trypanosomiasis. Bayer 2502 (nifurtimox), available from the CDC, is useful in some cases of acute infection.

MALARIA **Malaria** is believed to be the most common serious infectious disease of humans. One hundred million cases occur annually, and 1 million of them are fatal. Historically, it has been among the great scourges. The disease is naturally transmitted by the anopheles mosquito. Although the vector has been eliminated from most areas in the United

States, malaria continues to be transmitted through imported cases (soldiers, immigrants, and travelers) by alternative means, for example, by transfusion of contaminated blood. Drug users often acquire the disease by sharing contaminated needles. A fetus may acquire malaria by transplacental passage. In 1980 over 2000 cases of malaria were reported in the United States.

Four species of *Plasmodium* cause malaria in humans, *Plasmodium malariae*, *P. vivax*, *P. ovale*, and *P. falciparum*. The protozoa are injected into the bloodstream from the saliva of the female anopheles mosquito during a blood meal. The infectious forms of the organism, called *sporozoites*, invade the liver, where they divide by multiple fission into cells called *merozoites*. A single sporozoite may produce as many as 40,000 merozoites. These cells are released into the bloodstream, invade erythrocytes, and undergo a series of morphological changes into ring-shaped forms called *trophozoites*. The trophozoites reproduce inside the red blood cells, forming more merozoites. Ultimately the red blood cell is destroyed by lysis, and the released merozoites are free to infect more red blood cells. Rupture of the erythrocytes occurs in a synchronized periodic manner—every 48 to 72 hours depending on the multiplication rate of the particular *Plasmodium* species. Rupture is accompanied by the release of toxic materials into the blood, producing the cycles of chills and fever that are typical of malaria.

Some cells released from the erythrocytes differentiate into male and female *gametocytes*. These cells do not mature in the human, but must be ingested by the female anopheles mosquito in order to complete their development. Gametocytes fuse in the insect, and the resulting zygotes divide into asexual sporozoites. These then make their way to the insect's salivary glands and are introduced into a person while the mosquito feeds. Additional details of the *Plasmodium* life cycle are presented in Figure 23-12.

The most characteristic symptoms of malaria are the periodic cycles of chills, fever, and sweating, accompanied by headache and nausea. Additional manifestations may be serious, especially those caused by *Plasmodium falciparum* (jaundice, anemia, gastrointestinal disturbances, coagulation disorders, shock, and coma). As many as 60 percent of the circulating erythrocytes may be parasitized. Their lysis often causes *blackwater fever*, characterized by excretion of large amounts of hemoglobin in urine. Falciparum malaria is fatal in more than 10 percent of untreated cases. The other malarias are rarely life-threatening, but without appropriate treatment the parasites persist in the liver and may cause relapses. Relapses never occur when the disease has been acquired by contaminated blood or instruments because only the sporozoite, which develops only in the mosquito, can infect the liver. Laboratory diagnosis of malaria is established by detecting the trophozoite in red blood cells.

Most cases of malaria respond to therapy with chloroquine or related compounds, although drug resistance in *P. falciparum* is becoming more common. These resistant strains can be treated with quinine. Malaria relapses are prevented by combining chloroquine, which kills the circulat-

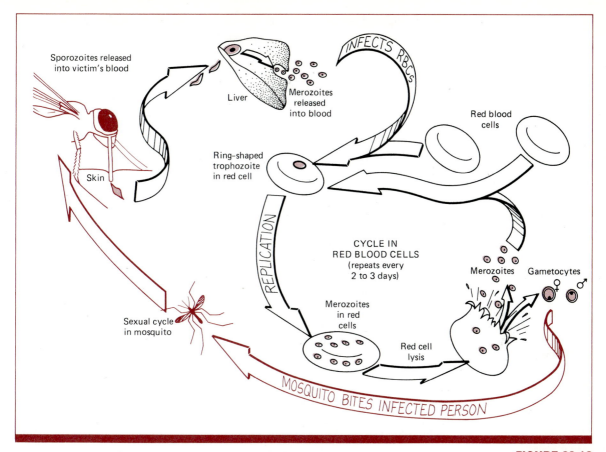

FIGURE 23-12

Life cycle of the malaria parasite requires two hosts, a human (shown in black) and a mosquito (shown in color). When an infected mosquito bites a person, it releases sporozoites into the blood. The parasite replicates in cells of the liver. (This is the *preerythrocyte* phase of malaria.) Days later, the bloodstream is showered with merozoites released from the liver. (This may also occur months later and cause a relapse of malaria.) Merozoites released from the liver infect red blood cells and begin the second, or *erythrocytic,* phase of malaria. Some merozoites in red blood cells mature into gametocytes. When a mosquito draws blood from an infected person these gametocytes fertilize in the new host, where they complete the pathogen's life cycle. Infectious sporozoites are shed in the mosquito's saliva.

ing form of the pathogen, and primaquine, which destroys the parasites in the liver. Prophylactic drug therapy is recommended for persons traveling to areas at high risk for malaria. Since protection is virtually assured, it is imperative that travelers be informed of the serious consequences of disease and of the overwhelming success of preventive measures.

In the United States infection by transfusion is prevented by screening blood donors and not accepting blood from persons who have returned from an area endemic for malaria until they are proved uninfected by remaining free from symptoms for at least 6 months.

625

Wound Inoculations

The characteristics of diseases acquired by wound inoculation are summarized in Table 23-9.

Tetanus Endospores of *Clostridium tetani* are common contaminants of soil, dust, or feces, and are frequently found on objects that cause accidental injuries. These spores germinate in the oxygen-starved environment provided by the necrotic tissues of deep wounds. Vegetative cells growing in the injured tissue produce a potent neurotoxin called *tetanospasm that is spread throughout the body by the bloodstream.* The toxin prevents relaxation of motor neurons controlling voluntary muscles, which become paralyzed in a state of tetanic contraction. *Lockjaw,* the common name for the disease, refers to the toxin's predilection for jaw and neck muscles, which become rigidly locked. The toxin eventually affects most of the voluntary muscles in the body, causing rigid, uncontrolled contractions that are sometimes strong enough to break the patient's spine (Fig. 23-13a). Without treatment the fatality rate may reach 70 percent, usually from respiratory failure.

Laboratory diagnosis confirms the clinical suspicion of tetanus. The pathogen is isolated only from the local lesion and only under anaerobic transport and culture conditions. *Clostridium tetani* is identified by its characteristic terminal, swollen endospores (Fig. 23-13b). Because of the urgent need for immediate treatment, therapy should not be delayed while waiting for laboratory confirmation when clinical symptoms suggest tetanus. Patients with histories of tetanus immunization are treated with tetanus toxoid to quickly boost their immunity against the effects of the toxin. This is usually sufficient to prevent any further manifestations of

TABLE 23-9

Characteristics of Diseases Acquired through Wound Inoculation

Disease	Pathogen	Reservoir	Incubation Period	Diagnosis	Antimicrobial Treatment
Tetanus	*Clostridium tetani*	Humans, animals, spores in soil	4–21 days	History of wound and clinical picture	Penicillin
Gas gangrene	*Clostridium perfringens,* other clostridia	Humans, animals, spores in soil	1–3 days	Isolation and identification of pathogen from wound	Penicillin
Sporotrichosis	*Sporothrix schenckii*	Soil, vegetation, wood	1 week to 3 months	Cultivation and identification of fungus from lesion	Iodides
Mycetoma	*Madurella mycetomi, Petriellidium boydii*	Soil, decaying vegetation	Months	Microscopic identification of fungus in lesion or isolation in culture	None
Hepatitis B	Hepatitis B virus	Humans	60–90 days	Serologic identification of viral antigens in blood	None
Non-A non-B hepatitis	Unknown virus	Humans	40–60 days	Evidence of viral hepatitis, negative tests for hepatitis A and B	None

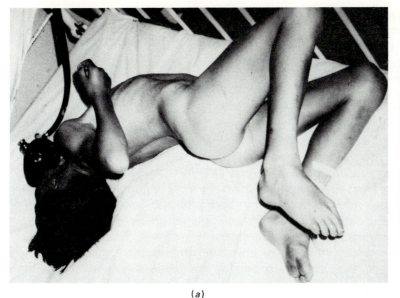

(a)

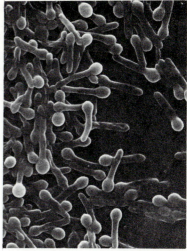

(b)

FIGURE 23-13

Tetanus: (a) Tetanic paralysis characteristic of advanced disease. (b) Scanning electron micrograph of *Clostridium tetani* spores showing the swollen terminal endospore.

disease. Patients who have never been immunized should receive passive immunization with tetanus antitoxin to provide immediate protection. These measures neutralize toxin before it attaches to nerve cells. Once it is fixed to neurons, the effect cannot be reversed by immunotherapy, and paralysis persists until the patient slowly metabolizes the toxin. Penicillin is administered to kill the pathogens in the wound. Sedatives, muscle relaxants, and respiratory support systems help maintain the patient until neurological functions normalize.

The best protection against tetanus is routine immunization with tetanus toxoid. A series of four DPT (diphtheria, pertussis, and tetanus) shots is recommended for all children 2, 4, 6, and 18 months of age. Booster shots at 10-year intervals maintain immunity. Because of the danger of hypersensitivity to the antigens in the vaccine, toxoid boosters are no longer given as a part of routine wound treatment unless either 10 years have elapsed since the patient last received toxoid or there are signs of developing tetanus. In areas of the world where immunization is not common and sanitation is poor, tetanus causes several hundred thousand deaths a year. In the United States the total number of reported cases of tetanus in 1981 was 72. Most cases were in persons over 60 years old who had presumably failed to maintain adequate immunizations.

Gas Gangrene Gas gangrene is a rapidly advancing muscle necrosis caused by several species of *Clostridium*, most notably *C. perfringens*, *C. novyi*, and *C. septicum*. These clostridia grow especially well in anaerobic tissues of severe traumatic injuries such as gunshot wounds, compound fractures, septic abortions, and surgical incisions. Postsurgical gas gangrene is an occasional complication when, because of poor asepsis, surgical wounds are contaminated with endospores during the operation or during subsequent therapy. Tissues insufficiently supplied with blood—for example, the poorly oxygenated tissues found in the extremities of diabetics—are also prone to infection. The clostridia release toxins and histolytic enzymes that kill the tissue surrounding the infected site. As the area of necrosis advances, enzymes dissolve muscle and connective tissue, facilitating the spread of infection into the liquefied areas and accounting for most of the clinical symptoms (Fig. 23-14). The enzymes produced by *C. perfringens*, the most common agent of gas gangrene, include collagenase, protease, and *lecithinase* (also called *alpha toxin*). The latter enzyme dissolves the cell membranes of muscle cells (it is the same poison found in rattlesnake venom) and appears to be the most important virulence factor in gas gangrene. The pathogens also produce gaseous metabolic by-products, CO_2 and H_2, that can distend tissues until blood flow through the area is obstructed, thus enhancing tissue necrosis. If not halted within hours, the disease may claim an entire limb or the patient's life. Gas gangrene of the thoracic or abdominal region is particularly life-threatening.

Gas gangrene often requires immediate therapeutic action to prevent irreversible tissue damage or death of the patient. A Gram stain of infected tissue provides a rapid preliminary diagnosis by differentiating between the major gram-positive and gram-negative agents of anaerobic wound infections (the clostridia are gram-positive rods). This helps in selecting the therapy most likely to arrest microbial growth and halt further damage. An important step in treating gas gangrene is the removal of necrotic tissue, a procedure called **debridement**. Sometimes debridement requires surgical removal of gangrenous muscle or even amputation of limbs. Simultaneous treatment with penicillin helps control the growth of pathogens not removed by debridement. This combination of treatments has reduced mortality rates to about 20 percent. *Hyperbaric oxygen treatment* also appears

FIGURE 23-14

Gas gangrene at amputation site.

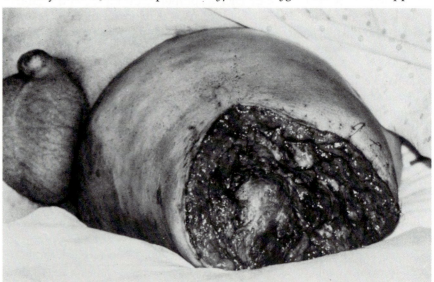

to reduce tissue loss and mortality rates. The patient is placed in an oxygen-filled high-pressure chamber that forces elevated concentrations of oxygen into tissues. This increase in oxygen concentration is lethal to many obligate anaerobic bacteria.

Other Anaerobic Wound Infections   In addition to gas gangrene, many other anaerobic infections are caused by microbes introduced into wounds. These infections are usually caused by a mixture of bacteria that may include anaerobic streptococci and species of *Bacteroides*, *Fusobacterium*, or *Clostridium*. Although most of these organisms are sensitive to antibiotics, the lack of blood flow to the necrotic region may prevent effective concentrations of the drug from reaching the infected site. Thorough wound cleansing and prompt surgical debridement are important therapeutic measures.

Wound botulism is a rare infection in which *Clostridium botulinum* multiplies in the anaerobic surroundings of a deep wound. The pathogen releases *botulin*, a toxin which, when absorbed, causes classical symptoms of botulism.

Burn Wound Infections   The dead or dying tissues of a burn wound are an ideal environment for microbial growth as well as a portal of entry for invasion of the bloodstream. Burns also impair many major defenses of the immune response. Circulating IgG and complement are reduced, T-cell functions are altered, and chemotaxis of phagocytic white blood cells is inhibited. Each year in the United States between 10 and 30 percent of the approximately 60,000 people hospitalized for burn wounds die of infection, which is the major cause of death of persons hospitalized for burn wounds who have survived the first 48 hours.

Because the integrity of the skin is breached, colonization and subsequent infection of the wound are inevitable. Organisms are usually transmitted to the wound by autoinoculation with the patient's normal flora or by contact with colonized personnel or contaminated fomites. Air is generally not a major route of transmission. Before the availability of antibiotics, *Streptococcus pyogenes* was the most frequent cause of invasive burn wound infections. This bacterium could spread so rapidly that bacteremia could be fatal within a few hours. Currently, *Streptococcus pyogenes* is cultured from only 5 percent of infected burn wounds. It has been replaced as the most prevalent microbe by *Pseudomonas*. The presence of *Pseudomonas* is often recognizable by the pigment it produces, which imparts a green color to the wound and dressings. Enterobacteria, *Staphylococcus aureus*, and the yeast *Candida albicans* are other common burn wound contaminants, although any pathogen can probably infect areas of extensive damage.

Sterile wounds do not exist. However, the danger of invasion increases as the types and numbers of microbes in the wound increase. Thus, persons with extensive burns are usually kept in areas where access to the patient is restricted and where rigorous precautions are practiced, particularly the wearing of masks and gowns and mandatory washing of hands.

Topical antimicrobial agents, most commonly silver sulfadiazene, mafenide, or povidine-iodine, are used to inhibit microbial growth. Unfortunately, some of these drugs delay epithelial growth and wound repair as well. In addition, these drugs are sometimes ineffective because of the emergence of resistant pathogens. Appropriate therapy depends on determining sensitivities and using drugs that are effective. Debridement or surgical excision of the nonliving tissue decreases infection, especially if the wound is covered with an autograft (a piece of intact skin transplanted from another part of the patient's body).

Fungal Infections   Many opportunistic fungi in soil or on vegetation may be introduced through the skin by piercing wounds. Spores of these fungi are found on thorns, barbs, and splinters. **Sporotrichosis**, the most common parenterally acquired mycosis in the United States, begins as a local subcutaneous lesion where the organism enters the body. The entry site is often a wound made by the accidental prick of a plant, thorn, or barb contaminated with the pathogen, *Sporothrix schenckii*. The disease is common as an occupational hazard among gardeners, farmers, and horticulturalists (hence the common name "rose gardener's disease"). The initial lesion often progresses from the site of inoculation to the regional lymph nodes, forming a series of enlarged nodules that extend up the arm or leg. The nodules may become necrotic and ulcerate. Although invasion of lungs and other organs is an occasional severe complication, spreading beyond the local lymph nodes is rare.

Mycetomas are common parenterally acquired mycoses in tropical and subtropical regions, especially among persons who seldom wear shoes. The initial lesion develops at the site of entry, most often on the feet (Fig. 23-15a). A variety of environmental organisms may produce the swelling and the local suppurative lesions of the skin and subcutaneous tissues that are characteristic of mycetoma. The organisms include *Petriellidium boydii* (the most common agent of mycetoma in the United States), *Madurella mycetomi,* and numerous other molds (Fig. 23-15b). Lesions usually remain localized, but can cause crippling disfiguration unless they are surgically removed. Another mycosis acquired by puncture wounds to the extremities is **chromomycosis**, caused by fungi from the genera *Phialophora, Fonsecaea,* and *Cladosporium.* Lesions slowly develop into a cauliflowerlike mass on the inoculated limb. Rarely, the pathogens invade the bloodstream and brain.

Diagnosis of the parenterally acquired mycoses depends on microscopic observation of fungi in tissue preparations and isolation of the pathogen. Materials aspirated from the lesions are the best specimens for examination and culture. The lesions of mycetoma characteristically produce granules of packed hyphae growth that are readily observed microscopically. Pigmented segmented spores called *fission bodies* (or "sclerotic bodies") are indicative of chromomycosis. *Sporothrix schenckii,* however, is rarely detected by examining tissue. Positive identification of this dimorphic fungus depends on growing the mycelial phases at 25°C and the yeast phase at 37°C.

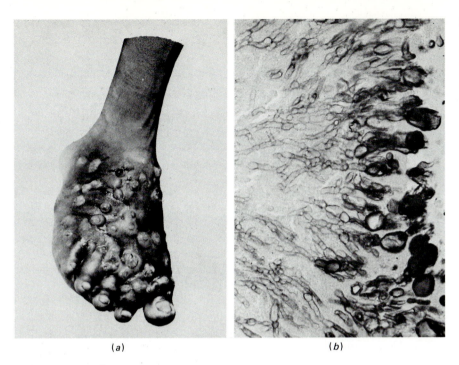

FIGURE 23-15

Mycetoma: (*a*) Clinical appearance and (*b*) microscopic appearance of the fungus *Madurella mycetomi* in tissue.

(*a*)　　　　　　　　　　　　(*b*)

Sporotrichosis is treated by oral administration of potassium iodide or intravenous therapy with amphotericin B. Chromomycoses also respond to amphotericin B. No drugs are effective against mycotic mycetomas, and surgical removal of lesions remains the best therapy.

Viral Hepatitis B   The clinical manifestations of B type hepatitis are similar to those associated with hepatitis A, discussed in Chapter 21. Together these viruses are among the most commonly reported infectious diseases and are believed to infect 300,000 people in the United States each year. **Hepatitis B** is usually acquired by parenteral inoculation with blood products or instruments (needles, syringes, or hemodialysis equipment) contaminated with the virus. The disease may also be spread by oral or sexual contact and is highly prevalent among male homosexuals. Women who acquire the disease during pregnancy may transmit the virus to the fetus. After an average incubation period of 90 days, the virus begins to cause nausea, vomiting, fever, and abdominal pain. Jaundice is a common symptom that usually disappears within 6 weeks. Hepatitis B is most common in persons aged 15 to 29, especially among drug addicts and workers in blood banks, renal dialysis units, and medical or dental laboratories. Mortality is approximately 1 percent, and recovery from infection provides permanent immunity to the virus. However, approximately 5 to 10 percent of the individuals who have recovered from hepatitis B become chronic carriers. More than 120 million people in the world currently are carriers and continue as potential disseminators of disease.

TABLE 23-10

Characteristic Features of
Hepatitis Viruses

	Hepatitis B	Hepatitis A
Virus nucleic acid	DNA	RNA
Average incubation period	90 days	30 days
Common vehicles of transmission	Blood products, contaminated needles and syringes	Food, water, shellfish
Major portal of entry	Parenteral	Oral
Major portal of exit	Wound, usually created by a needle	Fecal
Carriers	Yes	No
Mortality in untreated cases	1%	<0.1%

Major differences exist between the hepatitis B and hepatitis A viruses (Table 23-10). Hepatitis A is a RNA virus, whereas hepatitis B is a DNA virus. In addition to distinctions in physical properties, the viruses differ in the length of their incubation periods and in their primary modes of transmission.

Electron microscopy of serum containing hepatitis B virus shows the complete virus (called the *Dane particle*) and two other particles that are viral surface antigens (Fig. 23-16). Hepatitis B is diagnosed by immunoprecipitation, immunoelectrophoresis, or radioimmunoassay to detect these viral antigens circulating in the patient's serum. There is no specific antiviral treatment for hepatitis B, although interferon may someday be used to treat the disease. Passive immunization with sera from immune patients may prevent disease if administered shortly after exposure to the virus. Prevention of hepatitis depends largely on avoiding exposure to contaminated products. Blood products and donors are screened for hepatitis B antigen to reduce transmission of virus by transfusion. Using disposable needles and syringes further reduces the danger. Hepatitis B

FIGURE 23-16

Hepatitis B viral antigens in blood. The arrow indicates the infectious virus (Dane particle). The other forms are noninfectious aggregates of surface antigens.

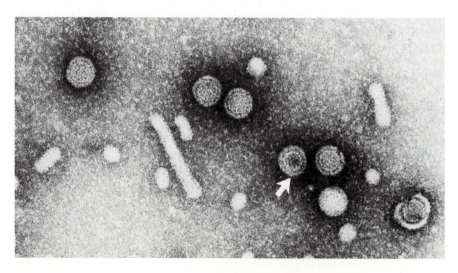

virus is especially resistant to many chemical disinfectants, but can be destroyed by solutions of sodium hypochlorite, formaldehyde, or glutaraldehyde. In 1982 a vaccine containing viral surface antigen was licensed for use in the United States and is recommended for protection of persons with high risk of exposure to the hepatitis B virus. Persons who work in blood banks, medical laboratories, and other high-risk areas should be well informed of the mode of transmission and methods for preventing infection.

Non-A Non-B Hepatitis Screening donors and monitoring blood for the presence of hepatitis B virus has lowered the incidence of posttransfusion hepatitis. Many transfusion recipients, however, still acquire classical viral hepatitis from blood that has been verified free of hepatitis B virus. These cases are presumably caused by another parenterally transmitted hepatitis virus (or viruses) designated the **non-A non-B hepatitis** (**NANA**) agent. The agent has not been isolated, but is antigenically distinct from hepatitis A or hepatitis B virus. NANB hepatitis has an incubation period of 50 days, intermediate between that of hepatitis A and hepatitis B. It causes approximately 90 percent of posttransfusion hepatitis cases. There is currently (1984) no way to specifically detect the agent in blood.

Inoculations through Superficial Breaks in the Skin
Table 23-11 presents the major features of diseases acquired by inoculation through superficial breaks in the skin.

Leprosy (Hansen's Disease) Throughout history **leprosy** has been an especially alarming disease. It tends to slowly disfigure its victims, who were often confined to remote asylums or made to carry conspicuous alarms such as bells to warn others of their approach. These abuses supposedly helped protect uninfected people from contracting the disease. Today we know that leprosy is one of the *least* communicable of all infectious diseases. Most people can safely live in the same household with a leprosy patient. For unidentified reasons, however, a few people are susceptible to contracting the disease, perhaps by direct intimate contact with leprosy lesions. The details of transmission are still poorly understood, but the pathogen may enter the body through otherwise inconsequential skin abrasions.

Leprosy has been an especially difficult disease to study because the causative agent, *Mycobacterium leprae,* cannot be grown on laboratory media. *M. leprae* can be cultured only in the nine-banded armadillo or in the foot pads of mice, where the pathogen grows extremely slowly, about one generation every 12 days. (Compare this to *Escherichia coli*'s generation time of 20 minutes.) Each experiment may require months or years to obtain even the earliest results. This slow growth probably accounts for leprosy's long incubation period of 2 to 10 years, and the slowly progressive nature of the disease.

The tissue affinity of *Mycobacterium leprae* for peripheral nerve cells triggers chronic neurological inflammation, often causing loss of feeling in

TABLE 23-11

Characteristics of Diseases Acquired through Superficial Breaks in Skin

Disease	Pathogen	Characteristic of Pathogen	Reservoir	Incubation Period	Diagnosis	Anti-Microbial Treatment
Leprosy	*Mycobacterium leprae*	Acid-fast rod	Humans	3–6 years	Microscopic identification of pathogen in lesion	Dapsone
Yaws	*Treponema pertenue*	Spirochete	Humans	14–90 days	Microscopic identification of pathogen in lesion	Penicillin
Pinta	*Treponema carateum*	Spirochete	Humans	3–60 days	Microscopic identification of pathogen in lesion	Penicillin
Anthrax	*Bacillus anthracis*	Gram-positive spore-forming rod	Domestic farm animals, spores in soil	2–5 days	Demonstration of pathogen in blood or lesions by serology or culture	Penicillin
Tularemia	*Francisella tularensis*	Gram-negative rod	Rabbits, muskrats, ticks	2–10 days	Isolation and identification of pathogen from lesion or blood	Streptomycin
Leptospirosis	*Leptospira interrogans*	Spirochete	Farm and pet animals, rodents	4–19 days	Serologic tests or isolation of pathogen from blood or urine	Penicillin

the region. This anesthetizing effect promotes physical injury and secondary infections of numb extremities, especially hands and fingers. The primary symptoms, however, are largely due to a delayed hypersensitivity response to the pathogen, which ultimately produces the characteristic cutaneous lesions and disabling neurological inflammation. This form of the disease, called *tuberculoid leprosy*, is infrequently fatal, and patients may recover without treatment. In persons with defective cell-mediated immunity, on the other hand, the pathogen may invade every organ in the body. If untreated, this progressive form, called *lepromatous leprosy*, is usually fatal. Its victims may die because of secondary bacterial infection, often tuberculosis, or because of waxy deposits that accumulate in and injure the liver, kidney, and spleen.

In some persons symptoms alone are characteristic enough to diagnose leprosy. Otherwise the disease can be diagnosed only by microbiological findings, such as demonstrating acid-fast bacilli in cells from the

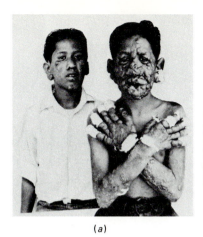
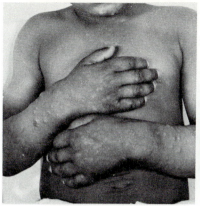
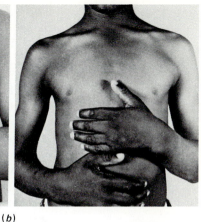

(a)   (b)

FIGURE 23-17

The effect of sulfone on the severity of leprosy. (a) Before sulfone was available, the disease in this patient followed a relentless course within two years. (b) Four years of sulfone treatment halted the progress of the disease in this patient.

lesions. Because the disease is poorly transmitted, isolation of leprosy patients is unnecessary. Most patients eventually recover after years of treatment with sulfone drugs (Fig. 23-17). An estimated 11 million persons are believed to currently have leprosy. Most of these are in Africa, Asia, and Latin America. There are about 4200 cases of leprosy in the United States.

Yaws and Pinta Yaws and pinta are treponemal infections that are primarily transmitted by nonvenereal routes among persons living in the tropics. These diseases are usually contracted by skin-to-skin contact among children or by flies that feed on the open lesions. *Treponema pertenue*, the etiologic agent of **yaws**, causes a primary skin lesion, called a "mother yaw," that releases bacteria into the bloodstream. If untreated, the pathogen spreads to bone, lymph, and skin sites over the body. The skin lesions are infectious if they become ulcerated. **Pinta** is caused by *Treponema carateum*. The hyperpigmented lesions later become nonpigmented and hyperkeratotic (thickening of the skin). They usually appear on the hands, feet, and scalp.

The microscopic appearance and serologic responses of *Treponema pertenue* and *Treponema carateum* are identical to those of *Treponema pallidum*. The three agents are so similar that persons who naturally recover from yaws and pinta acquire some cross-protection against syphilis. Ironically, successful control of yaws and pinta in the tropics has been accompanied by an increased incidence of syphilis in these regions.

Anthrax Anthrax is a disease of animals that spreads to people primarily through minor breaks in the skin or mucous membranes. Spores of *Bacillus anthracis* can survive on soil or on articles for years. Livestock is often infected while feeding on contaminated soil or contaminated feed products. Most human infections occur in workers whose occupations expose them to infected animals or their products. In addition to direct inoculation through broken skin (cutaneous anthrax), the bacterium may be acquired

635

by inhalation or more rarely by ingestion of contaminated, insufficiently cooked meat.

Cutaneous anthrax first appears as a papule that develops into a vesicle and in 2 to 6 days into a *black eschar* (a hard crust or scab) which is surrounded by small vesicles. If untreated, *Bacillus anthracis* may spread through the lymphatic system to the bloodstream. Between 5 and 20 percent of untreated cases are fatal.

Diagnosis is confirmed by direct microscopic identification of the pathogen or by isolation of *B. anthracis* from lesions or blood. The infection responds to penicillin and probably induces immunity. The incidence of anthrax is minimized by immunizing susceptible animals and persons most likely to be exposed to the pathogen. Additional control measures depend on disinfection and sterilization of animal products, as well as destruction and safe disposal of infected animals.

Tularemia   Many wild animals, domestic animals, and various ticks harbor *Francisella tularensis,* which causes **tularemia** in humans. Disease symptoms vary with the route of infection of the gram-negative bacterium. It is most commonly acquired either by inoculation of minor skin wounds while handling infected animals or by the bite of an infected tick. An ulcer usually forms at the inoculation site, and regional lymph nodes become enlarged. Hematogenous spread may produce infection of other organs and lead to meningitis or pneumonia. Occasionally the disease may be acquired by eating undercooked meat from an infected animal, resulting in gastrointestinal symptoms. When inhaled in contaminated dust, the bacteria may cause pneumonia.

Tularemia occurs throughout the United States, Canada, the European continent, Russia, and Japan. It is most prevalent among hunters who fail to wear gloves when skinning or handling animals. No direct person-to-person transmission occurs.

Diagnosis depends on isolating and identifying the organism from lesions. Most cases respond to streptomycin, but even without treatment fatalities are rare.

Leptospirosis   Animals that are infected with *Leptospira interrogans* shed the spirochetes in their urine, thereby contaminating water or soil. Humans contacting infected animals or bathing in contaminated water may become infected, especially if their skin is abraded. The disease may also be acquired by ingesting food or water contaminated by the urine of an infected reservoir. Most cases of **leptospirosis** result from recreational or occupational exposures. Rice and sugarcane field workers, farmers, sewer workers, veterinarians, and slaughterhouse workers are at high risk of exposure.

Symptoms usually appear within 10 days and characteristically include fever, severe headache, chills, and vomiting. Often the liver is affected and jaundice develops. Occasionally a rash, skin hemorrhages, or meningitis occurs.

DISEASES ACQUIRED
THROUGH THE SKIN

Diagnosis is confirmed by isolating and identifying the spirochete in blood or urine or by immunological tests. Penicillin and tetracycline are effective if administered early. In untreated patients with jaundice the mortality rate is about 20 percent.

OVERVIEW

Intact skin is an inhospitable environment for most microbes and effectively prevents their access to underlying tissues. Only a few pathogens can multiply on the surface or in the follicles of undamaged skin. Cutaneous infections are most likely to occur in areas of skin where moisture accumulates or in persons with decreased surface shedding. Underlying diseases also predispose for cutaneous infections. Most infections, however, are acquired through breaks in the skin's physical integrity and are not confined to the superficial surfaces.

Otherwise inconsequential and often unnoticeable abrasions, nicks, and cuts are portals of entry for several pathogens. Transmission of these microbes is usually by autoinoculation or by direct contact with an infected reservoir or contaminated vehicle.

Wounds that penetrate beyond the superficial layers of skin are portals for some dangerous pathogens, most of which gain access to the subcutaneous tissues at the time that the wound is inflicted. The microbes may then disseminate via the blood, lymph, or nervous system. Rabies and ratbite fever are transmitted by the bite of an infected animal. Arthropods are vectors of diseases caused by bacteria (plague and relapsing fever), rickettsias (spotted fevers and typhus), viruses (benign and hemorrhagic fevers and encephalitis), and protozoa (leishmaniasis, trypanosomiasis, and malaria). Controlling these diseases depends on reducing the animal reservoir and arthropod vector populations.

Extensive wounds and burns expose underlying tissue to pathogens in the environment. The threat of infection can often be minimized by proper cleansing and care of the wound. Clostridia and other anaerobic pathogens survive in the oxygen-poor environment of deep wounds. *Clostridium tetani* releases a neurotoxin from its localized infection site in deep wounds. *Clostridium perfringens,* on the other hand, releases enzymes that destroy surrounding host tissues and facilitate bacterial spread and the development of gas gangrene.

KEY WORDS

parenteral inoculation	sebum
epidermis	cutaneous infection
dermis	subcutaneous infection
keratin	tinea

pyogenic

bubo

Weil-Felix reaction

hemorrhagic fever

debridement

REVIEW QUESTIONS

1. Discuss four reasons why the skin is inhospitable for the growth of most microbes.

2. What factors predispose for each of the following?
   (a) Acne
   (b) Dermatophytoses
   (c) Cutaneous infection
   (d) Candidiasis
   (e) Hepatitis B?

3. How are the following used in disease diagnosis?
   (a) Negri bodies
   (b) Weil-Felix reaction
   (c) Hepatitis B surface antigens

4. What evidence suggests that humans are probably not the reservoirs of infections for viral encephalitis?

5. Describe three routes by which malaria can be transmitted among humans. In each case, how can transmission best be interrupted?

6. Compare the pathogenic agents, modes of transmission, and methods of treatment and control of rabies, plague, tetanus, and leprosy.

7. Why are wounds and burns so susceptible to infection? Discuss the role of enzymes in the development of gas gangrene. Why is this disease difficult to cure with antibiotics alone?

8. Distinguish between:
   (a) Bubonic and pneumonic plague
   (b) Jungle and urban yellow fever
   (c) Benign and hemorrhagic fever
   (d) Tinea unguium and tinea capitis
   (e) African and American trypanosomiasis
   (f) Hepatitis B and non-A non-B hepatitis

638

DISEASES ACQUIRED
THROUGH THE SKIN

Nosocomial Infections

24

The medical resources in hospitals are, in some ways, double-edged swords. Many of the very agents used to combat disease also contribute to one of the major problems in contemporary medicine: the alarming numbers of persons who contract infectious diseases as a direct result of exposure to the hospital environment. These hospital-acquired diseases are called **nosocomial infections** (*nosokomeion,* hospital).

Each year in the United States about 1.5 million people reportedly contract an infectious disease during their hospital stay. This reported nosocomial infection rate of 5 percent is considered by many experts to be lower than the actual incidence. Nosocomial rates in some American hospitals may run as high as 15 percent. In other words, one in every seven patients may acquire an additional infectious disease after admission to the hospital. On the average, the victim of nosocomial disease stays in the hospital an additional 7 days. At a cost of $300 per day, nosocomial diseases add an annual 3 billion dollars to hospitalization expenses. The cost in terms of human lives and suffering is even more important. More than half of all infectious diseases seen in hospitals of industrialized countries are nosocomial. Each year in the United States at least 15,000 patients die as a direct result of infections they acquire while hospitalized.

INFECTIONS IN THE HOSPITAL

Persons who develop diseases within their first 2 days in the hospital were probably in the incubation phase of a *community-acquired infection* at the time they were admitted. Only diseases that develop in patients who have been hospitalized for longer than 48 hours are considered true nosocomial infections. In these cases, it is assumed that exposure to the pathogens occurred during and as a direct result of hospitalization. These hospital infections pose a potential hazard to the general community since symptoms often don't develop until after the patient is discharged, and many of these patients harbor virulent pathogens that may cause infectious disease in the community.

EPIDEMIOLOGY OF NOSOCOMIAL INFECTIONS

Prevention of hospital-acquired infections depends on cooperation among hospital personnel, especially those in contact with patients. It requires an understanding of the special factors in the hospital that contribute to the problem—the types of reservoirs, the unusual opportunities for transmission, and the nature of the susceptible population combine to create a unique epidemiological setting. Because of this, patterns of nosocomial infections are different from those of infections acquired in the community (Fig. 24-1).

Reservoirs of Infection

Hospitals are unavoidably reservoirs of virulent and opportunistic pathogens. Some nosocomial diseases are acquired from infected patients, visitors, or hospital personnel. Persons with active disease shed infectious

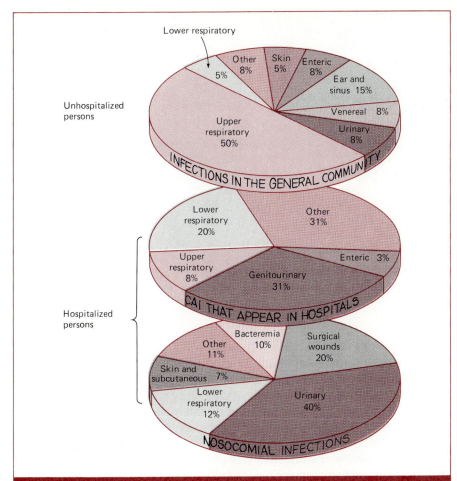

FIGURE 24-1

Estimated frequencies of
infections in the United
States.
CAI=Community-acquired
infections. *(Based on
information from the National
Health Survey and from CDC
hospital-based surveillance
studies.)*

agents that, unless confined, may contaminate the environment or directly infect patients, employees, or visitors. Healthy carriers also shed pathogens, usually with no knowledge of their asymptomatic infections. Most of these carriers probably acquired the pathogens from infected patients, usually because of failure to wash hands after contact with the patient, inadequate aseptic technique during treatment, or improper isolation of infectious patients. Persons who fail to take these precautions need not become infected to transfer organisms from an infected patient to other susceptible persons. Hands that harbor transient contaminants are common mechanical vectors of infection in the hospital. Surveys have discovered that one in five medical professionals carries potentially pathogenic antibiotic-resistant pathogens on his or her hands. Hand washing by medical professionals in some hospitals has been found to occur at only 30 percent of the ideal rate. Infected or colonized medical professionals (including physicians) are prevalent sources of epidemics within the hospital (Fig. 24-2). Failure to wash one's hands before and after each

FIGURE 24-2

Bacteria grown from the
hands of a nurse who had
just engaged in
cardiopulmonary
resuscitation.

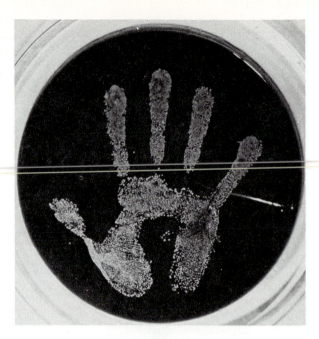

642

patient contact is probably the most important contributor to the spread of these infections.

Nosocomial infections are also acquired by contact with contaminated inanimate materials. For example, in one nationwide outbreak, 400 cases of *Enterobacter* bacteremia followed intravenous therapy with contaminated fluid distributed by a large manufacturer. Fifty-two of these patients died before the source of the infection was identified. Contaminated respiratory equipment has been responsible for epidemics of severe lower respiratory tract infections. Intestinal infections have been caused by contaminated endoscopes (used for examining internal mucous membranes), and infant diarrhea is often due to feeding contaminated bottled milk. Some opportunistic pathogens grow in soaps or lotions. In some instances, the disinfectants and antiseptics employed to reduce microbial contamination become heavily populated with resistant bacteria. Flower vases are no longer allowed in surgical wards and burn units because of their potential for harboring dangerous bacteria. Some opportunistic pathogens proliferate in the ventilation systems and are subsequently dispersed throughout the hospital.

Occasionally the source of nosocomial pathogens is the patient's own normal flora. These endogenous infections usually occur when a predisposing factor compromises the patient's normal defense mechanisms or when the indigenous flora are transferred to susceptible body sites by invasive procedures.

The epidemiology of nosocomial infections is depicted in Figure 24-3.

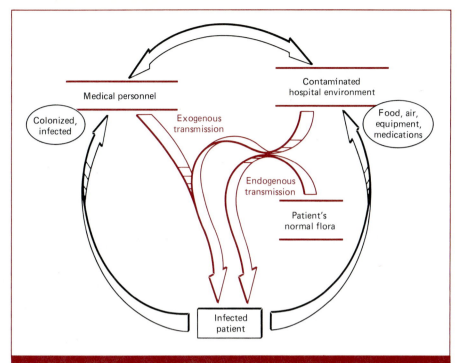

FIGURE 24-3

Epidemiology of
nosocomial infections.
Transmission occurs chiefly
by contact with the
reservoir, vehicle, or
mechanical vector, and to a
lesser extent by the
airborne route.

Mechanisms of Transmission

Procedures that bypass physical host defenses are often associated with
high rates of infection (Table 24-1). Of these *invasive procedures*, catheterization is an especially prevalent cause of nosocomial infections. Catheters are
hollow tubes that remain inserted in veins to deliver intravenous fluids and
medication or in the bladder through the urethra to relieve urinary
retention. Catheterization, inhalation therapy, and renal dialysis cause an
estimated 850,000 cases of nosocomial infections each year. This is approximately 45 percent of all reported hospital-acquired diseases. Even relatively noninvasive procedures can cause serious disease in susceptible persons

Procedure	Type of Disease	Common Pathogens
Urinary catheterization	Cystitis, pyelonephritis	Gram-negative bacilli, enterococci
Surgery	Infected wounds, septicemia	Staphylococci, gram-negative bacilli, *Bacteroides*
Intravenous therapy	Infection at injection site, septicemia	Staphylococci, *Klebsiella, Enterobacter, Serratia, Candida*
Respiratory intubation	Pneumonia	*Pseudomonas, Klebsiella, Serratia*
Renal dialysis	Sepsis, pyrogenic reactions	Hepatitis B virus, *Staphylococcus aureus, Pseudomonas*

TABLE 24-1

Medical Procedures Most
Commonly Associated with
Nosocomial Infections

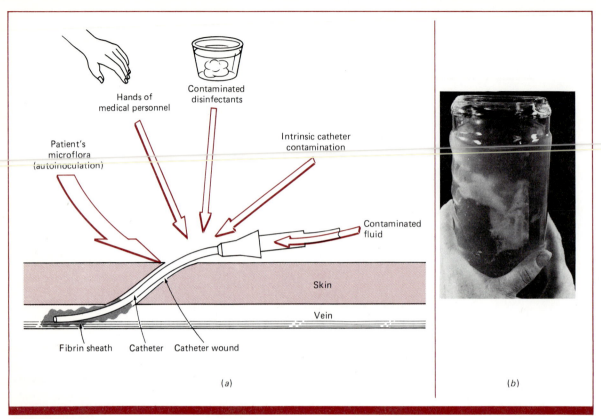

(a)

(b)

FIGURE 24-4

(a) Some possible routes for entry of infectious agents during intravenous therapy. A fibrin clot may develop on the portion of the needle within the blood vessel and may protect contaminating microbes as they reproduce. The infected site may shower the bloodstream with microbes. (b) The fungus contaminant entered the fluid through a hairline crack in the glass.

In the figure: Hands of medical personnel; Contaminated disinfectants; Patient's microflora (autoinoculation); Intrinsic catheter contamination; Contaminated fluid; Skin; Vein; Fibrin sheath; Catheter; Catheter wound

if the equipment is contaminated. The use of a contaminated blood pressure cuff on a newborn exposes this highly susceptible patient to the risk of infection. Contaminated bandages can cause serious wound infection.

Invasive procedures often transfer normal skin flora into susceptible body sites, either during the procedure or by allowing microbes to enter later—for example, where an indwelling intravenous catheter penetrates the skin, sometimes days after the device was inserted. Intravenous therapy may cause nosocomial infections when the needle is contaminated, the injection site is improperly decontaminated, or the injected material contains microbes (Fig. 24-4). Many nosocomial diseases could have been avoided had intravenous bottles with cracks or visible precipitates and turbidity been discarded rather than infused into unfortunate patients. As the duration of catheterization increases, so does the likelihood of nosocomial infection. These infections sometimes spontaneously subside if the catheter is removed from the vein.

Microorganisms growing in urine-collecting containers may enter the bladder by migrating directly up a urinary catheter (Fig. 24-5). Prior to 1960, when urinary catheters drained into easily contaminated open bottles, bladder infections were virtually inevitable in catheterized pa-

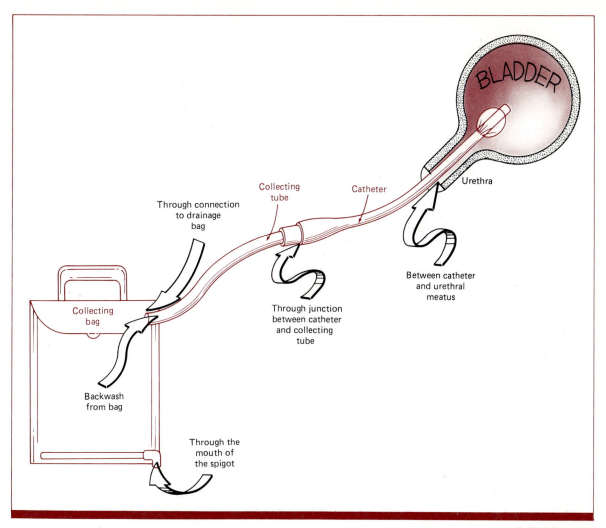

Through connection
to drainage
bag

Collecting
tube

Catheter

BLADDER

Urethra

Collecting
bag

Between catheter
and urethral
meatus

Backwash
from bag

Through junction
between catheter
and collecting
tube

Through the
mouth of
the spigot

FIGURE 24-5

Some possible entry points
for microbes during
urinary catheterization.

tients, causing more than 30,000 deaths per year from ascending kidney disease. Adoption of the closed-catheter system, which uses a sealed collecting bag, has reduced the risk of urinary tract infection to less than 20 percent. Unfortunately, because of improper aseptic technique during catheter insertion and inadequate care while the catheter is in place, urinary tract infections continue to be the most common nosocomial infection reported. Infection of the urinary tract accounts for 40 percent or more of all nosocomial infections.

The introduction of invasive procedures or devices creates new sources of hospital-acquired infections. Tracheostomy (surgical opening of the windpipe) and endotracheal intubation (inserting an air tube through the trachea) bypass mechanical respiratory defenses, sometimes depositing pathogens directly into the lower airways where they cause pneumonia. Pulmonary infections are also associated with the use of contaminated respirators and humidifiers. Although respiration equipment has been the source of nosocomial outbreaks in many hospitals, with care and attention to aseptic technique, these infection rates appear to be no higher than the rates prior to the development of respiratory therapy.

Each year an estimated 40,000 persons undergo renal dialysis, a procedure used in persons with malfunctioning kidneys to rid the body of its diffusible wastes. Some of these persons suffer pyrogenic reactions or serious systemic infections due to this invasive procedure. (Hepatitis B is the most common dialysis-associated infection.) Pathogens are usually introduced by insertion of the intravenous needle, but pyrogenic reactions are often related to the growth of microbes in the water and fluids used in the dialysis machine (Fig. 24-6). Although cells are too large to cross the dialysis membrane and enter the patient's blood, fragments of endotoxin released from gram-negative bacteria readily diffuse across the membrane and cause pyrogenic reactions. Bacterial concentrations in the water and dialysis fluid must be maintained at low levels to minimize the risk of endotoxic reactions in the dialysis patient. The dialysis procedure also increases the risk of infection among staff working in the dialysis unit. These persons are in frequent contact with potentially infectious blood. Infection may follow inadvertent inoculation or accidental ingestion.

Infected surgical wounds are the second most common type of nosocomial disease. The Centers for Disease Control have estimated that perhaps more than one in ten surgical patients develops postoperative infections after extensive surgical procedures. This number is even higher for surgical sites that are unavoidably populated by huge numbers of indigenous microbes. For example, the rate of postoperative infections following colon surgery may be as high as 30 percent.

Most agents of surgical wound infections are introduced during surgery. Microorganisms that cause these infections are usually gram-negative members of the patient's body flora, microbes shed from operating room personnel, or microbes released through punctured surgical gloves. The risk of infection increases when prosthetic devices or other foreign bodies are surgically implanted. The local tissue irritation that develops around an implant may protect microbes from host defenses after they gain entry into the surgical site. Surgical infection rates also increase as the duration of surgery increases, doubling for each 30 to 60 minutes of surgery.

Use of the term "sterile" to indicate personal decontamination following scrubbing fosters a false sense of security that leads to reduced emphasis on aseptic technique. This attitude is as dangerous as it is erroneous because the living human body is impossible to sterilize. To illustrate, resident staphylococci persist in the pores of the skin even after 6 minutes of proper scrubbing. The skin flora on the hands of surgeons wearing plastic gloves during lengthy operations are rapidly replenished in the warm, moist environment of the gloves. The bacterial population on the gloved hands soon exceeds the concentrations on the hands before

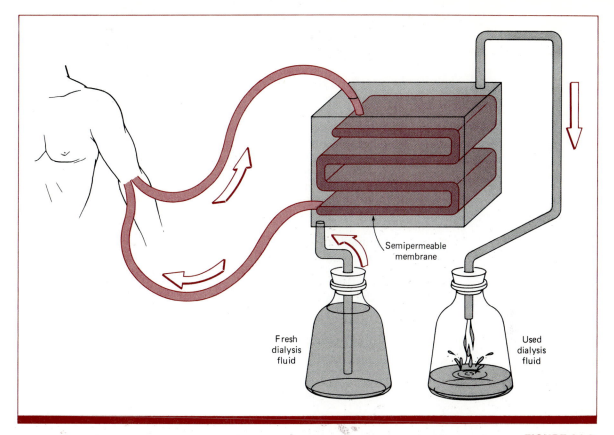

FIGURE 24-6

Amplified diagram of an artificial kidney (renal dialysis machine). All the blood in the body is pumped through a flattened chamber that is physically separated from the surrounding dialyzing surface by cellophane sheets. The total diffusing area is about 20,000 cm^2 per 500 ml of blood. Cells and proteins are too large to escape from the blood, but wastes diffuse out through the semipermeable cellophane membrane and flow away in the effluent. Although the membrane does not allow the passage of microbes from the diaysis fluid into the blood, it is no barrier for endotoxins and other molecules small enough to slip through the membrane's pores.

scrubbing. A careless nick in the gloves, often making a hole too small to observe, will allow these accumulated bacteria to pour into the surgical wound, leading to postsurgical infection. Glove punctures occur in as many as 60 percent of operations. Surgical wounds can also become contaminated during postoperative care, usually by *Staphylococcus aureus* shed from hands during wound inspection and care.

Characteristics of the Susceptible Host

The hospital contains a group of people who are unusually susceptible to infectious disease because of illness and medical procedures that reduce host resistance. Most infectious diseases acquired in the hospital are caused by opportunistic pathogens that pose no danger to normally

647

healthy persons or to persons not receiving treatments that increase the risk of infection. The following factors commonly found in hospitalized patients create an especially susceptible population:

■ Some infectious diseases (viral respiratory tract infections, for example) reduce resistance and may be followed by secondary bacterial infection. Secondary bacterial infection is responsible for most fatal cases of influenza.

■ The generous use of antibiotics disrupts the protective normal flora, leaving the patient susceptible to colonization by hospital strains of microorganisms resistant to the antibiotic used. Some antibiotics also damage organs essential to the host's defense against infection—the bone marrow, for example.

■ Large wounds, especially burns and surgical wounds, provide access to tissues that are ideal for bacterial growth.

■ Invasive procedures may introduce hospital strains of pathogens directly into the target area, bypassing host defenses.

■ Uncontrolled diabetes and other constitutional diseases increase susceptibility to infection by the opportunistic hospital strains.

■ Immunosuppression to accommodate organ transplants leaves a patient vulnerable to infectious disease by most microorganisms in the hospital. Infections in these people are commonly fatal.

■ Newborn infants and very old persons are especially susceptible to serious infections by organisms of low virulence.

■ Treatment of cancer patients with cytostatic or cytotoxic drugs not only kills tumor cells, but also the lymphatic cells needed for phagocytosis and acquired immunity. Cancer chemotherapy thus lowers the patient's ability to fight infectious disease.

Because so many hospitalized patients are suffering from at least one of these predisposing conditions, persons who are most susceptible to serious infections by opportunistic pathogens are often concentrated in hospitals.

Overuse of Antibiotics

The severity of the nosocomial problem is aggravated by the widespread and often irresponsible use of antibiotics in the hospital. With the discovery of penicillin and sulfa, a new era of medicine and medical education began to evolve. Not only were drugs available to treat existing bacterial diseases or to help prevent infections, but antibiotics were believed to be capable of preventing infection when proper asepsis was breached. (This myth is still prevalent in the medical community.) While the positive attributes of antibiotics should not be underrated, reliance on drugs rather

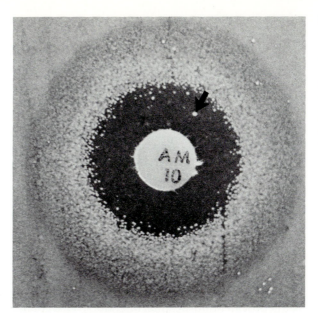

FIGURE 24-7

The colony of ampicillin-resistant *Haemophilus influenzae* (arrow) growing in the usual zone of inhibition reveals a reservoir of resistant bacteria that may become predominant in an infected patient treated with ampicillin.

than on medical asepsis has led to relaxed aseptic technique and to an alarming increase in nosocomial infections.

The overuse of antibiotics in the hospital has created another serious problem—a preponderance of drug-resistant pathogens (Fig. 24-7). While antibiotics reduced the prevalence of many formerly common pathogens in hospitals, the "microbial vacuum" left by these organisms has been rapidly filled by microbes that survive the most frequently used chemotherapeutic agents. This new breed of hospital pathogen often causes infections that are difficult to cure. As a consequence, many patients die before an effective chemotherapeutic regimen can be initiated. Furthermore, multiple-drug-resistant plasmids may be readily transferred from drug-resistant strains to antibiotic-susceptible bacteria (see Genetic Transfer in Bacteria, Chap. 12). The number of drug-resistant organisms therefore increases in proportion to antibiotic usage. Laboratory reports and antibiograms reduce reliance on the shotgun approach to chemotherapy and consequently discourage the emergence of drug-resistant pathogens. Although new antibiotics may provide successful chemotherapy against multiple-drug-resistant pathogens, bacterial populations become resistant to these drugs when they are excessively utilized.

Concern about the problems associated with antibiotics has prompted hospital personnel to examine the reasons for prescribing antibiotic therapy. Data obtained from several hospitals on antibiotic use have indicated that more than 50 percent of the patients received unneeded or inappropriate antimicrobial therapy. For example, although antibiotics have been shown to be of no value in preventing catheter-associated urinary tract infections or bacterial infections secondary to influenza and common colds, they continue to be frequently prescribed for these purposes. Such unjustifiable use of antibiotics not only increases the risk of nosocomial infections, but substantially adds to the cost of medical care. It is a task of the hospital infection control committee to establish and enforce prudent guidelines for antibiotic use.

Etiology of Nosocomial Infections

Hospitals are no longer pesthouses where people are constantly exposed to other persons with highly infectious diseases. Patients known to be shedding highly virulent pathogens are kept well isolated from susceptible patients. It is not the highly virulent pathogens that present the major nosocomial problems in modern hospitals, but microorganisms of surprisingly low virulence, those to which we are all frequently exposed. Many are free-living saprophytes.

Nearly any strain of bacterium residing in the hospital can cause nosocomial infection, but the most common of these opportunistic pathogens are bacteria in the enterobacteria group (Table 24-2). These and several other genera of gram-negative rods have replaced the staphylococci as the prevalent nosocomial offenders. This change may be partially due to

TABLE 24-2

Common Agents of
Nosocomial Infections

Organism	Percentage of Nosocomial Infections	Common Sites from which the Pathogen is Isolated
Gram-Negative Bacteria		
Escherichia coli	18.3	Urinary tract, surgical wounds, blood
Pseudomonas sp.	9.4	Urinary tract, lower respiratory tract, burn wounds
Klebsiella sp.	7.9	Urinary tract, lower respiratory tract, surgical wounds
Proteus sp.	6.6	Urinary tract, surgical wounds
Enterobacter sp.	4.2	Urinary tract, lower respiratory tract, surgical wounds
Serratia sp.	2.2	Urinary tract, lower respiratory tract, surgical wounds
Bacteroides fragilis	1.1	Surgical wounds, bacteremia
Gram-Positive Bacteria		
Streptococcus sp.	12.0	Urinary tract, surgical wounds, lower respiratory tract
Staphylococcus aureus	9.9	Skin, surgical wounds, blood
Staphylococcus epidermidis	3.8	Skin, surgical wounds, blood
Fungi		
Candida albicans	3.4	Urinary tract, blood
Other fungi	1.2	Urinary tract, blood, lower respiratory tract

increased recognition that the "harmless" saprophytes are not merely contaminants but cause serious disease.

Staphylococci These gram-positive cocci have historically been the major nosocomial problem, primarily because of the prevalence of healthy carriers. (About 20 percent of the adult population harbors *Staphylococcus aureus* in the nasopharynx or on the skin.) Most of the strains common to the hospital are resistant to antibiotics, especially penicillin. The major danger areas for severe staphylococcal infections are the nursery, the delivery room, the operating room, and the burn unit, although transfer of infections among patients in hospital wards is not uncommon. Patients with open wounds infected by staphylococci should be separated from highly susceptible individuals such as postsurgical patients.

Streptococci The two groups of streptococci most frequently associated with nosocomial infections are groups B and D. Of these, group D (the enterococci) causes the greatest number of streptococcal infections in the hospital. These cocci are members of the normal bowel flora and are second only to *Escherichia coli* as etiologic agents of nosocomial urinary tract infections. They are the third most common cause of surgical wound infections. These infections are generally treated with penicillin. Group B streptococci are responsible for infections of the uterus following childbirth and for septicemia and meningitis in newborns. Group B and D streptococcal infections are generally treated with penicillin. Although they are prevalent as pathogens of community-acquired infections, group A streptococci (*Streptococcus pyogenes*) and the pneumococci (*S. pneumoniae*) are infrequent causes of nosocomial infections.

Enterobacteria Before 1960, disease caused by *Serratia marcescens*, a common saprophyte, was unusual. Now this "harmless" bacterium, together with *Klebsiella*, *Escherichia*, and *Proteus*, is among the most common pathogens in the hospital. Each causes disease that is often incurable by the commonly used antibiotics. The mortality rate of bacteremia with any of these gram-negative organisms is 20 to 50 percent among hospitalized patients. *Escherichia coli* is the hospital's most troublesome pathogen, causing nearly twice the number of nosocomial infections as any other microbe. It is the most common agent of bacteremia and causes one-third of all hospital-acquired urinary tract infections. This microbe, as well as other members of the intestinal bacteria (*Klebsiella*, *Enterobacter*, and *Serratia*), readily grows in glucose solutions used for intravenous therapy. (Most bacteria either die or do not proliferate in these solutions.) *Klebsiella* readily colonizes the oropharynx of chronically and severely ill persons, who may develop pneumonia following direct extension of bacteria to the lungs. Most lower respiratory tract nosocomial infections are caused by *Klebsiella*.

Pseudomonas Pseudomonas is the prototype of saprophyte-turned-pathogen and is a major cause of nosocomial infections in American hospitals. A particularly hazardous hospital pathogen, *Pseudomonas aeruginosa* is a free-living gram-negative bacterium found in the bowels of 5 percent of healthy persons in the community. In the hospital this figure increases to 40 percent. *Pseudomonas* is one of the major causes of nosocomial infections, partially because of its widespread distribution within hospitals. This organism is one of a group of microbes collectively called "water bacteria," that present a special medical problem—their ability to grow in distilled and deionized water allows them to proliferate in the moist chambers of respirators, mist therapy units, renal dialysis units, and improperly dried plastic tubing used for intravenous therapy. When these contaminated items are used in a patient, high concentrations of bacteria are introduced directly into the bloodstream or respiratory system, and severe nosocomial disease is virtually inescapable. The concentration of *Pseudomonas* is often not enough to make water appear turbid, so contamination cannot be visually detected. Contaminated water looks as clear as cell-free water.

This small gram-negative bacillus is especially problematic in burn wounds, where the organism becomes easily established (Fig. 24-8). From an infected burn wound, the bacteria may cause septicemia, pneumonia, meningitis, and several other serious diseases. Infected burns often infect other burn patients who are present in the same ward. Burned persons isolated in burn units or located with patients who are not burned are less likely to contract serious infections than those confined to self-defeating burn wards. Sometimes the source of infection is the patient's own gastrointestinal tract, especially following treatment with antibiotics to

652

NOSOCOMIAL
INFECTIONS

FIGURE 24-8

Pseudomonas infection of a
burn wound.

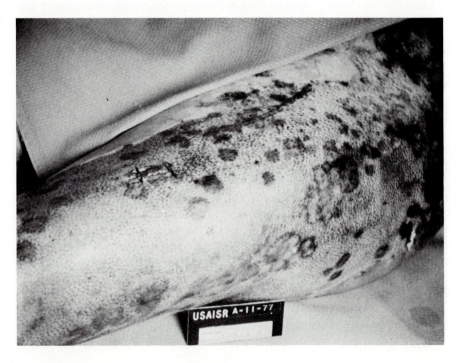

which the strain is resistant. These antibiotics may actually encourage growth of *Pseudomonas* in the bowel by reducing competition from other members of the normal microbial flora. Poorly designed wash basins that are difficult to disinfect are a common source of *Pseudomonas* infections in newborns. The organism is sometimes introduced into body orifices on contaminated catheters, cystoscopes, or other invasive instruments. One startling source of infection is contaminated disinfectants and medications used to prevent infection. Medications packaged in single dosages are less likely to be contaminated, especially if vials are dated and sterilized after preparation.

Fungi   In addition to opportunistic bacteria, molds and yeasts of normally low virulence may also establish nosocomial infections (Fig. 24-9). Antibiotic therapy or severe immunosuppression often precedes extensive invasion by *Candida albicans* and fatal lung infections with either *Aspergillus* or the molds in the order Mucorales. Sixty percent of patients dying with acute leukemia have fungal infections, as do 45 percent of those who die after renal transplant.

Viruses   Because of difficulty in detecting and propagating viruses, knowledge of their role as agents of nosocomial infection is very limited. Indeed, we have yet to even determine what viruses can normally be expected to be isolated from healthy humans. We do know that patients treated with immunosuppressive agents to prevent rejection of organ transplants often die of overwhelming herpesvirus infection. The hepatitis B virus, on the other hand, poses a threat to all people in the hospital,

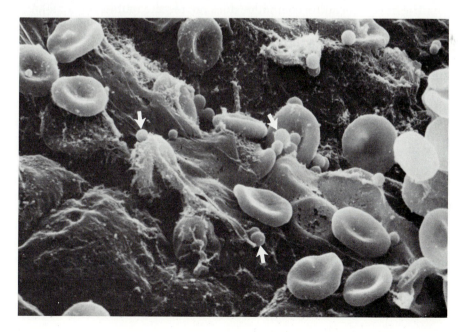

FIGURE 24-9

Candida albicans (arrows) colonizing a fibrin clot. This yeast causes opportunistic infections following cardiac surgery often colonizing prosthetic heart valves and catheters.

regardless of immunological competence. Hepatitis B viruses, usually spread by the parenteral route through contaminated blood or invasive equipment, also may enter the body through the mucous membrane surfaces. Inapparent carriers of the virus increase the danger of hospital-acquired hepatitis, especially to surgical personnel, persons performing renal dialysis, and personnel in blood banks.

Infection Control Committees

Hospital surveillance has shown that most nosocomial infections result from failure to adhere to established procedures for handling patients and infectious materials. Many of these procedures are established by an **infection control committee**, which, since 1971, is required in every accredited hospital in the United States.

The primary responsibility of the hospital's infection control committee is to identify and control hospital-acquired infections. The committee is usually composed of representatives from each major clinical department, the laboratory, hospital administration, and nursing. Housekeeping, pharmacy, dietary, and blood bank personnel may also sit on the committee. The major responsibilities for surveillance are assigned to the chief infection-control practitioner, usually a nurse epidemiologist. This individual collects data on all infections that occur in the hospital. These data include the body site, causative agents, a record of the responsible pathogen's antibiotic susceptibilities, and any host factors that may contribute to development of disease. This information alerts the committee when an increase in reported infections suggests the possibility of an epidemic. These data are also used to assess the effectiveness of measures implemented to prevent nosocomial infections.

The infection control committee does not require routine microbiological surveillance of possible environmental reservoirs of infection, nor are staff members routinely screened to detect carriers. During an outbreak, however, both of these measures are implemented to detect the source. Some critical items, on the other hand, are periodically monitored. The effectiveness of autoclaves and other sterilization systems is assessed to ensure that instruments, equipment, and fluids for direct use with patients are effectively treated and are sterile. Hospital-prepared injectable fluids or infant formulas are monitored for the presence of microbial contaminants. Disinfectants and antiseptics are periodically cultured to make sure that microbes are not growing in the chemical agents being used for decontamination.

LOWERING THE INFECTION RATE

Even under ideal conditions, some nosocomial infections are inevitable. For example, severely compromised patients cannot be entirely protected from endogenous infections caused by their own normal flora microbes. The nosocomial infection rate that now exists, however, can be significantly reduced. Isolation of patients infected with highly communicable diseases usually confines the pathogen

TABLE 24-3

Isolation Procedures

Strict Isolation

1. Gowns, gloves, and masks must be worn by all persons entering room and discarded before leaving room.
2. Hands must be washed thoroughly on entering and before leaving room.
3. Articles must be discarded or washed and wrapped before being taken from room for disinfection or sterilization.
4. Confinement to a private room (or with others who have same proven diagnosis) is necessary.

Recommended for patients with anthrax (inhalation); plague (pneumonic); smallpox; vaccinia; major burn, skin, or wound infections with *Staphylococcus aureus* or *Streptococcus pyogenes*; pneumonia (*S. aureus* or *S. pyogenes*); diphtheria; chickenpox; disseminated herpes zoster; congenital rubella syndrome; disseminated neonatal herpes simplex; rabies

Respiratory Isolation

1. Masks must be worn by all susceptible persons entering room.
2. Hands must be washed thoroughly on entering and before leaving room.
3. Articles coming in contact with patient's secretions must be discarded or washed and wrapped before being taken from room for disinfection or sterilization.
4. A private room is necessary.

Recommended for patients with tuberculosis; meningococcal meningitis; meningococcemia; measles; mumps; rubella; pertussis

Enteric Isolation

1. Gowns and gloves must be worn by persons having direct contact with excretions of patient.
2. Hands must be washed thoroughly on entering and before leaving room.
3. Articles coming in contact with patient's excretions must be discarded or washed and wrapped before being taken from room for disinfection or sterilization.
4. A private room is desirable, especially for children.

Recommended for patients with cholera; *Escherichia coli* gastroenteritis; salmonellosis; shigellosis; *Yersinia enterocolytica* gastroenteritis; typhoid fever; acute diarrhea with suspected infectious cause; viral hepatitis

Wound and Skin Isolation

1. Gowns and gloves must be worn by persons having direct contact with patient and must be changed between patients.
2. Hands must be washed thoroughly on entering and before leaving room.
3. Articles must be discarded or washed and wrapped before being taken from room for disinfection or sterilization.
4. A private room is desirable.

Recommended for patients with *Clostridium perfringens* gas gangrene; localized herpes zoster; limited burn, skin, or wound infections; plague (bubonic)

Reverse (Protective) Isolation

1. Gowns, masks, and gloves must be worn by all persons entering room and discarded outside of room.
2. Hands must be washed thoroughly *before* entering room.
3. A private room is necessary.

Recommended for patients with agranulocytosis; severe and extensive noninfected vesicular, bullous, or eczematous dermatitis; certain patients receiving immunosuppressive therapy; certain patients with lymphomas and leukemias

Source: Adapted from Centers for Disease Control, *Isolation Techniques for Use in Hospitals*, 2d ed., DHEW publ. no. (CDC) 76-8314, U.S. Government Printing Office, Washington, D.C., 1975.

655

LOWERING THE
INFECTION RATE

to a restricted area. The degree of isolation depends on how readily the disease is transmitted (Table 24-3). **Strict isolation** for highly infectious and dangerous diseases that are spread either by contact or by airborne routes requires a private room and the use of masks, gowns, and gloves by everyone entering the room. Patients who are exceedingly susceptible to infection are isolated for protection against exposure to opportunistic pathogens. This procedure is referred to as **reverse** or **protective isolation**.

Chemoprophylaxis has been shown to reduce the incidence of surgical wound infection following some types of surgery, particularly of the gastrointestinal and genitourinary tracts. (These drugs should be discontinued by the end of the operative day.) The prophylactic use of antibiotics should be minimized to reduce the likelihood of infection by antibiotic-resistant pathogens. The reflex response in prescribing antibiotics should be replaced by a more rational approach that determines if the pathogen is sensitive, if the drug interferes with other medications the patient is taking, and if the antibiotic is truly needed.

Invasive and immunosuppressive therapies should be used only when necessary and when noninvasive alternatives are unavailable. Invasive equipment should be monitored for contamination, changed at defined intervals, and removed from the patient's body as soon as possible.

TABLE 24-4
Inanimate Sources and Vehicles of Nosocomial Infections

Source or Vehicle	Precautions*
Water	Use sterile water for nebulizers and humidifiers. Replace drinking water carafes daily.
Mops	Do not allow mops to stand in water. Use mophead once (at most for 1 day), then autoclave or wash at 70°C.
Respiratory therapy equipment	Sterilize or properly disinfect before use and change every 24 hours when in continuous use.
Renal dialysis machine	Disinfect distribution system immediately before use; disinfect machine overnight; treat water by reverse osmosis.
Anesthesia equipment	Disinfect artificial airways and breathing bags before each use.
Instruments and fluids for injection into patients	Use disposable needles and syringes; open packages immediately before using; prepare skin properly; use single dose vials when possible; store multiple dose vials in refrigerator.
Thermometers	Use disposable thermometers; otherwise, sterilize with ethylene oxide between patients and store in fresh 70% ethyl alcohol.
Ventilation systems and air conditioners	Disinfect water reservoirs and monitor for contamination.
Linen	Gather soiled linen with minimum amount of turbulence; place immediately in bags; wash linen (and bag) *before* sorting; use sterile (autoclaved) linen for compromised patients and in operating and delivery rooms.
Patient rooms	Wet-vacuum floors after flooding with water containing germicide; regularly wash all surfaces (walls, ceilings, etc.); disinfect following discharge of patient with a communicable disease; wash furniture, lamps, and other fixtures daily; wash plumbing fixtures twice daily with a detergent-germicide; scrub tubs and showers immediately after use; change curtains after patient leaves.

*Precautions may vary according to the hospital.

Controlling nosocomial infections also depends on educating all hospital personnel as to the importance of excellent medical asepsis, the potential sources of infection, and how they can be controlled (Table 24-4). Each health care professional has a personal responsibility for adhering to infection-control policies and exercising adequate precautions. For example, if there is doubt about the sterility of any object, it should be treated as if it were unsterile. A sterile object, when being removed from its package, should not be allowed to touch the outside surface of the package, which is unsterile. Any sterile object that has contacted an unsterile object is contaminated and should be discarded. If the package is paper or cardboard, it must be dry to prevent contamination of the enclosed object. Objects in moist paper packaging should not be considered sterile.

Aseptic techniques should be strictly adhered to at all times. This is especially true for physicians, since many nurses and therapists model their behavior according to that displayed by the medical doctor. Personnel should never consider themselves "sterile," and their behavior should reflect awareness that potential pathogens are constantly being shed from their bodies. This danger is best combated by maintaining excellent hygiene of the total body (hands, hair, and clothing). Conscientious hand washing is believed to be the most effective single strategy in reducing the incidence of nosocomial infections. Simple rinsing or superficial washing with soap will not significantly reduce the resident organisms on the hands (see The Mechanical Scrub, Chap. 13). Using elbow, knee, or foot controls helps prevent recontamination of hands while turning off the water after washing (Fig. 24-10). In critical care areas, such as the intensive care unit and the nursery, as well as before surgery, special antiseptics should be used on the hands. Hand washing should be performed by everyone in contact with patients immediately *before and after* each contact.

FIGURE 24-10
Proper hand washing is perhaps the major element in infection control.

Ignaz Semmelweiss developed a system of disinfecting instruments and hands that dramatically reduced the infection rate following childbirth. Yet he was chastised by his colleagues for his radical ideas, and they resisted the implementation of these practices. One hundred years later many medical professionals still fail to perform the simple, life-saving task of hand washing before and after contact with each patient. Studies by infection control researchers reveal that, although most of these persons are aware of the importance of hand washing in preventing nosocomial disease, they are deterred by the "inconvenience" of the procedure. Some persons neglect to wash their hands frequently because they believe it will have a detrimental effect on their skin. Neither of these reasons is justified.

The alarming rate of nosocomial infections has helped maintain community mistrust of the hospital environment. The persisting stigma of the "pesthouse" will disappear only after practice of asepsis by medical personnel is elevated to a level required to prevent the transmission of infectious disease among patients and staff alike.

OVERVIEW

The prevalence of nosocomial infections is due to the concentration of highly susceptible persons in the hospital, the ubiquity of opportunistic pathogens, medical procedures that encourage infection, and breaches in sound aseptic practices. Hospital-acquired infections are often difficult to treat because widespread use of antimicrobial chemotherapy in the hospital encourages the emergence of drug-resistant microbes. Many of these strains possess R factors that confer multiple-antibiotic resistance. These drug-resistant hospital strains colonize the hospital staff, who then transmit them to patients.

The compromised patient is the most common victim of nosocomial infection. Illness and medical procedures combine to increase susceptibility to the many opportunistic pathogens in the hospital. Identifying these susceptible persons is one step in controlling nosocomial infections so that additional precautions can be employed to protect them. Controlling hospital-acquired infections is the task of all personnel, but is coordinated by the infection control committee. This committee detects cases of nosocomial infection and collects data on their occurrence. The committee also conducts educational programs to encourage awareness and practice of aseptic procedures by the hospital staff, monitors the use of antimicrobial chemotherapy, and coordinates surveillance of possible reservoirs. Hand washing before and after each patient contact is the single most important precaution for lowering the incidence of nosocomial infections.

KEY WORDS

nosocomial infection

infection control committee

strict isolation

reverse isolation

REVIEW QUESTIONS

1. Why are infections that emerge within the first 48 hours of hospitalization not considered nosocomial?

2. List four factors that contribute to the high incidence of hospital-acquired infection.

3. Why do the following increase the risk of infection to the hospitalized patients?
    (a) Antibiotic therapy
    (b) Respiratory therapy
    (c) Cancer and its treatment
    (d) Urinary catheterization

4. What pathogens are the most frequent causes of nosocomial infections? Describe three ways they gain entry into the body.

5. Why is proper hand washing considered the most important element in controlling nosocomial infections?

Microbial Biotechnology

25

USING MICROBES FOR FOOD PRODUCTION
Food Traditionally Produced by Microbes
Dairy Products / Nondairy Foods / Alcoholic Beverages
Single-Cell Protein
Food Additives

INDUSTRIAL MICROBIOLOGY
Industrial Fermentations
Pharmaceutical Products
Products of Genetic Engineering

AGRICULTURAL MICROBIOLOGY
Biofertilization
Microbial Inoculants That Increase Crop Yield
Biological Control of Agricultural Pests

WATER TREATMENT AND WASTE DISPOSAL
Water Treatment
Sewage Disposal
Evaluating Water Quality

BIODEGRADATION OF ENVIRONMENTAL POLLUTANTS

MICROBES AS SUPPLIERS OF NATURAL RESOURCES
Fuel Production
Biomining Techniques

OVERVIEW

Since the dawn of history, people have been putting microbes to work. Processes discovered hundreds or even thousands of years ago are still used today to produce bread, beer, wine, cheeses, yogurt, and many oriental foods. Moreover, recent developments in molecular biology are expanding the potential usefulness of microbes into new areas of industry, agriculture, medicine, and food production. The new field of microbial biotechnology has sprung from the primitive roots of early microbial manipulations. Sophisticated techniques are providing new and exciting ways of tapping the technological capacities of microorganisms. This technology is already a well-established factor in world economy. In the United States alone, microorganisms add tens of billions of dollars to our annual income.

Some of the more advanced techniques available today would be considered miraculous by many early scientists—for example, the ability to manufacture "new" organisms by genetic manipulation. So far, however, the applied fields of microbiology are still dominated by traditional applications of microbial activities—industrial fermentations, treatment of wastewater, and production of alcoholic beverages, foods, antibiotics, and vaccines. One of the goals of the new biotechnology is to improve the efficiency of the microorganisms performing these established techniques. Another objective is to create strains of organisms that will readily provide resources currently either unavailable or difficult to obtain from conventional sources. In this chapter we will discuss the traditional applications of microbiology in conjunction with the anticipated benefits of the emerging technology.

USING MICROBES FOR FOOD PRODUCTION

Microbes continue to be used in the food industry primarily in three ways.

■ Specific metabolic activities, usually fermentation reactions, generate organic compounds that accumulate and transform some edible substances into foods with more desirable characteristics. These altered properties usually help preserve foods and often enhance their flavor, texture, or digestibility. Dairy products, breads, soy sauce, pickled vegetables, alcoholic beverages, and vinegar are produced by microbial fermentation.

■ Microbial cells, cultivated in large quantities, are used as protein supplements in feed for livestock. Such single-cell proteins have also been suggested as alternative food resources for humans.

■ Certain microbes produce metabolic by-products that have nutritional or flavor-enhancing properties when added to foods and feeds. Enzymes isolated from microorganisms are also instrumental in food production.

Foods Traditionally Produced by Microbes

Dairy Products   Milk is fermented to cheese, yogurt, sour cream, or buttermilk by lactic acid bacteria and streptococci. Fermented milk products taste different, usually have a reduced moisture content, and are more resistant to spoilage than milk. Some cheeses, for example, can be stored for months at room temperatures, whereas pasteurized whole milk spoils quickly even when refrigerated. Dairy fermentations can best be controlled if milk is pasteurized to inactivate contaminating organisms and enzymes before processing. Starter cultures of desired microbes are then added. The final product depends on the kinds of microbes inoculated, the type of milk (cow's, sheep's, or goat's), and the subsequent conditions of processing.

Cheese production (Fig. 25-1) begins with *souring*, the bacterial conversion of lactose (the sugar in milk) to lactic acid. The resulting drop in pH causes protein in the soured milk to coagulate. This process may be facilitated by the addition of the milk-curdling enzyme, *rennin*. These coagulated solids form a *curd*, which, when separated from *whey* (the watery liquid), becomes cottage cheese. When cream rather than milk is curdled, the solid product is cream cheese. The curd may also be used to produce other types of cheese. It is first heated to reduce its moisture content and allowed to ripen through additional microbial activity. Depending on the residual moisture content, cheeses are classified as soft, semisoft, or hard. The microbes used for ripening break down proteins and fats or produce fermentation products that generate the characteristic flavors, aromas, and appearances of cheeses. Brie and Camembert are soft

FIGURE 25-1
Blue cheese is made by inoculating the curd with *Penicillium roqueforti*, the fungus that gives the ripe cheese its aesthetic characteristics.

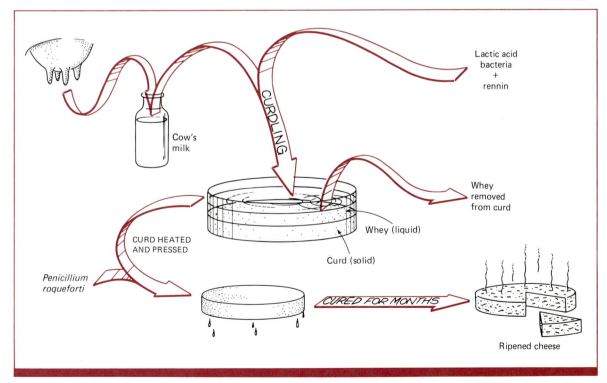

Lactic acid bacteria + rennin

Cow's milk

CURDLING

Whey removed from curd

Whey (liquid)

Curd (solid)

CURD HEATED AND PRESSED

Penicillium roqueforti

CURED FOR MONTHS

Ripened cheese

cheeses ripened by fungi growing on the surface of the pressed curd. The semisoft Roquefort cheese, on the other hand, is ripened by a blue-green mold that can be seen growing throughout the curd. Bacteria are responsible for ripening most hard cheeses. For example, propionibacteria transform a curd to Swiss cheese by producing propionic acid, which imparts flavor, and carbon dioxide, which accumulates into large bubbles, the holes of Swiss cheese.

Milk may also be acidified and flavored by controlled fermentation to produce yogurt. Milk solids are added prior to fermentation, and two bacteria, *Lactobacillus bulgaricus* and *Streptococcus thermophilus*, convert lactose to lactic acid and acetaldehyde, which impart the characteristic flavor to yogurt. Kefir, a dairy drink popular in eastern Europe, contains about 1% alcohol. It is formed by the combined fermentations of lactic-acid-producing bacteria and yeast that convert lactose in milk to alcohol. Buttermilk is usually made by fermenting pasteurized skim milk (although originally it was the liquid remaining after the fat was removed in churning cream to butter). Sour cream, as the name implies, is fermented cream. Lactic acid bacteria in cream produce a chemical, called *diacetyl*, responsible for the characteristic flavor of the butter churned from the cream. Thus, although butter is not a product of fermentation, its flavor is.

Nondairy Foods Vegetables, particularly cucumbers, cabbages, and olives, can be preserved by the fermenting activities of lactic acid bacteria and yeast that naturally reside on their surfaces. The growth of these microbes is selectively encouraged by placing the vegetables in salt solutions called *brines*. Lactic acid production usually continues until no fermentable carbohydrates remain. The salt, low pH, and absence of carbohydrates so effectively prevent the growth of spoilage organisms that pickles, sauerkraut, and olives can usually be kept indefinitely.

Soy sauce originated in China centuries ago. Its production requires two microbial transformations. Cooked soybeans and crushed wheat are first incubated with the mold *Aspergillus oryzae*, which converts much of the starch to fermentable sugars. The mash is then fermented for 6 to 12 months with a mixture of lactic acid bacteria and yeast. The resulting liquid is soy sauce; the remaining solids are used as livestock feed.

Bread production by fermentation is another ancient process. Leavened breads are made by adding baker's yeast, *Saccharomyces cerevisiae*, to dough. The yeast ferments sugar in the dough to alcohol and carbon dioxide. Bubbles of carbon dioxide gas become trapped in the dough, which rises and acquires a light texture. The alcohol evaporates during baking. Additional yeast or bacteria may be used to produce the characteristic flavors of sourdough and various other breads.

Alcoholic Beverages The term "fermentation," from the Latin word for "boil," was originally applied to the production of alcoholic beverages because the generation of carbon dioxide gas gives the fermenting liquid a frothy appearance. Most alcoholic beverages are manufactured by the metabolic activity of *Saccharomyces cerevisiae* or of closely related yeasts. Wines are prepared by the direct action of yeasts on sugars in fruit juices. Beers, on the other hand, are made from the starch in barley or other grain extracts. Since starch is a nonfermentable substrate, beer production begins with *malting*. During this process the barley germinates and produces enzymes that digest starch to the fermentable sugars glucose and maltose. These sugars are dissolved in the aqueous extract of the malted barley, called *wort*, which is then boiled with hops, a flower that imparts a characteristic bitter flavor to the brew. Boiling stops the enzymatic digestion of starch and therefore limits the amount of alcohol that can be produced by subsequent fermentation to between 3 and 6%. After boiling, the hops are removed and yeasts are added to ferment the sugars to alcohol and flavor-enhancing compounds. Because the undigested starch in the wort cannot be fermented by *Saccharomyces cerevisiae*, most beers have a high carbohydrate content, containing about 4% unfermented starch (each gram of which adds about four calories to the beverage). Yeasts that digest this starch have been mated with brewing yeasts to yield hybrid organisms that produce beers containing less than 1.5% undigested carbohydrate. These yeasts are used to manufacture some brands of low-calorie ("light") beer. Other light beers depend on predigestion of starch in the wort using enzymes instead of whole organisms.

The starch in rice used to make sake must also be digested to sugar before it can be fermented. This process is accomplished by *Aspergillus oryzae* inoculated into steamed rice. The resulting sugar is fermented to alcohol by yeast and to lactic acid by bacteria. Most distilled spirits are also made from fermented grain products in a process similar to beer brewing. The malted mash, however, is not boiled, and enzymes continue to hydrolyze starch to sugar during the fermentation process, resulting in higher alcohol production. The amount of alcohol produced is limited by the inability of yeast to thrive in ethanol concentrations higher than 14 to 18 percent. Thus, products of natural fermentation never exceed these concentrations of alcohol. Brandies, whiskeys, and other spirits have higher alcoholic contents as a result of *distillation*, the vaporizing of the volatile alcohol by heating the brew and collecting the vapor as it condenses back to liquid.

Grapes and other fruits used to make wine naturally contain large numbers of yeasts on their surface (up to 10^5 per grape). Originally, the sugar in crushed grapes was fermented to wine solely by these yeasts. However, a consistently better product is obtained when the natural flora are inhibited, after which a starter culture of *Saccharomyces cerevisiae* is added to the grape juice. The resulting wine usually contains about 12% alcohol. Red wines are fermented from crushed red grapes. The entire grape, including juice, pulp, skin, and seeds, is present during fermentation. Pigment in the grape's skin provides the wine's red color. Juice that is separated from the solids after 1 day and then fermented becomes rosé wine. White wines are fermented from either red or white grapes; however, only the juice is used. Fortified wines such as sherries, ports, and vermouths contain 15 to 20% alcohol and are produced by adding brandy to wine. The increased alcohol content of fortified wines inhibits microbial growth, thereby increasing the stability of the wines.

Vinegar is literally sour wine, originally manufactured by intentionally exposing wine to airborne bacteria that grow in the wine and convert ethanol to acetic acid, the chemical that gives vinegar its characteristic odor and taste. The modern method for vinegar production is not very different from the original "let-alone" process. It consists of inoculating wine or hard apple cider with *Acetobacter*, which, under highly aerated conditions, converts the ethanol to acetic acid.

Single-Cell Protein

Most microbes reproduce faster than either plants or animals. Many bacteria, for example, double their numbers in less than 1 hour when supplied with a favorable environment. The generation time of yeasts is usually 1 to 3 hours, of algae 2 to 6 hours, and even the slower-growing molds reproduce in 4 to 12 hours. Microbial cells are generally rich in proteins; some are as much as 80 percent protein. In reproducing themselves, these cells can quickly generate huge quantities of proteins. Thus compared to a 1000-pound steer, which produces 1 pound in new protein daily, or 1000 pounds of soybeans, which manufacture 80 pounds of protein per day, 1000 pounds of yeast may yield 50 tons of protein, and 1000 pounds of bacteria may increase their protein 10^{10}-fold in a day. Such rapid protein production makes microbes attractive as sources of **single-cell proteins** (**SCP**), alternative or supplementary protein sources to relieve shortages caused by the limited production of plants and animals.

The most useful single-cell proteins supply all essential amino acids. Several bacteria and yeasts, for example, have protein values comparable to animal products and in many cases are better protein sources than plants, which tend to be deficient in the amino acid lysine. Often these single-cell proteins can be produced inexpensively by growing microbes on inorganic nitrogen (ammonium salts or nitrates), using cheap sources of carbon and energy. Petroleum hydrocarbons were among the first and most efficient substrates for producing single-cell proteins but are no longer used because of their current demand as fuel. Agricultural and industrial waste products provide several inexpensive substrates. Such wastes include molasses, fruit pulp, milk whey, sulphite liquor (wastes from the paper industry), and methane (or methanol) generated from sewage treatment. When grown with these substrates, microbes not only supply a valuable resource but simultaneously help dispose of huge amounts of garbage.

Algae and photosynthetic procaryotes utilize light energy and carbon dioxide for growth. Because they need not be supplied with a prepared carbon and energy source, they are potentially inexpensive sources of SCP. Unfortunately, these organisms are difficult to cultivate. They must be grown in shallow ponds or lagoons and will produce high yields only if they are provided adequate sunlight and controlled incubation temperatures. Often it is impossible to prevent contamination, and it is difficult to harvest the organisms, particularly the unicellular forms. Despite these problems, some algae, notably *Spirulina,* are being used in dried forms as human food.

Most microbes cultivated as food sources are used primarily as animal feeds. For example, over 150,000 tons of yeast are harvested annually as fodder for the world's livestock. Microbes have yet to gain widespread acceptance as dietary supplements for humans. Many organisms are difficult to digest because of their cell walls or have a disagreeable appearance, smell, or taste. Some microbes produce endotoxins or other products that make them unsafe for use as food. Furthermore, most rapidly dividing cells contain high nucleic acid concentrations, which, when ingested in amounts that the body cannot effectively excrete, pose a threat to human health. Degradation of excess nucleic acids generates uric acid, a compound that forms kidney stones and accumulates in joints and tissues, causing symptoms of gout. Unfortunately, the cost of removing nucleic acids reduces the economic feasibility of these products.

Food Additives

Many vitamins, amino acids, nucleotides, and enzymes that are commercially valuable to the food industry are obtained in high concentrations from microbial cultures. The microbe's metabolic control mechanisms that prevent it from overproducing these compounds must be circumvented before large quantities of the desired by-product can be synthesized and harvested. Thus, it is desirable to obtain microbes that have lost their sensitivity to metabolic regulation. For example, some mutants of *Corynebacterium glutamicum* can produce more than 200 times the amount of lysine required for growth. The amino acid extracted from these cultures is used for supplementing lysine-deficient plant proteins in many animal feeds. More than 300,000 tons of glutamic acid, another amino acid manufactured commercially by microbes, is used each year, primarily as the flavor-enhancing salt, monosodium glutamate (MSG). Vitamin B_{12} is manufactured by commercial cultures of propionibacteria in amounts 50,000 times that required for its own growth. Other microbially produced vitamins include riboflavin and precursors of vitamins A and C. Microbes also produce huge quantities of nucleotides that have taste-enhancing properties.

Many enzymes instrumental in food production are also derived from microbes (Table 25-1). *Amylases* are used in the initial steps in several food fermentation processes to convert starch to fermentable sugars. They are also used to partially predigest foods for young children, to clarify fruit juices, and in the manufacture of corn and chocolate syrups. The enzyme *invertase,* obtained from *Saccharomyces cerevisiae,* prevents the crystallization of sugars by converting sucrose to the more soluble glucose and fructose. It is injected into some candies to liquefy the centers. Other fungal enzymes, *pectinases,* dissolve pectin, a naturally occurring thickening agent in ripe fruit. These enzymes are used to clarify fruit juices and allow them to be

TABLE 25-1

Enzymes of Microbial
Origin Used in Food
Production

Enzyme	Major Source	Reaction Catalyzed	Applications
Amylase	Aspergillus	Starch → sugar	Brewing, syrup production
Invertase	Saccharomyces cerevisiae	Sucrose → glucose + fructose	Manufacture of candies with liquid centers
Pectinase	Aspergillus	Pectin → sugars + galacturonic acid	Preparation of concentrated fruit juices
Rennin	Endothia, Mucor	Coagulation of casein	Production of cheese (helps form curd)
Proteases	Bacillus, Aspergillus	Protein hydrolysis	Tenderizing of meat
Diacetyl reductase	Enterobacter aerogenes	Diacetyl removal	Prevention of certain off-flavors in beer and fruit juices
Lactase	Kluyveromyces fragilis	Lactose → galactose + glucose	Reduction of lactose in milk (for people with lactose intolerance); prevention of lactose crystallization in ice cream
Naringinase	Aspergillus niger	Elimination of naringin	Removal of bitter taste from orange juice
Glucose oxidase	Aspergillus niger	Glucose → gluconic acid	Prevention of browning in dried eggs
Glucose isomerase	Bacillus, Arthrobacter	Glucose → fructose	Preparation of very sweet syrups

concentrated without solidifying, and to remove the gummy material around coffee beans prior to roasting. *Rennins* are enzymes derived from molds that are used to coagulate milk for cheese production. Not long ago, these curdling enzymes were obtained solely from the gastric juice of cattle and other animals.

INDUSTRIAL MICROBIOLOGY

Some microbes are metabolic factories that can manufacture many industrially important substances not readily or economically obtained from other sources. Industrial microbiologists maximize production of these materials by employing high-yielding strains and then manipulating their growth environment. For example, substrates that induce the formation of the desired enzymes may be included in the medium. Feedback inhibition can be overcome by provoking rapid excretion of a product before it can accumulate inside the cell and shut down its own production. In this way, detergents and other substances that cause nonlethal leaks in the cell membrane promote high yields of end products by reducing their intracellular concentrations. Ultraviolet irradiation and chemical mutagenesis help generate mutant strains that overproduce desired materials. The loss of regulation may be at the level of enzyme activity or enzyme synthesis (Fig. 25-2). The combination of

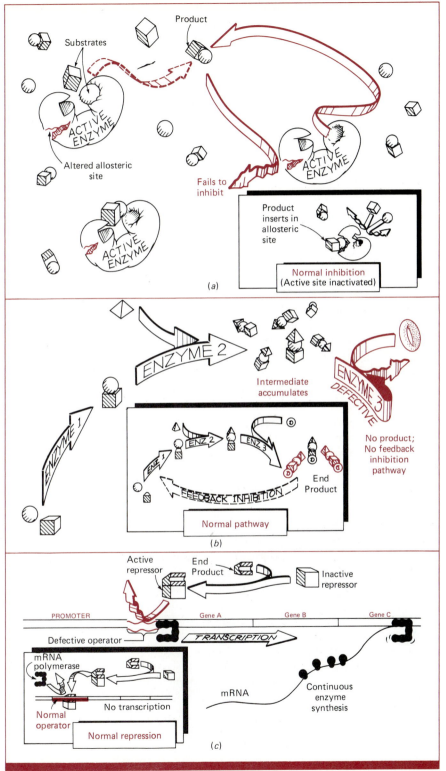

FIGURE 25-2

Genetic alterations
that lead to increased
production of a
specific compound.
(Normal regulation is
shown in the insets.)
(*a*) Enzyme, no longer
sensitive to feedback
inhibition, continues to
catalyze product formation
even when cytoplasm is
saturated with end
product. (*b*) Defective
enzyme, incapable of
converting the desired
compound to the next
chemical , blocks the
normal metabolic pathway;
the result is accumulation
of the intermediate.
(*c*) Operons in which the
repressor protein can no
longer bind to the operator
continuously synthesize
enzymes. The defect can be
in the operator (as shown
here) or in the repressor
protein.

669

mutagenesis and environmental manipulations to improve product yield has been exploited with great success, as evidenced by the 10,000-fold increase in today's penicillin production over that of Fleming's original strain.

Genetic recombination technology has created new frontiers in industrial microbiology. Recombination can occur following *fusion* of genetically distinct protoplasts to form a single hybrid cell with combined characteristics. This technique is valuable for incorporating into a single organism desirable traits that ordinarily occur in separate cells—for example, fusing a cell that produces high yields with one that grows rapidly. Recombinant DNA technology has unleashed the potential for creating organisms with even greater production capabilities. For example, *gene amplification* increases the number of copies of a gene in the cell, thereby increasing the amount of the corresponding product. Productivity is also increased by manipulating the operon that regulates a cluster of genes—for example, inserting structural genes and a promoter (the site to which messenger RNA polymerase binds) directly adjacent to each other in a plasmid. With the operator region eliminated, structural genes cannot be repressed, and huge quantities of enzymes are produced.

Microbes can also be turned into factories that manufacture useful substances they would not naturally synthesize. Plasmids may be used to introduce DNA from virtually any organism into bacteria or yeast (Fig. 25-3). The acquired genes then direct their new host to synthesize the corresponding product.

Three categories of microbial products are of industrial importance: (1) primary metabolites, (2) secondary metabolites, and (3) enzymes. **Primary metabolites** are compounds that are either intermediate or end products of biochemical pathways essential for the growth of the microbe. Sugars, amino acids, vitamins, nucleotides, and some organic acids and alcohols are examples of primary metabolites. They are usually produced in greatest abundance during logarithmic growth. **Secondary metabolites**, on

670

MICROBIAL
BIOTECHNOLOGY

FIGURE 25-3

Plasmids, such as this one from *Escherichia coli*, can be cleaved with endonucleases, joined with foreign genes, and introduced into bacteria that subsequently manufacture the foreign gene product.

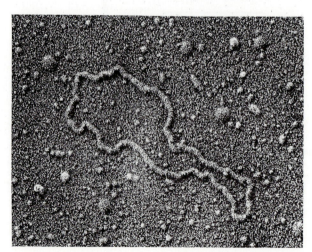

the other hand, are not essential for growth, and in many cases their role is unknown. Most secondary metabolites are low-molecular-weight compounds that accumulate during the stationary phase. Many antibiotics are secondary metabolites.

Industrial Fermentations

Some primary metabolites and enzymes of industrial importance that are produced by microbes are summarized in Table 25-2. Previously, the organic solvents glycerol, ethanol, acetone, and butanol were also derived from microbes, but they are currently produced more economically by chemical synthesis from petroleum products. However, microbes are becoming economically competitive with these chemical processes as the cost of petroleum substrates increases and as recombinant DNA technology produces microbes that use cheap organic wastes as substrates for solvent production.

Microbial fermentations manufacture more than 175,000 tons of organic acids annually throughout the world. Much of this is in the form of citric acid, a metabolic intermediate which accumulates when the Krebs cycle is disrupted. Citric acid is the principal acid added to jams, candies, fruit drinks, and other foods. It is also used to adjust the pH of cosmetic lotions and pharmaceutical products. Citric acid has metal binding properties that make it useful in shampoos, electroplating, leather tanning, and in clearing metal-clogged pipes in the oil industry. Itaconic acid is another compound derived from the Krebs cycle. Its major application is in the manufacture of plastics and synthetic fibers. Other organic acids are employed in the preparation of perfumes, plastics, dyes, and pharmaceuticals.

Dextrans and xanthans, the principal polysaccharides of industrial importance, are produced by bacteria as components of their capsules or slime layers. *Dextrans* have adsorbent properties that aid in chemical

TABLE 25-2
Some Industrial Products and Enzymes Obtained from Microbes

Product	Major Source	Application
Citric acid	*Aspergillus niger*	Food acid, cosmetics
Itaconic acid	*Aspergillus terreus*	Plastics
Dextrans	*Leuconostoc mesenteroides*	Blood plasma expander, adsorbent
Xanthans	*Xanthomonas campestrans*	Drilling muds, stabilizers and emulsifiers
Cellulose	*Acetobacter xylinum*	Filters, fiber production
Amino acids	*Corynebacterium glutamicum*	Food supplement and additive
Amylase	*Bacillus*	Textile industry, detergents
Proteases	*Bacillus, Streptomyces*	Detergents
Lipase	*Rhizopus, Saccharomycopsis*	Degreasing wool, digestive aid
Streptokinase	*Streptococcus*	Dissolution of blood clots
Uric oxidase	*Aspergillus*	Treatment of gout
Penicillin acylase	*Escherichia coli*	Production of semi-synthetic penicillin

separation and purification procedures. They are also used to expand the volume of blood plasma for transfusion. *Xanthans* are the most widely used microbial polysaccharide and are the only exopolysaccharide regarded by the Food and Drug Administration as a safe food additive. They are produced by *Xanthomonas campestrans,* a bacterium originally isolated from a rutabaga plant. The gelling and emulsifying properties of xanthans make them ideal solidifying agents in foods and pharmaceutical and cosmetic products, as well as stabilizers for suspending insoluble components of dairy products, salad dressings, and fruit drinks. Xanthans are also emulsifiers in paints and ceramic glazes, and thickeners in sauces and syrups, cosmetics, pharmaceuticals, textiles, and glue. When dissolved, xanthans produce highly viscous solutions that are used in the petroleum industry to lubricate drills, to counterbalance the pressure of oil in the well, or to recover otherwise unobtainable oil.

Another bacterium, *Acetobacter xylinum,* manufactures long fibers of cellulose. Currently most of the world's cellulosic textiles and paper come from cotton and trees. Although these sources are still abundant enough for our needs, commercial production of cellulose by *Acetobacter* may reduce the destruction of trees and allow the land currently used for cotton production to be planted with food crops.

Some microbial enzymes that can be harvested in high concentrations have industrial application, predominantly in the food industry. Enzymes may also be included in some commercial laundry detergents. These are proteinases that dissolve blood, mucus, chocolate, and other proteinaceous stains. Proteinases also soften and prepare hides for tanning. Lipases are used to degrease wool, amylases to reduce starch sizes in textiles. Amylases and lipases also serve as digestive aids in preparing foods for persons who produce deficient amounts of these enzymes.

Enzymes are employed to catalyze the production of several substances of industrial or medical value, thereby eliminating use of whole cells to accomplish these reactions. This approach has several advantages: it eliminates the need for growth medium; it reduces the likelihood of contaminants growing in the nutritionally deficient system; and it reduces subsequent steps needed for purification of the product. The efficiency of these processes may be augmented by attaching the enzyme to a solid carrier (Fig. 25-4). *Immobilized enzymes* remain in the reaction vessel as the medium flows through, carrying the substrate in and the products out. This approach is especially useful when the enzyme is very expensive or is available only in very low quantities.

Pharmaceutical Products

Each year over 100,000 tons of antibiotics are produced in the world, primarily penicillins, cephalosporins, and tetracyclines. These drugs are secondary metabolites, or their derivatives, that are synthesized by bacteria or fungi during the stationary phase of growth. (Chap. 14 details the source organisms for the major antibiotics.) Antibiotic-producing microbes are usually grown for several days in aerated batch cultures in huge tanks,

MICROBIAL
BIOTECHNOLOGY

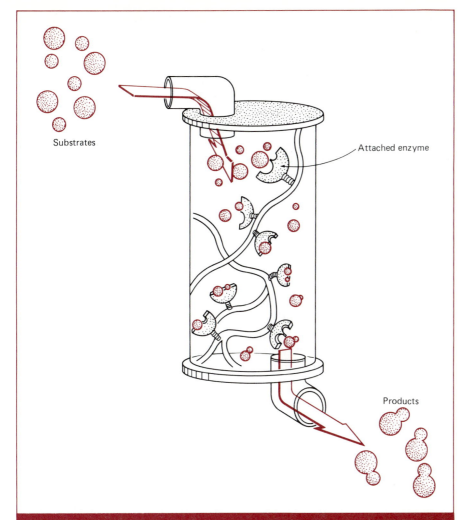

Substrates

Attached enzyme

Products

FIGURE 25-4

Immobilized enzymes
catalyze formation of
products in a
continuous-flow reaction
vessel. The active enzymes
remain attached to the
cellulose fibers, preventing
their loss in the effluent.

673

INDUSTRIAL
MICROBIOLOGY

some with volumes in excess of 100,000 liters. Under these conditions, penicillin production takes approximately 8 days. Conversion of penicillin G to semisynthetic derivatives usually requires a second microbial process. Penicillin G is incubated with bacteria that produce an enzyme that hydrolyzes the antibiotic to 6-aminopenicillanic acid, the core molecule needed to produce derivatives.

Microbial bioconversions are important in the manufacture of certain steroid hormones. For example, the use of microbes to synthesize the anti-inflammatory agent, cortisone, reduces the number of chemical reactions from the 37 needed for chemical synthesis to 11, some of which are chemical and some of which are microbial. By 1980, the aid of microbes reduced the cost of cortisone 400-fold. Other steroids produced by microbial transformation include hydrocortisone and prednisone.

The manufacture of vaccines depends on the ability to culture the organisms to obtain high yields of either the microbe itself or of specific immunogens—toxins, capsular material, ribosomes, or surface antigens. The development of vaccines against several pathogens has been hampered by the lack of technology for safe and inexpensive growth of these organisms. New vaccines that contain only the antigenic component(s) to which immunity is directed are being developed by genetic engineering of microbes (see explanation below).

Products of Genetic Engineering

Recombinant DNA technology has created "improved" microorganisms that are highly efficient at their natural metabolic activities as well as "superbugs" that are capable of doing things that they (or any other microorganism) had not done before. (This technology is discussed in Chap. 12.) Vaccine production is facilitated by extracting genes that direct the synthesis of surface antigens of virulent pathogens and transferring them to harmless bacteria. Transplanting a viral gene for capsomere production may create a microbe that pumps out large quantities of a single viral antigen. Pure preparations of the immunogen may be readily harvested from the growth medium for use as a vaccine that carries no risk of infection by viable pathogens or of deleterious side effects from noncritical viral antigens, endotoxins, or other extraneous materials.

Bacteria and yeasts have been genetically manipulated to produce human growth hormones, interferon, insulin, and pure viral antigens for vaccines. Practical quantities of these products are virtually unattainable from their natural sources. For example, it would take 100 sheep brains to obtain the amount of somatostatin (a growth-regulating hormone) produced in a 2-liter culture of an engineered *Escherichia coli*. Large-scale production of interferon for clinical studies is also made possible by genetically manipulated microbes that have acquired human interferon genes. Products of microbial origin are often safer than those derived from traditional sources. Human insulin produced by microbes, for example, does not cause the allergic reaction associated with commercial insulin obtained from its traditional source, cows and pigs.

Application of recombinant DNA techniques is limited by the accessibility of genes. Considering the tremendous amount of DNA in a human cell (1000 times that found in *Escherichia coli*), the likelihood of being able to isolate any given gene seems remote at best. The use of synthetic genes may soon overcome this limitation. The nucleotide sequence of a gene can be deduced by determining the amino acid sequence of the corresponding protein. Once this sequence is known, it may be possible to chemically synthesize the gene in vitro. This approach has been greatly helped by the invention of "gene machines" controlled by computers hooked up to four reservoirs, each containing a different deoxyribonucleotide. The nucleotides are sequentially added to the reaction tube and assembled on the growing end of the synthetic DNA chain. Gene synthesis is accelerated more than 300 times by the use of such machines, theoretically providing a practical source of virtually any gene.

AGRICULTURAL MICROBIOLOGY

Soil is a complex of inorganic and organic compounds. Inorganic compounds arise from the hard mineral layers that lie below the earth's surface. As the mineral layer gets closer to the surface, it mixes with organic matter, primarily from decomposing plants, to form topsoil, the fertile zone that supports terrestrial life. The organic matter, called humus, softens the soil, increases its porosity so it holds moisture better, makes it more permeable to penetration by plant roots, and provides the matrix and many nutrients for the vast array of microorganisms that help maintain the earth's growth cycle.

Each gram of organically rich soil may contain more than 2 billion bacteria. These microbes, along with saprophytic fungi, have several roles in soil fertility. They decompose dead organisms, freeing their molecular constituents for recycling by plants (see Microbial Activities, Chap. 1). Other soil organisms help plants by enriching the soil with essential nutrients and enhancing their ability to obtain water.

Biofertilization

All plants, indeed all organisms, depend on a source of nitrogen. This essential nutrient must be provided to plants in a reduced form (Fig. 25-5). Molecular nitrogen in the atmosphere is converted to biologically usable forms by nitrogen-fixing microorganisms, by natural processes such as lightning discharge, and by industrial production of synthetic fertilizers. Although chemical nitrogenous fertilizers have revolutionized world agriculture, their production depends on an expensive and vanishing resource, petroleum. In the United States farmers spend over a billion dollars each year on nitrogenous fertilizers for the corn crop alone. Scientists are attempting to manipulate the relationship between plants and nitrogen-fixing microorganisms to reduce our dependence on chemical fertilizers.

The problem has been partially solved by nature. Leguminous plants grow well in nitrogen-poor soil, as though manufacturing their own fertilizers. Actually, these plants form root nodules that contain populations of *Rhizobium,* a nitrogen-fixing bacterium. The more root nodules that are formed by the plant, the greater the crop yield. The symbiotic relationship is initiated when bacteria in the soil attach to root hairs of the plant (Fig. 25-6). This highly specific attachment process is mediated by plant proteins called *lectins* that bind the bacteria to the surface of the root hair, which is then penetrated by the microbes. The infected root hair provides the bacteria access to plant cells inside the root itself. These infected root cells divide and form a nitrogen-fixing nodule.

This association has been exploited since the time of the Roman Empire to enhance agricultural yields. Fields to be planted for the first time with legumes were seeded with soil from fields where these plants had already been grown. This provided an inoculum of *Rhizobium* for nodulation of the plants in the new field and subsequent nitrogen fixation. (The Romans, of course, were aware of neither the microbiological nor the chemical explanations for their agricultural success.) Nitrogen fixation by

FIGURE 25-5

The nitrogen cycle. The vast reservoir of molecular nitrogen in the atmosphere is available to plants and animals (and most microbes) only after it has been incorporated into its reduced forms (organic nitrogen, ammonia, and nitrate). Usable nitrogen is constantly being lost by the action of denitrifying bacteria and must be constantly replaced by nitrogen fixation for life to continue. Nitrogen is fixed from the atmosphere by bacteria such as *Rhizobium* and *Azotobacter*, in the heterocysts of cyanobacteria, by discharges of lightning, and by industrial manufacture of synthetic fertilizers.

676

MICROBIAL
BIOTECHNOLOGY

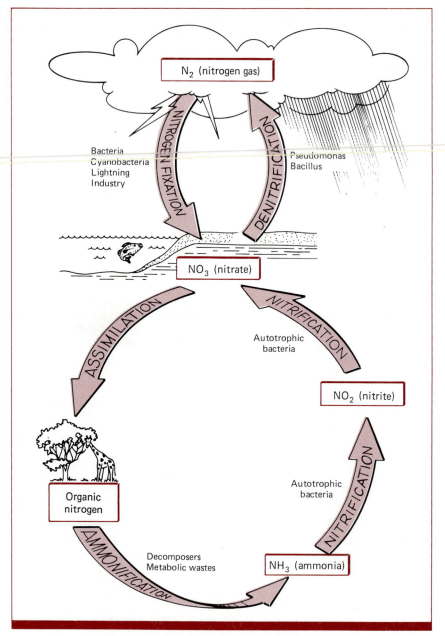

legumes also forms the basis for crop rotation to restore soil fertility. Nitrogen left in the soil by a single crop of soybeans, peanuts, peas, alfalfa, or even clover (these are all legumes) can be absorbed and utilized by subsequent crops of nonleguminous grain plants.

Several modern approaches seek to improve on the natural nitrogen-fixation process. Mutant strains of *Rhizobium* have been developed with superior nitrogen-fixing capacities. In the laboratory, these strains improve

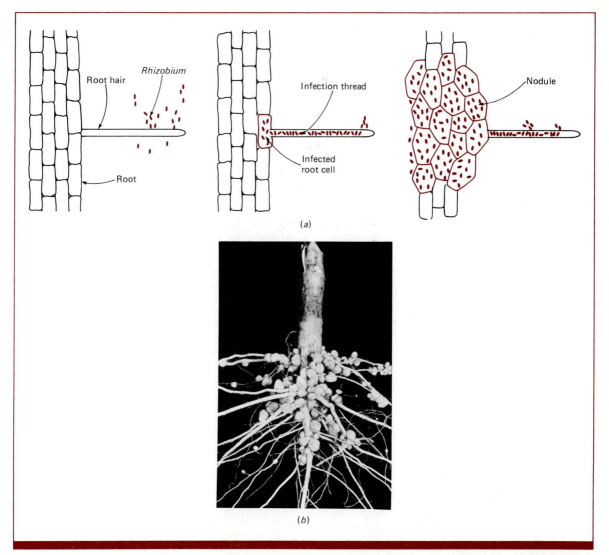

FIGURE 25-6

Symbiotic relationship between the nitrogen-fixing bacterium *Rhizobium* and leguminous plant. *(a)* The infection process is initiated by attachment of bacteria in soil to root hairs. The attachment is specific; recognition is facilitated by lectins produced by the plant. Bacteria penetrate the root hair and enter the root through an "infection thread." They proliferate in the cells of the root, causing them to swell and form a nodule. *(b)* Root nodules on the legume.

soybean yields by 50 percent. (At the time of this writing, however, this mutant had failed to flourish in field tests.) Scientists are also attempting to persuade *Azotobacter* (a free-living nitrogen-fixing bacterium) to stick tightly to the roots of corn, a nonleguminous plant, thereby freeing corn of its requirement for nitrogenous fertilizer.

The most sophisticated approach to biofertilization is to create plants that possess the genetic capacity for nitrogen fixation. Several groups of scientists are attempting to transfer genes for nitrogen fixation from bacteria to plants. The most promising source of these genes is the free-living bacterium *Klebsiella pneumoniae*, which contains all 17 nitrogen-fixation (*nif*) genes in a single large plasmid. The plasmid has been cloned and transferred to the yeast *Saccharomyces cerevisiae*. Although the introduction of genes from a procaryote into a eucaryote is a substantial accomplishment, the newly acquired genes were not expressed in the eucaryotic recipient, and the *Saccharomyces* remained dependent on an external source of usable nitrogen. This problem may be overcome by combining the *nif* genes with different cloning vehicles (see color box).

Another problem that will delay or perhaps prohibit the development of nitrogen-fixing plants is the extreme oxygen-sensitivity of *nitrogenase*, the key enzyme in reducing diatomic nitrogen (N_2) to ammonia (NH_3). Molecular oxygen generated by photosynthesis could inactivate the enzyme before usable levels of nitrogen have been fixed. Engineering a nitrogen-fixing plant therefore depends on the prior development of a mechanism for protecting nitrogenase from oxidation.

Microbial Inoculants That Increase Crop Yield

Plants may also benefit from association with several types of non-nitrogen-fixing bacteria and fungi. Certain strains of *Pseudomonas putida* and *Pseudomonas fluorescens* when inoculated into soil can promote growth of crop plants. These bacteria attach to root hairs and secrete extracellular substances called *siderophores*, compounds that tightly bind iron from the soil. This suppresses the growth of other microorganisms near the roots by depriving them of this essential nutrient. Siderophore production therefore reduces the number of fungi and soil bacteria that are either directly harmful to the root or that compete with the plant for other essential nutrients and water. The plant gets more of what it needs, resulting in higher yields.

Crop yields may also be increased by inoculating either soil or plants

A TAXI FOR FOREIGN GENES

■ Foreign genes are more readily transferred into and expressed by plant cells when the DNA is incorporated in a vehicle that not only introduces the nucleic acid into the recipient cell, but facilitates the stable integration of the newly acquired genes into the chromosome. One such method exploits the natural infection process of the plant pathogen *Agrobacterium tumefaciens*. Normally this bacterium infects plants through damaged tissue and causes crown gall disease. The bacterium contains a plasmid, a portion of which is inserted into the infected plant's chromosome. This integrated plasmid segment, called **transfer DNA (tDNA)**, triggers rapid growth of infected cells, forming a tumor called a crown gall. At least some of the tDNA is transcribed by the plant cell. Using recombinant DNA technology, foreign genes (nif genes, for example) may be inserted into the tDNA of *Agrobacterium tumefaciens*. The resulting recombinant plasmid might be readily incorporated into the cells of plants infected with *A. tumefaciens*. Cells containing the integrated plasmid and presumably the foreign genes could be isolated from crown gall tumors and cultured. Theoretically an entire plant may be regenerated from one of these genetically modified cells. Succeeding gene-rations would acquire the new trait by normal inheritance. (See figure.)

This approach has been used successfully to transfer genes from bean plants to sunflowers, creating a genetically engineered hybrid called the "sunbean." Unfortunately, the sunbean was covered with tumors that prevented the plants from growing to full size. The search for gene taxis continues.

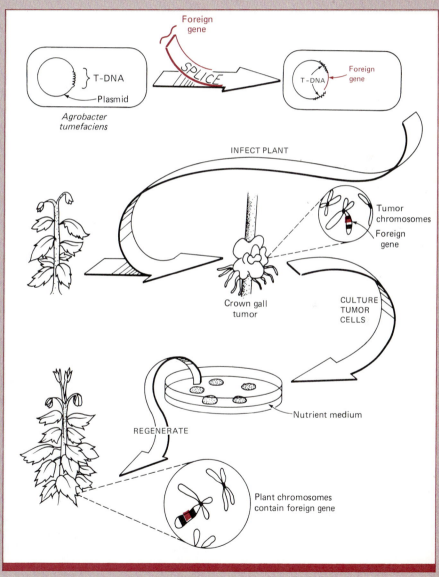

Foreign gene

SPLICE

T-DNA

Plasmid

Agrobacter tumefaciens

T-DNA

Foreign gene

INFECT PLANT

Tumor chromosomes

Foreign gene

Crown gall tumor

CULTURE TUMOR CELLS

Nutrient medium

REGENERATE

Plant chromosomes contain foreign gene

themselves with fungi that form **mycorrhizae**, literally, "fungus roots." These fungi colonize the fine root hairs and functionally increase the area of interface between plant roots and soil. Mycorrhiza formation facilitates more efficient uptake of water and nutrients, thereby increasing plant vigor, growth rates, and maximum yields. The beneficial effect is especially pronounced in marginally fertile soil where nutrients and water are scarce (Fig. 25-7). The symbiotic relationship is gaining in importance as the cost of irrigating and supplementing soil with synthetic fertilizers increases. Mycorrhizae associations reduce dependency on these supplements and may help convert otherwise marginal land into agriculturally productive fields.

Biological Control of Agricultural Pests

The cost of insect damage to crops is staggering. Insects substantially increase world starvation by annually consuming nearly one-third of the human food supplies on earth. These pests have traditionally been fought with synthetic organic chemical insecticides. Many of these insecticides have long-term toxicity, pollute the environment, and accumulate in the food chain. Their lethal effect is also fairly nonspecific, killing beneficial insects and spiders, predators that naturally control the pest population. In

FIGURE 25-7
Growth of plants augmented by association with the mycorrhiza-forming fungus. The effect is most pronounced in nutrient-poor soil (left). Colonies of mycorrhizae may extend into regions of the soil not accessible to the root system, thereby functionally increasing the volume of soil tapped by the plant.

(a)

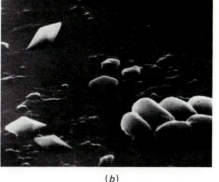
(b)

FIGURE 25-8
(a) This cabbage looper larva shows symptoms of *Bacillus thuringiensis* poisoning and will soon die. (b) Crystals of the fatal poison produced by the bacterium during sporulation.

a predator-free environment, pesticide-resistant insects quickly repopulate to overwhelming proportions that result in even greater crop damage than would have occurred without the use of the pesticide.

Several insect-specific pathogens are being produced commercially for use as **microbial pesticides**. Microbial pesticides have several advantages over chemical insecticides. They are specific for a small number of species of insects, sparing plants, animals, and many beneficial arthropods. They are also much cheaper than organic pesticides, which rely on petroleum for their production. In addition, some of these pesticides are self-perpetuating in the field. *Baculoviruses*, for example, are spread throughout the insect population by inoculating male insects with the virus (the males are trapped by using sex attractants, called pheromones, extracted from female insects). The inoculated males are then released. The virus does not kill immediately. The male insects have time to mate with and inoculate females with the virus. Eggs laid by infected females are also infected. Soon all members of the insect population have received a lethal dose of baculovirus.

During endospore formation, *Bacillus thuringiensis* produces toxic crystals that make it an excellent pesticide. Cultures of sporulated bacteria are dried and dusted onto crop plants. When ingested by gypsy moth larvae, cabbage worms, or tent caterpillars, the toxic protein lethally damages the insects' intestinal tracts (Fig. 25-8). Most other insects, plants, and animals are unaffected by the protein.

Now that the safety of many of these agents, as well as their profit potential, has been demonstrated, such easy-to-produce pesticides will likely play a role in reducing world starvation.

WATER TREATMENT AND WASTE DISPOSAL

One of the most significant uses for microorganisms is the destruction of hazardous or undesirable wastes, especially in water that has been used for domestic purposes. *Sewage* includes all such wastewater, plus storm runoff and water discharged from industry. Sewage also contains solids from these various sources as well as dissolved organic compounds that impart an offensive odor. The potential hazard of raw (untreated) sewage is due primarily to the presence of waterborne pathogens discharged from human intestinal tracts. Development of technologies that prevent ingestion or personal utilization of water containing these pathogens has been a dominant influence in the industrialization and economic growth of the world's leading societies. In developing countries, however, water treatment and sewage disposal is largely uncontrolled. According to a 1980 study by the World Health Organization (WHO), at least 30,000 people die *every day* in these countries because of unsanitary water supplies.

Ideally, water treatment technology has two facets: (1) treatment of public water supplies and (2) processing of raw sewage (Fig. 25-9).

Water Treatment

Water is often badly polluted by the time it collects in reservoirs. It may contain human and animal excrement, industrial wastes, and a variety of potential pathogens and noxious chemicals. Suspended particulate materials tend to increase water turbidity (cloudiness). The larger particles settle out when water is allowed to sit undisturbed in a holding reservoir. The colloidal solids, however, remain suspended in the water unless precipitated with a chemical, such as alum (aluminum potassium sulfate), that complexes the particles into *floccules,* large aggregates that quickly settle to the bottom of the tank. Flocculation also removes many viruses and bacteria. The clarified water is then *filtered* through sand or activated charcoal, a process that physically removes most bacteria, protozoa, and other cellular microbes, as well as many viruses. Poliovirus and other small viruses, however, may escape filtration. The addition of *chlorine* is the final step in assuring a safe public water supply. Even in low concentrations, chlorine inactivates most of the pathogens in water. (Some protozoan cysts survive chlorination, but they are readily removed from water by filtration.) Although in higher concentrations chlorine and its by-products are deleterious to people and the environment, it is unlikely that the amounts used to render public water safe exceed the danger level.

Sewage Disposal

There are several approaches to disposal of domestic wastewater. One is to return the raw sewage directly to the environment, burying it or dumping it into oceans, lakes, or rivers. Microorganisms, the decomposers, eventually digest the organic load and completely *mineralize* it, that is, convert it to the inorganic state. This is simply completion of the biogeochemical cycle. Decomposers also eliminate the pathogens in sewage by outcompeting them, rendering the material noninfective. After a period of time, the waste is safely free of pathogens, noxious chemicals, and offensive odors. Unfortunately, this process requires so much time that sewage-contaminated water will likely be used before the pathogens are eliminated, resulting in epidemics of serious waterborne disease. The natural decomposition cycle is not rapid enough to keep up with the volumes of sewage generated by communities, especially large municipalities.

MICROBIAL
BIOTECHNOLOGY

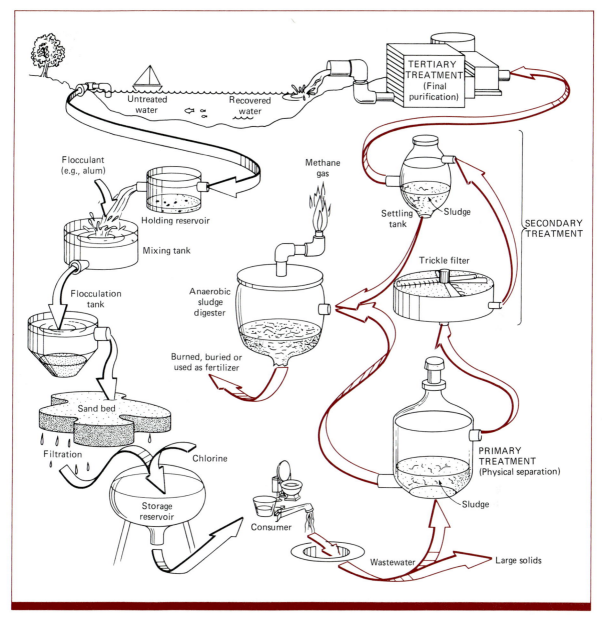

FIGURE 25-9

Modern approach to maintaining the safety of water. Treatment of municipal water supply is shown with black arrows. The steps in wastewater (sewage) treatment (colored arrows) convert sewage back to clean water. (See text for explanation.)

Families may safely dispose of their sewage in *septic tanks*. These are containers for storing sewage while solids settle out and decomposers break down organic compounds. Complete mineralization occurs only in the presence of molecular oxygen. The biological activity in the tank, however, quickly consumes oxygen and produces anaerobic conditions. Thus, decomposition is incomplete and the liquid discharged from septic tanks contains solublized organic compounds. In most systems, the liquid leaving the tank trickles along pipes or troughs lined with gravel. In this way, the **effluent** (discharged liquid) is aerated, and mineralization is completed by the decomposers attached to the surfaces over which the liquid flows. The treated effluent is discharged into the surrounding soil.

Larger communities and cities in the United States rely on large-scale sewage treatment plants for disposal of wastewater (see colored portion of Fig. 25-9). In these systems, removal of solids and microbial decomposition is speeded up to keep pace with the huge volumes of wastewater generated. Sewage is first subjected to *primary treatment*. Large solids are screened or skimmed off for burning and burying. The liquid is then transferred to a sedimentation tank where more solids settle out to form **sludge**. Primary treatment is therefore more a process of physical separation than one of microbial decomposition.

Both the effluent from the primary treatment and the sludge contain potentially dangerous pathogens and a high **biochemical oxygen demand (BOD)**. The BOD is a way of expressing the amount of organic compounds in sewage as measured by the volume of oxygen required by bacteria to metabolize it. The more dissolved organic matter in sewage, the more oxygen will be utilized by bacteria to mineralize it. Dumping sewage that contains high BOD elevates the concentration of soluble organic compounds in environments where it is discarded. Digestion of these organic compounds in natural ecosystems, such as lakes, can deplete available oxygen and result in asphyxiation of fish. This type of environmental disruption is minimized by treating sewage to reduce the BOD prior to its return to the environment. Primary treatment removes approximately one-fourth of the BOD from sewage.

Secondary treatment is mostly a biological process that depends on complete aeration of the system. The BOD of sewage may be reduced by another 80 to 90 percent in oxygen excess during secondary treatment. One of two systems is usually employed to saturate the sewage with oxygen. The most common is called a *trickle filter* (although no filtering actually occurs). Slowly moving sprinkler arms trickle the effluent from the settling tank over a bed of gravel. As the sewage seeps through the gravel, microbes attached to the rocks rapidly oxidize the dissolved organic compounds and much of the suspended solid material. The trickling action aerates the liquid and circulates the sewage. The effluent is then collected in a sedimentation tank for separation of remaining solids. The *activated sludge system* is slightly more efficient than the trickle-filter system. In this system, the effluent from the primary system is held in a tank, and air is vigorously pumped through the liquid. Resident microbes in the sewage decompose much of the soluble organic material and suspended solids. The sewage is then transferred to a settling tank for separation of liquid from the sludge. About a fifth of the sludge is recycled to the aeration tank as a starter inoculum for the incoming sewage.

Sludge from both primary and secondary sedimentation tanks is digested by anaerobic microorganisms. This process decomposes much of the organic material to methane, a flammable gas that is often recovered and used as fuel in the sewage facility. The remaining undigested (stabilized) sludge presents the most serious disposal problem. The 20,000 wastewater treatment plants in the United States generate 5 to 10 million dry tons of sludge each year. This material is usually burned or buried, but some is also dried and applied to soil as a fertilizer. The public safety of this practice is being evaluated, since a few viruses may remain following anaerobic digestion.

By the end of secondary treatment, the infectious potential of sewage is virtually eliminated. Pathogens, indeed the majority of microorganisms from the human alimentary tract, are poorly adapted for growth outside the body, especially when confronted by a vast number of microbes that are perfectly suited for exploiting the rich resources in sewage. These harmless microbes will quickly prevail at the expense of the pathogens. Before long the predominant organisms released in feces are completely replaced by the saprophytes favored by natural selection in sewage.

Although the products of secondary treatment may be safely discarded, too much organic material remains in the effluent for its direct use by people. In addition, nitrogen and phosphorus are poorly removed by primary and secondary treatments, and elevated concentrations of these compounds can be detrimental to the environment (see Algae and Human Disease, Chap. 8). Secondary effluent is being used in some communities to irrigate crops. The nitrogen and phosphorous compounds fertilize the plants while environmental microbes in the soil digest chemical contaminants. Some modern sewage facilities, however, further process wastewater by *tertiary treatment* to completely remove the BOD and all remaining nitrogen and phosphorous contaminants. This stage of processing is a combination of physical and chemical treatments that yields water safe for drinking. After tertiary treatment, water may be pumped directly into the public reservoir for recycling or returned to the environment with no significant disturbance. The process is extremely expensive, however, and most often the effluent from secondary processing is returned to the environment, where natural processes accomplish the equivalent of tertiary treatment.

Some communities utilize high levels of chlorine to oxidize BOD and eliminate pathogens from sewage. (This is much diferent from using chlorine to treat public water supplies.) The potential hazards of this approach are discussed in the color box.

Evaluating Water Quality

Water that has been contaminated by feces may contain virulent pathogens and transmit serious diseases. Fecal contamination is most commonly detected by assaying water for the presence of fecal coliforms, predominantly *Escherichia coli*. The only natural source of these bacteria is the intestines of humans and other mammals. Although coliforms are usually not pathogenic when ingested by healthy persons, their presence in water indicates fecal contamination and thus the possible presence of waterborne pathogens. In the United States, water is considered safe for drinking only if it contains fewer than four fecal coliforms per 100 milliliters, as deter-

MISUSE OF CHLORINATION: A NEW ENVIRONMENTAL HAZARD

■ Before chlorination of drinking water became a routine practice in most urban centers, epidemics of typhoid, dysentery, and other diseases acquired by contaminated water were common in the United States. New Jersey pioneered the use of chlorine to disinfect a public water supply in 1908. Other states soon followed suit, dramatically reducing the incidence of waterborne diseases. The number of persons dying each year from typhoid, for example, dropped from over seventy in 1908 to five by 1930. Undoubtedly water chlorination, coupled with other treatments, has contributed to public health. Chlorine also oxidizes organic compounds and removes objectionable tastes and odors. Seventy-five years of experience with using low-level chlorine to disinfect municipal water supplies have proved the safety of this approach.

Unfortunately, chlorine is being used indiscriminately, with increasing frequency, in a practice that magnifies its inherent hazards. This practice is the use of high levels of chlorine to treat wastewater discharged from municipal sewage treatment plants. This use of chlorine gained popularity because it oxidizes the heavy BOD and odors of wastewater much more quickly

and inexpensively than traditional microbial treatment. Unlike the low concentrations needed to prevent the spread of pathogens in drinking water, however, much higher levels of chlorine must be added to wastewater because of the water's high organic load. Much of this chlorine is discharged into the environment, where it may react with several chemical impurities in human excreta, fertilizers, industrial waste, and other substances common in sewage. Some of these chlorinated by-products are health hazards to persons who consume the water, even after it has been subsequently processed by water treatment plants. For example, one group of by-products, the chloramines, produces hemolytic anemia in uremic patients undergoing kidney dialysis. Because these chlorinated organic derivatives are more hazardous than the BOD, the chlorination of wastewater to reduce BOD is probably not justified. It is much safer (although more expensive) to depend on microbiological and physical treatments to decrease the organic load of wastewater. Unlike this practice, the use of low concentrations of chlorine in public water supplies to kill waterborne pathogens is a sound public health practice and should be continued.

mined by the membrane filter method (Fig. 25-10). Coliforms may also be detected by a multiple-tube fermentation test, a procedure that is detailed in water-microbiology laboratory manuals.

BIODEGRADATION OF ENVIRONMENTAL POLLUTANTS

Modern technology generates many organic materials that cannot be readily disposed of by natural biogeochemical cycles. Plastics, for example, are synthetic compounds that have appeared in the biosphere so recently that microbes have not had time to evolve the ability to decompose them. Many low-molecular-weight compounds, such as solvents, refrigerants, propellants, flame retardants, pesticides, and herbicides, also persist in the environment—a phenomenon called *recalcitrance*—because of the inability of natural soil and water microbes to metabolize them at significant rates. Many of these compounds are toxic to humans, and their endurance

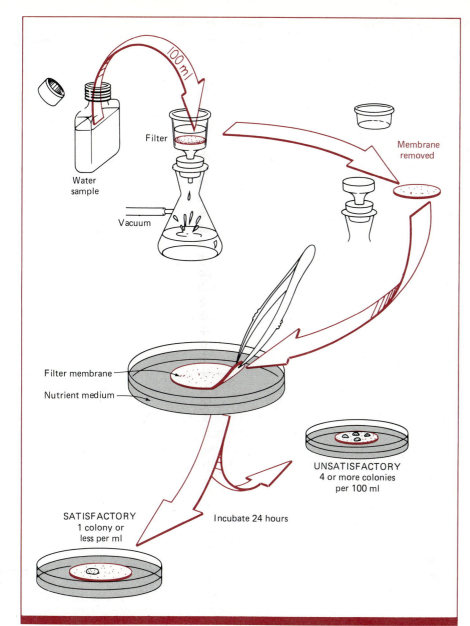

FIGURE 25-10

Detection of coliforms to determine water purity. A known quantitiy of water is passed through a sterile bacteriological membrane filter. The bacteria in the sample are thus collected on the membrane, which is then aseptically transferred to a differential nutrient medium and incubated. The number of coliform colonies that develop indicates the concentration in the sample. Acceptable water contains fewer than four colony-forming units per 100 ml.

Filter

Water sample

Vacuum

100 ml

Membrane removed

Filter membrane

Nutrient medium

UNSATISFACTORY
4 or more colonies
per 100 ml

SATISFACTORY
1 colony or
less per ml

Incubate 24 hours

687

BIODEGRADATION OF
ENVIRONMENTAL
POLLUTANTS

presents a hazard significant enough to require curtailment of their use. The pesticide DDT is one of the most notable examples of a toxic product that is banned because it is considered too environmentally persistent.

The large quantities of herbicides, pesticides, and other recalcitrant chemicals released into the environment have exerted enough selective pressure that organisms capable of metabolizing a few of these compounds have emerged. Scientists are using microbiological techniques to speed up this slow evolutionary process. One approach is a continuous-culture enrichment technique that starts with an organism that can utilize a naturally occurring compound that is molecularly related to the nonbiodegradable synthetic chemical. The microbiologist grows the microbe in continuous culture, gradually substituting compounds with increasing similarity to the target substance. Other investigators are attempting to genetically engineer bacteria that can convert recalcitrant pollutants, especially those in industrial wastes, to useful products.

Petroleum is another environmental pollutant that, in nature, is poorly degraded. Although no single organism degrades all the hydrocarbons in petroleum, degradation by mixed cultures has been accomplished. Hydrocarbon-digesting bacteria are seeded into activated sludge to degrade refined oil in sewage. Microbes are also routinely used by the United States Navy for degreasing tanks of ships; this procedure was adopted after the successful degreasing of the Queen Mary in 1967. When this ship was docked in Long Beach harbor, it contained 800,000 gallons of oily wastewater. About 150 pounds of dried microbes were poured into the bilge compartments. Six weeks later the water was pure enough to discharge into the harbor. Petroleum-digesting bacteria that have been created by genetic manipulation are being evaluated as tools for quickly eliminating oil spills and reducing subsequent environmental damage. Recombinant DNA technology has produced plasmid-carrying strains of marine bacteria that rapidly metabolize crude oil even in cold ocean water where oil spills are likely to occur.

Microbial biodegradation is being used for many tasks that are traditionally done mechanically. For example, municipal sewer lines eventually become clogged with a buildup of insoluble materials on the inside of the pipe. In Washington, D.C., 3000 miles of sewer pipe were effectively cleared by drip-feeding bacterial cultures into manholes at the extremes of the system. The project cost about one-third that of mechanically removing the buildup, saving the city nearly 3 million dollars. The same process is also being used on a smaller scale to clean drainpipes, grease traps, and septic tanks.

MICROBES AS SUPPLIERS OF NATURAL RESOURCES

Fuel Production

Dependence on petroleum and other nonrenewable energy sources has led to a fuel crisis as the world runs out of its reserves. Available oil supplies will probably be exhausted within the next 50 years. Microbes may provide a partial solution to this problem by converting waste materials and sewage into usable fuel. Already methane gas released as a by-product of sewage treatment is harvested and used to generate electricity. Most of the possibilities for using microorganisms in future energy technology, however, are yet unexplored. Biotransformation of sunlight into usable fuel is perhaps the most exciting, since the source is inexhaustible. A number of photosynthetic microorganisms generate hydrogen gas, a clean fuel that, when burned, produces only one waste product, water. Another promising possibility is the use of microbes or microbial enzymes to generate electricity directly. In these systems electrons released by microbial oxidations are concentrated on one side of a membrane. If an external circuit is provided, these electrons automatically flow through it, producing electricity.

Another virtually inexhaustible source of energy is cellulose, the most abundant of the earth's biologically renewable resources. Each year the production of textiles, paper, and other materials manufactured from cellulose generates about 15 tons of products *per person* in the United States, of which more than 900 million metric tons ends up as refuse. Ethanol and methane are two fuels that can be collected as a by-product of biodegradation of these huge cellulose wastes. Tapping these resources would have the triple advantage of energy production, waste disposal, and simultaneous production of single-cell protein. Small-scale digesters may help reduce the demands on our limited reserves of fossil fuels. These systems can be used by individual families to produce usable energy from animal manure or from mixtures of sewage, refuse, and weeds.

Genetic engineering may eventually create an oil-producing bacterium; indeed, microbial hydrocarbon production has already been enhanced by genetic transformation. However, it seems likelier that microbes will first be used to aid in oil recovery processes in the near future. As of 1984, approximately two-thirds of our current known oil reserves are not recoverable. Microbes could theoretically be used to degrade the mineral complexes that prevent access to shale-oil deposits.

Biomining Techniques

For nearly 10 centuries, microorganisms have assisted in the recovery of copper dissolved in drainage water from mines. Only in modern times, however, has the role of bacteria in this process been discovered. In modern copper and uranium mines, metal from low-grade ore is dissolved in a leaching solution with the aid of microbes, predominantly bacteria in the genus *Thiobacillus*. These microbes are *chemolithotrophs,* literally, "rock eaters." They break down sulfide minerals to sulfate (SO_4^{2-}) and cause the conversion of copper in the ore to the cupric (Cu^{2+}) state. These ions spontaneously react and yield soluble copper sulfate ($CuSO_4$). Dissolving the copper out of ore occurs in huge leaching dumps (see color photo section). Solutions containing the microbes percolate down through millions of tons of ore and are collected in catch basins at the bottom. The solution is then processed so that the dissolved copper can be harvested in its pure elemental form. Microbial mining accounts for more than 10 percent of the copper produced in the United States. Advances in biomining technology may soon make it applicable to many other metals in addition to copper and uranium.

Microbes may also be used to extract metals from industrial wastewater, reducing environmental contamination while helping to recover valuable resources that would otherwise be discarded.

From its ancient roots, microbial biotechnology has mushroomed into one of the most exciting fields of human endeavor. Microbes are used to protect our food from spoilage organisms and to impart different aesthetic characteristics to edible substances. Milk and cream are converted into cheese, yogurt, sour cream, or buttermilk by bacteria and fungi. Vegetables are soaked in brine solutions, then fermented to produce ripe olives, sauerkraut, pickles, and other foods that are much more resistant to spoilage than the original, untreated vegetable. Soy sauce is produced by fermenting soybean mash with several organisms over a period of about 1 year.

Bread is leavened when carbon dioxide is released by *Saccharomyces cerevisiae*, fermenting glucose in the dough. This yeast, and others closely related to it, also ferment sugars in fruit juices and grain mash, producing wine, beer, and other alcoholic beverages. Microbes, in the form of single-cell protein (SCP), are potential sources of nutrition for protein-starved people in many developing nations. Currently SCP is used primarily as a livestock feed to increase meat and dairy production. Microbes are sources of vitamins, amino acids, nucleotides, and other food supplements, as well as of enzymes employed for a variety of purposes in food preparation.

Microorganisms are metabolic factories that produce many industrially valuable products. Production of these compounds is increased by selecting deregulated mutants that overproduce the desired product. Recombinant DNA technology generates strains that give greater yields or that manufacture useful substances not normally synthesized by microbes. The products of industrial fermentation are either primary metabolites, such as vitamins, organic acids, and alcohols, or such secondary metabolites as antibiotics. In addition to antibiotics, the pharmaceutical industry depends on microbes for the manufacture of vaccines and some steroid hormones—cortisone, for example. Improved vaccines that contain only those antigens against which immunity is directed may soon be manufactured by genetically engineered bacteria and yeast. Genetic engineering has already yielded microbes that synthesize several medically valuable substances (interferon and human growth hormone) that are difficult to obtain from traditional sources. The use of synthetic genes manufactured in vitro promises to contribute to the potential of genetic engineering.

Several properties of agriculturally important microbes are being explored to enhance crop yields and help cultivate marginally fertile soil. Free-living nitrogen-fixing microbes, as well as those living in root nodules of legumes, provide nitrogenous plant fertilizers. These bacteria are exploited to reduce the need for expensive chemical fertilizers. These approaches include crop rotation, genetic manipulation to enhance bacterial nitrogen fixation, and methods to encourage free-living nitrogen fixers to colonize the roots of nonleguminous plants. Scientists are also attempting to transfer the nif gene from bacteria to plant cells. Other microbes increase crop yields by inhibiting soil organisms that compete with the plant for water and nutrients. Still others form mycorrhizae (fungus roots) that increase the functional surface area of the plant's root system. Agriculture also benefits from biological control of insect pests by using insect pathogens as microbial pesticides.

Sewage and other wastes are disposed of by microbial digestion. Treatment of domestic wastewater (sewage) depends on microorganisms to mineralize the BOD and eliminate pathogens. This may be accomplished in septic tanks or by large-scale municipal sewage treatment plants.

Unlike wastewater, municipal water supplies are first subjected to mechanical treatment to clarify the water and filter out impurities, then treated with chlorine to kill pathogens. The purity of drinking water is evaluated by testing for the presence of coliforms as evidence of fecal contamination.

Recalcitrant pollutants are poorly degraded by common environmental bacteria and fungi. Microbes capable of efficiently decomposing these compounds are being developed by continuous-culture enrichment technique and genetic engineering. These techniques are also being used to obtain hydrocarbon-digesting bacteria to degrade unwanted petroleum contaminants and oil spills. Other bacteria may help alleviate the energy shortage by producing usable fuel. The most attractive of these technologies is one that converts industrial waste and garbage into combustible hydrocarbons. Bacteria already aid in recovering hard-to-obtain petroleum deposits trapped in shale. Other microbes are used to extract copper and uranium from low-grade ore.

KEY WORDS

single-cell protein (SCP)   microbial pesticide
primary metabolite   effluent
secondary metabolite   sludge
transfer DNA (tDNA)   biochemical oxygen demand (BOD)
mycorrhiza

REVIEW QUESTIONS

1. List four advantages of using single-cell proteins over traditional food sources. Discuss four problems that currently limit the use of single-cell proteins.

2. Describe how microbes participate in the production of the following:
   (a) Swiss cheese   (c) Light beer
   (b) Soy sauce   (d) Sauerkraut

3. How can you increase production of industrially valuable metabolic by-products that are normally regulated by (a) feedback inhibition and (b) end-product repression?

4. List five enzymes of microbial origin and their industrial application.

5. Distinguish between:
   (a) Primary and secondary treatment of wastewater.
   (b) Mycorrhizae and root nodules.

6. Describe three ways by which genetic engineering contributes to new uses of microorganisms.

7. What are the biotechnological applications of each of the following microbes?
   (a) *Rhizobium*
   (b) *Bacillus thuringiensis*
   (c) *Thiobacillus*
   (d) *Xanthomonas*

MICROBIAL
BIOTECHNOLOGY

Appendix 1
Bergey's Manual: Identification of an Unknown Bacterium

A. Bergey's Manual of Determinative Bacteriology[*]

This single volume is a classification of bacteria that contains descriptions of all species known at the time of its publication. It serves as a practical guide for the identification of bacteria which are assigned to one of 19 Parts based primarily on (a) energy and carbon source, (b) mode of locomotion, (c) morphology and Gram stain reaction, (d) gaseous requirements, and (e) ability to produce endospores. Each Part contains keys and tables describing additional criteria for the diagnosis of genus and species.

Part	Distinguishing Features
1. Phototrophic bacteria	Photosynthetic without the production of oxygen
2. Gliding bacteria	Capable of slow, gliding movement but lacking locomotor organelles
3. Sheathed bacteria	Covered with a sheath
4. Budding or appendaged bacteria	Reproduce by budding and/or produce appendages (other than flagella or pili)
5. Spirochetes	Flexible, coiled morphology; motility by axial filament
6. Spiral and curved bacteria	Rigid, coiled morphology
7. Gram-negative aerobic rods and cocci	Capable of growth only in presence of oxygen *a.* Rods motile by polar flagella *b.* Coccobacilli of uncertain affiliation
8. Gram-negative facultatively anaerobic rods	Capable of growth in presence or absence of oxygen *a.* Motile by peritrichous flagella *b.* Motile by polar flagella *c.* Uncertain affiliation

[*]*Bergey's Manual of Determinative Bacteriology*, 8th ed., Buchanan and Gibbons (Editors), Williams and Wilkins Co., Baltimore, 1974.

9.	Gram-negative anaerobic bacteria	Capable of growth only in absence of oxygen; nonspore formers; pleomorphic
10.	Gram-negative cocci and coccoba-cilli	Generally occurring as diplococci; non-flagellated
11.	Gram-negative anaerobic cooci	Capable of growth only in absence of oxygen
12.	Gram-negative chemolithotrophic bacteria	Derive energy from inorganic compounds
13.	Methane-producing bacteria	Obtain energy for growth via formation of methane; anaerobic
14.	Gram-positive cocci	Nonspore formers
15.	Endospore-forming rods and cocci	a. Obligate aerobes b. Obligate anaerobes
16.	Gram-positive asporogenous rods	Nonspore formers
17.	Actinomycetes and related organisms	Rods, pleomorphic rods, or filaments; some acid-fast
18.	Rickettsias	Obligate intracellular parasites of eucaryotic cells
19.	Mycoplasmas	Lack cell wall; highly pleomorphic

Using *Bergey's Manual*

When an unknown bacterium is isolated in the laboratory, it is usually identified by a combination of information from microscopic observations, cultural (growth) characteristics on agar and in broth, and biochemical tests. The following information describing an unknown bacterium can be readily obtained in most laboratories.

MORPHOLOGICAL
    Shape: coccus
    Arrangement: pairs, short chains, clusters
    Motility: nonmotile
    Gram stain: positive
    Spore stain: nonspore former
    Buds or sheaths: negative

CULTURAL
    Nutrition: heterotroph
    Agar colonies: golden yellow, glistening
    Gelatin stab: liquefaction
    Temperature: 30–37°C
    Oxygen tolerance: facultative anaerobe
    Novobiocin sensitivity: sensitive

BIOCHEMICAL TESTS
    Hydrogen sulfide: trace
    Voges-Proskauer test: positive
    Indole: negative
    Nitrate: reduced

Starch: not hydrolyzed
Catalase: positive
Coagulase: strongly positive
DNAase: positive
Glucose fermentation
     Aerobic: acid
     Anaerobic: acid
Mannitol fermentation
     Aerobic: acid
     Anaerobic: acid

We can now identify the unknown bacterium using *Bergey's Manual*. Parts 1 through 13 can be immediately eliminated because:

1. The unknown is heterotrophic not phototrophic (eliminate Part 1).

2. The organism does not move by gliding, it is not sheathed, and it does not bud or possess appendages (eliminate Parts 2, 3, and 4).

3. It is not spiral, curved, or filamentous (eliminate Parts 5, 6, and 17).

4. It is neither Gram-negative nor a Gram-positive rod (eliminate Parts 7, 8, 9, 10, 11, 12, 13, and 16).

5. It does not produce endospores (eliminate Part 15).

6. It is not an obligate intracellular parasite (eliminate Part 18).

7. It possesses a cell wall (eliminate Part 19).

BERGEY'S MANUAL OF
DETERMINATIVE
BACTERIOLOGY

Thus the organism must belong to Part 14. Three families are classified in this Part.

Family—Micrococcaceae
Cocci, packets, or clusters
Aerobic or facultatively anaerobic
May or may not produce acid from carbohydrates
Catalase positive

Family—Streptococcaceae
Cocci, pairs, chains, or tetrads
Facultatively anaerobic
Produce acid from carbohydrates
Catalase negative

Family—Peptococcaceae
Cocci, singly, pairs, tetrads, irregular masses, packets, and chains
Strictly anaerobic

Immediately, Streptococcaceae and Peptococcaceae are eliminated because the organism is not a strict anaerobe and it is catalase positive. This leaves the family Micrococcaceae, which contains three genera:

Genus—*Micrococcus*
Heterotrophs, strict aerobes, indole not produced

Genus—*Staphylococcus*
Heterotrophs, facultative anaerobes, acid from carbohydrates, indole not produced

Genus—*Planococcus*
Heterotrophs, strict aerobes, motile

The organism is neither a strict aerobe nor is it motile. Therefore, the unknown is in the genus *Staphylococcus*. In this genus, there are three species: *Staphylococcus aureus*, *Staphylococcus epidermidis*, and *Staphylococcus saprophyticus*. The unknown bacterium produces acid anaerobically from glucose and mannitol and is coagulase positive. These properties are characteristic of *Staphylococcus aureus*. In addition, most of the other characteristics of the unknown correspond well with those listed for *Staphylococcus aureus* in *Bergey's Manual*. Therefore the unknown is *Staphylococcus aureus*.

B. Bergey's Manual of Systematic Bacteriology*

The 1984 revision of *Bergey's Manual* is an expanded four-volume edition (with a new name), containing the most recent and exact procedures for organism identification including data reflecting genetic relationships. The additional information gathered since 1974 has resulted in reclassifying the bacteria from the 19 Parts and assigning related organisms to one of 30 sections.

Volume I (1984)
Section 1 The Spirochaetes
2 Aerobic/Microaerophilic, Motile, Helical/Vibroid Gram-Negative Bacteria
3 Non-Motile (Or Rarely Motile), Gram-Negative Curved Bacteria
4 Aerobic Gram-Negative Rods and Cocci
5 Facultatively Anaerobic Gram-Negative Rods
6 Anaerobic Gram-Negative Straight, Curved, and Helical Rods
7 Dissimilatory Sulfate- or Sulfur-Reducing Bacteria
8 Anaerobic Gram-Negative Cocci
9 The Rickettsias and Chlamydias
10 The Mycoplasmas
11 Endosymbionts

Volumes II, III, and IV (Tentative Outlines)
12 Gram-Positive Cocci
13 Endospore-Forming Gram-Positive Rods and Cocci
14 Regular, Non-Sporing, Gram-Positive Rods
15 Irregular, Non-Sporing, Gram-Positive Rods

BERGEY'S MANUAL:
IDENTIFICATION OF AN
UNKNOWN BACTERIUM

*Williams and Wilkins Co., Baltimore, 1984

697

BERGEY'S MANUAL OF
SYSTEMATIC
BACTERIOLOGY

Appendix 2
Atomic Theory

All forms of matter* are composed of small units called **atoms**. Each atom has a central nucleus that contains subatomic particles. These nuclear particles are **protons** (particles with a positive electric charge) and **neutrons** (particles with no electric charge). The atomic nucleus therefore carries a positive charge, the magnitude of which depends on its number of protons. Each proton carries one unit of charge.

Smaller, negatively charged particles, called electrons, orbit the nucleus at a speed that generates enough centrifugal force to counter the pull of the positively charged nucleus (opposite charges attract one another). Atoms are electrically neutral particles that have one electron for each proton.

Atoms that have gained or lost one or more electrons have either a positive or negative electric charge and are called **ions**.

The number of protons in an atom determines its **atomic number**, a value that is unique for each type of atom. It is the number of protons that distinguishes each type of atom from all other types and allows their classification as **elements**, substances composed of a single kind of atom. In other words, all atoms with the same atomic number are considered to be the same element. For example, any atom with eight protons is the element oxygen.

The highest atomic number for any known element is 106. In other words we have discovered the existence of just over 100 elements, only 92 of which occur naturally. The others were created by scientists using extraordinary laboratory procedures. Living tissue is built from fewer than one-fourth of the 106 elements, the most abundant of which are hydrogen, carbon, nitrogen, and oxygen.

The total number of protons and neutrons in an atom's nucleus constitute its **atomic weight** (also called "atomic mass"), with each proton and neutron arbitrarily assigned a value of 1. Because electrons weigh only

---

*With the exception of subatomic particles.

1/1840 as much as a single proton or neutron, their contribution is often ignored when discussing atomic weight.

Unlike atomic number, the identity of an element cannot always be deduced from its atomic weight. This is because the number of neutrons may vary in a single element, thereby changing the weight of the atom without altering its identity (since the number of protons remains unchanged.). These different forms of the same element are called **isotopes**. A common example is carbon-12 (^{12}C), the most commonly occurring form, and its other isotopes, ^{13}C, ^{14}C, and ^{15}C. Most elements have several isotopes.

Some isotopes are unstable and tend to decay, losing neutrons in the process (some may lose protons or electrons). Isotopes that discharge such particles are called *radioisotopes* (or radioactive isotopes), and their discharged particles often contain enormous amounts of energy.

The number of electrons possessed by an atom determines the chemical behavior of that element, especially its tendency to react with other atoms or **molecules** (particles composed of two or more atoms). Electrons spin around the nucleus, forming an electron **shell**. Each shell has a maximum number of electrons it can accept. When an inner shell becomes saturated, additional electrons occupy the next shell away from the nucleus.

The first shell holds a maximum of two electrons; the next two shells can accept eight each. Thus the oxygen atom depicted in Fig. A2-1 has its eight electrons distributed in two shells—two in the interior shell and six in the outermost layer. (See Table A2-1 for a description of electron distribution in several types of atoms.)

The number of electrons in the outer shell of an atom imparts to the element its characteristic chemical behavior, such as the tendency to lose

700

ATOMIC THEORY

Figure A2-1

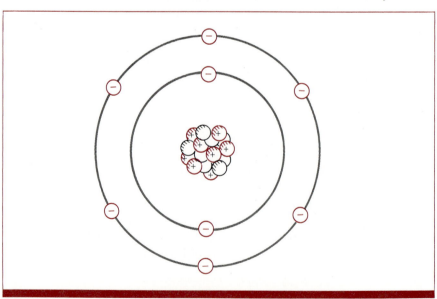

Element	Atomic Number	Number of Electrons in			Electrons in Outermost Shell	Number of Electrons to Complete Outermost Shell
		Shell 1	Shell 2	Shell 3		
Hydrogen	1	1	—	—	1	Gain or lose 1
Carbon	6	2	4	—	4	Gain 4
Nitrogen	7	2	5	—	5	Gain 3
Oxygen	8	2	6	—	6	Gain 2
Sodium	11	2	8	1	1	Lose 1
Magnesium	12	2	8	2	2	Lose 2
Phosphorus	15	2	8	5	5	Gain 3
Sulfur	16	2	8	6	6	Gain 2
Chlorine	17	2	8	7	7	Gain 1

electrons or gain them. This property of atoms is the underlying motive force that determines how an element interacts with other elements, in other words, its *chemical reactivity* (the subject of Appendix 3).

Appendix 3
Chemical Reactions

Most atoms tend to chemically react with other atoms and form larger, multiatomic units called **molecules.** These complexes may consist of two or more of the same elements, or they may be **compounds,** that is, molecules that contain different types of atoms.

Reactions follow a fundamental principle that determines whether atomic partnerships are created or destroyed, as well as the nature of the partnerships formed. This principle can be stated very simply: *All atoms are chemically "unsatisfied" until their outermost shell is filled with its maximum number of electrons.* Chlorine (atomic number 17) is just one electron shy of atomic fulfillment (how many electrons does it have in its outer shell?). Sodium, on the other hand, has only one electron in its outer shell. Sacrificing this extra electron would make its already saturated second shell the outermost orbital and create a chemically satisfied atom.

Elements join forces with each other to achieve their ideal states, forming partnerships of atoms held together by **chemical bonds.** This alliance may be an ionic bond or a covalent bond.

IONIC BONDS

When atoms that strongly attract electrons meet atoms that have only one or two electrons in their outer shell, the allure of the former usually proves irresistible. The electrons are kidnapped, a transaction that produces two electron-satisfied atoms, each with a saturated outer shell. Such a transaction, however, creates an electrical imbalance because the electron number no longer corresponds to the proton number. In other words, both atoms have acquired an electric charge and are now *ions.* (The electron appropriator has a negative charge, the electron donor has a positive charge.)

Oppositely charged ions attract one another in much the same way protons and electrons do. Two atoms that have entered into an electron transaction for their mutual satisfaction tend to be drawn together by their opposite electrical charges. If the attraction is strong enough, a chemical linkage called an ionic bond is formed (see Fig. A3-1).

703

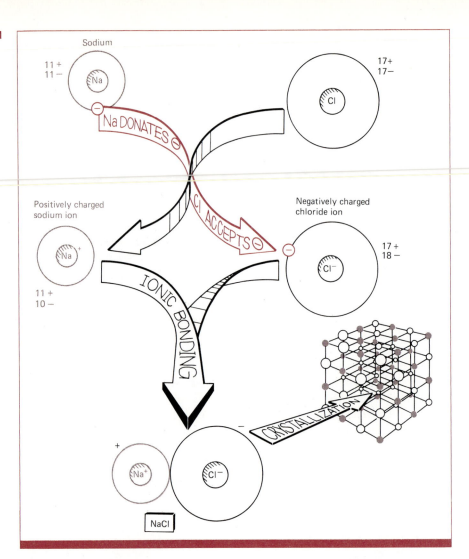

CHEMICAL REACTIONS

COVALENT BONDS

To achieve mutual electron satisfaction in their outer shells, some atoms form a **covalent bond** with each other. In covalent bonds, the outer shells of participating atoms *share* the same electrons, which spin alternately around each nucleus, as illustrated in Fig. A3-2.

Single covalent bonds join two atoms that share one pair of electrons, each atom donating an electron to the affiliation, as between the hydrogens and oxygen in water (H_2O; see Appendix 4). Oxygen can also satisfy its own requirement for two electrons by sharing two electron pairs with another oxygen, forming a *double bond*. Molecular nitrogen illustrates *triple bond* formation, each nitrogen in the duplex donating three electrons to be shared with the other atom of nitrogen.

No atom can form a quadruple bond with another atom—three is the maximum number of linkages between any two atoms. Carbon, with its

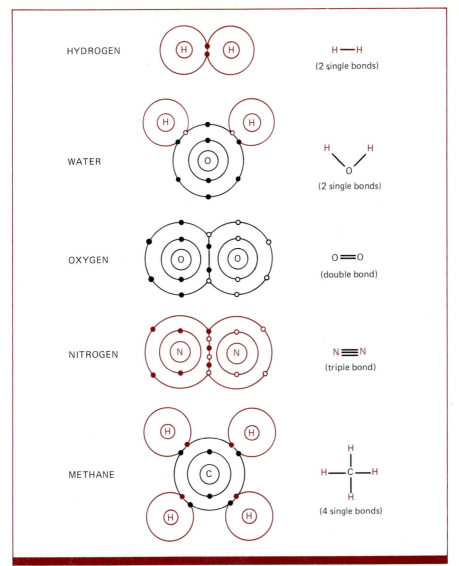

HYDROGEN — H —H (2 single bonds)

WATER — (2 single bonds)

OXYGEN — O ═ O (double bond)

NITROGEN — N ≡ N (triple bond)

METHANE — H—C—H (4 single bonds)

surplus (or deficit) of four outer-shell electrons, must combine with at least two other atoms to achieve its ideal state. Methane, for example, is the product of such multiple covalent bonding, carbon eliciting the cooperation of four hydrogens to generate saturated shells for all five participating atoms. (Methane is natural gas, a flammable organic compound often used for heating and cooking.)

ORGANIC MOLECULES

Carbon's ability to form four covalent bonds imparts special importance to its role in the chemistry of life. It enables carbon to form long chains by linking with itself. These carbon chains provide the chemical backbone of the enormously complex molecules needed to organize and direct the life processes, as well as those that comprise the fabric of life itself. These carbon backbones may be linear, branched, or cyclic. Equally important, every carbon in the chain has at least two free bonding sites to which other elements and chemical groups can attach. These accessory chemicals influence the property of the compound. With the exception of CO_2 and a few other simple compounds, the presence of carbon (and usually hydrogen) distinguishes a compound as *organic*. In some organic compounds, carbon forms double or triple bonds with other carbon atoms.

Hydrogen Bonds

Another type of adhesive force between molecules is formed when two molecules share an atom of hydrogen. This occurs only between **polar molecules.** Polar molecules have their electrons unequally distributed and therefore have slight positive and negative charges concentrated in different parts of the molecule. The single oxygen atom in water, for example, tends to attract electrons more efficiently than do its two hydrogens. The shared electrons therefore spend more time at the oxygen end of the molecule. This imparts an unequal electron distribution to the molecule. The oxygen end becomes slightly more negatively charged than the two hydrogen ends, which are left with a small positive charge. As a result, water molecules are drawn together, each positively charged end of the molecule attracted to the negatively charged portion of another water molecule. (The attraction is similar to that between the opposite poles of a magnet.) In effect, the oxygen atoms of two water molecules share a common hydrogen atom. The sharing is not equal, of course, since the shared hydrogen is covalently bonded to one of the oxygens; nonetheless, the effect is to generate a weak association between the two participating molecules, an attraction called a **hydrogen bond.** The strength of these bonds keeps water molecules closely associated as a liquid rather than breaking apart and vaporizing (unless it is heated to the boiling point).

Hydrogen bonding is not limited to water but is common to many polar molecules. Some large polar molecules even form hydrogen bonds with other atoms within the same molecule. For example, parts of a single protein molecule form hydrogen bonds with each other, bonds that help stabilize the giant molecule. Although a hydrogen bond is only about 1/20 as strong as a typical covalent bond, a single large molecule may contain several hundred of these polar linkages, creating considerable combined strength.

Appendix 4
Life-Supporting Properties of Water, Solvents, and Solutes

Life as we know it on earth depends on water. About 70 to 80 percent of an organism's living material is composed of water. In addition to being a reactant in or product of some essential chemical reactions, water provides the matrix in which nearly all life's biochemical processes occur. The peculiar properties of water that make it invaluable to life are largely the product of its polar nature.

MOLECULAR STRUCTURE

Water's single oxygen tends to attract electrons more efficiently than do its two hydrogens. Such polarity creates interactions between individual water molecules that dictate its properties. These interactions result from the attraction between the oppositely charged ends of adjacent molecules, weakly pulling water molecules together by hydrogen bonding. Each molecule of water can form hydrogen bonds with four other water molecules (Fig. A4-1).

The polar nature of water makes it earth's most efficient solvent, gives it surface tension, propels it upward against the pull of gravity (capillary action), and makes it ideal for temperature regulation in organisms. Each of these special properties of water is discussed in this appendix.

WATER AS A SOLVENT

More types of substances dissolve in greater quantities in water than in any other medium. Many biologically active molecules, large and small, perform their functions only when dissolved in water. Virtually any substance that is either ionic (a salt, for example) or polar can dissolve in water, forming a solution.

SOLUTIONS AND SUSPENSIONS

When a substance dissolves into individual molecules in a fluid, the resulting mixture is called a **solution.** The fluid matrix in which the substance is dissolved is the *solvent;* the

707

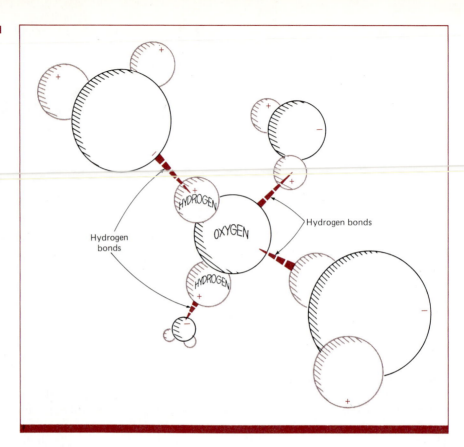

Hydrogen
bonds

Hydrogen bonds

HYDROGEN

OXYGEN

HYDROGEN

708

material that goes into solution is the *solute*. Prior to being dissolved, a solute can be a solid, liquid, or gas. Water is such an efficient solvent that rainwater becomes a solution containing the dissolved gases oxygen, nitrogen, and carbon dioxide, even before it reaches the ground.

Some substances that don't dissolve into small molecules become suspended as particles in a fluid matrix, forming a *suspension*. Particles that are 10^{-4} μm or larger in diameter eventually settle out of suspension, collecting at the bottom of the fluid. If the particles are small enough, however, they resist gravity's pull and remain in suspension, forming a **colloidal suspension.** Common examples of colloids include milk, Jell-O, and fog (water droplets suspended in air). Cytoplasm is also a colloidal suspension, a semifluid substance in which most of the life processes take place.

SURFACE TENSION

The attraction between the water molecules imparts to the liquid a high surface tension, that is, a tendency of the surface to resist being easily broken. Position a leaf broadside on top of a bowlfull of water and see what happens. It not only floats, it seems to be supported by a membrane across the water's surface (you can even see indentations in the water's surface around the edge of the leaf). This "membrane" is the product of the cohesive force generated by hydrogen bonding between adjacent water molecules. It takes considerable force to overcome the attraction and break the water's surface. The same effect is demonstrated by filling a glass with water just beyond the top of the container. The water bulges above the top of the glass rather than spilling out because of hydrogen bonding with the underlying water molecules.

Water molecules are not only attracted to each other, but also to any polar (hydrophilic) substance. Many surfaces, such as glass, are hydrophilic and attract water molecules with enough force to overcome the effects of gravity. The results of this type of attraction are apparent when one dips a narrow glass tube into water. The water seems to crawl up the sides of the tube, pulled along by the attraction to the hydrophilic (wettable) surface (Fig. A4-2). The water molecules adjacent to the surface pull nearby molecules up by the strength of hydrogen bonds. This tendency of water to be pulled into the tube is called **capillary action**.

TEMPERATURE REGULATION

Most organisms are remarkably flexible in their ability to withstand changes in the amount of heat to which they are exposed. This flexibility is largely the product of the thermal stability of water, its ability to absorb or lose large amounts of heat without changing temperature.

Figure A4-2

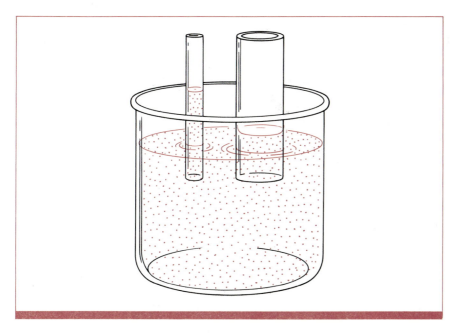

ICE FORMATION

As the temperature of water decreases, the molecules tend to slide in closer to one another. In other words, as water gets colder it becomes heavier, at least until it reaches 4°C. At this temperature water is at its heaviest state. As the temperature approaches 0°C, however, this trend reverses itself, and the molecules orient themselves in an arrangement that takes full advantage of hydrogen bonding. Below 0°C, water freezes to a solid, forming a rigid molecular lattice in which each molecule is firmly hydrogen bonded to four other water molecules (see Fig. A4-3). Since this configuration prevents the molecules from slipping into closer proximity with each other, the intramolecular space is greater in ice than in liquid water. Consequently, water expands as it freezes and is therefore lighter as ice than as a liquid. Freezing is fatal to many organisms because the expansion of water tends to damage cell membranes and tissues, especially if large crystals with sharp edges from. Rapid freezing with extremely cold temperatures tends to reduce the formation of such destructive crystals.

FIGURE A4-3

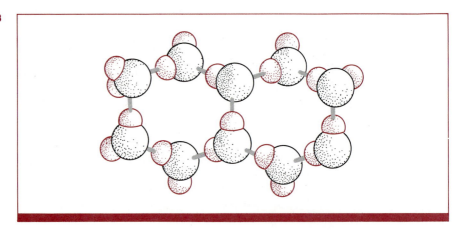

Glossary

abscess Localized accumulation of pus in tissue.

acetyl coenzyme A (acetyl CoA) Two-carbon fragment coupled to coenzyme A that introduces carbon from pyruvate metabolism into the Krebs cycle. Acetyl CoA is also a key intermediate in many metabolic reactions.

acid Substance with a pH below 7.0.

acid-fast stain Differential stain for detecting bacteria that retain carbolfuchsin when treated with acid alcohol.

acquired immunity Body's resistance to specific pathogens or toxins; protection is due to antibodies or immune cells acquired during a person's lifetime.

acquired immunodeficiency disease (AIDS) Immunological defect caused by transmissible agent. AIDS is associated with fatal cases of Kaposi's sarcoma and opportunistic infections.

actinomycete Filamentous bacterium; many actinomycetes superficially resemble fungi.

activated sludge Sewage solids coated with microbes that decompose sewage during secondary treatment.

activation energy Minimum energy needed to initiate a chemical reaction.

active immunity Production of antibodies or lymphocytes by the host in response to the presence of an antigen.

active site Area on an enzyme that reacts with the substrate.

active transport Energy-dependent movement of molecules across a cell membrane against a concentration gradient.

acute disease Disease with a rapid onset and short duration.

acyclovir Drug successful for treating some herpesvirus infections.

adenine Purine nucleotide base found in nucleic acids.

adenosine triphosphate (ATP) Compound which stores energy in phosphate bonds. This energy is released by converting ATP to ADP and phosphate.

adenyl cyclase Enzyme that converts ATP to cyclic AMP (cAMP).

adjuvant Substance that enhances the immune response when injected along with an antigen.

aerial hypha Fungal filament (hypha) that protrudes above the surface and produces asexual spores.

aerobe Organism that requires molecular oxygen for growth.

aerobic respiration Energy production that depends on molecular oxygen (O_2) as the terminal electron acceptor.

aerotolerant anaerobe Organism that neither requires nor is inhibited in the presence of molecular oxygen.

aflatoxin Poisonous substance produced by some fungi that damages the liver and may cause cancer.

711

agammaglobulinemia Absence of antibody production due to a defect in the B-cell system.

agar Extract of seaweed used as a solidifying agent in media.

agglutination Clumping of cells or particles in the presence of specific antibody.

agglutinin Antibody that reacts with particulate antigens, causing agglutination.

agranulocyte White blood cell that lacks observable intracellular granules. Lymphocytes and monocytes are examples.

alga Photosynthetic eucaryote that lacks true roots, stems, and leaves.

alkaline Characterized by a pH above 7.0.

allergen Antigen that elicits an allergic response.

allergy Antigen-triggered immune-mediated response that injures the host.

allergy Introducing small doses of allergen to eliminate IgE-mediated allergy.

allosteric enzyme Enzyme whose activity is regulated by the binding of an effector molecule to a site other than the active site.

allosteric site Site on an allosteric enzyme to which an effector molecule binds.

alpha hemolysis Greenish zone of partial clearing around colonies of certain microbes growing on blood agar.

alveolar macrophage Phagocytic cell in the lungs; also called "dust cells."

alveolus Terminal portion of the respiratory tract where gas exchange occurs.

amantadine Antiviral agent used to prevent influenza.

Ames test Laboratory procedure using bacteria to screen compounds for mutagenic activity.

amination Addition of an amino group (NH_2) to a compound.

amino acid Organic building block of protein. It contains an amino group (NH_2) and a carboxyl group (COOH).

aminoglycoside Any of a group of related antibiotics that attach to ribosomes and interfere with protein synthesis. Streptomycin, gentamicin, and neomycin are examples.

amino group $—NH_2$.

ammonification Chemical release of ammonia (NH_3) from nitrogen-containing organic compounds.

amoeba Protozoan in the class Sarcodina that moves and feeds by pseudopods.

amphitrichous Form of polar flagellation having tufts of flagella at each end of the cell.

amphotericin B Polyene antibiotic used for treating systemic fungal infections.

amylase Enzyme that hydrolyzes starch.

anabolism Biosynthesis of complex molecules from simple compounds.

anaerobe Organism that grows in the absence of molecular oxygen.

anaerobic Energy production that depends on an inorganic molecule other than oxygen as the terminal electron acceptor.

anamnestic response Immunological memory that accounts for rapid immune reactions following antigen exposure in sensitized individuals.

anaphylaxis Generalized IgE-mediated allergic response; may be fatal.

anemia Deficiency in erythrocytes or hemoglobin (or both).

angstrom 10^{-10} meter.

anionic Possessing a negative charge.

antagonism Reduction in the effectiveness of drugs when used in combination.

antibiogram Spectrum of antibiotic susceptibility of a microbe.

antibiotic Chemical produced by microorganisms that in low concentrations selectively inhibits or kills other microorganisms.

antibiotic resistance Ability of a microbe to survive in the presence of an antibiotic.

antibody Protein produced in the body in response to an antigen and capable of specifically binding with the antigen that stimulated its formation.

antibody titer Concentration of an antibody in the blood.

anticodon Three-nucleotide region on transfer RNA that binds to the codon of messenger RNA.

antigen Substance that stimulates an immune response when introduced into the body.

antigenic determinant Small site on the antigen that determines the specificity of the immune response.

antigenic drift Minor structural changes in the virus surface due to mutation in the viral genome.

antigenic shift Sudden major structural

change in the virus surface resulting in immunologically distinct strains; usually due to recombination.

antihistamine Chemical that inhibits the effects of histamine in the body.

anti-inflammatory agent Substance that reduces the intensity of the inflammatory response.

antimetabolite Chemical that competitively inhibits specific microbial metabolic reactions.

antimicrobial agent Substance that kills microbes or inhibits their growth.

antisepsis Inhibition or destruction of microorganisms on the surface of living tissue.

antiseptic Chemical used to destroy or inhibit microbes on body surfaces.

antiserum Blood serum that contains antibodies of known specificity.

antitoxin Antibody that neutralizes a toxin.

archaebacterium Any of a group of microbes that, like other procaryotes, lack a true nucleus and membrane-bound organelles but have some chemical and metabolic properties not characteristic of procaryotes.

arthrospore Fungal spore formed by segmentation of hyphae into thick-walled cells.

artifact Artificial characteristic that develops in a specimen as a result of its manipulation.

artificial active immunity Production of antibodies or lymphocytes by the host in response to vaccination.

artificial passive immunity Immunity acquired by injecting presynthesized antibody into a nonimmune host.

Ascomycetes Class of fungi characterized by production of sexual ascospores.

ascospore Sexual fungal spore produced in an ascus.

ascus Sac that contains ascospores.

aseptic technique Procedure that prevents the introduction of microbial contaminants.

asexual reproduction Reproduction without fusion of haploid sex cells.

asexual spore Spore formed without fusion of nuclear material.

asthma Allergic response characterized by narrowing of the trachea or bronchi.

asymptomatic Exhibiting no symptoms.

atom Smallest unit of an element that retains the characteristic properties of that element; it is composed of protons, neutrons, and electrons.

attachment site Location on the viral surface of the molecules that adhere to the cell's receptor site.

attenuation Reduction in a pathogen's capacity to cause disease.

autoclave Instrument that sterilizes with steam under pressure.

autocytolysis Killing of an infected host cell by its own lytic enzymes.

autoimmune disease Disease caused by an immune response against one's body tissues.

autoinoculation Transfer of microbes indigenous to one area of the body to another area where they may cause disease.

autotroph Organism that uses carbon dioxide as the sole source of carbon.

auxotroph Mutant that has acquired an additional nutritional requirement.

avirulent Lacking the capacity to produce disease.

axial filament structure responsible for spirochete motility.

bacillus Rod-shaped bacterium.

bacitracin Antibiotic active against the cell wall of many gram-positive bacteria.

bacteremia Presence of bacteria in the bloodstream.

bactericidal agent Substance or process that is able to kill bacteria.

bacteriochlorophyll Light-absorbing pigment in photosynthetic bacteria.

bacteriocin Substance produced by some bacteria that inhibits or kills closely related bacteria.

bacteriological membrane filter Filter that contains pores small enough to prevent the passage of bacteria.

bacteriophage Virus that infects bacteria.

bacteriostatic Substance or process that is able to inhibit the growth of bacteria.

bacterium Procaryotic cell that is not a cyanobacterium.

bacteriuria Presence of bacteria in the urine.

baking Hot-air sterilization process requiring exposure to temperatures between 150 and 180°C.

base Substance with a pH above 7.0.

base analog Chemical that resembles one of the purine or pyrimidine bases normally found in nucleic acids; some may cause mutations by being incorporated into DNA in place of the normal base.

base pairing Mutual attraction between adenine and thymine (or uracil) and between guanine and cytosine.

Basidiomycetes Class of fungi characterized by production of sexual basidiospores.

basidiospore Sexual fungal spore produced by a basidium.

basidium Structure which produces basidiospores.

basophil Nonphagocytic granulocyte that stains with basic dyes.

batch culture Culture system containing a medium to which nutrients are neither added nor removed during the growth of the culture.

BCG vaccine Vaccine to provide immunity against tuberculosis; made from attenuated bovine tubercle bacillus.

Bergey's Manual A published classification of bacteria; used as a practical aid for their identification.

beta hemolysis Zone of complete clearing around colonies of certain microbes growing on blood agar.

beta lysin Soluble blood chemical that disrupts bacterial membranes.

binary fission Process of asexual reproduction whereby one cell divides into two cells of equal size.

binomial nomenclature System of naming organisms with two terms, the first representing the genus, the second the species.

bioassay Method of measuring the concentration or activity of a substance by determining its effect on a living organism.

biochemical Compound that is a component of living organisms.

biochemical oxygen demand (BOD) Amount of oxygen consumed by aerobic organisms; used to determine the concentration of biodegradable organic matter in water.

biofertilization Nutrient enrichment of soil as a result of microbial activity.

biogenesis Doctrine stating that living organisms arise only from other living organisms.

biogeochemical cycle Series of conversions by which nutrients are recycled in nature.

biological indicator Living organism whose viability or activity is used as a measure of a specific process.

biological vector A type of vector in which the pathogen proliferates, as opposed to a mechanical vector, which is a carrier only.

biomass Weight of all organisms that inhabit a defined area.

biomining Use of microorganisms to recover metals from ores.

biopsy Surgical removal of tissues for diagnosis.

biosynthesis Construction of chemical components in the growing cell.

biotype Subgroup of organisms within a species that possess biochemical properties distinguishing them from other members of the species.

bivalent Property of an immunoglobulin that enables it to combine with two identical antigens at the same time.

blastospore Asexual fungal spore produced by budding.

blocking antibody Antibody able to bind with an allergen and prevent an allergic response.

blood agar Solid-culture medium containing whole blood.

blood-brain barrier Barrier that determines which substances will pass between brain capillaries and brain tissue.

bloom Visible accumulation of algae or cyanobacteria in a body of water.

B lymphocyte Cell that has been programmed to participate in the humoral immune response; when stimulated by antigen it proliferates and differentiates into antibody-producing plasma cells and memory cells.

booster Immunization given to enhance the memory response to an antigen.

brightfield microscope Microscope that uses a direct light source for illumination.

broad-spectrum antibiotic Antibiotic affecting a wide variety of microorganisms.

bubo Swollen lymph node due to inflammation.

bubonic plague Form of plague transmitted from rats by fleas and characterized by the development of bubos.

budding Asexual reproduction in which progeny arise from protuberances on the surface of a parent cell; in virology, the release of an enveloped virus from the surface of an infected cell.

buffer Chemical that helps resist changes in the pH of a solution.

cancer Uncontrolled proliferation of abnormal host cells.

candle jar Container used to create microaerophilic or capneic conditions.

capneic Requiring increased levels of CO_2.

capsid Protein coat surrounding the nucleic acid of a virus.

capsomere Repeating protein unit of which a capsid is composed.

capsule Structure composed of polysaccharide or polypeptide that surrounds the cell wall in some bacteria and fungi.

carbohydrate Organic molecule that contains carbon, hydrogen, and oxygen, with a hydrogen-to-oxygen ratio of 2:1.

carbon cycle Series of reactions in nature in which carbon dioxide is converted to organic carbon and ultimately metabolized back to carbon dioxide.

carboxyl group An organic acid group; −COOH.

carbuncle Extensive suppurative inflammation of skin and underlying tissues caused by bacterial infection.

carcinogen Any agent that causes cancer.

cardiolipin Component of beef-heart extract used in rapid screening tests for syphilis.

carrier Person with an inapparent infection who sheds pathogens and can therefore transmit disease.

caseation Dead tissue with a cheeselike appearance, typically seen in tuberculosis.

catabolism Degradation of complex molecules to simpler molecules.

catalase Enzyme that decomposes hydrogen peroxide to water and oxygen.

catalyst Substance that increases the rate of a chemical reaction without being consumed in the reaction.

catheter Hollow tube used to gain access to narrow body canals in order to drain fluids from body cavities or to introduce material into the body.

cationic Possessing a positive charge.

cell Structural and functional unit of all living organisms.

cell culture Growth of eucaryotic cells in vitro.

cell-mediated immunity (CMI) Immune response mediated by T lymphocytes.

cell membrane Selectively permeable structure enclosing the cytoplasmic contents of a cell; composed of phospholipids and proteins.

cellulose Polysaccharide that comprises the cell wall of plants and most algae.

cell wall Rigid structure surrounding the cell membrane in most plant, algal, fungal, and bacterial cells.

Centers for Disease Control (CDC) Federal agency that gathers and analyzes data on notifiable diseases.

cephalosporin Any of a group of antibiotics produced by *Cephalosporium* or a semisynthetic derivative; inhibits cell-wall synthesis in gram-positive bacteria.

chancre Ulcer with a hard, rubbery base; the initial lesion of syphilis.

chemical bond Force that holds atoms together in a molecule.

chemical energy Energy stored within a chemical bond and released when that bond is broken.

chemically defined medium See synthetic medium.

chemoautotroph Organism that uses inorganic chemicals as a source of energy and carbon dioxide as the sole source of carbon.

chemoheterotroph Organism that uses organic chemicals as a source of energy and carbon.

chemolithotroph *See* chemoautotroph.

chemoprophylaxis Administering drugs to prevent infectious disease.

chemostat Device used to maintain continuous cultures.

chemotactic factor Substance that attracts phagocytic white blood cells.

chemotaxis Movement of an organism toward or away from a chemical stimulus.

chemotherapeutic agent Chemical administered in treatment of disease.

chemotherapy Use of chemicals in the body to treat disease.

chemotroph Organism that uses chemicals to secure energy.

chitin Structural polysaccharide of fungal cell walls.

chlamydium Obligate intracellular parasitic bacterium that lacks a functional ATP-generating system.

chlamydospore Thick-walled asexual fungal spore formed within a vegetative hypha.

chloramphenicol Antibacterial chemotherapeutic agent that inhibits protein synthesis.

chlorine Halogen used in water purification and as a disinfectant.

chlorophyll Light-absorbing pigment used by cyanobacteria, algae, and plants for photosynthesis.

chloroplast Cellular site of eucaryotic photosynthesis.

chloroquine Antimalarial drug.

chocolate agar Solid medium containing blood that has been heated.

chromatophore Invagination of the cell membrane in photosynthetic bacteria that contains enzymes and pigments for photosynthesis.

chromosome Structure that carries genes and is responsible for heredity.

chronic disease Slowly progressing disease of long and often indeterminable duration.

cilia Short, hairlike processes on the eucaryotic cell surface that provide motility or allow a cell to move the surrounding liquids.

ciliated mucosa Lining of the respiratory tract that traps microbes and other particles in a layer of mucus and propels them toward the pharynx.

Ciliophora Class of protozoa that possess cilia.

citric acid cycle See Krebs cycle.

clinical specimen Fluid or tissue removed from the human body for laboratory analysis.

clone One of a group of genetically identical cells that have descended from a common parent cell.

coagulase Enzyme produced by staphylococci that causes the clotting of blood plasma.

coagulation Formation of a blood clot.

coccobacillus Oval-shaped bacterium.

coccus Spherical bacterium.

codon Triplet sequence of nucleotides in messenger RNA that specifies a particular amino acid to be inserted in the protein chain.

coenzyme Small organic molecule required by some enzymes to function.

coenzyme A (CoA) Coenzyme that transfers acetic groups.

cofactor Metallic ion required by some enzymes to function.

coinfection Infection caused by the presence of more than one kind of pathogen.

cold agglutinin Antibody formed in response to some infections that clumps type 0 erythrocytes at 4°C but not at 37°C.

cold sterilization Sterilization method that does not rely on heat for killing.

coliform Colon bacillus; *Escherichia coli* and similar gram-negative rods that normally inhabit the colon.

collagenase Enzyme that hydrolyzes collagen.

colonization Establishment of microbes on the skin or mucous membranes.

colony Visible aggregate of cells on a solid medium, all descended from the same parent organism.

colony-forming unit Single cell or cluster of cells that generates an isolated colony on a solid medium.

commensalism Association between organisms in which one is benefitted and the other is neither benefitted nor harmed.

common-source epidemic Outbreak in which the agent is acquired from a contaminated source rather than from an infected individual.

communicable disease Disease that may be transmitted from an infected host to uninfected individuals.

community-acquired infection Infection that develops as a result of exposure to a pathogen outside of a hospital environment.

competitive inhibition Inhibition of enzyme activity due to competition from a chemical that is a structural analog of the substrate.

complement Group of serum proteins that facilitate bacterial phagocytosis and lysis.

complement fixation Reaction of complement with an antigen-antibody complex.

complex medium Medium in which the exact identity and concentration of each chemical compound is unknown.

compound Chemical composed of two or more elements bound to one another.

compound microscope Microscope containing at least two lenses.

concentration gradient Existence of a higher concentration of a substance in one region of a system than in another.

condenser Part of the microscope that focuses light on the specimen.

congenital infection Infection acquired in utero and present at birth.

conidium Asexual fungal spore formed at the tip of aerial hypha.

conjugation Transfer of genetic material requiring cell-to-cell contact.

constitutive enzyme Enzyme whose production is not sensitive to regulation.

contagious disease Infectious disease that can be readily transmitted from person to person.

continuous culture System in which microbial populations are maintained in exponential growth.

convalescence Recovery from disease.

core Nucleocapsid of an enveloped virus.

corepressor Molecule that activates the repressor in a repressible enzyme system.

coupled reactions Chemical reactions that are linked to one another by a common intermediate.

covalent bond Chemical bond in which atoms share electrons.

cresol Any of a group of disinfectants derived from phenol.

cross reaction Reaction between an antibody and a different antigen than that which stimulated its formation; this phenomenon is due to similarities between the two antigens.

cross resistance Insensitivity of a microbe to several related antibiotics.

cubic symmetry Virus morphology that resembles a spherelike structure.

culture Microorganisms grown in an artificial medium.

culture medium Nutrient material that supports cell growth.

curd Solid portion of milk after it coagulates.

cutaneous mycosis Fungal infection of the skin, nails, or hair.

cyanobacterium Procaryote that contains chlorophyll A and performs O_2-generating photosynthesis.

cyclic AMP (cAMP) Regulatory molecule synthesized in cells from ATP.

cyst Protective, dormant structure formed by some protozoa.

cystitis Inflammation of the urinary bladder.

cytochrome Iron-containing compound that functions in electron transport in respiration and photosynthesis.

cytopathic effect (CPE) Virus-induced change in cell cultures.

cytoplasm Internal contents of the cell other than the nucleus.

cytoplasmic inclusion Intracellular storage granule in procaryotes; examples are metachromatic granules and poly-β-hydroxybutyric acid granules.

cytoplasmic streaming Constant motion of cytoplasm in the eucaryotic cell.

cytosine Pyrimidine nucleotide base found in nucleic acids.

cytostome Oral opening of some protozoa.

cytotoxicity Death of cells.

Dane particle Infectious form of hepatitis-B virus in serum.

darkfield microscope Compound microscope in which specimens appear light against a dark background.

dead-end infection Infection that is not normally transmissible to another host.

deamination Removal of an amino group (NH_2) from a compound.

death phase Stage in the growth curve where organisms are dying at an exponential rate.

debridement Removal of necrotic tissue from a wound.

decarboxylation Removal of carbon dioxide from an organic molecule.

decomposer Microorganism that degrades organic material into inorganic molecules.

decontamination Process of removing or destroying harmful microorganisms or their toxic products.

degradation Conversion of an organic compound to a smaller organic compound.

degranulation Discharge of lysosomal contents into the phagocytic vacuole.

dehydration Removal of water.

delayed hypersensitivity lymphocyte (DHS) T cell that enhances immunity by secreting lymphokines.

delayed-type hypersensitivity Allergic response produced by the cell-mediated immune system.

deletion mutation Loss of a section of DNA; may be loss of a single nucleotide or of entire genes.

denaturation Change in the configuration of a macromolecule, often resulting in a loss of function.

denitrification Metabolic conversion of nitrate to nitrite or nitrogen gas.

dental caries Localized disintegration of teeth due to acids produced by oral bacteria adhering to the tooth surface.

dental plaque Combination of bacteria and dextran adhering to teeth.

deoxyribonucleic acid (DNA) Macromolecule that encodes the genetic information of the cell.

dermatophyte Fungus that invades the superficial keratinized areas of the body such as skin, hair, and nails.

dermis Layer of skin below the epidermis.

detergent Surface-acting agent that often has antibacterial activity.

Deuteromycetes Class of fungi characterized by the absence of sexual reproduction.

dextran Type of polysaccharide composed of glucose.

diacetyl Bacterial fermentation product responsible for the characteristic flavor of butter.

diapedesis Process by which white blood cells pass through blood vessels into tissues.

diarrhea Increased frequency of movement and fluid consistency of stools.

diatom Unicellular alga having a cell wall composed of silica.

diatomaceous earth Sediment composed of the silica remnants of the cell walls of diatoms.

differential blood count Percentage of each kind of leukocyte in a blood sample.

differential medium Culture medium used to distinguish microorganisms based on their characteristic macroscopic appearance.

differential stain Stain used to differentiate bacteria, their structures, or their tissue components.

diffusion Tendency of molecules to move from areas where they are in higher concentration to areas of lower concentration.

dimorphic fungus Fungus that exists as either a yeast or mold, depending on environmental conditions.

dimorphism The characteristic of existing in two structurally distinct forms.

diphtheroid Any of a group of bacteria (morphologically) resembling the diphtheria bacillus.

diplobacilli Bacilli that occur in pairs.

diplococci Cocci that occur in pairs.

diploid Containing two copies of each type of chromosome.

direct count Determination of the number of microorganisms by microscopic observation.

direct transmission Immediate transfer of an infectious agent from the reservoir to a host with no intervening intermediate.

disaccharide Sugar consisting of two monosaccharides.

disc diffusion method Method for determining the antibiotic sensitivity of a microbe; performed by placing filter-paper discs impregnated with the drugs onto a plate seeded with the test organism and measuring any zone of inhibition that develops.

disease Injury or damage to the body.

disinfectant Agent used on inanimate objects that destroys or inhibits pathogenic microbes.

disinfection Elimination from an inanimate object of the vegetative forms of most pathogenic organisms.

DNA homology The degree of similarity between the nucleotide sequences in different DNAs.

Donovan body Intracellular inclusion found in cells of persons infected with *Calymmatobacterium granulomatis*.

DPT vaccine An immunizing mixture against diphtheria, tetanus, and pertussis.

droplet Expelled particle of moisture that can transmit microbes between hosts.

droplet nucleus Dried droplet containing microorganisms; it is smaller than a droplet and may remain suspended in air; if inhaled it may enter the lower respiratory tract.

dysentery Inflammation of the intestine characterized by diarrhea with blood and mucus.

early protein Protein synthesized by a virus during the early period of replication; most early proteins are enzymes.

eclipse phase Time during a virus infection of a host cell when mature virions cannot be found.

edema Accumulation of fluid in tissues.

effector Compound that inhibits an allosteric enzyme.

effector cell T lymphocyte that provides protection from an antigen.

effluent Fluid discharged for sewage treatment.

electron Negatively charged subatomic particle.

electron microscope Instrument that uses electron beams instead of light to produce images.

electron transport system (ETS) Series of compounds that generate energy by the transfer of electrons.

element Chemical composed of only one kind of atom.

elementary body Extracellular, infectious form of *Chlamydia*.

ELISA (enzyme-linked immunosorbant assay) Immunological test that uses an enzyme-mediated reaction as an indicator of an antigen-antibody reaction.

encephalitis Inflammation of the brain.

encystment Process of cyst formation.

endemic Occurring with a constant frequency in the population.

endergonic Pertaining to a chemical reaction that consumes energy.

endocarditis Inflammation of the lining of the heart.

endogenous disease Opportunistic disease caused by normally harmless microorganisms inhabiting the body.

endoplasmic reticulum Internal membrane network in the eucaryotic cell extending from the outer cell membrane to the nucleus.

endospore Highly resistant bacterial spore formed within the parent cell.

endotoxin Lipopolysaccharide component of gram-negative cell walls associated with fever and shock in humans.

enriched medium Medium supplemented with nutrients that support the growth of many fastidious bacteria.

enrichment broth Liquid medium in which organisms in mixed culture will grow at different rates, encouraging the isolation of the favored organism.

enteric Pertaining to the intestine.

enteric fever Syndrome characterized by fever usually produced by disseminated infection with strains of *Salmonella*.

enterotoxin Exotoxin that affects the intestinal mucosa.

enterovirus Member of a group of small, nonenveloped, single-stranded RNA viruses that may be isolated from the alimentary tract.

envelope Outer membrane layer which characteristically surrounds certain viruses.

enzyme Protein that catalyzes chemical reactions.

enzyme induction Process by which a substance stimulates synthesis of enzymes.

enzyme repression Process by which a substance, such as an end product, prevents the synthesis of enzymes.

enzyme-substrate complex Temporary interaction between an enzyme and its substrate during a chemical reaction.

eosinophil Nonphagocytic granulocyte that stains with acid dyes.

epidemic Significant increase in the usual number of cases of a disease.

epidemiological typing Determining strains of an organism for the purpose of identifying its source.

epidemiology Study of the factors that contribute to the occurrence of diseases.

epidermis Outer portion of the skin.

epiglottis Structure that covers the entrance to trachea during swallowing.

epizootic Epidemic among animals.

ergosterol Sterol found in cell membranes of fungi.

erythema Redness of the skin.

erythrocyte Red blood cell.

erythromycin Antibacterial antibiotic that inhibits protein synthesis.

ethambutol Tuberculostatic drug.

ethylene oxide Gas used as a sterilizing agent.

etiologic agent Specific cause of a disease.

eucaryote Cell that contains a membrane-bound nucleus.

eutrophication Accumulation of nutrients in an aquatic environment and subsequent increase in microorganisms; this usually results in depletion of oxygen available for fish and other aerobic organisms.

excretion Discharge of waste products.

exergonic Pertaining to a reaction that releases energy.

exogenous disease Disease acquired by exposure to pathogens from sources external to the body.

exotoxin Soluble toxin excreted by microorganisms into their surroundings.

extracellular enzyme Enzyme secreted by a microorganism into its surroundings, often to break down large molecules for transport into the cell.

extracellular pathogen A pathogen that multiplies in the cavities, fluids, or tissue spaces of the body.

exudate White blood cells and other substances that have passed from blood into tissues during inflammation; pus.

facultative anaerobe Organism that grows in the presence or absence of molecular oxygen.

FAD Flavin adenine denucleotide; one type of coenzyme that transfers electrons (and hydrogens) during metabolism.

fastidious Having exacting nutritional or physical requirements for growth.

fat Organic compound composed of glycerol and fatty acids.

fatty acid Hydrocarbon chain terminating with a carboxyl group.

F⁺ cell Donor cell in bacterial conjugation; contains the fertility factor.

F⁻ cell Recipient cell in bacterial conjugation.

feedback inhibition Inhibition of the activity of an enzyme in a metabolic pathway by an end product of that pathway.

fermentation Energy production that depends on an organic compound as final electron acceptor.

fertility (F) pilus Pilus on a donor capable of connecting bacterial cell to a recipient during conjugation.

fertilization Fusion of two haploid nuclei to form a diploid nucleus.

fertilization (F) factor Plasmid that codes for the F pilus and converts a bacterium to a donor in conjugation.

fibrinolysin Enzyme that dissolves fibrin clots.

filamentous Tending to form long strands.

filtration Passage of liquid or gases through a material that retains microorganisms.

fimbria Pilus.

fission Asexual reproduction by division into two or more cells.

fixed macrophage Macrophage attached to a specific tissue or lining certain organs.

flagellin Noncontractile protein from which procaryotic flagella are composed.

flagellum Whiplike appendage used for locomotion.

flaming Dry-heat sterilization procedure used primarily for inoculating loops or needles.

flocculation Precipitation of colloidal solids in water treatment.

fluorescence microscope Compound microscope that uses ultraviolet light to illuminate a specimen that will fluoresce.

fluorescent antibody technique Method to detect an antigen by treating it with specific antibody conjugated to a fluorescent dye.

focal infection Localized infection from which microbes spread to distant body sites.

folliculitis Infection of a hair follicle.

fomite Contaminated inanimate object that transmits infectious disease.

food vacuole Intracellular vesicle in protozoa containing ingested material to be degraded as a source of nutrients.

formaldehyde A chemical used for sterilization, disinfection, or antisepsis, depending on its concentration.

F-prime cell Bacterial cell with a plasmid that contains some information from the host chromosome and some from the F factor.

fragmentation Process of asexual reproduction in which a filament breaks into smaller portions, each of which can develop into a new organism.

frame-shift mutation Addition or deletion of nucleotides that alter the reading frame.

fruiting body Structure that bears sexual spores in fungi.

FTA-ABS test Fluorescent antibody test specific for syphilis.

fungicidal agent Chemical or process capable of killing fungi.

fungi imperfecti Deuteromycetes.

fungistatic agent Chemical or process capable of inhibiting the growth of fungi.

fungus Heterotrophic eucaryote with a cell wall.

furuncle Infection of a hair follicle that develops into a solitary abscess.

G + C content Percentage of guanine and cytosine nucleotides in DNA; used to determine similarities between organisms.

gametocyte Male or female reproductive cell of *Plasmodium*.

gamma hemolysis Lack of hemolysis.

gamma ray Ionizing radiation emitted by radioactive elements, for example, cobalt 60.

gangrene Tissue death resulting from loss of a blood supply.

gastroenteritis Inflammation of the stomach and intestines.

gene Segment of a DNA molecule that contains information directing a cell to synthesize a specific protein product.

gene amplification Increase in the number of copies of a specific gene in the cell.

generation time Time it takes for a cell or population to double.

genetic code Triplet nucleotide sequences, each of which specifies a particular amino acid.

genetic engineering Techniques for manipulating a cell's genetic composition and its expression.

genetics Study of heredity.

genotype Entire complement of genes possessed by an organism.

genus Taxonomic group that contains closely related species.

germicidal agent Chemical or process capable of destroying microorganisms.

germination Formation of a vegetative cell from a spore.

germ theory Theory that infectious diseases are caused by microbes.

gingivostomatitis Inflammation of the gums and oral mucosa.

glomerulonephritis Inflammation of the glomeruli in the kidneys, often due to deposits of immune complexes.

glottis Opening between the upper and lower respiratory tracts.

glutaraldehyde Chemical used in sterilization or disinfection, depending on its concentration.

glycerol A three-carbon alcohol; found in fatty acids and as an intermediate in glycolysis.

glycocalyx Tangled mass of thin polysaccharide fibers that extend from the surface of some bacteria.

glycogen Polysaccharide used as a storage form of glucose in some cells.

glycolipid Compound composed of a carbohydrate and a lipid.

glycolysis Series of reactions by which one glucose is split into molecules of pyruvic acid.

glycoprotein Compound composed of a carbohydrate and a protein.

glycosidic bond Chemical link between the monomers in a polysaccharide.

Golgi complex Intracellular membrane system functioning in protein secretion and intracellular vesicle formation.

gram-negative In the Gram stain, bacteria that do not retain the crystal violet dye after decolorization with alcohol.

gram-positive In the Gram stain, bacteria that retain the crystal violet dye after decolorization with alcohol.

Gram's stain Differential stain that forms the basis for dividing bacteria into two groups, gram-positive and gram-negative.

granulocyte White blood cell characterized by granules in its cytoplasm. Eosinophils, basophils, and neutrophils are examples of granulocytes.

griseofulvin Fungistatic antibiotic used to treat dermatophyte infections.

growth curve Graphic expression of the changes in microbial population size in a batch culture.

growth factor Organic substance required for growth.

growth rate Number of doublings per hour.

guanine Purine nucleotide base found in nucleic acids.

gumma Rubbery necrotic lesion characteristic of tertiary syphilis.

halogen Any one of the elements chlorine, iodine, bromine, or fluorine.

halophile Organism that requires a high concentration of salt for growth.

hand washing A mechanical method extremely successful in reducing the population of pathogens that can be transmitted by the hands.

haploid Containing a single copy of each type of chromosome.

hapten Small molecule that reacts with a specific antibody but is nonantigenic unless coupled to a large carrier molecule.

hay fever Allergic rhinitis; antibody-mediated upper respiratory response to inhaled allergens.

heavy metal High-molecular-weight metal that kills cells by denaturing proteins.

helical symmetry Virus morphology that resembles a long rod.

helper T cell T lymphocyte that cooperates with B cells to initiate an antibody response.

hemadsorption Attachment of red blood cells to the surface of virus-infected cells.

hemagglutination Clumping of red blood cells caused by antibodies or by some enveloped viruses.

hemagglutination inhibition One type of test to determine the amount of antiviral antibody in a person's blood.

hemagglutinin Substance that causes red blood cells to clump.

hematogenous Relating to the blood; disseminated by the bloodstream.

hemolysin Substance that destroys red blood cells.

hemolytic Having the ability to lyse red blood cells (see alpha and beta hymolysis).

hemorrhagic fever Acute febrile disease caused by certain viruses and characterized by fever and hemorrhages into the skin and internal organs.

hepatitis Inflammation of the liver.

herd immunity Protection of individuals in a group from infection by virtue of extensive acquired immunity that decreases the number of infected individuals and therefore the likelihood of exposure of nonimmune persons to the pathogen.

heterocyst Structure in some cyanobacteria specialized for nitrogen fixation.

heterophil antigen Antigen from plant or animal that coincidentally cross-reacts with antibody to specific microbes.

heterotroph Organism that requires carbon in the form of organic molecules.

hexachlorophene An antiseptic derivative of phenol.

Hfr cell Donor bacterium containing an F factor integrated into the chromosome; it can transfer chromosomal information to the F^- cell by conjugation.

high-level germicide Chemical that can be used for sterilization.

histamine Chemical released from tissue cells that triggers changes in capillaries and smooth muscles.

histiocyte Fixed macrophage; a macrophage attached to tissue.

horizontal transmission Spread of disease from person to person within a group.

host The organism on or in which a parasite lives.

humoral immunity Immunity due to the production of antibodies.

humus Organic matter in soil from decayed plants.

hyaluronidase Enzyme that hydrolyzes hyaluronic acid in connective tissue, facilitating the spread of pathogens in tissues.

hybridoma Cell resulting from the fusion of a tumor cell with an antibody-producing cell; it produces abundant quantities of monoclonal antibody.

hydrogen bond Weak attraction between a slightly positive-charged hydrogen of one compound with a slightly negative-charged portion of another molecule or a different part of the same molecule.

hydrogen ion Positively charged nucleus of a hydrogen atom (a proton).

hydrolysis Breaking a molecular bond by the addition of water.

hydrophilic Having an affinity for water molecules.

hydrophobic Lacking an affinity for water molecules.

hyperhidrosis Excessive sweating.

hyperimmune serum Serum that contains specific antibodies from persons who have been immunized with a vaccine.

hypersensitivity (allergy) Immune-mediated response to antigen that injures the host.

hypertonic Having a solute concentration greater than another solution against which it is compared.

hypha Filament of a mold.

hypochlorite Chlorine-containing compound with disinfectant and bleaching properties.

hypogammaglobulinemia Abnormally low amounts of antibody in blood.

hypotonic Having a solute concentration lower than another solution against which it is compared.

icosahedral Virus morphology resembling a spherical structure with 20 triangular sides.

ID$_{50}$ Number of microbes required to cause disease in 50 percent of laboratory animals experimentally infected with the pathogen.

IgA Principal antibody found in body secretions.

IgD Immunoglobulin found on the surface of B cells; may function during fetal development of the immune system.

IgE Antibody associated with allergic reactions.

IgG Most abundant antibody found in blood.

IgM First antibody to appear after initial exposure to an antigen.

imidazole Antifungal drug that acts on the cell membrane.

immediate-type hypersensitivity Allergic response mediated by IgE antibodies; symptoms appear within minutes of exposure to antigen.

immobilized enzymes Enzymes attached to a solid carrier.

immune-complex disease Damage caused by the deposit of antigen-antibody complexes on capillary or renal membranes.

immune interferon Interferon produced by DHS cells.

immunity Body's resistance to invasion by microorganisms and damage by foreign substances.

immunization Process rendering a host immune.

immunodeficiency Deficit in either T-cell or B-cell immunity.

immunodiffusion test Laboratory technique that uses immunoprecipitation in agar to indicate the presence of a specific antigen or antibody.

immunogen An antigen.

immunoglobulin (Ig) An antibody.

immunological tolerance Lack of immunological activity against certain antigens.

immunology Study of antibodies and immune cells and their interactions with antigens.

immunopotentiator Substance that enhances the natural immune response.

immunosuppressant Substance that inhibits the immune response.

inapparent infection Infection that causes no clinically apparent symptoms.

incidence Rate of occurrence of a disease within a defined population.

incidental host Host that is not essential to the life cycle of a parasite.

incineration Sterilization procedure by which contaminated materials are burned to ash.

inclusion body Microscopically visible site within the cell containing aggregates of developing viruses.

incubation period Time interval between exposure to a pathogen and appearance of disease symptoms.

indigenous Belonging to a certain location or environment.

indirect transmission Transfer of an infectious agent from an infected host by an intermediate, either a vector or vehicle.

inducer Substance that stimulates synthesis of an enzyme.

inducible enzyme Enzyme that is synthesized only when the appropriate inducer is present.

induction In virology, the onset of virus production (lytic cycle) in a lysogenic cell (see also enzyme induction).

infant botulism A disease caused by *Clostridium botulinum* infection in children less than 8 months old.

infection Growth of microbes in the body or on its surface.

infection control committee Epidemiological unit in the hospital responsible for monitoring nosocomial infections.

infectious disease Injury of host tissue due to infection.

infectious dose Number of organisms needed to initiate infection in a host.

infectivity Ability of a microbe to infect a host.

inflammation Nonspecific host response to tissue injury characterized by swelling, pain, redness, and heat.

initiator codon Nucleotide triplet on messenger RNA that specifies the site where protein synthesis begins.

inoculum Microorganisms introduced to culture media or into a host.

inorganic Most compounds that do not contain carbon.

interferon Group of proteins released by animal cells in response to antigens and certain other triggers; some interferons induce antiviral activity; others enhance the immune response.

intermediate-level germicide Chemical that kills most vegetative cells and viruses but not endospores.

intoxication Food poisoning following consumption of foods containing preformed microbial toxins.

intracellular pathogen Microbe that invades and multiplies within the cells of the host.

intrauterine infection Infection acquired by the fetus as it develops within the uterus.

in-use test Evaluation of disinfectants under actual conditions of use.

invasive pathogen Microbe capable of penetrating into deep tissues and disseminating to secondary sites in the body.

invasive procedure Any technique that introduces an instrument into the body.

in vitro (Latin, "on glass") Occurring in a test tube or culture dish.

in vivo (Latin, "in life") Occurring in a living organism.

iodine Halogen used as an antiseptic.

iodophor Antiseptic composed of a complex of iodine and surfactant that slowly releases the iodine.

ion Atom or group of atoms having either a positive or negative charge.

ionic bond Bond formed when electrons are transferred from one atom to another.

ionizing radiation High-energy radiation that causes molecules to be ionized.

isolation Separation of an individual from the community to prevent the transfer of infectious disease.

isoniazid (INH) Tuberculostatic drug.

jaundice Yellowing of the skin due to increased bile pigments in the blood; a common symptom of hepatitis.

keratin Protein found in skin, hair, and nails.

killer T cell *See* T killer cell.

Kirby-Bauer technique Disc diffusion method for determining antibiotic susceptibility of bacteria.

Koch's postulates Criteria used to establish the etiology of an infectious disease.

Koplik's spots Oral rash characteristic of measles.

Krebs cycle Metabolic pathway that oxidizes two-carbon compounds to CO_2 and water, with the release of energy-rich electrons that yield ATP when processed by the electron transport system.

lag phase Initial stage in the bacterial growth curve. The microbes may be metabolically active but are not increasing in number.

laminar flow hood Large box for performing microbial manipulations with minimum chance of contamination; sterile air is circulated through the instrument and passed through a series of filters.

Lancefield group Classification of streptococci into groups A through O, based on antigenic differences in a surface carbohydrate.

latent infection Stage of an infection in which a pathogen remains in a host without producing disease for long periods of time.

late protein Protein synthesized by a virus during the late period of replication; most are structural components of the virion.

lecithinase Enzyme that degrades lecithin (in cell membranes).

lectin Surface protein of plant cells to which microbes, such as nitrogen-fixing bacteria, can attach.

legume Plant with root nodules containing nitrogen-fixing bacteria.

leukemia Uncontrolled proliferation of leukocytes and their immature precursors.

leukocidin Substance that kills white blood cells.

leukocyte White blood cell.

leukocytosis Increase in the number of circulating leukocytes.

leukopenia Decrease in the number of circulating leukocytes.

L form Bacterium that has lost the ability to synthesize peptodoglycan and therefore lacks a cell wall.

lichen Symbiotic association between a fungus and an alga or cyanobacterium.

ligase Enzyme that seals gaps in the sugar-phosphate backbone of DNA.

limulus amoebocyte lysate assay (LAL) In-vitro test for endotoxin; also employed as a rapid screening test for diagnosing gonorrhea.

lipase Enzyme that hydrolyzes lipids.

lipid Organic molecule characterized by water insolubility; lipids include fats, oils, waxes, phospholipids, and steroids.

lipid A Component of lipopolysaccharide in the gram-negative bacterial cell wall that possesses endotoxin activity.

lipopolysaccharide Compound composed of lipid and polysaccharide.

lipoprotein Compound composed of lipid and protein.

local infection Infection limited to a single body site.

localized invasive disease Disease in which pathogens invade tissues in one body site but do not disseminate to other body sites.

logarithmic (log) phase Stage in the bacterial growth curve characterized by increases in cell numbers at an exponential rate.

lophotrichous Arrangement of polar flagella in tufts at one end of a cell.

low-level germicide Chemical that kills only a few types of bacteria, fungi, and viruses.

lymphocyte Type of agranulocyte; B cells and T cells are examples.

lymphokine Soluble substance released by T cells in response to antigenic stimulation; lymphokines include lymphotoxin, macrophage-activating factor (MAF), specific-macrophage-arming factor (SMAF), migration-inhibition factor (MIF), and immune interferon.

lymphotoxin Lymphokine that damages or lyses many types of cells.

lyophilization Freeze-drying; rapid dehydration of organisms while they are in a frozen state.

lysis Bursting of a cell; also splitting of a chemical bond, for example, proteolysis.

lysogenic cell Bacterium that contains a prophage.

lysogenic conversion Acquisition of new properties by a cell following lysogeny.

lysogeny Integration of the DNA of a temperate bacteriophage into the bacterial chromosome.

lysosome Intracellular vesicle containing hydrolytic enzymes and some powerful oxidizing chemicals.

lysozyme Enzyme that degrades bacterial cell walls by disassembling peptidoglycan; found in many types of body secretions.

lytic cycle Virus infection that results in cell lysis.

macromolecule Protein, polysaccharide, nucleic acid, or other large molecule composed of subunits assembled in polymeric chains.

macronucleus The large nucleus in some protozoa that controls vegetative, but not sexual, reproduction.

macrophage Active phagocytic white blood cell that develops from a monocyte.

macrophage-activating factor (MAF) Lympnokine that enhances intracellular killing efficiency of macrophages.

macule Small red spot on the skin.

magnification Enlargement of an object's image.

maintenance medium Medium used to store microbes in which they remain viable but grow very slowly.

malignant Progressively growing tumor that tends to metastasize.

malting Digestion of starch in grain to fermentable sugars.

mast cell One type of cell that binds IgE and, in the presence of specific allergen, releases histamines and other chemicals that elicit symptoms of allergy.

Mastigophora Class of protozoa that use flagella for motility.

mechanical vector Living organism, especially an arthropod, that transfers pathogens from one host to another but is not required for the multiplication or development of the microorganism.

meiosis Process by which the number of

chromosomes in a diploid nucleus is reduced in half.

memory cell Antigen-sensitive B or T lymphocyte that persists in the host long after the immunological response to that antigen has waned. Subsequent antigen exposures trigger the cell to differentiate into immunologically active T cells or plasma cells.

meningitis Inflammation of the membranes of the brain or spinal cord.

mercurial Compound that contains mercury or its salt, often used as an antiseptic.

merozoite Product of asexual reproduction of *Plasmodium* in humans.

mesophile Organism whose optimal growth temperature is between 20 and 40°C.

mesosome Cytoplasmic invagination of the cell membrane in procaryotes.

messenger RNA (mRNA) Single-stranded RNA that determines the sequence in which amino acids are assembled during protein synthesis.

metabolic intermediate Compound in a metabolic pathway that is neither the initial substrate nor the final product of that pathway.

metabolic pathway Series of reactions that sequentially alter a starting compound to a final end product.

metabolism The sum of all the chemical reactions performed by an organism.

metachromatic granule Storage form of phosphate in some bacteria (*see* volutin granule).

metastasis Spread of disease from its primary site to another part of the body.

metronidazole Antiprotozoal chemotherapeutic agent; also used to treat infections caused by anaerobic bacteria.

microaerophile Organism that requires molecular oxygen at concentrations less than that in normal air for growth.

microbial pesticide Microorganism used to control the growth of an insect by virtue of its specific pathogenicity for the undesirable arthropod.

microbicidal Capable of killing microbes.

microbistatic Capable of inhibiting microbial growth.

micrometer 10^{-6} meter.

micronucleus Smaller of two nuclei in some ciliated protozoa; required for sexual reproduction but has no role in vegetative reproduction.

microorganism Life form too small to be seen with the naked eye.

microtubule Hollow protein filaments in eucaryotic flagella and cilia.

midstream-catch technique Procedure for obtaining a urine sample which reduces contamination from the resident microorganisms on the genitals.

migration-inhibition factor (MIF) Lymphokine that retains macrophages at the site of activated T cells.

mineralization Conversion of organic material to an inorganic state.

minimum bactericidal concentration (MBC) Smallest concentration of a drug that kills the pathogen against which the drug is used.

minimum inhibitory concentration (MIC) The smallest concentration of a drug that inhibits the multiplication of the pathogen against which the drug is to be used.

minimum medium Medium that supplies only a source of carbon, nitrogen, inorganic salts, and energy.

mitochondrion Organelle in eucaryotic cells that generates ATPs by respiration.

mitosis Duplication of chromosomes and division of the nucleus in eucaryotic cells; usually accompanied by cell division.

mixed culture Culture containing more than one species of microorganism.

mixed infection Infection with more than one type of organism.

mold Fungus that forms large, multicellular aggregates of long, branching filaments.

molecule Smallest quantity of a compound that retains its characteristic properties.

monera Kingdom of procaryotes; one of the five kingdoms defined by Whittaker; the other kingdoms are those of protists, fungi, plant, and animal.

monoclonal antibody Immunoglobulin, specific for a single antigenic determinant, produced in vitro by hybridomas (lymphocytes fused with tumor cells).

monocyte Mononuclear phagocyte from which macrophages differentiate.

monolayer Uniform single layer of cells growing on the surface of a culture container.

monomer Subunit from which a polymer is made.

mononuclear phagocytic system Defense system that consists of monocytes and macrophages.

monosaccharide Sugar that cannot be hydrolyzed to a simpler sugar.

monospecific Capable of reacting with only one kind of antigenic determinant.

monotrichous Arrangement of a single polar flagellum at the end of the cell.

morbidity Number of cases of a disease in a population.

Morbidity and Mortality Weekly Report (MMWR) Publication from the Centers for Disease Control reporting statistics on notifiable disease and current information on public health issues.

mordant Substance that forms an insoluble complex with a dye in the cell, increasing the intensity or tenacity of staining.

morphology Physical form of an organism or its parts.

mortality Number of people in a defined population who die from a specific disease.

motility Ability to move by oneself.

M protein Antiphagocytic antigen on the surface of *Streptococcus pyogenes*.

mucormycosis Infection, usually of a compromised host and usually severe, caused by fungi in the group Mucorales.

multiple fission Asexual reproduction characterized by a series of nuclear divisions, followed by cytoplasmic division into as many parts as there are nuclei.

mutagen Any physical or chemical agent that increases the rate of mutation.

mutant Organism in which a mutation has occurred.

mutation Alteration in the nucleotide sequence in DNA.

mutualism Symbiotic relationship that benefits both members of the relationship.

mycelium Filaments of fungi forming an interwoven mass.

mycology Study of fungi.

mycoplasma Group of bacteria that lack cell walls.

mycorrhiza Growth of fungi on the roots of plants in a relationship that benefits both the plant and the fungus.

mycosis Disease caused by a fungus.

mycotoxicosis Disease caused by ingestion of mycotoxin.

mycotoxin Poisonous substance produced by a fungus.

NAD (nicotinamide adenine denucleotide) Coenzyme that transfers hydrogen atoms between metabolic reactions.

NADP (nicotinamide adenine denucleotide phosphate) Coenzyme that transfers hydrogen atoms between metabolic reactions.

nanometer 10^{-9} meter.

narrow-spectrum antibiotic Drug with high degree of specificity against a small number of microbial types.

natural active immunity Active immunity induced by natural infection.

natural passive immunity Immunity acquired by transfer of maternal antibodies across the placenta to the fetus or in breast milk to an infant.

natural killer (NK) cell Protective cell that lyses virus-infected cells and tumor cells.

nebulization Dispersal of a liquid in a fine spray; used to loosen and liquefy respiratory secretions.

necrosis Localized tissue death that occurs in response to infection or injury.

negative stain Procedure that stains the background and leaves cells colorless.

negri body Cytoplasmic inclusion body that develops in nerve cells infected with rabies virus.

neuraminidase Enzyme that digests neuraminic acid, a component of mucus and cell membranes.

neurotoxin Toxin that damages nerve tissue.

neutralizing antibody Antibody that inactivates the harmful effects of a virus or toxin.

neutron An electrically neutral subatomic particle.

neutrophil Highly phagocytic granulocyte that does not stain with either basic or acid dyes.

nitrification Conversion of nitrogen in ammonia to nitrites and nitrates.

nitrogenase Enzyme that converts nitrogen (N_2) to ammonia (NH_3).

nitrogen cycle Series of reactions in nature

in which atmospheric nitrogen (N_2) and inorganic nitrogen are converted to organic nitrogen and ultimately degraded back to inorganic and atmospheric forms.

nitrogen fixation Conversion of molecular nitrogen (N_2) into a biologically useful form.

nodule See root nodule.

nongonococcal urethritis (NGU) Inflammation of the urethra caused by an organism other than *Neisseria gonorrhoeae,* most often a chlamydia or mycoplasma.

noninvasive disease Infectious disease restricted to the primary site of infection.

nonsense codon Codon for which no amino acid is specified; it signals protein chain termination.

nonseptate hypha Hypha that lacks partitions.

normal flora Those microorganisms which live on the body surfaces in a harmonious relationship with their host.

nosocomial infection Infection caused by exposure to a pathogen within the hospital environment.

notifiable disease Disease that by law must be reported to public health officials whenever diagnosed.

nuclear membrane Membrane separating the nucleus from the cytoplasm in eucaryotes.

nucleic acid Macromolecule composed of nucleotides; RNA and DNA.

nucleocapsid Virus structure composed of the capsid and the enclosed nucleic acid.

nucleoid Region in a procaryotic cell that contains the chromosome.

nucleolus Structure within the cell nucleus where ribosomal RNA is synthesized.

nucleotide Monomer from which nucleic acids are constructed; composed of a purine or pyrimidine base, a five-carbon sugar, and a phosphate group.

nucleus Membrane-bound structure in eucaryotes that contains the genetic material.

nutrient agar Nutrient broth that has been solidified by the addition of agar.

nutrient broth Complex medium containing beef extract and peptone.

nystatin Polyene antibiotic used topically for treating localized yeast infections.

O antigen A surface antigen that is a component of lipopolysaccharide in gram-negative bacteria and is useful for their identification.

objective lens Lens located closest to the specimen in the compound microscope.

obligate anaerobe Organism that grows only in the absence of oxygen.

obligate intracellular parasite Microorganism that grows only in the cytoplasm or nucleoplasm of a host cell.

oligodynamic action Antimicrobial activity of low concentrations of heavy metals.

oncogene Gene responsible for transforming normal cells to malignant cells.

oncogenesis Formation of a tumor.

oncogenic virus Virus that can induce tumor production.

oocyst Sexual stage of many protozoa, especially Sporozoa.

oospore Sexual fungal spore produced by many Zygomycetes.

operator Region in the operon that can bind a specific repressor protein that prevents transcription of the operon's structural genes.

operon Genetic unit consisting of an operator site and the adjacent structural genes, which are regulated as a unit.

ophthalmia neonatorum Eye infection acquired by newborns during passage through an infected birth canal.

opportunistic pathogen Microbe that can cause disease only in an injured, debilitated, or immunologically compromised host.

opsonin Serum protein, such as an antibody, that promotes phagocytosis of the antigen with which it reacts.

opsonization Enhancement of phagocytosis by coating a cell with opsonin.

optical density See turbidity.

optocin Chemical used to distinguish *Streptococcus pneumoniae* from other alpha-hemolytic streptococci by virtue of its lethal effect on the pathogen.

organelle Specialized structure bound by membranes in eucaryotic cells.

organic compound Compound that contains carbon.

osmosis Movement of a solvent across a semipermeable membrane from a dilute solution into a more concentrated one.

osmotic pressure Pressure exerted against a membrane when solutions of different concentrations are present on opposite sides of the membrane.

outer membrane Lipid-protein bilayer that surrounds the peptidoglycan layer in the cell walls of gram-negative bacteria.

oxidation Loss of electrons.

palisade arrangement Orientation of bacilli in parallel rows resembling a picket fence.

pandemic Epidemic that occurs on several continents.

papilloma Benign cutaneous tumor; warts are papillomas.

papule Small, elevated lesion of the skin.

para-aminosalicylic acid (PAS) Antituberculosis chemotherapeutic agent.

parasite Organism that lives in or on a host organism from which it secures some advantage.

parenteral Entering the body by means other than through the alimentary canal.

passive immunity Transfer of antibodies from an immune host to another individual.

Pasteur effect Preference of facultative anaerobes to use aerobic pathways when molecular oxygen is available.

pasteurization Mild heating process used to destroy spoilage microorganisms and most types of pathogens.

pathogen Microbe that causes disease.

pathogenesis Sequence of events during the development of disease and mechanism(s) by which tissues are injured.

pathogenicity Ability to cause disease.

penicillin Any of a group of antibiotics produced by fungi of the genus *Penicillium* that inhibit bacterial peptidoglycan synthesis. The natural antibiotic is often chemically modified to generate semisynthetic penicillins.

penicillinase Enzyme that hydrolyzes penicillin to an inactive form.

peptide Short chain of two or more amino acids.

peptide bond Chemical linkage between amino acids, forming the fundamental backbone of proteins.

peptidoglycan Rigid matrix of bacterial cell walls; consists of N-acetyl glucosamine and N-acetyl muramic acid, arranged in strands cross-linked to one another by amino-acid side chains.

peptone Mixture of short amino-acid chains produced by partially hydrolyzing protein; used as a nutrient supplement in culture media.

periodontal disease Disease of the tissues surrounding teeth.

peristalsis Rhythmic muscular contractions that force substances to move through a tube, such as food through the digestive tract.

peritrichous Arrangement of flagella characterized by even distribution of the appendages over the entire surface of a bacterium.

petri dish Shallow container used for solid-culture media.

pH Measure of the relative hydrogen-ion concentration of a solution; pH values below 7.0 indicate acidic solutions and above 7.0 indicate basic (alkaline) conditions.

phage typing Identification of bacterial strains using bacteriophage susceptibility as the specific indicator.

phagocyte White blood cell that is actively phagocytic.

phagocytosis Cellular engulfment of solids.

phagolysosome Vacuole formed within the phagocyte by the fusion of a phagocytic vacuole and a lysosome.

phase-contrast microscope Compound microscope with optics that detect density differences and highlight details of structures within a cell.

phenol (carbolic acid) Caustic chemical used as an antimicrobial agent.

phenol coefficient Number that describes a chemical's antimicrobial effectiveness relative to that of phenol.

phenotype Total of the genetically controlled characteristics that are expressed at a given time.

phospholipid Complex lipid containing two fatty acids and phosphate attached to a single glycerol molecule.

phosphorylation Coupling of inorganic phosphate (PO_4) to an organic molecule; the term is often used to describe ATP production.

photoautotroph Organism that utilizes light

as its energy source and CO_2 as its principal carbon source.

photoheterotroph Organism that uses light as its energy source and organic molecules as its principal carbon source.

photolysis Use of light energy to split water into hydrogen and oxygen.

photophosphorylation Photosynthetic transformation of light energy into chemical energy of ATP.

photosynthesis Process by which cells convert energy of sunlight into chemical energy.

phototroph An organism that relies on sunlight as its principal energy source.

phycology Study of algae.

pilosebaceous unit Complex of a sebaceous gland and a hair follicle.

pilus Tube extending from the surface of some bacteria that is used for attachment and conjugation.

pinocytosis Engulfment of liquids by a cell.

plaque Zone of clearing in a layer of susceptible cells due to virus replication.

plasma cell Antibody-producing cell derived from a B lymphocyte after antigen activation.

plasma membrane See cell membrane.

plasmid Small, circular piece of extrachromosomal DNA in bacteria. Some plasmids can integrate into a bacterial chromosome.

plate count *See* viable count.

pleomorphic Exhibiting several different shapes.

pneumonia Inflammation of the lung.

pneumonic plague Form of plague that is transmitted directly from person to person by respiratory droplets from someone infected with *Yersinia pestis*.

pneumovax Multivalent vaccine that contains capsule antigens of the 14 most common strains of *Streptococcus pneumoniae*.

point mutation Change in DNA sequence by substituting one nucleotide for another.

polar flagellum Flagellum located on one or both ends of a bacillus or vibrio.

poly-beta-hydroxybutyric acid Storage granule for lipids found in some bacteria.

polyene Any of a group of antibiotics that react with sterols in eucaryotic membranes.

polymer Large molecule composed of simpler molecules (monomers) repeating in a linear or branched arrangement.

polymorphic Changes in form associated with different stages of a complex life cycle.

polymorphonuclear leukocyte (PMN) Phagocytic white blood cell with a multilobed nucleus. See neutrophil.

polymyxin Antibiotic that alters the permeability of the cell membrane by disrupting phospholipid.

polysaccharide Macromolecule composed of repeating sugar subunits.

polysome Chain of ribosomes held together by mRNA; the functional protein synthesis complex.

portal of entry Site in the host that provides a microorganism access to tissues in which initial infection is established.

portal of exit Site from which pathogens are shed from an infected individual.

pour plate Bacterial culture prepared by suspending organisms in melted agar medium and poured into a petri dish to solidify.

precipitation Solidification of soluble substances, often used to detect antigen-antibody reactions.

precipitin Antibody that reacts with soluble antigens and converts them to a solid precipitate.

predisposing factor Condition that reduces a person's resistance to infectious disease.

preservation Prevention of microbial growth to retard spoilage of prepared products or to maintain viability of laboratory cultures.

preservative Compound that retards microbial proliferation in prepared products.

prevalence Number of persons with a specific disease present in a defined population at a given time.

primary immune response Initial production of antibodies following the first exposure to a specific antigen.

primary infection Acute infection that causes the initial illness in a complex disease process.

primary metabolite Intermediate or end product of a biochemical pathway essential for growth of the microbe.

primary stain Initial dye used in a differential staining process.

primary treatment Mechanical removal of most solids from sewage during the first stages of its treatment.

prions Infectious protein complexes that appear to lack nucleic acid.

probenecid Adjunct that, when injected with penicillin or cephalosporins, prolongs therapeutic effectiveness by delaying excretion of the drug.

procaine Adjunct that is often injected with penicillin to delay absorption of the antibiotic and prolong therapeutic effectiveness.

procaryote Cell whose genetic material is not surrounded by a nuclear membrane.

prodromal period Earliest stage of a developing condition or disease.

product Substance formed by a chemical reaction.

promoter Site on DNA to which messenger RNA polymerase binds.

propagated epidemic Outbreak in which the infectious agent can be directly transmitted by infected individuals rather than through a common source.

prophage DNA of a temperate bacteriophage that has integrated into a bacterial chromosome and established lysogeny.

prosthetic group Nonprotein portion of an enzyme.

protein Macromolecule composed of a linear sequence of amino acids that folds into a specific shape.

proteinase Enzyme that hydrolyzes protein to peptides or amino acids.

protists Kingdom of single-cell eucaryotic organisms; one of the five kingdoms defined by Whittaker; in an older, three-kingdom classification scheme, the kingdom of all bacteria, algae, fungi, and protozoa.

proton Positively charged subatomic particle.

protoplast Gram-positive bacterium from which the cell wall has been completely removed.

protozoa Unicellular, nonphotosynthetic eucaryote that lacks polysaccharide cell walls.

pseudohypha Chain of elongated yeast cells that resembles a mold filament.

pseudopod Extension of an ameboid cell's surface for motility and phagocytosis.

psychrophile Organism that grows optimally at temperatures below 20°C.

pure culture Culture that contains a single species of microorganism.

purified protein derivative (PPD) Purified protein derived from *Mycobacterium tuberculosis* cultures and used for skin testing.

purine Category of nucleotide bases; adenine and guanine are purines.

purulent Pus-producing.

pus Accumulation of dead white blood cells, bacteria, and serous fluid in tissues.

pustule Elevated surface lesion that contains pus.

pyelonephritis Inflammation of the kidney.

pyrimidine Category of nucleotide bases; thymine, cytosine, and uracil are pyrimidines.

pyrogen Fever-inducing substance.

pyruvic acid A three-carbon compound that is the end product of glycolysis.

quarantine Isolation of individuals having (or suspected of having) communicable disease.

quarternary ammonium compound Cationic surfactant used as a disinfectant and antiseptic.

quellung reaction Appearance of swollen bacterial capsules following reaction between bacteria and capsule-specific antibody.

quinine Antimalarial chemotherapeutic agent.

radiation Energy that travels as particles or in waves; includes energy emitted from a radioactive source and light, such as ultraviolet radiation.

radioimmune assay (RIA) Immunological test that uses a radioactively labeled antigen or antibody to detect minute amounts of its complementary substance.

reading frame Synchrony in which nucleotide triplets are read during translation (protein synthesis).

reagin Immunoglobulin E antibody that mediates immediate-type allergy (Type I hypersensitivity); also a substance produced in response to treponemal infections. Its presence in the blood of persons with syphilis, easily detected by the VDRL test and other syphilis-screening procedures, provides evidence of this disease.

recalcitrance Persistence of compounds in the environment because of the inability of microbes to metabolize them at significant rates.

receptor site　Location of molecules on the cell's surface to which a virus attaches.

recombinant DNA technology　Genetic engineering techniques that restructure cellular DNA by recombination, producing genetically unique organisms.

recombination　Joining of two or more pieces of DNA from different sources.

red tide　Bloom of dinoflagellate algae in seawater; often associated with paralytic shellfish poisoning.

reducing agent　Chemical that reacts with molecular oxygen; used in media for growing anaerobes.

reduction　Gain in electrons.

refractive index　The amount light is bent when it travels from air to a medium with a different density; measures the ability of microscope lenses to magnify images.

regulatory gene　Gene that specifies the production of a repressor protein.

relapse　Recurrence of disease after apparent recovery.

relative risk　Likelihood that one group of people will acquire a disease as compared with another group.

rennin　Enzyme obtained from a mold; coagulates milk and is used for cheese production.

repression　Inhibition of enzyme production in the presence of adequate amounts of the pathway's end products.

repressor　Protein that binds to DNA and prevents transcription.

reservoir of infection　Place where a microbe normally lives and multiplies.

resident flora　*See* normal flora.

resistance　Body's nonspecific and specific defenses against disease; lack of sensitivity of a microbe to an antimicrobial agent.

resistance (R) factor　Transmissible plasmid that provides bacteria with resistance to one or more antimicrobial drugs.

resistance transfer factor (RTF)　Portion of the R factor that allows the plasmid to be replicated and transferred.

resolution　Ability to distinguish two adjacent objects as separate and distinct images; clarity of image.

respiration　Production of ATP by oxidation of chemical compounds, using oxygen (or some other inorganic molecule) as the terminal electron acceptor.

respiratory droplet　Moist particle discharged into the environment from the respiratory tract.

restriction analysis　Comparison of nucleic acid pieces produced by digestion with restriction endonucleases.

restriction endonuclease　Enzyme that cleaves DNA at specific nucleotide sequences, often leaving single-stranded regions ("sticky ends").

reticulate body　Vegetative form of chlamydia that actively multiplies in host cells.

reticuloendothelial system (RES)　See mononuclear phagocytic system.

reverse (protective) isolation　Protecting persons who are especially susceptible to infection by minimizing their exposure to microbes.

reverse transcriptase　Enzyme that polymerizes a molecule of DNA using RNA as the template for determining sequence.

Reye's syndrome　Acute, frequently fatal, condition of the brain and liver; occurrence has been linked to treating viral-induced fevers with aspirin.

R factor　*See* resistance factor.

ribonucleic acid (RNA)　Macromolecule built of ribonucleotides that contain the bases adenine, uracil, guanine, and cytosine (see messenger RNA, transfer RNA, and ribosomal RNA).

ribosomal RNA　Ribonucleic acid from which ribosomes are constructed.

ribosome　Site of protein synthesis.

rickettsia　Organism in the family Rickettsiaceae.

rifampin　Chemotherapeutic agent that inhibits bacterial RNA synthesis; used for treating tuberculosis and carriers of gonorrhea.

ringworm　*See* tinea.

RNA polymerase　Enzyme that synthesizes molecules of RNA.

RODAC plate　Petri plate that elevates the surface of the solid-culture medium above the edges of the container so the medium can be pressed directly against objects to be sampled for microbial contamination.

root nodule Swollen area that contains symbiotic nitrogen-fixing bacteria; develops on the roots of leguminous plants.

rose spot Tiny red lesion that occurs on the skin of persons with typhoid or paratyphoid fever.

rumen Stomach chamber in ruminants (cows, sheep, etc.) that contains cellulose-digesting microorganisms.

Sabin vaccine Attenuated trivalent preparation of poliovirus administered orally to induce immunity against poliomyelitis.

Sabouraud's agar Culture medium usually employed for isolating and culturing fungi.

saliva Watery secretions of the mouth.

Salk vaccine Formalin-inactivated poliovirus preparation administered by injection to induce immunity against poliomyelitis.

salpingitis Inflammation of the fallopian tubes.

saprophyte Organism that grows on dead organic material.

sarcina Eight cocci grouped together, often in a cubic arrangement.

Sarcodina Class of protozoa that are motile by pseudopods.

satellitism Growth of bacteria around colonies of another kind of microbe that supplies a nutrient needed by the satellite organism.

saturated steam Water vapor at the temperature at which it is produced.

scanning electron microscope Microscope that uses an electron beam reflected from the surface of the specimen to create a three-dimensional image of the object.

sebum Oil secretion of sebaceous glands.

secondary immune response Rapid production of high antibody concentrations that follows the primary response.

secondary infection An infection that develops because of reduced host resistance induced by a primary infection.

secondary metabolite Compound produced by a microorganism but not essential for the microorganism's growth.

secondary treatment Biological degradation of organic matter in sewage after primary treatment.

secretion Release of a substance from the cell into its surroundings.

selective medium Culture medium that inhibits the growth of contaminating microorganisms while allowing proliferation of the desired type of cells.

selective toxicity Ability of a drug to destroy the pathogen while leaving the host unharmed.

selective transport Ability of the selectively permeable cell membrane to determine which molecules may travel between the external medium and the interior of the cell.

semipermeable membrane Membrane that allows the passage of small molecules but not larger ones.

semisynthetic antibiotic Antibiotic that is chemically modified after its biological production by microbes.

sensitization Production of immunological memory following initial contact with an antigen; often denotes the exposure that confers hypersensitivity (*see* sensitizing dose).

sensitized memory cell Antigen-stimulated lymphocyte that can be rapidly activated upon exposure to the corresponding antigen.

sensitizing dose Initial exposure to an allergen that stimulates the development of hypersensitivity. There are no symptoms, however, until subsequent contact with the same allergen.

sepsis Presence of pathogenic microorganisms in the blood or other tissues of the body.

septate Presence of partitions (septa) between adjacent cells in a fungal hypha.

septicemia Presence of actively proliferating microorganisms in blood.

septic tank Underground container for storing sewage for primary treatment and partial biological decomposition.

sequela Disease that develops only after the initial infection is resolved.

serial dilution Series of dilutions, each being more dilute than the preceding one.

serological test Any diagnostic technique that depends on specific reactions between antigen and antibody.

serology Study of antigen-antibody reactions in vitro.

serotype Strain of microbe that is immunologically distinct from other members of the same species.

serum Liquid portion of the blood that remains after removal of erythrocytes and clotting factors.

serum sickness Immune complex disease due to production of antibodies against the foreign antibodies in serum used for passive immunization.

sewage Domestic wastewater; contains surface runoff, human excreta, and other domestic wastes.

sex pilus *See* fertility (F) pilus.

sexually transmitted disease (STD) Disease that is transmitted directly from person to person primarily by sexual contact; some STDs may also be transmitted by nonsexual routes.

sexual reproduction Reproduction by the union of two haploid sex cells.

sexual spore Spore formed by the fusion of two nuclei.

shocking dose Allergen exposure that elicits symptoms of hypersensitivity in a sensitized host.

siderophore Compound that tightly binds iron.

simple stain Use of a single dye to stain the cells in a smear for microscopic examination.

single-cell protein Food produced by large-scale cultivation of microorganisms.

skin test Intradermal injection of antigen to determine immunological sensitivity to a pathogen or its products.

slant Solid-culture medium that solidifies in a tilted test tube to produce a slanted surface for microbial inoculation.

slime layer Loosely adhering surface layer surrounding some types of bacterial cells, but not as distinct as a capsule.

slow virus Unidentified infectious agent that causes several slowly progressive neurological diseases such as kuru and Creutzfeldt-Jakob disease.

sludge Solid matter that settles during sewage treatment.

smear Material on a glass slide for microscopic examination.

solubility Substance's tendency to dissolve in a solute, usually water.

specialized transduction Transfer of bacterial genes to a recipient by a temperate bacteriophage; the transferred genes lie adjacent to the prophage integration site.

species An organism's ultimate designation in the taxonomic hierarchy; members of the same species share common characteristics that distinguish them from members of other species.

specific-macrophage-arming factor (SMAF) Lymphokine that enhances the ability of macrophages to kill specific antigenic target cells.

spectrophotometer Instrument that measures the amount of light transmitted through a sample.

spike Protein projection on the surface of some enveloped viruses.

spirillum A rigid spiral or corkscrew-shaped bacterium.

spirochete A motile, corkscrew-shaped bacterium that possesses an axial filament.

spontaneous generation Theory that nonliving substances can be converted into living organisms in the absence of preexisting cells.

sporangiospore Asexual fungal spore contained in a sporangium.

sporangium Sac containing one or more sporangiospores borne on the tip of an aerial hypha.

spore Reproductive structure produced by some bacteria, fungi, and a few primitive plants; some spores are more resistant to adverse conditions than vegetative cells are.

spore strip Filter paper impregnated with bacterial endospores and used to determine the effectiveness of sterilization processes.

sporicidal Lethal to spores.

Sporozoa Class of protozoa characterized by the absence of motility in the adult forms.

sporozoite Infectious stage in the complex life cycle of the malaria parasite (*Plasmodium* sp.); these trophozoites develop in the mosquito and are transmitted by its bite.

sporulation Production of one or more spores by vegetative cells.

spread-plate method Technique of obtaining isolated colonies by spreading microbes in a liquid sample on the surface of solid medium.

sputum Secretion of the lower respiratory tract.

standard curve Graphic plot that shows the

relationship between two characteristics, such as bacterial concentration and turbidity.

starch Linear polysaccharide composed of glucose subunits linked together by alpha-glycoside bonds.

starter culture Inoculum for initiating fermentations or other useful microbial processes.

stationary phase Stage in the bacterial growth curve when total bacterial numbers are neither increasing nor decreasing.

sterile Free from all living microorganisms.

sterilization Elimination of all microbial forms of life, including spores and viruses.

steroids Group of lipids characterized by a four-ring structure.

sterol Type of lipid found in the membranes of eucaryotes and mycoplasma; cholesterol and ergosterol are examples.

sticky ends Single-stranded ends of a linear piece of double-stranded DNA; they base-pair with complementary sticky ends.

strain Subgroup of individuals within a single species that possess several properties distinct from the other members of the species.

streak-plate method Technique for obtaining isolated colonies by inoculating a sample over a large area of the surface of a solid-culture medium.

streptobacilli Rod-shaped bacteria that form chains.

streptococci Spherical bacteria that form chains.

strict isolation Restriction of an infectious patient to a private hospital room which no one enters without wearing a mask, gown, and gloves.

structural analog Chemical that closely resembles another compound in molecular structure.

structural gene Gene that specifies the production of an enzyme.

sty Infected eyelash follicle.

subacute sclerosing panencephalitis (SSPE) Rare form of brain inflammation that occurs in children, usually with fatal consequences; a sequela of measles.

subclinical infection See inapparent infection.

subculture Microbial culture that results from the transfer of microbes from one culture to a fresh medium.

subcutaneous In the tissues beneath the skin.

subcutaneous mycosis Fungal infection usually restricted to the skin, the tissue beneath the skin, and the lymphatics.

substrate Reactant in an enzyme-mediated reaction.

substrate-level phosphorylation Production of ATP by direct transfer of a high-energy phosphate group from an organic substrate to ADP.

sulfa drugs Synthetic sulfur-containing antimicrobial compounds that are structural analogs of para-aminobenzoic acid (see sulfonamide).

sulfonamide Synthetic chemotherapeutic agent that competitively inhibits microbial metabolism by interfering with the synthesis of folic acid from para-aminobenzoic acid.

superficial mycosis Fungal infection that affects only the outer dead layers of the skin.

superinfection Secondary infection that develops during the course of chemotherapy against another infectious disease.

superoxide dismutase Enzyme that degrades the superoxide (O^-) radical.

surfactant (surface-active agent) Wetting agent, detergent, or any other compound that interferes with the interaction between the cell's surface and its aqueous environment.

susceptibility Lack of resistance to disease.

sylvatic plague Bubonic plague acquired from a wild animal reservoir rather than from domestic rats.

symbiosis Close association of two dissimilar organisms living together.

symptom Disease-induced change in condition that is perceived by the person suffering the disease.

syndrome Complex of signs and symptoms that accompany a specific disease.

synergy Enhancement of effectiveness of chemicals when used in combination; the increased activity is greater than the sum of the activity of the two agents used alone.

synthetic medium Nutrient growth medium in which each chemical compound is added separately so that the exact chemical nature of the solution is known.

systemic infection Active proliferation of microorganisms throughout the body.

systemic mycosis Fungal disease that may affect the brain, bone, viscera, skin, or any area of the body; also called "deep mycosis."

target organ Body site most commonly attacked by a particular pathogen or its by-products.

taxonomy Science of classifying organisms into categories that contain individuals with similar characteristics.

teichoic acid Macromolecules composed of sugars and phosphates found in cell walls of gram-negative bacteria.

temperate bacteriophage Virus that establishes lysogeny with its host bacterium; its genome integrates into the host chromosome and replicates with the bacterial DNA.

tertiary treatment Final treatment of sewage; use of chemical and physical means to remove BOD, nitrogen, and phosphorus from wastewater, rendering it suitable for drinking.

tetanospasm Neurotoxin, produced by *Clostridium tetani*, responsible for the paralytic symptoms of the disease tetanus.

tetracycline Broad-spectrum antibiotic isolated from *Streptomyces* and used for treatment of bacterial and amebic infections; tetracyclines interfere with protein synthesis.

tetrad Arrangement of cocci in packets of four.

thermophile Organism that grows best at temperatures above 40°C.

thrombocyte (platelet) Smallest of the cells in the blood; essential for coagulation.

thylakoid Folded membrane structure that contains the photosynthetic pigments of cyanobacteria and chloroplasts.

thymine Pyrimidine nucleotide base found in DNA but not RNA.

thymus Organ that programs lymphocytes to become T cells.

tincture Solution that uses alcohol or a water-alcohol combination as the solvent.

tinea Cutaneous infection by dermatophytes; ringworm.

tissue tropism Affinity of a pathogen for a particular tissue.

T killer lymphocyte T cell that physically attaches to cells and destroys them by membrane disruption and lysis.

T lymphocyte Lymphocyte that has been programmed by the thymus for participation in cell-mediated immunity.

toxemia Presence of toxins in the blood.

toxic shock syndrome Severe acute disease caused by strains of *Staphylococcus aureus*, phage group I, that produce an exotoxin that is distributed by the blood from a site of local infection; occurs most often in menstruating women using highly absorbent tampons.

toxigenic Exotoxin-producing.

toxoid Inactivated form of toxin that is antigenically identical to the active toxin; used for immunization and skin testing.

trace element Element essential for growth but in extremely small amounts.

tracheostomy Surgical opening into the trachea to maintain an airway to the lungs.

transcriptase Enzyme responsible for transcribing mRNA from DNA; DNA-dependent mRNA polymerase.

transcription Process of assembling a molecule of messenger RNA with a nucleotide sequence complementary to a corresponding segment of DNA.

transduction Bacteriophage-mediated gene transfer from one bacterium to another.

transfer DNA (tDNA) Plasmid transferred from the bacterium *Agrobacterium tumefaciens* to infected plant cells, where it integrates into the eucaryote's chromosome.

transfer factor Soluble component extracted from immune T lymphocytes that can sensitize other T cells to an antigen.

transferrins Iron-binding proteins that reduce free-iron levels in the body.

transfer RNA (tRNA) Class of RNA molecules that carry specific amino acids and insert them into growing protein chains.

transformation Transfer of genetic information by free DNA released from disrupted bacteria.

transient flora Microbes that temporarily colonize the surface of a person or animal.

translation Process of protein synthesis using the nucleotide sequence in mRNA to determine the amino acid sequence in the corresponding protein.

transmission electron microscope Microscope that transmits an electron beam through thin sections of a specimen.

transport medium
tain the viability of microorganisms in a specimen being transported to the laboratory.

transtracheal aspiration Suction of sputum into a sterile needle passed through the skin into the trachea.

***Treponema pallidum* immobilization (TPI) test** Diagnostic test for syphilis that uses darkfield microscopy to observe whether spirochetes from a lesion are immobilized by *T. pallidum*-specific antibodies.

tricarboxylic acid (TCA) cycle *See* Krebs cycle.

trickle filter Apparatus for secondary treatment of sewage; wastewater is sprayed over a gravel bed seeded with microbes that digest the organic compounds in sewage.

trophozoite Vegetative stage of a protozoan.

tubercle Characteristic lesion produced by *Mycobacterium tuberculosis* infection.

turbidity Cloudiness of a liquid.

ultramicrotome Knife that cuts ultrathin sections of a specimen, usually in preparation for transmission electron microscopy.

ultraviolet (UV) radiation Radiation at wavelengths between 180 and 390 nm.

uncoating Release of virus nucleic acid from its capsid.

undulating membrane Flexible sheet of material that joins the flagella of certain protozoa to the cell surface.

uracil Pyrimidine nucleotide base found in RNA but not DNA.

urethritis Inflammation of the urethra.

use-dilution method Laboratory evaluation of a disinfectant; determines the highest dilution that kills a standard number of test bacteria.

vaccination Inducing active immunity against disease by introducing a vaccine into the individual; artificial active immunity.

vaccine Preparation of microbes or toxoid that can no longer induce severe disease but can still stimulate immunity against the corresponding pathogen or toxin.

vacuole Membrane-bound intracellular inclusion in eucaryotic cells; a hollow, gas-filled chamber in some procaryotes.

vaginitis Inflammation of the vagina.

VDRL (venereal disease research laboratory) test Rapid screening test for syphilis that detects the presence of reagin as an indicator of *Treponema pallidum* infection.

vector Living organism, such as an arthropod, that transmits disease from one individual to another (*see* biological vector and mechanical vector).

vegetative cell Actively growing, feeding, and proliferating stage of an organism, as opposed to dormant forms such as spores or cysts.

vehicle Nonliving material or object that can transmit infectious disease.

venereal disease Infectious disease acquired by sexual intercourse or genital contact.

venipuncture Inserting a needle into a vein.

vertical transmission Spread of disease from parent to offspring by an infected sperm or egg, by passage across the placenta, or during the birth process.

vesicle Fluid-filled blister.

viable count Laboratory process of determining the microbial concentration in a sample by plating samples on (or in) solid media and counting the resulting colonies.

vibrio Comma-shaped bacterium.

viremia Presence of viruses in the blood.

virion Infectious virus particle.

viroid Group of small RNA molecules that cause some infectious diseases of plants.

virology Study of viruses.

viropexis Engulfment of viruses by a host cell using the process of phagocytosis.

virucidal Capable of destroying viruses.

virulence Degree of pathogenicity of a microorganism.

virulence factors Microbial attributes that increase either infectivity or severity of disease.

virus Submicroscopic, noncellular infectious entity that is an obligate intracellular parasite; consists of nucleic acid surrounded by a protein coat and sometimes an envelope.

virustatic Capable of inhibiting viral proliferation without destroying the virus.

vitamin Nutrient supplement; some vitamins are coenzyme precursors.

volutin granule Intracellular reservoir of phosphate characteristic of some bacteria; the granules have a marked affinity for basic dyes (see metachromatic granule).

vulnerable food Food that can support the proliferation of microbes that cause foodborne illness.

wandering macrophage Actively phagocytic cell that can travel to sites of infection or inflammation.

Weil-Felix reaction Serological test for diagnosing rickettsial disease; uses strains of *Proteus* OX as the indicator of positive reactions.

whey Fluid portion of milk after curdling.

wild type Organism with the nonmutant genotype.

wort Aqueous extract of malted barley produced during beer production.

xanthan Microbial exopolysaccharide used as a solidifying and stabilizing agent in cosmetics and many other commercial products.

yeast Unicellular fungus cell.

zone of inhibition Clear area within a lawn of bacteria in which growth fails to occur because of the presence of an inhibitor.

zoonosis Disease transmitted from a vertebrate animal to human.

Zygomycetes Class of fungi characterized by nonseptate hyphae, asexual sporangiospores, and sexual zygospores.

zygospore Sexual spore produced by Zygomycetes fungi.

Chapter 1

1-1   (a) Courtesy of Rijksmuseum.

     (b) *Philosophical Transactions of the Royal Society of London* **14**, No. 159:568 (1684).

1-3   (a) Courtesy of S. Hay.

     (b) Courtesy of Fleischmann's Yeast Nabisco Brands Inc.

     (d) E. Kaneshiro, L. Buschel, S. Merkel, and D. Rhoads, *J. Protozool.* **26**: 147 (1979).

     (e) Courtesy of S. Murray.

     (f) R. W. Castenholz, *Microbial Ecol.* **3**: 79 (1977).

     (g) Courtesy of J. D. Almeida.

1-5   Bettmann Archives.

1-6   (a) Bettmann Archives.

     (b) Courtesy of J. Higbee and R. Marchino.

1-7   Bettmann Archives.

Chapter 2

2-1   With permission from J. D. Dodge, *The Fine Structure of Algae*, Academic Press, London, 1973.

2-2   I. D. J. Burdett and L. G. E. Murray, *J. Bacteriol.* **119**: 1039 (1974).

2-3   M. R. Edwards, M. A. Gordon, E. W. Lapa, and W. C. Ghiorsi, *J. Bacteriol.* **94**: 766 (1967).

2-6   Courtesy of P. Ma.

2-7   (a) K. J. Aufderheide, P. M. Daggett, and T. A. Nerad, *J. Protozool.* **30**: 128 (1983).

     (b) Courtesy of R. Dute.

2-8   (a) J. W. Rippon, *Medical Mycology*, W. B. Saunders, Philadelphia, 1974.

     (b) Courtesy of J. Moore.

2-9   Courtesy of D. Diener.

2-10   Courtesy of S. Murray.

2-11   Courtesy of D. C. Birdsell.

2-11   Courtesy of S. Murray.

2-13   Courtesy of B. Larsen.

2-14   (a) Courtesy of F. Eiserling.

     (b) A. K. Kleinschmidt, *Biochim. Biophys. Acta* **61**: 861 (1962).

Chapter 3

3-1   From *Cell Ultrastructure* by William A. Jensen and Roderic B. Park. © 1967 by Wadsworth Publishing Company, Inc. Reprinted by permission of Wadsworth Publishing Company, Belmont, California.

3-2   (a) Courtesy of American Optical.

3-3   (b) Courtesy of R. T. Kelley. From J. B. Henry (Ed.), *Clinical Diagnosis and Management by Laboratory Methods*, Vol. II, W. B. Saunders, Philadelphia, 1979.

3-4   J. Wolfe and G. W. Grimes, *J. Protozool.* **26**: 82 (1979).

3-5   Courtesy of Centers for Disease Control.

3-6   Courtesy of J. P. Delafield.

3-7   I. D. J. Burdett and R. G. E. Murray, *J. Bacteriol.* **119**: 303 (1974).

3-8   Z. Yoshii, J. Tokunaga, and J. Tawara. *Atlas of Scanning Electron Microscopy in Microbiology*, Igaku-Shoin Ltd., Tokyo, 1976.

3-10   (b) Courtesy of J. Higbee and R. Marchino.

Chapter 4

4-1   (a) D. L. Shungu, J. B. Cornett, and G. D. Schockman, *J. Bacteriol.* **138**: 598 (1979).

     (b) L. F. Ellis, D. K. Herron, D. A. Preston, L. K. Simmons, and R. A. Schlegel, *Antimicrob. Agents Chemother.* **9**: 334 (1976).

     (c) N. S. Hayes, K. E. Muse, A. M. Collier, and J. B. Baseman, *Infect. Immun.* **17**: 174 (1977).

4-2   (a) Courtesy of J. M. Slack and M. H. Gerenscer.

(b) Courtesy of E. S. Boatman. From M. H. Barile and S. Razin (Eds.), *The Mycoplasma*, Vol. I, Academic Press, New York, 1979.

4-4   (b) Courtesy of G. Cohen-Bazire.

4-5   Courtesy of G. B. Chapman.

4-7   T. Kawata, K. Asaki, and A. Tagaki, *J. Bacteriol.* **81:** 160 (1961).

4-9   T. J. Beveridge, *Can. J. Microbiol.* **26:** 643 (1980).

4-10   J. P. Robinson, R. A. Robinson, and J. H. Hash, *J. Bacteriol.* **117:** 900 (1974).

4-11   (a) T. J. Beveridge, *Can. J. Microbiol.* **26:** 643 (1980).

4-12   E. Juni and W. Taylor, *J. Bacteriol.* **81:** 688 (1961).

4-13   Courtesy of J. W. Costerton.

4-14   Courtesy of R. J. Gibbons. From E. H. Beachey (Ed.), *Bacterial Adherence*, Chapman and Hall, London, 1980.

4-15   Courtesy of E. Leifson.

4-17   Department of Health and Human Services, Centers for Disease Control.

4-18   J. P. Duguid, *J. Gen. Microbiol.* **21:** 271 (1959).

4-20   A. Umeda and K. Amako, *J. Gen. Microbiol.* **118:** 215 (1980).

Insert photo: W. Stoeckenius, *J. Bacteriol.* **148:** 352 (1981).

Chapter 5

5-2   Courtesy of S. Murray.

5-3   M. R. Gambrill and C. Wisseman, *Infec. Immun.* **8:** 519 (1973).

5-4   (a) J. W. Moulder, in D. Schlessinger (Ed.), *Microbiology 1979*, American Society for Microbiology, Washington, D.C., 1979.

(b) Courtesy of J. W. Moulder.

5-5   D. F. Kohn, L. S. Magill, and N. Chinookoswong, *Infect. Immun.* **35:** 730 (1984).

5-6   R. M. Putzrath and J. Maniloff, *J. Virol.* **28:** 254 (1978).

Chapter 6

6-1   S. A. Morse, in D. Schlessinger (Ed.), *Microbiology 1976*, American Society for Microbiology, Washington, D.C., 1976.

6-3   (a) Z. Yoshii, J. Tokunaga, and J. Tawara, *Atlas of Scanning Electron Microscopy in Microbiology*, Igaku-Shoin, Ltd., Tokyo, 1976.

6-7   Courtesy of Forma Scientific.

6-12   Courtesy of Gelman Sciences.

Chapter 7

7-1   (b) J. W. Rippon, *Medical Mycology*, W. B. Saunders Co., Philadelphia, 1974.

7-2   (a) Courtesy of Fleischmann's Yeast Nabisco Brands, Inc.

7-4   (a) Z. Yoshii, J. Tokunaga, and J. Tawara, *Atlas of Scanning Electron Microscopy in Microbiology*, Igaku-Shoin Ltd., Tokyo, 1976.

(b) E. J. Bottone, I. Weitzman, and B. A. Hanna, *J. Clin. Microbiol.* **9:** 530 (1979).

(c) Courtesy of E. S. Beneke.

(d) D. J. Bibel, D. A. Crumrine, K. Yee, and R. D. King, *Infect. Immun.* **15:** 959 (1977).

7-7   (a) C. W. Emmons, C. H. Binford, J. P. Utz, and K. J. Kwon-Chung, *Medical Mycology*, Lea and Febiger, Philadelphia, 1977.

(b) R. C. Jones and R. A. Goodwin, *Amer. J. Med.*, **70**, 864, 1981.

7-8   Courtesy of Armed Forces Institute of Pathology. (a) Neg. No. 56-2004-15; (b) Neg. No. 75-5301.

7-9   (a) Courtesy of Armed Forces Institute of Pathology, Neg. No. 57-6860.

(b) From J. Sinski, *Dermatophytes in Human Skin, Hair, and Nails*, 1974. Courtesy of Charles C. Thomas, Publisher, Springfield, Illinois

7-10   (a) Courtesy of K. Hunter.

(b) Courtesy of C. W. Emmons.

7-11   (a) Courtesy of G. P. Segal.

(b,c) Courtesy of S. H. Sun.

Chapter 8

8-1   Courtesy of Armed Forces Institute of Pathology. (a) Neg. No. 75-9300; (b) Neg. No. 75-9301.

8-6   Courtesy of T. Watanabe.

8-7   K. W. Jeon and M. S. Jeon, *J. Protozool.* **23:** 83 (1976).

8-9   Courtesy of Armed Forces Institute of Pathology. (a) Neg. No. 74-19698; (b) Neg. No. 75-14469; (c) Neg. No. 219371-45.

8-10   Courtesy of World Health Organization.

8-11   Courtesy of the Centers for Disease Control.

8-12   Courtesy of G. Fryxell.

8-13   Courtesy of R. R. Davies and the Liverpool School of Tropical Medicine.

8-14   J. M. Sieburth, H. L. Pratt, P. W. Johnson, and D. Scales, *Microbial Seascapes*, University Park Press, Baltimore, 1975.

Chapter 9

9-1   Courtesy of E. Kellenberger.

9-2   R. L. Steere and T. L. Schaffer, *Biochim. Biophys. Acta* **28:** 241 (1958).

9-3   (c) Courtesy of H. Frank. Modified from *Z. Naturforsch.* **33C:** 124 (1978).

9-4   (a) Courtesy of A. Knight.

    (b) C. R. Madeley, *Virus Morphology*, Churchill Livingstone, London, 1972.

    (c) Courtesy of J. T. Finch.

    (d) C. R. Madeley, *Virus Morphology*, Churchill Livingstone, London, 1972.

    (e) Courtesy of C. J. Smale.

9-8   (a,b) P.K.Y. Wong, *Cell* **10:** 245 (1977).

9-9   (b) Courtesy of E. Kellenberger.

9-10 Courtesy of G. Kellenberger and E. Kellenberger.

9-12 C. H. Zierdt, *J. Virol.* **14:** 1268 (1974).

9-13 Courtesy of N. Sharon.

9-14 E. M. Morgan and F. Rapp, *Bacterial Rev.* **41:** 636 (1977).

Chapter 10

10-4 G. A. Peters and R. A. Cellarius, *Bioenergetics* **3:** 345 (1972).

10-12 P. A. Sokol, D. E. Ohman, and B. H. Iglewski, *J. Clin. Microbiol.* **9:** 538 (1979).

10-15 (a) Courtesy of L. J. LeBeau.

    (b) A. T. Willis, in A. Balows (Ed.), *Anaerobic Bacteria: Role in Disease*, 1974. Courtesy of Charles C. Thomas, Publisher, Springfield, Illinois

Chapter 11

11-4 (a) C. C. Remsen, S. W. Watson, J. N. Waterbury, and H. G. Truper, *J. Bacteriol.* **95:** 2374 (1968).

    (b) G. A. Peters and R. A. Cellarius, *Bioenergetics* **3:** 345 (1972).

Chapter 12

12-8 O. L. Miller, *Sci. Amer.* **228:** 34 (1973).

12-15 R. Curtiss, L. G. Caro, D. P. Allison, and D. R. Stallions, *J. Bacteriol.* **100:** 1091 (1969).

12-17 Courtesy of J. D. Griffith.

Insert photo: B. N. Ames, J. McCann, and E. Yamasaki, *Mutation Res.* **31:** 347 (1975).

Chapter 13

13-5 J. G. Franklin and D. E. Bradley, *J. Appl. Bacteriol.* **20:** 467 (1957).

13-6 Courtesy of Millipore Corp.

13-8 Courtesy of Loma Linda University.

13-9 Courtesy of A. T. I.

13-11 Courtesy of L. J. Morse.

Chapter 14

14-5 E. W. Goodell, M. Fuzio, and A. Tomasz, *Antimicrob. Agents Chemother.* **13:** 514 (1978).

14-7 Courtesy of C. H. Kirkpatrick.

14-8 C. Witkop and R. Wolf, *J. Amer. Med. Assoc.* **185:** 1008 (1963). Copyright © 1963, American Medical Association.

14-10 Courtesy of J. H. Carter.

14-11 (b) Courtesy of Micro-Media Systems.

Insert photo: National Library of Medicine.

Chapter 15

15-4 Courtesy of W. Fluegel.

15-5 (a) T-Arai, Y, Mikami, and K. Yokoyama, *Sabouraudia* **15:** 171 (1977).

Chapter 16

16-1 Modified from L. S. Young, *Ann. Intern. Med.* **86:** 456 (1977).

16-4 (a) Courtesy of B. L. Wasilauskas. Reproduced with permission from *Amer. J. Clin. Pathol.* **72,** 463 (1979).

16-5 Courtesy of G. Koshi, Reproduced with permission from *Amer. J. Clin. Pathol.* **71:** 709 (1979).

16-6 Courtesy of A. Esser.

16-8, From R. G. Kessel and R. H. Kardon, *Tissues and Organs: A Text-Atlas of Scanning Electron Microscopy*, W. H. Freeman and Co. Copyright © 1979.
16-11

16-14 P. Rohlich, *J. Cell Biol.* **51:** 465 (1971).

16-16 Courtesy of Armed Forces Institute of Pathology Neg. No. 53-1280.

16-18 Courtesy of S. Sell.

Chapter 17

17-1 Courtesy of J. P. Delafield.

17-2 Courtesy of G. Segal.

17-3 (a) M. Ward, J. N. Robertson, P. M. Englefield, and P. J. Watt, in D. Schlessinger (Ed.), *Microbiology 1975*, and American Society for Microbiology, Washington, D.C., 1975.

    (b) C. S. Eden, B. Eriksson, and L. A. Hanson, *Infect. Immun.* **18:** 767 (1977).

17-5 Courtesy of T. Stanley.

Chapter 18

18-2 F.R. Moulton. AAAS Publication No. 17, p. 118. Copyright © 1942 by the American Association for the Advancement of Science.

18-3 Courtesy of C. J. Webb.

18-4 Centers for Disease Control, "Annual Summary 1981," *Morbidity Mortality Weekly Report*. **30**:54 (1982).

18-5 From Snow on Cholera, The Commonwealth Fund, New York, 1936.

18-6 Courtesy of Centers for Disease Control.

18-7 Centers for Disease Control, "Annual Summary 1981," *Morbidity Mortality Weekly Report* **30**:54 (1982).

18-8 Courtesy of Central Office of Information, London.

18-9 Courtesy of the World Health Organization.

Chapter 19

19-1 Charles Gardner.

19-4 Courtesy of R. C. Bartlett.

Insert: Courtesy of the American Society for Microbiology.

Chapter 20

20-2 K. Tsai, *Infect. Immun.* **18**: 780 (1977).

20-6 From T. Nasemann, *Viral Diseases of the Skin, Mucous Membranes, and Genitals*, Georg Thieme Verlag, Stuttgart, 1977.

20-7 E. M. Morgan, and F. Rapp, *Bacteriol. Rev.* **41**: 636 (1977).

20-8 (a) Courtesy of I. S. Snyder.

20-9 J. Walter, *Introduction to the Principles of Disease*, W. B. Saunders, Philadelphia, 1977.

20-10 (a) Courtesy of L. J. LeBeau.
(b) Courtesy of R. Facklam.

20-11 M. R. Dische, P. A. Quinn, E. Czegledy-Nagy, and J. N. Sturgess. Reprinted with permission from *Amer. J. Clin. Pathol.* **71**: 167 (1979).

20-12 Courtesy of F. Chandler. Reproduced with permission from *Amer. J. Clin. Pathol.* **71**: 43 (1979).

20-14 Courtesy of Armed Forces Institute of Pathology. Neg. No. 59-11164.

20-15 (a) N. F. Conant, D. T. Smith, R. D. Baka, J. L. Callaway, and D. S. Martin, *Manual of Clinical Mycology*, W. B. Saunders, Philadelphia, 1954.
(b) E. W. Koneman, G. D. Roberts, and S. F. Wright, *Practical Laboratory Mycology*, Williams and Wilkins, Baltimore, 1978.

Insert: Courtesy of R. Facklam.

Chapter 21

21-3 (a) Courtesy of S. Jones. Reprinted with permission from *Arch. Oral Biol.* **17**: 613 (1972). Copyright © Pergamon Press, Ltd.
(b) Courtesy of R. J. Gibbons. From E. H. Beachey (Ed.), *Bacterial Adherence*, Chapman and Hall, London, 1980.

21-4 T. Nasemann, *Viral Diseases of the Skin, Mucous Membranes, and Genitals*, Georg Thieme Verlag, Stuttgart, 1977.

21-5 (a) Courtesy of Armed Forces Institute of Pathology. Neg. No. 74-12823.
(b) E. S. Beneke, *Human Mycoses*, The Upjohn Co., Kalamazoo, MI, 1972.

21-7 Courtesy of World Health Organization.

21-8 J. H. L. Watson, J. Goodwin, and K. W. Rajan, *Micron*, **10**: 61.

21-10 Courtesy of Centers for Disease Control.

21-12 Courtesy of National Library of Medicine.

Chapter 22

22-3 (a) Courtesy of F. J. Fleury.
(b) J. W. Rippon, *Medical Mycology*, W. B. Saunders, Philadelphia, 1974.

22-4 Courtesy of F. J. Fleury.

22-5 Sexually Transmitted Disease Laboratory Program, Center for Infectious Diseases, CDC.

22-6 N. S. Hayes, K. E. Muse, A. M. Collier, and J. B. Baseman, *Infect. Immun.* **17**: 174 (1977).

22-7 (a) Courtesy of L. J. LeBeau.
(b,c) Technical Information Services.

22-8 (b) Sexually Transmitted Disease Laboratory Program, Center for Infectious Diseases, CDC.

22-10 T. Nasemann, *Viral Diseases of the Skin, Mucous Membranes, and Genitals*, Georg Thieme Verlag, Stuttgart, 1977.

22-11 Courtesy of F. J. Fleury.

Chapter 23

23-2 Courtesy of W. C. Noble.

23-3 (a) J. W. Rippon, *Medical Mycology*, W. B. Saunders, Philadelphia, 1974.
(b) Courtesy of Armed Forces Institute of Pathology. Neg. No. 75-1175.

23-4 (a) R. J. Schlegel, *Pediatrics* **45**: 926 (1970). Copyright © American Academy of Pediatrics, 1970.

23-5 Courtesy of L. J. LeBeau.

23-6 (a) K. Hummeler and H. Koprowski, *Nature* **221**: 418 (1969).
(b) M. Fekadu, F. W. Chandler, and A. K. Harrison, *Arch. Virol.* **71**: 109 (1982).

23-7 (a) Courtesy of Armed Forces Institute of Pathology. Neg. No. 21990-7.
(b) Courtesy of L. J. LeBeau.

23-8   Courtesy of Armed Forces Institute of Pathology. Neg. No. N67987-3.

23-9   D. H. Walker, W. T. Firth, and C. S. Edgell, *Infect. Immun.* **37**: 301 (1982).

23-10   Courtesy of Armed Forces Institute of Pathology. Neg. No. 55-17580.

23-11   Courtesy of Eastman Kodak Co.

23-13   *(a)* Courtesy of A. T. Willis. From A. Balows (Ed.), *Anaerobic Bacteria*, 1974. By permission of Charles C. Thomas, Publisher, Springfield, Illinois.

        (b) A. Yoshii, J. Tokunaga, and J. Tawara, *Atlas of Scanning Electron Microscopy in Microbiology*, Igaku-Shoin, Ltd., Tokyo, 1976.

23-14   Courtesy of R. Fass, Ohio State Department of Medicine.

23-15   Courtesy of Armed Forces Institute of Pathology. *(a)* Neg. No. N39280; *(b)* Neg. No. 62-1295.

23-16   Courtesy of J. D. Almeida.

23-17   *(a)* Courtesy of Armed Forces Institute of Pathology. Neg. No. 75-2479-1.

        (b) Courtesy of National Hansen's Disease Center, Carville, Louisiana.

Chapter 24

24-2   Courtesy of L. J. Morse.

24-4   *(a)* Courtesy of D. Maki. From I. Phillips, P. D. Meers, and D. F. D'Arcy (Eds.), *Microbiological Hazards of Infusion Therapy*, John Wright - PSG, Littleton, Mass., 1976.

        (b) Courtesy of A. Plumer. From *Principles and Practice of Intravenous Therapy*, Little Brown, Boston, 1975.

24-5   Adapted from C. Kunin, *Detection, Prevention and Management of Urinary Tract Infections*, Lea and Febiger, Philadelphia, 1979.

24-7   B. M. Gray, C. Hubbell, and H. C. Dillon, *Antimicrob. Agents Chemother.* **11**:102 (1977).

24-8   Courtesy of B. A. Pruitt. From L D. Sabath (Ed.), *Pseudomonas aeruginosa: The Organism, Diseases it Causes, and Their Treatment*, Hans Huber, Bern, 1980.

24-9   Courtesy of R. Calderone.

24-10   Courtesy of Loma Linda University.

Chapter 25

25-3   Courtesy of S. N. Cohen.

25-6   *(b)* Courtesy of Nitragin Co.

25-7   Courtesy of J. Menge.

25-8   Courtesy of C. M. Ignaffo.

Color Plates

1b	S. Murray
1c	S. Hay
1d	Nikon Corporation
2a	S. Murray
2b,c	Nikon Corporation
3a	H. Morton
4a,b	N. Furjanick
5	Nikon Corporation
6	R. Williams
7	J. Higbee and R. Marchino
8	R. Williams
9	J. Higbee and R. Marchino
10,11	H. Morton
12	Analytab Products
13	Syva Company
14	TAP Pharmaceuticals
15	Computer Graphics Laboratory, University of California, San Francisco
16	E. Bottone
17	Florida Department of Natural Resources, Bureau of Marine Research
18	Zoological Society of San Diego
19	D. Roberts
20	Kennecott Company

Index

Page numbers in *italic* indicate illustrations or tables.

750

INDEX

756

INDEX

Eye, conjunctiva of, bacterial flora indigenous to, *476*

Eye specimens, collection of, *461*, 463

F factor (fertility factor), 280–283
F pilus, 83
F-prime (F') cells, 282
F⁺ donors, 281
Facultative anaerobes, 93, 114
FAD (flavin adenine dinucleotide), 233–234
FADH₂ (flavin adenine dinucleotide, reduced), 234
   in respiration, 242, *243*
Fastidious microbes, 109
Fat, 213
   construction of, *214*
Favus, 600
Feedback inhibition and metabolism, 252–253
Fermentation, 240–241
   and alcoholic beverages, 664–666
   industrial, 671–672
   of milk, 662–664
   for nondairy foods, 664
Fertility factor (F factor), 280–283
Fetus, intrauterine infections and, 588, *590*–591
"Fever blisters," 532
Fibrinolysin, 353–*354*
Filamentation, 102
Filamentous bacteria, *61*
Filtration:
   of clarified water, 682
   for removal of microbes, 299–*300*, *302*
Finlay, Carlos, 618
Firmicutes, 92
Fission, 103–106, 160, *161*
Fission bodies (sclerotic bodies), 630
Flagellation:
   amphitrichous, 80, *81*
   lophotrichous, 80
   monotrichous, 80, *81*
   peritrichous, 80, *81*
Flagellin, 31
Flagellum (flagella):
   of bacteria, 80–*81*
     structure of, *81*
   of eucaryotic cell, *26–27*
   polar, 80, *81*
   in procaryotic cells, 31
Flagyl (*see* Metronidazole)
Flavin adenine dinucleotide (FAD), 233–234
   reduced (FADH₂), 234, 242, *243*
Flea-borne typhus (endemic typhus), *613*, 615
Fleas, and plague transmission, 610–612
Fleming, Alexander, and penicillin, 13–14, 317
Fleshy fungi, 134, 142
Flocculation, 682
Flu (*see* Influenza)

Flucytosine, 331, 512
Fluorescence microscope (microscopy), 44–45, *46*, *49*
Fluorescent antibody technique, 45, *46*, 55–56, *403*, *404*
Fluorescent treponema antibody (FTA) test, *404*, 579, 580
Folic acid inhibitors, *332–333*
Folliculitis, 603, *607*
Fomites, 434
Fonsecaea, 630
Food:
   canning of, 295
   spoilage of, 8–9
Food additives, microbes and production of, 667–668
Food allergy, 394
Food poisonings (intoxications), 526–529, *530*
Food production, microbes used for, 662–668
Food vacuoles of protozoa, 159
Foodborne diseases, *525*–526
Foraminifera, 164
Formaldehyde, 303
Frame-shift mutation, 274
Francisella tularensis, 501, *634*, 636
Freezing, 710
Fruiting bodies, 137, *138*
FTA (fluorescent treponema antibody) test, *404*, 579, 580
Fuchsin, 49
Fuel production, microbes and, 688–689
Fungal diseases, 143–151
   control and treatment of, 150–151
   from wound inoculation, 630–631
   [*See also* Mycosis (mycoses)]
Fungi Imperfecti (deuteromycetes), *141*, 142
Fungus (fungi), 134–152
   classification of, *141–142*
   dimorphism in, 27, 140–141
   diseases caused by, 143–151
     control and treatment of, 150–151
     from wound inoculation, 630–631
     [*See also* Mycosis (mycoses)]
   drugs for use against, *341*
   fertilization in, 136
   fleshy, 134, 142
   fruiting bodies of, 137, *138*
   as microbes, 4, 27
   morphology of, 134–141
   and nosocomial infections, 653
   notifiable diseases caused by, *439*
   nutrition and cultivation of, 142–143
   opportunistic, 148
   pathogenic, in soil, 432
   saprophytic, 142
   (*See also specific fungi*)
Furuncle (boil), 603
Fusobacterium, 629

758

INDEX